Biology

Concepts and Applications
without Physiology 8e

Biology

Concepts and Applications
without Physiology 8e

Cecie Starr

Christine A. Evers

Lisa Starr

BROOKS/COLE
CENGAGE Learning

Australia • Brazil • Japan • Korea • Mexico • Singapore • Spain • United Kingdom • United States

BROOKS/COLE
CENGAGE Learning

Biology: Concepts and Applications without Physiology, Eighth Edition
Cecie Starr, Christine A. Evers, Lisa Starr

Senior Acquisitions Editor, Life Sciences:
Peggy Williams

Publisher, Life Sciences: Yolanda Cossio

Editor in Chief: Michelle Julet

Assistant Editor: Elizabeth Momb

Editorial Assistant: Alexis Glubka

Managing Media Editor: Shelley Ryan

Media Editor: Lauren Oliveira

Senior Marketing Manager: Tom Ziolkowski

Marketing Coordinator: Elizabeth Wong

Senior Marketing Communications Manager:
Linda Yip

Content Project Manager: Hal Humphrey

Senior Art Director: John Walker

Senior Print Buyer: Karen Hunt

Permissions Account Manager, Text: Bob Kauser

Senior Permissions Account Manager, Images:
Dean Dauphinais

Production Service:
Grace Davidson & Associates, Inc.

Text Designer: John Walker

Photo Researcher: Paul Forkner,
Myrna Engler Photo Research Inc.

Copy Editor: Anita Wagner

Illustrators:
ScEYEnce Studios, Lisa Starr,
Precision Graphics

Cover Image: A pair of Japanese or red
crowned cranes engage in a courtship
dance, Hokkaido, Japan (TIM LAMAN/
National Geographic Stock).

Compositor: Lachina Publishing Services

For product information and technology assistance, contact us at
Cengage Learning Customer & Sales Support, 1-800-354-9706.
For permission to use material from this text or product,
submit all requests online at **cengage.com/permissions**.
Further permissions questions can be emailed to
permissionrequest@cengage.com.

ExamView® and *ExamView Pro*® are registered trademarks of FSCreations, Inc. Windows is a registered trademark of the Microsoft Corporation used herein under license. Macintosh and Power Macintosh are registered trademarks of Apple Computer, Inc. Used herein under license.

Library of Congress Control Number: 2010926064

International Edition:

ISBN-13: 978-0-538-73618-3

ISBN-10: 0-538-73618-6

Cengage Learning International Offices

Asia
www.cengageasia.com
tel: (65) 6410 1200

Australia/New Zealand
www.cengage.com.au
tel: (61) 3 9685 4111

Brazil
www.cengage.com.br
tel: (55) 11 3665 9900

India
www.cengage.co.in
tel: (91) 11 4364 1111

Latin America
www.cengage.com.mx
tel: (52) 55 1500 6000

UK/Europe/Middle East/Africa
www.cengage.co.uk
tel: (44) 0 1264 332 424

Represented in Canada by
Nelson Education, Ltd.
tel: (416) 752 9100 / (800) 668 0671
www.nelson.com

Cengage Learning is a leading provider of customized learning solutions with office locations around the globe, including Singapore, the United Kingdom, Australia, Mexico, Brazil, and Japan. Locate your local office at:
www.cengage.com/global

For product information: **www.cengage.com/international**
Visit your local office: **www.cengage.com/global**
Visit our corporate website: **www.cengage.com**

AVAILABILITY OF RESOURCES MAY DIFFER BY REGION. Check with your local Cengage Learning representative for details.

Printed in the United States of America
1 2 3 4 5 6 7 14 13 12 11 10

CONTENTS IN BRIEF

DETAILED CONTENTS

UNIT II GENETICS

UNIT V PRINCIPLES OF ECOLOGY

25 Animal Behavior

26 Population Ecology

30 Human Effects on the Biosphere 8.5

486—501

PREFACE

Aptostichus stephencolberti

Since the last edition of this book was published, researchers have discovered ice on Mars and have documented rapid glacial melting here on Earth. They have identified thousands of new species, including a spider (*left*) now named after a television show host. They have also uncovered a wealth of information about extinct species, including some of our own close relatives. Biologists have created a vaccine that protects against cervical cancer, and can now make adult cells behave like embryonic stem cells.

In short, what we know about biological systems changes very rapidly. Given the pace of the changes, we would do well to help our students become lifelong learners. Most of the students who use this book will not be biologists, and many will never take another science course. Yet, as voting members of our society, all will face decisions that require a basic understanding of the principles of biology and the process of science in general. This book provides a foundation for such decisions. It teaches students how to draw connections between abstract ideas and everyday experiences, between recent discoveries and long-standing biological theories.

New To This Edition

Graphically Enhanced Key Concepts
The opening pages of each chapter introduces **Key Concepts**, concise statements that set the stage for the more detailed discussion that follows. In this edition, each key concept is accompanied by a photo or graphic that catches the student's eye. The student will encounter these images again within the chapter and in the **Illustrated Chapter Summary** as part of a visual message that threads through each chapter.

Improved Integration Among Chapters
We retained the **Links to Earlier Concepts** on the opening spread of each chapter. This brief paragraph reminds students of relevant information that has been covered in previous chapters. We have also improved the usefulness of the in-chapter Links references by including a brief description of the relevant material. For example, the heading of the section that covers membrane potential in neurons notes that this material links to "Potential energy 5.2, Transport proteins 5.7."

Section-Based Glossary
In addition to a full glossary of terms at the end of the book, each section now has a **Section-Based Glossary**. Important new terms that are boldfaced and defined in the text of a section are redefined in the Section-Based Glossary. A student can simply glance at this glossary for a quick check of a term's definition. As an additional aid, the Chapter Summary includes all glossary terms used in context. Here again, the terms appear in boldface.

Figure-Based Self-Assessment Questions
Many figure captions now include a **Figure-It-Out Question** that allows the student to check whether he or she understands the figure's content. For example, Figure 12.5, which shows the stages of meiosis, is accompanied by a question that asks the student to identify the stage of meiosis in which the chromosome number becomes reduced. Answers to Figure-It-Out questions are provided on the same page. Additional self-assessment material is provided in the Self-Quiz and Critical Thinking Questions at the end of the chapter.

Emphasis on Analyzing Scientific Data

A new feature in the chapter end matter, **Data Analysis Activities**, serves two important purposes. First, most familiarize a student with the process of science by describing an actual scientific study or experiment related to the chapter material. Second, it sharpens a student's analytical skills by asking questions that require the interpretation of data in a graph or table. For example, the Data Analysis Activity in Chapter 6, Where It Starts—Photosynthesis, describes a study that compares the efficiencies of deriving biofuels from native perennial grasses, corn, or soybeans. It then asks students to analyze the data generated by the study.

Chapter-Specific Changes
This new edition contains 140 new photographs and almost 400 new or updated illustrations. In addition, the text of every chapter has been updated and revised for clarity. A page-by-page guide to new content and figures is available upon request, but we summarize the highlights here. Note that we have added a new final chapter, *Human Effects on the Biosphere*. We have also deleted a chapter (*Plants and Animals—Common Challenges*) that appeared in the previous edition. Material that had been in the *Common Challenges* chapter is now integrated into other chapters.

• *Chapter 1, Invitation to Biology*   Revised and expanded coverage of critical thinking and the process of science; levels of life's organization now illustrated with the same organism.

• *Chapter 2, Life's Chemical Basis*   New opening essay discusses mercury toxicity and prevalence; revisions emphasize electron behavior in atoms as it relates to ions and bonding.

• *Chapter 3, Molecules of Life*   New opening essay discusses health impacts of *trans* fats;. importance of protein structure now exemplified by prions.

• *Chapter 4, Cell Structure*   New opening essay about *E. coli* O157:H7 contamination of food; expanded discussion and new photos of archaeans.

• *Chapter 5, Ground Rules of Metabolism*   Example in opening essay is now ADH, to tie in with revisited section about ethanol metabolism, defects in the pathway that affect drinking behavior, and metabolic effects of alcohol abuse.

• *Chapter 6, Where It Starts—Photosynthesis*   New opening essay about biofuels ties in to discussion regarding photosynthesis, carbon dioxide, and global climate change.

• *Chapter 7, How Cells Release Chemical Energy*   Introductory section now discusses the relationship between the evolution of oxygenic photosynthesis and that of aerobic respiration.

• *Chapter 8, DNA Structure and Function*   Opening essay now discusses the cloning of 9/11 rescue dog Trackr; discussion of animal cloning techniques and applications expanded.

• *Chapter 9, From DNA to Protein*   Expanded introductory material now includes an overview of genetic information and gene expression; all new illustrations of translation.

• *Chapter 10, Controls Over Genes*   Theme of evolutionary connections strengthened throughout; for example, in the comparison of effects of mutations in the same homeotic gene in humans and flies.

• *Chapter 11, How Cells Reproduce*   Mitosis panel now compares micrographs of plant and animal cells; new section on cell cycle controls and neoplasms.

• *Chapter 12, Meiosis and Sexual Reproduction*   Opening essay updated; new photo shows pollen grains germinating on stigma; meiosis and segregation graphics updated.

• *Chapter 13, Observing Patterns in Inherited Traits*   Updated essay with current model of CF pathogenesis and greater emphasis on genetics behind the ubiquity of the allele.

• *Chapter 14, Human Inheritance*   New opening essay explores the genetics of skin color variation and discusses evolutionary advantages that reinforce small allelic differences.

• *Chapter 15, Biotechnology*   New opening essay discusses personal genetic testing; updated, expanded genomics section.

• *Chapter 16, Evidence of Evolution*   New graphics include a sheet from Darwin's evolution journal and "missing links" in cetacean evolutionary history; updated geologic time scale now graphically correlated with Grand Canyon stratigraphy.

• *Chapter 17, Processes of Evolution*   New illustrated examples include allopatric speciation in snapping shrimp; sympatric speciation in Lake Victoria cichlids; adaptive radiation of honeycreepers; coevolution of the large blue butterfly and *Maculinea arion* ant. Stasis and exaptation added.

• *Chapter 18, Life's Origin and Early Evolution*   New opening section now focuses on astrobiology; added information about biomarkers as historical evidence.

• *Chapter 19, Viruses, Bacteria, and Archaeans*   New opening section discusses evolution of HIV; new art of HIV replication cycle; new information about influenzas H5N1 and H1N1.

• *Chapter 20, The Protists*   Opening section now covers harmful algal blooms; new graphic showing relationships of protists to other groups; new art of *Plasmodium* life cycle.

• *Chapter 21, Plant Evolution*   Opening section now focuses on Nobel Prize winner Wangari Maathai's efforts on behalf of tropical forests; new section focusing on the ecological and economic importance of angiosperms.

• *Chapter 22, Fungi*   Opening section describes the threat wheat stem rust poses to world food supplies; more extensive coverage of fungi as plant and human pathogens.

• *Chapter 23, Animals I: Major Invertebrate Groups*   Updated evolutionary tree showing relationships among animal groups; major reorganization of coverage of insects and improved discussion of their ecological, economic, and health impacts.

• *Chapter 24, Animals II: The Chordates*   New evolutionary tree diagrams for chordates and for primates.

• *Chapter 25, Animal Behavior*   New material on behavioral genetics and expanded coverage of mechanisms of learning.

• *Chapter 26, Population Ecology*   Opening essay now focuses on soaring numbers of Canada goose; human life table and age-structure diagrams updated.

• *Chapter 27, Community Ecology*   New description and photo of interspecific competition; more focused, concise coverage of exotic species; new subsection about herbivory.

• *Chapter 28, Ecosystems*   Opening section focuses on phosphate pollution of waterways; new diagrams of water, carbon, nitrogen, and phosphorus cycles.

• *Chapter 29, The Biosphere*   Discussion of deserts expanded; soil profiles integrated into biome descriptions.

• *Chapter 30, Human Effects on the Biosphere*   Includes material about declining biodiversity, desertification, deforestation, water shortages and pollution, biological accumulation and magnification, effects of trash on marine ecosystems, depletion of the ozone layer, ground level ozone pollution, effects of climate change, conservation biology and sustainable uses of resources.

Student and Instructor Resources

Test Bank   Nearly 4,000 test items, ranked according to difficulty and consisting of multiple-choice (organized by section heading), selecting the exception, matching, labeling, and short answer exercises. Includes selected images from the text. Also included in Microsoft® Word format on the PowerLecture DVD.

ExamView®   Create, deliver, and customize tests (both print and online) in minutes with this easy-to-use assessment and tutorial system. Each chapter's end-of-chapter material is also included.

Instructor's Resource Manual   Includes chapter outlines, objectives, key terms, lecture outlines, suggestions for presenting the material, classroom and lab enrichment ideas, discussion topics, paper topics, possible answers to critical thinking questions, answers to data analysis activities, and more. Also included in Microsoft® Word format on the PowerLecture DVD.

Resource Integration Guide   A chapter-by-chapter guide to help you use the book's resources effectively. Each chapter includes applied readings, all chapter animations, specific BBC and ABC video segments, hyperlink examples, a listing of the chapter's How Would You Vote? question, and pop-up tutor videos.

Student Interactive Workbook   Labeling exercises, self-quizzes, review questions, and critical thinking exercises help students with retention and better test results.

PowerLecture   This convenient tool makes it easy for you to create customized lectures. Each chapter includes the following features, all organized by chapter: lecture slides, all chapter art and photos, bonus photos, animations, videos, Instructor's Manual, Test Bank, ExamView testing software, and JoinIn polling and quizzing slides. This single disc places all the media resources at your fingertips.

The Brooks/Cole Biology Video Library 2009 featuring BBC Motion Gallery   Looking for an engaging way to launch your lectures? The Brooks/Cole series features short high-interest segments: Pesticides: Will More Restrictions Help or Hinder?; A Reduction in Biodiversity; Are Biofuels as Green as They Claim?; Bone Marrow as a New Source for the Creation of Sperm; Repairing Damaged Hearts with Patients' Own Stem Cells; Genetically Modified Virus Used to Fight Cancer; Seed Banks Helping to Save Our Fragile Ecosystem; The Vanishing Honeybee's Impact on Our Food Supply.

CengageNOW   Save time, learn more, and succeed in the course with CengageNOW, an online set of resources (including Personalized Study Plans) that give you the choices and tools you need to study smarter and get the grade. You will have access to hundreds of animations that clarify the illustrations in the text, videos, and quizzing to test your knowledge. You can also access live online tutoring from an experienced biology instructor. New to this edition are pop-up tutors, which help explain key topics with short video explanations. Get started today!

Webtutors for WebCT and BlackBoard   Jump-start your course with customizable, rich, text-specific content. Whether you want to Web-enable your class or put an entire course online, WebTutor delivers. WebTutor offers a wide array of resources including media assets, quizzing, web links, exercises, flashcards, and more. Visit webtutor.cengage.com to learn more. New to this edition are pop-up tutors, which help explain key topics with short video explanations.

Biology CourseMate   Cengage Learning's Biology Course-Mate brings course concepts to life with interactive learning, study, and exam preparation tools that support the printed textbook, or the included eBook. With CourseMate, professors can use the included Engagement Tracker to assess student preparation and engagement. Use the tracking tools to see progress for the class as a whole or for individual students.

Premium eBook   This complete online version of the text is integrated with multimedia resources and special study features, providing the motivation that so many students need to study and the interactivity they need to learn. New to this edition are pop-up tutors, which help explain key topics with short video explanations.

Acknowledgments

Thanks to our academic advisors for their ongoing impact on the book's content. We are especially grateful to Jean deSaix, David Rintoul, and Michael Plotkin for their ongoing advice and constructive criticism. This edition also reflects influential contributions of the instructors, listed on the following page, who helped shape our thinking. *Key Concepts, Data Analysis Activities,* custom videos—such features are direct responses to their suggestions.

Cengage Learning continues to prove why it is one of the world's foremost publishers. Michelle Julet and Yolanda Cossio, thank you again for allowing us to maintain our ideals and to express our creativity. Peggy Williams, we are, as always, grateful for your continuing guidance and encouragement; thank you for giving us the freedom to improve this edition while updating it. Producing this book would not have been possible without the organizational wizardry and unfailing patience of Grace Davidson. The talented John Walker deserves primary credit for this book's visual appeal; his inspiring design, together with Paul Forkner's dedicated photo research, were critical to our ongoing efforts to create the perfect marriage of text and illustration. Copyeditor Anita Wagner and proofreader Kathleen Dragolich helped us keep our text clear, concise, and correct; tireless editorial assistant Alexis Glubka organized meetings, reviews, and paperwork. Elizabeth Momb managed production of the book's many print supplements and Lauren Oliveira created a world-class technology package for both students and instructors.

LISA STARR, CHRIS EVERS, AND CECIE STARR   *2010*

Contributors to This Edition: Influential Class Tests and Reviews

Brenda Alston-Mills
North Carolina State University

Norris Armstrong
University of Georgia

Dave Bachoon
Georgia College & State University

Neil R. Baker
The Ohio State University

Andrew Baldwin
Mesa Community College

David Bass
University of Central Oklahoma

Lisa Lynn Boggs
Southwestern Oklahoma State University

Gail Breen
University of Texas at Dallas

Marguerite "Peggy" Brickman
University of Georgia

David William Bryan
Cincinnati State College

Uriel Buitrago-Suarez
Harper College

Sharon King Bullock
Virginia Commonwealth University

John Capehart
University of Houston - Downtown

Daniel Ceccoli
American InterContinental University

Tom Clark
Indiana University South Bend

Heather Collins
Greenville Technical College

Deborah Dardis
Southeastern Louisiana University

Cynthia Lynn Dassler
The Ohio State University

Carole Davis
Kellogg Community College

Lewis E. Deaton
University of Louisiana - Lafayette

Jean Swaim DeSaix
University of North Carolina - Chapel Hill

(Joan) Lee Edwards
Greenville Technical College

Hamid M. Elhag
Clayton State University

Patrick Enderle
East Carolina University

Daniel J. Fairbanks
Brigham Young University

Amy Fenster
Virginia Western Community College

Kathy E. Ferrell
Greenville Technical College

Rosa Gambier
Suffok Community College - Ammerman

Tim D. Gaskin
Cuyahoga Community College - Metropolitan

Stephen J. Gould
Johns Hopkins University

Marcella Hackney
Baton Rouge Community College

Gale R. Haigh
McNeese State University

John Hamilton
Gainesville State

Richard Hanke
Rose State Community College

Chris Haynes
Shelton St. Community College

Kendra M. Hill
South Dakota State University

Juliana Guillory Hinton
McNeese State University

W. Wyatt Hoback
University of Nebraska, Kearney

Kelly Hogan
University of North Carolina

Norma Hollebeke
Sinclair Community College

Robert Hunter
Trident Technical College

John Ireland
Jackson Community College

Thomas M. Justice
McLennan College

Timothy Owen Koneval
Laredo Community College

Sherry Krayesky
University of Louisiana - Lafayette

Dubear Kroening
University of Wisconsin - Fox Valley

Jerome Krueger
South Dakota State University

Jim Krupa
University of Kentucky

Mary Lynn LaMantia
Golden West College

Kevin T. Lampe
Bucks County Community College

Susanne W. Lindgren
Sacramento State University

Madeline Love
New River Community College

Dr. Kevin C. McGarry
Kaiser College - Melbourne

Jeanne Mitchell
Truman State University

Alice J. Monroe
St. Petersburg College - Clearwater

Brenda Moore
Truman State University

Erin L. G. Morrey
Georgia Perimeter College

Rajkumar "Raj" Nathaniel
Nicholls State University

Francine Natalie Norflus
Clayton State University

Alexander E. Olvido
Virginia State University

John C. Osterman
University of Nebraska, Lincoln

Bob Patterson
North Carolina State University

Shelley Penrod
North Harris College

Mary A. (Molly) Perry
Kaiser College - Corporate

John S. Peters
College of Charleston

Michael Plotkin
Mt. San Jacinto College

Ron Porter
Penn State University

Karen Raines
Colorado State University

Larry A. Reichard
Metropolitan Community College - Maplewood

Jill D. Reid
Virginia Commonwealth University

Robert Reinswold
University of Northern Colorado

Ashley E. Rhodes
Kansas State University

David Rintoul
Kansas State University

Darryl Ritter
Northwest Florida State College

Amy Wolf Rollins
Clayton State University

Sydha Salihu
West Virginia University

Jon W. Sandridge
University of Nebraska

Robin Searles-Adenegan
Morgan State University

Julie Shepker
Kaiser College - Melbourne

Rainy Shorey
Illinois Central College

Eric Sikorski
University of South Florida

Robert (Bob) Speed
Wallace Junior College

Tony Stancampiano
Oklahoma City Community College

Jon R. Stoltzfus
Michigan State University

Peter Svensson
West Valley College

Jeffrey L. Travis
University at Albany

Nels H. Troelstrup, Jr.
South Dakota State University

Allen Adair Tubbs
Troy University

Will Unsell
University of Central Oklahoma

Rani Vajravelu
University of Central Florida

Jack Waber
West Chester University of Pennsylvania

Kathy Webb
Bucks County Community College

Amy Stinnett White
Virginia Western Community College

Virginia White
Riverside Community College

Robert S. Whyte
California University of Pennsylvania

Kathleen Lucy Wilsenn
University of Northern Colorado

Penni Jo Wilso
Cleveland State Community College

Michael L. Womack
Macon State College

Maury Wrightson
Germanna Community College

Mark L. Wygoda
McNeese State University

Lan Xu
South Dakota State University

Poksyn ("Grace") Yoon
Johnson and Wales University

Muriel Zimmermann
Chaffey College

Biology

Concepts and Applications
without Physiology 8e

The organization of topics in this book parallels life's levels of organization. In both cases, each level builds on the last. At the beginning of each chapter, we will remind you of concepts in previous chapters that will help you understand the material presented in the current chapter. Within chapters, cross-references will link you to the relevant sections in previous chapters.

Key Concepts

The Science of Nature

We understand life by studying it at different levels of organization, which extend from atoms and molecules to the biosphere. The quality we call "life" emerges at the level of cells.

Life's Unity

All organisms consist of one or more cells that take in energy and raw materials to stay alive; all sense and respond to stimuli; and all function and reproduce with the help of DNA.

1 Invitation to Biology

1.1 The Secret Life of Earth

In this era of Google Earth and global positioning systems, could there possibly be any unexplored places left on Earth? Well, yes, actually. In 2005, for instance, helicopters dropped a team of scientists into the middle of a vast and otherwise inaccessible Indonesian cloud forest (Figure 1.1). Within minutes, the explorers realized that their landing site, a dripping, moss-covered swamp, was home to plants and animals that had been previously unknown to science. Over the next month, they discovered dozens of new species there, including a plant with flowers the size of dinner plates and a frog the size of a pea. They also came across hundreds of species that are on the brink of extinction in other parts of the world, some that supposedly were extinct, and one that had not been seen for so many years that scientists had forgotten about it.

The animals in the forest had never learned to be afraid of humans, so they could be approached and even picked up. A few new species were discovered as they casually wandered through the campsite. Team member Bruce Beehler remarked, "Everywhere we looked, we saw amazing things we had never seen before. I was shouting. This trip was a once-in-a-lifetime series of shouting experiences."

New species are discovered all the time, often in places much more mundane than Indonesian cloud forests. How do we know what species a particular organism belongs to? What is a species, anyway, and why should discovering a new one matter to anyone other than a scientist? You will find the answers to such questions in this book. They are part of the scientific study of life, **biology**, which is one of many ways we humans try to make sense of the world around us.

Trying to understand the immense scope of life on Earth gives us some perspective on where we fit into it. For example, we routinely discover hundreds of species every year, but about

---

biology The scientific study of life.

20 species become extinct *every minute* in rain forests alone—and those are only the ones we know about. The current rate of extinctions is about 1,000 times faster than normal. Human activities are responsible for the acceleration. At this rate, we

Figure 1.1 A peek inside the cloud forest of New Guinea's Foja Mountains, (*opposite*). Explorers recently discovered dozens of very rare species—and some new ones—in this forest. *Above*, a jungle hawk-owl (*Ninox theomacha*). This species is a not-so-rare resident of Indonesia, including the Foja Mountains.

will never know about most of the species that are alive on Earth today. Does that matter? Biologists think so. Whether or not we are aware of it, we are intimately connected with the world around us. Our activities are profoundly changing the entire fabric of life on Earth. The changes are, in turn, affecting us in ways we are only beginning to understand.

Ironically, the more we learn about the natural world, the more we realize we have yet to learn. But don't take our word for it. Find out what biologists know, and what they do not, and you will have a solid foundation upon which to base your own opinions about our place in this world. By reading this book, you are choosing to learn about the human connection—your connection—with all life on Earth.

Life's Diversity
Observable characteristics vary tremendously among organisms. Various classification systems help us keep track of the differences.

The Nature of Science
Science helps us be objective about our observations by addressing only the observable. It involves making, testing, and evaluating hypotheses.

Experiments and Research
Researchers carefully design and carry out experiments in order to unravel cause-and-effect relationships in complex natural systems.

> Biologists study life by thinking about it at different levels of organization.

1 atom
Atoms are fundamental units of all substances, living or not. This image shows a model of a single atom.

2 molecule
Atoms joined in chemical bonds. This is a model of a water molecule. The molecules of life are much larger and more complex than water.

3 cell
The cell is the smallest unit of life. Some, like this plant cell, live and reproduce as part of a multicelled organism; others do so on their own.

4 tissue
Organized array of cells and substances that interact in a collective task. This is epidermal tissue on the outer surface of a flower petal.

5 organ
Structural unit of interacting tissues. Flowers are the reproductive organs of many plants.

6 organ system
A set of interacting organs. The shoot system of this poppy plant includes its aboveground parts: leaves, flowers, and stems.

Figure 1.3 **Animated** Levels of life's organization.

Life Is More Than the Sum of Its Parts

What, exactly, is the property we call "life"? We may never actually come up with a good definition. Living things are too diverse, and they consist of the same basic components as nonliving things. When we try to define life, we end up only identifying the properties that differentiate living from nonliving things.

Complex properties, including life, often emerge from the interactions of much simpler parts. For an example, take a look at the drawings in Figure 1.2. The property of "roundness" emerges when the parts are organized one way, but not the other ways. Characteristics of a system that do not appear in any of the system's components are called **emergent properties**. The idea that structures with emergent properties can be assembled from the same basic building blocks is a recurring theme in our world, and also in biology.

A Pattern in Life's Organization

Biologists study all aspects of life, past and present. Through their work, we are beginning to understand a great pattern in life's organization. That organization occurs in successive levels, with new emergent properties appearing at each level (Figure 1.3).

Life's organization starts when atoms interact. **Atoms** are fundamental building blocks of all substances, living and nonliving **1**. There are no atoms unique to life, but there are unique molecules. **Molecules** are atoms joined

Figure 1.2 **Animated**
Example of how different objects can be assembled from the same parts. Roundness is an emergent property of the *rightmost* object.

atom Fundamental building block of all matter.
biosphere All regions of Earth where organisms live.
cell Smallest unit of life.
community All populations of all species in a given area.
ecosystem A community interacting with its environment.
emergent property A characteristic of a system that does not appear in any of the system's component parts.
molecule An association of two or more atoms.
organ In multicelled organisms, a grouping of tissues engaged in a collective task.
organism Individual that consists of one or more cells.
organ system In multicelled organisms, set of organs engaged in a collective task that keeps the body functioning properly.
population Group of individuals of the same species that live in a given area.
tissue In multicelled organisms, specialized cells organized in a pattern that allows them to perform a collective function.

in chemical bonds **2**. In today's world, only living things make the "molecules of life," which are complex carbohydrates and lipids, proteins, DNA, and RNA.

The emergent property of "life" appears at the next level, when many molecules of life become organized as a cell **3**. A **cell** is the smallest unit of life that can survive and reproduce on its own, given information in DNA, energy, and raw materials. Some cells live and reproduce independently. Others do so as part of a multicelled organism. An **organism** is an individual that consists of one or more cells. A poppy plant is an example of a multicelled organism **7**.

In most multicelled organisms, cells make up tissues **4**. Cells of a **tissue** are specialized and organized in a particular pattern. The arrangement allows the cells to collectively perform a special function such as movement (muscle tissue), fat storage (adipose tissue), and so on.

An **organ** is an organized array of tissues that collectively carry out a particular task or set of tasks **5**. For example, a flower is an organ of reproduction in plants; a heart, an organ that pumps blood in animals. An **organ system** is a set of organs and tissues that interact to keep the individual's body working properly **6**. Examples of organ systems include the aboveground parts of a plant (the shoot system), and the heart and blood vessels of an animal (the circulatory system).

A **population** is a group of individuals of the same type, or species, living in a given area **8**. An example would be all of the California poppies in California's Antelope Valley Poppy Reserve. At the next level, a **community** consists of all populations of all species in a given area. The Antelope Valley Reserve community includes California poppies and many other organisms such as microorganisms, animals, and other plants **9**. Communities may be large or small, depending on the area defined.

The next level of organization is the **ecosystem**, or a community interacting with its physical and chemical environment **10**. The most inclusive level, the **biosphere**, encompasses all regions of Earth's crust, waters, and atmosphere in which organisms live **11**.

7 multicelled organism
Individual that consists of different types of cells. The cells of this California poppy plant are part of its two organ systems: aboveground shoots and belowground roots.

8 population
Group of single-celled or multicelled individuals of a species in a given area. This population of California poppy plants is in California's Antelope Valley Poppy Reserve.

9 community
All populations of all species in a specified area. These flowering plants are part of the Antelope Valley Poppy Reserve community.

10 ecosystem
A community interacting with its physical environment through the transfer of energy and materials. Sunlight and water sustain the natural community in the Antelope Valley.

11 biosphere
The sum of all ecosystems: every region of Earth's waters, crust, and atmosphere in which organisms live. The biosphere is a finite system, so no ecosystem in it can be truly isolated from any other.

Take-Home Message How does "life" differ from "nonlife"?

❯ All things, living or not, consist of the same building blocks—atoms. Atoms join as molecules.

❯ The unique properties of life emerge as certain kinds of molecules become organized into cells.

❯ Higher levels of life's organization include multicelled organisms, populations, communities, ecosystems, and the biosphere.

❯ Emergent properties occur at each successive level of life's organization.

> Continual inputs of energy and the cycling of materials maintain life's complex organization.
> Organisms sense and respond to change.
> All organisms use information in the DNA they inherited from parents to function and to reproduce.

A Producers harvest energy from the environment. Some of that energy flows from producers to consumers.

sunlight energy

Producers
plants and other self-feeding organisms

B Nutrients that become incorporated into the cells of producers and consumers are eventually released by decomposition. Some cycle back to producers.

Consumers
animals, most fungi, many protists, bacteria

C All of the energy that enters the world of life eventually flows out of it, mainly as heat released back to the environment.

Figure 1.4 Animated The one-way flow of energy and the cycling of materials in the world of life. The photo shows a producer acquiring energy and nutrients from the environment, and consumers acquiring energy and nutrients by eating the producer.

Even though we cannot define "life," we can intuitively understand what it means because all living things share some key features. All require ongoing inputs of energy and raw materials; all sense and respond to change; and all have DNA that guides their functioning.

Organisms Require Energy and Nutrients

Not all living things eat, but all require energy and nutrients on an ongoing basis. Both are essential to maintain life's organization and functioning. **Energy** is the capacity to do work. A **nutrient** is a substance that an organism needs for growth and survival but cannot make for itself.

Organisms spend a lot of time acquiring energy and nutrients. However, what type of energy and nutrients are acquired varies considerably depending on the type of organism. The differences allow us to classify all living things into two categories: producers and consumers. **Producers** make their own food using energy and simple raw materials they get directly from their environment. Plants are producers that use the energy of sunlight to make sugars from water and carbon dioxide (a gas in air), a process called **photosynthesis**. By contrast, **consumers** cannot make their own food. They get energy and nutrients by feeding on other organisms. Animals are consumers. So are decomposers, which feed on the wastes or remains of other organisms. The leftovers of consumers' meals end up in the environment, where they serve as nutrients for producers. Said another way, nutrients cycle between producers and consumers.

Energy, however, is not cycled. It flows through the world of life in one direction: from the environment, through organisms, and back to the environment. The flow of energy maintains the organization of individual organisms, and it is the basis of how organisms interact with one another and their environment. It is also a one-

consumer Organism that gets energy and nutrients by feeding on tissues, wastes, or remains of other organisms.
development Multistep process by which the first cell of a new individual becomes a multicelled adult.
DNA Deoxyribonucleic acid; carries hereditary information that guides growth and development.
energy The capacity to do work.
growth In multicelled species, an increase in the number, size, and volume of cells.
homeostasis Set of processes by which an organism keeps its internal conditions within tolerable ranges.
inheritance Transmission of DNA from parents to offspring.
nutrient Substance that an organism needs for growth and survival, but cannot make for itself.
photosynthesis Process by which producers use light energy to make sugars from carbon dioxide and water.
producer Organism that makes its own food using energy and simple raw materials from the environment.
reproduction Processes by which parents produce offspring.

Figure 1.5 Organisms sense and respond to stimulation. This baby orangutan is laughing in response to being tickled. Apes and humans make different sounds when being tickled, but the airflow patterns are so similar that we can say apes really do laugh.

way flow, because with each transfer, some energy escapes as heat. Cells do not use heat to do work. Thus, all of the energy that enters the world of life eventually leaves it, permanently (Figure 1.4).

Organisms Sense and Respond to Change

An organism cannot survive for very long in a changing environment unless it adapts to the changes. Thus, every living thing has the ability to sense and respond to conditions both inside and outside of itself (Figure 1.5). For example, after you eat, the sugars from your meal enter your bloodstream. The added sugars set in motion a series of events that causes cells throughout the body to take up sugar faster, so the sugar level in your blood quickly falls. This response keeps your blood sugar level within a certain range, which in turn helps keep your cells alive and your body functioning.

The fluid in your blood is part of your body's internal environment, which consists of all body fluids outside of cells. Unless that internal environment is kept within certain ranges of composition, temperature, and other conditions, your body cells will die. By sensing and adjusting to change, you and all other organisms keep conditions in the internal environment within a range that favors cell survival. **Homeostasis** is the name for this process, and it is a defining feature of life.

Organisms Use DNA

With little variation, the same types of molecules perform the same basic functions in every organism. For example, information encoded in an organism's **DNA** (deoxyribonucleic acid) guides the ongoing metabolic activities that

sustain the individual through its lifetime. Such activities include **growth**: increases in cell number, size, and volume; **development**: the process by which the first cell of a new individual becomes a multicelled adult; and **reproduction**: processes by which parents produce offspring.

Individuals of every natural population are alike in certain aspects of their body form and behavior, an outcome of shared information encoded in DNA. Orangutans look like orangutans and not like caterpillars because they inherited orangutan DNA, which differs from caterpillar DNA in the information it carries. **Inheritance** refers to the transmission of DNA from parents to offspring. All organisms receive their DNA from parents.

Thus, DNA is the basis of similarities in form and function among organisms. However, the details of DNA molecules differ, and herein lies the source of life's diversity. Small variations in the details of DNA's structure give rise to differences among individuals, and among types of organisms. As you will see in later chapters, these differences are the raw material of evolution.

> There is great variation in the details of appearance and other observable characteristics of living things.

Living things differ tremendously in their observable characteristics. Various classification schemes help us organize what we understand about this variation, which we call Earth's **biodiversity**.

For example, organisms can be classified into broad groups depending on whether they have a **nucleus**, a sac with two membranes that encloses and protects a cell's DNA. **Bacteria** (singular, bacterium) and **archaeans** are organisms whose DNA is not contained within a nucleus. All bacteria and archaeans are single-celled, which means each organism consists of one cell (Figure 1.6A,B). As a group, they are also the most diverse organisms: Different kinds are producers or consumers in nearly all regions of the biosphere. Some inhabit such extreme environments as frozen desert rocks, boiling sulfurous lakes, and nuclear reactor waste. The first cells on Earth may have faced similarly hostile challenges to survival.

Traditionally, organisms without a nucleus have been called prokaryotes, but this designation is an informal one. Despite their similar appearance, bacteria and archaeans are less related to one another than we had once thought. Archaeans are actually more closely related to **eukaryotes**, organisms whose DNA is contained within a nucleus. Some eukaryotes live as individual cells; others are multicelled (Figure 1.6C). Eukaryotic cells are typically larger and more complex than bacteria or archaeans.

Structurally, **protists** are the simplest eukaryotes, but as a group they vary dramatically. Protists range from single-celled amoebas to giant, multicelled seaweeds.

Most **fungi** (singular, fungus), such as the types that form mushrooms, are multicelled eukaryotes. Many are decomposers. All secrete enzymes that digest food outside the body, then their cells absorb the released nutrients.

Plants are multicelled eukaryotes that live on land or in freshwater environments. Most are photosynthetic producers. Besides feeding themselves, plants and other photosynthesizers also serve as food for most of the other organisms in the biosphere.

Animals are multicelled consumers that ingest tissues or juices of other organisms. Herbivores graze, carnivores eat meat, scavengers eat remains of other organisms, parasites withdraw nutrients from the tissues of a host, and so on. Animals grow and develop through a series of stages that lead to the adult form, and all kinds actively move about during at least part of their lives.

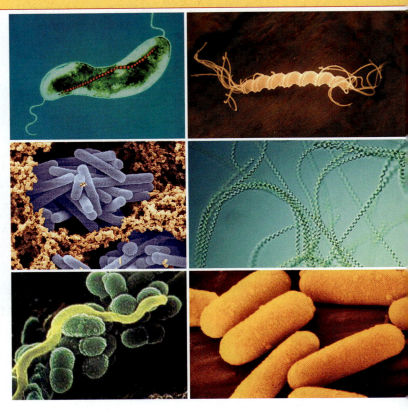

A **Bacteria** are the most numerous organisms on the planet. All are single-celled, but different types vary in shape and size. *Clockwise from upper left*, a bacterium with a row of iron crystals that acts like a tiny compass; *Helicobacter*, a common resident of cat and dog stomachs; spiral cyanobacteria; *E. coli*, a beneficial resident of human intestines; types found in dental plaque; *Lactobacillus* cells in yogurt.

B **Archaeans** resemble bacteria, but are more closely related to eukaryotes. *Left*, an archaean from volcanic ocean sediments. *Right*, two types of archaeans from a hydrothermal vent on the sea floor.

Figure 1.6 **Animated** Representatives of life's diversity.

---

animal Multicelled consumer that develops through a series of stages and moves about during part or all of its life cycle.
archaean Member of a group of single-celled organisms that differ from bacteria.
bacterium Member of a large group of single-celled organisms.
biodiversity Variation among living organisms.
eukaryote Organism whose cells characteristically have a nucleus.
fungus Type of eukaryotic consumer that obtains nutrients by digestion and absorption outside the body.
nucleus Double-membraned sac that encloses a cell's DNA.
plant A multicelled, typically photosynthetic producer.
protist Member of a diverse group of simple eukaryotes.

Take-Home Message **How do organisms differ from one another?**

> Different types of organisms vary tremendously in observable characteristics. For example, some organisms have a nucleus; others do not.

protists are a group of extremely diverse eukaryotes that range from giant multicelled seaweeds to microscopic single cells. Many biologists are now viewing "protists" as several major groups.

plants are multicelled eukaryotes, most of which are photosynthetic. Nearly all have roots, stems, and leaves. Plants are the primary producers in land ecosystems.

fungi are eukaryotes. Most are multicelled. Different kinds are parasites, pathogens, or decomposers. Without decomposers such as fungi, communities would be buried in their own wastes.

animals are multicelled eukaryotes that ingest tissues or juices of other organisms. All actively move about during at least part of their life.

C Eukaryotes are single-celled or multicelled organisms whose DNA is contained within a nucleus.

> Each type of organism, or species, is given a unique name.
> We define and group species based on shared traits.

Each time we discover a new **species**, or kind of organism, we name it. **Taxonomy**, a system of naming and classifying species, began thousands of years ago. However, doing it in a consistent way did not become a priority until the eighteenth century. At that time, European explorers and naturalists were beginning to discover the scope of life's diversity. They started having more and more trouble communicating with one another because species often had multiple names.

For example, one type of plant native to Europe, Africa, and Asia was alternately known as the dog rose, briar rose, witch's briar, herb patience, sweet briar, wild briar, dog briar, dog berry, briar hip, eglantine gall, hep tree, hip fruit, hip rose, hip tree, hop fruit, and hogseed—and those are only the English names! Species often had multiple scientific names too, Latin names that were descriptive but often cumbersome. For example, the scientific name of the dog rose was *Rosa sylvestris inodora seu canina* (odorless woodland dog rose), and also *Rosa sylvestris alba cum rubore, folio glabro* (pinkish white woodland rose with smooth leaves).

An eighteenth-century naturalist, Carolus Linnaeus, came up with a much simpler naming system that we still use. By the Linnaean system, every species is given a unique two-part scientific name. The first part is the name of the **genus** (plural, genera), a group of species that share a unique set of features. The second part is the **specific epithet**. Together, the genus name plus the specific epithet designate one species. Thus, the dog rose now has one official name: *Rosa canina*.

Genus and species names are always italicized. For example, *Panthera* is a genus of big cats. Lions belong to the species *Panthera leo*. Tigers belong to a different species in the same genus (*Panthera tigris*), and so do leopards (*P. pardus*). Note how the genus name may be abbreviated after it has been spelled out once.

Linnaeus ranked species into ever more inclusive categories. Each Linnaean category, or **taxon** (plural, taxa), is a group of organisms. The categories above species—genus, family, order, class, phylum, kingdom, and domain—are the higher taxa (Figure 1.7). Each higher taxon consists of a group of the next lower taxon. Using this system, we can sort all life into a few categories (Figure 1.8 and Table 1.1).

A Rose by Any Other Name . . .

The individuals of a species share a unique set of features, or traits. For example, giraffes normally have very long necks, brown spots on white coats, and so on. These are examples of morphological traits (*morpho*– means form). Individuals of a species also share physiological traits, such as metabolic activities, and they respond the same way to certain stimuli, as when hungry giraffes feed on tree leaves. These are behavioral traits.

DOMAIN Eukarya	Eukarya	Eukarya	Eukarya	Eukarya
KINGDOM Plantae	Plantae	Plantae	Plantae	Plantae
PHYLUM Magnoliophyta	Magnoliophyta	Magnoliophyta	Magnoliophyta	Magnoliophyta
CLASS Magnoliopsida	Magnoliopsida	Magnoliopsida	Magnoliopsida	Magnoliopsida
ORDER Apiales	Rosales	Rosales	Rosales	Rosales
FAMILY Apiaceae	Cannabaceae	Rosaceae	Rosaceae	Rosaceae
GENUS *Daucus*	*Cannabis*	*Malus*	*Rosa*	*Rosa*
SPECIES *carota*	*sativa*	*domesticus*	*acicularis*	*canina*
COMMON NAME carrot	marijuana	apple	arctic rose	dog rose

Figure 1.7 Linnaean classification of five species that are related at different levels. Each species has been assigned to ever more inclusive groups, or taxa: in this case, from genus to domain.

>> **Figure It Out** Which of the plants shown here are in the same order?

Answer: Marijuana, apple, arctic rose, and dog rose are all in the order Rosales.

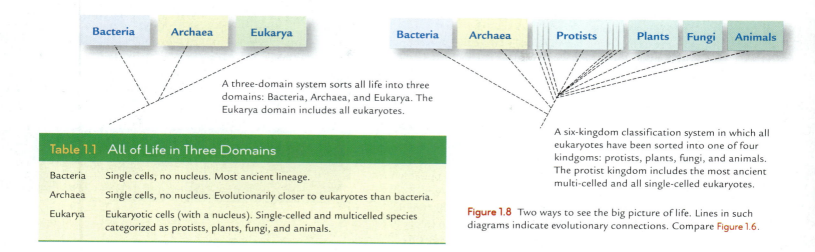

A three-domain system sorts all life into three domains: Bacteria, Archaea, and Eukarya. The Eukarya domain includes all eukaryotes.

Table 1.1	All of Life in Three Domains
Bacteria	Single cells, no nucleus. Most ancient lineage.
Archaea	Single cells, no nucleus. Evolutionarily closer to eukaryotes than bacteria.
Eukarya	Eukaryotic cells (with a nucleus). Single-celled and multicelled species categorized as protists, plants, fungi, and animals.

A six-kingdom classification system in which all eukaryotes have been sorted into one of four kindgoms: protists, plants, fungi, and animals. The protist kingdom includes the most ancient multi-celled and all single-celled eukaryotes.

Figure 1.8 Two ways to see the big picture of life. Lines in such diagrams indicate evolutionary connections. Compare Figure 1.6.

A species is assigned to higher taxa based on some subset of traits it shares with other species. That assignment may change as we discover more about the species and the traits involved. For example, Linnaeus grouped plants by the number and arrangement of reproductive parts, a scheme that resulted in odd pairings such as castor-oil plants with pine trees. Having more information today, we place these plants in separate phyla.

Traits vary a bit within a species, such as eye color does in people. However, there are often tremendous differences between species. Think of petunias and whales, beetles and emus, and so on. Such species look very different, so it is easy to tell them apart. Species that share a more recent ancestor may be much more difficult to distinguish (Figure 1.9).

How do we know if similar-looking organisms belong to different species or not? The short answer is that we rely on whatever information we have. For example, early naturalists studied anatomy and distribution—essentially the only techniques available at the time. Thus, species were named and classified according to what they looked like and where they lived. Today's biologists have at their disposal an array of techniques and machinery much more sophisticated than those of the eighteenth century. They are able to study traits that the early naturalists did not even know about—biochemistry, for example.

Evolutionary biologist Ernst Mayr defined a species as one or more groups of individuals that potentially can interbreed, produce fertile offspring, and do not interbreed with other groups. This "biological species concept"

is useful but it is not universally applicable. For example, not all populations of a species actually continue to interbreed. In many cases, we may never know whether populations separated by a great distance could interbreed successfully even if they did get together. Also, populations often continue to interbreed even as they become different, so the exact moment at which they become separate species is often impossible to pinpoint. We return to speciation and how it occurs in Chapter 17, but for now it is useful to remember that a "species" is a convenient but artificial construct of the human mind.

Figure 1.9 Four butterflies, two species: Which are which? Two forms of the species *Heliconius melpomene* are on the top row; two of *H. erato* are on the bottom row. These two species never cross-breed. Their alternate but similar patterns of coloration evolved as a shared warning signal to local birds that these butterflies taste terrible.

genus A group of species that share a unique set of traits; also the first part of a species name.
species A type of organism.
specific epithet Second part of a species name.
taxon Linnaean category; a grouping of organisms.
taxonomy The science of naming and classifying species.

Take-Home Message How do we keep track of all the species we know about?

❯ Each species has a unique, two-part scientific name.

❯ Various classification systems group species on the basis of shared traits.

> Critical thinking is judging the quality of information before accepting it.
> Scientists make and test potentially falsifiable predictions about how the natural world works.
> Science addresses only what is observable.

Thinking About Thinking

Most of us assume that we do our own thinking, but do we, really? You might be surprised to find out just how often we let others think for us. For instance, a school's job, which is to impart as much information as possible to students, meshes perfectly with a student's job, which is to acquire as much knowledge as possible. In this rapid-fire exchange of information, it is all too easy to forget about the quality of what is being exchanged. Any time you accept information without question, you allow someone else to think for you.

Critical thinking is the deliberate process of judging the quality of information before accepting it. "Critical" comes from the Greek *kriticos* (discerning judgment). When you use critical thinking, you move beyond the content of new information to consider supporting evidence, bias, and alternative interpretations. How does the busy student manage this? Critical thinking does not require extra time, just a bit of extra awareness. There are many ways to do it. For example, you might ask yourself some of the following questions while you are learning something new:

> What message am I being asked to accept?
> Is the message based on opinion or evidence?
> Is there a different way to interpret the evidence?
> What biases might the presenter have?
> How do my own biases affect what I'm learning?

Such questions are simply a way of being conscious about learning. They will help you to decide whether to allow new information to guide your beliefs and actions.

How Science Works

Critical thinking is a big part of **science**, the systematic study of the observable world and how it works (Figure 1.10). A scientific line of inquiry usually begins with curiosity about something observable, such as, say, a decrease in the number of birds in a particular area. Typically, a scientist will read about what others have discovered before making a **hypothesis**, a testable explanation for a natural phenomenon. An example of a hypothesis would be, "The number of birds is decreasing because the number of cats is increasing." Making a hypothesis this way is an example of **inductive reasoning**, which means arriving at a conclusion based on one's observations. Inductive reasoning is the way we come up with new ideas about groups of objects or events.

A **prediction**, or statement of some condition that should exist if the hypothesis is correct, comes next. Making predictions is called the if–then process, in which the "if" part is the hypothesis, and the "then" part is the prediction. Using a hypothesis to make a prediction is a form of **deductive reasoning**, or logical process of using a general premise to draw a conclusion about a specific case.

Table 1.2   Example of the Scientific Method

1. Form a hypothesis	Observe some aspect of nature	Hangover symptoms vary in severity.
2. Test the hypothesis	Think of an explanation for the observation (a hypothesis)	Eating artichokes reduces the severity of hangover symptoms.
	Make a prediction based on the hypothesis	If eating artichokes reduces the severity of hangover symptoms, then taking artichoke extract will reduce the severity of a hangover after drinking alcohol.
	Test the prediction (experiments or surveys)	During a party at which alcohol is served, administer artichoke extract to half of the people who are drinking. The next day, ask everyone who drank alcohol at the party to rate the severity of their hangover symptoms.
3. Evaluate the hypothesis	Analyze the results of the tests (data) and make conclusions	See if there is a correlation between taking the artichoke extract and reduced hangover symptoms. Submit your results and conclusions to a peer-reviewed journal for publication.

Figure 1.10 Scientists doing research. *From left to right*, surveying wildlife in New Guinea; sequencing the human genome; looking for fungi in atmospheric dust collected in Cape Verde; improving the efficiency of biofuel production from agricultural wastes; studying the benefits of weedy buffer zones on farms.

Next, a scientist will devise ways to test a prediction. Tests may be performed on a **model**, or analogous system, if working with an object or event directly is not possible. For example, animal diseases are often used as models to investigate similar human diseases. Careful observations are one way to test predictions that flow from a hypothesis. So are **experiments**: tests designed to support or falsify a prediction. A typical experiment explores a cause and effect relationship.

Researchers investigate causal relationships by changing or observing **variables**, which are characteristics or events that can differ among individuals or over time. An **independent variable** is defined or controlled by the person doing the experiment. A **dependent variable** is an observed result that is supposed to be influenced by the independent variable. For example, an independent

variable in an investigation of hangover preventions may be the administration of artichoke extract before alcohol consumption. The dependent variable in this experiment would be the severity of the forthcoming hangover.

Biological systems are complex, with many interacting variables. It can be difficult to study one variable separately from the rest. Thus, biology researchers often test two groups of individuals simultaneously. An **experimental group** is a set of individuals that have a certain characteristic or receive a certain treatment. This group is tested side by side with a **control group**, which is identical to the experimental group except for one independent variable—the characteristic or the treatment being tested. Any differences in experimental results between the two groups should be an effect of changing the variable.

Test results—**data**—that are consistent with the prediction are evidence in support of the hypothesis. Data that are inconsistent with the prediction are evidence that the hypothesis is flawed and should be revised.

A necessary part of science is reporting one's results and conclusions in a standard way, such as in a peer-reviewed journal article. The communication gives other scientists an opportunity to check and confirm the work.

Forming, testing, and evaluating hypotheses are collectively called the **scientific method** (Table 1.2).

---

control group In an experiment, group of individuals who are not exposed to the independent variable that is being tested.
critical thinking Judging information before accepting it.
data Experimental results.
deductive reasoning Using a general idea to make a conclusion about a specific case.
dependent variable In an experiment, variable that is presumably affected by the independent variable being tested.
experiment A test designed to support or falsify a prediction.
experimental group In an experiment, group of individuals who are exposed to an independent variable.
hypothesis Testable explanation of a natural phenomenon.
independent variable Variable that is controlled by an experimenter in order to explore its relationship to a dependent variable.
inductive reasoning Drawing a conclusion based on obervation.
model Analogous system used for testing hypotheses.
prediction Statement, based on a hypothesis, about a condition that should exist if the hypothesis is correct.
science Systematic study of the observable world.
scientific method Making, testing, and evaluating hypotheses.
variable In an experiment, a characteristic or event that differs among individuals or over time.

Take-Home Message What is science?

> Science is concerned only with the observable—those objects or events for which objective evidence can be gathered.

> The scientific method consists of making, testing, and evaluating hypotheses. It is a way of critical thinking, or systematically judging the quality of information before allowing it to guide one's beliefs and actions.

> Experiments measure how changing an independent variable affects a dependent variable.

> Researchers unravel cause-and-effect relationships in complex natural processes by changing one variable at a time.

There are different ways to do research, particularly in biology. Some biologists do surveys; they observe without making hypotheses. Some make hypotheses and leave the experimentation to others. However, despite a broad range of subject matter, scientific experiments are typically designed in a consistent way. Experimenters try to change one independent variable at a time, and see what happens to a dependent variable.

To give you a sense of how biology experiments work, we summarize two published studies here.

Potato Chips and Stomachaches

In 1996 the FDA approved Olestra®, a fat replacement manufactured from sugar and vegetable oil, as a food additive. Potato chips were the first Olestra-containing food product on the market in the United States.

Controversy about the food additive soon raged. Many people complained of intestinal problems after eating the chips, and thought that the Olestra was at fault. Two years later, researchers at Johns Hopkins University School of Medicine designed an experiment to test whether Olestra causes cramps.

The researchers predicted that if Olestra indeed causes cramps, then people who eat Olestra will be more likely to get cramps than people who do not. To test the prediction, they used a Chicago theater as a "laboratory." They asked 1,100 people between the ages of thirteen and thirty-eight to watch a movie and eat their fill of potato chips. Each person got an unmarked bag that contained 13 ounces of chips.

In this experiment, the individuals who got Olestra-containing potato chips constituted the experimental group, and individuals who got regular chips were the control group. The independent variable was the presence or absence of Olestra in the chips.

A few days after the experiment was finished, the researchers contacted all of the people and collected any reports of post-movie gastrointestinal problems. Of 563 people making up the experimental group, 89 (15.8 percent) complained about cramps. However, so did 93 of the 529 people (17.6 percent) making up the control group—who had munched on the regular chips.

People were about as likely to get cramps whether or not they ate chips made with Olestra. These results did not support the prediction, so the researchers concluded that eating Olestra does not cause cramps (Figure 1.11).

Butterflies and Birds

Consider the peacock butterfly, a winged insect that was named for the large, colorful spots on its wings. In 2005, researchers published a report on their tests to identify factors that help peacock butterflies defend themselves against insect-eating birds. The researchers made two observations. First, when a peacock butterfly rests, it folds its wings, so only the dark underside shows (Figure 1.12A). Second, when a butterfly sees a predator approaching, it repeatedly flicks its wings open and closed, while also moving the hindwings in a way that produces a hissing sound and a series of clicks.

A Hypothesis
Olestra® causes intestinal cramps.

B Prediction
People who eat potato chips made with Olestra will be more likely to get intestinal cramps than those who eat potato chips made without Olestra.

C Experiment

	Control Group	Experimental Group
	Eats regular potato chips	Eats Olestra potato chips

D Results

	Control Group	Experimental Group
	93 of 529 people get cramps later (17.6%)	89 of 563 people get cramps later (15.8%)

E Conclusion
Percentages are about equal. People who eat potato chips made with Olestra are just as likely to get intestinal cramps as those who eat potato chips made without Olestra. These results do not support the hypothesis.

Figure 1.11 The steps in a scientific experiment to determine if Olestra causes cramps. A report of this study was published in the *Journal of the American Medical Association* in January 1998.

>> **Figure It Out** What was the dependent variable in this experiment?

Answer: Whether or not a person got cramps

A

B

C

Figure 1.12 Peacock butterfly defenses against predatory birds. **A** With wings folded, a resting peacock butterfly looks a bit like a dead leaf. **B** When a bird approaches, the butterfly repeatedly flicks its wings open and closed, a behavior that exposes brilliant spots and produces hissing and clicking sounds.

Researchers tested whether the butterfly's behavior deters blue tits **C**. They painted over the spots of some butterflies, cut the sound-making part of the wings on other butterflies, and did both to a third group; then the biologists exposed each butterfly to a hungry bird.

The results, listed in Table 1.3, support the hypotheses that peacock butterfly spots and sounds can deter predatory birds.

›› Figure It Out What percentage of butterflies with no spots and no sound survived the test?

Answer: 20 percent.

Table 1.3 Results of Peacock Butterfly Experiment*

Wing Spots	Wing Sound	Total Number of Butterflies	Number Eaten	Number Survived
Spots	Sound	9	0	9 (100%)
No spots	Sound	10	5	5 (50%)
Spots	No sound	8	0	8 (100%)
No spots	No sound	10	8	2 (20%)

* Proceedings of the Royal Society of London, Series B (2005) 272: 1203–1207.

The researchers were curious about why the peacock butterfly flicks its wings. After they reviewed earlier studies, they came up with two hypotheses that might explain the wing-flicking behavior:

1. Although wing-flicking probably attracts predatory birds, it also exposes brilliant spots that resemble owl eyes (Figure 1.12B). Anything that looks like owl eyes is known to startle small, butterfly-eating birds, so exposing the wing spots might scare off predators.
2. The hissing and clicking sounds produced when the peacock butterfly moves its hindwings may be an additional defense that deters predatory birds.

The researchers used their hypotheses to make the following predictions:

1. If peacock butterflies startle predatory birds by exposing their brilliant wing spots, then individuals with wing spots will be less likely to get eaten by predatory birds than those without wing spots.
2. If peacock butterfly sounds deter predatory birds, then sound-producing individuals will be less likely to get eaten by predatory birds than silent individuals.

The next step was the experiment. The researchers used a marker to paint the wing spots of some butterflies black, and scissors to cut off the sound-making part of the hindwings of others. A third group had their wing spots painted and their hindwings cut. The researchers then put each butterfly into a large cage with a hungry blue tit (Figure 1.12C) and watched the pair for thirty minutes.

Table 1.3 lists the results of the experiment. All of the butterflies with unmodified wing spots survived, regardless of whether they made sounds. By contrast, only half of the butterflies that had spots painted out but could make sounds survived. Most of the butterflies with neither spots nor sound structures were eaten quickly. The test results confirmed both predictions, so they support the hypotheses. Birds are deterred by peacock butterfly sounds, and even more so by wing spots.

Take-Home Message **Why do biologists perform experiments?**

> Natural processes are often influenced by many interacting variables.

> Experiments help researchers unravel causes of complex natural processes by focusing on the effects of changing a single variable.

> Science is, ideally, a self-correcting process because scientists check one another's work.

1 Natalie, blindfolded, randomly plucks a jelly bean from a jar. There are 120 green and 280 black jelly beans in that jar, so 30 percent of the jelly beans in the jar are green, and 70 percent are black.

2 The jar is hidden from Natalie's view before she removes her blindfold. She sees only one green jelly bean in her hand and assumes that the jar must hold only green jelly beans.

3 Still blindfolded, Natalie randomly picks out 50 jelly beans from the jar. She ends up picking out 10 green and 40 black ones.

4 The larger sample leads Natalie to assume that one-fifth of the jar's jelly beans are green (20 percent) and four-fifths are black (80 percent). The sample more closely approximates the jar's actual green-to-black ratio of 30 percent to 70 percent. The more times Natalie repeats the sampling, the greater the chance she will come close to knowing the actual ratio.

Figure 1.13 Animated Demonstration of sampling error.

The Trouble With Trends

Researchers can rarely observe all individuals of a group. For example, the explorers you read about in Section 1.1 did not—and could not—survey every uninhabited part of New Guinea. The cloud forest itself cloaks more than 2 million acres of the Foja Mountains, so surveying all of it would take unrealistic amounts of time and effort. Besides, tromping about even in a small area can damage delicate forest ecosystems.

Given such limitations, researchers often look at subsets of an area, a population, an event, or some other aspect of nature. They test or survey the subset, then use the results to make generalizations. However, generalizing from a subset is risky because a subset may not be representative of the whole.

For example, the golden-mantled tree kangaroo pictured on the *right* was first discovered in 1993 on a single forested mountaintop in New Guinea. For more than a decade, the species was never seen outside of that habitat, which is getting smaller every year because of human activities. Thus, the golden-mantled tree kangaroo was considered to be one of the most endangered animals on the planet. Then, in 2005, the New Guinea explorers discovered that this kangaroo species is fairly common in the Foja Mountain cloud forest. As a result, biologists now believe its future is secure, at least for the moment.

Problems With Probability

Making generalizations from testing or surveying a subset is risky because of sampling error. **Sampling error** is a difference between results obtained from a subset, and results from the whole (Figure 1.13).

Sampling error may be unavoidable, as illustrated by the example of the golden-mantled tree kangaroo. However, knowing how it can occur helps researchers design their experiments to minimize it. For example, sampling error can be a substantial problem with a small subset, so experimenters try to start with a relatively large sample, and they repeat their experiments.

To understand why such practices reduce the risk of sampling error, think about what happens when you flip a coin. There are two possible outcomes: The coin lands heads up, or it lands tails up. Thus, with each flip, the chance that the coin will land heads up is one in two (1/2), which is a proportion of 50 percent. However, when you flip a coin repeatedly, it often lands heads up, or tails up, several times in a row. With just 3 flips, the proportion of times that heads actually land up may not even be close

Figure 1.14 Example of error bars in a graph. This particular graph was adapted from the peacock butterfly research described in Section 1.7.

The researchers recorded the number of times each butterfly flicked its wings in response to an attack by a bird.

The squares represent average frequency of wing flicking for each sample set of butterflies. The error bars that extend above and below the squares indicate the range of values—the sampling error.

>> Figure It Out What was the fastest rate at which a butterfly with no spots or sound flicked its wings? Answer: 22 times per minute.

(graph axis) Wing-flicks per minute — 0, 4, 8, 12, 16, 20, 24

− spots / + sound   − spots / − sound   + spots / − sound

to 50 percent. With 1,000 flips, the overall proportion of times that the coin lands heads up is likely to be close to 50 percent.

In cases like flipping a coin, it is possible to calculate **probability**: the measure, expressed as a percentage, of the chance that a particular outcome will occur. That chance depends on the total number of possible outcomes. For instance, if 10 million people enter a drawing, each has the same probability of winning: 1 in 10 million, or (an extremely improbable) 0.00001 percent.

Analysis of experimental data often includes calculations of probability. If a result is very unlikely to have occurred by chance alone, it is said to be **statistically significant**. In this context, the word "significant" does not refer to the result's importance. It means that the result has been subjected to a rigorous statistical analysis that shows it has a very low probability (usually 5 percent or less) of being skewed by sampling error.

Variation in data is often shown as error bars on a graph (Figure 1.14). Depending on the graph, error bars may indicate variation around an average for one sample set, or the difference between two sample sets.

Bothering With Bias

Particularly when studying humans, experimenting with a single variable apart from all others is not often pos-

sible. For example, remember that the people who participated in the Olestra experiment were chosen randomly. That means the study was not controlled for gender, age, weight, medications taken, and so on. Such variables may well have influenced the results.

Human beings are by nature subjective, and scientists are no exception. Experimenters risk interpreting their results in terms of what they want to find out. That is why they often design experiments to yield quantitative results, which are counts or some other data that can be measured or gathered objectively. Such results minimize the potential for bias, and also give other scientists an opportunity to repeat the experiments and check the conclusions drawn from them.

This last point gets us back to the role of critical thinking in science. Scientists expect one another to recognize and put aside bias in order to test their hypotheses in ways that may prove them wrong. If a scientist does not, then others will, because exposing errors is just as useful as applauding insights. The scientific community consists of critically thinking people trying to poke holes in one another's ideas. Their collective efforts make science a self-correcting endeavor.

probability The chance that a particular outcome of an event will occur; depends on the total number of outcomes possible.
sampling error Difference between results derived from testing an entire group of events or individuals, and results derived from testing a susbet of the group.
statistically significant Refers to a result that is statistically unlikely to have occurred by chance.

> Scientific theories are our best descriptions of reality.
> Science helps us to be objective about our observations, in part because it is limited to the observable.

About the Word "Theory"

Suppose a hypothesis stands even after years of tests. It is consistent with all data ever gathered, and it has helped us make successful predictions about other phenomena. When a hypothesis meets these criteria, it is considered to be a **scientific theory** (Table 1.4).

To give an example, all observations to date have been consistent with the hypothesis that matter consists of atoms. Scientists no longer spend time testing this hypothesis for the compelling reason that, since we started looking 200 years ago, no one has discovered matter that doesn't consist of atoms. Thus, scientists use the hypothesis, now called atomic theory, to make other hypotheses about matter and the way it behaves.

Scientific theories are our best descriptions of reality. However, they can never be proven absolutely, because to do so would necessitate testing under every possible circumstance. For example, in order to prove atomic theory, the atomic composition of all matter in the universe would have to be checked—an impossible task even if someone wanted to try.

Like all hypotheses, a scientific theory can be disproven by a single observation or result that is inconsistent with it. For example, if someone discovers a form of matter that does not consist of atoms, atomic theory would have to be revised. The potentially falsifiable nature of scientific theories means that science has a

built-in system of checks and balances. A theory is revised until no one can prove it to be incorrect. For example, the theory of evolution, which states that change occurs in a line of descent over time, still holds after a century of observations and testing. As with all other scientific theories, no one can be absolutely sure that it will hold under all possible conditions, but it has a very high probability of not being wrong. Few other theories have withstood as much scrutiny.

You may hear people apply the word "theory" to a speculative idea, as in the phrase "It's just a theory." This everyday usage of the word differs from the way it is used in science. Speculation is an opinion, belief, or personal conviction that is not necessarily supported by evidence. A scientific theory is different. By definition, it is supported by a large body of evidence, and it is consistent with all known facts.

A scientific theory also differs from a **law of nature**, which describes a phenomenon that has been observed to occur in every circumstance without fail, but for which we currently do not have a complete scientific explanation. The laws of thermodynamics, which describe energy, are examples. We know how energy behaves, but not exactly why it behaves the way it does.

The Limits of Science

Science helps us be objective about our observations in part because of its limitations. For example, science does not address many questions, such as "Why do I exist?" Answers to such questions can only come from within as an integration of the personal experiences and mental

Table 1.4 Examples of Scientific Theories

Atomic theory	All substances consist of atoms.
Big bang	The universe originated with an explosion and continues to expand.
Cell theory	All organisms consist of one or more cells, the cell is the basic unit of life, and all cells arise from existing cells.
Evolution	Change occurs in the inherited traits of a population over generations.
Global warming	Human activities are causing Earth's average temperature to increase.
Plate tectonics	Earth's crust is cracked into pieces that move in relation to one another.

connections that shape our consciousness. This is not to say subjective answers have no value, because no human society can function for long unless its individuals share standards for making judgments, even if they are subjective. Moral, aesthetic, and philosophical standards vary from one society to the next, but all help people decide what is important and good. All give meaning to our lives.

Neither does science address the supernatural, or anything that is "beyond nature." Science neither assumes or denies that supernatural phenomena occur, but scientists often cause controversy when they discover a natural explanation for something that was thought to have none. Such controversy often arises when a society's moral standards are interwoven with its understanding of nature.

For example, Nicolaus Copernicus concluded in 1540 that Earth orbits the sun. Today that idea is generally accepted, but during Copernicus's time the prevailing belief system had Earth as the immovable center of the universe. In 1610, astronomer Galileo Galilei published evidence for the Copernican model of the solar system, an act that resulted in his imprisonment. He was publicly forced to recant his work, spent the rest of his life under house arrest, and was never allowed to publish again.

As Galileo's story illustrates, exploring a traditional view of the natural world from a scientific perspective might be misinterpreted as a violation of morality, even though the two are not the same. As a group, scientists are no less moral than anyone else. However, they follow a particular set of rules that do not necessarily apply to others: Their work concerns only the natural world, and their ideas must be testable in ways others can repeat.

Science helps us communicate our experiences without bias. As such, it may be as close as we can get to a universal language. We are fairly sure, for example, that the laws of gravity apply everywhere in the universe. Intelligent beings on a distant planet would likely understand the concept of gravity. We might well use gravity or another scientific concept to communicate with them, or anyone, anywhere. The point of science, however, is not to communicate with aliens. It is to find common ground here on Earth.

---

law of nature Generalization that describes a consistent natural phenomenon for which there is incomplete scientific explanation.
scientific theory Hypothesis that has not been disproven after many years of rigorous testing.

Take-Home Message Why does science work?

> Science has built-in checks and balances that help us to be objective about our observations.

> Because a scientific theory is revised until no one can prove it wrong, it is our best way of describing reality.

The Secret Life of Earth (revisited)

> We have discovered only a small fraction of the species that share Earth with us.

Of an estimated 100 billion species that have ever lived, at least 100 million are still with us today. That number is only an estimate because we are still discovering them—by the thousands every year. For example, a return expedition to New Guinea's Foja Mountains turned up a mouse-sized opossum and a cat-sized rat. Other surveys revealed lemurs (Figure 1.15) and sucker-footed bats in Madagascar; birds in the

Figure 1.15 It is traditional for the discoverer of a new species to have the honor of naming it. *Top*, this tiny mouse lemur, discovered in Madagascar in 2005, was named *Microcebus lehilahytsara* in honor of primatologist Steve Goodman (*lehilahytsara* is a combination of the Malagasy words for "good" and "man"). *Bottom*, Dr. Jason Bond holding a spider he discovered in California in 2008. Bond named the spider *Aptostichus stephencolberti*, after TV personality Stephen Colbert.

Philippines; monkeys in Tanzania, Brazil, and India; cave-dwelling spiders and insects in two of California's national parks; carnivorous sponges near Antarctica; whales, sharks, giant jellylike animals, fishes, and other aquatic wildlife; and scores of plants and single-celled organisms. Most were discovered by biologists who were simply trying to find out what lives where.

Biologists make discoveries every day, though we may never hear of them. Each new species they discover is another reminder that we do not yet know all of the organisms on our own planet. We don't even know how many to look for. The vast information about the 1.8 million species we do know about changes so quickly that collating it has been impossible—until recently. A new web site, titled the Encyclopedia of Life, is intended to be an online reference source and database of species information that is maintained by collaborative effort. See its progress at www.eol.org.

How Would You Vote? There is a possibility that substantial populations of some species currently listed as endangered may exist in unexplored areas. Should we wait to protect endangered species until all of Earth has been surveyed? See Cengage-Now for details, then vote online (cengagenow.com).

Summary

Section 1.1 Biology is the systematic study of life. We have encountered only a fraction of the organisms that live on Earth, in part because we have explored only a fraction of its inhabited regions.

Section 1.2 Biologists think about life at different levels of organization. **Emergent properties** appear at successively higher levels. Life emerges at the cellular level. All matter consists of **atoms**, which combine as **molecules**. **Organisms** are individuals that consist of one or more **cells**. Cells of larger multicelled organisms are organized as **tissues**, **organs**, and **organ systems**. A **population** is a group of individuals of a species in a given area; a **community** is all populations of all species in a given area. An **ecosystem** is a community interacting with its environment. The **biosphere** includes all regions of Earth that hold life.

Section 1.3 Life has underlying unity in that all living things have similar characteristics. (1) All organisms require **energy** and **nutrients** to sustain themselves: **producers** harvest energy from the environment to make their own food by processes such as **photosynthesis**; **consumers** eat other organisms, or their wastes and remains. (2) Organisms keep the conditions in their internal environment within ranges that their cells tolerate—a process called **homeostasis**. (3) **DNA** contains information that guides all of an organism's metabolic activities, including **growth**, **development**, and **reproduction**. The passage of DNA from parents to offspring is **inheritance**.

Section 1.4 The different types of organisms that currently exist on Earth differ greatly in details of body form and function. **Biodiversity** is the sum of differences among living things. **Bacteria** and **archaeans** are all single-celled, and their DNA is not contained within a **nucleus**. **Eukaryotes** (**protists**, **plants**, **fungi**, and **animals**) can be single-celled or multicelled. Their DNA is contained within a nucleus.

Section 1.5 Each type of organism has a two-part name. The first part is the **genus** name. When combined with the **specific epithet**, it designates a particular **species**. Linnean **taxonomy** ranks all species into successive **taxa** on the basis of shared traits.

Section 1.6 Critical thinking, the self-directed act of judging the quality of information as one learns, is an important part of **science**. Generally, a researcher observes something in nature, uses **inductive reasoning** to form a **hypothesis** (testable explanation) for it, then uses **deductive reasoning** to make a **prediction** about what might occur if the hypothesis is not wrong. Predictions are tested with observations, **experiments**, or both. Experiments typically are performed on an **experimental group** as compared with a **control group**, and sometimes on **models**. Conclusions are drawn from experimental results, or **data**. A hypothesis that is not consistent with data is modified. Making, testing, and evaluating hypotheses is the **scientific method**.

Biological systems are usually influenced by many interacting **variables**. An **independent variable** influences a **dependent variable**.

Section 1.7 Scientific approaches differ, but experiments are typically designed in a consistent way. A researcher changes an independent variable, then observes the effects of the change on a dependent variable. This practice allows the researcher to unravel a cause-and-effect relationship in a complex natural system.

Section 1.8 Small sample size increases the potential for **sampling error** in experimental results. In such cases, a subset may be tested that is not representative of the whole. Researchers design experiments carefully to minimize sampling error and bias, and they use **probability** rules to check the **statistical significance** of their results. Science is ideally a self-correcting process because scientists check and test one another's ideas.

Section 1.9 Science helps us be objective about our observations because it is only concerned with testable ideas about observable aspects of nature. Opinion and belief have value in human culture, but they are not addressed by science. A **scientific theory** is a long-standing hypothesis that is useful for making predictions about other phenomena. It is our best way of describing reality. A **law of nature** describes something that occurs without fail, but for which we do not have a complete scientific explanation.

Self-Quiz   Answers in Appendix III

1. _____ are fundamental building blocks of all matter.

2. The smallest unit of life is the _____ .

3. _____ move around for at least part of their life.

4. Organisms require _____ and _____ to maintain themselves, grow, and reproduce.

5. _____ is a process that maintains conditions in the internal environment within ranges that cells can tolerate.

6. DNA _____ .
   a. guides growth   c. is transmitted from
      and development   parents to offspring
   b. is the basis of traits   d. all of the above

7. A process by which an organism produces offspring is called _____ .

Data Analysis Activities

Peacock Butterfly Predator Defenses

The photographs on the *right* represent the actual experimental and control groups used in the peacock butterfly experiment discussed in Section 1.7.

See if you can identify the experimental groups, and match them up with the relevant control group(s). *Hint:* Identify which variable is being tested in each group (each variable has a control).

A Wing spots painted out

B Wing spots visible; wings silenced

C Wing spots painted out; wings silenced

D Wings painted but spots visible

E Wings cut but not silenced

F Wings painted, spots visible; wings cut, not silenced

---

8. _____ is the transmission of DNA to offspring.
   - a. Reproduction
   - b. Development
   - c. Homeostasis
   - d. Inheritance

9. An animal is a(n) _____ (choose all that apply).
   - a. organism
   - b. domain
   - c. species
   - d. eukaryote
   - e. consumer
   - f. producer
   - g. hypothesis
   - h. trait

10. Plants are _____ (choose all that apply).
    - a. organisms
    - b. a domain
    - c. a species
    - d. eukaryotes
    - e. consumers
    - f. producers
    - g. hypotheses
    - h. traits

11. Science only addresses that which is _____ .
    - a. alive
    - b. observable
    - c. variable
    - d. indisputable

12. A control group is _____ .
    - a. a set of individuals that have a certain characteristic or receive a certain treatment
    - b. the standard against which an experimental group is compared
    - c. the experiment that gives conclusive results

13. Match the terms with the most suitable description.
    - ____ emergent property
    - ____ species
    - ____ scientific theory
    - ____ hypothesis
    - ____ prediction
    - ____ probability
    - a. statement of what a hypothesis leads you to expect
    - b. type of organism
    - c. occurs at a higher organizational level
    - d. time-tested hypothesis
    - e. testable explanation
    - f. measure of chance

Additional questions are available on **CENGAGENOW**.

Critical Thinking

1. A person is declared to be dead upon the irreversible cessation of spontaneous body functions: brain activity, or blood circulation and respiration. However, only about 1% of a person's cells have to die in order for all of these things to happen. How can someone be dead when 99% of his or her cells are still alive?

2. Why would you think twice about ordering from a cafe menu that lists the genus name but not the specific epithet of its offerings? *Hint:* Look up *Homarus americanus*, *Ursus americanus*, *Ceanothus americanus*, *Bufo americanus*, *Lepus americanus*, and *Nicrophorus americanus*.

3. Once there was a highly intelligent turkey that had nothing to do but reflect on the world's regularities. Morning always started out with the sky turning light, followed by the master's footsteps, which were always followed by the appearance of food. Other things varied, but food always followed footsteps. The sequence of events was so predictable that it eventually became the basis of the turkey's theory about the goodness of the world. One morning, after more than 100 confirmations of the goodness theory, the turkey listened for the master's footsteps, heard them, and had its head chopped off.

   Any scientific theory is modified or discarded upon discovery of contradictory evidence. The absence of absolute certainty has led some people to conclude that "facts are irrelevant—facts change." If that is so, should we stop doing scientific research? Why or why not?

4. In 2005, researcher Woo-suk Hwang reported that he had made immortal stem cells from human patients. His research was hailed as a breakthrough for people affected by degenerative diseases, because stem cells may be used to repair a person's own damaged tissues. Hwang published his results in a peer-reviewed journal. In 2006, the journal retracted his paper after other scientists discovered that Hwang's group had faked their data. Does the incident show that results of scientific studies cannot be trusted? Or does it confirm the usefulness of a scientific approach, because other scientists discovered and exposed the fraud?

Animations and Interactions on **CENGAGENOW**:
› Life's building blocks; Life's levels of organization; Energy flow and materials cycling; Life's diversity; Three domains of life; Sampling error.

See an annotated scientific paper in Appendix II.

❮ **Links to Earlier Concepts**

Take a moment to review Section 1.2 because life's organization starts with atoms. Life's organization requires continuous inputs of energy (1.3). Organisms store that energy in chemical bonds between atoms. You will see examples of how the body's built-in mechanisms maintain homeostasis (1.3), and how scientists make major discoveries (1.6).

Key Concepts

Atoms and Elements
Atoms, the building blocks of all matter, differ in their numbers of protons, neutrons, and electrons.

Why Electrons Matter
How an atom interacts with other atoms depends on the number and arrangement of its electrons.

2 Life's Chemical Basis

2.1 ## Mercury Rising

Actor Jeremy Piven, best known for his Emmy-winning role on the television series *Entourage*, began starring in a Broadway play in 2008. He quit suddenly after two shows, citing medical problems. Piven said he was suffering from mercury poisoning caused by eating too much sushi. The play's producers and his co-actors were skeptical. The playwright ridiculed Piven, saying he was leaving to pursue a career as a thermometer. But mercury poisoning is no laughing matter.

Mercury is a naturally occurring metal. Most of it is safely locked away in rocky minerals, but volcanic activity and other geologic processes release it into the atmosphere. So do human activities, especially burning coal (Figure 2.1). Once airborne, mercury can drift long distances before settling to Earth's surface. There, microbes combine it with carbon to form a substance called methylmercury.

Unlike mercury alone, methylmercury easily crosses skin and mucous membranes. In water, it ends up in the tissues of aquatic organisms. All fish and shellfish contain methylmercury. Humans contain it too, mainly as a result of eating seafood.

When mercury enters the body, it damages the nervous system, brain, kidneys, and other organs. A dose as low as 3 micrograms per kilogram of body weight (about 200 micrograms for an average-sized adult) can cause tremors, itching or burning sensations, and loss of coordination. Exposure to larger amounts can result in thought and memory impairment, coma, and death.

The developing brain is particularly sensitive to mercury because the metal interferes with nerve formation. Thus, mercury is acutely toxic in infants, and it causes long-term neurological effects in children. Methylmercury in a pregnant woman's blood passes to her unborn child, along with a legacy of permanent developmental problems.

The U.S. Food and Drug Administration requires that foods contain less than 1 part per million of mercury, and for the most part they do. However, it takes months or even years for mer-

Figure 2.1 Atmospheric fallout from coal-fired power plant emissions is now the biggest cause of mercury pollution. *Opposite*, mercury can accumulate to toxic levels in the tissues of tuna and other large predatory fish.

cury to be cleared from the body, so it can build up to high levels if even small amounts are ingested on a regular basis. That is why large predatory fish have the most mercury in their tissues. It is also why the U.S. Environmental Protection Agency recommends that adults ingest less than 0.1 microgram of mercury per kilogram of body weight per day. For an average-sized person, that limit works out to be about 7 micrograms per day, which is not a big amount if you eat seafood. A two-ounce piece of sushi tuna typically contains about 40 micrograms of mercury, and the occasional piece has many times that amount. It doesn't matter if the fish is raw, grilled, or canned, because mercury is unaffected by cooking. Eat a medium-sized tuna steak, and you could be getting more than 700 micrograms of mercury along with it.

With this chapter, we turn to the first of life's levels of organization: atoms. Interactions between atoms make the molecules that sustain life, and also some that destroy it.

Atoms Bond
Atoms of many elements interact by acquiring, sharing, and giving up electrons. Interacting atoms may form ionic, covalent, or hydrogen bonds.

Water of Life
Water stabilizes temperature. It also has cohesion, and it can act as a solvent for many other substances. These properties make life possible.

The Power of Hydrogen
Most of the chemistry of life occurs in a narrow range of pH, so the fluids inside organisms are buffered to stay within that range.

> The behavior of elements, which make up all living things, starts with the structure of individual atoms.
> The number of protons in the atomic nucleus defines the element, and the number of neutrons defines the isotope.
< Link to Atoms 1.2

The idea that different structures can be assembled from the same basic building blocks is a recurring theme in our world. The theme is apparent in all levels of nature's organization, including atomic structure. Life's unique characteristics start with the properties of different **atoms**, tiny particles that are building blocks of all substances. Even though they are about 20 million times smaller than a grain of sand, atoms consist of even smaller subatomic particles: positively charged **protons** (p^+), uncharged **neutrons**, and negatively charged **electrons** (e^-). **Charge** is an electrical property. Opposite charges attract, and like charges repel. Protons and neutrons cluster in an atom's central core, or **nucleus**, and electrons move around the nucleus (Figure 2.2).

Atoms differ in their number of subatomic particles. The number of protons in an atom's nucleus is called the **atomic number**, and it determines the type of atom, or element. **Elements** are pure substances, each consisting only of atoms that have the same number of protons in their nucleus. For example, the atomic number of carbon is 6, so all atoms with six protons in their nucleus are carbon atoms, no matter how many electrons or neutrons they have. A chunk of carbon consists only of carbon atoms, and all of those atoms have six protons.

The same elements that make up a living body also occur in nonliving things, but their proportions differ. For example, a human body contains a much larger pro-

Figure 2.2 Atoms consist of electrons moving around a core, or nucleus, of protons and neutrons.

Models such as this diagram cannot show what atoms really look like. Electrons zoom around in fuzzy, three-dimensional spaces about 10,000 times bigger than the nucleus.

- ● proton
- ● neutron
- ● electron

portion of carbon atoms than do rocks or seawater. Why? Unlike rocks or seawater, a body consists of a very high proportion of the molecules of life, which in turn consist of a high proportion of carbon atoms.

Knowing about the numbers of electrons, protons, and neutrons helps us predict how elements will behave. For example, elements in each vertical column of the periodic table behave in similar ways. The **periodic table** shows all of the known elements arranged in order of atomic number (Figure 2.3). Each of the 117 elements is listed in the table by a symbol that is usually an abbreviation of the element's Latin or Greek name. For instance, Pb (lead) is short for *plumbum;* the word "plumbing" is related—ancient Romans made their water pipes with lead. Carbon's symbol, C, is from *carbo*, the Latin word for coal (which is mostly carbon).

The first ninety-four elements occur in nature. We know about the rest because we have synthesized them a few atoms at a time. An atomic nucleus cannot be altered by heat or any other ordinary means, so synthesizing an element requires a particle accelerator and other equipment found only in nuclear physics laboratories.

Figure 2.3 Animated The periodic table. The table was created in 1869 by chemist Dmitry Mendeleyev (*left*), who arranged the known elements by chemical properties. The arrangement turned out to be by atomic number. Until he came up with the table, Mendeleyev had been known mainly for his extravagant hair (he cut it only once per year).

atom Particle that is a fundamental building block of all matter.
atomic number Number of protons in the atomic nucleus; determines the element.
charge Electrical property. Opposite charges attract, and like charges repel.
electron Negatively charged subatomic particle that occupies orbitals around an atomic nucleus.
element A pure substance that consists only of atoms with the same number of protons.
isotopes Forms of an element that differ in the number of neutrons their atoms carry.
mass number Total number of protons and neutrons in the nucleus of an element's atoms.
neutron Uncharged subatomic particle in the atomic nucleus.
nucleus Core of an atom; occupied by protons and neutrons.
periodic table Tabular arrangement of the known elements by atomic number.
proton Positively charged subatomic particle that occurs in the nucleus of all atoms.
radioactive decay Process by which atoms of a radioisotope emit energy and/or subatomic particles when their nucleus spontaneously disintegrates.
radioisotope Isotope with an unstable nucleus.
tracer A molecule labelled with a detectable substance.

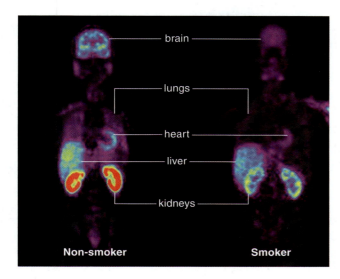

Figure 2.4 **Animated** PET scans. The result of a PET scan is a digital image of a process in the body's interior. These PET scans show the activity of a molecule called MAO-B in the body of a non-smoker (*left*) and a smoker (*right*). The activity is color coded from *red* (highest activity) to *purple* (lowest). Low MAO-B activity is associated with violence, impulsiveness, and other behavioral problems.

Isotopes and Radioisotopes

Atoms of an element that differ in their number of neutrons are called **isotopes**. All elements have isotopes. We define isotopes by their **mass number**, which is the total number of protons and neutrons in their nucleus. Mass number is written as a superscript to the left of an element's symbol. For example, the most common isotope of carbon has six protons and six neutrons, so it is designated ^{12}C, or carbon 12. The other naturally occurring isotopes of carbon are ^{13}C (six protons, seven neutrons), and ^{14}C (six protons, eight neutrons).

In 1896, physicist Henri Becquerel made a discovery after leaving some crystals of a uranium salt in a desk drawer, on top of a metal screen. Under the screen was an exposed film wrapped tightly in black paper. When Becquerel developed the film, he was surprised to see a negative image of the screen. He realized that "invisible radiations" coming from the uranium salts had passed through the paper and exposed the film around the screen.

Uranium, like many other elements, has radioactive isotopes, or **radioisotopes**. The atoms of radioisotopes spontaneously emit subatomic particles or energy (radiation) when their nucleus breaks down. This process, **radioactive decay**, can transform one element into another. For example, carbon 14 is a radioisotope that decays when one of its neutrons splits into a proton and an electron. The nucleus emits the electron, so an atom with eight neutrons and six protons (^{14}C) becomes an atom with seven neutrons and seven protons, which is nitrogen (^{14}N).

Each radioisotope decays at a constant rate into certain products. This process occurs independently of external factors such as temperature, pressure, or whether the atoms are part of molecules, so it is very predictable. For example, we can reliably predict that about half of the atoms of ^{14}C in any sample will be ^{14}N atoms after 5,730 years. Thus, we can use the isotope content of a rock or fossil to estimate its age (we will return to this topic in Section 16.5).

Researchers and clinicians also use radioisotopes to track biological processes inside living organisms. Remember, isotopes are atoms of the same element. All isotopes of an element generally have the same chemical properties regardless of the number of neutrons in their atoms. This consistent chemical behavior means that organisms use atoms of one isotope (such as ^{14}C) the same way that they use atoms of another (such as ^{12}C). Thus, radioisotopes can be used in tracers. A **tracer** is any molecule with a detectable substance attached.

A typical radioactive tracer is a molecule in which radioisotopes have been swapped for one or more atoms. Researchers deliver radioactive tracers into a biological system such as a cell or a multicelled body. Instruments that can detect radioactivity let researchers follow the tracer as it moves through the system. For example, Melvin Calvin and his colleagues used a radioactive tracer to identify specific reaction steps of photosynthesis. The researchers made carbon dioxide with ^{14}C, then let green algae (simple aquatic organisms) take up the radioactive gas. Using instruments that detected the radioactive decay of ^{14}C, they tracked carbon through steps by which the algae—and all plants—make sugars.

Radioisotopes have medical applications as well. For example, PET (short for positron-emission tomography) helps us "see" a functional process inside the body. By this procedure, a radioactive sugar or other tracer is injected into a patient, who is then moved into a PET scanner. Inside the patient's body, cells with differing rates of activity take up the tracer at different rates. The scanner detects radioactive decay wherever the tracer is, then translates that data into an image (Figure 2.4).

Take-Home Message **What are the basic building blocks of all matter?**

> All matter consists of atoms, tiny particles that in turn consist of electrons moving around a nucleus of protons and neutrons.

> An element is a pure substance that consists only of atoms with the same number of protons. All elements occur as isotopes, which are forms of an element that have different numbers of neutrons.

> The unstable nuclei of radioisotopes disintegrate spontaneously (decay) at a predictable rate to form predictable products.

> Atoms acquire, share, and donate electrons.
> Whether an atom will interact with other atoms depends on how many electrons it has.
‹ Link to Atomic interactions 1.2

Energy Levels

Electrons are really, really small. How small are they? If they were as big as apples, you would be about 3.5 times taller than our solar system is wide. Simple physics explains the motion of, say, an apple falling from a tree, but electrons are so tiny that such everyday physics cannot explain their behavior. However, that behavior underlies atomic interactions.

A typical atom has about the same number of electrons and protons. In larger atoms, a lot of electrons may be zipping around one nucleus. Those electrons move at nearly the speed of light (300,000 kilometers per second, or 670 million miles per hour), but they never collide. Why not? Electrons avoid one another because they travel in different orbitals, which are defined volumes of space around the nucleus.

Imagine that an atom is a multilevel apartment building with a nucleus in the basement. Each "floor" of the building corresponds to a certain energy level, and each has a certain number of "rooms" (orbitals) available for rent to one or two electrons at a time. Electrons populate rooms from the ground floor up; in other words, they fill orbitals from lower to higher energy levels. The farther an electron is from the nucleus in the basement, the greater its energy. An electron can move to a room on a higher floor if an energy input gives it a boost, but it immediately emits its extra energy and moves back down.

Shell models help us visualize how electrons populate atoms. In this model, nested "shells" correspond to successively higher energy levels. Thus, a shell includes all of the rooms on one floor of our atomic apartment building.

We draw a shell model of an atom by filling its shells with electrons (represented as balls or dots), from the innermost shell out, until there are as many electrons as the atom has protons. For example, there is only one room on the first floor, one orbital at the lowest energy level. It fills up first. In hydrogen, the simplest atom, a single electron occupies that room (Figure 2.5A). Helium, with two protons, has two electrons that fill its first shell. In larger atoms, more electrons rent the second-floor rooms (Figure 2.5B). When the second floor fills, more electrons rent third-floor rooms (Figure 2.3C), and so on.

A The first shell corresponds to the first energy level, and it can hold up to 2 electrons. Hydrogen has one proton, so it has 1 electron and 1 vacancy. A helium atom has 2 protons, 2 electrons, and no vacancies. The number of protons in each model is shown.

first shell

1 proton ⟶ 1
1 electron ⟶

hydrogen (H)

helium (He)

B The second shell corresponds to the second energy level, and it can hold up to 8 electrons. Carbon has 6 protons, so its first shell is full. Its second shell has 4 electrons, and four vacancies. Oxygen has 8 protons and two vacancies. Neon has 10 protons and no vacancies.

second shell

carbon (C)

oxygen (O)

neon (Ne)

C The third shell, which corresponds to the third energy level, can hold up to 8 electrons. A sodium atom has 11 protons, so its first two shells are full; the third shell has one electron. Thus, sodium has seven vacancies. Chlorine has 17 protons and one vacancy. Argon has 18 protons and no vacancies.

third shell

sodium (Na)

chlorine (Cl)

argon (Ar)

Figure 2.5 Animated Shell models. Each circle (shell) represents all orbitals at one energy level. A model is filled with electrons from the innermost shell out, until there are as many electrons as protons. Atoms with vacancies (room for additional electrons) in their outermost shell tend to interact with other atoms.

If an atom's outermost shell is full of electrons, we say that it has no vacancies. Helium is an example. Atoms of such elements are chemically inactive, which means they are most stable as single atoms. By contrast, if an atom's outermost shell has room for another electron, it has a vacancy. Hydrogen has one vacancy. Atoms with vacancies tend to interact with other atoms: They give up, acquire, or share electrons until they have no vacancies in their outermost shell. Any atom is in its most stable state when it has no vacancies.

Swapping Electrons

The negative charge of an electron is the same magnitude as the positive charge of a proton, so the two charges cancel one another. Thus, an atom with as many electrons as protons carries no overall (net) charge. An atom with different numbers of electrons and protons carries a charge. When an atom is charged, we call it an **ion**.

An atom acquires a positive charge by losing an electron. Conversely, an atom acquires a negative charge by pulling an electron away from another atom (Figure 2.6). **Electronegativity** is a measure of an atom's ability to pull electrons away from other atoms. Electronegativity is not the same as charge. Rather, an atom's electronegativity depends on its size and how many vacancies it has. For example, when a chlorine atom is uncharged, it has 17 protons and 17 electrons. Seven electrons are in its outer (third) shell, which can hold eight, so this atom has one vacancy. An uncharged chlorine atom is highly electronegative—it can easily pull an electron away from another atom to fill its third shell. When that happens, the atom becomes a chloride ion (Cl^-) with 17 protons, 18 electrons, and a net negative charge (Figure 2.6A).

As another example, an uncharged sodium atom has 11 protons and 11 electrons. This atom has one electron in its outer (third) shell, which can hold eight. It has seven vacancies. An uncharged sodium atom is weakly electronegative, so it cannot pull seven electrons from other atoms to fill its third shell. Instead, it tends to lose the single electron in its third shell. When that happens, two full shells—and no vacancies—remain. The atom has now become a sodium ion (Na^+), with 11 protons, 10 electrons, and a net positive charge (Figure 2.6B).

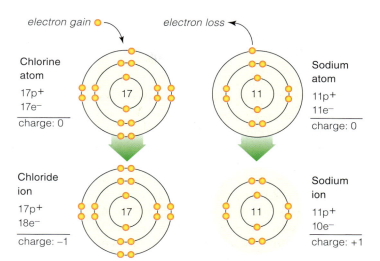

electron gain ○   *electron loss* ◀

Chlorine atom
17p+
17e−
charge: 0

Sodium atom
11p+
11e−
charge: 0

Chloride ion
17p+
18e−
charge: −1

Sodium ion
11p+
10e−
charge: +1

A A chlorine atom (Cl) becomes a negatively charged chloride ion (Cl^-) when it gains an electron and fills the vacancy in its third, outermost shell.

B A sodium atom (Na) becomes a positively charged sodium ion (Na^+) when it loses the electron in its third shell. The atom's full second shell is now its outermost, so it has no vacancies.

Figure 2.6 Animated Ion formation.

Sharing Electrons

An atom can get rid of vacancies by participating in a **chemical bond**, which is an attractive force that arises between two atoms when their electrons interact. A **molecule** forms when two or more atoms of the same or different elements join in chemical bonds. The next section explains the main types of bonds in biological molecules.

Compounds are molecules that consist of two or more different elements. The proportions of elements in a molecular substance do not vary. For example, all water molecules have one oxygen atom bonded to two hydrogen atoms. By contrast, a **mixture** is an intermingling of two or more substances. The proportions of substances in a mixture can vary because chemical bonds do not form. For example, sugar may dissolve in water in variable amounts because no chemical bonds form between the two substances. A liquid mixture is called a solution.

---

chemical bond An attractive force that arises between two atoms when their electrons interact.
compound Type of molecule that has atoms of more than one element.
electronegativity Measure of the ability of an atom to pull electrons away from other atoms.
ion Charged atom.
mixture An intermingling of two or more types of molecules.
molecule Group of two or more atoms joined by chemical bonds.
shell model Model of electron distribution in an atom.

Take-Home Message Why do atoms interact?

❯ An atom's electrons are the basis of its chemical behavior.

❯ Shells represent all electron orbitals at one energy level in an atom. When the outermost shell is not full of electrons, the atom has a vacancy.

❯ Atoms tend to get rid of vacancies by gaining or losing electrons (thereby becoming ions), or by sharing electrons with other atoms.

❯ Atoms with vacancies can form chemical bonds. Chemical bonds connect atoms into molecules.

Why Atoms Interact

> The characteristics of a bond arise from the properties of atoms that take part in it.
‹ Link to Molecules 1.2

Arranged in different ways, the same atomic building blocks make different molecules. For example, carbon atoms bonded one way form layered sheets of a soft, slippery substance known as graphite. The same carbon atoms bonded a different way form the rigid crystal lattice of the hardest substance, which is diamond. Carbon bonded to oxygen and hydrogen atoms make sugar.

Although bonding applies to a range of interactions among atoms, we can categorize most bonds into distinct types based on their different properties. Which type forms—an ionic, covalent, or hydrogen bond—depends on the atoms that take part in it.

Ionic Bonds

Remember from the last section that a strongly electronegative atom tends to gain electrons until its outermost shell is full. At that point it will be a negatively charged ion. A weakly electronegative atom tends to lose electrons until its outermost shell is full. At that point it will be a positively charged ion.

An **ionic bond** is a strong mutual attraction of oppositely charged ions. Such bonds do not usually form by the direct transfer of an electron from one atom to another; rather, atoms that have already become ions stay close together because of their opposite charges. Common table salt offers an example. Each crystal of this substance consists of a lattice of sodium and chloride ions interacting in ionic bonds (Figure 2.7).

A Each crystal of table salt is a cubic lattice of many sodium and chloride ions locked in ionic bonds.

ionic bond

Sodium ion
$11p^+$, $10e^-$

Chloride ion
$17p^+$, $18e^-$

B The mutual attraction of opposite charges holds the two kinds of ions together in the lattice.

Figure 2.7 Animated Ionic bonds.

Covalent Bonds

In a **covalent bond**, two atoms share a pair of electrons. Such bonds typically form between atoms with similar electronegativity and unpaired electrons. By sharing their electrons, each atom's vacancy becomes partially filled (Figure 2.8). Covalent bonds can be stronger than ionic bonds, but they are not always so.

Structural formulas show how covalent bonds connect atoms. A line between two atoms represents a single covalent bond, in which two atoms share one pair of electrons. A simple example is molecular hydrogen (H_2), with one covalent bond between hydrogen atoms (H—H).

Many atoms participate in more than one covalent bond at the same time. The oxygen atom in a water molecule (H—O—H) is one example (Table 2.1). Two, three, or even four covalent bonds may form between two atoms when they share multiple pairs of electrons. For example, two atoms sharing two pairs of electrons are connected by two covalent bonds. Such double bonds are represented by a double line between the atoms. A double bond links the two oxygen atoms in molecular oxygen (O=O). Three lines indicate a triple bond, in which two atoms share three pairs of electrons. A triple covalent bond links the two nitrogen atoms in molecular nitrogen (N≡N).

Some covalent bonds are nonpolar, meaning that the atoms participating in the bond are sharing electrons equally. There is no difference in charge between the two ends of such bonds. The molecular hydrogen (H_2), oxygen

Table 2.1 Different Ways To Represent the Same Molecule		
Common name	Water	Familiar term.
Chemical name	Dihydrogen monoxide	Systematically describes elemental composition.
Chemical formula	H_2O	Indicates unvarying proportions of elements. Subscripts show number of atoms of an element per molecule. The absence of a subscript means one atom.
Structural formula	H—O—H H∕O∖H	Represents each covalent bond as a single line between atoms. Bond angles also may be represented.
Structural model		Shows the positions and relative sizes of atoms.
Shell model		Shows how pairs of electrons are shared in covalent bonds.

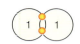

Molecular hydrogen (H — H)
Two hydrogen atoms, each with one proton, share two electrons in a single nonpolar covalent bond.

Molecular oxygen (O = O)
Two oxygen atoms, each with eight protons, share four electrons in a double covalent bond.

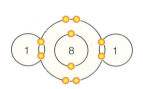

Water molecule (H — O — H)
Two hydrogen atoms share electrons with an oxygen atom in two polar covalent bonds. The oxygen exerts a greater pull on the shared electrons, so it has a slight negative charge. Each hydrogen has a slight positive charge.

Figure 2.8 Animated Covalent bonds, in which atoms with unpaired electrons in their outermost shell become more stable by sharing electrons. Two electrons are shared in each covalent bond. When sharing is equal, the bond is nonpolar. When one atom exerts a greater pull on the electrons, the bond is polar.

water molecule   water molecule

A A hydrogen (H) bond is an attraction between an electronegative atom and a hydrogen atom taking part in a separate polar covalent bond.

B Hydrogen bonds are individually weak, but many of them form. Collectively, they are strong enough to stabilize the structures of large biological molecules such as DNA, shown here.

Figure 2.9 Animated Hydrogen bonds. Hydrogen bonds form at a hydrogen atom taking part in a polar covalent bond. The hydrogen atom's slight positive charge weakly attracts an electronegative atom. As shown here, hydrogen (H) bonds can form between molecules or between different parts of the same molecule.

(O_2), and nitrogen (N_2) mentioned earlier are examples. These molecules are some of the gases that make up air.

Atoms participating in polar covalent bonds do not share electrons equally. Such bonds can form between atoms with a difference in electronegativity. The atom that is more electronegative pulls the electrons a little more toward its "end" of the bond, so that atom bears a slightly negative charge. The atom at the other end of the bond bears a slightly positive charge. The greater the difference in electronegativity between the atoms, the more polar is the covalent bond that forms between them.

For example, a water molecule has two polar covalent bonds. The oxygen atom carries a slight negative charge, and each of the hydrogen atoms carries a slight positive charge. Any such separation of charge into distinct positive and negative regions is called **polarity**. As you will see in the next section, the polarity of the water molecule is very important for the world of life.

Hydrogen Bonds

Hydrogen bonds form between polar regions of two molecules, or between regions of the same molecule.

covalent bond Chemical bond in which two atoms share a pair of electrons.
hydrogen bond Attraction that forms between a covalently bonded hydrogen atom and another atom taking part in a separate covalent bond.
ionic bond Type of chemical bond in which a strong mutual attraction forms between ions of opposite charge.
polarity Any separation of charge into distinct positive and negative regions.

A **hydrogen bond** is a weak attraction between a covalently bonded hydrogen atom and another atom taking part in a separate polar covalent bond. In a hydrogen bond, the atom interacting with the hydrogen is typically an oxygen, nitrogen, or other highly electronegative atom.

Like ionic bonds, hydrogen bonds form by the mutual attraction of opposite charges: The hydrogen atom carries a slight positive charge and the other atom carries a slight negative charge. However, unlike ionic bonds, hydrogen bonds do not make molecules out of atoms, so they are not chemical bonds.

Hydrogen bonds are individually weak. They form and break much more easily than covalent or ionic bonds. Even so, many of them form between molecules, or between different parts of a large one. Collectively, they are strong enough to stabilize the characteristic structures of large biological molecules such as DNA (Figure 2.9).

> Water is essential to life because of its unique properties.
> The unique properties of water are a result of the extensive hydrogen bonding among water molecules.

Life evolved in water. All living organisms are mostly water, many of them still live in it, and all of the chemical reactions of life are carried out in water. What is so special about water?

Each Water Molecule Is Polar

Water's special properties as a liquid begin with the two polar covalent bonds in each water molecule. Overall, the molecule has no charge, but the oxygen pulls the shared electrons a bit more than the hydrogen atoms do. Thus, each of the atoms in a water molecule carries a slight charge. The oxygen atom is slightly negative, and the hydrogen atoms are slightly positive:

slight negative charge

slight positive charge   slight positive charge

The separation of charge means that the water molecule itself is polar. The polarity is very attractive to other water molecules, and hydrogen bonds form between them in tremendous numbers (Figure 2.10). Extensive hydrogen bonding between water molecules imparts unique properties to liquid water that make life possible.

Figure 2.10 Animated Water. Many hydrogen bonds (dashed lines) that quickly form and break keep water molecules clustered together tightly. The extensive hydrogen bonding in liquid water gives it unique properties.

Figure 2.11 Animated Water molecules that surround an ionic solid pull its atoms apart, thereby dissolving them.

Water Is an Excellent Solvent

Water is a **solvent**—a substance, usually a liquid, that can dissolve other substances. When a substance dissolves, its individual molecules or ions become **solutes** as they disperse. Salts, sugars, and other compounds that dissolve easily in water are polar, so many hydrogen bonds form between them and water molecules.

A **salt** is a compound that dissolves easily in water and releases ions other than H^+ and OH^- when it does. Sodium chloride (NaCl) is an example of a salt. Water easily dissolves salts and other **hydrophilic** (water-loving) substances. Hydrogen bonds form between water and the polar molecules of such substances. These bonds dissolve solutes by pulling their molecules away from one another and keeping them apart (Figure 2.11).

You can see how water interacts with **hydrophobic** (water-dreading) substances if you shake a bottle filled with water and salad oil, then set it on a table and watch what happens. Salad oil consists of nonpolar molecules, and hydrogen bonds do not form between nonpolar molecules and water. Shaking breaks some of the hydrogen bonds that keep water molecules together, so the water breaks into small droplets that mix with the oil. However, the water quickly begins to cluster into larger and larger drops as new hydrogen bonds form among its molecules. The bonding excludes molecules of oil and pushes them together into drops that rise to the surface of the water. The same interactions occur at the thin, oily membrane

---

cohesion Tendency of molecules to resist separating from one another.
evaporation Transition of a liquid to a gas.
hydrophilic Describes a substance that dissolves easily in water.
hydrophobic Describes a substance that resists dissolving in water.
salt Compound that releases ions other than H^+ and OH^- when it dissolves in water.
solute A dissolved substance.
solvent Liquid that can dissolve other substances.
temperature Measure of molecular motion.

Figure 2.13 Ice floats on water. *Left*, a covering of ice can insulate water underneath it, thus keeping aquatic organisms from freezing during harsh winters. *Right*, ice forms below about 0°C (32°F), as hydrogen bonds lock water molecules in a rigid, three-dimensional lattice. It floats because the molecules pack less densely than in water.

Figure 2.12 Visible effect of cohesion: a wasp drinking, not sinking. Cohesion imparts surface tension to liquid water, which means that the surface of liquid water behaves a bit like a sheet of elastic.

that separates the watery fluid inside of cells from the watery fluid outside of them. The organization of membranes—and of life—starts with such interactions (you will read more about membranes in Chapter 4).

Cohesion

Another life-sustaining property of water is **cohesion**, which means that water molecules resist separating from one another. Hydrogen bonds collectively exert a continuous pull on individual water molecules. You can see the effect of cohesion as surface tension (Figure 2.12).

Cohesion is an important component of many processes that sustain multicelled bodies. As one example, water molecules constantly escape from the surface of liquid water as vapor, a process called **evaporation**. Evaporation is resisted by the hydrogen bonding that keeps water molecules together. In other words, overcoming water's cohesion takes energy. Thus, evaporation sucks energy in the form of heat from liquid water, lowering the water's surface temperature. Evaporative water loss can help you and some other mammals cool off when you sweat in hot, dry weather. Sweat, which is about 99 percent water, cools the skin as it evaporates.

Cohesion works inside organisms, too. For instance, plants continually absorb water from soil as they grow. Water molecules evaporate from leaves, and replacements are pulled upward from roots. Cohesion makes it possible for columns of liquid water to rise from roots to leaves inside narrow pipelines of vascular tissue. In some trees, these pipelines extend straight up for hundreds of feet.

Water Stabilizes Temperature

Temperature is a way to measure the energy of molecular motion. All molecules jiggle nonstop, and they jiggle faster as they absorb heat. However, extensive hydrogen bonding restricts the movement of water molecules—it keeps them from jiggling as much as they would otherwise. Thus, it takes more heat to raise the temperature of water compared with other liquids. Temperature stability is an important component of homeostasis, because most of the molecules of life function properly only within a certain range of temperature.

Below 0°C (32°F), water molecules do not jiggle enough to break hydrogen bonds, and they become locked in the rigid, lattice-like bonding pattern of ice. Individual water molecules pack less densely in ice than they do in water, so ice floats on water. Sheets of ice that form near the surface of ponds, lakes, and streams can insulate the water under them from subfreezing air temperatures. Such "ice blankets" protect aquatic organisms during extremely cold winters (Figure 2.13).

> pH is a measure of the concentration of hydrogen ions.
> Most biological processes occur within a narrow range of pH, typically around pH 7.
< Link to Homeostasis 1.3

In liquid water, water molecules spontaneously separate into hydrogen ions (H^+) and hydroxide ions (OH^-). These ions can combine again to form water:

$$H_2O \longrightarrow H^+ + OH^- \longrightarrow H_2O$$

water   hydrogen ions   hydroxide ions   water

Concentration refers to the amount of a particular solute that is dissolved in a given volume of fluid. Hydrogen ion concentration is a special case. We measure the number of hydrogen ions in a solution using a value called **pH**. When the number of H^+ ions is the same as the number of OH^- ions, the pH of the solution is 7, or neutral. The pH of pure water (but not rainwater or seawater) is like this. The higher the number of hydrogen ions, the lower the pH. A one-unit decrease in pH corresponds to a tenfold increase in the number of H^+ ions, and a one-unit increase corresponds to a tenfold decrease in the number of H^+ ions (Figure 2.14).

One way to get a sense of pH is to taste dissolved baking soda (pH 9), distilled water (pH 7), and lemon juice (pH 2). Nearly all of life's chemistry occurs near pH 7. Most of your body's internal environment (tissue fluids and blood) stays between pH 7.3 and 7.5.

Substances called **acids** give up hydrogen ions when they dissolve in water, so they lower the pH of fluids and make them acidic (below pH 7). **Bases** accept hydrogen ions, so they can raise the pH of fluids and make them basic, or alkaline (above pH 7).

Acids and bases can be weak or strong. Weak acids are stingy H^+ donors. Strong acids give up more H^+ ions. Hydrochloric acid (HCl) is a strong acid that, when added to water, very easily separates into H^+ and Cl^-:

HCl H^+ Cl^-
hydrochloric acid   hydrogen ions   chloride ions

Inside your stomach, the H^+ from HCl makes gastric fluid acidic (pH 1–2). The acidity activates enzymes that digest proteins in your food.

Most biological molecules can function properly only within a narrow range of pH. Even a slight deviation from that range can halt cellular processes. Under normal circumstances, the fluids inside cells and bodies stay within

Figure 2.14 Animated A pH scale. Here, *red* dots signify hydrogen ions (H^+) and *blue* dots signify hydroxyl ions (OH^-). Also shown are approximate pH values for some common solutions.

This pH scale ranges from 0 (most acidic) to 14 (most basic). A change of one unit on the scale corresponds to a tenfold change in the amount of H^+ ions.

>> **Figure It Out** What is the approximate pH of cola?

Answer: 2.5

acid Substance that releases hydrogen ions in water.
base Substance that accepts hydrogen ions in water.
buffer Set of chemicals that can keep the pH of a solution stable by alternately donating and accepting ions that contribute to pH.
concentration The number of molecules or ions per unit volume of a solution.
pH Measure of the number of hydrogen ions in a fluid.

Figure 2.15 Corrosive effect of acid rain. Airborne pollutants dissolve in water vapor and form compounds that change the pH of rain.

a consistent range of pH because they are buffered. A **buffer** is a set of chemicals, often a weak acid or base and a salt, that can keep the pH of a solution stable. The two chemicals can alternately donate and accept ions that contribute to pH changes.

For example, when a base is added to an unbuffered fluid, the number of OH^- ions increases, so the pH rises. However, if the fluid is buffered, the addition of base causes the buffer to release H^+ ions. These combine with OH^- ions to form water, which has no effect on pH. So a buffered fluid's pH stays the same when base is added.

Carbon dioxide gas becomes a weak acid when it dissolves in the fluid portion of human blood:

$$H_2O \ + \ CO_2 \longrightarrow H_2CO_3$$
$$\text{carbon dioxide} \qquad \text{carbonic acid}$$

Carbonic acid can separate into hydrogen ions and bicarbonate ions, which can in turn recombine to form carbonic acid:

$$H_2CO_3 \longrightarrow H^+ \ + \ HCO_3^- \longrightarrow H_2CO_3$$
$$\text{carbonic acid} \qquad \text{bicarbonate} \qquad \text{carbonic acid}$$

Together, carbonic acid and bicarbonate constitute a buffer. Any excess OH^- in blood combines with the H^+ to form water, which does not contribute to pH. Any excess H^+ in blood combines with the bicarbonate. Thus bonded, the hydrogen does not affect pH. The exchange of ions between carbonic acid and bicarbonate keeps the blood pH between 7.3 and 7.5, but only up to a point. A buffer can neutralize only so many ions. Even slightly more than that limit and the pH of the fluid changes dramatically.

Buffer failure can be catastrophic in a biological system. For example, too much carbonic acid forms in blood when breathing is impaired suddenly. The resulting decline in blood pH may cause an individual to enter a coma, which is a dangerous level of unconsciousness. Hyperventilation (sustained rapid breathing) causes the body to lose too much CO_2. The loss can result in a rise in

Mercury Rising (revisited)

Today, all ecosystems on Earth have detectable effects of air pollution. However, such effects are understudied, so they are probably underestimated. We do know that the amount of mercury in Earth's waters is rising (*inset*). The concentration of mercury in the Pacific Ocean is predicted to double within forty years. We also know that this rise is occurring as a consequence of human activities that release mercury into the atmosphere. In 2005 alone, our activities released more than 2,000 tons of mercury worldwide.

All human bodies, too, now have detectable amounts of mercury. Some of it comes from dental fillings, particularly when bleached. Imported skin-bleaching cosmetics and broken fluorescent lamps also contribute. However, most comes from eating contaminated fish and shellfish. Tuna harvested from the Pacific Ocean accounts for almost half of the mercury in human bodies.

How Would You Vote? A U.S. Food and Drug Administration advisory warns children, and women who are pregnant, nursing, or planning to conceive, to avoid eating tuna because of high mercury levels. The California attorney general requires products that contain dangerous chemicals from non–natural sources to carry warning labels. A typical 6–ounce can of albacore tuna contains about 60 micrograms of mercury. Canned tuna companies say that most of that mercury comes from natural sources, and consider the federal advisory to be sufficient. Do you think cans of tuna should carry labels warning of mercury content? See CengageNow for details, then vote online (cengagenow.com).

blood pH. If blood pH rises too much, prolonged muscle spasm (tetany) or coma may occur.

Burning fossil fuels such as coal releases sulfur and nitrogen compounds that affect the pH of rain and other forms of precipitation. Water is not buffered, so the addition of acids or bases has a dramatic effect. In places with a lot of fossil fuel emissions, the rain and fog can be more acidic than vinegar (Figure 2.15). This acid rain drastically changes the pH of water in soil, lakes, and streams. Such changes can overwhelm the buffering capacity of fluids inside organisms, with lethal effects. We will return to the topic of acid rain in Section 28.9.

Take-Home Message **Why are hydrogen ions important in biological systems?**

> pH reflects the number of hydrogen ions in a fluid. Most biological systems function properly only within a narrow range of pH.

> Acids release hydrogen ions in water; bases accept them. Salts release ions other than H^+ and OH^-.

> Buffers help keep pH stable. Inside organisms, they are part of homeostasis.

Summary

Section 2.1 At life's first level of organization, atoms interact with other atoms to form molecules. The properties of molecules depend on, but differ from, those of their atomic components.

Section 2.2 Atoms consist of **electrons**, which carry a negative **charge**, moving about a **nucleus** of positively charged **protons** and (except for hydrogen) uncharged **neutrons** (Table 2.2). The **periodic table** lists **elements** in order of their **atomic number**. **Isotopes** are atoms of an element that differ in the number of neutrons, so they also differ in **mass number**. Researchers can make **tracers** with **radioisotopes**, which spontaneously emit particles and energy by the process of **radioactive decay**.

Table 2.2 Players in the Chemistry of Life

Atoms	Particles that are basic building blocks of all matter; the smallest unit that retains an element's properties.
Proton (p^+)	Positively charged particle of an atom's nucleus.
Electron (e^-)	Negatively charged particle that can occupy a volume of space (orbital) around an atom's nucleus.
Neutron	Uncharged particle of an atom's nucleus.
Element	Pure substance that consists entirely of atoms with the same, characteristic number of protons.
Isotopes	Atoms of an element that differ in the number of neutrons.
Radioisotope	Unstable isotope that emits particles and energy when its nucleus disintegrates.
Tracer	Molecule that has a detectable substance (such as a radioisotope) attached. Used to track the movement or destination of the molecule in a biological system.
Ion	Atom that carries a charge after it has gained or lost one or more electrons. A single proton without an electron is a hydrogen ion (H^+).
Molecule	Two or more atoms joined in a chemical bond.
Compound	Molecule of two or more different elements in unvarying proportions (for example, water).
Mixture	Intermingling of two or more elements or compounds in proportions that can vary.
Solute	Molecule or ion dissolved in some solvent.
Hydrophilic	Refers to a substance that dissolves easily in water. Such substances consist of polar molecules.
Hydrophobic	Refers to a substance that resists dissolving in water. Such substances consist of nonpolar molecules.
Acid	Compound that releases H^+ when dissolved in water.
Base	Compound that accepts H^+ when dissolved in water.
Salt	Compound that releases ions other than H^+ or OH^- when dissolved in water.
Solvent	Substance that can dissolve other substances.

Section 2.3 Up to two electrons occupy each orbital (volume of space) around a nucleus. Which orbital an electron occupies depends on its energy. A **shell model** represents successive energy levels as concentric circles.

Atoms fill vacancies by gaining or losing electrons, or by sharing electrons with other atoms. **Electronegativity** is a measure of how strongly an atom attracts electrons from other atoms. **Ions** are charged atoms. A **chemical bond** is an attractive force that unites two atoms as a **molecule**. A **compound** is a molecule that consists of two or more elements. A **mixture** is an intermingling of substances.

Section 2.4 Atoms form different types of bonds depending on their electronegativity. An **ionic bond** is a strong association between oppositely charged ions; it arises from the mutual attraction of opposite charges. Atoms share a pair of electrons in a **covalent bond**, which is nonpolar if the sharing is equal, and polar if it is not. A molecule that has a separation of charge is said to show **polarity**. **Hydrogen bonds** collectively stabilize the structures of large molecules.

Section 2.5 Polar covalent bonds join two hydrogen atoms to one oxygen atom in each water molecule. The polarity invites extensive hydrogen bonding between water molecules, and this bonding is the basis of unique properties that sustain life: a capacity to act as a **solvent** for **salts** and other polar **solutes**; resistance to **temperature** changes; and **cohesion**. **Hydrophilic** substances dissolve easily in water; **hydrophobic** substances do not. **Evaporation** is the transition of a liquid to a gas.

Section 2.6 A solute's **concentration** refers to the amount of solute in a given volume of fluid. **pH** reflects the number of hydrogen ions (H^+) in a fluid. At neutral pH (7), the amounts of H^+ and OH^- ions are the same. **Acids** release hydrogen ions in water; **bases** accept them. A **buffer** keeps a solution within a consistent range of pH. Most cell and body fluids are buffered because most molecules of life work only within a narrow range of pH.

Self-Quiz   Answers in Appendix III

1. Is this statement true or false? All atoms consist of electrons, protons, and neutrons.

2. In the periodic table, symbols for the elements are arranged according to _____ .
   a. size
   b. charge
   c. mass number
   d. atomic number

3. A(n) _____ is a molecule into which a radioisotope has been incorporated.
   a. compound
   b. tracer
   c. salt
   d. acid

Data Analysis Activities

Mercury Emissions By weight, coal does not contain much mercury, but we burn a lot of it. In addition to coal-fired power plants, several other industries contribute substantially to atmospheric mercury pollution. Figure 2.16 shows mercury emissions from different regions of the world in the year 2006.

1. About how many tons of mercury were released worldwide in 2006?
2. Which industry tops the list of mercury-pollution offenders? Which industry is next on the list?
3. Which region emits the most mercury from producing cement?
4. About how many tons of mercury were released from gold production in South America?

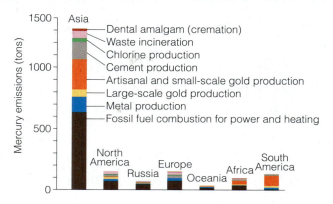

Figure 2.16 Global mercury emissions, 2006. *Source: Global Atmospheric Mercury Assessment: Sources, Emissions and Transport. United Nations Environmental Programme, Chemicals Branch. 2008*

---

4. An ion is an atom that has _____ .
   a. the same number of electrons and protons
   b. different numbers of electrons and protons

5. The measure of an atom's ability to pull electrons away from another atom is called _____ .
   a. electronegativity   c. charge
   b. polarity   d. concentration

6. The mutual attraction of opposite charges holds atoms together as molecules in a(n) _____ bond.
   a. ionic   c. polar covalent
   b. hydrogen   d. nonpolar covalent

7. Atoms share electrons unequally in a(n) _____ bond.

8. A(n) _____ substance repels water.
   a. acidic   c. hydrophobic
   b. basic   d. polar

9. A salt releases ions other than _____ in water.

10. Hydrogen ions (H^+) are _____ .
    a. indicated by a pH scale c. in blood
    b. unbound protons   d. all of the above

11. When dissolved in water, a(n) _____ donates H^+; a(n) _____ accepts H^+.
    a. acid; base   c. buffer; solute
    b. base; acid   d. base; buffer

12. A(n) _____ is a chemical partnership between a weak acid or base and its salt.
    a. covalent bond   c. buffer
    b. hydrogen bond   d. pH

13. A(n) _____ is dissolved in a solvent.
    a. molecule   c. salt
    b. solute   d. acid

14. Match the terms with their most suitable description.
    ___ hydrophilic   a. protons > electrons
    ___ atomic number   b. number of protons in nucleus
    ___ charged atom   c. polar; easily dissolves in water
    ___ mass number   d. ion
    ___ temperature   e. protons < electrons
    ___ uncharged   f. protons = electrons
    ___ negative charge g. measure of molecular motion
    ___ positive charge h. number of protons and
                            neutrons in atomic nucleus

Additional questions are available on **CENGAGENOW**.

Critical Thinking

1. Alchemists were medieval scholars and philosophers who were the forerunners of modern-day chemists. Many spent their lives trying to transform lead (atomic number 82) into gold (atomic number 79). Explain why they never did succeed in that endeavor.

2. Draw a shell model of an uncharged nitrogen atom (nitrogen has 7 protons).

3. Polonium is a rare element with 33 radioisotopes. The most common one, ^{210}Po, has 82 protons and 128 neutrons. When ^{210}Po decays, it emits an alpha particle, which is a helium nucleus (2 protons and 2 neutrons). ^{210}Po decay is tricky to detect because alpha particles do not carry very much energy compared to other forms of radiation. They can be stopped by, for example, a sheet of paper or a few inches of air. That is one reason that authorities failed to discover toxic amounts of ^{210}Po in the body of former KGB agent Alexander Litvinenko until after he died suddenly and mysteriously in 2006. What element does an atom of ^{210}Po change into after it emits an alpha particle?

4. Some undiluted acids are not as corrosive as when they are diluted with water. That is why lab workers are told to wipe off splashes with a towel before washing. Explain.

Animations and Interactions on **CENGAGENOW**:
❯ Isotopes; PET scans; Shell models; Types of bonds; Ions and ionic bonds; Hydrogen bonds; Water; Dissolution; pH.

❮ Links to Earlier Concepts

Having learned about atomic interactions (Section 2.3), you are now in a position to understand the structure of the molecules of life. Keep the big picture in mind by reviewing Section 1.2. You will be building on your knowledge of the nature of covalent bonding (2.4), acids and bases (2.6), and the effects of hydrogen bonds (2.4) as you learn about how the molecules of life are put together.

Key Concepts

Structure Dictates Function

We define cells partly by their capacity to build complex carbohydrates and lipids, proteins, and nucleic acids. All of these organic compounds have functional groups attached to a backbone of carbon atoms.

Carbohydrates

Carbohydrates are the most abundant biological molecules. They function as energy reservoirs and structural materials. Different types of carbohydrates are built from the same sugars, bonded in different patterns.

3 Molecules of Life

3.1 Fear of Frying

The human body requires only about a tablespoon of fat each day to stay healthy, but most people in developed countries eat far more than that. The average American eats about 70 pounds of fat per year, which may be part of the reason why the average American is overweight. Being overweight increases one's risk for many health conditions. However, the total quantity of fat may be less important than the kinds of fats we eat. Fats are more than just inert molecules that accumulate in strategic areas of our bodies. They are major constituents of cell membranes, and as such they have powerful effects on cell function.

The typical fat molecule has three fatty acids, each with a long chain of carbon atoms. Different fats consist of different fatty acids. Fats with a certain arrangement of hydrogen atoms around that carbon chain are called *trans* fats (Figure 3.1). Small amounts of *trans* fats occur naturally in red meat and dairy products, but most of the *trans* fats that humans eat come from partially hydrogenated vegetable oil, an artificial food product.

Hydrogenation, a manufacturing process that adds hydrogen atoms to a substance, changes liquid vegetable oils into solid fats. Procter & Gamble Co. developed partially hydrogenated soybean oil in 1908 as a substitute for the more expensive solid animal fats they had been using to make candles. However, the demand for candles began to wane as more households in the United States became wired for electricity, and P & G began to look for another way to sell its proprietary fat. Partially hydrogenated vegetable oil looks a lot like lard, so in 1911 the company began aggressively marketing it as a revolutionary new food: a solid cooking fat with a long shelf life, mild flavor, and lower cost than lard or butter.

By the mid-1950s, hydrogenated vegetable oil had become a major part of the American diet. It was (and still is) found in many manufactured and fast foods: french fries, butter substitutes, cookies, crackers, cakes and pancakes, peanut butter, pies, doughnuts, muffins, chips, granola bars, breakfast bars, choco-late, microwave popcorn, pizzas, burritos, chicken nuggets, fish sticks, and so on.

For decades, hydrogenated vegetable oil was considered more healthy than animal fats because it was made from plants, but we now know otherwise. The *trans* fats in hydrogenated vegetable oils raise the level of cholesterol in our blood more than any other fat, and they directly alter the function of our arteries and veins. The effects of such changes are quite serious. Eating as little as 2 grams a day of hydrogenated vegetable oils increases a person's risk of atherosclerosis (hardening of the arteries), heart attack, and diabetes. A small serving of french fries made with hydrogenated vegetable oil contains about 5 grams of *trans* fat.

All organisms consist of the same kinds of molecules, but small differences in the way those molecules are put together can have big effects in a living organism. With this concept, we introduce you to the chemistry of life. This is your chemistry. It makes you far more than the sum of your body's molecules.

oleic acid (a *trans* fatty acid)

elaidic acid (a *cis* fatty acid)

Figure 3.1 *Trans* fats, an unhealthy food. The arrangement of hydrogen atoms around a double bond (in *red*) is what makes a fat *trans* (*top*) or *cis* (*bottom*). This seemingly small difference in structure makes a big difference in our bodies.

Lipids
Lipids function as energy reservoirs and as waterproofing or lubricating substances. Some are remodeled into other compounds such as vitamins. Lipids are the main structural component of all cell membranes.

Proteins
Structurally and functionally, proteins are the most diverse molecules of life. They include enzymes and structural materials. A protein's function arises from and depends on its structure.

Nucleic Acids
Nucleotides are the building blocks of nucleic acids; some have additional roles in metabolism. DNA and RNA are part of a cell's system of storing and retrieving heritable information.

> All of the molecules of life are built with carbon atoms.
> The function of organic molecules in biological systems begins with their structure.
< Links to Elements 2.2, Ions 2.3, Covalent bonds 2.4, Polarity 2.4, Acids and bases 2.6

Carbon accounts for a high proportion of the elements in living things, mainly because the molecules of life—complex carbohydrates and lipids, proteins, and nucleic acids—are organic. **Organic** compounds consist primarily of carbon and hydrogen atoms. The term is a holdover from a time when such molecules were thought to be made only by living things, as opposed to "inorganic" molecules that formed by nonliving processes. The term persists, even though we now know that organic compounds were present on Earth long before organisms were.

Carbon's importance to life starts with its versatile bonding behavior: Each carbon atom can form covalent bonds with up to four other atoms. Most organic compounds have a backbone, or chain, of carbon atoms, that may include rings. Using different models to represent organic compounds allows us to visualize different aspects of their structure.

Structural formulas show how all of the atoms connect (Figure 3.2A). Some atoms or bonds may be implied but not shown (Figure 3.2B). For further simplification, carbon ring structures are often represented as polygons that imply atoms at their corners (Figure 3.2C).

Molecular models show the positions of atoms in three dimensions. Atoms in such models are typically represented by colored balls, the size of which reflects relative atomic size. Ball-and-stick models show covalent bonds as a stick (Figure 3.2D). Space-filling models show overall shape (Figure 3.2E). To reduce visual complexity, other types of models omit individual atoms. Such models can reveal large-scale features, such as folds or pockets, that can be difficult to see when individual atoms are shown. For example, very large molecules are often shown as ribbons, which highlight structural features such as coils (see Figure 3.14 for an example).

Group	Character	Location	Structure		
hydroxyl	polar	amino acids; sugars, alcohols	—OH		
methyl	nonpolar	fatty acids, some amino acids	H | —C—H | H		
carbonyl	polar, reactive	sugars, amino acids, nucleotides	—C—H || O (aldehyde)	—C— || O (ketone)	
carboxyl	acidic	fatty acids, amino acids, carbohydrates	—C—OH || O	—C—O⁻ || O (ionized)	
amine	basic	amino acids, some nucleotide bases	—N—H | H	H | —NH⁺ | H (ionized)	
phosphate	high energy, polar	nucleotides (e.g., ATP); DNA, RNA; phospholipids, many proteins	O⁻ | —O—P—O⁻ || O	Ⓟ icon	
sulfhydryl	forms disulfide bridges	cysteine (an amino acid)	—SH	—S—S— (disulfide bridge)	

Figure 3.3 Animated Common functional groups. Such groups impart specific chemical characteristics to organic compounds.

Functional Groups

An organic molecule that consists only of hydrogen and carbon atoms is called a **hydrocarbon**. Methane, the simplest hydrocarbon, is one carbon atom bonded to four hydrogen atoms. Most of the molecules of life have at least one **functional group**, which is a cluster of atoms covalently bonded to a carbon atom of an organic molecule. Functional groups impart specific chemical proper-

A glucose   **B** glucose   **C** glucose

D glucose   **E** glucose

Figure 3.2 Modeling an organic molecule. **A** A structural formula shows atoms and bonds. **B,C** Structural formulas are often abbreviated to omit labels for some atoms such as the carbons at the corners of ring structures. **D** A ball-and-stick model shows the arrangement of atoms in three dimensions. **E** A space-filling model shows a molecule's overall shape. *Right*, typical color code for elements in molecular models.

carbon   hydrogen   oxygen   nitrogen   phosphorus

ties such as polarity or acidity (Figure 3.3). The chemical behavior of the molecules of life arises largely from the number, kind, and arrangement of their functional groups. For example, hydroxyl groups (—OH) impart polarity to alcohols. Thus, alcohols (at least the small ones) dissolve quickly in water. Larger alcohols do not dissolve as easily because their long, hydrophobic hydrocarbon chains repel water.

What Cells Do to Organic Compounds

All biological systems are based on the same organic molecules, a similarity that is one of many legacies of life's common origin. However, the details of those molecules differ among organisms. Simple organic building blocks bonded in different numbers and arrangements form different versions of the molecules of life, just as atoms bonded in different numbers and arrangements form different molecules. Cells maintain reserves of small organic molecules that they can assemble into complex carbohydrates, lipids, proteins, and nucleic acids. When used as subunits of larger molecules, the small organic molecules (simple sugars, fatty acids, amino acids, and nucleotides) are called **monomers**. Molecules that consist of multiple monomers are called **polymers**.

Cells build polymers from monomers, and break down polymers to release monomers. **Metabolism** refers to activities by which cells acquire and use energy as they make and break apart organic compounds. These activities help cells stay alive, grow, and reproduce. Metabolism also requires **enzymes**, which are organic molecules that speed up reactions without being changed by them.

Table 3.1 lists some common metabolic reactions. For now, start thinking about two types of reactions. Large organic molecules are often built from smaller ones by **condensation**, a process in which an enzyme covalently bonds two molecules together. Water usually forms as a product of condensation when a hydroxyl group (—OH)

---

condensation Process by which enzymes build large molecules from smaller subunits; water also forms.
enzyme Compound (usually a protein) that speeds a reaction without being changed by it.
functional group A group of atoms bonded to a carbon of an organic compound; imparts a specific chemical property.
hydrocarbon Compound or region of one that consists only of carbon and hydrogen atoms.
hydrolysis Process by which an enzyme breaks a molecule into smaller subunits by attaching a hydroxyl group to one part and a hydrogen atom to the other.
metabolism All the enzyme-mediated chemical reactions by which cells acquire and use energy as they build and break down organic molecules.
monomers Molecules that are subunits of polymers.
organic Type of compound that consists primarily of carbon and hydrogen atoms.
polymer Molecule that consists of multiple monomers.

Table 3.1 What Cells Do to Organic Compounds

Type of Reaction	What Happens
Condensation	Two molecules covalently bond and become a larger molecule.
Hydrolysis	A molecule splits into two smaller molecules.
Functional group transfer	A functional group is transferred from one molecule to another.
Electron transfer	One molecule accepts electrons from another.
Rearrangement	Juggling of covalent bonds converts one organic compound into another.

A Condensation. Cells build a large molecule from smaller ones by this reaction. An enzyme removes a hydroxyl group from one molecule and a hydrogen atom from another. A covalent bond forms between the two molecules, and water also forms.

B Hydrolysis. Cells split a large molecule into smaller ones by this water-requiring reaction. An enzyme attaches a hydroxyl group and a hydrogen atom (both from water) at the cleavage site.

Figure 3.4 Animated Two common metabolic processes by which cells build and break down organic molecules.

from one of the molecules combines with a hydrogen atom from the other molecule (Figure 3.4A). **Hydrolysis**, which is the reverse of condensation, breaks apart large organic molecules into smaller ones (Figure 3.4B). Hydrolysis enzymes break a bond by attaching a hydroxyl group to one atom and a hydrogen atom to the other. The —OH and —H come from a water molecule, so this reaction requires water.

Take-Home Message How do organic molecules function in living systems?

> The molecules of life are organic, which means they consist mainly of carbon and hydrogen atoms.

> Functional groups impart certain chemical characteristics to organic molecules. Such groups contribute to the particular function of a biological molecule.

> Cells assemble large polymers from smaller monomers. They also break apart polymers into component monomers.

> Carbohydrates are the most plentiful biological molecules.
> Cells use some carbohydrates as structural materials; they use others for fuel, or to store or transport energy.
< Link to Hydrogen bonds 2.4

Carbohydrates are organic compounds that consist of carbon, hydrogen, and oxygen in a 1:2:1 ratio. Cells use different kinds as structural materials, for fuel, and for storing and transporting energy. The three main types of carbohydrates in living systems are monosaccharides, oligosaccharides, and polysaccharides.

Simple Sugars

"Saccharide" is from *sacchar*, a Greek word that means sugar. Monosaccharides (one sugar unit) are the simplest type of carbohydrate, but they have extremely important roles as components of larger molecules. Common monosaccharides have a backbone of five or six carbon atoms, one carbonyl group, and two or more hydroxyl groups. Enzymes can easily break the bonds of monosaccharides to release energy (we will return to carbohydrate metabolism in Chapter 7). The solubility of these molecules also means that they move easily throughout the water-based internal environments of all organisms.

Monosaccharides that are components of the nucleic acids DNA and RNA have five carbon atoms. Glucose (at *left*) has six. Glucose can be used as a fuel to drive cellular processes, or as a structural material to build larger molecules. It can also be used as a precursor, or parent molecule, that is remodeled into other molecules. For example, cells of plants and many animals make vitamin C from glucose (human cells cannot, so we need to get our vitamin C from food).

Short–Chain Carbohydrates

An oligosaccharide is a short chain of covalently bonded monosaccharides (*oligo*– means a few). Disaccharides consist of two sugar monomers. The lactose in milk is a disaccharide, with one glucose and one galactose unit. Sucrose, the most plentiful sugar in nature, has a glucose and a fructose unit (Figure 3.5). Sucrose extracted from sugarcane or sugar beets is our table sugar. Oligosaccharides with three or more sugar units are often attached to lipids or proteins that have important functions in immunity.

Complex Carbohydrates

The "complex" carbohydrates, or polysaccharides, are straight or branched chains of many sugar monomers—often hundreds or thousands of them. There may be one type or many types of monomers in a polysaccharide.

The most common polysaccharides are cellulose, glycogen, and starch. All consist of glucose monomers, but they differ dramatically in their chemical properties. Why? The answer begins with differences in patterns of covalent bonding that link their glucose monomers.

Cellulose, the major structural material of plants, is the most abundant biological molecule in the biosphere. It consists of long, straight chains of glucose monomers. Hydrogen bonds lock the chains into tight, sturdy bundles (Figure 3.6A). In plants, these tough cellulose fibers act like reinforcing rods that help stems resist wind and other forms of mechanical stress.

Cellulose does not dissolve in water, and it is not easily broken down. Some bacteria and fungi make enzymes that break it apart into its component sugars, but humans and other mammals do not. When we talk about dietary fiber, or "roughage," we are usually referring to the cellulose and other indigestible polysaccharides in our vegetable foods. Bacteria that live in the gut of termites and grazers such as cattle and sheep help these animals digest the cellulose in plants.

In starch, a different covalent bonding pattern between glucose monomers makes a chain that coils up into a spiral (Figure 3.6B). Starch is not as stable as cellulose, and it does not dissolve easily in water. Both properties make the molecule ideal for storing chemical energy in the

glucose   +   fructose   ⟶   sucrose   +   water

Figure 3.5 Animated The synthesis of a sucrose molecule is an example of a condensation reaction. You are already familiar with sucrose—it is common table sugar.

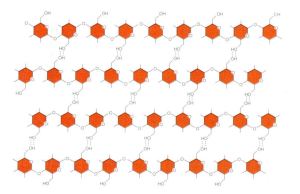

A Cellulose, a structural component of plants. Chains of glucose units stretch side by side and hydrogen bond at many —OH groups. The hydrogen bonds stabilize the chains in tight bundles that form long fibers. Very few types of organisms can digest this tough, insoluble material.

B In amylose, one type of starch, a series of glucose units form a chain that coils. Starch is the main energy reserve in plants, which store it in their roots, stems, leaves, fruits, and seeds (such as coconuts).

C Glycogen. In humans and other animals, this polysaccharide functions as an energy reservoir. It is stored in muscles and in the liver.

Figure 3.6 Structure of **A** cellulose, **B** starch, and **C** glycogen, and their typical locations in a few organisms. All three carbohydrates consist only of glucose units, but the different bonding patterns that link the subunits result in substances with very different properties.

watery, enzyme-filled interior of plant cells. Most plants make much more glucose than they can use. The excess is stored as starch inside cells that make up roots, stems, and leaves. However, because it is insoluble, starch cannot be transported out of the cells and distributed to other parts of the plant. When sugars are in short supply, hydrolysis enzymes break the bonds between starch's monomers to release glucose subunits. Humans also have enzymes that hydrolyze starch, so this carbohydrate is an important component of our food.

The covalent bonding pattern in glycogen forms highly branched chains of glucose monomers (Figure 3.6C). In animals, glycogen is the sugar-storage equivalent of starch in plants. Muscle and liver cells store it in reserve to meet a sudden need for glucose. When the sugar level in blood falls, liver cells break down stored glycogen, and the released glucose subunits enter the blood.

---

carbohydrate Molecule that consists primarily of carbon, hydrogen, and oxygen atoms in a 1:2:1 ratio.

Figure 3.7 Chitin. This polysaccharide strengthens the hard parts of many small animals, such as crabs.

Chitin is a polysaccharide similar to cellulose. Its monomers are glucose with a nitrogen-containing carbonyl group (Figure 3.7). Long, unbranching chains of these monomers are linked by hydrogen bonds. As a structural material, chitin is durable, translucent, and flexible. It strengthens hard parts of many animals, including the outer cuticle of crabs, beetles, and ticks, and it reinforces the cell wall of many fungi.

Take-Home Message What are carbohydrates?

> Subunits of simple carbohydrates (sugars), arranged in different ways, form various types of complex carbohydrates.

> Cells use carbohydrates for energy or as structural materials.

> Lipids are hydrophobic organic compounds.
> Common lipids include triglycerides, phospholipids, waxes, and steroids.

Lipids are fatty, oily, or waxy organic compounds. They vary in structure, but all are hydrophobic. Many lipids incorporate **fatty acids**, which are small organic molecules that consist of a hydrocarbon "tail" topped with a carboxyl group "head" (Figure 3.8). The tail of a fatty acid is hydrophobic (or fatty, hence the name), but the carboxyl group (the "acid" part of the name) makes the head hydrophilic. You are already familiar with the properties of fatty acids because these molecules are the main component of soap. The hydrophobic tails attract oily dirt, and the hydrophilic heads dissolve the dirt in water.

Fatty acids can be saturated or unsaturated. Saturated types have only single bonds in their tails. In other words, their carbon chains are fully saturated with hydrogen atoms (Figure 3.8A). Saturated fatty acid tails are flexible and they wiggle freely. The tails of unsaturated fatty acids have one or more double bonds that limit their flexibility (Figure 3.8B,C). These bonds are termed *cis* or *trans*, depending on the way the hydrogens are arranged around them. You can see in Figure 3.1 how a *cis* bond kinks the tail, and a *trans* bond keeps it straight.

Fats

The carboxyl group of a fatty acid easily forms bonds with other molecules. **Fats** are lipids with one, two, or three fatty acids bonded to a small alcohol called glycerol. A fatty acid attaches to a glycerol via its carboxyl group

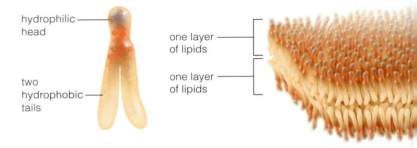

Figure 3.10 Phospholipids as components of cell membranes. *Left*, the head of a phospholipid is hydrophilic, and the tails are hydrophobic. *Right*, a double layer of phospholipids—the lipid bilayer—is the structural foundation of all cell membranes.

head. When it does, the fatty acid loses its hydrophilic character. When three fatty acids attach to a glycerol, the resulting molecule, which is called a **triglyceride**, is entirely hydrophobic (Figure 3.9A).

Because they are hydrophobic, triglycerides do not dissolve easily in water. Most "neutral" fats, such as butter and vegetable oils, are like this. Triglycerides are the most abundant and richest energy source in vertebrate bodies. They are concentrated in adipose tissue that insulates and cushions body parts.

Animal fats are saturated, which means they consist mainly of triglycerides with three saturated fatty acid tails. Saturated fats tend to remain solid at room temperature because their floppy saturated tails can pack

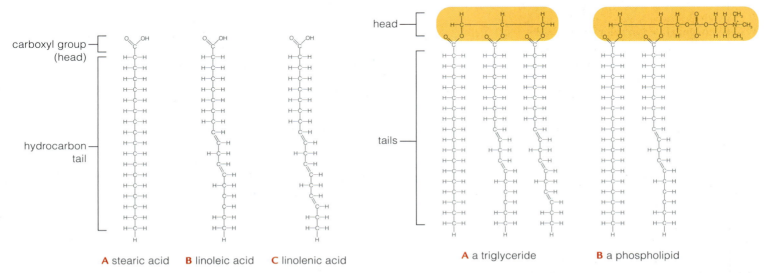

Figure 3.8 Fatty acids. **A** The tail of stearic acid is fully saturated with hydrogen atoms. **B** Linoleic acid, with two double bonds, is unsaturated. The first double bond occurs at the sixth carbon from the end, so linoleic acid is called an omega-6 fatty acid. Omega-6 and **C** omega-3 fatty acids are "essential fatty acids." Your body does not make them, so they must come from food.

Figure 3.9 Animated Lipids with fatty acid tails. **A** Fatty acid tails of a triglyceride are attached to a glycerol head. **B** Fatty acid tails of a phospholipid are attached to a phosphate-containing head.

>> **Figure It Out** Is the triglyceride saturated or unsaturated?

Answer: Unsaturated

an estrogen   testosterone

female wood duck   male wood duck

Figure 3.11 Estrogen and testosterone, steroid hormones that cause different traits to arise in males and females of many species such as wood ducks (*Aix sponsa*), pictured at *right*.

tightly together. Most vegetable oils are unsaturated, which means these fats consist mainly of triglycerides with one or more unsaturated fatty acid tails. Kinky tails do not pack tightly, so unsaturated fats are typically liquid at room temperature. The partially hydrogenated vegetable oils that you learned about in Section 3.1 are an exception. They are solid at room temperature. The special *trans* double bond that keeps fatty acid tails straight allows them to pack tightly, just like saturated fats do.

Phospholipids

A **phospholipid** has two fatty acid tails and a head that contains a phosphate group (Figure 3.9B). The tails are hydrophobic, but the highly polar phosphate group makes the head very hydrophilic. The opposing properties of a phospholipid molecule give rise to cell membrane structure. Phospholipids are the most abundant lipids in cell membranes, which have two layers of lipids (Figure 3.10). The heads of one layer are dissolved in the cell's watery interior, and the heads of the other layer are dissolved in the cell's fluid surroundings. In such lipid bilayers, all of the hydrophobic tails are sandwiched between the hydrophilic heads. You will read more about the structure of cell membranes in Chapter 4.

fat Lipid that consists of a glycerol molecule with one, two, or three fatty acid tails.
fatty acid Organic compound that consists of a chain of carbon atoms with an acidic carboxyl group at one end. Carbon chain of saturated types has single bonds only; that of unsaturated types has one or more double bonds.
lipid Fatty, oily, or waxy organic compound.
phospholipid A lipid with a phosphate group in its hydrophilic head, and two nonpolar fatty acid tails; main constituent of eukaryotic cell membranes.
steroid Type of lipid with four carbon rings and no fatty acid tails.
triglyceride A fat with three fatty acid tails.
wax Water-repellent mixture of lipids with long fatty acid tails bonded to long-chain alcohols or carbon rings.

Waxes

A **wax** is a complex, varying mixture of lipids with long fatty acid tails bonded to long-chain alcohols or carbon rings. The molecules pack tightly, so the resulting substance is firm and water-repellent. A layer of secreted waxes that covers the exposed surfaces of plants helps restrict water loss and keep out parasites and other pests. Other types of waxes protect, lubricate, and soften skin

and hair. Waxes, together with fats and fatty acids, make feathers waterproof. Bees store honey and raise new generations of bees inside honeycomb made from wax that they secrete.

Steroids

Steroids are lipids with a rigid backbone of four carbon rings and no fatty acid tails. All eukaryotic cell membranes contain them. Cholesterol, the most common steroid in animal tissue, is also a starting material that cells remodel into many molecules, such as bile salts (which help digest fats) and vitamin D (required to keep teeth and bones strong). Steroid hormones are also derived from cholesterol. Estrogens and testosterone, hormones that govern reproduction and secondary sexual traits, are steroid hormones (Figure 3.11).

Take-Home Message **What are lipids?**

> Lipids are fatty, waxy, or oily organic compounds. Common types include fats, phospholipids, waxes, and steroids.

> Triglycerides are lipids that serve as energy reservoirs in vertebrate animals.

> Phospholipids are the main lipid component of cell membranes.

> Waxes are lipid components of water-repelling and lubricating secretions.

> Steroids are lipids that occur in cell membranes. Some are remodeled into other molecules.

> ❯ Proteins are the most diverse biological molecule. All cellular processes involve them.
> ❯ Cells build thousands of different types of proteins by stringing together amino acids in different orders.
> ❮ Link to Covalent bonding 2.4

Of all biological molecules, proteins are the most diverse in both structure and function. Structural proteins support cell parts and, as part of tissues, multicelled bodies. Spiderwebs, feathers, hooves, and hair, as well as bones and other body parts, consist mainly of structural proteins. A tremendous number of different proteins, including some structural types, actively participate in all processes that sustain life. Most enzymes that drive metabolic reactions are proteins. Proteins move substances, help cells communicate, and defend the body.

Amino Acids

Amazingly, cells can make all of the thousands of different kinds of proteins they need from only twenty kinds of monomers called amino acids. **Proteins** are polymers of amino acids. An **amino acid** is a small organic compound with an amine group, a carboxyl group (the acid), and one or more atoms called an "R group." In most amino acids, all three groups are attached to the same carbon atom (Figure 3.12). The structures of the twenty amino acids used in eukaryotic proteins are shown in Appendix V.

Figure 3.12 Generalized structure of amino acids. Appendix V has models of all twenty of the amino acids used in eukaryotic proteins.

Building Proteins

Protein synthesis involves covalently bonding amino acids into a chain. For each type of protein, instructions coded in DNA specify the order in which any of the twenty kinds of amino acids will occur at every place in the chain.

During protein synthesis, the amine group of one amino acid becomes bonded to the carboxyl group of the next. The bond that forms between the two amino acids is called a **peptide bond** (Figure 3.13). Enzymes repeat this bonding process hundreds or thousands of times, so a long chain of amino acids (a **polypeptide**) forms (Figure 3.14). The linear sequence of amino acids in the polypeptide is called the protein's primary structure ❶. You will learn more about protein synthesis in Chapter 9.

Protein Structure

You and all other organisms depend on working proteins: enzymes that speed metabolic processes, receptors that receive signals, hemoglobin that carries oxygen in your red blood cells, and so on. One of the fundamental ideas in biology is that structure dictates function. This idea is particularly appropriate as applied to proteins, because the shape of a protein defines its biological activity.

There are several levels of protein structure beyond amino acid sequence. Even before a polypeptide is finished being synthesized, it begins to twist and fold as hydrogen bonds form among the amino acids of the chain. This hydrogen bonding may cause parts of the polypeptide to form flat sheets or coils (helices), patterns that constitute a protein's secondary structure ❷. The primary structure of each type of protein is unique, but most proteins have sheets and coils.

Much as an overly twisted rubber band coils back upon itself, hydrogen bonding between different parts of a protein make it fold up even more into compact domains. A domain is a part of a protein that is organized

Figure 3.13 Animated Polypeptide formation. Chapter 9 offers a closer look at protein synthesis.

A Two amino acids (here, methionine and serine) are joined by condensation. A peptide bond forms between the carboxyl group of the methionine and the amine group of the serine.

B One by one, additional amino acids are added to the carboxyl end of the chain. The resulting polypeptide can be thousands of amino acids long.

| lysine | – | glycine | – | glycine | – | arginine |

1 A protein's primary structure consists of a linear sequence of amino acids (a polypeptide chain). Each type of protein has a unique primary structure.

Figure 3.14 Animated Protein structure.

2 Secondary structure arises as a polypeptide chain twists into a coil (helix) or sheet held in place by hydrogen bonds between different parts of the molecule. The same patterns of secondary structure occur in many different proteins.

3 Tertiary structure occurs when a chain's coils and sheets fold up into a functional domain such as a barrel or pocket. In this example, the coils of a globin chain form a pocket.

4 Some proteins have quaternary structure, in which two or more polypeptide chains associate as one molecule. Hemoglobin, shown here, consists of four globin chains (*green* and *blue*). Each globin pocket now holds a heme group (*red*).

as a structurally stable unit. Such units are part of a protein's overall three-dimensional shape, or tertiary structure **3**.

Tertiary structure is what makes a protein a working molecule. For example, the sheets of some proteins curl up into a barrel shape. A barrel domain often functions as a tunnel for small molecules, allowing them to pass, for example, through a cell membrane. Globular domains of enzymes form chemically active pockets that can make or break bonds of other molecules.

Many proteins also have quaternary structure, which means they consist of two or more polypeptide chains that are in close association or covalently bonded together. Most enzymes and many other proteins consist of two or more polypeptide chains that collectively form a roughly spherical shape **4**.

Some proteins aggregate by many thousands into much larger structures, with their polypeptide chains organized into strands or sheets. The keratin in your hair is an example **5**. Some fibrous proteins contribute to the structure and organization of cells and tissues. Others, such as the actin and myosin filaments in muscle cells, are part of the mechanisms that help cells, cell parts, and bodies move.

Enzymes often attach sugars or lipids to proteins. A glycoprotein forms when oligosaccharides are attached to

5 Many proteins aggregate by the thousands into much larger structures, such as the keratin filaments that make up hair.

a polypeptide. The molecules that allow a tissue or a body to recognize its own cells are glycoproteins, as are other molecules that help cells interact in immunity.

Some lipoproteins form when enzymes covalently bond lipids to a protein. Other lipoproteins are aggregate structures that consist of variable amounts and types of proteins and lipids. These aggregate molecules carry fats and cholesterol through the bloodstream. Low-density lipoprotein, or LDL, transports cholesterol out of the liver and into cells. High-density lipoprotein, or HDL, ferries the cholesterol that is released from dead cells back to the liver.

protein   lipid

an HDL particle

---

amino acid Small organic compound that is a subunit of proteins. Consists of a carboxyl group, an amine group, and a characteristic side group (R), all typically bonded to the same carbon atom.
peptide bond A bond between the amine group of one amino acid and the carboxyl group of another. Joins amino acids in proteins.
polypeptide Chain of amino acids linked by peptide bonds.
protein Organic compound that consists of one or more chains of amino acids (polypeptides).

Take-Home Message **What are proteins?**

❯ Proteins are chains of amino acids. The order of amino acids in a polypeptide chain dictates the type of protein.

❯ Polypeptide chains twist and fold into coils, sheets, and loops, which fold and pack further into functional domains.

❯ A protein's shape is the source of its function.

The Importance of Protein Structure

> Changes in a protein's shape may have drastic consequences to health.

Protein shape depends on many hydrogen bonds and other interactions that heat, some salts, shifts in pH, or detergents can disrupt. At such times, proteins **denature**, which means they unwind and otherwise lose their shape. Once a protein's shape unravels, so does its function.

You can see denaturation in action when you cook an egg. A protein called albumin is a major component of egg white. Cooking does not disrupt the covalent bonds of albumin's primary structure, but it does destroy the weaker hydrogen bonds that maintain the protein's shape. When a translucent egg white turns opaque, the albumin has been denatured. For very few proteins, denaturation is reversible if normal conditions return, but albumin is not one of them. There is no way to uncook an egg.

Prion diseases, including mad cow disease (bovine spongiform encephalitis, or BSE) in cattle, Creutzfeldt–Jakob disease in humans, and scrapie in sheep, are the dire aftermath of a protein that changes shape. These

Figure 3.15 Variant Creutzfeldt–Jakob disease (vCJD).

A Charlene Singh, here being cared for by her mother, was one of three people who developed symptoms of vCJD disease while living in the United States. Like the others, Singh most likely contracted the disease elsewhere; she spent her childhood in Britain. She was diagnosed in 2001, and she died in 2004.

B Slice of brain tissue from a person with vCJD. Characteristic holes and prion protein fibers radiating from several deposits are visible.

infectious diseases may be inherited, but more often they arise spontaneously. All are characterized by relentless deterioration of mental and physical abilities that eventually causes the individual to die (Figure 3.15A).

All prion diseases begin with a protein that occurs normally in mammals. One such protein, PrP^C, is found in cell membranes throughout the body. This copper-binding protein is especially abundant in brain cells, but we still know very little about what it does. Very rarely, a PrP^C protein spontaneously misfolds so that it loses some of its

---

denature To unravel the shape of a protein or other large biological molecule.
prion Infectious protein.

Figure 3.16 The PrP^C protein becomes a prion when it misfolds into an as yet unknown conformation. Prions cause other PrP^C proteins to misfold, and the misfolded proteins aggregate into long fibers.

coils. In itself, a single misfolded protein molecule would not pose much of a threat. However, when this particular protein misfolds, it becomes a **prion**, or infectious protein (Figure 3.16). The altered shape of a misfolded PrP^C protein somehow causes normally folded PrP^C proteins to misfold too. Because each protein that misfolds becomes infectious, the number of prions increases exponentially.

The shape of misfolded PrP^C proteins allows them to align tightly into long fibers. Fibers composed of aggregated prion proteins begin to accumulate in the brain as large, water-repellent patches (Figure 3.15B). The patches grow as more prions form, and they begin to disrupt brain cell function, causing symptoms such as confusion, memory loss, and lack of coordination. Tiny holes form in the brain as its cells die. Eventually, the brain becomes so riddled with holes that it looks like a sponge.

In the mid-1980s, an epidemic of mad cow disease in Britain was followed by an outbreak of a new variant of Creutzfeldt–Jakob disease (vCJD) in humans. Researchers isolated a prion similar to the one in scrapie-infected sheep from cows with BSE, and also from humans affected by the new type of Creutzfeldt–Jakob disease. How did the prion get from sheep to cattle to people? Prions are not denatured by cooking or typical treatments that inactivate other types of infectious agents. The cattle became infected by the prion after eating feed prepared from the remains of scrapie-infected sheep, and people became infected by eating beef from the infected cattle.

Two hundred people have died from vCJD since 1990. The use of animal parts in livestock feed is now banned, and the number of cases of BSE and vCJD has since declined. Cattle with BSE still turn up, but so rarely that they pose little threat to human populations.

Take-Home Message **Why is protein structure important?**

> A protein's function depends on its structure, so changes in a protein's structure may also alter its function.

Nucleic Acids

> Nucleotides are subunits of DNA and RNA. Some have roles in metabolism.

< Links to Inheritance 1.3, Diversity 1.4, Hydrogen bonds 2.4

Nucleotides are small organic molecules that function as energy carriers, enzyme helpers, chemical messengers, and subunits of DNA and RNA. Each consists of a sugar with a five-carbon ring, bonded to a nitrogen-containing base and one or more phosphate groups. The nucleotide **ATP** (adenosine triphosphate) has a row of three phosphate groups attached to its ribose sugar (Figure 3.17A). When the outer phosphate group of an ATP is transferred to another molecule, energy is transferred along with it. You will read about such phosphate-group transfers and their important metabolic role in Chapter 5.

base: adenine (A)

3 phosphate groups

sugar: ribose

A ATP, a nucleotide monomer of RNA, and also an essential participant in many metabolic processes.

B A chain of nucleotides is a nucleic acid. The sugar of one nucleotide is covalently bonded to the phosphate group of the next, forming a sugar–phosphate backbone.

C DNA consists of two chains of nucleotides, twisted into a double helix. Hydrogen bonding maintains the three-dimensional structure of this nucleic acid.

Figure 3.17 Animated Nucleic acid structure.

Fear of Frying (revisited)

Trans fatty acids are relatively rare in unprocessed foods, so it makes sense from an evolutionary standpoint that our bodies may not have enzymes to deal with them efficiently. The enzymes that hydrolyze *cis* fatty acids have difficulty breaking down *trans* fatty acids, a problem that may be a factor in the ill effects of *trans* fats on our health.

How Would You Vote? All prepackaged foods in the United States are now required to list *trans* fat content, but may be marked "zero grams of *trans* fats" even when a single serving contains up to half a gram. Should hydrogenated oils be banned from all food? See CengageNow for details, then vote online (cengagenow.com).

Nucleic acids are polymers, chains of nucleotides in which the sugar of one nucleotide is joined to the phosphate group of the next (Figure 3.17B). An example is **RNA**, or ribonucleic acid, named after the ribose sugar of its component nucleotides. RNA consists of four kinds of nucleotide monomers, one of which is ATP. RNA molecules carry out protein synthesis, which we discuss in detail in Chapter 9.

DNA, or deoxyribonucleic acid, is a nucleic acid named after the deoxyribose sugar of its component nucleotides. A DNA molecule consists of two chains of nucleotides twisted into a double helix (Figure 3.17C). Hydrogen bonds between the nucleotides hold the two chains together.

Each cell starts out life with DNA inherited from a parent cell. That DNA contains all of the information necessary to build a new cell and, in the case of multicelled organisms, an entire individual. The cell uses the order of nucleotide bases in DNA—the DNA sequence—to guide production of RNA and proteins. Parts of the sequence are identical or nearly so in all organisms, but most is unique to a species or an individual (Chapter 8 returns to the topic of DNA structure and function).

ATP Adenosine triphosphate. Nucleotide that consists of an adenine base, a five-carbon ribose sugar, and three phosphate groups.
DNA Deoxyribonucleic acid. Nucleic acid that carries hereditary information about traits; consists of two nucleotide chains twisted in a double helix.
nucleic acid Single- or double-stranded chain of nucleotides joined by sugar–phosphate bonds; for example, DNA, RNA.
nucleotide Monomer of nucleic acids; has five-carbon sugar, nitrogen-containing base, and phosphate groups.
RNA Ribonucleic acid. Some types have roles in protein synthesis.

Take-Home Message What are nucleic acids?

> Nucleotides are monomers of the nucleic acids DNA and RNA. Many kinds, such as ATP, have other functions in metabolism.

> DNA's nucleotide sequence encodes heritable information.

> RNA molecules have roles in the processes by which a cell uses the information in its DNA.

Summary

Section 3.1 All organisms consist of the same kinds of molecules. Seemingly small differences in the way those molecules are put together can have big effects inside a living organism.

Section 3.2 Under present-day conditions in nature, only living things make the molecules of life—complex carbohydrates and lipids, proteins, and nucleic acids. All of these molecules are **organic**, which means they consist primarily of carbon and hydrogen atoms. **Hydrocarbons** have only carbon and hydrogen atoms.

Carbon chains or rings form the backbone of the molecules of life. **Functional groups** attached to the backbone influence the function of these compounds.

Metabolism includes all processes by which cells acquire and use energy as they make and break the bonds of organic compounds. By metabolic reactions such as **condensation**, **enzymes** build **polymers** from **monomers** of simple sugars, fatty acids, amino acids, and nucleotides. Reactions such as **hydrolysis** release the monomers by breaking apart the polymers.

Section 3.3 Enzymes assemble complex **carbohydrates** such as cellulose, glycogen, and starch from simple carbohydrate (sugar) subunits. Cells use carbohydrates for energy, and as structural materials.

Section 3.4 **Lipids** are fatty, oily, or waxy compounds. All are nonpolar. **Fats** and some other lipids have **fatty acid** tails; **triglycerides** have three.

Cells use lipids as major sources of energy and as structural materials. **Phospholipids** are the main structural component of cell membranes. **Waxes** are lipids that are part of water-repellent and lubricating secretions. **Steroids** occur in cell membranes, and some are remodeled into other molecules.

Section 3.5 Structurally and functionally, **proteins** are the most diverse molecules of life. The shape of a protein is the source of its function.

Protein structure begins as a linear sequence of **amino acids** linked by **peptide bonds** into a **polypeptide** (primary structure). Polypeptides twist into loops, sheets, and coils (secondary structure) that can pack further into functional domains (tertiary structure).

Many proteins, including most enzymes, consist of two or more polypeptides (quaternary structure). Fibrous proteins aggregate into much larger structures.

Section 3.6 A protein's structure dictates its function, so changes in a protein's structure may also alter its function. Shifts in pH or temperature, and exposure to detergent or some salts may disrupt hydrogen bonds and other molecular interactions that are responsible for the protein's shape. If that happens, the protein unravels, or **denatures**, and so loses its function. **Prion** diseases are a consequence of misfolded proteins.

Section 3.7 **Nucleotides** are small organic molecules consisting of a sugar, a phosphate group, and a nitrogen-containing base. Nucleotides are monomers of **DNA** and **RNA**, which are **nucleic acids**. Some nucleotides have additional functions. For example, **ATP** energizes many kinds of molecules by phosphate-group transfers. DNA encodes heritable information that guides the synthesis of RNA and proteins. RNAs interact with DNA and with one another to carry out protein synthesis.

Self-Quiz   Answers in Appendix III

1. Organic molecules consist mainly of _____ atoms.
   - a. carbon
   - b. carbon and oxygen
   - c. carbon and hydrogen
   - d. carbon and nitrogen

2. Each carbon atom can share pairs of electrons with as many as _____ other atom(s).

3. _____ groups impart polarity to alcohols.
   - a. Hydroxyl ($-OH^-$)
   - b. Phosphate ($-PO_4$)
   - c. Methyl ($-CH_3$)
   - d. Sulfhydryl ($-SH$)

4. _____ is a simple sugar (a monosaccharide).
   - a. Glucose
   - b. Sucrose
   - c. Ribose
   - d. Starch
   - e. both a and c
   - f. a, b, and c

5. Unlike saturated fats, the fatty acid tails of unsaturated fats incorporate one or more _____ .
   - a. phosphate groups
   - b. glycerols
   - c. double bonds
   - d. single bonds

6. Is this statement true or false? Unlike saturated fats, all unsaturated fats are beneficial to health because their fatty acid tails kink and do not pack together.

7. Steroids are among the lipids with no _____ .
   - a. double bonds
   - b. fatty acid tails
   - c. hydrogens
   - d. carbons

8. Name three kinds of carbohydrates that can be built using only glucose monomers.

9. Which of the following is a class of molecules that encompasses all of the other molecules listed?
   - a. triglycerides
   - b. fatty acids
   - c. waxes
   - d. steroids
   - e. lipids
   - f. phospholipids

10. _____ are to proteins as _____ are to nucleic acids.
   - a. Sugars; lipids
   - b. Sugars; proteins
   - c. Amino acids; hydrogen bonds
   - d. Amino acids; nucleotides

Data Analysis Activities

Effects of Dietary Fats on Lipoprotein Levels Cholesterol that is made by the liver or that enters the body from food does not dissolve in blood, so it is carried through the bloodstream by lipoproteins. Low-density lipoprotein (LDL) carries cholesterol to body tissues such as artery walls, where it can form deposits associated with cardiovascular disease. Thus, LDL is often called "bad" cholesterol. High-density lipoprotein (HDL) carries cholesterol away from tissues to the liver for disposal, so HDL is often called "good" cholesterol.

In 1990, Ronald Mensink and Martijn Katan published a study that tested the effects of different dietary fats on blood lipoprotein levels. Their results are shown in Figure 3.18.

1. In which group was the level of LDL ("bad" cholesterol) highest?
2. In which group was the level of HDL ("good" cholesterol) lowest?
3. An elevated risk of heart disease has been correlated with increasing LDL-to-HDL ratios. Which group had the highest LDL-to-HDL ratio?
4. Rank the three diets from best to worst according to their potential effect on heart disease.

Main Dietary Fats			
	cis fatty acids	*trans* fatty acids	saturated fats	optimal level
LDL	103	117	121	<100
HDL	55	48	55	>40
ratio	1.87	2.44	2.2	<2

Figure 3.18 Effect of diet on lipoprotein levels. Researchers placed 59 men and women on a diet in which 10 percent of their daily energy intake consisted of *cis* fatty acids, *trans* fatty acids, or saturated fats.

Blood LDL and HDL levels were measured after three weeks on the diet; averaged results are shown in mg/dL (milligrams per deciliter of blood). All subjects were tested on each of the diets. The ratio of LDL to HDL is also shown.

---

11. A denatured protein has lost its _____ .
    a. hydrogen bonds   c. function
    b. shape   d. all of the above

12. _____ consists of nucleotides.
    a. Sugars   b. DNA   c. RNA   d. b and c

13. Which of the following is not found in DNA?
    a. amino acids   c. nucleotides
    b. sugars   d. phosphate groups

14. In the following list, identify the carbohydrate, the fatty acid, the amino acid, and the polypeptide:
    a. NH_2—CHR—COOH   c. (methionine)$_{20}$
    b. $C_6H_{12}O_6$ d. $CH_3(CH_2)_{16}COOH$

15. Match the molecules with the best description.
    ___ wax   a. protein primary structure
    ___ starch   b. an energy carrier
    ___ triglyceride c. water-repellent secretions
    ___ DNA   d. carries heritable information
    ___ polypeptide e. sugar storage in plants
    ___ ATP   f. richest energy source

16. Match each polymer with the most appropriate set of component monomers.
    ___ protein   a. glycerol, fatty acids, phosphate
    ___ phospholipid   b. amino acids, sugars
    ___ glycoprotein   c. glycerol, fatty acids
    ___ fat   d. nucleotides
    ___ nucleic acid   e. sugars
    ___ carbohydrate   f. sugar, phosphate, base
    ___ nucleotide   g. amino acids
    ___ lipoprotein   h. glucose, fructose
    ___ sucrose   i. lipids, amino acids

Additional questions are available on **CENGAGENOW**.

Critical Thinking

1. In 1976, a team of chemists in the United Kingdom was developing new insecticides by modifying sugars with chlorine (Cl_2), phosgene (Cl_2CO), and other toxic gases. One young member of the team misunderstood his verbal instructions to "test" a new molecule. He thought he had been told to "taste" it. Luckily, the molecule was not toxic, but it was very sweet. It became the food additive sucralose.

Sucralose has three chlorine atoms substituted for three hydroxyl groups of sucrose. It binds so strongly to the sweet-taste receptors on the tongue that the human brain perceives it as 600 times sweeter than sucrose (table sugar). Sucralose was originally marketed as an artificial sweetener called Splenda®, but it is now available under several other brand names.

Researchers proved that the body does not recognize sucralose as a carbohydrate by feeding sucralose labeled with ^{14}C to volunteers. Analysis of the radioactive molecules in the volunteers' urine and feces showed that 92.8 percent of the sucralose passed through the body without being altered. Many people are worried that the chlorine atoms impart toxicity to sucralose. How would you respond to that concern?

sucrose

sucralose

Animations and Interactions on **CENGAGENOW**:
❯ Functional groups; Condensation and hydrolysis; Sucrose synthesis; Fatty acids; Triglyceride formation; Amino acids; Peptide bond formation.

Links to Earlier Concepts

Reflect on the Section 1.2 overview of life's levels of organization. In this chapter, you will see how the properties of lipids (3.4) give rise to cell membranes; consider the location of DNA (3.7) and the sites where carbohydrates are built and broken apart (3.2, 3.3); and expand your understanding of the vital roles of proteins in cell function (3.5, 3.6). You will also see an application of tracers (2.2), and revisit scientific philosophy (1.6, 1.9).

Key Concepts

What Is a Cell?
A cell is the smallest unit of life. Each has a plasma membrane that separates its interior from the exterior environment. A cell's interior contains cytoplasm and DNA.

Microscopes
Most cells are too small to see with the naked eye. We use different types of microscopes to reveal different details of their structure.

4 Cell Structure

We find bacteria at the bottom of the ocean, high up in the atmosphere, miles underground—essentially anywhere we look. Mammalian intestines typically harbor fantastic numbers of them, but bacteria are not just stowaways there. Intestinal bacteria make vitamins that mammals cannot, and they crowd out more dangerous germs. Cell for cell, bacteria that live in and on a human body outnumber the person's own cells by about ten to one.

Escherichia coli is one of the most common intestinal bacteria of warm-blooded animals. Only a few of the hundreds of types, or strains, of *E. coli*, are harmful. One that is called O157:H7 (Figure 4.1) makes a potent toxin that can severely damage the lining of the human intestine. After ingesting as few as ten O157:H7 cells, a person may become ill with severe cramps and bloody diarrhea that lasts up to ten days. In some people, complications of O157:H7 infection result in kidney failure, blindness, paralysis, and death. About 73,000 people in the United States become infected with *E. coli* O157:H7 each year, and more than 60 of them die.

E. coli O157:H7 can live in the intestines of other animals—mainly cattle, deer, goats, and sheep—apparently without sickening them. Humans are exposed to the bacteria when they come into contact with feces of animals that harbor it, for example, by eating contaminated ground beef. During slaughter, meat can come into contact with feces. Bacteria in the feces stick to the meat, then get thoroughly mixed into it during the grinding process. Unless contaminated meat is cooked to at least 71°C (160°F), live bacteria will enter the digestive tract of whoever eats it.

People also become infected by eating fresh fruits and vegetables that have come into contact with animal feces. Washing contaminated produce with water does not remove *E. coli* O157:H7, because the bacteria are sticky. In 2006, more than 200 people became ill and 3 died after eating fresh spinach grown in a field close to a cattle pasture. Water contaminated with manure may have been used to irrigate the field.

Figure 4.1 *E. coli* O157:H7 bacteria (*red*) clustering on intestinal cells of a small child. This type of bacteria can cause a serious intestinal illness in people who eat foods contaminated with it.

The economic impact of such outbreaks, which occur with some regularity, extends beyond the casualties. Growers lost $50–100 million recalling fresh spinach after the 2006 outbreak. In 2007, about 5.7 million pounds of ground beef were recalled after 14 people were sickened. More than 2.8 million pounds of meat products were recalled between November, 2009 and January, 2010. Food growers and processors are now using procedures that they hope will reduce *E. coli* O157:H7 outbreaks. Some meats and produce are now tested for pathogens before sale, and improved documentation should allow a source of contamination to be pinpointed more quickly.

What makes bacteria sticky? Why do people but not cows get sick with *E. coli* O157:H7? You will begin to find answers to these and many more questions that affect your health in this chapter, as you learn about cells and how they work.

Cell Membranes
All cell membranes consist mainly of a lipid bilayer and different types of proteins. The proteins carry out various tasks, including control over which substances cross the membrane.

Bacteria and Archaeans
Archaeans and bacteria have few internal membrane-enclosed compartments. In general, they are the smallest and structurally the simplest cells, but they are also the most numerous.

Eukaryotic Cells
Cells of protists, plants, fungi, and animals are eukaryotic; they have a nucleus and other membrane-enclosed compartments. Cells differ in internal parts and surface specializations.

> All cells have a plasma membrane and cytoplasm, and all start out life with DNA.
< Links to Lipid structure 3.4, DNA 3.7

Traits Common to All Cells

A **cell** is the smallest unit that shows the properties of life. Cells vary dramatically in shape and in function (Figure 4.2). Despite their differences, however, all cells share certain organizational and functional features. Every cell has a **plasma membrane**, an outer membrane that separates the cell's contents from its environment. A plasma membrane is selectively permeable, which means it allows only certain materials to cross. Thus, it controls exchanges between the cell and its environment. All cell membranes, including the plasma membrane, consist mainly of lipids.

The plasma membrane encloses a fluid or jellylike mixture of water, sugars, ions, and proteins called **cytoplasm**. Some or all of a cell's metabolism occurs in the cytoplasm, and the cell's internal components, including organelles, are suspended in it. **Organelles** are structures that carry out special metabolic functions inside a cell.

All cells start out life with DNA, although a few types of cells lose it as they mature. We categorize cells based on whether their DNA is housed in a nucleus or not (Figure 4.3). Only eukaryotic cells have a **nucleus** (plural, nuclei), an organelle with a double membrane that contains the cell's DNA. In most bacteria and archaeans, the DNA is suspended directly in the cytoplasm.

Constraints on Cell Size

Almost all cells are too small to see with the naked eye. Why? The answer begins with the processes that keep a cell alive. A living cell must exchange substances with

Figure 4.2 Examples of cells. Each one of the cells pictured here is an individual organism; all are protists.

its environment at a rate that keeps pace with its metabolism. These exchanges occur across the plasma membrane, which can handle only so many exchanges at a time. Thus, cell size is limited by a physical relationship called the **surface-to-volume ratio**. By this ratio, an object's volume increases with the cube of its diameter, but its surface area increases only with the square.

Apply the surface-to-volume ratio to a round cell. As Figure 4.4 shows, when a cell expands in diameter, its volume increases faster than its surface area does. Imagine that a round cell expands until it is four times its original diameter. The volume of the cell has increased 64 times (4^3), but its surface area has increased only 16 times (4^2). Each unit of plasma membrane must now handle exchanges with four times as much cytoplasm ($64 = 16 \times 4$). If the cell gets too big, the inward flow of nutrients and the outward flow of wastes across that membrane will not be fast enough to keep the cell alive.

Surface-to-volume limits also affect the body plans of multicelled species. For example, small cells attach end

A Bacterial cell   — cytoplasm — DNA — plasma membrane

B Plant cell   cytoplasm — DNA in nucleus — plasma membrane

C Animal cell   cytoplasm — DNA in nucleus — plasma membrane

Figure 4.3 Animated Overview of the general organization of bacteria and eukaryotic cells. Archaeans are similar to bacteria in overall structure. The three examples are not drawn to the same scale.

>> **Figure It Out:** Which of these cells is/are eukaryotic?

Answer: The plant and animal cells are eukaryotic.

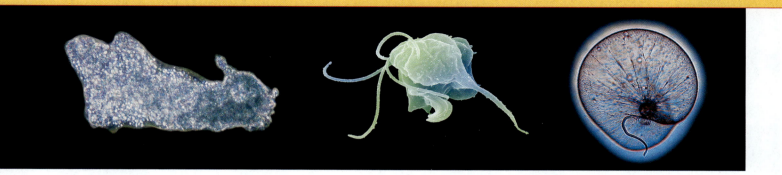

	Diameter (cm)	2	3	6
	Surface area (cm²)	12.6	28.2	113
	Volume (cm³)	4.2	14.1	113
	Surface-to-volume ratio	3:1	2:1	1:1

Figure 4.4 Animated Three examples of the surface-to-volume ratio. This physical relationship between increases in volume and surface area constrains cell size and shape.

to end to form strandlike algae, so that each can interact directly with its surroundings. Muscle cells in your thighs are as long as the muscle in which they occur, but each is thin, so it exchanges substances efficiently with fluids in the tissue surrounding it.

Cell Theory

Nearly all cells are so small that no one even knew they existed until after the first microscopes were invented. In 1665, Antoni van Leeuwenhoek, a Dutch draper, wrote about the tiny moving organisms he spied in rainwater, insects, fabric, sperm, feces, and other samples. In scrapings of tartar from his teeth, Leeuwenhoek saw "many very small animalcules, the motions of which were very pleasing to behold." He (incorrectly) assumed that movement defined life, and (correctly) concluded that the moving "beasties" he saw were alive. Perhaps Leeuwenhoek

cell Smallest unit that has the properties of life.
cell theory Theory that all organisms consist of one or more cells, which are the basic unit of life.
cytoplasm Semifluid substance enclosed by a cell's plasma membrane.
nucleus Organelle with two membranes that holds a eukaryotic cell's DNA.
organelle Structure that carries out a specialized metabolic function inside a cell.
plasma membrane A cell's outermost membrane.
surface–to–volume ratio A relationship in which the volume of an object increases with the cube of the diameter, but the surface area increases with the square.

was so pleased to behold his animalcules because he did not grasp the implications of what he was seeing: Our world, and our bodies, teem with microbial life. The term cell was coined when Robert Hooke, a contemporary of Leeuwenhoek, magnified a piece of thinly sliced cork. Hooke named the tiny compartments he observed "cellae," a Latin word for the small chambers that monks lived in.

In the 1820s, botanist Robert Brown was the first to identify a cell nucleus. Matthias Schleiden, another botanist, hypothesized that a plant cell is an independent living unit even when it is part of a plant. Schleiden compared notes with the zoologist Theodor Schwann, and both concluded that the tissues of animals as well as plants are composed of cells and their products. Together, the two scientists recognized that cells have a life of their own even when they are part of a multicelled body.

Later, physiologist Rudolf Virchow realized that all cells he studied descended from another living cell. These and many other observations yielded four generalizations that today constitute the **cell theory**:

1. Every living organism consists of one or more cells.
2. The cell is the structural and functional unit of all organisms. A cell is the smallest unit of life, individually alive even as part of a multicelled organism.
3. All living cells come from division of preexisting cells.
4. Cells contain hereditary material, which they pass to their offspring during division.

The cell theory, first articulated in 1839 by Schwann and Schleiden and later revised, remains an important foundation of modern biology.

Take-Home Message How are all cells alike?

> All cells start life with a plasma membrane, cytoplasm, and a region of DNA, which, in eukaryotic cells only, is enclosed by a nucleus.

> The surface-to-volume ratio limits cell size and influences cell shape.

> Observations of cells led to the cell theory: All organisms consist of one or more cells; the cell is the smallest unit of life; each new cell arises from another cell; and a cell passes hereditary material to its offspring.

path of light rays (bottom to top) to eye

prism that directs rays to ocular lens

ocular lens

objective lenses

focusing knob

specimen stage

condenser lens

illuminator

A

light source (in base)

incoming electron beam

condenser lens

specimen on grid

objective lens

projective lens

B

phosphor screen

Figure 4.5 Animated Examples of microscopes. **A** A compound light microscope. **B** A transmission electron microscope (TEM).

> We use different types of microscopes to study different aspects of organisms, from the smallest to the largest.
< Link to Tracers 2.2

Microscopes allow us to study cells in detail. The ones that use visible light to illuminate objects are called light microscopes (Figure 4.5). All light travels in waves. This property makes light bend when it passes through curved glass lenses. Inside a light microscope, such lenses focus light that passes through a specimen, or bounces off of one, into a magnified image. Photographs of images enlarged with any microscope are called micrographs, and you will find many of them in this book.

Phase-contrast microscopes shine light through specimens. Most cells are nearly transparent, so their internal details may not be visible unless they are first stained, or exposed to dyes that only some cell parts soak up. Parts that absorb the most dye appear darkest. Staining results

in an increase in contrast (the difference between light and dark) that allows us to see a greater range of detail (Figure 4.6A). Surface details can be revealed by reflected light (Figure 4.6B).

With a fluorescence microscope, a cell or a molecule is the light source; it fluoresces, or emits energy in the form of light, when a laser beam is focused on it. Some molecules fluoresce naturally (Figure 4.6C). More typically, researchers attach a light-emitting tracer (Section 2.2) to the cell or molecule of interest.

Other types of microscopes can reveal finer details. For example, electron microscopes use electrons instead of visible light to illuminate samples. Because electrons travel in wavelengths that are much shorter than those

light microscopes

electron microscopes

molecules of life

complex carbohydrates

viruses

mitochondria, chloroplasts

most bacteria

most eukaryotic cells

proteins

lipids

DNA (width)

small molecules

| 0.1 nm | 1 nm | 10 nm | 100 nm | 1 μm | 10 μm |

A Light micrograph. A phase-contrast microscope yields high-contrast images of transparent specimens. Dark areas have taken up dye.

B Light micrograph. A reflected light microscope captures light reflected from specimens.

C Fluorescence micrograph. This image shows fluorescent light emitted by chlorophyll molecules in the cells.

D A transmission electron micrograph reveals fantastically detailed images of internal structures.

E A scanning electron micrograph shows surface details. SEMs may be artificially colored to highlight specific details.

Figure 4.6 Different microscopes reveal different characteristics of the same organism, a green alga (*Scenedesmus*).

of visible light, electron microscopes can resolve details that are much smaller than you can see with light microscopes. Electron microscopes use magnetic fields to focus beams of electrons onto a sample. Transmission electron microscopes beam electrons through a thin specimen. The specimen's internal details appear on the resulting image as shadows (Figure 4.6D). Scanning electron microscopes direct a beam of electrons back and forth across a surface of a specimen, which has been coated with a thin layer of gold or another metal. The metal emits both electrons and x-rays, which are converted into an image of the surface (Figure 4.6E). Both types of electron microscopes can resolve structures as small as 0.2 nanometer.

Figure 4.7 compares the resolving power of light and electron microscopes with that of the unaided human eye.

Take-Home Message How do we see cells?

> Most cells are visible only with the help of microscopes.

> Different types of microscopes reveal different aspects of cell structure.

Figure 4.7 Relative sizes. *Below*, the diameter of most cells is in the range of 1 to 100 micrometers. *Right*, converting among units of length; also see Units of Measure, Appendix IX.

>> **Figure It Out:** Which one is smallest: a protein, a lipid, or a water molecule?

Answer: A water molecule

1 centimeter (cm)	= 1/100 meter, or 0.4 inch
1 millimeter (mm)	= 1/1000 meter, or 0.04 inch
1 micrometer (µm)	= 1/1,000,000 meter, or 0.00004 inch
1 nanometer (nm)	= 1/1,000,000,000 meter, or 0.00000004 inch
1 meter	= 10^2 cm = 10^3 mm = 10^6 µm = 10^9 nm

human eye (no microscope)

largest organisms

small animals

frog eggs

humans

| 100 µm | 1 mm | 1 cm | 10 cm | 1 m | 10 m | 100 m |

> A cell membrane is organized as a lipid bilayer with many proteins embedded in it and attached to its surfaces.

< Links to Emergent properties 1.2, Enzyme 3.2, Lipids 3.4, Proteins 3.5

Lipids—mainly phospholipids—make up the bulk of a cell membrane. A phospholipid consists of a phosphate-containing head and two fatty acid tails (Section 3.4). The polar head is hydrophilic, which means that it interacts with water molecules. The nonpolar tails are hydrophobic, so they do not interact with water molecules, but they do interact with the tails of other phospholipids. As a result of these influences, phospholipids swirled into water will spontaneously organize themselves into a **lipid bilayer**, two layers of lipids with nonpolar tails sandwiched between polar heads. Lipid bilayers are the basic structural and functional framework of all cell membranes (Figure 4.8A). In water, a lipid bilayer spontaneously shapes itself into a sheet or bubble. A cell is essentially a lipid bilayer bubble filled with fluid (Figure 4.9).

fluid

Figure 4.9 At its most basic, a cell is a lipid bilayer bubble filled with fluid.

A In a watery fluid, phospholipids spontaneously line up into two layers, tails to tails. This lipid bilayer spontaneously shapes itself into a sheet or a bubble. It is the basic structural and functional framework of all cell membranes. Many types of proteins intermingle among the lipids.

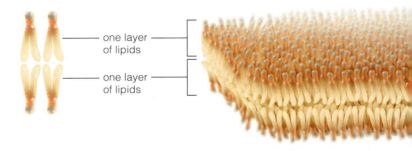

one layer of lipids

one layer of lipids

Figure 4.8 Animated Cell membrane structure. **A** Organization of phospholipids in cell membranes. **B–E** Examples of common membrane proteins.

Other molecules, including steroids and proteins, are embedded in or associated with the lipid bilayer of every cell membrane. Most of these molecules move around the membrane more or less freely. A cell membrane behaves like a two-dimensional liquid of mixed composition, so we describe it as a **fluid mosaic**. The "mosaic" part of the name comes from a cell membrane's mixed composition of lipids and proteins. The fluidity occurs because the phospholipids in a cell membrane are not bonded to one another. They stay organized as a bilayer as a result of collective hydrophobic and hydrophilic attractions, which, on an individual basis, are relatively weak. Thus, phospholipids in a bilayer can drift sideways and spin around their long axis, and their tails can wiggle.

Membrane Proteins

A cell membrane physically separates an external environment from an internal one, but that is not its only task. Many types of proteins are associated with a cell membrane, and each type adds a specific function to it (Table 4.1). Thus, different cell membranes can have different characteristics depending on which proteins are associated with them.

For example, a plasma membrane has certain proteins that no internal cell membrane has. Many plasma membrane proteins are enzymes. Others are **adhesion proteins**, which fasten cells together in animal tissues. **Recognition proteins** function as identity tags for a cell type, individual, or species (Figure 4.8B). Being able to

Table 4.1 Common Types of Membrane Proteins

Category	Function	Examples
Passive transporters	Allow ions or small molecules to cross a membrane to the side where they are less concentrated. Open or gated channels.	Porins; glucose transporter
Active transporters	Pump ions or molecules through membranes to the side where they are more concentrated. Require energy input, as from ATP.	Calcium pump; serotonin transporter
Receptors	Initiate change in a cell activity by responding to an outside signal (e.g., by binding a signaling molecule or absorbing light energy).	Insulin receptor; B cell receptor
Adhesion proteins	Help cells stick to one another and to protein matrixes that are part of tissues.	Integrins; cadherins
Recognition proteins	Identify cells as self (belonging to one's own body or tissue) or nonself (foreign to the body).	MHC molecules
Enzyme	Speeds a specific reaction. Membranes provide a relatively stable reaction site for enzymes that work in series with other molecules.	Cytochrome *c* oxidase of mitochondria

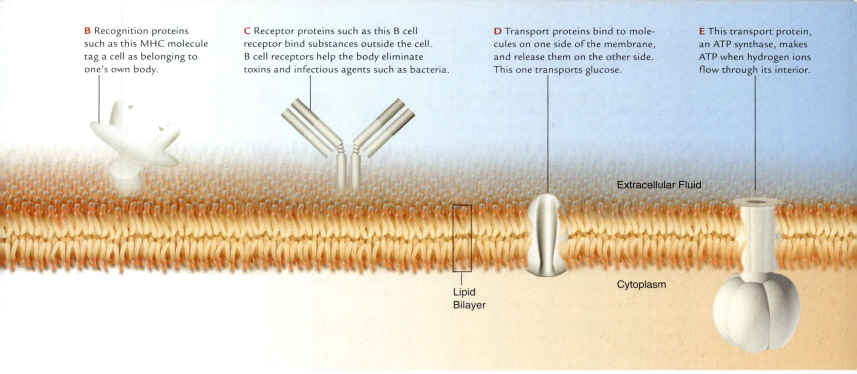

B Recognition proteins such as this MHC molecule tag a cell as belonging to one's own body.

C Receptor proteins such as this B cell receptor bind substances outside the cell. B cell receptors help the body eliminate toxins and infectious agents such as bacteria.

D Transport proteins bind to molecules on one side of the membrane, and release them on the other side. This one transports glucose.

E This transport protein, an ATP synthase, makes ATP when hydrogen ions flow through its interior.

Extracellular Fluid

Cytoplasm

Lipid Bilayer

recognize "self" means that foreign cells (harmful ones, in particular) can also be recognized. **Receptor proteins** bind to a particular substance outside of the cell, such as a hormone or toxin (Figure 4.8C). Binding triggers a change in the cell's activities that may involve metabolism, movement, division, or even cell death. Receptors for different types of substances occur on different cells, but all are critical for homeostasis.

Additional proteins occur on all cell membranes. **Transport proteins** move specific substances across a membrane, typically by forming a channel through it. These proteins are important because lipid bilayers are impermeable to most substances, including ions and polar molecules. Some transport proteins are open channels through which a substance moves on its own across a membrane (Figure 4.8D,E). Others use energy to actively pump a substance across. We return to the topic of membrane transport in the next chapter.

---

adhesion protein Membrane protein that helps cells stick together in tissues.

fluid mosaic Model of a cell membrane as a two-dimensional fluid of mixed composition.

lipid bilayer Structural foundation of cell membranes; double layer of lipids arranged tail-to-tail.

receptor protein Plasma membrane protein that binds to a particular substance outside of the cell.

recognition protein Plasma membrane protein that identifies a cell as belonging to self (one's own body).

transport protein Protein that passively or actively assists specific ions or molecules across a membrane.

Variations on the Model

Membranes differ in composition. For example, different kinds of cells may have different kinds of membrane phospholipids. Fatty acid tails of membrane phospholipids vary in length and saturation. Usually, at least one of the two tails is unsaturated (Section 3.4).

Some proteins stay put, such as those that cluster as rigid pores. Protein filaments inside the cell lock these and other proteins in place. For example, plasma membrane adhesion proteins connect the internal filaments of adjacent cells in animal tissues. This arrangement strengthens a tissue, and it can constrain certain membrane proteins to an upper or lower surface of the cells.

Archaeans do not build their phospholipids with fatty acids. Instead, they use molecules that have reactive side chains, so the tails of archaean phospholipids form covalent bonds with one another. As a result of this rigid crosslinking, archaean phospholipids do not drift, spin, or wiggle in a bilayer. Thus, the membranes of archaeans are far more rigid than those of bacteria or eukaryotes, a characteristic that may help these cells survive in extreme habitats that destroy other cells.

Take-Home Message What is the function of a cell membrane?

› A cell membrane selectively controls exchanges between the cell and its surroundings. It is a mosaic of different kinds of lipids and proteins.

› The foundation of cell membranes is the lipid bilayer—two layers of lipids (mainly phospholipids), with tails sandwiched between heads.

❯ All bacteria and archaeans are single-celled organisms with no nucleus.
❮ Links to Domains 1.4, Polysaccharides 3.3, ATP 3.7

All bacteria and archaeans are single-celled, and none have a nucleus. Outwardly, they appear so similar that archaeans were once thought to be an unusual group of bacteria. Both were classified as prokaryotes, a word that means "before the nucleus." By 1977, it had become clear that archaeans are more closely related to eukaryotes than to bacteria, so they were given their own separate domain. The term "prokaryote" is now being retired.

As a group, bacteria and archaeans are the smallest and most metabolically diverse forms of life that we know about. They inhabit nearly all of Earth's environments, including some very hostile places. The two kinds of cells differ in structure and metabolic details (Figure 4.10). Chapter 19 revisits them in more detail; here we present an overview of their structure (Figure 4.11).

Most bacteria and archaeans are not much bigger than a few micrometers. None has a complex internal framework, but protein filaments under the plasma membrane reinforce the cell's shape. Such filaments also act as scaffolding for internal structures. The cytoplasm of these cells contains many **ribosomes** (organelles upon which polypeptides are assembled), and in some species, additional organelles. **Plasmids** also occur in cytoplasm. These small circles of DNA carry a few genes (units of inheritance) that can provide advantages, such as resistance to antibiotics. The cell's single chromosome, a circular DNA molecule, is located in an irregularly shaped region of cytoplasm called the **nucleoid**. Some nucleoids are enclosed by a membrane, but most are not.

Many bacteria and archaeans have one or more flagella projecting from their surface. **Flagella** (singular, flagellum) are long, slender cellular structures used for motion. A bacterial flagellum rotates like a propeller that drives

A Protein filaments, or pili, anchor bacterial cells to one another and to surfaces. Here, *Salmonella typhimurium* cells (*red*) use their pili to invade human cells.

B Ball-shaped *Nostoc* cells are a type of freshwater photosynthetic bacteria. The cells in each strand stick together in a sheath of their own jellylike secretions.

Figure 4.10 A sampling of bacteria (*this page*) and archaeans (*facing page*).

the cell through fluid habitats, such as an animal's body fluids. Some bacteria also have protein filaments called **pili** (singular, pilus) projecting from their surface. Pili help cells cling to or move across surfaces (Figure 4.10A). One kind, a "sex" pilus, attaches to another bacterium and then shortens. The attached cell is reeled in, and a plasmid is transferred from one cell to the other through the pilus.

A durable **cell wall** surrounds the plasma membrane of nearly all bacteria and archaeans. Dissolved substances easily cross this permeable layer on the way to and from the plasma membrane. The cell wall of most archaeans consists of proteins. The wall of most bacteria consists of a polymer of peptides and polysaccharides.

Sticky polysaccharides form a slime layer, or capsule, around the wall of many types of bacteria. The sticky capsule helps these cells adhere to many types of surfaces

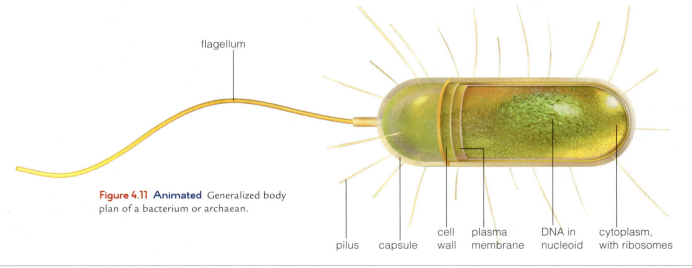

flagellum

Figure 4.11 Animated Generalized body plan of a bacterium or archaean.

pilus   capsule   cell wall   plasma membrane   DNA in nucleoid   cytoplasm, with ribosomes

C The archaean *Pyrococcus furiosus* was discovered in ocean sediments near an active volcano. It lives best at 100°C (212°F), and it makes a rare kind of enzyme that contains tungsten atoms.

D *Ferroglobus placidus* prefers superheated water spewing from the ocean floor. The durable composition of archaean lipid bilayers (note the gridlike texture) keeps their membranes intact at extreme heat and pH.

E *Metallosphaera prunae*, an archaean discovered in a smoking pile of ore at a uranium mine, prefers high temperatures and low pH. (*White* shadows are an artifact of electron microscopy.)

(such as spinach leaves and meat), and it also offers some protection against predators and toxins.

The plasma membrane of all bacteria and archaeans selectively controls which substances move into and out of the cell, as it does for eukaryotic cells. The plasma membrane bristles with transporters and receptors, and it also incorporates proteins that carry out important metabolic processes. For example, part of the plasma membrane of cyanobacteria (Figure 4.10B) folds into the cytoplasm. Molecules that carry out photosynthesis are embedded in this membrane, just as they are in the inner membrane of chloroplasts, which are organelles specialized for photosynthesis in eukaryotic cells (we return to chloroplasts in Section 4.9).

Figure 4.12 A biofilm. A single species of bacteria, *Bacillus subtilis*, formed this biofilm. Note the distinct "neighborhoods."

biofilm Community of microorganisms living within a shared mass of slime.
cell wall Semirigid but permeable structure that surrounds the plasma membrane of some cells.
flagellum Long, slender cellular structure used for motility.
nucleoid Region of cytoplasm where the DNA is concentrated inside a bacterium or archaean.
pilus A protein filament that projects from the surface of some bacterial cells.
plasmid Small circle of DNA in some bacteria and archaeans.
ribosome Organelle of protein synthesis.

Biofilms

Bacterial cells often live so close together that an entire community shares a layer of secreted polysaccharides and proteins. A communal living arrangement in which single-celled organisms live in a shared mass of slime is called a **biofilm**. In nature, a biofilm typically consists of multiple species, all entangled in their own mingled secretions. It may include bacteria, algae, fungi, protists, and archaeans. Participating in a biofilm allows the cells to linger in a favorable spot rather than be swept away by fluid currents, and to reap the benefits of living communally. For example, rigid or netlike secretions of some species serve as permanent scaffolding for others; species that break down toxic chemicals allow more sensitive ones to thrive in polluted habitats that they could not withstand on their own; and waste products of some serve as raw materials for others.

Like a bustling metropolitan city, a biofilm organizes itself into "neighborhoods," each with a distinct microenvironment that stems from its location within the biofilm and the particular species that inhabit it (Figure 4.12).

Take-Home Message How are bacteria and archaeans alike?

❯ Bacteria and archaeans do not have a nucleus. Most kinds have a cell wall around their plasma membrane. The wall is permeable, and it reinforces and imparts shape to the cell body.

❯ The structure of bacteria and archaeans is relatively simple, but as a group these organisms are the most diverse forms of life. They inhabit nearly all regions of the biosphere.

❯ Some metabolic processes occur at the plasma membrane of bacteria and archaeans. They are similar to complex processes that occur at certain internal membranes of eukaryotic cells.

> Eukaryotic cells carry out much of their metabolism inside organelles enclosed by membranes.

All protists, fungi, plants, and animals are eukaryotes. Some of these organisms are independent, free-living cells; others consist of many cells working together as a body. By definition, a eukaryotic cell starts out life with a nucleus (*eu–* means true; *karyon* means nut, or kernel).

Like many other organelles, a nucleus has a membrane. An organelle's outer membrane controls the types and amounts of substances that cross it. Such control maintains a special internal environment that allows the organelle to carry out its particular function. That function may be isolating toxic or sensitive substances from the rest of the cell, transporting substances through cytoplasm, maintaining fluid balance, or providing a favorable environment for a special process. For example, a mitochondrion makes ATP after concentrating hydrogen ions inside its membrane system.

Much as interactions among organ systems keep an animal body running, interactions among organelles keep a cell running. Substances shuttle from one kind of organelle to another, and to and from the plasma membrane. Some metabolic pathways take place in a series of different organelles.

Figure 4.13 shows two typical eukaryotic cells. In addition to a nucleus, these and most other eukaryotic cells contain an endomembrane system (ER, vesicles, and Golgi bodies), mitochondria, and a cytoskeleton, which is a dynamic "skeleton" of proteins (*cyto–* means cell). Certain cell types have additional organelles that allow them to do particular tasks (Table 4.2). We will discuss the main components of eukaryotic cells in the remaining sections of this chapter.

Figure 4.13
Transmission electron micrographs of eukaryotic cells.

A Human white blood cell.

B Photosynthetic cell from a blade of timothy grass.

Table 4.2	Components of Eukaryotic Cells
Organelles with membranes	
Nucleus	Protecting and controlling access to DNA
Endoplasmic reticulum (ER)	Routing, modifying new polypeptide chains; synthesizing lipids; other tasks
Golgi body	Modifying new polypeptide chains; sorting, shipping proteins and lipids
Vesicles	Transporting, storing, or digesting substances in a cell; other functions
Mitochondrion	Making ATP by glucose breakdown
Chloroplast	Making sugars in plants, some protists
Lysosome	Intracellular digestion
Peroxisome	Inactivating toxins
Vacuole	Storage
Organelles without membranes	
Ribosomes	Assembling polypeptide chains
Centriole	Anchor for cytoskeleton
Other components	
Cytoskeleton	Contributes to cell shape, internal organization, movement

Take-Home Message What do all eukaryotic cells have in common?

> All eukaryotic cells start life with a nucleus and other membrane-enclosed organelles.

cell wall

central vacuole

vacuole

plasma membrane

chloroplast

mitochondrion

nucleus

1 µm

1 µm

A

B

Figure 4.14 Animated The cell nucleus. TEM at *right*, nucleus of a mouse pancreas cell.

❯ A nucleus protects and controls access to a eukaryotic cell's DNA.

A eukaryotic cell's nucleus serves two important functions. First, it keeps the cell's genetic material—its DNA—safe and sound. Isolated in its own compartment, the cell's DNA stays separated from the bustling activity of the cytoplasm, and from metabolic processes that might damage it.

The second function of a nucleus is to control the passage of certain molecules between the nucleus and the cytoplasm. The nuclear membrane, which is called the **nuclear envelope**, carries out this function. A nuclear envelope consists of two lipid bilayers folded together as a single membrane. As Figure 4.14 shows, the outer bilayer of the membrane is continuous with the membrane of another organelle, the ER.

Receptors and transporters stud both sides of the nuclear envelope; other proteins cluster to form tiny pores that span it (Figure 4.15). These molecules and structures work as a system that selectively transports various molecules across the nuclear membrane.

Cells access their DNA when they make RNA and proteins, so the molecules involved in this process must pass into the nucleus and out of it. Control over their transport through the nuclear membrane is one way the cell regulates the amount of RNA and proteins it makes.

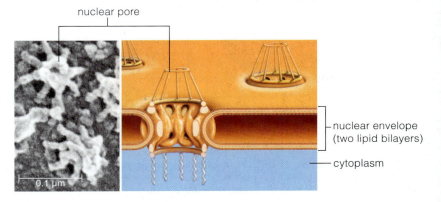

Figure 4.15 Animated Nuclear pores. Each is an organized cluster of membrane proteins that selectively allows substances to cross the nuclear membrane.

The nuclear envelope encloses **nucleoplasm**, a viscous fluid that is similar to cytoplasm. The nucleus also contains at least one **nucleolus** (plural, nucleoli), a dense, irregularly shaped region where subunits of ribosomes are assembled from proteins and RNA. The subunits pass through nuclear pores into the cytoplasm, where they join and become active in protein synthesis.

nuclear envelope A double membrane that constitutes the outer boundary of the nucleus.
nucleolus In a cell nucleus, a dense, irregularly shaped region where ribosomal subunits are assembled.
nucleoplasm Viscous fluid enclosed by the nuclear envelope.

Take-Home Message **What is the function of a nucleus?**
❯ A nucleus protects and controls access to a eukaryotic cell's DNA.
❯ The nuclear envelope is a double lipid bilayer. Proteins embedded in it control the passage of molecules between the nucleus and cytoplasm.

> The endomembrane system is a set of organelles that makes, modifies, and transports proteins and lipids.
< Links to Lipids 3.4, Proteins 3.5

The **endomembrane system** is a series of interacting organelles between the nucleus and the plasma membrane. Its main function is to make lipids, enzymes, and proteins for secretion, or for insertion into cell membranes. It also destroys toxins, recycles wastes, and has other specialized functions. The system's components vary among different types of cells, but here we present the most common ones (Figure 4.16).

Rough ER
Modifies proteins made by ribosomes attached to it

Smooth ER
Makes lipids, breaks down carbohydrates and fats, inactivates toxins

Golgi Body
Finishes, sorts, ships lipids, enzymes, and proteins

Lysosome
Digests, recycles materials

Figure 4.16 Animated Endomembrane system, where many proteins are modified and lipids are built. These molecules are sorted and shipped to cellular destinations or to the plasma membrane for export.

Endoplasmic Reticulum

Part of the endomembrane system is an extension of the nuclear envelope called **endoplasmic reticulum**, or **ER**. ER forms a continuous compartment that folds into flattened sacs and tubes. The space inside the compartment is the site where many new polypeptide chains are modified. Two kinds of ER, rough and smooth, are named for their appearance in electron micrographs. Thousands of ribosomes are attached to the outer surface of rough ER. These ribosomes make polypeptides that thread into the interior of the ER as they are assembled ❶. Inside the ER, the polypeptides fold and take on their tertiary structure. Some of them become part of the ER membrane itself.

Cells that make, store, and secrete proteins have a lot of rough ER. For example, ER-rich gland cells in the pancreas make and secrete enzymes that help digest meals in the small intestine.

Smooth ER does not make protein, so it has no ribosomes ❷. Some of the polypeptides made in the rough ER end up as enzymes in the smooth ER. These enzymes assemble most of the lipids that form the cell's membranes. They also break down carbohydrates, fatty acids, and some drugs and poisons. In skeletal muscle cells, one type of smooth ER stores calcium ions and has a role in muscle contraction.

A Variety of Vesicles

Small, membrane-enclosed, saclike **vesicles** form in great numbers, in a variety of types, either on their own or by budding ❸. Many vesicles transport substances from one organelle to another, or to and from the plasma mem-

polypeptide

RNA

❶ **Rough ER**
Some of the RNA in the cytoplasm is translated into polypeptide chains by ribosomes attached to the rough ER. The chains enter the rough ER, where they are modified into final form.

❸ **Vesicles**
Vesicles that bud from the rough ER carry some of the new proteins to Golgi bodies. Other proteins migrate through the interior of the rough ER, and end up in the smooth ER.

nucleus

ribosome attached to ER

vesicle budding from ER

brane. *Endocytic* vesicles form as a patch of plasma membrane sinks into the cytoplasm. *Exocytic* vesicles bud from the ER or Golgi membranes and transport substances to the plasma membrane for export. Still other vesicles form on their own in the cytoplasm. Eukaryotic cells also contain **vacuoles**, which are vesicles that appear empty under a microscope. Amino acids, sugars, toxins, and ions accumulate in the fluid-filled interior of a plant cell's large **central vacuole** (Figure 4.13B). Fluid pressure in a central vacuole keeps plant cells plump, so stems, leaves, and other structures stay firm.

Other types of vesicles are a bit like trash cans that collect waste, debris, or toxins. These vesicles dispose of their contents by fusing with other vesicles called lysosomes. **Lysosomes** that bud from Golgi bodies take part in intracellular digestion. They contain powerful enzymes that

---

central vacuole Fluid-filled vesicle in many plant cells.
endomembrane system Series of interacting organelles (endoplasmic reticulum, Golgi bodies, vesicles) between nucleus and plasma membrane; produces lipids, proteins.
endoplasmic reticulum (ER) Organelle that is a continuous system of sacs and tubes; extension of the nuclear envelope. Rough ER is studded with ribosomes; smooth ER is not.
Golgi body Organelle that modifies polypeptides and lipids; also sorts and packages the finished products into vesicles.
lysosome Enzyme-filled vesicle that functions in intracellular digestion.
peroxisome Enzyme-filled vesicle that breaks down amino acids, fatty acids, and toxic substances.
vacuole A fluid-filled organelle that isolates or disposes of waste, debris, or toxic materials.
vesicle Small, membrane-enclosed, saclike organelle; different kinds store, transport, or degrade their contents.

can break down carbohydrates, proteins, nucleic acids, and lipids. Vesicles inside white blood cells or amoebas deliver ingested bacteria, cell parts, and other debris to lysosomes for destruction.

In plants and animals, vesicles called **peroxisomes** form and divide on their own, so they are not part of the endomembrane system. Peroxisomes contain enzymes that digest fatty acids and amino acids. They also break down hydrogen peroxide, a toxic by-product of fatty acid metabolism. Peroxisome enzymes convert hydrogen peroxide to water and oxygen, or use it in reactions that break down alcohol and other toxins.

Golgi Bodies

Some vesicles fuse with and empty their contents into a **Golgi body**. This organelle has a folded membrane that typically looks like a stack of pancakes ❹. Enzymes in a Golgi body put finishing touches on proteins and lipids that have been delivered from the ER. They attach phosphate groups or oligosaccharides, and cut certain polypeptides. The finished products are sorted and packaged into new vesicles that carry them to lysosomes or to the plasma membrane ❺.

Take-Home Message What is the endomembrane system?

❭ The endomembrane system includes rough and smooth endoplasmic reticulum, vesicles, and Golgi bodies.

❭ This series of organelles works together mainly to make and modify cell membrane proteins and lipids.

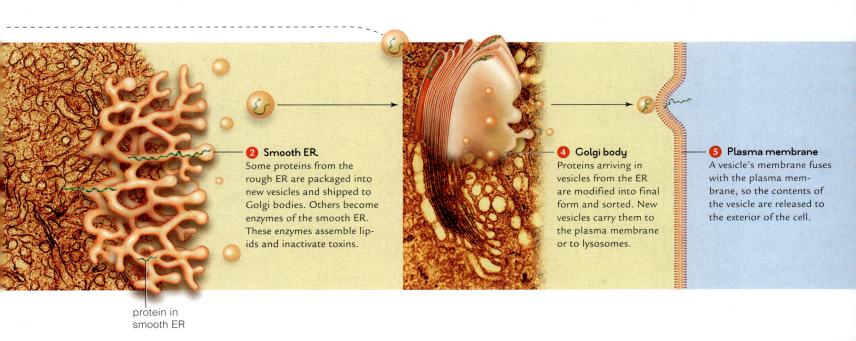

❷ **Smooth ER**
Some proteins from the rough ER are packaged into new vesicles and shipped to Golgi bodies. Others become enzymes of the smooth ER. These enzymes assemble lipids and inactivate toxins.

protein in smooth ER

❹ **Golgi body**
Proteins arriving in vesicles from the ER are modified into final form and sorted. New vesicles carry them to the plasma membrane or to lysosomes.

❺ **Plasma membrane**
A vesicle's membrane fuses with the plasma membrane, so the contents of the vesicle are released to the exterior of the cell.

❯ Eukaryotic cells make most of their ATP in mitochondria.
❯ Organelles called plastids function in storage and photosynthesis in plants and some types of algae.
❮ Links to Metabolism 3.2, ATP 3.7

Mitochondria

The **mitochondrion** (plural, mitochondria) is a type of organelle that specializes in making ATP (Figure 4.17). Aerobic respiration, an oxygen-requiring series of reactions that proceeds inside mitochondria, can extract more energy from organic compounds than any other metabolic pathway. With each breath, you are taking in oxygen mainly for mitochondria in your trillions of aerobically respiring cells.

outer membrane
outer compartment
inner compartment
inner membrane

Typical mitochondria are between 1 and 4 micrometers in length; a few are as long as 10 micrometers. Some are branched. Mitochondria can change shape, split in two, and fuse together. Each has two membranes, one highly folded inside the other. This arrangement creates two compartments. During aerobic respiration, hydrogen ions accumulate between the two membranes. The buildup causes the ions to flow across the inner mitochondrial membrane, through the interior of membrane transport proteins. That flow drives the formation of ATP.

Bacteria and archaeans have no mitochondria; they make ATP in their cell walls and cytoplasm. Nearly all eukaryotic cells have mitochondria, but the number varies by the type of cell and by the type of organism. For example, a single-celled yeast (a type of fungus) might have only one mitochondrion, but a human skeletal muscle cell may have a thousand or more. Cells that have a very high demand for energy tend to have many mitochondria.

Mitochondria resemble bacteria in size, form, and biochemistry. They have their own DNA, which is similar to bacterial DNA. They divide independently of the cell, and have their own ribosomes. Such clues led to a theory that mitochondria evolved from aerobic bacteria that took up permanent residence inside a host cell. By the theory of endosymbiosis, one cell was engulfed by another cell, or entered it as a parasite, but escaped digestion. The ingested cell kept its plasma membrane and reproduced inside its host. In time, the cell's descendants became permanent residents that offered their hosts the benefit of extra ATP. Structures and functions once required for independent life were no longer needed and were lost over time. Later descendants evolved into mitochondria. We will explore evidence for the theory of endosymbiosis in Section 18.6.

Mitochondrion
Energy powerhouse; produces many ATP by aerobic respiration

0.5 μm

Figure 4.17 Animated The mitochondrion. This organelle specializes in producing large quantities of ATP.

❯❯ **Figure It Out:** What organelle is visible in the upper right–hand corner of the TEM?   Answer: Rough ER

Chloroplasts and Other Plastids

Plastids are a category of membrane-enclosed organelles that function in photosynthesis or storage in plant and algal cells. Plastids called **chloroplasts** are organelles specialized for photosynthesis. In many ways, chloroplasts resemble photosynthetic bacteria, and like mitochondria they may have evolved by endosymbiosis.

Most chloroplasts are oval or disk-shaped (Figure 4.18). Each has two outer membranes enclosing a semifluid interior, the stroma, that contains enzymes and the chloroplast's own DNA. Inside the stroma, a third, highly folded membrane forms a single, continuous compartment. The folded membrane resembles stacks of flattened disks. The stacks are called grana (singular, granum). Photosynthesis takes place at this membrane, which is called the thylakoid membrane.

The thylakoid membrane incorporates many pigments. The most common ones are chlorophylls, which are green. The abundance of chlorophylls in plant cell chloroplasts is the reason most plants are green.

By the process of photosynthesis, chlorophylls and other molecules in the thylakoid membrane harness the energy in sunlight to drive the synthesis of ATP. The ATP is then used inside the stroma to build carbohydrates from carbon dioxide and water. (Chapter 6 describes the process of photosynthesis in more detail.)

Chromoplasts are plastids that make and store pigments other than chlorophylls. They typically have red, orange, and yellow pigments that color many flowers, leaves, fruits, and roots. For example, as a tomato ripens, its chlorophyll-containing (green) chloroplasts are converted to lycopene-containing (red) chromoplasts, so the color of the fruit changes.

Amyloplasts are unpigmented plastids. Typical amyloplasts store starch grains, and are notably abundant in starch-storing cells of stems, tubers (underground stems), and seeds. Starch-packed amyloplasts are dense and heavy compared to cytoplasm. In some plant cells, they function as gravity-sensing organelles.

---

chloroplast Organelle of photosynthesis in the cells of plants and many protists.
mitochondrion Double-membraned organelle that produces ATP by aerobic respiration in eukaryotes.
plastid An organelle that functions in photosynthesis or storage; for example, chloroplast, amyloplast.

Chloroplast
Specializes in photosynthesis

Mitochondrion

Figure 4.18 Animated The chloroplast, a defining character of photosynthetic eukaryotic cells. *Bottom*, transmission electron micrograph of a chloroplast from a tobacco leaf (*Nicotiana tabacum*). The lighter patches are nucleoids where the chloroplast's own DNA is stored.

two outer membranes

stroma

thylakoids (inner membrane system folded into flattened disks)

1 µm

> Mitochondria are eukaryotic organelles that produce ATP from organic compounds in reactions that require oxygen.

> Chloroplasts are plastids that carry out photosynthesis.

❯ Eukaryotic cells have an extensive and dynamic internal framework called a cytoskeleton.
❮ Links to Protein structure and function 3.5, 3.6

Between the nucleus and plasma membrane of all eukaryotic cells is a system of interconnected protein filaments collectively called the **cytoskeleton**. Elements of the cytoskeleton reinforce, organize, and move cell structures, and often the whole cell (Figure 4.19). Some are permanent; others form only at certain times.

Microtubules are long, hollow cylinders that consist of subunits of the protein tubulin. They form a dynamic scaffolding for many cellular processes, rapidly assem-

bling when they are needed and then disassembling when they are not. For example, before a eukaryotic cell divides, microtubules assemble, separate the cell's duplicated chromosomes, then disassemble. As another example, microtubules that form in the growing end of a young nerve cell support and guide its lengthening in a particular direction.

Microfilaments are fibers that consist primarily of subunits of the globular protein actin. They strengthen or change the shape of eukaryotic cells. Crosslinked, bundled, or gel-like arrays of them make up the **cell cortex**, which is a reinforcing mesh under the plasma membrane. Actin microfilaments that form at the edge of a cell drag or extend it in a certain direction. Myosin and actin microfilaments interact to bring about contraction of muscle cells.

Intermediate filaments that support cells and tissues are the most stable elements of the cytoskeleton. These filaments form a framework that lends structure and resilience to cells and tissues. Some kinds underlie and reinforce membranes. The nuclear envelope, for example, is supported by an inner layer of intermediate filaments called lamins. Other kinds connect to structures that lock cell membranes together in tissues.

Among the many accessory molecules associated with cytoskeletal elements are **motor proteins**, which move cell parts when energized by a phosphate-group transfer from ATP. A cell is like a bustling train station, with molecules and structures being transported continuously throughout its interior. Motor proteins are like freight trains, dragging their cellular cargo along tracks of dynamically assembled microtubules and microfilaments (Figure 4.20).

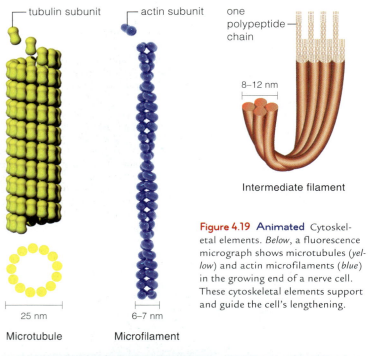

tubulin subunit · actin subunit · one polypeptide chain

8–12 nm

Intermediate filament

25 nm · 6–7 nm

Microtubule · Microfilament

Figure 4.19 Animated Cytoskeletal elements. *Below*, a fluorescence micrograph shows microtubules (*yellow*) and actin microfilaments (*blue*) in the growing end of a nerve cell. These cytoskeletal elements support and guide the cell's lengthening.

10 μm

Figure 4.20 Animated Motor proteins. Here, kinesin (*tan*) drags a *pink* vesicle as it inches along a microtubule.

basal body Organelle that develops from a centriole.
cell cortex Reinforcing mesh of cytoskeletal elements under a plasma membrane.
centriole Barrel-shaped organelle from which microtubules grow.
cilium Short, movable structure that projects from the plasma membrane of some eukaryotic cells.
cytoskeleton Dynamic framework of protein filaments that support, organize, and move eukaryotic cells and their internal structures.
intermediate filament Stable cytoskeletal element that structurally supports cells and tissues.
microfilament Reinforcing cytoskeletal element; a fiber of actin subunits.
microtubule Cytoskeletal element involved in cellular movement; hollow filament of tubulin subunits.
motor protein Type of energy-using protein that interacts with cytoskeletal elements to move the cell's parts or the whole cell.

Figure 4.21 Examples of motile structures in cells.
A Flagellum of a human sperm, which is about to penetrate an egg.
B A predatory amoeba (*Chaos carolinense*) extending two pseudopods around its hapless meal: a single-celled green alga (*Pandorina*).

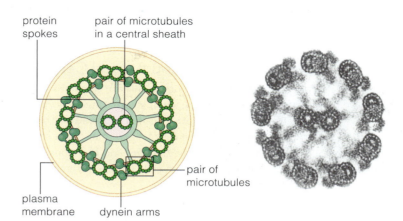

A Sketch and micrograph of one eukaryotic flagellum, cross-section. Like a cilium, it contains a 9+2 array: a ring of nine pairs of microtubules plus one pair at its core. Stabilizing spokes and linking elements that connect to the microtubules keep them aligned in this radial pattern.

A motor protein called dynein interacts with organized arrays of microtubules to bring about movement of eukaryotic flagella and cilia. Eukaryotic flagella are structures that whip back and forth to propel cells such as sperm through fluid (Figure 4.21A). They have a different internal structure and type of motion than flagella of bacteria and archaeans. **Cilia** (singular, cilium) are short, hairlike structures that project from the surface of some cells. Cilia are usually more profuse than flagella. The coordinated waving of many cilia propels cells through fluid, and stirs fluid around stationary cells.

A special array of microtubules extends lengthwise through a flagellum or cilium. This 9+2 array consists of nine pairs of microtubules ringing another pair in the center (Figure 4.22). Protein spokes and links stabilize the array. The microtubules grow from a barrel-shaped organelle called the **centriole**, which remains below the finished array as a structure called a **basal body**.

Amoebas and other types of eukaryotic cells form pseudopods, or "false feet." As these temporary, irregular lobes bulge outward, they move the cell and engulf a target such as prey (Figure 4.21B). Elongating microfilaments force the lobe to advance in a steady direction. Motor proteins that are attached to the microfilaments drag the plasma membrane along with them.

B Projecting from each pair of microtubules in the outer ring are "arms" of dynein, a motor protein that has ATPase activity. Phosphate-group transfers from ATP cause the dynein arms to repeatedly bind the adjacent pair of microtubules, bend, and then disengage. The dynein arms "walk" along the microtubules. Their motion causes adjacent microtubule pairs to slide past one another

C Short, sliding strokes occur in a coordinated sequence around the ring, down the length of each microtubule pair. The flagellum bends as the array inside bends:

basal body, a microtubule organizing center that gives rise to the 9+2 array and then remains beneath it, inside the cytoplasm

Take-Home Message What is a cytoskeleton?

> A cytoskeleton of protein filaments is the basis of eukaryotic cell shape, internal structure, and movement.

> Microtubules organize the cell and help move its parts. Networks of microfilaments reinforce the cell surface. Intermediate filaments strengthen and maintain the shape of cells and tissues.

> When energized by ATP, motor proteins move along tracks of microtubules and microfilaments. As part of cilia, flagella, and pseudopods, they can move the whole cell.

Figure 4.22 Animated Mechanism of movement of eukaryotic flagella and cilia.

> Many cells secrete materials that form a covering or matrix outside their plasma membrane.
< Links to Tissue 1.2, Chitin 3.3

Matrixes Between and Around Cells

Most cells of multicelled organisms are surrounded and organized by a nonliving, complex mixture of fibrous proteins and polysaccharides called **extracellular matrix**, or **ECM**. Secreted by the cells it surrounds, ECM supports and anchors cells, separates tissues, and functions in cell signaling. Different types of cells secrete different kinds of ECM. For example, a waxy ECM secreted by plant cells forms a **cuticle**, or covering, that protects the plant's exposed surfaces and limits water loss (Figure 4.23). The cuticle of crabs, spiders, and other arthropods is mainly chitin, a polysaccharide. ECM in animals typically consists of various kinds of carbohydrates and proteins; it is the basis of tissue organization, and it provides structural support. Bone is mostly an extracellular matrix composed of collagen, a fibrous protein, hardened by mineral deposits.

The cell wall around the plasma membrane of plant cells is a type of ECM that is structurally different from the cell wall of bacteria and archaeans. Both types of wall protect, support, and impart shape to a cell. Both are also porous: Water and solutes easily cross it on the way to and from the plasma membrane. Cells could not live without exchanging these substances with their environment. A plant cell wall forms as a young cell secretes pectin and

Figure 4.23 A plant ECM. Section through a plant leaf showing cuticle, a protective covering of deposits secreted by living cells.

thick, waxy cuticle at leaf surface
cell of leaf epidermis
photosynthetic cell inside leaf

other polysaccharides onto the outer surface of its plasma membrane. The sticky coating is shared between adjacent cells, and it cements them together. Each cell then forms a **primary wall** by secreting strands of cellulose into the coating. Some of the coating remains as the middle lamella, a sticky layer in between the primary walls of abutting plant cells (Figure 4.24A).

Being thin and pliable, a primary wall allows a growing plant cell to enlarge. Cells with only a thin primary wall can change shape as they develop. At maturity, cells in some plant tissues stop enlarging and begin to secrete material onto the primary wall's inner surface. These deposits form a firm **secondary wall** (Figure 4.24B). One of the materials deposited is **lignin**, an organic compound that makes up as much as 25 percent of the secondary

A Plant cell secretions form the middle lamella, a layer that cements adjoining cells together.

middle lamella
plasma membrane
cytoplasm
primary cell wall

B In many plant tissues, cells also secrete materials that are deposited in layers on the inner surface of their primary wall. These layers strengthen the wall and maintain its shape. They remain after the cells die, and become part of pipelines that carry water through the plant.

secondary cell wall (added in layers)
primary cell wall

pipeline made of abutting cell walls

middle lamella

C Plasmodesmata are channels across the cell walls and the plasma membranes of living cells that are pressed against one another in tissues.

plasmodesma
middle lamella

Figure 4.24 Animated Some characteristics of plant cell walls.

Figure 4.25 **Animated** Three types of cell junctions in animal tissues: tight junctions, gap junctions, and adhering junctions.

In the micrograph *above*, a profusion of tight junctions (*green*) seals abutting surfaces of kidney cell membranes and forms a waterproof tissue. The DNA in each cell nucleus appears *red*.

free surface of epithelial tissue

tight junctions

gap junction

basement membrane (extracellular matrix)

adhering junction

wall of cells in older stems and roots. Lignified plant parts are stronger, more waterproof, and less susceptible to plant-attacking organisms than younger tissues.

Cell Junctions

ECM does not prevent a cell from interacting with other cells or the surroundings. In multicelled species, such interaction occurs by way of **cell junctions**, which are structures that connect a cell to other cells and to the environment. Cells send and receive ions, molecules, or signals through some junctions. Other kinds help cells recognize and stick to each other and to extracellular matrix.

In most animal tissues, cells are connected to their neighbors and to ECM by cell junctions (Figure 4.25). In epithelial tissues that line body surfaces and internal cavities, rows of proteins that form **tight junctions** between plasma membranes prevent body fluids from seeping between adjacent cells. To cross these tissues, fluid must pass directly through the cells. Thus, transport proteins

embedded in the cell membranes help control which ions and molecules cross the tissue. For example, an abundance of tight junctions in the lining of the stomach normally keeps acidic fluid from leaking out. If a bacterial infection damages this lining, acid and enzymes can erode the underlying layers. The result is a painful peptic ulcer.

Adhering junctions composed of adhesion proteins snap cells to each other and anchor them to ECM. Skin and other tissues that are subject to abrasion or stretching have a lot of adhering junctions. These cell junctions also strengthen contractile tissues such as heart muscle. **Gap junctions** form channels that connect the cytoplasm of adjoining cells, thus permitting ions and small molecules to pass directly from the cytoplasm of one cell to another. By opening or closing, these channels allow entire regions of cells to respond to a single stimulus. Heart muscle and other tissues in which the cells perform a coordinated action have many of these communication channels.

In plants, open channels called **plasmodesmata** (singular, plasmodesma) extend across cell walls, connecting the cytoplasm of adjoining cells (Figure 4.24C). Substances such as water, nutrients, and signaling molecules can flow quickly from cell to cell through these cell junctions.

adhering junction Cell junction composed of adhesion proteins; anchors cells to each other and extracellular matrix.
cell junction Structure that connects a cell to another cell or to extracellular matrix.
cuticle Secreted covering at a body surface.
extracellular matrix (ECM) Complex mixture of cell secretions; supports cells and tissues; has roles in cell signaling.
gap junction Cell junction that forms a channel across the plasma membranes of adjoining animal cells.
lignin Material that stiffens cell walls of vascular plants.
plasmodesmata Cell junctions that connect the cytoplasm of adjacent plant cells.
primary wall The first cell wall of young plant cells.
secondary wall Lignin-reinforced wall that forms inside the primary wall of a plant cell.
tight junctions Arrays of fibrous proteins; join epithelial cells and collectively prevent fluids from leaking between them.

Take-Home Message **What structures form on the outside of eukaryotic cells?**

> Cells of many protists, nearly all fungi, and all plants have a porous wall around the plasma membrane. Animal cells do not have walls.

> Plant cell secretions form a waxy cuticle that helps protect the exposed surfaces of soft plant parts.

> Cell secretions form extracellular matrixes between cells in many tissues.

> Cells make structural and functional connections with one another and with extracellular matrix in tissues.

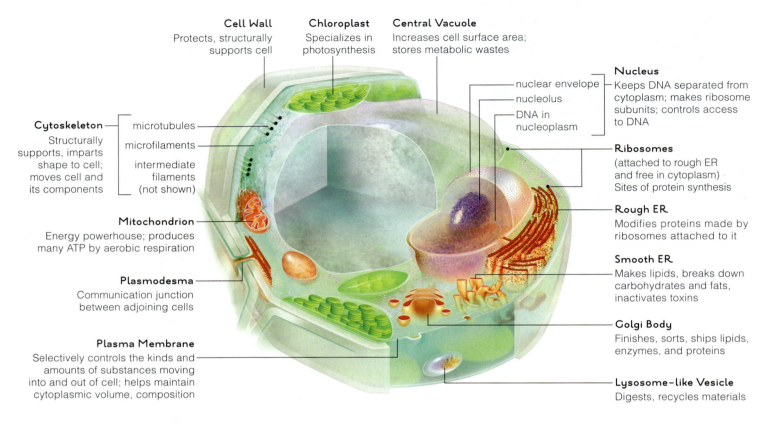

Cell Wall
Protects, structurally supports cell

Chloroplast
Specializes in photosynthesis

Central Vacuole
Increases cell surface area; stores metabolic wastes

Nucleus
Keeps DNA separated from cytoplasm; makes ribosome subunits; controls access to DNA
- nuclear envelope
- nucleolus
- DNA in nucleoplasm

Cytoskeleton
Structurally supports, imparts shape to cell; moves cell and its components
- microtubules
- microfilaments
- intermediate filaments (not shown)

Ribosomes
(attached to rough ER and free in cytoplasm) Sites of protein synthesis

Rough ER
Modifies proteins made by ribosomes attached to it

Mitochondrion
Energy powerhouse; produces many ATP by aerobic respiration

Smooth ER
Makes lipids, breaks down carbohydrates and fats, inactivates toxins

Plasmodesma
Communication junction between adjoining cells

Golgi Body
Finishes, sorts, ships lipids, enzymes, and proteins

Plasma Membrane
Selectively controls the kinds and amounts of substances moving into and out of cell; helps maintain cytoplasmic volume, composition

Lysosome-like Vesicle
Digests, recycles materials

A Typical plant cell components.

Nucleus
Keeps DNA separated from cytoplasm; makes ribosome subunits; controls access to DNA
- nuclear envelope
- nucleolus
- DNA in nucleoplasm

Cytoskeleton
Structurally supports, imparts shape to cell; moves cell and its components
- microtubules
- microfilaments
- intermediate filaments

Ribosomes
(attached to rough ER and free in cytoplasm) Sites of protein synthesis

Rough ER
Modifies proteins made by ribosomes attached to it

Mitochondrion
Energy powerhouse; produces many ATP by aerobic respiration

Smooth ER
Makes lipids, breaks down carbohydrates and fats, inactivates toxins

Centrioles
Special centers that produce and organize microtubules

Golgi Body
Finishes, sorts, ships lipids, enzymes, and proteins

Plasma Membrane
Selectively controls the kinds and amounts of substances moving into and out of cell; helps maintain cytoplasmic volume, composition

Lysosome
Digests, recycles materials

B Typical animal cell components.

Figure 4.26 Animated Organelles and structures typical of **A** plant cells and **B** animal cells.

4.13 The Nature of Life

> We define life by describing the set of properties that is unique to living things.

< Links to Life's levels of organization 1.2, Homeostasis 1.3, Philosophy of science 1.9

In this chapter, you learned about the structure of cells, which have at their minimum a plasma membrane, cytoplasm, and a region of DNA. Most cells have many other components. But what, exactly, makes a cell, or an organism that consists of them, alive? We say that life is a property that emerges from cellular components, but a collection of those components in the right amounts and proportions is not necessarily alive.

We know intuitively what "life" is, but defining it unambiguously is challenging, if not impossible. We can more easily describe what sets living things apart from nonliving things, but even that is tricky. For example, living organisms have a high proportion of the organic molecules of life, but so do the remains of dead organisms such as coal and other fossil fuels. Living things use energy to reproduce themselves, but computer viruses, which are arguably not alive, can do that too.

So how do biologists, who study life as a profession, describe it? The short answer is that their best description of life is very long. It consists of a list of properties associated with things we know to be alive. You already know about two of these properties of organisms:

1. They make and use the organic molecules of life.
2. They consist of one or more cells.

The remainder of this book details the other properties of living things:

3. They engage in self-sustaining biological processes such as metabolism and homeostasis.
4. They change over their lifetime, for example by growing, maturing, and aging.
5. They use DNA as their hereditary material.
6. They have the collective capacity to change over successive generations, for example by adapting to environmental pressures.

Collectively, these properties characterize living things as different from nonliving things.

Take-Home Message What is life?

> We describe the characteristic of "life" in terms of a set of properties. The set is unique to living things.

> In living things, the molecules of life are organized as one or more cells that engage in self-sustaining biological processes.

> Organisms use DNA as their hereditary material.

> Living things change over lifetimes, and over generations.

Food for Thought (revisited)

One food safety measure involves sterilization, which kills *E. coli* O157:H7 and other bacteria. Recalled, contaminated ground beef is typically cooked or otherwise sterilized, then processed into ready-to-eat products. Raw beef trimmings are effectively sterilized when sprayed with ammonia and ground to a paste. The resulting meat product is now routinely added to hamburger patties, fresh ground beef, hot dogs, lunch meats, sausages, frozen entrees, canned foods, and other items sold to quick service restaurants, hotel and restaurant chains, institutions, and school lunch programs. Meat, poultry, milk, and fruits sterilized by exposure to radiation are available in supermarkets. By law, irradiated foods must be marked with the symbol on the *left*, but foods sterilized with chemicals are not currently required to carry any disclosure. Some worry that sterilization may alter food or leave harmful chemicals in it. Whether any health risks are associated with consuming sterilized foods is unknown.

How Would You Vote? Some think the safest way to protect consumers from food poisoning is by sterilizing food with chemicals or radiation to kill any bacteria that may be in it. Others think we should make sure our food does not get contaminated in the first place. Would you choose sterilized food? See CengageNow for details, then vote online (cengagenow.com).

Summary

 Sections 4.1 and 4.2 The **cell** is the smallest unit of life. Although cells differ in size, shape, and function, each starts out life with a **plasma membrane**, **cytoplasm**, and a region of DNA—in eukaryotic cells, a **nucleus**. A lipid bilayer is the structural foundation of cell membranes, including **organelle** membranes.

The **surface-to-volume ratio** limits cell size. By the **cell theory**, all organisms consist of one or more cells; the cell is the smallest unit of life; each new cell arises from another, preexisting cell; and a cell passes hereditary material to its offspring.

 Section 4.3 Most cells are far too small to see with the naked eye. We use different types of microscopes and techniques to reveal cells and their internal and external details.

 Section 4.4 A cell membrane is a mosaic of lipids (mainly phospholipids) and proteins. It functions as a selectively permeable barrier that separates an internal environment from an external one. The lipids are organized as a **lipid bilayer**: a double layer of lipids in which the nonpolar tails of both layers are sandwiched between the polar heads.

The membranes of most cells can be described as a **fluid mosaic**. Proteins associated with a membrane carry out most membrane functions. All membranes have **transport proteins**. Plasma membranes also have **receptor proteins**, **adhesion proteins**, enzymes, and **recognition proteins**.

Summary

Section 4.5 Bacteria and archaeans have no nucleus, but they have **nucleoids** and **ribosomes**. Many have a **cell wall**, **flagella** or **pili**, and **plasmids**. Bacteria and other microbial organisms often share living arrangements in **biofilms**.

Sections 4.6, 4.7 Eukaryotic cells start out life with membrane-enclosed organelles, including a nucleus. A nucleus has a double-membraned **nuclear envelope** surrounding **nucleoplasm**. In the nucleus, ribosome subunits are assembled in dense regions called **nucleoli**.

Section 4.8 The **endomembrane system** includes rough and smooth **endoplasmic reticulum (ER)**, **vesicles**, and **Golgi bodies**. This system makes and modifies lipids and proteins; it also recycles and disposes of molecules and particles. **Peroxisomes**, **lysosomes**, and **vacuoles** (including **central vacuoles**) are vesicles.

Section 4.9 Mitochondria make ATP by breaking down organic compounds in the oxygen-requiring pathway of aerobic respiration. **Chloroplasts** are **plastids** that produce sugars by photosynthesis.

Section 4.10 A **cytoskeleton** includes **microtubules**, **microfilaments**, and **intermediate filaments**. A microfilament **cell cortex** reinforces plasma membranes. Interactions between **motor proteins** and microtubules move **cilia** and eukaryotic flagella. **Centrioles** give rise to microtubules, then become **basal bodies**.

Section 4.11 Cuticle is an example of **extracellular matrix (ECM)**. Pliable **primary walls** enclose **secondary walls** strengthened with **lignin**. **Cell junctions** connect cells to one another and to ECM. **Adhering junctions**, **gap junctions**, and **tight junctions** connect animal cells. **Plasmodesmata** connect plant cells.

Sections 4.12, 4.13 We describe the characteristic of life as a set of properties. The set is unique to living things.

Table 4.3 Summary of Typical Components of Cells

Cell Component	Main Functions	Bacteria, Archaeans	Eukaryotes			
			Protists	Fungi	Plants	Animals
Cell wall	Protection, structural support	✔*	✔*	✔	✔	None
Plasma membrane	Control of substances moving into and out of cell	✔	✔	✔	✔	✔
Nucleus	Physical separation of DNA from cytoplasm	None	✔	✔	✔	✔
DNA	Encoding of hereditary information	✔	✔	✔	✔	✔
RNA	Transcription, translation	✔	✔	✔	✔	✔
Nucleolus	Assembly of ribosome subunits	None	✔	✔	✔	✔
Ribosome	Protein synthesis	✔	✔	✔	✔	✔
Endoplasmic reticulum (ER)	Initial modification of polypeptide chains; lipid synthesis	None	✔	✔	✔	✔
Golgi body	Final modification of proteins, lipid assembly, and packaging of both for use inside cell or export	None	✔	✔	✔	✔
Lysosome	Intracellular digestion	None	✔	✔*	✔*	✔
Mitochondrion	ATP formation	**	✔	✔	✔	✔
Photosynthetic pigments	Light-to-energy conversion	✔*	✔*	None	✔	None
Chloroplast	Photosynthesis; some starch storage	None	✔*	None	✔	None
Central vacuole	Increasing cell surface area; storage	None	None	✔*	✔	None
Bacterial flagellum	Locomotion through fluid surroundings	✔*	None	None	None	None
Eukaryotic flagellum or cilium	Locomotion through or motion within fluid surroundings	None	✔*	✔*	✔*	✔
Cytoskeleton	Cell shape; internal organization; basis of cell movement and, in many cells, locomotion	Rudimentary***	✔*	✔*	✔*	✔

 * Known to be present in cells of at least some groups.
 ** Many groups use oxygen-requiring (aerobic) pathways of ATP formation, but mitochondria are not involved.
 *** Protein filaments form a simple scaffold that helps support the cell wall in at least some species.

Data Analysis Activities

Abnormal Motor Proteins Cause Kartagener Syndrome An abnormal form of a motor protein called dynein causes Kartagener syndrome, a genetic disorder characterized by chronic sinus and lung infections. Biofilms form in the thick mucus that collects in the airways, and the resulting bacterial activities and inflammation cause recurring respiratory infections that damage tissues.

Affected people are usually infertile. Men can produce sperm (Figure 4.27), and some have become fathers after a doctor injects their sperm cells directly into eggs. Review Figure 4.22, then explain how abnormal dynein could cause infertility in males.

Figure 4.27 Cross-section of the flagellum of a sperm cell from **A** a human male affected by Kartagener syndrome and **B** an unaffected male.

1. The _____ is the smallest unit of life.

2. Every cell is descended from another cell. This idea is part of _____ .
   a. evolution
   b. the theory of heredity
   c. the cell theory
   d. cell biology

3. True or false? Some protists have no nucleus.

4. Cell membranes consist mainly of a _____ .
   a. carbohydrate bilayer and proteins
   b. protein bilayer and phospholipids
   c. lipid bilayer and proteins

5. Unlike eukaryotic cells, bacterial cells _____ .
   a. have no plasma membrane   c. have no nucleus
   b. have RNA but not DNA   d. a and c

6. In a lipid bilayer, _____ of all the lipid molecules are sandwiched between all the _____ .
   a. hydrophilic tails; hydrophobic heads
   b. hydrophilic heads; hydrophilic tails
   c. hydrophobic tails; hydrophilic heads
   d. hydrophobic heads; hydrophilic tails

7. Enzymes contained in _____ break down worn-out organelles, bacteria, and other particles.

8. Put the following structures in order according to the pathway of a secreted protein:
   a. plasma membrane   c. endoplasmic reticulum
   b. Golgi bodies   d. post-Golgi vesicles

9. The main function of the endomembrane system is building and modifying _____ and _____ .

10. Is this statement true or false? The plasma membrane is the outermost component of all cells. Explain.

11. Most membrane functions are performed by _____ .
    a. proteins   c. nucleic acids
    b. phospholipids   d. hormones

12. No animal cell has a _____ .
    a. plasma membrane   c. lysosome
    b. flagellum   d. cell wall

13. _____ connect the cytoplasm of plant cells.
    a. Plasmodesmata   c. Tight junctions
    b. Adhering junctions   d. a and b

14. Match each cell component with its function.
    ___ mitochondrion   a. protein synthesis
    ___ chloroplast   b. associates with
    ___ ribosome   ribosomes
    ___ smooth ER   c. ATP production
    ___ Golgi body   d. sorts and ships
    ___ rough ER   e. assembles lipids
                             f. photosynthesis

Additional questions are available on **CENGAGENOW**.

Critical Thinking

1. In a classic episode of *Star Trek*, a gigantic amoeba engulfs an entire starship. Spock blows the cell to bits before it reproduces. Think of at least one problem a biologist would have with this particular scenario.

2. A student is examining different samples with a transmission electron microscope. She discovers a single-celled organism swimming in a freshwater pond (*below*). Which of this organism's structures can you identify? What type of cell do you think this is? Be as specific as you can. Look ahead to Section 20.3 to check your answers.

Animations and Interactions on **CENGAGENOW**:
❯ Overview of cells; Surface-to-volume ratio; Microscopes; Lipid bilayer; Membrane proteins; Bacterial structure; Eukaryotic organelles; The nuclear envelope; Endomembrane system; Mitochondria; Chloroplasts; Cytoskeletal elements; Motor proteins; Eukaryotic flagella and cilia; Plant cell walls; Animal cell junctions.

‹ Links to Earlier Concepts

Here, we revisit electrons (Section 2.3), chemical bonds (2.4), carbohydrates (3.3), ATP (3.7), motor proteins (4.10), solutes (2.5), and plant cell walls (4.11), as we discuss how organisms use energy to maintain their organization (1.2, 1.3). Protein structure and function (3.5, 3.6), membrane proteins (4.4), and vesicles (4.8) reappear in the context of metabolism (3.2). You will learn about how enzymes work, and how factors such as temperature (2.5) and pH (2.6) affect them.

Key Concepts

Energy Flow
Organisms maintain their organization only by continually harvesting energy from their environment. ATP couples reactions that release usable energy with reactions that require it.

How Enzymes Work
Enzymes tremendously increase the rate of metabolic reactions. Cofactors assist enzymes, and environmental factors such as temperature, salt, and pH can influence enzyme function.

5 Ground Rules of Metabolism

5.1 A Toast to Alcohol Dehydrogenase

The next time someone asks you to have a drink, stop for a moment and think about the cells in your body that detoxify alcohol. It makes no difference whether a person drinks a bottle of beer, a glass of wine, or 1–1/2 ounces of vodka. Each holds the same amount of alcohol or, more precisely, ethanol. Ethanol molecules move quickly from the stomach and small intestine into the bloodstream. Almost all of the ethanol ends up in the liver, which is a large organ in the abdomen. The liver has impressive numbers of enzymes. One of them, alcohol dehydrogenase, helps rid the body of ethanol and other toxins (Figure 5.1).

Ethanol harms the liver. For one thing, breaking it down produces molecules that directly damage liver cells, so the more a person drinks, the fewer liver cells are left to do the breaking down. Ethanol also interferes with normal processes of metabolism. For example, oxygen that would normally take part in breaking down fatty acids is diverted to breaking down ethanol, so fats tend to accumulate as large globules in the tissues of heavy drinkers. A common outcome of such interference is alcoholic hepatitis, a disease characterized by inflammation and destruction of liver tissue. Long-term, heavy drinking can also lead to cirrhosis, or scarring of the liver. (The term cirrhosis is from the Greek *kirros*, meaning orange-colored, after the abnormal skin color of people with the disease.)

Eventually, the liver of a heavy drinker may just quit working, with dire health consequences. The liver is the largest gland in the human body, and it has many important functions. In addition to breaking down fats and toxins, it helps regulate the body's blood sugar level, and it makes proteins that are essential for blood clotting, immune function, and maintaining the solute balance of body fluids.

Heavy drinking can be dangerous in the short term too. Binge drinking, a self-destructive behavior that involves consuming large amounts of alcohol in a brief period of time, is currently the most serious drug problem on college campuses in the United States. Tens of thousands of undergraduate students have been polled about their drinking habits in recent surveys. More than half of them reported that they regularly drink five or more alcoholic beverages within a two-hour period.

Figure 5.1 Alcohol metabolism. Alcohol dehydrogenase helps the body break down toxic alcohols such as ethanol. This enzyme makes it possible for humans to drink beer, wine, and other alcoholic beverages.

Binge drinking does far more than damage one's liver. Aside from the related 500,000 injuries from accidents, the 600,000 assaults by intoxicated students, 100,000 cases of date rape, and 400,000 incidences of unprotected sex among students, binge drinking is responsible for killing or causing the death of more than 1,700 college students every year. Ethanol is toxic: If you put more of it into your body than your enzymes can deal with, then you will die. With this sobering example, we invite you to learn about how and why your cells break down organic compounds, including toxic molecules such as ethanol.

The Nature of Metabolism
Metabolic pathways are energy-driven sequences of enzyme-mediated reactions. They build, convert, and dispose of materials in cells. Controls that govern steps in metabolic pathways can quickly shift cell activities.

Movement of Fluids
Gradients drive the directional movements of substances across membranes. Water tends to diffuse across selectively permeable membranes, including cell membranes, to regions where solute concentration is higher.

Membrane Trafficking
Transport proteins that work with or against gradients adjust or maintain solute concentrations. Large packets of substances move across the plasma membrane by processes of endocytosis and exocytosis.

> Energy input into living cells drives assembly of the molecules of life.

‹ Links to Life's organization 1.2, Energy 1.3, Laws of nature 1.9, Chemical bonding 2.4, ATP 3.7

Energy Disperses

We define **energy** as the capacity to do work, but this definition is not very satisfying. Even the brilliant physicists who study it cannot say what energy is, exactly. However, even without a perfect definition, we have an intuitive understanding of energy just by thinking about familiar forms of it, such as light, heat, electricity, and motion. We also understand intuitively that one form of energy can be converted to another. Think about how a light-bulb changes electricity into light, or how an automobile changes gasoline into the energy of motion, which is also called **kinetic energy** (Figure 5.2).

The formal study of heat and other forms of energy is called thermodynamics (*therm* is a Greek word for heat; *dynam* means energy). By making careful measurements, thermodynamics researchers discovered that the total amount of energy before and after every conversion is always the same. In other words, energy cannot be created or destroyed—a phenomenon that we call the **first law of thermodynamics**. Remember, a law of nature describes something that occurs without fail, but our explanation of why it occurs is incomplete (Section 1.9).

Energy also tends to spread out, or disperse, until no part of a system holds more than another part. In a kitchen, for example, heat always flows from a hot pan to cool air until the temperature of both is the same. We

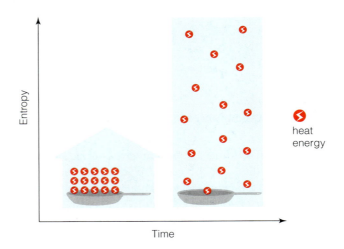

Figure 5.3 Entropy. Entropy tends to increase, but the total amount of energy in any system always stays the same.

never see cool air raising the temperature of a hot pan. **Entropy** is a measure of how much the energy of a particular system has become dispersed. Let's use the hot pan in a cool kitchen as an example of a system. As heat flows from the pan into the air, the entropy of the system increases (Figure 5.3). Entropy continues to increase until the heat is evenly distributed throughout the kitchen, and there is no longer a net (or overall) flow of heat from one area to another. Our system has now reached its maximum entropy with respect to heat. The tendency of entropy to increase is the **second law of thermodynamics**. This is just a fancy way of saying that energy tends to spread out spontaneously.

Biologists use the concept of entropy as it applies to chemical bonding, because energy flow in living things occurs mainly by the making and breaking of chemical bonds. How is entropy related to chemical bonding? Think about it just in terms of motion. Two unbound atoms can vibrate, spin, and rotate in every direction, so they are at high entropy with respect to motion. A covalent bond between the atoms restricts their movement, so they are able to move in fewer ways than they did before bonding. Thus, the entropy of two atoms decreases when a bond forms between them. Such entropy changes are part of the reason why some reactions occur spontaneously and others require an energy input, as you will see in the next section.

Energy's One-Way Flow

Work occurs as a result of an energy transfer. For example, making ATP is work, so it requires energy. A plant cell can do this work by transferring energy from the environment (light) to molecules that use the energy to

Figure 5.2 Demonstration of a familiar type of energy: motion, or kinetic energy.

Figure 5.4 It takes more than 10,000 pounds of soybeans and corn to raise a 1,000-pound steer. Where do the other 9,000 pounds go? About half of the steer's food is indigestible. The animal's body breaks down molecules in the remaining half to access energy stored in chemical bonds. Only about 15% of that energy goes toward building body mass. The rest is lost during energy conversions, as heat.

build ATP. This particular transfer involves an energy conversion (light energy to chemical energy). Most other types of cellular work occur by the transfer of chemical energy from one molecule to another. For example, cells make glucose by transferring chemical energy from ATP to other molecules.

As you learn about such processes, remember that every time energy is transferred, a bit of it disperses. The energy lost from the transfer is usually in the form of heat. As a simple example, a typical incandescent light-bulb converts about 5 percent of the energy of electricity into light. The remaining 95 percent of the energy ends up as heat that radiates from the bulb.

Dispersed heat is not very useful for doing work, and it is not easily converted to a more useful form of energy (such as electricity). Because some of the energy in every transfer disperses as heat, and heat is not useful for doing work, we can say that the total amount of energy available for doing work in the universe is always decreasing.

Is life an exception to this inevitable flow? An organized body is hardly dispersed. Energy becomes concentrated in each new organism as the molecules of life organize into cells. Even so, living things constantly use energy to grow, to move, to acquire nutrients, to reproduce, and so on. Inevitable losses occur during the energy transfers that maintain life (Figure 5.4). Unless those losses are replenished with energy from another source, the complex organization of life will end.

Most of the energy that fuels life on Earth comes from the sun. In our world, energy flows from the sun, through

sunlight energy

Producers
plants and other self–feeding organisms

nutrient cycling

Consumers
animals, most fungi, many protists, bacteria

Energy In
Sunlight energy reaches environments on Earth. Producers in those environments capture some of the energy and convert it to other forms that can drive cellular work.

Energy Out
With each conversion, some energy escapes into the environment, mainly as heat. Living things do not use heat to drive cellular work, so the energy flows only in one direction.

Figure 5.5 Animated Energy flows from the environment into living organisms, and then back to the environment. The flow drives a cycling of materials among producers and consumers.

producers, then consumers (Figure 5.5). During this journey, the energy is transferred many times. With each transfer, some energy escapes as heat until, eventually, all of it is permanently dispersed. However, the second law of thermodynamics does not say how quickly the dispersal has to happen. Energy's spontaneous dispersal is resisted by chemical bonds. The energy in chemical bonds is a type of **potential energy**, because it can be stored. Think of all the bonds in the countless molecules that make up your skin, heart, liver, fluids, and other body parts. Those bonds hold the molecules, and you, together—at least for the time being.

energy The capacity to do work.
entropy Measure of how much the energy of a system is dispersed.
first law of thermodynamics Energy cannot be created or destroyed.
kinetic energy The energy of motion.
potential energy Stored energy.
second law of thermodynamics Energy tends to disperse spontaneously.

Take-Home Message What is energy?

❯ Energy is the capacity to do work. It can be transferred between systems or converted from one form to another, but it cannot be created or destroyed.

❯ Energy disperses spontaneously.

❯ Some energy is lost during every transfer or conversion.

❯ Organisms can maintain their complex organization only as long as they replenish themselves with energy they harvest from someplace else.

> All cells store and retrieve energy in chemical bonds of the molecules of life.
> ATP functions as an energy carrier in cellular reactions.
< Links to Bonding 2.4, Carbohydrates 3.3, Nucleotides 3.7

Energy In, Energy Out

Cells store energy in chemical bonds, and access the energy stored in chemical bonds by breaking them. Both processes change molecules. Any process by which such chemical change occurs is called a **reaction**.

During a reaction, one or more **reactants** (molecules that enter a reaction) become one or more **products** (molecules that remain at the reaction's end). Intermediate molecules may form between reactants and products. We show a chemical reaction as an equation in which an arrow points from reactants to products (Figure 5.6). A number before a chemical formula in the equation indicates the number of molecules. Atoms shuffle around in a reaction, but they never disappear: The same number of atoms that enter a reaction remain at the reaction's end.

Every chemical bond holds energy, and the amount of energy depends on which elements are taking part in the bond. For example, the covalent bond between an oxygen and hydrogen atom in any water molecule always holds the same amount of energy. That is the amount of energy required to break the bond, and it is also the amount of energy released when the bond forms. Bond energy and entropy both contribute to a molecule's free energy, which is the amount of energy that is available (free) to do work.

2H₂ (hydrogen)   O₂ (oxygen)   →   2H₂O (water)

Reactants	Products
4 hydrogen atoms + 2 oxygen atoms	4 hydrogen atoms + 2 oxygen atoms

Figure 5.6 Animated Chemical bookkeeping. In equations that represent chemical reactions, reactants are written to the left of an arrow that points to the products. A number before a formula indicates the number of molecules. Atoms shuffle around in a reaction, but they never disappear: The same number of atoms that enter a reaction remain at the reaction's end.

In most reactions, the free energy of reactants differs from the free energy of products (Figure 5.7). Reactions in which reactants have less free energy than products will not proceed without a net energy input. Such reactions are **endergonic**, which means "energy in" ❶. In other reactions, reactants have greater free energy than products. Such reactions are **exergonic**, which means "energy out," because they end with a net release of free energy ❷. Cells access the free energy of molecules by running exergonic reactions. An example is the overall process of aerobic respiration, which converts glucose and oxygen to carbon dioxide and water for a net energy output.

Why Earth Does Not Go Up in Flames

The molecules of life release energy when they combine with oxygen. For example, think of how a spark ignites tinder-dry wood in a campfire. Wood is mostly cellulose, which is a carbohydrate that consists of long chains of repeating glucose units. The spark initiates a reaction that converts cellulose in the wood and oxygen in air to water and carbon dioxide. The reaction is exergonic, and it releases enough energy to initiate the same reaction with other cellulose and oxygen molecules. That is why a campfire keeps burning once it has been lit.

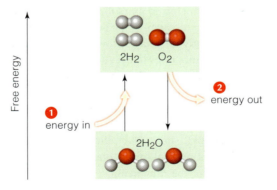

Free energy

2H₂   O₂

❶ energy in

❷ energy out

2H₂O

Figure 5.7 Energy inputs and outputs in chemical reactions.

❶ Endergonic reactions convert molecules with lower energy to molecules with higher energy, so they require a net energy input in order to proceed.

❷ Exergonic reactions convert molecules with higher energy to molecules with lower energy, so they end with a net energy output.

>> **Figure It Out** Which law of thermodynamics explains energy inputs and outputs in chemical reactions?

Answer: The first law

activation energy Minimum amount of energy required to start a reaction.
ATP Adenosine triposphate; an energy carrier that couples endergonic with exergonic reactions in cells.
ATP/ADP cycle Process by which cells regenerate ATP. ADP forms when ATP loses a phosphate group, then ATP forms again as ADP gains a phosphate group.
endergonic Type of reaction that requires a net input of free energy to proceed.
exergonic Type of reaction that ends with a net release of free energy.
phosphorylation Addition of a phosphate group to a molecule; occurs by the transfer of a phosphate group from a donor molecule such as ATP.
product A molecule that remains at the end of a reaction.
reactant Molecule that enters a reaction.
reaction Process of chemical change.

Figure 5.8 **Animated** Activation energy. Most reactions will not begin without an input of activation energy, which is shown here as a bump in an energy hill. In this example, the reactants have more energy than the products. Activation energy keeps this and other exergonic reactions from starting spontaneously.

A Structure of ATP.

B After ATP loses one phosphate group, the nucleotide is ADP (adenosine diphosphate); after losing two phosphate groups, it is AMP (adenosine monophosphate).

C ATP forms by endergonic reactions. ADP forms again when ATP energy is transferred to another molecule along with a phosphate group. Energy from such transfers drives cellular work.

Figure 5.9 ATP, the energy currency of cells.

Earth is rich in oxygen—and in potential exergonic reactions. Why doesn't it burst into flames? Luckily, energy is required to break the chemical bonds of reactants, even in an exergonic reaction. **Activation energy** is the minimum amount of energy that will get a chemical reaction started (Figure 5.8). Activation energy is a bit like a hill that reactants must climb before they can coast down the other side to products.

Both endergonic and exergonic reactions have activation energy, but the amount varies with the reaction. Consider guncotton (nitrocellulose), a highly explosive derivative of cellulose. Christian Schönbein accidentally discovered a way to manufacture it when he used his wife's cotton apron to wipe up a nitric acid spill on his kitchen table, then hung it up to dry next to the oven. The apron exploded. Being a chemist in the 1800s, Schönbein immediately thought of marketing guncotton as a firearm explosive, but it proved to be too unstable to manufacture. So little activation energy is needed to make guncotton react with oxygen that it tends to explode unexpectedly. The substitute? Gunpowder, which has a higher activation energy for a reaction with oxygen.

ATP—The Cell's Energy Currency

ATP Cells pair reactions that require energy with reactions that release energy. ATP and other nucleotides are often part of that process. **ATP**, or adenosine triphosphate, functions as an energy carrier by accepting energy released by exergonic reactions, and delivering energy to endergonic reactions. ATP is the main currency in a cell's energy economy, so we use a cartoon coin to symbolize it.

ATP is a nucleotide with three phosphate groups (Figure 5.9A). The bonds that link those phosphate groups hold a lot of energy. When a phosphate group is trans-

ferred from ATP to another molecule, energy is transferred along with the phosphate. That energy contributes to the "energy in" part of an endergonic reaction. The transfer of a phosphate group is called **phosphorylation**.

Cells constantly use up ATP to drive endergonic reactions, so they constantly replenish it. When ATP loses a phosphate, ADP (adenosine diphosphate) forms (Figure 5.9B). ATP forms again when ADP binds phosphate in an endergonic reaction. The cycle of using and replenishing ATP is called the **ATP/ADP cycle** (Figure 5.9C).

Take-Home Message How do cells use energy?

❯ Cells store and retrieve energy by making and breaking chemical bonds.

❯ Activation energy is the minimum amount of energy required to start a chemical reaction.

❯ Endergonic reactions cannot run without a net input of energy. Exergonic reactions end with a net release of energy.

❯ Energy carriers such as ATP couple exergonic reactions with endergonic ones.

❯ Enzymes make specific reactions occur much faster than they would on their own.

❮ Links to Temperature 2.5, pH 2.6, Protein structure 3.5, Denaturation 3.6

Centuries might pass before sugar would break down to carbon dioxide and water on its own, yet that same conversion takes just seconds inside your cells. Enzymes make the difference. In a process called **catalysis**, an enzyme makes a reaction run much faster than it would on its own. The enzyme is unchanged by participating in the reaction, so it can work again and again.

Most enzymes are proteins, but some are RNAs. Each kind recognizes specific reactants, or **substrates**, and alters them in a specific way. For instance, the enzyme hexokinase adds a phosphate group to the hydroxyl group on the sixth carbon of glucose. Such specificity occurs because an enzyme's polypeptide chains fold up into one or more **active sites**, which are pockets where substrates bind and where reactions proceed (Figure 5.10). An active

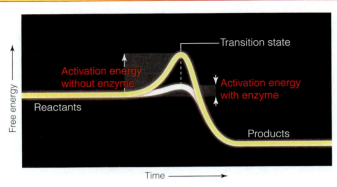

Figure 5.11 Animated An enzyme enhances the rate of a reaction by lowering its activation energy.

❯❯ **Figure It Out** Is this reaction endergonic or exergonic?

Answer: Exergonic

site is complementary in shape, size, polarity, and charge to the enzyme's substrate. This fit is the reason why each enzyme acts in a specific way on specific substrates.

When we talk about activation energy, we are really talking about the energy required to break the bonds of the reactants. Depending on the reaction, that energy may force substrates close together, redistribute their charge, or cause some other change. The change brings on the transition state, when substrate bonds reach their breaking point and the reaction will run spontaneously to product. Enzymes can help bring on the transition state by lowering activation energy (Figure 5.11). They do this by the following four mechanisms, which work alone or in combination.

Helping Substrates Get Together Binding at an active site brings two or more substrates close together. The closer the substrates are to one another, the more likely they are to react.

Orienting Substrates in Positions That Favor Reaction On their own, substrates collide from random directions. By contrast, binding at an active site positions substrates so they align appropriately for a reaction.

Inducing a Fit Between Enzyme and Substrate By the **induced–fit model**, an enzyme's active site is not quite complementary to its substrate. Interacting with a sub-

active site
enzyme

A Like other enzymes, hexokinase's active sites bind and alter specific substrates. A model of the whole enzyme is shown to the *left*.

— reactant(s)

B A close-up shows glucose and phosphate meeting inside the enzyme's active site. The microenvironment of the site favors a reaction between the two substrate molecules.

— product(s)

C Here, the glucose has bonded with the phosphate. The product of this reaction, glucose-6-phosphate, is shown leaving the active site.

Figure 5.10 An example of an active site. This one is in a hexokinase, an enzyme that phosphorylates glucose and other six-carbon sugars.

active site Pocket in an enzyme where substrates bind and a reaction occurs.
antioxidant Substance that prevents molecules from reacting with oxygen.
catalysis The acceleration of a reaction rate by a molecule that is unchanged by participating in the reaction.
coenzyme An organic molecule that is a cofactor.
cofactor A metal ion or a coenzyme that associates with an enzyme and is necessary for its function.
induced–fit model Substrate binding to an active site improves the fit between the two.
substrate A molecule that is specifically acted upon by an enzyme.

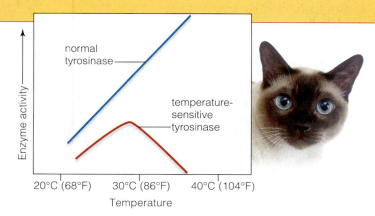

Figure 5.12 **Animated** Enzymes and temperature. Tyrosinase is involved in the production of melanin, a black pigment in skin cells. The form of this enzyme in Siamese cats is inactive above about 30°C (86°F), so the warmer parts of the cat's body end up with less melanin, and lighter fur.

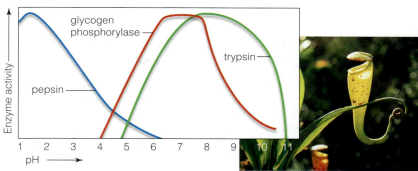

Figure 5.13 Enzymes and pH. *Left*, how pH affects three enzymes. *Right*, carnivorous plants of the genus *Nepenthes* grow in nitrogen-poor habitats. They secrete acids and protein-digesting enzymes into a fluid–filled cup that consists of a modified leaf. The enzymes release nitrogen from insects that are attracted to odors from the fluid and then drown in it. One of these enzymes functions best at pH 2.6.

strate molecule causes the enzyme to change shape so that the fit between them improves. The improved fit may result in a stronger bond between enzyme and substrate, or it may better bring on the substrate's transition state.

Shutting Out Water Molecules Metabolism occurs in water-based fluids, but water molecules can interfere with certain reactions. The active sites of some enzymes repel water, and keep it away from the reactions.

Effects of Temperature, pH, and Salinity

Adding energy in the form of heat boosts free energy, which is one reason why molecular motion increases with temperature. The greater the free energy of reactants, the closer a reaction is to its activation energy. Thus, the rate of an enzymatic reaction typically increases with temperature. However, that rule holds only up to a point. An enzyme denatures above a characteristic temperature. Then, the reaction rate falls sharply as the shape of the enzyme changes and it stops functioning (Figure 5.12).

The pH tolerance of enzymes varies. Most enzymes in the human body have an optimal pH between 6 and 8. For instance, the hexokinase molecule shown in Figure 5.10A is most active in areas of the small intestine where the pH is around 8. Some enzymes work outside the typical range of pH. The enzyme pepsin functions only in stomach fluid, where it breaks down proteins in food. The fluid is very acidic, with a pH of about 2 (Figure 5.13).

An enzyme's activity is also influenced by the amount of salt in the surrounding fluid. Too much or too little salt can disrupt the hydrogen bonding that holds an enzyme in its three-dimensional shape.

Help From Cofactors

Cofactors are atoms or molecules (other than proteins) that associate with enzymes and are necessary for their function. Some are metal ions. Organic molecules that are cofactors are called **coenzymes**. Vitamin C is a coenzyme, and many B vitamins are remodeled into coenzymes.

Coenzymes in some reactions are tightly bound to an enzyme. In others, they participate as separate molecules. Unlike enzymes, many coenzymes are modified by taking part in a reaction. They typically become regenerated in other reactions. For example, the coenzyme NAD^+ (nicotinamide adenine dinucleotide) becomes NADH by accepting electrons and hydrogen atoms in a reaction. NAD^+ is regenerated when the NADH gives up the electrons and hydrogen atoms in a different reaction.

We can use an enzyme called catalase as an example of how cofactors work. Like hemoglobin, catalase has four hemes. The iron atom at the center of each heme (*right*) is a cofactor. Iron is a metal, and metal atoms affect nearby electrons. Catalase works by holding a substrate molecule close to one of its iron atoms. The iron pulls on the substrate's electrons, which brings on the transition state.

iron atom

a heme

Catalase is an **antioxidant**, which means its actions stop other molecules from reacting with oxygen. Oxygen reactions often produce free radicals: atoms or molecules with unpaired electrons. Free radicals form during normal metabolic reactions, but they are dangerous because they attack the structure of biological molecules. Antioxidants such as catalase are critical to health because they reduce the amount of free radicals that form in the body.

Take-Home Message How do enzymes work?

❯ Enzymes greatly enhance the rate of specific reactions. Binding at an enzyme's active site causes a substrate to reach its transition state. In this state, the substrate's bonds are at the breaking point, so the reaction can run spontaneously.

❯ Each enzyme works best at certain temperatures, pH, and salt concentration.

❯ Cofactors associate with enzymes and assist their function.

> ATP, enzymes, and other molecules interact in organized pathways of metabolism.
< Links to Electrons 2.3, Metabolism 3.2, Amino acids 3.5, Membrane proteins 4.4

Types of Metabolic Pathways

Metabolism, remember, refers to the activities by which cells acquire and use energy as they build and break down organic molecules. Building, rearranging, or breaking down an organic substance often occurs stepwise, in a series of reactions called a **metabolic pathway**.

Some metabolic pathways are linear, meaning that the reactions run straight from reactant to product:

Other metabolic pathways are cyclic. In a cyclic pathway, the last step regenerates a reactant for the first step:

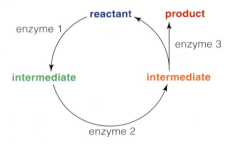

Controls Over Metabolism

Cells conserve energy and resources by making only what they need—no more, no less—at any given moment. How does a cell adjust the types and amounts of molecules it produces? Several mechanisms help a cell maintain, raise,

Figure 5.14 Animated
Feedback inhibition. In this example, three kinds of enzymes act in sequence to convert a substrate to a product, which inhibits the activity of the first enzyme.

>> **Figure It Out** Is this an example of a cyclic or a linear metabolic pathway?

Answer: This is a linear pathway.

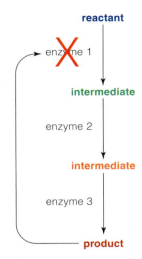

or lower its production of thousands of different substances. For example, reactions do not only run from reactants to products. Many also run in reverse at the same time, with some of the products being converted back to reactants. The rates of the forward and reverse reactions often depend on the concentrations of reactants and products: A high concentration of reactants pushes the reaction in the forward direction, and a high concentration of products pushes it in the reverse direction.

Other mechanisms more actively regulate enzymes. Certain molecules in a cell govern how fast enzyme molecules are made, or influence the activity of enzymes that have already been built. For example, the end product of a series of enzymatic reactions may inhibit the activity of one of the enzymes in the series, an effect called **feedback inhibition** (Figure 5.14).

Some regulatory molecules or ions activate or inhibit an enzyme by binding directly to its active site. Others bind to **allosteric** sites, which are regions of an enzyme (other than the active site) where regulatory molecules

A inactive form   **B** active form

Figure 5.15 Animated Allosteric effects.

A Pyruvate kinase is an enzyme that consists of four identical polypeptide chains. Each chain has an active site and a binding site for a regulatory molecule.

B When regulatory molecules (*red*) bind to the allosteric sites, the overall shape of the enzyme changes. The change makes the enzyme functional. Here, substrate molecules (*yellow*) are bound to the active sites. Small green balls are metal cofactors.

A Glucose and oxygen react (burn) when exposed to a spark. Energy is released all at once as light and heat when CO_2 and water form.

B The same overall reaction occurs in small steps with an electron transfer chain. Energy is released in amounts that cells can harness for cellular work.

① Energy input splits glucose into carbon dioxide, electrons, and hydrogen ions (H^+).

② Electrons lose energy as they move through an electron transfer chain.

③ Energy released by electrons is harnessed for cellular work.

④ Electrons, hydrogen ions, and oxygen combine to form water.

Figure 5.16 Animated Comparing uncontrolled and controlled energy release.

bind. *Allo–* means other, and *steric* means structure. Binding of an allosteric regulator alters the shape of the enzyme in a way that enhances or inhibits its function (Figure 5.15).

Redox Reactions

The bonds of organic molecules hold a lot of energy that can be released by a reaction with oxygen. One type of reaction with oxygen—burning—releases the energy of organic molecules all at once, explosively (Figure 5.16A).

Cells use oxygen to break the bonds of organic molecules. However, they have no way to harvest an explosive burst of energy. Instead, they break the molecules apart in several steps that release the energy in small, manageable increments. Most of these steps are oxidation–reduction reactions, or redox reactions for short. In a **redox reaction**, one molecule accepts electrons (it becomes reduced) from another molecule (which becomes oxidized). Redox

reactions are also called electron transfers. To remember what reduced means, think of how the negative charge of an electron "reduces" the charge of a recipient molecule.

In the next two chapters, you will learn about the importance of redox reactions in electron transfer chains. An **electron transfer chain** is an organized series of reaction steps in which membrane-bound arrays of enzymes and other molecules give up and accept electrons in turn. Electrons are at a higher energy level when they enter a chain than when they leave. Electron transfer chains can harvest the energy given off by an electron as it drops to a lower energy level (Figure 5.16B).

Many coenzymes deliver electrons to electron transfer chains in photosynthesis and aerobic respiration. Energy released at certain steps in those chains helps drive the synthesis of ATP. These pathways will occupy our attention in chapters to come.

allosteric Describes a region of an enzyme other than the active site that can bind regulatory molecules.
electron transfer chain Array of enzymes and other molecules that accept and give up electrons in sequence, thus releasing the energy of the electrons in usable increments.
feedback inhibition Mechanism by which a change that results from some activity decreases or stops the activity.
metabolic pathway Series of enzyme-mediated reactions by which cells build, remodel, or break down an organic molecule.
redox reaction Oxidation–reduction reaction in which one molecule accepts electrons (it becomes reduced) from another molecule (which becomes oxidized). Also called electron transfer.

Take-Home Message What is a metabolic pathway?

❯ A metabolic pathway is a sequence of enzyme-mediated reactions that builds, rearranges, or breaks down an organic molecule.

❯ Control mechanisms enhance or inhibit the activity of many enzymes. The adjustments help cells produce only what they require in any given interval.

❯ Many metabolic pathways involve oxidation–reduction (redox) reactions, which are also called electron transfers.

❯ Redox reactions occur in electron transfer chains, which are important processes of energy exchange in photosynthesis and aerobic respiration.

> Ions and molecules tend to move from one region to another in response to concentration gradients.
> Water diffuses across cell membranes by osmosis.
< Links to Homeostasis 1.3, Ions and molecules 2.3, Solutes 2.5, Cell membrane structure 4.2, Plant cell walls 4.11

Understanding how metabolism works in cells begins with the behavior of solutions. How much of a solute is dissolved in a given amount of fluid is the solute's **concentration**. A difference in solute concentration between adjacent regions of solution is a **concentration gradient**.

Solute molecules or ions tend to move "down" their concentration gradient, from a region of higher concentration to one of lower concentration. Why? Molecules and ions are always in motion; they collide at random and bounce off one another millions of times each second. The more crowded they are, the more often they collide. Rebounds from the collisions propel molecules away from one another. Thus, during any given interval, more molecules get bumped out of a region of higher concentration than get bumped into it.

Diffusion (*left*) is the net movement of molecules or ions in response to a concentration gradient. It is an essential way in which substances move into, through, and out of cells. The rate of diffusion depends on five factors:

1. *Size.* It takes more energy to move a large molecule than it does to move a small one. Thus, smaller molecules diffuse more quickly than larger ones.
2. *Temperature.* Molecules move faster at higher temperature, so they collide more often. Thus, the higher the temperature, the faster the diffusion.
3. *Steepness of the concentration gradient.* The rate of diffusion is higher with steeper gradients, because molecules collide more often in a region of greater concentration.
4. *Charge* can affect the rate and direction of diffusion between two regions. Each ion or charged molecule in a fluid contributes to the fluid's overall electric charge. A difference in charge between two regions can affect the rate and direction of diffusion between them. For exam-

ple, positively charged substances, such as sodium ions, diffuse toward a region with an overall negative charge.
5. *Pressure.* Diffusion may be affected by a difference in pressure between two adjoining regions. Pressure squeezes molecules together, and molecules that are more crowded collide and rebound more frequently. Thus, diffusion occurs faster at higher pressures.

Diffusion Across Membranes

Tonicity refers to the total concentration of solutes in fluids separated by a selectively permeable membrane (such membranes allow some substances, but not others, to cross). When the overall solute concentrations of the two fluids differ, the fluid with the lower overall concentration of solutes is said to be **hypotonic** (*hypo–*, under). The other one, with the higher overall solute concentration, is **hypertonic** (*hyper–*, over). Fluids that are **isotonic** have the same overall solute concentration.

A typical selectively permeable membrane allows water to cross it. When this type of membrane separates two fluids that are not isotonic, water will diffuse across the membrane, and move from the hypotonic fluid into the hypertonic one (Figure 5.17). The diffusion will continue until the two fluids are isotonic, or until some

Figure 5.18 Animated Selective permeability of lipid bilayers.
A Hydrophobic molecules, gases, and water molecules can cross a lipid bilayer on their own. **B** Ions in particular and most polar molecules such as glucose cannot; they cross cell membranes only with the help of transport proteins in the bilayer.

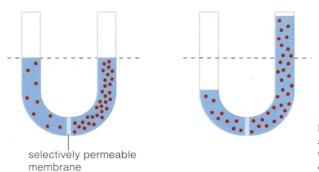

selectively permeable membrane

Figure 5.17 Animated Osmosis. Water moves across a selectively permeable membrane that separates two fluids with differing concentration. The fluid volume changes in the two compartments as water follows its gradient and diffuses across the membrane.

A

2% sucrose   10% sucrose   water

Figure 5.19 Animated Tonicity.

A Experiment that shows what happens when a selectively permeable membrane bag is immersed in solutions of different tonicity.

B–D The micrographs show human red blood cells immersed in fluids of different tonicity.

›› **Figure It Out** Which of the three solutions in **A** is hypertonic with respect to the fluid in the bag?

Answer: The 10% sucrose solution

B Red blood cells immersed in an isotonic solution do not change in volume. The fluid portion of blood is typically isotonic with cytoplasm.

C Red blood cells immersed in a hypertonic solution shrivel up because more water diffuses out of the cells than into them.

D Red blood cells immersed in a hypotonic solution swell up because more water diffuses into the cells than out of them.

pressure against the hypertonic fluid counters it. The diffusion of water across a membrane is so important in biology that it is given a special name: **osmosis**.

A lipid bilayer is one type of selectively permeable membrane that ions and most polar molecules cannot cross (Figure 5.18). If a cell's cytoplasm is hypertonic with respect to the fluid outside of its plasma membrane, water diffuses into it. If the cytoplasm is hypotonic with respect to the fluid on the outside, water diffuses out. In either case, the solute concentration of the cytoplasm may change. If it changes enough, the cell's enzymes will stop working, with lethal results. Most free-living cells, and some that are part of multicelled organisms, have built-in mechanisms that compensate for differences in tonicity between cytoplasm and external fluid. In cells with no such mechanism, the volume—and solute concentration—of cytoplasm will change as water diffuses into or out of the cell (Figure 5.19).

concentration Number of molecules or ions per unit volume.
concentration gradient Difference in concentration between adjoining regions of fluid.
diffusion Net movement of molecules or ions from a region where they are more concentrated to a region where they are less so.
hypertonic Describes a fluid that has a high overall solute concentration relative to another fluid.
hypotonic Describes a fluid that has a low overall solute concentration relative to another fluid.
isotonic Describes two fluids with identical solute concentrations.
osmosis The diffusion of water across a selectively permeable membrane in response to a differing overall solute concentration.
osmotic pressure Amount of turgor that prevents osmosis into cytoplasm or other hypertonic fluid.
turgor Pressure that a fluid exerts against a wall, membrane, or other structure that contains it.

Turgor

Cell walls of plants and many protists, fungi, and bacteria can resist an increase in the volume of cytoplasm even in hypotonic environments. In the case of plant cells, cytoplasm usually contains more solutes than soil water does. Thus, water usually diffuses from soil into a plant—but only up to a point. Rigid walls keep plant cells from expanding very much. Thus, osmosis causes pressure to build up inside of these cells. Pressure that a fluid exerts against a structure that contains it is called **turgor**. When enough pressure builds up inside a plant cell, water stops diffusing into its cytoplasm. The amount of turgor that stops osmosis is called **osmotic pressure**.

Osmotic pressure keeps walled cells plump, just as high air pressure inside a tire keeps it inflated. A young land plant can resist gravity to stay erect because its cells are plump with cytoplasm. When soil dries out, it loses water but not solutes, so the concentration of solutes increases in whatever water remains in it. If soil water becomes hypertonic with respect to cytoplasm, water diffuses out of the plant's cells, so their cytoplasm shrinks. As turgor inside the cells decreases, the plant wilts.

Take-Home Message What influences the movement of ions and molecules?

› Molecules or ions tend to diffuse into an adjoining region of fluid in which they are not as concentrated.

› The steepness of a concentration gradient as well as temperature, molecular size, charge, and pressure affect the rate of diffusion.

› Osmosis is a net diffusion of water between two fluids that differ in water concentration and are separated by a selectively permeable membrane.

› Fluid pressure that a solution exerts against a membrane or wall influences the osmotic movement of water.

> Many types of molecules and ions diffuse across a lipid bilayer only with the help of transport proteins.

< Link to Transport proteins 4.4

You learned in the previous section that gases, water, and small nonpolar molecules can diffuse directly across a lipid bilayer. Most other molecules, and ions in particular, cross only with the help of membrane transport proteins. Each type of transport protein can move a specific ion or molecule. Glucose transporters only transport glucose; calcium pumps only pump calcium; and so on. The specificity of transport proteins means that the amounts and types of substances that cross a membrane depend on which transport proteins are embedded in it. It also allows cells to control the volume and composition of

their fluid interior by moving particular solutes one way or the other across their membranes.

For example, a glucose transporter in a plasma membrane can bind to a molecule of glucose, but not to a molecule of phosphorylated glucose. Enzymes in cytoplasm phosphorylate glucose as soon as it enters the cell. Phosphorylation prevents the molecule from moving back through the glucose transporter and leaving the cell.

Passive Transport

In **passive transport**, the movement of a solute (and the direction of the movement) through a transport protein is driven entirely by the solute's concentration gradient. For this reason, passive transport is also called facilitated diffusion. The solute simply binds to the passive transport protein, and the protein releases it to the other side of the membrane (Figure 5.20).

A glucose transporter is an example of a passive transport protein ❶. This protein changes shape when it binds to a molecule of glucose ❷. The shape change moves the solute to the opposite side of the membrane, where it detaches. Then, the transporter reverts to its original shape ❸. Some passive transporters do not change shape; they form permanently open channels through a membrane. Others are gated, which means they open and close in response to a stimulus such as a shift in electric charge or binding to a signaling molecule.

Active Transport

Solute concentrations may shift in extracellular fluid or in cytoplasm. Maintaining a particular solute's concentration at a certain level often means transporting the solute against its gradient, to the side of a membrane where it is more concentrated. Pumping a solute against its gradient

Figure 5.20 Animated Passive transport. This model shows a glucose transporter.

❶ A glucose molecule (here, in extracellular fluid) binds to a glucose transporter in the plasma membrane.

❷ Binding causes the transport protein to change shape.

❸ The glucose molecule detaches from the transport protein on the other side of the membrane (here, in cytoplasm), and the protein resumes its original shape.

>> **Figure It Out** In this example, which fluid is hypotonic: extracellular fluid or the cytoplasm?   Answer: Cytoplasm

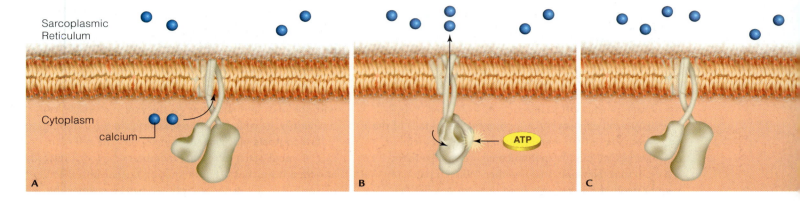

Figure 5.21 Animated Active transport. This model shows a calcium pump embedded in a muscle cell membrane. **A** Two calcium ions bind to the transport protein. **B** Energy in the form of a phosphate group is transferred from ATP to the protein. The transfer causes the protein to change shape so that it ejects the calcium ions to the opposite side of the membrane. **C** After it loses the calcium ions, the transport protein resumes its original shape.

takes energy. In **active transport**, a transport protein uses energy to pump a solute against its gradient across a cell membrane. After a solute binds to an active transporter, an energy input (often in the form of a phosphate-group transfer from ATP) changes the shape of the protein. The change causes the transporter to release the solute to the other side of the membrane.

A calcium pump is an example of an active transporter. This protein moves calcium ions across cell membranes (Figure 5.21). Calcium ions act as potent messengers inside cells, and many enzymes have allosteric sites that bind these ions. Thus, their presence in cytoplasm is tightly regulated. Calcium pumps in the plasma membrane of all eukaryotic cells keep the concentration of calcium in cytoplasm 10,000 times higher than it is in extracellular fluid.

Cotransporters are active transport proteins that move two substances at the same time, in the same or opposite directions across a membrane. Nearly all of the cells in your body have cotransporters called sodium–potassium pumps (Figure 5.22). Sodium ions (Na^+) in the cytoplasm diffuse into the pump's open channel and bind to its interior. A phosphate-group transfer from ATP causes the pump to change shape. Its channel opens to extracellular fluid, where it releases the Na^+. Then, potassium ions (K^+) from extracellular fluid diffuse into the channel and bind to its interior. The transporter releases the phosphate group and reverts to its original shape. The channel opens to the cytoplasm, where it releases the K^+.

Bear in mind, the membranes of all cells, not just those of animals, have active transporters. For example, active transporters in plant leaf cells pump sugars into tubes that distribute them throughout the plant body.

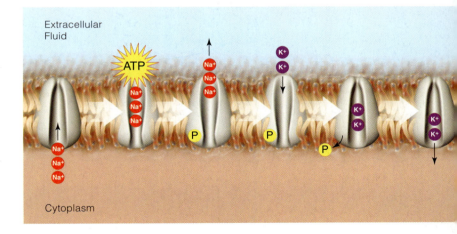

Figure 5.22 Cotransport. This model shows how a sodium–potassium pump transports sodium ions (Na^+, *red*) from cytoplasm to extracellular fluid, and potassium ions (K^+, *purple*) in the other direction across the plasma membrane. A phosphate-group transfer from ATP provides energy for the transport.

---

active transport Energy-requiring mechanism by which a transport protein pumps a solute across a cell membrane against its concentration gradient.
passive transport Mechanism by which a concentration gradient drives the movement of a solute across a cell membrane through a transport protein. Requires no energy input.

> By processes of exocytosis and endocytosis, cells take in and expel particles that are too big for transport proteins, as well as substances in bulk.

< Links to Lipoproteins 3.5, Vesicles 4.8, Motor proteins 4.10

Think back on the structure of a lipid bilayer. When a bilayer is disrupted, such as when part of the plasma membrane pinches off as a vesicle, it seals itself. Why? The disruption exposes the nonpolar fatty acid tails of the phospholipids to their watery surroundings. Remember, in water, phospholipids spontaneously rearrange themselves so that their tails stay together. When a patch of membrane buds, its phospholipid tails are repelled by water on both sides. The water "pushes" the phospholipid tails together, which helps round off the bud as a vesicle, and also seals the rupture in the membrane.

As part of vesicles, patches of membrane constantly move to and from the cell surface (Figure 5.23). The formation and movement of vesicles, which is called membrane trafficking, involves motor proteins and requires ATP.

By **exocytosis**, a vesicle moves to the cell surface, and the protein-studded lipid bilayer of its membrane fuses with the plasma membrane. As the exocytic vesicle loses its identity, its contents are released to the surroundings.

Endocytosis takes up substances near the cell's surface. A small patch of plasma membrane balloons inward, and then it pinches off after sinking farther into the cytoplasm (Figure 5.24). The membrane patch becomes the outer boundary of an endocytic vesicle, which delivers its contents to an organelle or stores them in cytoplasm.

As long as a cell is alive, exocytosis and endocytosis are continually replacing and withdrawing patches of its plasma membrane.

Phagocytosis ("cell eating") is an endocytic pathway in which phagocytic cells such as amoebas engulf microorganisms, cellular debris, or other particles. In animals, macrophages and other white blood cells engulf and digest pathogenic viruses and bacteria, cancerous body cells, and other threats. During phagocytosis, microfilaments form a mesh under the plasma membrane. When they contract, they force some cytoplasm and plasma membrane above it to bulge outward as a lobe called a pseudopod (Figure 5.25). Pseudopods engulf a target and merge as a vesicle that sinks into the cytoplasm and fuses with a lysosome. Enzymes in the lysosome break down the vesicle's contents. The resulting molecular bits may be recycled by the cell, or expelled by exocytosis.

Endocytosis   Exocytosis

A Molecules or particles enter pits in the plasma membrane.

pit

B The pits sink inward and become endocytic vesicles.

C Vesicle contents are sorted.

D Many of the sorted molecules cycle back to the plasma membrane.

E Some vesicles are routed to the nuclear envelope or ER membrane. Others fuse with Golgi bodies.

F Some vesicles and their contents are delivered to lysosomes.

Figure 5.23 Animated Endocytosis and exocytosis.

plasma membrane

aggregates of lipoproteins

Figure 5.24 Endocytosis of lipoproteins.

A Pseudopods surround a pathogen (*brown*).

B Endocytic vesicle forms.

C Lysosome fuses with vesicle; enzymes digest pathogen.

D Cell uses the digested material or expels it.

Figure 5.25 Animated Phagocytosis. A phagocytic cell's pseudopods (extending lobes of cytoplasm) surround a pathogen. The plasma membrane above the bulging lobes fuses and forms an endocytic vesicle. Once inside the cytoplasm, the vesicle fuses with a lysosome, which digests its contents.

The composition of a plasma membrane begins in the ER. There, membrane proteins and lipids are made and modified, and both become part of vesicles that transport them to Golgi bodies for final modification. The finished proteins and lipids are repackaged as new vesicles that travel to the plasma membrane and fuse with it. The lipids and proteins of the vesicle membrane become part of the plasma membrane. This is the process by which new plasma membrane forms.

In a cell that is no longer growing, the total area of the plasma membrane remains more or less constant. Membrane is lost as a result of endocytosis, but it is replaced by membrane arriving as exocytic vesicles.

endocytosis Process by which a cell takes in a small amount of extracellular fluid by the ballooning inward of its plasma membrane.
exocytosis Process by which a cell expels a vesicle's contents to extracellular fluid.
phagocytosis "Cell eating"; an endocytic pathway by which a cell engulfs particles such as microbes or cellular debris.

Take-Home Message How do cells take in large particles and bulk substances?

> Exocytosis and endocytosis move materials in bulk across plasma membranes.

> In exocytosis, a cytoplasmic vesicle fuses with the plasma membrane and releases its contents to the outside of the cell.

> In endocytosis, a patch of plasma membrane sinks inward and forms a vesicle in the cytoplasm.

> Phagocytosis is an endocytic pathway by which cells engulf particles such as microorganisms.

A Toast to Alcohol Dehydrogenase (revisited)

In the human body, alcohol dehydrogenase (ADH) converts ethanol to acetaldehyde, an organic molecule even more toxic than ethanol and the most likely source of various hangover symptoms. A different enzyme, ALDH, very quickly converts acetaldehyde to nontoxic acetate. Thus, the overall pathway of ethanol metabolism in humans is:

$$\text{ethanol} \xrightarrow[\text{NAD}^+ \quad \text{NADH}]{\text{ADH}} \text{acetaldehyde} \xrightarrow[\text{NAD}^+ \quad \text{NADH}]{\text{ALDH}} \text{acetate}$$

In the average adult human body, this metabolic pathway can detoxify between 7 and 14 grams of ethanol per hour. The average alcoholic beverage contains between 10 and 20 grams of ethanol, which is why having more than one drink in any two-hour interval may result in a hangover.

Alcohol dehydrogenase detoxifies the tiny quantities of alcohols that form in some metabolic pathways. In animals, the enzyme also detoxifies alcohols made by gut-inhabiting bacteria, and those in foods such as ripe fruit.

Defects in ADH or ALDH affect alcohol metabolism. For example, if ADH is overactive, acetaldehyde accumulates faster than ALDH can detoxify it:

$$\text{ethanol} \xrightarrow{\text{ADH}} \begin{array}{c}\text{acetaldehyde}\\\text{acetaldehyde}\\\text{acetaldehyde}\end{array} \xrightarrow{\text{ALDH}} \text{acetate}$$

People with an overactive form of ADH become flushed and feel ill after drinking even a small amount of alcohol. The unpleasant experience may be part of the reason that these people are less likely to become alcoholic than others. Underactive ALDH also causes acetaldehyde to accumulate:

$$\text{ethanol} \xrightarrow{\text{ADH}} \begin{array}{c}\text{acetaldehyde}\\\text{acetaldehyde}\\\text{acetaldehyde}\end{array} \xrightarrow{\quad\text{✗}\quad} \text{acetate}$$

Underactive ALDH is associated with the same effect—and the same protection from alcoholism—as overactive ADH. Both types of variant enzymes are common in people of Asian descent. For this reason, the alcohol flushing reaction is informally called "Asian flush."

Having an underactive ADH enzyme has the opposite effect. It results in slowed alcohol metabolism, so people with an underactive ADH may not feel the ill effects of drinking alcoholic beverages as much as other people do. When these people drink alcohol, they have a tendency to become alcoholics. One-quarter of undergraduate students who binge also have signs of alcoholism.

Alcohol abuse is the leading cause of cirrhosis of the liver in the United States. The liver becomes so scarred, hardened, and filled with fat that it loses its function. It stops making the protein albumin, so the solute balance of body fluids is disrupted, and the legs and abdomen swell with watery fluid. It cannot remove drugs and other toxins from the blood, so they accumulate in the brain—which impairs mental functioning and alters personality. Restricted blood flow through the liver causes veins to enlarge and rupture, so internal bleeding is a risk. The damage to the body results in a heightened risk of diabetes and liver cancer. Once cirrhosis has been diagnosed, a person has about a 50 percent chance of death within 10 years.

How Would You Vote? Some people have damaged their liver because they drank too much alcohol; others have had liver–damaging infections. Because there are not enough liver donors, should life–style be a factor in deciding who gets a liver transplant? See CengageNow for details, then vote online (cengagenow.com).

Summary

Section 5.1 Metabolic processes build and break down organic molecules such as ethanol and other toxins. Currently the most serious drug problem on college campuses is binge drinking.

Section 5.2 **Kinetic energy**, **potential energy**, and other forms of **energy** cannot be created or destroyed (**first law of thermodynamics**). Energy can be converted from one form to another and transferred between objects or systems. Energy tends to disperse spontaneously (**second law of thermodynamics**). Some energy disperses with every transfer, usually as heat. **Entropy** is a measure of how much the energy of a system is dispersed.

Living things maintain their organization only as long as they harvest energy from someplace else. Energy flows in one direction through the biosphere, starting mainly from the sun, then into and out of ecosystems. Producers and then consumers use energy to assemble, rearrange, and break down organic molecules that cycle among organisms throughout ecosystems.

Section 5.3 Cells store and retrieve free energy by making and breaking chemical bonds in metabolic **reactions**, in which **reactants** are converted to **products**. **Activation energy** is the minimum energy required to start a reaction. Phosphate-group transfers (**phosphorylations**) to and from **ATP** couple **exergonic** reactions with **endergonic** ones. Cells regenerate ATP by the **ATP/ADP cycle**.

Section 5.4 Enzymes enhance the rate of reactions without being changed by them, a process called **catalysis**. Enzymes lower activation energy by boosting local concentrations of **substrates**, orienting them in positions that favor reaction, inducing the fit between a substrate and the enzyme's **active site** (**induced-fit model**), or excluding water.

Each type of enzyme works best within a characteristic range of temperature, salt concentration, and pH. Most enzymes require **cofactors**, which are metal ions or organic **coenzymes**. Cofactors in some **antioxidants** help them stop reactions with oxygen that produce free radicals.

Section 5.5 Cells build, convert, and dispose of most substances in enzyme-mediated reaction sequences called **metabolic pathways**. Controls over enzymes allow cells to conserve energy and resources by producing only what they require. **Allosteric** sites are points of control by which a cell adjusts the types and amounts of substances it makes. **Feedback inhibition** is an example of enzyme control. **Redox** (oxidation–reduction) **reactions** in **electron transfer chains** allow cells to harvest energy in manageable increments.

Section 5.6 A **concentration gradient** is a difference in the **concentration** of a substance between adjoining regions of fluid. Molecules or ions tend to follow their own gradient and move toward the region where they are less concentrated, a behavior called **diffusion**. The steepness of the gradient, temperature, solute size, charge, and pressure influence the diffusion rate.

Osmosis is the diffusion of water across a selectively permeable membrane, from the region with a lower solute concentration (**hypotonic**) toward the region with a higher solute concentration (**hypertonic**). There is no net movement of water between **isotonic** solutions. **Osmotic pressure** is the amount of **turgor** (fluid pressure against a cell membrane or wall) that stops osmosis.

Section 5.7 Gases, water, and small nonpolar molecules can diffuse across a lipid bilayer. Most other molecules, and ions, cross only with the help of transport proteins, which allow a cell or membrane-enclosed organelle to control which substances enter and exit. The types of transport proteins in a membrane determine which substances cross it. **Active transport** proteins such as calcium pumps use energy, usually from ATP, to pump a solute against its concentration gradient. **Passive transport** proteins work without an energy input; solute movement is driven by the concentration gradient.

Section 5.8 Bulk substances and large particles move across plasma membranes by endocytosis and exocytosis. With **exocytosis**, a cytoplasmic vesicle fuses with the plasma membrane, and its contents are released to the outside of the cell. The vesicle's membrane lipids and proteins become part of the plasma membrane. With **phagocytosis** and other processes of **endocytosis**, a patch of plasma membrane balloons into the cell, and forms a vesicle that sinks into the cytoplasm.

Self-Quiz   Answers in Appendix III

1. _____ is life's primary source of energy.
   a. Food   b. Water   c. Sunlight   d. ATP

2. If we liken a chemical reaction to an energy hill, then a(n) _____ reaction is an uphill run.
   a. endergonic   c. catalytic
   b. exergonic   d. both a and c

3. Which of the following statements is not correct?
   a. Energy cannot be created or destroyed.
   b. Energy cannot change from one form to another.
   c. Energy tends to disperse spontaneously.

4. Enzymes _____ .
   a. are proteins, except for a few RNAs
   b. lower the activation energy of a reaction
   c. are changed by the reactions they catalyze
   d. a and b

Effects of Artichoke Extract on Hangovers Ethanol is a toxin, so it makes sense that drinking it can cause various symptoms of poisoning: headache, stomachache, nausea, fatigue, impaired memory, dizziness, tremors, and diarrhea, among other ailments. All are symptoms of hangover, the common word for what happens as the body is recovering from a bout of heavy drinking.

The most effective treatment for a hangover is to avoid drinking in the first place. Folk remedies (such as aspirin, coffee, bananas, more alcohol, honey, barley grass, pizza, milkshakes, glutamine, raw eggs, charcoal tablets, asparagus, or cabbage) abound, but few have been studied scientifically. In 2003, Max Pittler and his colleagues tested one of them. The researchers gave 15 participants an unmarked pill containing either artichoke extract or a placebo (an inactive substance) just before or after drinking enough alcohol to cause a hangover. The results are shown in Figure 5.26.

1. How many participants experienced a hangover that was worse with the placebo than with the artichoke extract?
2. How many participants had a worse hangover with the artichoke extract?
3. Express the numbers you counted in questions 1 and 2 as a percentage of the total number of participants. How much difference is there between the percentages?
4. Do these results support the hypothesis that artichoke extract is an effective hangover treatment? Why or why not?

Participant (Age, Gender)	Severity of Hangover	
	Artichoke Extract	Placebo
1 (34, F)	1.9	3.8
2 (48, F)	5.0	0.6
3 (25, F)	7.7	3.2
4 (57, F)	2.4	4.4
5 (34, F)	5.4	1.6
6 (30, F)	1.5	3.9
7 (33, F)	1.4	0.1
8 (37, F)	0.7	3.6
9 (62, M)	4.5	0.9
10 (36, M)	3.7	5.9
11 (54, M)	1.6	0.2
12 (37, M)	2.6	5.6
13 (53, M)	4.1	6.3
14 (48, F)	0.5	0.4
15 (32, F)	1.3	2.5

Figure 5.26 Results of a study that tested artichoke extract as a hangover preventive. All participants were tested once with the placebo and once with the extract, with a week interval between. Each rated the severity of 20 hangover symptoms on a scale of 0 (not experienced) to 10 ("as bad as can be imagined"). The 20 ratings were averaged as an overall rating, which is listed here.

5. _____ are always changed by participating in a reaction. (Choose all that are correct.)
   a. Enzymes   c. Reactants
   b. Cofactors   d. Coenzymes

6. Name one environmental factor that typically influences enzyme function.

7. A metabolic pathway _____ .
   a. may build or break down molecules
   b. generates heat
   c. can include an electron transfer chain
   d. a and c
   e. all of the above

8. Diffusion is the movement of ions or molecules from a region where they are _____ (more/less) concentrated to another where they are _____ (more/less) concentrated.

9. Name one molecule that can readily diffuse across a lipid bilayer.

10. Transporters that require an energy boost help sodium ions across a cell membrane. This is a case of _____ .
    a. passive transport   c. facilitated diffusion
    b. active transport   d. a and c

11. Immerse a living human cell in a hypotonic solution, and water will tend to _____ .
    a. diffuse into the cell   c. show no net movement
    b. diffuse out of the cell d. move in by endocytosis

12. Vesicles form during _____ .
    a. endocytosis   d. halitosis
    b. exocytosis   e. a through c
    c. phagocytosis   f. all of the above

13. Match each term with its most suitable description.
    ___ reactant
    ___ enzyme
    ___ first law of thermodynamics
    ___ product
    ___ cofactor
    ___ gradient
    ___ passive transport
    ___ active transport
    ___ phagocytosis

    a. assists enzymes
    b. forms at reaction's end
    c. enters a reaction
    d. unchanged by participating in a reaction
    e. energy cannot be created or destroyed
    f. basis of diffusion
    g. no energy boost required
    h. one cell engulfs another
    i. requires energy boost

Additional questions are available on **CENGAGENOW**.

Critical Thinking

1. Often, beginning physics students are taught the basic concepts of thermodynamics with two phrases: First, you can't win. Second, you can't break even. Explain.

2. Water molecules tend to diffuse in response to their own concentration gradient. How can water be more or less concentrated?

Animations and Interactions on **CENGAGENOW**:
❯ Energy flow and materials cycling; Chemical bookkeeping; Activation energy; How enzymes work; Enzymes and temperature; Feedback inhibition; Allosteric effects; Controlling energy release; Diffusion; Osmosis; Selective permeability; Tonicity; Passive and active transport; Endocytosis and exocytosis; Phagocytosis.

‹ **Links to Earlier Concepts**

A review of how energy flows through the biosphere (Sections 5.2, 5.3) will be useful before starting this chapter, which revisits electron energy levels (2.3), chemical bonds (2.4), carbohydrates (3.3), membrane proteins (4.4), and antioxidants (5.4). Chloroplasts (4.9) and surface specializations of plant cells (4.11) support photosynthesis. You will see an example of how cells harvest energy with electron transfer chains (5.5).

Key Concepts

The Rainbow Catchers
The flow of energy through the biosphere starts when chlorophylls and other photosynthetic pigments absorb the energy of visible light. In plants, some bacteria, and many protists, that energy ultimately drives the synthesis of glucose and other carbohydrates.

What Is Photosynthesis?
Photosynthesis has two stages in the chloroplasts of plants and many types of protists. In the first stage, sunlight energy is converted to chemical energy. Molecules that form in the first stage of photosynthesis power the formation of sugars in the second stage.

6 Where It Starts—Photosynthesis

6.1 Green Energy

Today, the expression "food is fuel" is not just about eating. With fossil fuel prices soaring, there is an increasing demand for biofuels, which are oils, gases, or alcohols made from organic matter that is not fossilized. Much of the material currently used for biofuel production consists of food crops—mainly corn, soybeans, and sugarcane. Growing these crops in large quantities is typically expensive and damaging to the environment, and using them to make biofuel competes with our food supply. The diversion of food crops to biofuel production contributes to increases in food prices worldwide.

How did we end up competing with our vehicles for food? We both run on the same fuel: energy that plants have stored in chemical bonds. Fossil fuels such as petroleum, coal, and natural gas formed from the remains of ancient swamp forests that decayed and compacted over millions of years. They consist mainly of molecules originally assembled by ancient plants. Biofuels—and foods—consist mainly of molecules originally assembled by modern plants.

Autotrophs harvest energy directly from the environment, and obtain carbon from inorganic molecules (*auto–* means self; *–troph* refers to nourishment). Plants and most other autotrophs make their own food by **photosynthesis**, a process in which they use the energy of sunlight to assemble carbohydrates from carbon dioxide and water. Directly or indirectly, photosynthesis also feeds most other life on Earth. Animals and other **heterotrophs** get energy and carbon by breaking down organic molecules assembled by other organisms (*hetero–* means other). We and almost all other organisms sustain ourselves by extracting energy from organic molecules originally assembled by photosynthesizers.

A lot of energy is locked up in the chemical bonds of molecules made by plants. That energy can fuel heterotrophs, as when an animal cell powers ATP synthesis by breaking the bonds of sugars. It can also fuel our cars, which run on energy released by burning biofuels or fossil fuels. Both processes are fundamentally the same: They release energy by breaking the bonds of organic molecules. Both use oxygen to break those bonds, and both produce carbon dioxide.

Corn and other food crops are rich in oils, starches, or sugars that can be easily converted to biofuels. The starch in corn kernels, for example, can be enzymatically broken down to glucose, which is converted to ethanol by heterotrophic bacteria or yeast. Making biofuels from other types of plant matter requires additional steps, because these materials contain a higher proportion of cellulose. Breaking down this tough, insoluble carbohydrate to its glucose monomers adds substantial cost to the biofuel product. Researchers are currently trying to find cost-effective ways to break down the abundant cellulose in fast-growing weeds such as switchgrass (Figure 6.1), and agricultural wastes such as wood chips, wheat straw, cotton stalks, and rice hulls.

autotroph Organism that makes its own food using carbon from inorganic molecules such as CO_2, and energy from the environment.
heterotroph Organism that obtains energy and carbon from organic compounds assembled by other organisms.
photosynthesis Metabolic pathway by which most autotrophs capture light energy and use it to make sugars from CO_2 and water.

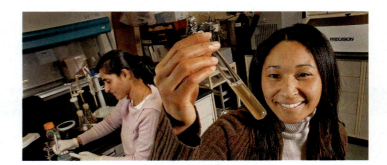

Figure 6.1 Biofuels. *Opposite*, switchgrass (*Panicum virgatum*), a weed that grows wild in North American prairies. *Above*, researchers Ratna Sharma and Mari Chinn of North Carolina State University working to reduce the costs of producing biofuel from biomass such as switchgrass and agricultural wastes.

Making ATP and NADPH
ATP forms in the first stage of photosynthesis, which is light-dependent because the reactions run on the energy of light. The coenzyme NADPH forms in a noncyclic pathway that releases oxygen. ATP also forms in a cyclic pathway that does not release oxygen.

Making Sugars
The second stage is the "synthesis" part of photosynthesis. Sugars are assembled with carbon and oxygen atoms from CO_2. The reactions run on the chemical bond energy of ATP, and electrons donated by NADPH—molecules that formed in the first stage of photosynthesis.

Alternate Pathways
Details of light-independent reactions that vary among organisms are evolutionary adaptations to different environmental conditions.

> Photosynthetic organisms use pigments to capture the energy of sunlight.
< Links to Electrons 2.3, Chemical bonds 2.4, Carbohydrates 3.3, Plastids 4.9, Energy 5.2, Antioxidants 5.4

Properties of Light

Energy flow through nearly all ecosystems on Earth begins when photosynthesizers intercept energy from the sun. Harnessing the energy of sunlight for work is complicated business, or we would have been able to do it in an economically sustainable way by now. Plants do it by converting light energy to chemical energy, which they and most other organisms use to drive cellular work. The first step involves capturing light. In order to understand how that happens, you have to understand a little about the nature of light.

Most of the energy that reaches Earth's surface is in the form of visible light, so it should not be surprising that visible light is the energy that drives photosynthesis. Visible light is a very small part of a large spectrum of electromagnetic energy radiating from the sun. Like all other forms of electromagnetic energy, light travels in waves, moving through space a bit like waves moving across an ocean. The distance between the crests of two successive waves is the light's **wavelength**, which we measure in nanometers (nm). About 25 million nanometers are equal to one inch.

Visible light occurs between wavelengths of 380 and 750 nanometers. We see light of particular wavelengths in this range as different colors, and all wavelengths combined as white. White light separates into its component colors when it passes through a prism. The prism bends the longer wavelengths more than it bends the shorter ones, so a rainbow of colors forms (Figure 6.2A).

Light travels in waves, but it is also organized in packets of energy called photons. A photon's energy and its wavelength are related, so all photons traveling at the same wavelength carry the same amount of energy. Photons that carry the least amount of energy travel in longer wavelengths; those that carry the most energy travel in shorter wavelengths (Figure 6.2B). Photons of wavelengths shorter than about 380 nanometers carry enough energy to alter or break the chemical bonds of DNA and other biological molecules. That is why UV (ultraviolet) light, x-rays, and gamma rays are a threat to life.

Pigments: The Rainbow Catchers

Photosynthesizers use pigments to capture light. A **pigment** is an organic molecule that selectively absorbs light of specific wavelengths. Wavelengths of light that are not absorbed are reflected, and that reflected light gives each pigment its characteristic color. For example, a pigment that absorbs violet, blue, and green light reflects the rest of the visible light spectrum—yellow, orange, and red light. This pigment appears orange to us.

Chlorophyll *a* is by far the most common photosynthetic pigment in plants, and also in photosynthetic protists and bacteria. Chlorophyll *a* absorbs violet and red light, so it appears green to us. Accessory pigments,

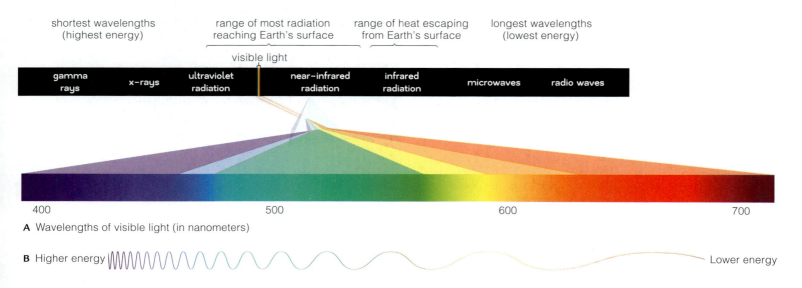

Figure 6.2 Animated Properties of light. **A** Electromagnetic spectrum of radiant energy, which undulates across space as waves that we measure in nanometers. Visible light makes up a very small part of the spectrum. Raindrops or a prism can separate its different wavelengths, which we see as different colors. **B** Light is organized as packets of energy called photons. The shorter a photon's wavelength, the greater its energy.

Pigment	Color	Occurrence			
		Plants	Protists	Bacteria	Archaeans
Chlorophyll *a*	green	●	●	●	
Other chlorophylls	green	●	●	●	
Phycobilins					
phycocyanobilin	blue		●	●	
phycoerythrobilin	red		●	●	
phycoviolobilin	violet		●	●	
Carotenoids					
beta-carotene	orange	●	●	●	
lycopene	red	●	●		
lutein	yellow	●	●	●	
zeaxanthin	yellow	●	●	●	
fucoxanthin	brown	●	●		
Anthocyanins	red to blue	●	●	●	
Retinal	violet				●

A

B

Figure 6.3 Examples of photosynthetic pigments.

A Collectively, photosynthetic pigments are capable of absorbing almost all visible light wavelengths.

B Structures of two photosynthetic pigments. The light-trapping ring structure of chlorophyll is almost identical to a heme group. Heme groups are part of hemoglobin, which is a red pigment.

chlorophyll *a*   beta-carotene

including other chlorophylls, work together with chlorophyll *a* to harvest a wide range of light wavelengths for photosynthesis. The colors of a few of the 600 or so known accessory pigments are shown in Figure 6.3A.

Accessory pigments are typically multipurpose molecules. Their antioxidant properties protect plants and other organisms from the damaging effects of UV light in the sun's rays; their appealing colors attract animals to ripening fruit or pollinators to flowers. You may already be familiar with some of these molecules because they color familiar roots, fruits, and flowers. For example, carrots are orange because they have a lot of beta-carotene (which is also spelled β-carotene). The yellow color of corn comes from zeaxanthin. The color of a tomato changes from green to red as it ripens because its chlorophyll-containing chloroplasts develop into lycopene-containing chromoplasts (Section 4.9). Roses are red and violets are blue because of their anthocyanin content.

Most photosynthetic organisms use a combination of pigments for photosynthesis. In plants, chlorophylls are usually so abundant that they mask the colors of the other pigments, so leaves typically appear green. The green leaves of many plants change color during autumn because they stop making pigments in preparation for a period of dormancy. Chlorophyll breaks down faster than the other pigments, so the leaves turn red, orange, yellow, or violet as their chlorophyll content declines and their accessory pigments become visible.

The light-trapping part of a pigment is an array of atoms in which single bonds alternate with double bonds (Figure 6.3B). Electrons in such arrays easily absorb photons, so pigment molecules function a bit like antennas that are specialized for receiving light energy of only certain wavelengths.

Absorbing a photon excites electrons. Remember, an energy input can boost an electron to a higher energy level (Section 2.3). The excited electron returns quickly to a lower energy level by emitting the extra energy. As you will see, photosynthetic cells can capture energy emitted from an electron returning to a lower energy level. Arrays of chlorophylls and other photosynthetic pigments in these cells hold on to the energy by passing it back and forth between them. When the energy reaches a special pair of chlorophylls, the reactions of photosynthesis begin.

---

chlorophyll *a* Main photosynthetic pigment in plants.
pigment An organic molecule that can absorb light of certain wavelengths.
wavelength Distance between the crests of two successive waves of light.

Take-Home Message How do photosynthesizers absorb light?

❭ Energy radiating from the sun travels through space in waves and is organized as packets called photons.

❭ The spectrum of radiant energy from the sun includes visible light. Humans perceive different wavelengths of visible light as different colors. The shorter the wavelength, the greater the energy.

❭ Pigments absorb light at specific wavelengths. Photosynthetic species use pigments such as chlorophyll *a* to harvest the energy of light for photosynthesis.

> Photosynthetic pigments work together to harvest light of different wavelengths.

In 1882, botanist Theodor Engelmann designed an experiment to test his hypothesis that the color of light affects photosynthesis. It had long been known that photosynthesis releases oxygen, so Engelmann used the amount of oxygen released by photosynthetic cells as a measure of how much photosynthesis was occurring in them. He used a prism to divide a ray of light into its component colors, then directed the resulting spectrum across a single strand of *Chladophora*, a photosynthetic alga (Figure 6.4A), suspended in a drop of water.

Oxygen-sensing equipment had not yet been invented, so Engelmann used oxygen-requiring bacteria to show him where the oxygen concentration in the water was highest. The bacteria moved through the water and gathered mainly where violet or red light fell across the strand of algae (Figure 6.4B). Engelmann concluded that the algal cells illuminated by light of these colors were releasing the most oxygen—a sign that violet and red light are the best for driving photosynthesis.

Engelmann's experiment allowed him to correctly identify the colors of light (red and violet) that are most efficient at driving photosynthesis in *Chladophora*. His results constituted an absorption spectrum, which is a graph that shows how efficiently the different wavelengths of light are absorbed by a substance. Peaks in the graph indicate wavelengths of light that the substance absorbs best (Figure 6.4C). Engelmann's results represent the combined spectra of all the photosynthetic pigments in *Chladophora*.

Most photosynthetic organisms use a combination of pigments to drive photosynthesis, and the combination differs by species. Why? Different proportions of wavelengths in sunlight reach different parts of Earth. The particular set of pigments in each species is an adaptation that allows an organism to absorb the particular wavelengths of light available in its habitat.

For example, water absorbs light between wavelengths of 500 and 600 nm less efficiently than other wavelengths. Algae that live deep underwater have pigments that absorb light in the range of 500–600 nm, which is the range that water does not absorb very well. Phycobilins are the most common pigments in deep-water algae.

A Light micrograph of photosynthetic cells in a strand of *Chladophora*. Engelmann used this green alga to show that certain colors of light are best for photosynthesis.

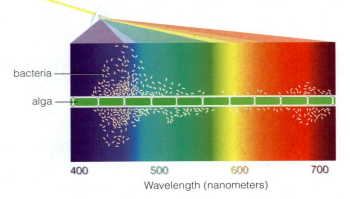

B Engelmann directed light through a prism so that bands of colors crossed a water droplet on a microscope slide. The water held a strand of *Chladophora* and oxygen-requiring bacteria. The bacteria clustered around the algal cells that were releasing the most oxygen—the ones that were most actively engaged in photosynthesis. Those cells were under red and violet light.

C Absorption spectra of a few photosynthetic pigments. Line color is the characteristic color of each pigment.

Figure 6.4 Animated Discovery that photosynthesis is driven by particular wavelengths of light. Theodor Engelmann used the green alga *Chladophora* **A** in an early photosynthesis experiment **B**. His results constituted one of the first absorption spectra.

C Absorption spectra of chlorophylls *a* and *b*, β-carotene, and two phycobilins reveal the efficiency with which these pigments absorb different wavelengths of visible light.

>> **Figure It Out** Of the five pigments represented in **C**, which three are the main photosynthetic pigments in *Chladophora*?

Answer: Chlorophyll *a*, chlorophyll *b*, and β-carotene

> Photosynthesis occurs in two stages in the chloroplasts of plants and other photosynthetic eukaryotes.

The **chloroplast** is an organelle that specializes in photosynthesis in plants and many protists (Figure 6.5). Plant chloroplasts have two outer membranes, and are filled with a semifluid matrix called **stroma**. Stroma contains the chloroplast's own DNA, some ribosomes, and an inner, much-folded **thylakoid membrane**. The folds of a thylakoid membrane typically form stacks of disks (thylakoids) that are connected by channels. The space inside all of the disks and channels is one continuous compartment.

Photosynthesis is often summarized by this equation:

$$6CO_2 \;+\; 6H_2O \;\xrightarrow{\textit{light energy}}\; C_6H_{12}O_6 \;+\; 6O_2$$

carbon dioxide   water   glucose   oxygen

However, photosynthesis is not one reaction. It is a series of many reactions that occur in two stages. The first stage occurs at the thylakoid membrane. It is driven by light, so the collective reactions of this stage are called the **light-dependent reactions**. Two different sets of light-dependent reactions constitute a noncyclic and a cyclic pathway. Both pathways convert light energy to chemical bond energy of ATP. The noncyclic pathway, which is the main one in chloroplasts, yields NADPH and O_2 in addition to ATP:

| ADP
NADP⁺
H₂O
energy | Light-dependent reactions | ATP
NADPH
O₂ |

ADP / NADP$^+$ / H$_2$O / energy → Light-dependent reactions → ATP / NADPH / O$_2$

The reactions of the second stage of photosynthesis, which run in the stroma, make glucose and other carbohydrates from carbon dioxide and water. Light energy does not power them, so they are collectively called the **light-independent reactions**. They run on energy delivered by ATP and coenzymes produced in the first stage:

ATP / NADPH / CO$_2$ / H$_2$O → Light-independent reactions (Calvin–Benson cycle) → ADP / NADP+ / glucose

---

chloroplast Organelle specialized for photosynthesis in plants and some protists.
light–dependent reactions First stage of photosynthesis; convert light energy to chemical energy of ATP and NADPH.
light–independent reactions Second stage of photosynthesis; use ATP and NADPH to assemble sugars from water and CO_2.
stroma Semifluid matrix between the thylakoid membrane and the two outer membranes of a chloroplast.
thylakoid membrane A chloroplast's highly folded inner membrane system; forms a continuous compartment in the stroma.

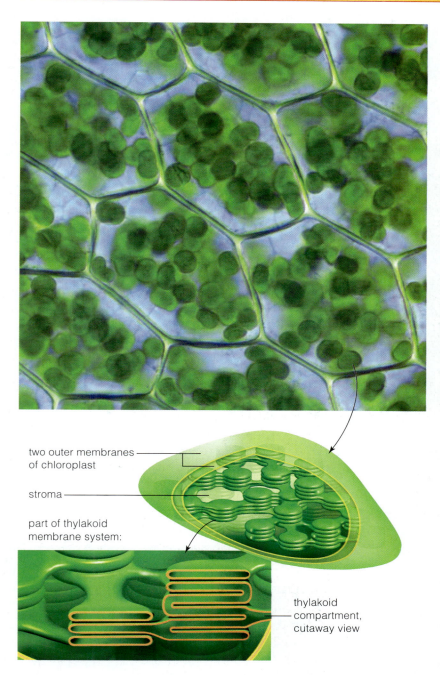

two outer membranes of chloroplast

stroma

part of thylakoid membrane system:

thylakoid compartment, cutaway view

Figure 6.5 Animated The chloroplast: site of photosynthesis in the cells of typical leafy plants. The micrograph shows chloroplast-stuffed cells of a moss, *Plagiomnium affine*.

Take-Home Message **What is photosynthesis and where in a eukaryotic cell does it take place?**

> In the first stage of photosynthesis, light energy drives the formation of ATP and NADPH, and oxygen is released. In eukaryotic cells, these light-dependent reactions occur at the thylakoid membrane of chloroplasts.

> The second stage of photosynthesis, the light-independent reactions, occur in the stroma. ATP and NADPH drive the synthesis of carbohydrates from water and carbon dioxide.

> The reactions of the first stage of photosynthesis convert the energy of light to the energy of chemical bonds.
< Links to Electrons and energy levels 2.3, Membrane proteins 4.4, Chloroplasts 4.9, Energy 5.2, Electron transfer chains 5.5, Gradients 5.6

Capturing Light for Photosynthesis

When a pigment in a thylakoid membrane absorbs a photon, the photon's energy boosts one of the pigment's electrons to a higher energy level. The electron quickly emits the extra energy and drops back to its unexcited state. That energy would be lost to the environment if nothing else were to happen, but in a thylakoid membrane the energy of excited electrons is not lost. Light-harvesting complexes keep it in play. Millions of these circular arrays of chlorophylls, accessory pigments, and proteins are embedded in each thylakoid membrane (Figure 6.6). Pigments that are part of a light-harvesting complex can hold on to energy by passing it back and forth, a bit like volleyball players pass a ball among team members. The energy gets volleyed from cluster to cluster until a photosystem absorbs it. **Photosystems** are groups of hundreds of chlorophylls, accessory pigments, and other molecules that work as a unit to begin the reactions of photosynthesis.

The Noncyclic Pathway

Thylakoid membranes contain two kinds of photosystems, type I and type II, which were named in the order of their discovery. They work together in a set of reactions called the noncyclic pathway of photosynthesis. These reactions begin when energy being passed among light-harvesting complexes reaches a photosystem II (Figure 6.7). At the center of each photosystem is a special pair of chlorophyll *a* molecules. When a photosystem absorbs energy, electrons are ejected from its special pair ❶. These electrons immediately enter an electron transfer chain in the thylakoid membrane.

Replacing Lost Electrons

A photosystem can donate only a few electrons to electron transfer chains before it must be restocked with more. Where do replacements come from? Photosystem II gets more electrons by pulling them off of water molecules in the thylakoid compartment. This reaction causes the water molecules to dissociate into hydrogen ions and oxygen ❷. The released oxygen diffuses out of the cell as O_2 gas. This and any other process by which a molecule is broken apart by light energy is called **photolysis**.

Harvesting Electron Energy

The actual conversion of light energy to chemical energy occurs when a photosystem donates electrons to an electron transfer chain ❸. Light does not take part in chemical reactions, but electrons do. In a series of redox reactions, electrons pass from one molecule of the chain to the next. With each reaction, the electrons release a bit of their extra energy.

The molecules of the electron transfer chain use the released energy to move hydrogen ions (H^+) across the membrane, from the stroma to the thylakoid compartment ❹. Thus, the flow of electrons through electron transfer chains sets up and maintains a hydrogen ion gradient across the thylakoid membrane. This gradient motivates hydrogen ions in the thylakoid compartment to move back into the stroma. However, ions cannot diffuse through lipid bilayers (Section 4.4). H^+ leaves the thylakoid compartment only by flowing through membrane transport proteins called ATP synthases.

Hydrogen ion flow through an ATP synthase causes this protein to attach a phosphate group to ADP ❼, so ATP forms in the stroma. The process by which the flow of electrons through electron transfer chains drives ATP formation is called **electron transfer phosphorylation**.

After the electrons have moved through the first electron transfer chain, they are accepted by a photosystem I. When this photosystem absorbs light energy, electrons

light-harvesting complex   photosystem

Figure 6.6 Artist's view of some of the components of the thylakoid membrane as seen from the stroma. Molecules of electron transfer chains and ATP synthases are also present, but not shown for clarity.

The Light–Dependent Reactions of Photosynthesis

Figure 6.7 Animated Light-dependent reactions of photosynthesis. This example shows the noncyclic reactions in a thylakoid membrane.

1 Light energy ejects electrons from a photosystem II.

2 The photosystem pulls replacement electrons from water molecules, which break apart into oxygen and hydrogen ions. The oxygen leaves the cell as O_2.

3 The electrons enter an electron transfer chain in the thylakoid membrane.

4 Energy lost by the electrons as they move through the transfer chain causes hydrogen ions to be pumped from the stroma into the thylakoid compartment. A hydrogen ion gradient forms across the thylakoid membrane.

5 Light energy ejects electrons from a photosystem I. Replacement electrons come from an electron transfer chain.

6 The electrons move through a second electron transfer chain, then combine with $NADP^+$ and H^+, so NADPH forms.

7 Hydrogen ions in the thylakoid compartment are propelled through the interior of ATP synthases by their gradient across the thylakoid membrane. Hydrogen ion flow causes ATP synthases to attach phosphate to ADP, so ATP forms in the stroma.

are ejected from its special pair of chlorophylls **5**. These electrons enter a second, different electron transfer chain. At the end of this chain, the coenzyme $NADP^+$ accepts the electrons along with H^+, so NADPH forms **6**:

$$NADP^+ + 2e^- + H^+ \longrightarrow NADPH$$

ATP and NADPH continue to form as long as electrons continue to flow through transfer chains in the thylakoid membrane, and electrons flow through the chains as long as water, $NADP^+$, and light are plentiful. The flow of electrons slows—and so does ATP and NADPH production—at night, or when water or $NADP^+$ is scarce.

The Cyclic Pathway

At high oxygen levels, or when NADPH accumulates in the stroma, the noncyclic pathway backs up and stalls. Even when the noncyclic pathway is not running, a cell can continue producing ATP by photosynthesis with the cyclic pathway. This pathway involves photosystem I and an electron transfer chain that cycles electrons back to it. The chain that acts in the cyclic pathway uses electron energy to move hydrogen ions into the thylakoid compartment. The resulting hydrogen ion gradient drives ATP formation, just as it does in the noncyclic pathway. However, NADPH does not form, because electrons at the end of this

chain are accepted by a photosystem I, not $NADP^+$. Oxygen (O_2) does not form either, because photosystem I does not rely on photolysis to resupply itself with electrons.

---

electron transfer phosphorylation Process in which electron flow through electron transfer chains sets up a hydrogen ion gradient that drives ATP formation.
photolysis Process by which light energy breaks down a molecule.
photosystem Cluster of pigments and proteins that converts light energy to chemical energy in photosynthesis.

Take-Home Message What happens during the light–dependent reactions of photosynthesis?

❯ In the light-dependent reactions of photosynthesis, chlorophylls and other pigments in the thylakoid membrane transfer the energy of light to photosystems.

❯ Absorbing energy causes electrons to leave photosystems and enter electron transfer chains in the membrane. The flow of electrons through the transfer chains sets up hydrogen ion gradients that drive ATP formation.

❯ In the noncyclic pathway, oxygen is released and electrons end up in NADPH.

❯ A cyclic pathway involving only photosystem I allows the cell to continue making ATP even when the noncyclic pathway is not running. NADPH does not form, and oxygen is not released.

> Energy flow in the light-dependent reactions is an example of how organisms harvest energy from the environment.
< Links to Energy in metabolism 5.3, Redox reactions 5.5

One of the recurring themes in biology is that organisms use energy harvested from the environment to drive cellular processes. Energy flow in the light-dependent reactions of photosynthesis is a classic example of how that happens. Figure 6.8 compares energy flow in the two pathways of light-dependent reactions.

The simpler cyclic pathway evolved first, and still operates in nearly all photosynthesizers. Later, the photosynthetic machinery in some organisms became modified so that photosystem II became part of it. That modification was the beginning of a combined sequence of reactions that removes electrons from water molecules, with the release of hydrogen ions and oxygen.

In the noncyclic reactions, electrons that leave photosystem II do not return to it. Instead, they end up in NADPH, a powerful reducing agent (electron donor). In the cyclic reactions, electrons lost from photosystem I are cycled back to it. No NADPH forms, and no oxygen is released.

In both the cyclic and noncyclic reactions, molecules in electron transfer chains use electron energy to shuttle H^+ across the thylakoid membrane. Hydrogen ions accumulate in the thylakoid compartment, forming a gradient that powers ATP synthesis.

The plasma membrane of different species of modern photosynthetic bacteria incorporates either type I or type II photosystems. Cyanobacteria, plants, and all photosynthetic protists use both types. Which of the two pathways predominates at any given time depends on the organism's immediate metabolic demands for ATP and NADPH.

Having the alternate pathways is efficient, because cells can direct energy to producing NADPH and ATP or to producing ATP alone. NADPH accumulates when it is not being used, such as when sugar production declines during cold snaps. The excess NADPH backs up the noncyclic pathway, so the cyclic pathway predominates. The cell still makes ATP, but not NADPH. When sugar production is in high gear, NADPH is being used quickly. It does not accumulate, and the noncyclic pathway is the predominant one.

Sunlight and water are essentially unlimited, free resources here on Earth. If we could use solar energy to split water molecules even a fraction as efficiently as photosystem II does, we would have a cheap source of hydrogen that we could use as clean-burning, renewable fuel.

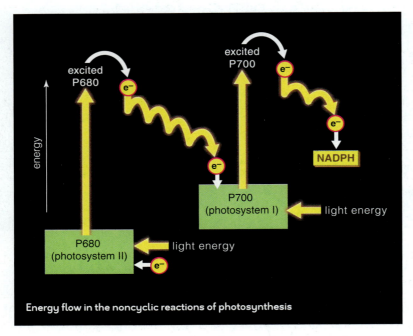

Energy flow in the noncyclic reactions of photosynthesis

A The noncyclic pathway is a one-way flow of electrons from water, to photosystem II, to photosystem I, to NADPH. As long as electrons continue to flow through the two electron transfer chains, H^+ continues to be carried across the thylakoid membrane, and ATP and NADPH keep forming. Light provides the energy boosts that keep the pathway going.

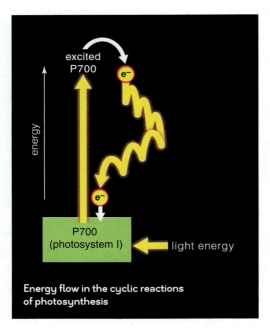

Energy flow in the cyclic reactions of photosynthesis

B In the cyclic pathway, electrons ejected from photosystem I are returned to it. As long as electrons continue to pass through its electron transfer chain, H^+ continues to be carried across the thylakoid membrane, and ATP continues to form. Light provides the energy boost that keeps the cycle going.

Figure 6.8 Energy flow in the light-dependent reactions of photosynthesis. An energy input of an appropriate wavelength ejects electrons from a photosystem's special pair of chlorophylls. The special pair in photosystem I absorbs photons of a 680-nanometer wavelength, so it is called P680. The special pair in photosystem II absorbs photons of a 700-nanometer wavelength, so it is called P700.

Take-Home Message How does energy flow during the reactions of photosynthesis?

> Light provides energy inputs that keep electrons flowing through electron transfer chains.

> Energy lost by electrons as they flow through the chains sets up a hydrogen ion gradient that drives the synthesis of ATP alone, or ATP and NADPH.

❯ The chloroplast is a sugar factory operated by enzymes of the Calvin–Benson cycle. The cyclic, light-independent reactions are the "synthesis" part of photosynthesis.

❮ Links to Carbohydrates 3.3, ATP energy transfers and phosphorylation 5.3

The enzyme-mediated reactions of the **Calvin–Benson cycle** build sugars in the stroma of chloroplasts (Figure 6.9). These reactions are light-independent because light energy does not power them. Instead, they run on ATP and NADPH that formed in the light-dependent reactions.

Light-independent reactions use carbon atoms from CO_2 to make sugars. Extracting carbon atoms from an inorganic source and incorporating them into an organic molecule is a process called **carbon fixation**. In most plants, photosynthetic protists, and some bacteria, the enzyme **rubisco** fixes carbon by attaching CO_2 to five-carbon RuBP (ribulose bisphosphate) **❶**.

The six-carbon intermediate that forms by this reaction is unstable, so it splits right away into two three-carbon molecules of PGA (phosphoglycerate). Each of the PGAs receives a phosphate group from ATP, and hydrogen and electrons from NADPH. Thus, ATP energy and the reducing power of NADPH convert each molecule of PGA into a molecule of PGAL (phosphoglyceraldehyde), a phosphorylated sugar **❷**.

glucose

In later reactions, two or more of the three-carbon PGAL molecules can be combined and rearranged to form larger carbohydrates. Glucose, remember, has six carbon atoms. To make one glucose molecule, six CO_2 must be attached to six RuBP molecules, so twelve PGAL form. Two PGAL combine to form one glucose molecule **❸**. The ten remaining PGAL regenerate the starting compound of the cycle, RuBP **❹**.

Plants can use the glucose they make in the light-independent reactions as building blocks for other organic molecules, or they can break it down to access the energy held in its bonds. However, most of the glucose is converted at once to sucrose or starch by other pathways that conclude the light-independent reactions. Excess glucose is stored in the form of starch grains inside the stroma of chloroplasts. When sugars are needed in other parts of the plant, the starch is broken down to sugar monomers and exported from the cell.

Calvin–Benson cycle Light-independent reactions of photosynthesis; cyclic carbon-fixing pathway that forms sugars from CO_2.
carbon fixation Process by which carbon from an inorganic source such as carbon dioxide gets incorporated into an organic molecule.
rubisco Ribulose bisphosphate carboxylase. Carbon-fixing enzyme of the Calvin–Benson cycle.

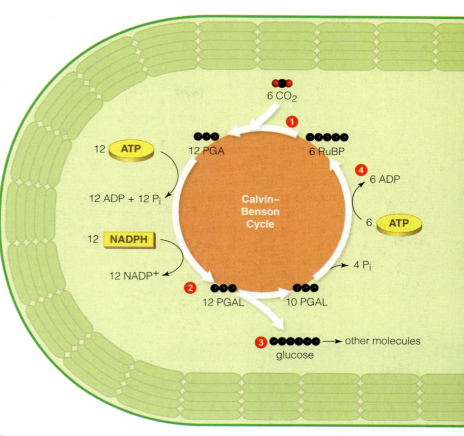

Figure 6.9 Animated Light-independent reactions of photosynthesis. The sketch shows a cross-section of a chloroplast with the light-independent reactions cycling in the stroma.

The steps shown are a summary of six cycles of the Calvin–Benson reactions. Black balls signify carbon atoms. Appendix VI details the reaction steps.

❶ Six CO_2 diffuse into a photosynthetic cell, and then into a chloroplast. Rubisco attaches each to a RuBP molecule. The resulting intermediates split, so twelve molecules of PGA form.

❷ Each PGA molecule gets a phosphate group from ATP, plus hydrogen and electrons from NADPH. Twelve PGAL form.

❸ Two PGAL combine to form one glucose molecule.

❹ The remaining ten PGAL receive phosphate groups from ATP. The transfer primes them for endergonic reactions that regenerate the 6 RuBP.

Take-Home Message **What happens during the light–independent reactions of photosynthesis?**

❯ The light-independent reactions of photosynthesis run on the bond energy of ATP and the energy of electrons donated by NADPH. Both molecules formed in the light-dependent reactions.

❯ Collectively called the Calvin–Benson cycle, these carbon-fixing reaction use hydrogen (from NADPH), and carbon and oxygen (from CO_2) to build sugars.

❯ Environments differ, and so do details of photosynthesis.
❮ Links to Surface specializations 4.11, Controls over metabolic reactions 5.5

Several adaptations allow plants to live where water is scarce or sporadically available. For example, a thin, waterproof coating called a cuticle prevents water loss by evaporation from aboveground plant parts. However, a cuticle also prevents gases from entering and exiting a plant by diffusing through cells at the surfaces of leaves and stems. Gases play a critical role in photosynthesis, so photosynthetic parts are often studded with tiny, closable gaps called **stomata** (singular, stoma). When stomata are open, carbon dioxide for the light-independent reactions can diffuse from air into the plant's photosynthetic tissues, and oxygen produced by the light-dependent reactions can diffuse from photosynthetic cells into the air.

Plants that use only the Calvin–Benson cycle are called **C3 plants**, because 3-carbon PGA is the first stable intermediate to form in the light-independent reactions. C3

plants typically conserve water on dry days by closing their stomata. However, when stomata are closed, oxygen produced by the light-dependent reactions cannot escape from the plant. Oxygen that accumulates in photosynthetic tissues limits sugar production. Why? At high oxygen levels, rubisco attaches oxygen (instead of carbon) to RuBP. This pathway, which is called **photorespiration**, produces carbon dioxide, so the plant loses carbon instead of fixing it (Figure 6.10A,B). In addition, ATP and NADPH are used to convert the pathway's intermediates to a molecule that can enter the Calvin–Benson cycle, so extra energy is required to make sugars on dry days. C3 plants compensate for rubisco's inefficiency by making a lot of it: Rubisco is the most abundant protein on Earth.

Some plants have adaptations that help them minimize photorespiration. A 4-carbon molecule is the first stable intermediate that forms in the light-independent reactions of corn, bamboo, and other **C4 plants** (Figure 6.10C). Such plants fix carbon twice, in two kinds of cells. In the first kind of cell, carbon is fixed by an enzyme that does not use oxygen even at high oxygen levels. The resulting intermediate is transported to another kind of cell, where it is converted to carbon dioxide that enters the Calvin–Benson cycle. The extra reactions keep the carbon dioxide level high near rubisco, thus minimizing photorespiration.

In cactuses and other **CAM plants**, the extra reactions run at a different time rather than in different cells: The C4 reactions run during the day, and the Calvin–Benson cycle runs at night. CAM stands for crassulacean acid metabolism, after the Crassulaceae family of plants in which this pathway was first studied (Figure 6.10D).

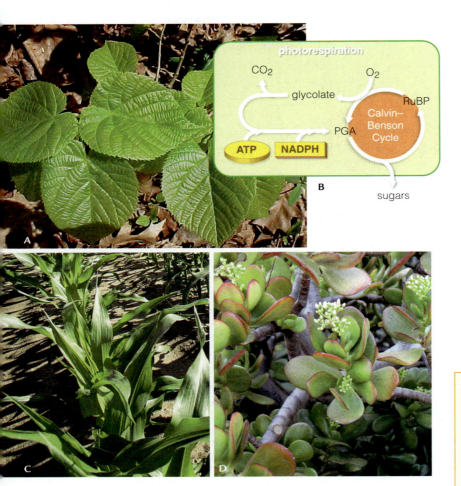

C3 plant Type of plant that uses only the Calvin–Benson cycle to fix carbon.
C4 plant Type of plant that minimizes photorespiration by fixing carbon twice, in two cell types.
CAM plant Type of C4 plant that conserves water by fixing carbon twice, at different times of day.
photorespiration Reaction in which rubisco attaches oxygen instead of carbon dioxide to ribulose bisphosphate.
stomata Gaps that open on plant surfaces; allow water vapor and gases to diffuse across the epidermis.

Figure 6.10 Animated Carbon-fixing adaptations. Most plants, including basswood (*Tilia americana* **A**), are C3 plants. **B** Photorespiration in C3 plants makes sugar production inefficient on dry days. Additional reactions minimize photorespiration in C4 plants such as corn (*Zea mays* **C**), and CAM plants such as jade plants (*Crassula argentea* **D**).

Take-Home Message How do carbon–fixing reactions vary?

❯ When stomata are closed, oxygen builds up inside leaves of C3 plants. Rubisco then can attach oxygen (instead of carbon dioxide) to RuBP. This reaction, photorespiration, reduces the efficiency of sugar production, so it can limit the plant's growth.

❯ Plants adapted to dry conditions limit photorespiration by fixing carbon twice. C4 plants separate the two sets of reactions in space; CAM plants separate them in time.

Green Energy (revisited)

Your body is about 9.5 percent carbon by weight, which means that you contain an enormous number of carbon atoms. Where did they all come from? You eat other organisms to get the carbon atoms your body uses for energy and for raw materials. Those atoms may have passed through other heterotrophs before you ate them, but at some point they were part of photosynthetic organisms. Photosynthesizers strip carbon from carbon dioxide, then use the atoms to build organic compounds. Your carbon atoms—and those of most other organisms—were recently part of Earth's atmosphere, in molecules of CO_2.

Photosynthesis removes carbon dioxide from the atmosphere, and locks its carbon atoms inside organic compounds. When photosynthesizers and other aerobic organisms break down the organic compounds for energy, carbon atoms are released in the form of CO_2, which then reenters the atmosphere. Since photosynthesis evolved, these two processes have constituted a balanced cycle of the biosphere. You will learn more about the carbon cycle in Section 28.8. For now, know that the amount of carbon dioxide that photosynthesis removes from the atmosphere is roughly the same amount that organisms release back into it. At least it was, until humans came along.

As early as 8,000 years ago, humans began burning forests to clear land for agriculture. When trees and other plants burn, most of the carbon locked in their tissues is released into the atmosphere as carbon dioxide. Fires that occur naturally release carbon dioxide the same way.

Today, we are burning a lot more than our ancestors ever did. In addition to wood, we are burning fossil fuels—coal, petroleum, and natural gas—to satisfy our greater and greater demands for energy (Figure 6.11). Fossil fuels are the organic remains of ancient organisms. When we burn these fuels, the carbon that has been locked in them for hundreds of millions of years is released back into the atmosphere, mainly as carbon dioxide.

Our activities have put Earth's atmospheric cycle of carbon dioxide out of balance. We are adding far more CO_2 to the atmosphere than photosynthetic organisms are removing from it. Today, we release about 28 billion tons of carbon dioxide into the atmosphere each year, more than ten times the amount we released in the year 1900. Most of it comes from burning fossil fuels. How do we know? Researchers can determine how long ago the carbon atoms in a sample of CO_2 were part of a living organism by measuring the ratio of different carbon isotopes in it (you will read more about radioisotope dating techniques in Section 16.5). These results are correlated with fossil fuel extraction, refining, and trade statistics.

Figure 6.11
Visible evidence of fossil fuel emissions in the atmosphere: the sky over New York City on a sunny day.

Researchers find pockets of our ancient atmosphere in Antarctica. Snow and ice have been accumulating in layers there, year after year, for the last 15 million years. Air and dust trapped in each layer reveal the composition of the atmosphere that prevailed when the layer formed. Thus, we now know that the atmospheric CO_2 level had been relatively stable for about 10,000 years before the industrial revolution. Since 1850, the CO_2 level has been steadily rising. In 2008, it was higher than it had been in *24 million years.*

The increase in atmospheric carbon dioxide is having dramatic effects on climate. CO_2 contributes to global climate change. We are seeing a warming trend that mirrors the increase in CO_2 levels: Earth is now the warmest it has been for 12,000 years. The trend is affecting biological systems everywhere. Life cycles are changing—birds are laying eggs earlier; plants are flowering earlier than usual; mammals are hibernating for shorter periods. Migration patterns and habitats are also changing. These changes may be too fast for many species, and the rate of extinctions is rising.

Under normal circumstances, extra carbon dioxide stimulates photosynthesis, which means extra carbon dioxide uptake. However, changes in temperature and moisture patterns as a result of global warming are offsetting this benefit because they are proving harmful to plants and other photosynthesizers.

Making biofuel production economically feasible is a high priority for today's energy researchers. Biofuels are a renewable source of energy: We can always make more of them simply by growing more biomass. Also, unlike fossil fuels, using plant matter for fuel recycles carbon that is already in the atmosphere, because plants remove carbon dioxide from the atmosphere as they grow.

How Would You Vote? Ethanol and other fuels manufactured from crops currently cost more than gasoline, but they are renewable energy sources and have fewer emissions. Would you pay a premium to drive a vehicle that runs on biofuels? See CengageNow for details, then vote online (cengagenow.com).

Summary

Section 6.1 Autotrophs make their own food using energy they get directly from the environment, and carbon from inorganic sources such as CO_2. By metabolic pathways of **photosynthesis**, plants and other autotrophs capture the energy of light and use it to build sugars from water and carbon dioxide. **Heterotrophs** get energy and carbon from molecules that other organisms have already assembled.

Sections 6.2, 6.3 Visible light is a very small part of the spectrum of electromagnetic energy radiating from the sun. That energy travels in waves, and it is organized as photons. Visible light drives photosynthesis, which begins when photons are absorbed by photosynthetic pigment molecules. **Pigments** are molecules that absorb light of particular **wavelengths** only; photons not captured by a pigment are reflected as its characteristic color. The main photosynthetic pigment—**chlorophyll** *a*—absorbs violet and red light, so it appears green. Accessory pigments absorb additional wavelengths.

Section 6.4 In **chloroplasts**, the **light-dependent reactions** of photosynthesis occur at a much-folded **thylakoid membrane**. The membrane forms a continuous compartment in the chloroplast's interior (**stroma**) where the **light-independent reactions** occur.

The following diagram summarizes the overall process of photosynthesis (with the noncyclic reactions):

Sections 6.5, 6.6 In the light reactions of photosynthesis, light-harvesting complexes in the thylakoid membrane absorb photons and pass the energy to **photosystems**, which then release electrons.

In the noncyclic pathway, electrons released from photosystem II flow through an electron transfer chain, then to photosystem I. Photon energy causes photosystem I to release electrons, which end up in NADPH. Photosystem II replaces lost electrons by pulling them from water, which then dissociates into H^+ and O_2 (an example of **photolysis**).

In the cyclic pathway, the electrons released from photosystem I enter an electron transfer chain, then cycle back to photosystem I. NADPH does not form.

ATP forms by **electron transfer phosphorylation** in both pathways. Electrons flowing through electron transfer chains cause H^+ to accumulate in the thylakoid compartment. The H^+ follows its gradient back across the membrane through ATP synthases, driving ATP synthesis.

Section 6.7 Carbon fixation occurs in light-independent reactions. Inside the stroma, the enzyme **rubisco** attaches a carbon from CO_2 to RuBP to start the **Calvin–Benson cycle**. This cyclic pathway uses energy from ATP, carbon and oxygen from CO_2, and hydrogen and electrons from NADPH to make sugars.

Section 6.8 Environments differ, and so do details of the light-independent reactions. On dry days, plants conserve water by closing their **stomata**. However, when stomata are closed, O_2 from photosynthesis cannot escape the plant, and CO_2 for photosynthesis cannot enter it. In **C3 plants**, the resulting high O_2 level in the plant's tissues causes rubisco to attach O_2 instead of CO_2 to RuBP. This pathway, which is called **photorespiration**, reduces the efficiency of sugar production on dry days. In **C4 plants**, carbon fixation occurs twice. The first reactions release CO_2 near rubisco, and thus limit photorespiration when stomata are closed. **CAM plants** minimize photorespiration by opening their stomata and fixing carbon at night.

Self-Quiz   Answers in Appendix III

1. A cat eats a bird, which ate a caterpillar that chewed on a weed. Which organisms are autotrophs? Which ones are heterotrophs?

2. Photosynthetic autotrophs use _____ from the air as a carbon source and _____ as their energy source.

3. Chlorophyll *a* appears green because it absorbs mainly _____ light.
   - a. violet and red
   - b. green
   - c. yellow
   - d. blue

4. Light-dependent reactions in plants proceed in the _____ .
   - a. thylakoid membrane
   - b. plasma membrane
   - c. stroma
   - d. cytoplasm

5. When a photosystem absorbs light, _____ .
   - a. sugar phosphates are produced
   - b. electrons are transferred to ATP
   - c. RuBP accepts electrons
   - d. electrons are ejected from its special pair

6. In the light-dependent reactions, _____ .
   - a. carbon dioxide is fixed
   - b. ATP forms
   - c. sugars form
   - d. CO_2 accepts electrons
   - e. b and c
   - f. a and c

7. What accumulates inside the thylakoid compartment during the light-dependent reactions?
   - a. glucose
   - b. hydrogen ions
   - c. O_2
   - d. CO_2

Energy Efficiency of Biofuel Production From Corn, Soy, and Prairie Grasses

Most corn is grown intensively in vast swaths, which means that farmers who grow it use fertilizers and pesticides, both of which are typically made from fossil fuels. Corn is an annual plant, and yearly harvests tend to cause runoff that depletes soil and pollutes rivers.

In 2006, David Tilman and his colleagues published the results of a 10-year study comparing the net energy output of various biofuels. The researchers grew a mixture of native perennial grasses without irrigation, fertilizer, pesticides, or herbicides, in sandy soil that was so depleted by intensive agriculture that it had been abandoned. They measured the usable energy in biofuels made from the grasses, from corn, and from soy. They also measured the energy it took to grow and produce each kind of biofuel. Some of their results are shown in Figure 6.12.

1. About how much energy did ethanol produced from one hectare of corn yield? How much energy did it take to grow the corn to make that ethanol?

2. Which of the biofuels tested had the highest ratio of energy output to energy input?

3. Which of the three crops would require the least amount of land to produce a given amount of biofuel energy?

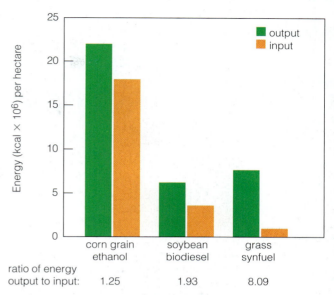

	corn grain ethanol	soybean biodiesel	grass synfuel
ratio of energy output to input:	1.25	1.93	8.09

Figure 6.12 Energy inputs and outputs of biofuels from corn and soy grown on fertile farmland, and grasses grown in infertile soil. One hectare is about 2.5 acres.

---

8. Light-independent reactions in plants proceed in the _____ .
   a. thylakoid membrane   c. stroma
   b. plasma membrane   d. cytoplasm

9. The Calvin–Benson cycle starts when _____ .
   a. light is available
   b. carbon dioxide is attached to RuBP
   c. electrons leave a photosystem II

10. Which of the following substances does *not* participate in the Calvin–Benson cycle?
   a. ATP   d. PGAL
   b. NADPH   e. O_2
   c. RuBP   f. CO_2

11. In the light-independent reactions, _____ .
   a. carbon dioxide is fixed   d. CO_2 accepts electrons
   b. ATP forms   e. b and c
   c. sugars form   f. a and c

12. Match each with its most suitable description.
   ___ PGAL formation
   ___ CO_2 fixation
   ___ photolysis
   ___ ATP forms; NADPH
        does not
   ___ photorespiration
   ___ photosynthesis
   ___ pigment
   ___ autotroph

   a. absorbs light
   b. converts light to
        chemical energy
   c. self-feeder
   d. electrons cycle back
        to photosystem I
   e. problem in C3 plants
   f. ATP, NADPH
        required
   g. water molecules split
   h. rubisco function

Additional questions are available on **CENGAGENOW**.

Critical Thinking

1. About 200 years ago, Jan Baptista van Helmont wanted to know where growing plants get the materials necessary for increases in size. He planted a tree seedling weighing 5 pounds in a barrel filled with 200 pounds of soil and then watered the tree regularly. After five years, the tree weighed 169 pounds, 3 ounces, and the soil weighed 199 pounds, 14 ounces. Because the tree had gained so much weight and the soil had lost so little, he concluded that the tree had gained all of its additional weight by absorbing the water he had added to the barrel, but of course he was incorrect. What really happened?

2. While gazing into an aquarium, you observe bubbles coming from an aquatic plant (*left*). What are the bubbles and where do they come from?

3. A C3 plant absorbs a carbon radio-isotope (as part of $^{14}CO_2$). In which stable, organic compound does the labeled carbon appear first? Which compound forms first if a C4 plant absorbs the same radioisotope?

Animations and Interactions on **CENGAGENOW**:
❯ Wavelengths of light; Englemann's experiment; Sites of photosynthesis in a plant; Chloroplast structure; The noncyclic pathway; Calvin–Benson cycle; Photosynthesis in C3 and C4 plants compared.

‹ Links to Earlier Concepts

This chapter expands the picture of energy flow through the world of life (Section 5.2). It focuses on metabolic reactions and pathways (3.2, 5.3, 5.5) that make ATP by degrading carbohydrates (3.3). These reactions occur either in cytoplasm or in mitochondria (4.9). You will revisit lipids (3.4) and proteins (3.5), membrane transport (4.4, 5.7), free radicals (5.4), electron transfer chains (5.5, 6.5), and photosynthesis pathways (6.4, 6.7).

Key Concepts

Energy From Carbohydrates

Various pathways convert the chemical energy of glucose and other organic compounds to the chemical energy of ATP. Aerobic respiration yields the most ATP from each glucose molecule. In eukaryotes, this pathway ends in mitochondria.

Glycolysis

Glycolysis, the first stage of aerobic respiration and of anaerobic fermentation pathways, occurs in cytoplasm. Enzymes of glycolysis convert one molecule of glucose to two molecules of pyruvate for a net yield of two ATP.

7 How Cells Release Chemical Energy

7.1 When Mitochondria Spin Their Wheels

In the early 1960s, a Swedish physician, Rolf Luft, mulled over a patient's odd symptoms. The young woman felt weak and hot all the time. Even on the coldest winter days she could not stop sweating, and her skin was always flushed. She was thin, yet had a huge appetite. Luft inferred that his patient's symptoms pointed to a metabolic disorder: Her cells were very active, but much of their activity was being lost as metabolic heat. Luft checked the patient's rate of metabolism, the amount of energy her body was expending. Even while resting, her oxygen consumption was the highest that had ever been recorded.

Examination of a tissue sample revealed that the patient's skeletal muscles had plenty of mitochondria, the cell's ATP-producing powerhouses. But there were too many of them, and they were abnormally shaped. The mitochondria were making very little ATP despite working at top speed.

The disorder, now called Luft's syndrome, was the first to be linked to defective mitochondria. The cells of someone with Luft's syndrome are like cities that are burning tons of coal in many power plants but not getting much usable energy output. Skeletal and heart muscles, the brain, and other hardworking body parts with high energy demands are most affected.

More than forty disorders related to defective mitochondria are now known. One called Friedreich's ataxia causes loss of coordination (ataxia), weak muscles, and heart problems (Figure 7.1). Many of those affected die when they are young adults.

Like the chloroplasts described in the previous chapter, mitochondria have an internal folded membrane system that allows them to make ATP. By the process of aerobic respiration, electron transfer chains in the mitochondrial membrane set up hydrogen ion gradients that power ATP formation.

In Luft's syndrome, electron transfer chains in mitochondria work overtime, but too little ATP forms. In Friedreich's ataxia, a protein called frataxin does not work properly. This protein helps build some of the iron-containing enzymes of electron

Figure 7.1 The mitochondrion, an ATP-producing powerhouse. *Above*, a mitochondrion's folded internal membrane is the source of its function.

Opposite, mitochondrial diseases such as Friedreich's ataxia cause serious health problems. Leah (*left*) started to lose her sense of balance and coordination at age five. Six years later she was in a wheelchair; now she is diabetic and partially deaf. Her brother Joshua (*right*) could not walk by the time he was eleven, and is now blind. Both have heart problems; both had spinal fusion surgery. Special equipment allows them to attend school and work part-time. Leah is a professional model.

transfer chains. When it malfunctions, iron atoms that were supposed to be incorporated into the enzymes accumulate inside mitochondria instead. Oxygen is present in mitochondria, and free radicals form when oxygen reacts with the iron atoms (Section 5.4). When too much iron accumulates in mitochondria, too many free radicals form. These free radicals destroy the molecules of life faster than they can be repaired or replaced. Eventually, the mitochondria stop working, and the cell dies.

You already have a sense of how cells harvest energy in electron transfer chains. Details of the reactions vary from one type of organism to the next, but all life relies on this ATP-forming machinery. When you consider mitochondria in this chapter, remember that without them, you would not make enough ATP to even read about how they do it.

Aerobic Respiration
The final stages of aerobic respiration break down pyruvate to CO_2. Many coenzymes that become reduced deliver electrons and hydrogen ions to electron transfer chains, where ATP forms by electron transfer phosphorylation.

Fermentation
Fermentation pathways start with glycolysis. Substances other than oxygen accept electrons at the end of the pathways. Compared with aerobic respiration, the net yield of ATP from fermentation is small.

Other Metabolic Pathways
Molecules other than carbohydrates are common energy sources in the animal body. Many different pathways can convert dietary lipids and proteins to molecules that may enter glycolysis or the Krebs cycle.

> Photoautotrophs use the ATP they produce by photosynthesis to make sugars.
> Most organisms, including photoautotrophs, make ATP by breaking down sugars and other organic compounds.
< Links to Energy flow 5.2, Free radicals 5.4, Metabolic pathways 5.5, Photosynthesis 6.4

Evolution of Earth's Atmosphere

The first cells on Earth did not use sunlight for energy. Like some modern archaeans, these ancient organisms extracted energy and carbon from simple molecules such as methane and hydrogen sulfide—gases that were plentiful in the nasty brew that constituted Earth's early atmosphere (Figure 7.2).

The first photosynthetic autotrophs, or **photoautotrophs**, evolved about 3.2 billion years ago, probably in shallow ocean waters. Sunlight offered these organisms an essentially unlimited supply of energy, and they were very successful. Oxygen gas (O_2) released from uncountable numbers of water molecules began seeping out of uncountable numbers of photosynthesizers, and it accumulated in the ocean and the atmosphere. From that time on, the world of life would never be the same.

Molecular oxygen had been a very small component of Earth's early atmosphere before photosynthesis evolved. The new abundance of atmospheric oxygen exerted tremendous selection pressure on all organisms. Why? Oxygen gas reacts with metals such as enzyme cofactors, and free radicals form during those reactions. Free radicals, remember, damage biological molecules, so they are dangerous to life (Section 5.4).

The ancient cells had no way to detoxify oxygen radicals, so most of them quickly died out. Only a few types persisted in deep water, muddy sediments, and other **anaerobic** (oxygen-free) habitats. Then, new metabolic pathways that detoxified oxygen radicals evolved in the survivors. Organisms with such pathways were the first

Figure 7.2 Then and now: An artist's conception of how Earth was permanently altered by the evolution of photosynthesis and aerobic respiration.

aerobic organisms—they could live in the presence of oxygen. One of the pathways, aerobic respiration, put the reactive properties of oxygen to good use. **Aerobic respiration** is one of several pathways by which organisms access the energy stored in carbohydrates. This equation summarizes aerobic respiration:

$$C_6H_{12}O_6 \ + \ O_2 \longrightarrow CO_2 \ + \ H_2O$$

glucose   oxygen   carbon dioxide   water

Note that aerobic respiration requires oxygen (a by-product of photosynthesis), and it produces carbon dioxide and water (the same raw materials from which photosynthesizers make sugars). With this connection, the cycling of carbon, hydrogen, and oxygen through the biosphere came full circle (Figure 7.3).

Carbohydrate Breakdown Pathways

Like those early organisms, photosynthetic autotrophs of today's world capture energy from the sun, and store it in the form of carbohydrates. They and most other organisms use energy stored in carbohydrates to run the diverse reactions that sustain life. However, carbohydrates rarely participate in such reactions, so how do cells harness their energy? In order to use the energy stored in carbohydrates, cells must first transfer it to energy-carriers such as ATP, which does participate in many of the energy-requiring reactions that a cell runs. The transfer occurs by breaking the bonds of the carbohydrates, which releases energy that drives ATP synthesis.

There are a few different pathways that break down carbohydrates, but aerobic respiration is the one that typical eukaryotic cells use at least most of the time. Aerobic respiration yields more ATP than other carbohydrate-breakdown pathways. You and other multicelled organisms could not live without its higher yield.

Figure 7.3 Animated
The connection between photosynthesis and aerobic respiration. Note the cycling of materials, and the one-way flow of energy (compare Figure 5.5).

Aerobic Respiration

In the Cytoplasm

A The first stage, glycolysis, occurs in the cell's cytoplasm. Enzymes convert a glucose molecule to 2 pyruvate for a net yield of 2 ATP. 2 NAD+ combine with electrons and hydrogen ions during the reactions, so 2 NADH also form.

In the Mitochondrion

B The second stage occurs in mitochondria. The 2 pyruvate are converted to a molecule that enters the Krebs cycle. CO_2 forms and leaves the cell. 2 ATP, 8 NADH, and 2 $FADH_2$ form during the reactions.

C The third and final stage, electron transfer phosphorylation, occurs inside mitochondria. 10 NADH and 2 $FADH_2$ give up electrons and hydrogen ions to electron transfer chains. Electron flow through the chains sets up hydrogen ion gradients that drive ATP formation. Oxygen accepts electrons at the end of the chains.

Figure 7.4 Animated Overview of aerobic respiration. The reactions start in the cytoplasm and end in mitochondria. **›› Figure It Out** What is aerobic respiration's typical net yield of ATP?

Answer: 38 – 2 = 36 ATP per glucose

Most types of eukaryotic cells either use aerobic respiration exclusively, or they use it most of the time. Many bacteria, archaeans, and protists use alternative pathways. Bacteria and single-celled protists that inhabit sea sediments, animal guts, improperly canned food, sewage treatment ponds, deep mud, and other anaerobic habitats use **fermentation**, a pathway by which cells harvest energy from carbohydrates anaerobically. Some of these organisms, including the bacteria that cause botulism, cannot tolerate aerobic conditions, and will die when they are exposed to oxygen.

Fermentation and aerobic respiration begin with the same reactions in the cytoplasm. These reactions, which are collectively called **glycolysis**, convert one six-carbon molecule of glucose into two molecules of **pyruvate**, an organic compound with a three-carbon backbone (Figure 7.4A). After glycolysis, the pathways of fermentation and aerobic respiration diverge. Aerobic respiration continues with two more stages that occur inside mitochondria (Figure 7.4B,C). It ends when oxygen accepts electrons at the end of electron transfer chains. Fermentation ends in the cytoplasm, where a molecule other than oxygen accepts electrons. Aerobic respiration is much more efficient than fermentation. You and other multicelled organisms could not live without its higher yield of ATP.

aerobic Involving or occurring in the presence of oxygen.
aerobic respiration Oxygen-requiring pathway that breaks down carbohydrates to produce ATP.
anaerobic Occurring in the absence of oxygen.
fermentation An anaerobic pathway by which cells harvest energy from carbohydrates to produce ATP.
glycolysis Set of reactions in which glucose or another sugar is broken down to two pyruvate for a net yield of two ATP.
photoautotroph Photosynthetic autotroph.
pyruvate Three-carbon end product of glycolysis.

Take-Home Message How do cells access the chemical energy in carbohydrates?

› Most cells convert the chemical energy of carbohydrates to the chemical energy of ATP by aerobic respiration or fermentation. Aerobic respiration and fermentation pathways start in cytoplasm, with glycolysis.

› Fermentation is anaerobic and ends in the cytoplasm.

› In eukaryotes, aerobic respiration requires oxygen and ends in mitochondria. This pathway evolved after a dramatic increase in oxygen content of the atmosphere brought about by the evolution of photosynthesis.

❯ The reactions of glycolysis convert one molecule of glucose to two molecules of pyruvate for a net yield of two ATP.

❯ An energy investment of ATP is required to start glycolysis.

❮ Links to Hydrolysis 3.2, Glucose 3.3, Endergonic reactions and phosphorylation 5.3, Redox reaction 5.5, Gradients 5.6, Glucose transporter 5.7, Calvin–Benson cycle 6.7

Glycolysis is a series of reactions that begins carbohydrate breakdown pathways. The reactions, which occur in the cytoplasm, convert one molecule of glucose to two molecules of pyruvate:

glucose   glycolysis →   pyruvate (2)

The word glycolysis (from the Greek words *glyk–*, sweet; and *–lysis*, loosening) refers to the release of chemical energy from sugars. Different sugars can enter glycolysis, but for clarity we focus here on glucose.

Glycolysis begins when a molecule of glucose enters a cell through a glucose transporter, a passive transport protein you encountered in Section 5.7. The cell invests two ATP in the endergonic reactions that begin the pathway (Figure 7.5). In the first reaction, a phosphate group is transferred from ATP to the glucose, thus forming glucose-6-phosphate ❶. A model of hexokinase, the enzyme that catalyzes this reaction, is pictured in Section 5.4.

Unlike glucose, glucose-6-phosphate does not pass through glucose transporters in the plasma membrane, so it is trapped inside the cell. Almost all of the glucose that enters a cell is immediately converted to glucose-6-phosphate. This phosphorylation keeps the glucose concentration in the cytoplasm lower than it is in the fluid outside of the cell. By maintaining this concentration gradient across the plasma membrane, the cell favors uptake of even more glucose.

Glycolysis continues as glucose-6-phosphate accepts a phosphate group from another ATP, then splits in two ❷, forming two PGAL (phosphoglyceraldehyde). This phosphorylated sugar also forms during the Calvin–Benson cycle (Section 6.7). A second phosphate group is attached to each PGAL, so two PGA (phosphoglycerate) form ❸. During the reaction, two electrons and a hydrogen ion are transferred from each PGAL to NAD+, so two NADH form.

Next, a phosphate group is transferred from each PGA to ADP, so two ATP form ❹. Two more ATP form when a phosphate group is transferred from another pair of inter-

mediates to two ADP ❺. This and any other reaction that transfers a phosphate group directly from a substrate to ADP is called a **substrate–level phosphorylation**.

Glycolysis ends with the formation of two three-carbon pyruvate molecules. These products may now enter the second-stage reactions of either aerobic respiration or fermentation. Remember, two ATP were invested to initiate the reactions of glycolysis. A total of four ATP form, so the net yield is two ATP per molecule of glucose that enters glycolysis ❻. Two NAD+ also pick up hydrogen ions and electrons, thereby becoming reduced to NADH:

Figure 7.5 Animated Glycolysis.

This first stage of carbohydrate breakdown starts and ends in the cytoplasm of all cells. *Opposite*, for clarity, we track only the six carbon atoms (*black* balls) that enter the reactions as part of glucose. Appendix VI has more details for interested students.

Cells invest two ATP to start glycolysis, so the net energy yield from one glucose molecule is two ATP. Two NADH also form, and two pyruvate molecules are the end products.

glucose		pyruvate
ADP + Pi	**Glycolysis**	ATP
NAD+		NADH

Take-Home Message What is glycolysis?

❯ Glycolysis is the first stage of carbohydrate breakdown in both aerobic respiration and fermentation.

❯ The reactions of glycolysis occur in the cytoplasm.

❯ Glycolysis converts one molecule of glucose to two molecules of pyruvate, with a net energy yield of two ATP. Two NADH also form.

---

substrate–level phosphorylation A reaction that transfers a phosphate group from a substrate directly to ADP, thus forming ATP.

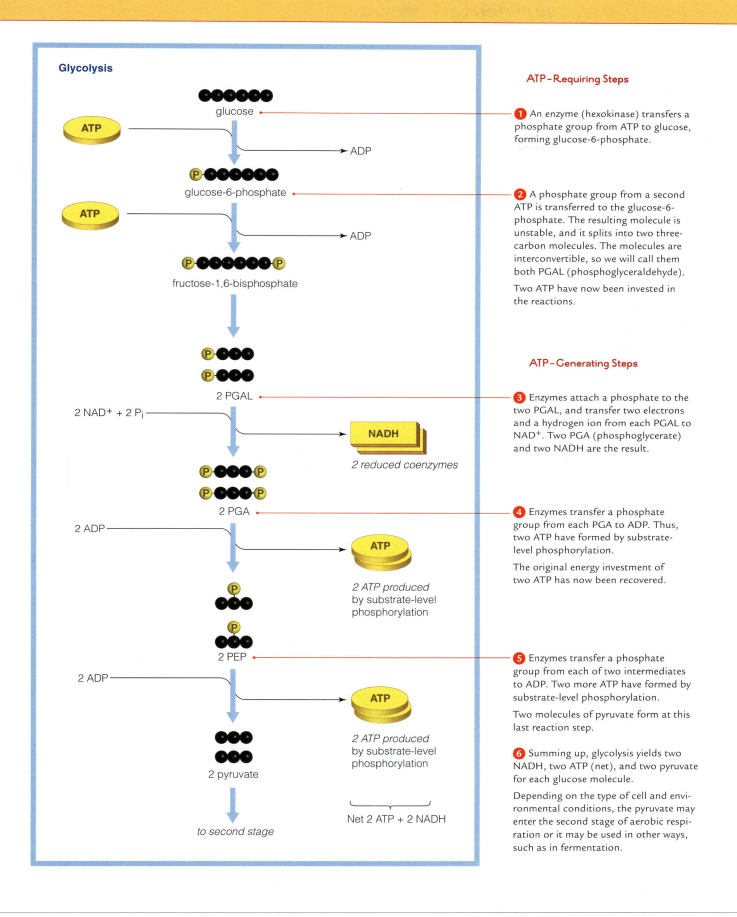

Glycolysis

glucose

ATP → ADP

1 An enzyme (hexokinase) transfers a phosphate group from ATP to glucose, forming glucose-6-phosphate.

glucose-6-phosphate

ATP → ADP

2 A phosphate group from a second ATP is transferred to the glucose-6-phosphate. The resulting molecule is unstable, and it splits into two three-carbon molecules. The molecules are interconvertible, so we will call them both PGAL (phosphoglyceraldehyde).

Two ATP have now been invested in the reactions.

fructose-1,6-bisphosphate

ATP–Generating Steps

2 PGAL

$2 NAD^+ + 2 P_i$

NADH

2 reduced coenzymes

3 Enzymes attach a phosphate to the two PGAL, and transfer two electrons and a hydrogen ion from each PGAL to NAD^+. Two PGA (phosphoglycerate) and two NADH are the result.

2 PGA

2 ADP

ATP

2 ATP produced by substrate-level phosphorylation

4 Enzymes transfer a phosphate group from each PGA to ADP. Thus, two ATP have formed by substrate-level phosphorylation.

The original energy investment of two ATP has now been recovered.

2 PEP

2 ADP

ATP

2 ATP produced by substrate-level phosphorylation

5 Enzymes transfer a phosphate group from each of two intermediates to ADP. Two more ATP have formed by substrate-level phosphorylation.

Two molecules of pyruvate form at this last reaction step.

2 pyruvate

Net 2 ATP + 2 NADH

to second stage

6 Summing up, glycolysis yields two NADH, two ATP (net), and two pyruvate for each glucose molecule.

Depending on the type of cell and environmental conditions, the pyruvate may enter the second stage of aerobic respiration or it may be used in other ways, such as in fermentation.

❭ The second stage of aerobic respiration finishes the break-down of glucose that began in glycolysis.
❬ Links to Mitochondria 4.9, Cyclic pathways 5.5

The second stage of aerobic respiration occurs inside mitochondria (Figure 7.6). It includes two sets of reactions, acetyl–CoA formation and the **Krebs cycle**, that break down pyruvate, the product of glycolysis. All of the carbon atoms that were once part of glucose end up in CO_2, which departs the cell. Only two ATP form. The big payoff is the formation of many reduced coenzymes that drive the third and final stage of aerobic respiration.

The second stage begins when the two pyruvate molecules formed by glycolysis enter a mitochondrion. Pyruvate is transported across the mitochondrion's inner membrane and into the inner compartment, which is called the matrix. Use Figure 7.7 to follow what happens next. In the first reaction, an enzyme splits each molecule of pyruvate into one molecule of CO_2 and a two-carbon acetyl group ❶. The CO_2 diffuses out of the cell, and the acetyl group combines with a molecule called coenzyme A (abbreviated CoA). The product of this reaction is acetyl–CoA. Electrons and hydrogen ions released by the reaction combine with NAD^+, so NADH also forms.

The Krebs Cycle

The Krebs cycle breaks down acetyl–CoA to CO_2. Remember from Section 5.5 that a cyclic pathway is not a physical object, such as a wheel. It is called a cycle because the last

Figure 7.7 Animated Aerobic respiration's second stage: formation of acetyl–CoA and the Krebs cycle. The reactions occur in the mitochondrion's matrix.

Opposite, it takes two cycles of Krebs reactions to break down two pyruvate molecules. After two cycles, all six carbons that entered glycolysis in one glucose molecule have left the cell, in six CO_2.

Two ATP, eight NADH, and two $FADH_2$ form during the two cycles. See Appendix VI for details of the reactions.

reaction in the pathway regenerates the substrate of the first. In the Krebs cycle, a substrate of the first reaction—and a product of the last—is four-carbon oxaloacetate.

During each cycle of Krebs reactions, two carbon atoms of acetyl–CoA are transferred to four-carbon

The breakdown of 2 pyruvate to 6 CO_2 yields 2 ATP and 10 reduced coenzymes (8 NAD^+, 2 FAD). The coenzymes will carry their cargo of electrons and hydrogen ions to the third stage of aerobic respiration.

a mitochondrion

Figure 7.6 Animated The second stage of aerobic respiration, acetyl–CoA formation and the Krebs cycle, occurs inside mitochondria. *Left*, an inner membrane divides a mitochondrion's interior into two fluid-filled compartments. *Right*, the second stage of aerobic respiration takes place in the mitochondrion's innermost compartment, or matrix.

Acetyl–CoA Formation and the Krebs Cycle

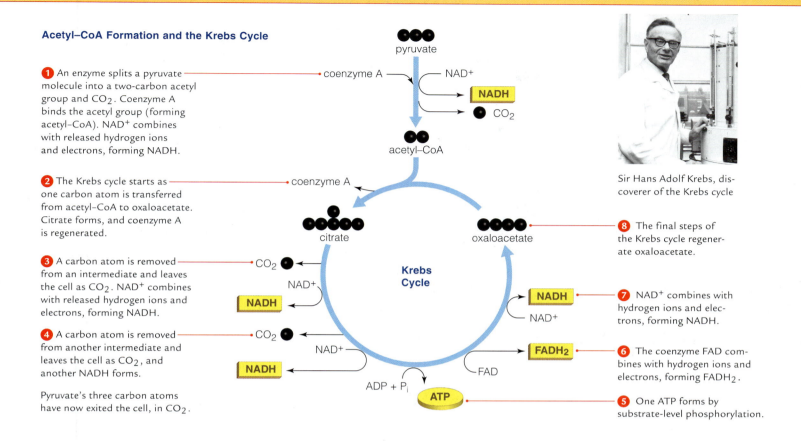

1 An enzyme splits a pyruvate molecule into a two-carbon acetyl group and CO_2. Coenzyme A binds the acetyl group (forming acetyl–CoA). NAD^+ combines with released hydrogen ions and electrons, forming NADH.

2 The Krebs cycle starts as one carbon atom is transferred from acetyl–CoA to oxaloacetate. Citrate forms, and coenzyme A is regenerated.

3 A carbon atom is removed from an intermediate and leaves the cell as CO_2. NAD^+ combines with released hydrogen ions and electrons, forming NADH.

4 A carbon atom is removed from another intermediate and leaves the cell as CO_2, and another NADH forms.

Pyruvate's three carbon atoms have now exited the cell, in CO_2.

pyruvate

coenzyme A NAD^+

NADH

CO_2

acetyl–CoA

coenzyme A

citrate

Krebs Cycle

oxaloacetate

CO_2

NAD^+

NADH

CO_2

NAD^+

NADH

NAD^+

NADH

NAD^+

FADH2

FAD

ADP + P_i

ATP

Sir Hans Adolf Krebs, discoverer of the Krebs cycle

8 The final steps of the Krebs cycle regenerate oxaloacetate.

7 NAD^+ combines with hydrogen ions and electrons, forming NADH.

6 The coenzyme FAD combines with hydrogen ions and electrons, forming $FADH_2$.

5 One ATP forms by substrate-level phosphorylation.

oxaloacetate, forming citrate, the ionized form of citric acid **2**. The Krebs cycle is also called the citric acid cycle after this first intermediate. In later reactions, two CO_2 form and depart the cell. Two NAD^+ are reduced when they accept hydrogen ions and electrons, so two NADH form **3** and **4**. ATP forms by substrate-level phosphorylation **5**, and FAD **6** and another NAD^+ **7** are reduced. The final steps of the pathway regenerate oxaloacetate **8**.

Remember, glycolysis converted one glucose molecule to two pyruvate, and these were converted to two acetyl–CoA when they entered the matrix of a mitochondrion. There, the second-stage reactions convert the two molecules of acetyl–CoA to six CO_2. At this point in aerobic respiration, one glucose molecule has been broken down completely: Six carbon atoms have left the cell, in six CO_2. Two ATP formed, which adds to the small net yield of glycolysis. However, six NAD^+ were reduced to six NADH, and two FAD were reduced to two $FADH_2$.

What is so important about reduced coenzymes? A molecule becomes reduced when it receives electrons (Section 5.5), and electrons carry energy that can be used to drive endergonic reactions. In this case, the electrons picked up by coenzymes during the first two stages of

aerobic respiration carry energy that drives the reactions of the third stage.

In total, two ATP form and ten coenzymes (eight NAD^+ and two FAD) are reduced during acetyl–CoA formation and the Krebs cycle. Add in the two NAD^+ reduced in glycolysis, and the full breakdown of each glucose molecule has a big potential payoff. Twelve reduced coenzymes will deliver electrons (and the energy they carry) to the third stage of aerobic respiration:

pyruvate
ADP + P_i
NAD^+
FAD

Acetyl–CoA formation and the Krebs cycle

CO_2
ATP
NADH
$FADH_2$

Krebs cycle Cyclic pathway that, along with acetyl–CoA formation, breaks down pyruvate to carbon dioxide.

Take-Home Message What happens during the second stage of aerobic respiration?

> The second stage of aerobic respiration, acetyl–CoA formation and the Krebs cycle, occurs in the inner compartment (matrix) of mitochondria.

> The pyruvate that formed in glycolysis is converted to acetyl–CoA and carbon dioxide. The acetyl–CoA enters the Krebs cycle, which breaks it down to CO_2.

> For each two pyruvate molecules broken down in the second-stage reactions, two ATP form, and ten coenzymes (eight NAD^+ and two FAD) are reduced.

> Many ATP are formed during the third and final stage of aerobic respiration.

< Links to Membrane proteins 4.4, Thermodynamics 5.2, Electron transfer chains 5.5, Selective permeability of cell membranes 5.6, Electron transfer phosphorylation 6.5

Electron Transfer Phosphorylation

Figure 7.8 The third and final stage of aerobic respiration, electron transfer phosphorylation, occurs at the inner mitochondrial membrane.

1 NADH and FADH$_2$ deliver electrons to electron transfer chains in the inner mitochondrial membrane.

2 Electron flow through the chains causes hydrogen ions (H$^+$) to be pumped from the matrix to the intermembrane space.

3 The activity of the electron transfer chains causes a hydrogen ion gradient to form across the inner mitochondrial membrane.

4 Hydrogen ion flow back to the matrix through ATP synthases drives the formation of ATP from ADP and phosphate (P$_i$).

5 Oxygen (O$_2$) accepts electrons and hydrogen ions at the end of mitochondrial electron transfer chains, so water forms.

>> **Figure It Out** Which other metabolic pathway that you have learned about involves electron transfer phosphorylation?

Answer: The light-dependent reactions of photosynthesis

Electron Transfer Phosphorylation

The third stage of aerobic respiration, electron transfer phosphorylation, also occurs inside mitochondria. Remember that electron transfer phosphorylation is a process in which the flow of electrons through electron transfer chains ultimately results in the attachment of phosphate to ADP (Section 6.5).

The third-stage reactions take place at the inner mitochondrial membrane (Figure 7.8). They begin with the coenzymes NADH and FADH$_2$, which became reduced in the first two stages of aerobic respiration. These coenzymes donate their cargo of electrons and hydrogen ions to electron transfer chains embedded in the inner mitochondrial membrane **1**. As the electrons pass through the chains, they give up energy little by little (Section 5.5). Some molecules of the transfer chains harness that energy to actively transport hydrogen ions across the inner membrane, from the matrix to the intermembrane space **2**. The ions that accumulate in the intermembrane space set up a hydrogen ion gradient across the inner mitochondrial membrane **3**.

This gradient attracts hydrogen ions back toward the matrix. However, ions cannot diffuse through a lipid bilayer on their own (Section 5.6). H$^+$ can only cross the inner mitochondrial membrane by flowing through the interior of ATP synthases **4**. The flow causes these membrane transport proteins to attach phosphate groups to ADP, so ATP forms. The twelve coenzymes that were reduced in the first two stages can drive the synthesis of about thirty-two ATP in the third stage.

Oxygen accepts electrons at the end of the mitochondrial electron transfer chains **5**. Aerobic respiration, which literally means "taking a breath of air," refers to oxygen as the final electron acceptor in this pathway. When oxygen accepts electrons, it combines with H$^+$ to form water, which is one product of the third stage:

A First stage: Glucose is converted to 2 pyruvate; 2 NADH and 4 ATP form. An energy investment of 2 ATP began the reactions, so the net yield is 2 ATP.

B Second stage: 10 more coenzymes accept electrons and hydrogen ions during the second-stage reactions. All six carbons of glucose leave the cell (as 6 CO_2), and 2 ATP form.

C Third stage: Coenzymes that were reduced in the first two stages give up electrons and hydrogen ions to electron transfer chains. Energy lost by the electrons as they flow through the chains is used to move H^+ across the membrane. The resulting gradient causes H^+ to flow through ATP synthases, driving synthesis of ATP.

glucose

2 ATP

2 NAD+

Glycolysis

4 ATP (2 net)

2 NADH

2 pyruvate

cytoplasm

outer membrane

intermembrane space

inner membrane

matrix

2 NADH

2 acetyl–CoA

2 NADH

2 CO_2

4 CO_2

6 NADH
2 FADH₂

Krebs Cycle

2 ATP

ATP ATP ATP

32 ATP

Electron Transfer Phosphorylation H^+ H^+ H^+ H^+ H^+ H^+

oxygen

Figure 7.9 **Animated** Summary of the three stages of aerobic respiration in a mitochondrion.

Summing Up: The Energy Harvest

The breakdown of one glucose molecule in aerobic respiration typically yields thirty-six ATP (Figure 7.9). Four ATP form in the first and second stages, and thirty-two form in the third stage. However, the overall yield varies. Factors such as cell type affect it. For example, the typical yield of aerobic respiration in brain and skeletal muscle cells is thirty-eight ATP, not thirty-six.

Remember that some energy dissipates with every transfer (Section 5.2). Even though aerobic respiration is a very efficient way of retrieving energy from carbohydrates, about 60 percent of the energy harvested in this pathway disperses as metabolic heat.

Take-Home Message **What happens during the third stage of aerobic respiration?**

> In aerobic respiration's third stage, electron transfer phosphorylation, energy released by electrons flowing through electron transfer chains is ultimately captured in the attachment of phosphate to ADP. A typical net yield of aerobic respiration is thirty-six ATP per glucose.

> The reactions begin when coenzymes that were reduced in the first and second stages of reactions deliver electrons and hydrogen ions to electron transfer chains in the inner mitochondrial membrane.

> Energy released by electrons as they pass through electron transfer chains is used to pump H^+ from the mitochondrial matrix to the intermembrane space.

> The H^+ gradient that forms across the inner mitochondrial membrane drives the flow of hydrogen ions through ATP synthases, which results in ATP formation.

❯ Fermentation pathways break down carbohydrates without using oxygen. The final steps in these pathways regenerate NAD⁺ but do not produce ATP.

Fermentation is a type of anaerobic pathway that harvests energy from carbohydrates. Aerobic respiration and fermentation begin with precisely the same set of reactions in the cytoplasm: glycolysis. Again, two pyruvate,

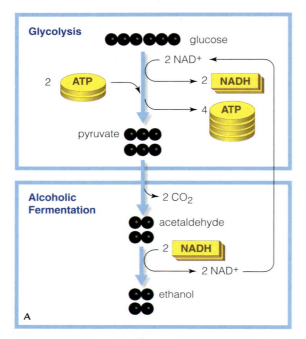

Glycolysis

glucose

2 ATP

2 NAD⁺

2 NADH

4 ATP

pyruvate

Alcoholic Fermentation

2 CO_2

acetaldehyde

2 NADH

2 NAD⁺

ethanol

A

two NADH, and two ATP form during the reactions of glycolysis. However, after that, fermentation and aerobic respiration pathways differ. The final steps of fermentation occur in the cytoplasm. In these reactions, pyruvate is converted to other molecules, but it is not fully broken down to carbon dioxide and water as occurs in aerobic respiration. Electrons do not flow through transfer chains, so no more ATP forms. However, electrons are removed from NADH, so NAD⁺ is regenerated. Regenerating this coenzyme allows glycolysis—along with the small ATP yield it offers—to continue. Thus, the net ATP yield of fermentation consists of the two ATP that form in glycolysis. That yield provides enough energy to sustain many single-celled anaerobic species. It also helps cells of aerobic species produce ATP under anaerobic conditions.

Alcoholic Fermentation

In **alcoholic fermentation**, the pyruvate from glycolysis is converted to ethyl alcohol, or ethanol. First, 3-carbon pyruvate is split into carbon dioxide and 2-carbon acetaldehyde. Then, electrons and hydrogen are transferred from NADH to the acetaldehyde, forming NAD⁺ and ethanol (Figure 7.10).

Bakers use the alcoholic fermentation capabilities of one species of yeast, *Saccharomyces cerevisiae*, to make bread. These cells break down carbohydrates in bread dough, and release CO_2 by alcoholic fermentation. The dough expands (rises) as CO_2 forms bubbles in it.

Some wild and cultivated strains of *Saccharomyces* are also used to produce wine. Crushed grapes are left in vats along with large populations of yeast cells, which convert the sugars in the juice to ethanol.

Lactate Fermentation

In **lactate fermentation**, the electrons and hydrogen ions carried by NADH are transferred directly to pyruvate. This reaction converts pyruvate to 3-carbon lactate (the ionized form of lactic acid), and also converts NADH to NAD⁺ (Figure 7.11A).

Some lactate fermenters spoil food, but we use others to preserve it. For instance, a bacterium called *Lactobacillus acidophilus* breaks down lactose in milk by fermentation. We use this bacteria to produce dairy products such

B

C

D

Figure 7.10 Animated Alcoholic fermentation. **A** Alcoholic fermentation begins with glycolysis, but the final steps do not produce ATP. They regenerate NAD⁺. The net yield of these reactions is two ATP per molecule of glucose (from glycolysis). **B** A vintner examines a product of alcoholic fermentation. **C** A commercial vat of yeast dough rising with the help of **D** live *Saccharomyces* cells.

Figure 7.11 Animated Lactate fermentation. **A** Like alcoholic fermentation, the final steps of lactate fermentation regenerate NAD⁺. The net yield is two ATP per molecule of glucose.

B Two types of muscle fibers, red and white, are visible in this cross-section through a human thigh muscle. The white fibers, which make ATP by lactate fermentation, support sprints and other intense bursts of activity **C**. The darker red fibers, which make ATP by aerobic respiration, sustain endurance activities **D**. Chickens spend most of their time walking, so their leg muscles consist mainly of red fibers.

as buttermilk, cheese, and yogurt. Yeast species ferment and preserve pickles, corned beef, sauerkraut, and kimchi.

Animal skeletal muscles, which move bones, consist of cells fused as long fibers. The fibers differ in how they make ATP. Red fibers have many mitochondria and produce ATP by aerobic respiration. These fibers sustain prolonged activity, such as marathon runs. They are red because they have an abundance of myoglobin, a protein that stores oxygen for aerobic respiration in these fibers (Figure 7.11B).

White muscle fibers contain few mitochondria and no myoglobin, so they do not carry out a lot of aerobic respiration. Instead, they make most of their ATP by lactate fermentation. This pathway makes ATP quickly but not for long, so it is useful for quick, strenuous activities such as sprinting or weight lifting. The low ATP yield does

not support prolonged activity. That is one reason why chickens cannot fly very far: Their flight muscles consist mostly of white fibers (thus, the "white" breast meat). Chickens fly only in short bursts. More often, a chicken walks or runs. Its leg muscles consist mostly of red muscle fibers, the "dark meat." Most human muscles consist of a mixture of white and red fibers, but the proportions vary among muscles and among individuals. Great sprinters tend to have more white fibers. Great marathon runners tend to have more red fibers.

---

alcoholic fermentation Anaerobic carbohydrate breakdown pathway that produces ATP and ethanol. Begins with glycolysis; end reactions regenerate NAD⁺ so glycolysis can continue.
lactate fermentation Anaerobic carbohydrate breakdown pathway that produces ATP and lactate.

Take-Home Message What is fermentation?

> ATP can form by carbohydrate breakdown in fermentation pathways, which are anaerobic.

> The end product of lactate fermentation is lactate. The end product of alcoholic fermentation is ethanol.

> Both pathways have a net yield of two ATP per glucose molecule. The ATP forms during glycolysis.

> Fermentation reactions regenerate the coenzyme NAD⁺, without which glycolysis (and ATP production) would stop.

❯ Aerobic respiration can produce ATP from the breakdown of fats and proteins.

❮ Links to Metabolic reactions 3.2, Carbohydrates 3.3, Lipids 3.4, Proteins 3.5

As you eat, glucose and other breakdown products of digestion are absorbed across the gut lining, and blood transports these small organic molecules throughout the body. The concentration of glucose in the bloodstream rises, and in response the pancreas (an organ) increases its rate of insulin secretion. Insulin is a hormone (a type of signaling molecule) that causes cells to take up glucose, so an increase in insulin production causes cells to take up glucose faster.

In cytoplasm, glucose is immediately converted to glucose-6-phosphate, an intermediate of glycolysis. Unless ATP is being used quickly, its concentration rises in the cytoplasm. A high concentration of ATP causes glucose-6-phosphate to be diverted away from glycolysis and into a pathway that forms glycogen. Liver and muscle cells espe-cially favor the conversion of glucose to glycogen, and these cells maintain the body's largest stores of it.

What happens if you eat too many carbohydrates? When the blood level of glucose gets too high, acetyl–CoA is diverted away from the Krebs cycle and into a pathway that makes fatty acids. That is why excess dietary carbohydrate ends up as fat.

Between meals, the level of glucose declines in the blood. The pancreas responds by secreting glucagon, a hormone that causes liver cells to convert stored glycogen to glucose. The cells release glucose into the bloodstream, so the blood glucose level rises. Thus, hormones control whether the body's cells use glucose as an energy source immediately or save it for use at a later time.

Glycogen makes up about 1 percent of an average adult's total energy reserves, which is the energy equivalent of about two cups of cooked pasta. Unless you eat regularly, you will completely deplete your liver's glycogen stores in less than twelve hours.

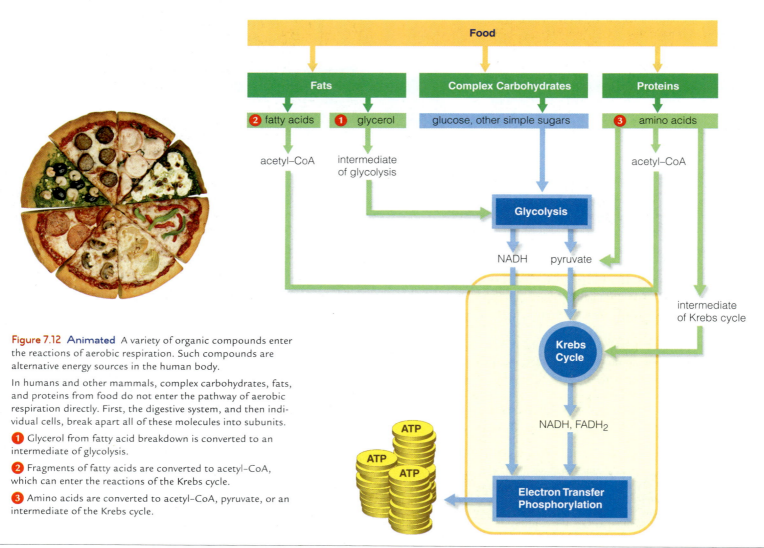

Figure 7.12 Animated A variety of organic compounds enter the reactions of aerobic respiration. Such compounds are alternative energy sources in the human body.

In humans and other mammals, complex carbohydrates, fats, and proteins from food do not enter the pathway of aerobic respiration directly. First, the digestive system, and then individual cells, break apart all of these molecules into subunits.

❶ Glycerol from fatty acid breakdown is converted to an intermediate of glycolysis.

❷ Fragments of fatty acids are converted to acetyl–CoA, which can enter the reactions of the Krebs cycle.

❸ Amino acids are converted to acetyl–CoA, pyruvate, or an intermediate of the Krebs cycle.

Energy From Fats

Of the total energy reserves in a typical adult who eats well, about 78 percent (about 10,000 kilocalories) is stored in body fat, and 21 percent in proteins. How does a body access its fat reservoir? A fat molecule has a glycerol head and one, two, or three fatty acid tails (Section 3.4). The body stores most fats as triglycerides, which have three fatty acid tails. Triglycerides accumulate in fat cells of adipose tissue. This tissue serves as an energy reservoir, and it also insulates and pads the buttocks and other strategic areas of the body.

When the blood glucose level falls, triglycerides are tapped to provide energy (Figure 7.12). Enzymes in fat cells break the bonds that connect glycerol with the fatty acids, and both are released into the bloodstream. Enzymes in liver cells convert the glycerol to another intermediate of glycolysis ❶. Nearly all cells of your body can take up the free fatty acids. Inside the cells, enzymes split the fatty acid backbones and convert the fragments to acetyl–CoA, which can enter the Krebs cycle ❷.

Compared to carbohydrate breakdown, fatty acid breakdown yields more ATP per carbon atom. Between meals or during steady, prolonged exercise, fatty acid breakdown supplies about half of the ATP that muscle, liver, and kidney cells require.

Diets that are extremely low in carbohydrates force the body to break down fats. The breakdown produces molecules called ketones, which most of your cells can use for energy instead of glucose. This metabolic state, which is normally associated with starvation, can raise the level of LDL ("bad" cholesterol) and damage the kidneys and liver.

Energy From Proteins

Some enzymes in your digestive system split dietary proteins into their amino acid subunits, which are then absorbed into the bloodstream. Cells use amino acids to build proteins or other molecules. Even so, when you eat more protein than your body needs, the amino acids become broken down further. The amino (NH_3^+) group is removed, and it becomes ammonia (NH_3), a waste product that the body eliminates in urine. Their carbon backbone is split, and acetyl–CoA, pyruvate, or an intermediate of the Krebs cycle forms, depending on the amino acid. Your cells can divert any of these organic molecules into the Krebs cycle ❸.

> **Take-Home Message** **How does nutrition influence the metabolism of organic compounds?**
>
> ❭ In humans and other organisms, the entrance of organic compounds into an energy-releasing pathway depends on the kinds and proportions of carbohydrates, fats, and proteins in the diet.

When Mitochondria Spin Their Wheels (revisited)

At least 83 proteins are directly involved in the electron transfer chains of electron transfer phosphorylation in mitochondria. A defect in any one of them—or in any of the thousands of other proteins used by mitochondria, such as frataxin (*left*)—can wreak havoc in the body. About one in 5,000 people suffer from a known mitochondrial disorder. New research is showing that mitochondrial defects may be involved in many other illnesses such as diabetes, hypertension, Alzheimer's and Parkinson's disease, and even aging.

How Would You Vote? Developing new drugs is costly, which is why pharmaceutical companies tend to ignore Friedreich's ataxia and other disorders that affect relatively few people. Should governments fund private companies that research potential treatments? See CengageNow for details, then vote online (cengagenow.com).

Summary

Section 7.1 Mitochondria are the ATP-producing powerhouses of eukaryotic cells. Defects in mitochondrial function, such as occurs in Friedreich's ataxia, cause serious health problems.

Section 7.2 Most organisms convert chemical energy of carbohydrates to the chemical energy of ATP. **Anaerobic** and **aerobic** pathways of carbohydrate breakdown start in the cytoplasm with the same set of reactions, **glycolysis**, which converts glucose and other sugars to **pyruvate**. Anaerobic **fermentation** pathways end in cytoplasm and yield two ATP per molecule of glucose. **Aerobic respiration**, which uses oxygen and yields much more ATP than fermentation, evolved after oxygen released by early **photoautotrophs** changed Earth's atmosphere. In modern eukaryotes, aerobic respiration is completed inside mitochondria.

Section 7.3 Glycolysis, the first stage of aerobic respiration and fermentation, occurs in cytoplasm. In the reactions, enzymes use two ATP to convert one molecule of glucose or another six-carbon sugar to two molecules of pyruvate. Electrons and hydrogen ions are transferred to two NAD^+, which are thereby reduced to NADH. Four ATP also form by **substrate–level phosphorylation**, the direct transfer of a phosphate group from a reaction intermediate to ADP.

The net yield of glycolysis is two pyruvate, two ATP, and two NADH per glucose molecule. The pyruvate may continue in fermentation in the cytoplasm, or it may enter mitochondria and the next steps of aerobic respiration.

Summary

Section 7.4 A mitochondrion's inner membrane divides its interior into two fluid-filled spaces: the inner compartment, or matrix, and the intermembrane space. The second stage of aerobic respiration, acetyl–CoA formation and the Krebs cycle, takes place in the matrix. The first steps convert two pyruvate from glycolysis to two acetyl–CoA and two CO_2. The acetyl–CoA enters the Krebs cycle.

It takes two cycles of Krebs reactions to dismantle the two acetyl–CoA. At this stage, all of the carbon atoms in the glucose molecule that entered glycolysis have left the cell in CO_2. During these reactions, electrons and hydrogen ions are transferred to NAD^+ and FAD, which are thereby reduced to NADH and $FADH_2$. ATP forms by substrate-level phosphorylation.

In total, the breakdown of two pyruvate molecules in the second stage of aerobic respiration yields ten reduced coenzymes and two ATP.

Section 7.5 Aerobic respiration ends in mitochondria. In the third stage of reactions, electron transfer phosphorylation, coenzymes that were reduced in the first two stages deliver their cargo of electrons and hydrogen ions to electron transfer chains in the inner mitochondrial membrane. Electrons moving through the chains release energy bit by bit; molecules of the chain use that energy to move H^+ from the matrix to the intermembrane space.

Hydrogen ions that accumulate in the intermembrane space form a gradient across the inner membrane. The ions follow the gradient back to the matrix through ATP synthases. H^+ flow through these transport proteins drives ATP synthesis.

Oxygen combines with electrons and H^+ at the end of the transfer chains, thus forming water.

Overall, aerobic respiration typically yields thirty-six ATP for each glucose molecule.

Section 7.6 Anaerobic fermentation pathways begin with glycolysis and finish in the cytoplasm. A molecule other than oxygen accepts electrons at the end of these reactions. The end product of **alcoholic fermentation** is ethyl alcohol, or ethanol. The end product of **lactate fermentation** is lactate. The final steps of fermentation serve to regenerate NAD^+, which is required for glycolysis to continue, but they produce no ATP. Thus, the breakdown of one glucose molecule in either alcoholic or lactate fermentation yields only the two ATP that form in glycolysis reactions.

Skeletal muscle has two types of fibers: red and white fibers. ATP is produced primarily by aerobic respiration in red muscle fibers, so these fibers sustain activities that require endurance. Lactate fermentation in white fibers supports activities that occur in short, intense bursts.

Section 7.7 In humans and other mammals, the simple sugars from carbohydrate breakdown, glycerol and fatty acids from fat breakdown, and carbon backbones of amino acids from protein breakdown may enter aerobic respiration at various reaction steps.

Self-Quiz   Answers in Appendix III

1. Is the following statement true or false? Unlike animals, which make many ATP by aerobic respiration, plants make all of their ATP by photosynthesis.

2. Glycolysis starts and ends in the _____ .
   a. nucleus   c. plasma membrane
   b. mitochondrion   d. cytoplasm

3. Which of the following metabolic pathways require(s) molecular oxygen (O_2)?
   a. aerobic respiration
   b. lactate fermentation
   c. alcoholic fermentation
   d. all of the above

4. Which molecule does not form during glycolysis?
   a. NADH   b. pyruvate   c. $FADH_2$ d. ATP

5. In eukaryotes, aerobic respiration is completed in the _____ .
   a. nucleus   c. plasma membrane
   b. mitochondrion   d. cytoplasm

6. Which of the following reaction pathways is not part of the second stage of aerobic respiration?
   a. electron transfer   c. Krebs cycle
      phosphorylation   d. glycolysis
   b. acetyl–CoA formation   e. a and d

7. After the Krebs reactions run through _____ cycle(s), one glucose molecule has been completely broken down to CO_2.
   a. one   b. two   c. three   d. six

8. In the third stage of aerobic respiration, _____ is the final acceptor of electrons.
   a. water   b. hydrogen   c. oxygen   d. NADH

9. In alcoholic fermentation, _____ is the final acceptor of electrons.
   a. oxygen   c. acetaldehyde
   b. pyruvate   d. sulfate

10. Fermentation pathways make no more ATP beyond the small yield from glycolysis. The remaining reactions serve to regenerate _____ .
   a. FAD   c. glucose
   b. NAD^+ d. oxygen

11. Most of the energy that is released by the full breakdown of glucose to CO_2 and water ends up in _____ .
   a. NADH   c. heat
   b. ATP   d. electrons

Data Analysis Activities

Mitochondrial Abnormalities in Tetralogy of Fallot Tetralogy of Fallot (TF) is a genetic disorder characterized by four major malformations of the heart. The circulation of blood is abnormal, so TF patients have too little oxygen in their blood. Inadequate oxygen levels result in damaged mitochondrial membranes, which in turn cause cells to self-destruct. In 2004, Sarah Kuruvilla and her colleages looked at abnormalities in the mitochondria of heart muscle in TF patients. Some of their results are shown in **Figure 7.13**.

1. Which abnormality was most strongly associated with TF?
2. Can you make any correlations between blood oxygen content and mitochondrial abnormalities in these patients?

Patient (age)	SPO$_2$ (%)	Mitochondrial Abnormalities in TF			
		Number	Shape	Size	Broken
1 (5)	55	+	+	−	−
2 (3)	69	+	+	−	−
3 (22)	72	+	+	−	−
4 (2)	74	+	+	−	−
5 (3)	76	+	+	−	+
6 (2.5)	78	+	+	−	+
7 (1)	79	+	+	−	−
8 (12)	80	+	−	+	−
9 (4)	80	+	+	−	−
10 (8)	83	+	−	+	−
11 (20)	85	+	+	−	−
12 (2.5)	89	+	−	+	−

Figure 7.13 Mitochondrial changes in tetralogy of Fallot (TF).

A Normal heart muscle. Many mitochondria between the fibers provide muscle cells with ATP for contraction. **B** Heart muscle from a person with TF shows swollen, broken mitochondria.

C Mitochondrial abnormalities in TF patients. SPO$_2$ is oxygen saturation of the blood. A normal value of SPO$_2$ is 96%. Abnormalities are marked "+".

C

---

12. Your body cells can use _____ as an alternative energy source when glucose is in short supply.
    a. fatty acids   c. amino acids
    b. glycerol   d. all of the above

13. Which of the following is not produced by an animal muscle cell operating under anaerobic conditions?
    a. heat   d. ATP
    b. pyruvate   e. lactate
    c. NAD$^+$ f. all are produced

14. Match the event with its most suitable description.
    ___ glycolysis   a. ATP, NADH, FADH$_2$,
    ___ fermentation   and CO$_2$ form
    ___ Krebs cycle   b. glucose to two pyruvate
    ___ electron transfer   c. NAD$^+$ regenerated, little ATP
           phosphorylation   d. H$^+$ flows via ATP synthases

15. Match the term with the best description.
    ___ matrix   a. needed for glycolysis
    ___ pyruvate   b. inner space
    ___ NAD$^+$ c. makes many ATP
    ___ mitochondrion   d. end of glycolysis
    ___ intermembrane   e. reduced coenzyme
          space   f. hydrogen ions
    ___ NADH   accumulate here

Additional questions are available on **CENGAGENOW**.

Animations and Interactions on **CENGAGENOW**:
❯ Links with photosynthesis; Overview of aerobic respiration; Glycolysis; Second stage of aerobic respiration; Oxidative phosphorylation; Fermentation pathways; Alternative energy sources.

Critical Thinking

1. At high altitudes, the oxygen level in air is low. Climbers of very tall mountains risk altitude sickness, which is characterized by shortness of breath, weakness, dizziness, and confusion.

    The early symptoms of cyanide poisoning are the same as those for altitude sickness. Cyanide binds tightly to cytochrome *c* oxidase, a protein complex that is the last component of mitochondrial electron transfer chains. Cytochrome *c* oxidase with bound cyanide can no longer transfer electrons. Explain why cyanide poisoning starts with the same symptoms as altitude sickness.

2. As you learned, membranes impermeable to hydrogen ions are required for electron transfer phosphorylation. Membranes in mitochondria serve this purpose in eukaryotes. Bacteria do not have this organelle, but they do make ATP by electron transfer phosphorylation. How do you think they do it, given that they have no mitochondria?

3. The bar-tailed godwit is a type of shorebird that makes an annual migration from Alaska to New Zealand and back. The birds make each 11,500-kilometer (7,145-mile) trip by flying over the Pacific Ocean in about nine days, depending on weather, wind speed, and direction of travel. One bird was observed to make the entire journey uninterrupted, a feat that is comparable to a human running a nonstop seven-day marathon at 70 kilometers per hour (43.5 miles per hour). Would you expect the flight (breast) muscles of bar-tailed godwits to be light or dark colored? Explain your answer.

‹ Links to Earlier Concepts

This chapter builds on your earlier introduction to nucleic acid structure (Section 3.7). You will revisit carbohydrates (3.3) as you explore the details of DNA's double helix structure, and phosphorylations (5.3) as you learn about DNA replication. You will also see another application of radioisotope tracers (2.2) in the research that led to the discovery that DNA, not protein (3.5), is the hereditary material of all organisms.

Key Concepts

Chromosomes
The DNA of a eukaryotic cell is divided among a characteristic number of chromosomes that differ in length and in shape. Sex chromosomes determine an individual's gender. Proteins associated with eukaryotic DNA help organize chromosomes so they can pack into a nucleus.

Discovery of DNA's Function
The work of many scientists over more than a century led to the discovery that DNA is the molecule that stores hereditary information.

8 DNA Structure and Function

8.1 A Hero Dog's Golden Clones

On September 11, 2001, an off-duty Canadian police officer, Constable James Symington, drove his search dog Trakr from Nova Scotia to Manhattan. Within hours of arriving, the dog led rescuers to the spot where the fifth and final survivor of the World Trade Center attacks was buried. She had been clinging to life, pinned under rubble from the building where she had worked. Symington and Trakr helped with the search and rescue efforts for three days nonstop, until Trakr collapsed from smoke and chemical inhalation, burns, and exhaustion.

Trakr survived the ordeal, but later lost the use of his limbs from a degenerative neurological disease probably linked to toxic smoke exposure at Ground Zero. The hero dog died in April 2009, but his DNA lives on in his genetic copies—his **clones**. Symington's essay about Trakr's superior nature and abilities as a search and rescue dog won the Golden Clone Giveaway, a contest to find the world's most clone-worthy dog. Trakr's DNA was shipped to Korea, where it was inserted into dog eggs, which were then implanted into surrogate mother dogs. Five puppies, all clones of Trakr, were delivered to Symington in July 2009 (Figure 8.1).

Many adult animals have been cloned besides Trakr, but cloning mammals is still unpredictable and far from routine. Typically, less than 2 percent of the implanted embryos result in a live birth. Of the clones that survive, many have serious health problems. Why the difficulty? Even though all cells of an individual inherit the same DNA, an adult cell uses only a fraction of it compared to an embryonic cell. To make a clone from an adult cell, researchers must reprogram its DNA to function like the DNA of an egg. Even though we are getting better at doing that, we still have a lot to learn.

So why do we keep trying? The potential benefits are enormous. Already, cells of cloned human embryos are helping researchers unravel the molecular mechanisms of human genetic diseases. Such cells may one day be induced to form replacement tissues or organs for people with incurable diseases. Endangered animals might be saved from extinction; extinct animals may be brought back. Livestock and pets are already being cloned commercially.

clone Genetically identical copy of an organism.

Figure 8.1 Animal cloning. *Opposite*, James Symington and his dog Trakr at Ground Zero in 2001. *Above*, Symington with Trakr's clones in 2009.

Perfecting the methods for cloning animals brings us closer to the possibility of cloning humans, both technically and ethically. For example, if cloning a lost cat for a grieving pet owner is acceptable, why would it not be acceptable to clone a lost child for a grieving parent? Different people have very different answers to such questions, so controversy over cloning continues to rage even as the techniques improve. Understanding the basis of heredity—what DNA is and how it works—will help inform your own opinions about the issues surrounding cloning.

Structure of DNA
A DNA molecule consists of two long chains of nucleotides coiled into a double helix. Four kinds of nucleotides make up the chains: adenine, thymine, guanine, and cytosine. The order of these bases in DNA differs among individuals and among species.

DNA Replication
Before a cell divides, it copies its DNA so that each of its descendants gets a full complement of hereditary information. Newly forming DNA is monitored for errors, most of which are corrected quickly. Uncorrected errors may be perpetuated as mutations.

Cloning Animals
Several methods are now commonly used to produce clones of adult animals for research and agriculture. The techniques are far from perfect, and the practice continues to raise serious ethical questions.

> The DNA in a eukaryotic cell nucleus is organized as one or more chromosomes.
< Link to DNA 3.7

All organisms pass DNA to offspring when they reproduce. Inside a cell, each DNA molecule is organized as a structure called a **chromosome** (Figure 8.2). Eukaryotic cells typically have a number of chromosomes ①. During most of a cell's life, each of its chromosomes consists of one DNA molecule. As it prepares to divide, the cell duplicates its chromosomes, so that both of its offspring get a full set. After the chromosomes are duplicated, each consists of two DNA molecules, which are then called **sister chromatids**. Sister chromatids attach to each other at a constricted region called a **centromere**:

a chromosome
(unduplicated)

a chromosome
(duplicated)

centromere
one chromatid
its sister chromatid

Stretched out end to end, the 46 chromosomes in a human cell would be about 2 meters (6.5 feet) long—a lot of DNA to fit into a nucleus that is typically less than 10 micrometers in diameter! Interactions between a DNA molecule and the proteins that associate with it structurally organize a chromosome and help it pack tightly.

At its most condensed, a duplicated chromosome consists of two long, tangled filaments (the sister chromatids) bunched into a characteristic X shape ②. A closer look reveals that each filament is actually a hollow tube formed by coils, like a phone cord ③. The coils them-

selves consist of a twisted fiber ④ of DNA wrapped twice at regular intervals around "spools" of proteins called **histones** ⑤. In micrographs, these DNA–histone spools look like beads on a string. Each "bead" is a **nucleosome**, the smallest unit of chromosomal organization in eukaryotes. As you will see Section 8.4, the DNA molecule consists of two strands twisted into a double helix ⑥.

Chromosome Number

The genetic information of each eukaryotic species is distributed among some number of chromosomes, which differ in length and shape. The sum of all chromosomes in a cell of a given type is called the **chromosome number**. Each species has a characteristic chromosome number. For example, the chromosome number of oak trees is 12, so the nucleus of a cell from an oak tree contains 12 chromosomes. The chromosome number of king crab cells is 208, so they have 208 chromosomes. That of human body cells is 46, so human body cells have 46 chromosomes.

Actually, human body cells have two of each type of chromosome, which means that their chromosome number is **diploid** ($2n$). The 23 pairs of chromosomes are like two sets of books numbered 1 to 23. There are two versions of each book: a pair. Except for a pairing of sex chromosomes (XY) in males, the two members of each pair have the same length and shape, and they hold information about the same traits. Think of them as two sets of books on how to build a house. Your father gave you one set. Your mother had her own ideas about wiring, plumbing, and so on. She gave you an alternate set that says slightly different things about many of those tasks.

Figure 8.2 Animated Zooming in on chromosome structure. Tight packing allows a lot of DNA to fit into a very small nucleus.

① The DNA inside the nucleus of a eukaryotic cell is typically divided up into a number of chromosomes. *Inset*: a duplicated human chromosome.

② At its most condensed, a duplicated chromosome is packed tightly into an X shape.

③ A chromosome unravels as a single fiber, a hollow cylinder formed by coiled coils.

④ The coiled coils consist of a long molecule of DNA (*blue*) and the proteins that are associated with it (*purple*).

⑤ At regular intervals, the DNA molecule is wrapped twice around a core of histone proteins. In this "beads-on-a-string" structure, the "string" is the DNA, and each "bead" is called a nucleosome.

⑥ The DNA molecule itself has two strands that are twisted into a double helix.

2 µm

Figure 8.3 Animated A karyotype is an image of a single cell's diploid set of chromosomes. This human karyotype shows 22 pairs of autosomes and a pair of X chromosomes.

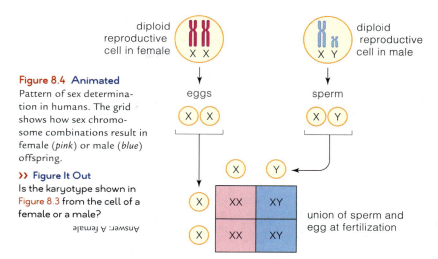

Figure 8.4 Animated
Pattern of sex determination in humans. The grid shows how sex chromosome combinations result in female (*pink*) or male (*blue*) offspring.

›› Figure It Out
Is the karyotype shown in Figure 8.3 from the cell of a female or a male?

Answer: A female

union of sperm and egg at fertilization

Types of Chromosomes

Karyotyping reveals an individual's diploid complement of chromosomes. With this technique, cells taken from an individual are treated to make their chromosomes condense, and stained so the chromosomes become visible under a microscope. The microscope reveals the chromosomes in every cell. A micrograph of a single cell is digitally rearranged so the images of the chromosomes are lined up by centromere location, and arranged according to size, shape, and length (Figure 8.3). The finished array constitutes the individual's **karyotype**, which is compared with a normal standard. A karyotype shows how many chromosomes are in the individual's cells. Comparison with a standard can also reveal any extra or missing chromosomes, and some structural abnormalities.

All except one pair of a diploid cell's chromosomes are **autosomes**, which are the same in both females and males. The two autosomes of a pair have the same length, shape, and centromere location. Members of a pair of **sex chromosomes** differ between females and males. The differences determine an individual's sex.

The sex chromosomes of humans are called X and Y. Body cells of human females contain two X chromosomes (XX); those of human males contain one X and one Y chromosome (XY). XX females and XY males are the rule among fruit flies, mammals, and many other animals, but there are other patterns. In butterflies, moths, birds, and certain fishes, males have two identical sex chromosomes, and females do not. Environmental factors (not sex chromosomes) determine sex in some species of invertebrates, turtles, and frogs. As an example, the temperature of the sand in which sea turtle eggs are buried determines the sex of the hatchlings.

In humans, a new individual inherits a combination of sex chromosomes that dictates whether it will become a male or a female. All eggs made by a human female have one X chromosome. One-half of the sperm cells made by a male carry an X chromosome; the other half carry a Y chromosome. If an X-bearing sperm fertilizes an X-bearing egg, the resulting individual will develop into a female. If the sperm carries a Y chromosome, the individual will develop into a male (Figure 8.4).

autosome Any chromosome other than a sex chromosome.
centromere Constricted region in a eukaryotic chromosome where sister chromatids are attached.
chromosome A structure that consists of DNA and associated proteins; carries part or all of a cell's genetic information.
chromosome number The sum of all chromosomes in a cell of a given type.
diploid Having two of each type of chromosome characteristic of the species (2*n*).
histone Type of protein that structurally organizes eukaryotic chromosomes.
karyotype Image of an individual's complement of chromosomes arranged by size, length, shape, and centromere location.
nucleosome A length of DNA wound around a spool of histone proteins.
sex chromosome Member of a pair of chromosomes that differs between males and females.
sister chromatid One of two attached DNA molecules of a duplicated eukaryotic chromosome.

Take-Home Message What are chromosomes?

› A eukaryotic cell's DNA is divided among some characteristic number of chromosomes, which differ in length and shape.

› Members of a pair of sex chromosomes differ between males and females. Other chromosomes are autosomes—the same in males and females.

› Proteins that associate with DNA structurally organize chromosomes and allow them to pack tightly.

> Investigations that led to our understanding that DNA is the molecule of inheritance reveal how science advances.
< Links to Radioisotopes 2.2, Proteins 3.5, DNA 3.7

Early and Puzzling Clues

In 1865, a Swiss medical student, Johannes Miescher, became ill with typhus. The typhus left Miescher partially deaf, so becoming a doctor was no longer an option for him. He switched to organic chemistry instead. By 1869, he was collecting white blood cells from pus-filled bandages and sperm from fish so he could study the composition of the nucleus. Such cells do not contain much cytoplasm, which made isolating the substances in their nucleus easy. Miescher found that nuclei contain an acidic substance composed mostly of nitrogen and phosphorus. Later, that substance would be called deoxyribonucleic acid, or DNA.

Sixty years later, a British medical officer, Frederick Griffith, was trying to make a vaccine for pneumonia. He isolated two strains (types) of *Streptococcus pneumoniae*, a bacteria that causes pneumonia. He named one strain R, because it grows in Rough colonies. He named the other strain S, because it grows in Smooth colonies. Griffith used both strains in a series of experiments that did not lead to the development of a vaccine, but did reveal a clue about inheritance (Figure 8.5).

First, he injected mice with live R cells ❶. The mice did not develop pneumonia, so the R strain was harmless.

Second, he injected other mice with live S cells ❷. The mice died. Blood samples from them teemed with live S cells. The S strain was pathogenic; it caused pneumonia.

Third, he killed S cells by exposing them to high temperature. Mice injected with dead S cells did not die ❸.

Fourth, he mixed live R cells with heat-killed S cells. Mice injected with the mixture died ❹. Blood samples drawn from them teemed with live S cells.

What happened in the fourth experiment? If heat-killed S cells in the mix were not really dead, then mice injected

with them in the third experiment would have died. If the harmless R cells had changed into killer cells, then mice injected with R cells in experiment 1 would have died.

The simplest explanation was that heat had killed the S cells, but had not destroyed their hereditary material, including whatever part specified "infect mice." Somehow, that material had been transferred from the dead S cells into the live R cells, which put it to use.

The transformation was permanent and heritable. Even after hundreds of generations, the descendants of transformed R cells were infectious. What had caused the transformation? Which substance encodes the information about traits that parents pass to offspring?

In 1940, Oswald Avery and Maclyn McCarty set out to identify that substance, which they termed the "transforming principle." They used a process of elimination that tested each type of molecular component of S cells. Avery and McCarty repeatedly froze and thawed S cells (ice crystals that form during freeze–thaw cycles disrupt membranes, thus releasing cell contents). The researchers then filtered any intact cells from the resulting slush. At the end of this process, the researchers had a fluid that contained the lipid, protein, and nucleic acid components of S cells.

The S cell extract could still transform R cells after it had been treated with lipid- and protein-destroying enzymes. Thus, the transforming principle could not be lipid or protein. Carbohydrates had been removed during the purification process, so Avery and McCarty realized that the substance they were seeking must be nucleic acid—RNA or DNA. The S cell extract could still transform R cells after treatment with RNA-degrading enzymes, but not after treatment with DNA-degrading enzymes. DNA had to be the transforming principle.

The result surprised Avery and McCarty, who, along with most other scientists, had assumed that proteins were the substance of heredity. After all, traits are diverse, and proteins were thought to be the most diverse biologi-

❶ Mice injected with live cells of harmless strain R do not die. Live R cells in their blood.

❷ Mice injected with live cells of killer strain S die. Live S cells in their blood.

❸ Mice injected with heat-killed S cells do not die. No live S cells in their blood.

❹ Mice injected with live R cells plus heat-killed S cells die. Live S cells in their blood.

Figure 8.5 Animated Summary of results from Fred Griffith's experiments. The hereditary material of harmful *Streptococcus pneumoniae* cells transformed harmless cells into killers.

DNA inside protein coat

tail fiber

hollow sheath

A *Top*, model of a bacteriophage. *Bottom*, micrograph of three viruses injecting DNA into an *E. coli* cell.

Virus particle coat proteins labeled with ^{35}S

DNA being injected into bacterium

^{35}S remains outside cells

B In one experiment, bacteria were infected with virus particles that had been labeled with a radioisotope of sulfur (^{35}S). The sulfur had labeled only viral proteins. The viruses were dislodged from the bacteria by whirling the mixture in a kitchen blender. Most of the radioactive sulfur was detected in the viruses, not in the bacterial cells. The viruses had not injected protein into the bacteria.

Virus DNA labeled with ^{32}P

Labeled DNA being injected into bacterium

^{32}P remains inside cells

C In another experiment, bacteria were infected with virus particles that had been labeled with a radioisotope of phosphorus (^{32}P). The phosphorus had labeled only viral DNA. When the viruses were dislodged from the bacteria, the radioactive phosphorus was detected mainly inside the bacterial cells. The viruses had injected DNA into the cells—evidence that DNA is the genetic material of this virus.

Figure 8.6 Animated The Hershey–Chase experiments. Alfred Hershey and Martha Chase tested whether the genetic material injected by bacteriophage into bacteria is DNA, protein, or both. The experiments were based on the knowledge that proteins contain more sulfur (S) than phosphorus (P), and DNA contains more phosphorus than sulfur.

cal molecules. Other molecules just seemed too uniform. The two scientists were so skeptical that they published their results only after they had convinced themselves, by years of painstaking experimentation, that DNA was indeed hereditary material. They were also careful to point out that they had not proven DNA was the only hereditary material.

Confirmation of DNA's Function

By 1950, researchers had discovered **bacteriophage**, a type of virus that infects bacteria (Figure 8.6A). Like all viruses, these infectious particles carry hereditary information about how to make new viruses. After a virus infects a cell, the cell starts making new virus particles. Bacteriophages inject genetic material into bacteria, but was that material DNA, protein, or both?

Alfred Hershey and Martha Chase found the answer to that question by exploiting the long-known properties of protein (high sulfur content) and DNA (high phosphorus content). They cultured bacteria in growth medium containing an isotope of sulfur, ^{35}S. In this medium, the protein (but not the DNA) of bacteriophage that infected the bacteria became labeled with the ^{35}S tracer.

Hershey and Chase allowed the labeled viruses to infect a fresh culture of unlabeled bacteria. They knew from electron micrographs that phages attach to bacteria by their slender tails. They reasoned it would be easy to break this precarious attachment, so they poured the virus–bacteria mixture into a blender and turned it on. (At the time, kitchen appliances were commonly used as laboratory equipment.)

After blending, the researchers separated the bacteria from the virus-containing fluid, and measured the ^{35}S content of each separately. The fluid contained most of the ^{35}S. Thus, the viruses had not injected protein into the bacteria (Figure 8.6B).

Hershey and Chase repeated the experiment using an isotope of phosphorus, ^{32}P, which labeled the DNA (but not the proteins) of the bacteriophage. This time, infected bacteria contained most of the isotope. The viruses had injected DNA into the bacteria (Figure 8.6C).

Both of these experiments—and many others after them—supported the hypothesis that DNA, not protein, is the material of heredity common to all life on Earth.

---

bacteriophage Virus that infects bacteria.

Take-Home Message What is the molecular basis of inheritance?

❯ DNA is the material of heredity common to all life on Earth.

> Watson and Crick's discovery of DNA's structure was based on almost fifty years of research by other scientists.
‹ Links to Carbohydrate rings 3.3, Protein structure 3.5, Nucleic acids 3.7

adenine (A)
deoxyadenosine triphosphate

guanine (G)
deoxyguanosine triphosphate

thymine (T)
deoxythymidine triphosphate

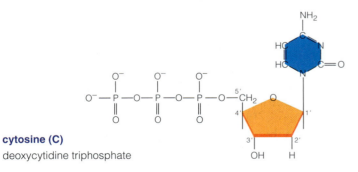

cytosine (C)
deoxycytidine triphosphate

Figure 8.7 Animated The four nucleotides in DNA. Each kind has three phosphate groups, a deoxyribose sugar (*orange*), and a nitrogen-containing base (*blue*) after which it is named. Biochemist Phoebus Levene identified the structure of these bases and how they are connected in DNA in the early 1900s. Levene worked with DNA for almost 40 years.

Numbering the carbons in the sugar rings (Section 3.3) allows us to keep track of the orientation of nucleotide chains, which is important in processes such as DNA replication. Compare Figure 8.8.

DNA's Building Blocks

Each strand of DNA is a polymer of nucleotides that have been linked into a chain. Even though a chain can be hundreds of millions of nucleotides long, it is composed of only four kinds of nucleotides. A DNA nucleotide has a five-carbon sugar, three phosphate groups, and one of four nitrogen-containing bases (Figure 8.7). Just how those four nucleotides—adenine (A), guanine (G), thymine (T), and cytosine (C)—are arranged in DNA was a puzzle that took over 50 years to solve. As molecules go, DNA is gigantic, and chromosomal DNA has a complex structural organization. Both factors made the molecule difficult to work with, given the laboratory methods of the time.

In 1950, Erwin Chargaff, one of many researchers who had been trying to solve the structure of DNA, made two discoveries. First, the amounts of thymine and adenine in all DNA are the same, as are the amounts of cytosine and guanine. We call this discovery Chargaff's first rule:

$$A = T \quad \text{and} \quad G = C$$

Chargaff's second discovery, or rule, is that the proportion of adenine and guanine differs among the DNA of different species.

Meanwhile, American biologist James Watson and British biophysicist Francis Crick, both at Cambridge University, had been sharing their ideas about the structure of DNA. The helical pattern of secondary structure that occurs in many proteins (Section 3.5) had just been discovered, and Watson and Crick suspected that the DNA molecule was also a helix. They had spent many hours arguing about the size, shape, and bonding requirements of the four kinds of nucleotides that make up DNA. They had pestered chemists to help them identify bonds they might have overlooked. They had fiddled with cardboard cutouts, and made models from scraps of metal connected by suitably angled "bonds" of wire.

At King's College in London, biochemist Rosalind Franklin had also been working on the structure of DNA. Like Crick, Franklin specialized in x-ray crystallography, a technique in which x-rays are directed through a purified and crystallized substance. Atoms in the substance's molecules scatter the x-rays in a pattern that can be captured as an image. Researchers can use the pattern to calculate the size, shape, and spacing between any repeating elements of the molecules—all of which are details of molecular structure.

DNA is a large molecule, and difficult to crystallize. Also, as Franklin discovered, "wet" and "dry" DNA samples have different shapes. She made the first clear x-ray diffraction image of "wet" DNA, the form that occurs in cells. From the information in that image, she calculated that DNA is very long compared to its 2-nanometer diameter. She also identified a repeating pattern every 0.34 nanometer along its length, and another every 3.4 nanometers.

Franklin's image and data came to the attention of Watson and Crick, who now had all the information they needed to build a model of the DNA helix—one with two sugar–phosphate chains running in opposite directions, and paired bases inside (Figure 8.8). Bonds between the sugar of one nucleotide and the phosphate of the next form the backbone of each chain. Hydrogen bonds between the internally positioned bases hold the two strands together. Only two kinds of base pairings form: A to T, and G to C, which explains the first of Chargaff's rules. Most scientists had assumed (incorrectly) that the bases had to be on the outside of the helix, because they would be more accessible to DNA-copying enzymes that way. You will see in Section 8.6 how DNA replication enzymes access the bases on the inside of a double helix.

DNA's Base Pair Sequence

Just two kinds of base pairings give rise to the incredible diversity of traits we see among living things. How? Even though DNA is composed of only four bases, the order in which one base pair follows the next varies tremendously among species (which explains Chargaff's second rule). For example, a small piece of DNA from a tulip, a human, or any other organism might be:

Notice how the two strands of DNA match up. They are complementary, which means each base on one strand is suitably paired with a partner base on the other. This bonding pattern (A to T, G to C) is the same in all molecules of DNA. However, which base pair follows the next differs among species, and among individuals of the same species. The information encoded by that sequence is the basis of visible traits that define species and distinguish individuals. Thus DNA, the molecule of inheritance in every cell, is the basis of life's unity. Variations in its base sequence are the foundation of life's diversity.

Take-Home Message What is the structure of DNA?

> A DNA molecule consists of two nucleotide chains (strands) running in opposite directions and coiled into a double helix. Internally positioned nucleotide bases hydrogen-bond between the two strands. A pairs with T, and G with C.

> The sequence of bases along a DNA strand is genetic information.

> DNA sequences vary among species and among individuals. This variation is the basis of life's diversity.

Figure 8.8 **Animated** Structure of DNA, as illustrated by a composite of different models.

Above, Watson and Crick with their model.

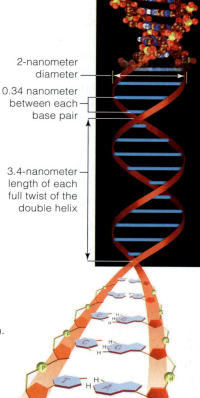

2-nanometer diameter

0.34 nanometer between each base pair

3.4-nanometer length of each full twist of the double helix

The numbers indicate the carbon of the deoxy-ribose sugars (compare Figure 8.7). The 3' carbon of each sugar is joined by the phosphate group to the 5' carbon of the next sugar. These links form each strand's sugar–phosphate backbone.

The two sugar–phosphate backbones run in parallel but opposite directions (*green* arrows). Think of one strand as upside down compared with the other.

> In science, as in other professions, public recognition
> for a discovery does not always include all contributors.

By the time she arrived at King's College, Rosalind Frank-lin was an expert x-ray crystallographer. She had solved the structure of coal, which is complex and unorganized (as are large biological molecules such as DNA). Like Paul-ing, she had built three-dimensional molecular models. Her assignment was to investigate DNA's structure.

Franklin had been told she would be the only one in the department working on the problem, so she did not know that Maurice Wilkins was already doing the same thing just down the hall. When Wilkins proposed a col-laboration with her, Franklin thought that Wilkins was oddly overinterested in her work and declined bluntly.

Wilkins and Franklin had been given identical samples of DNA carefully prepared by Rudolf Signer. Franklin's meticulous work with hers yielded the first clear x-ray diffraction image of DNA as it occurs inside cells (Figure 8.9). She gave a presentation on this work in 1952. DNA, she said, had two chains twisted into a double helix, with a backbone of phosphate groups on the outside, and bases arranged in an unknown way on the inside. She had cal-culated DNA's diameter, the distance between its chains and between its bases, the angle of the helix, and the number of bases in each coil. Crick, with his crystallogra-phy background, would have recognized the significance of the work—if he had been there. Watson was in the audience but he was not a crystallographer, and he did not understand the implications of Frank-lin's x-ray diffraction image or her calculations.

Franklin started to write a research paper on her findings. Meanwhile, and perhaps without her knowledge, Watson reviewed Franklin's x-ray diffraction image with Wilkins, and Watson and Crick read a report detailing Frank-lin's unpublished data. Crick, who had more experience with molecular modeling than Frank-lin, immediately understood what the image and the data meant. Watson and Crick used that information to build their model of DNA.

On April 25, 1953, Franklin's paper appeared third in a series of articles about the structure of DNA in the journal *Nature*. It supported with solid experimental evidence Watson and Crick's theoretical model, which appeared in the first article of the series.

Rosalind Franklin died at age 37, of ovarian cancer probably caused by extensive exposure to x-rays. Because the Nobel Prize is not given post-humously, she did not share in the 1962 honor that went to Watson, Crick, and Wilkins for the discovery of the structure of DNA.

Figure 8.9 Rosalind Frank-lin and her x-ray diffraction image of DNA.

> A cell copies its DNA before it reproduces.
> DNA repair mechanisms correct most replication errors.
‹ Link to Phosphate-group transfers 5.3

During most of its life, a cell contains only one set of chromosomes. However, when the cell reproduces, it must contain two sets of chromosomes: one for each of its future offspring. By a process called **DNA replication**, a cell copies its DNA before it divides (Figure 8.10). Before DNA replication begins, each chromosome consists of one molecule of DNA—one double helix ❶. During replica-tion, an enzyme called DNA helicase breaks the hydrogen bonds that hold the double helix together, so the two DNA strands unwind ❷. Another enzyme, **DNA polymerase**, assembles a complementary strand of DNA on each of the parent strands.

As each new DNA strand lengthens, it winds up with its "parent" strand into a double helix (Figure 8.11A). So, after replication, two double-stranded molecules of DNA have formed. One strand of each molecule is parental (old), and the other is new; hence the name of the process, semicon-servative replication (Figure 8.11B).

As you will see, the order of nucleotide bases in a strand of DNA—the **DNA sequence**—is genetic infor-mation. Descendant cells must get an exact copy of that information, or inheritance will go awry. Because each new strand of DNA is complementary in sequence to a par-ent strand, both double-stranded molecules that result from DNA replication are duplicates of the parent.

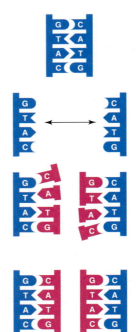

❶ The two strands of a DNA molecule are complementary: their nucleotides match up according to base-pairing rules (G to C, T to A).

❷ As replication starts, the two strands of DNA unwind at many sites along the length of the molecule.

❸ Each parent strand serves as a template for the assembly of a new DNA strand from nucleotides, according to base-pairing rules.

❹ DNA ligase seals any gaps that remain between bases of the "new" DNA, so a con-tinuous strand forms. The base sequence of each half-old, half-new DNA molecule is identical to that of the parent.

Figure 8.10 Animated DNA replication. Both strands of the double helix serve as templates, so two double-stranded DNA molecules result.

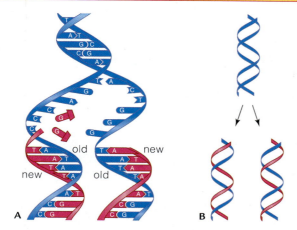

Figure 8.11 Semiconservative replication of DNA. **A** A parental DNA strand serves as a template for assembly of a new strand of DNA. The two parental DNA strands (*blue*) stay intact. A new strand (*magenta*) is assembled on each of the parental (old) strands. **B** One strand of each DNA molecule that forms is new.

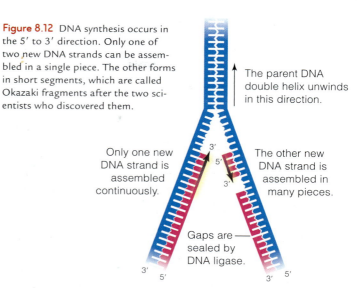

Figure 8.12 DNA synthesis occurs in the 5' to 3' direction. Only one of two new DNA strands can be assembled in a single piece. The other forms in short segments, which are called Okazaki fragments after the two scientists who discovered them.

The parent DNA double helix unwinds in this direction.

Only one new DNA strand is assembled continuously.

The other new DNA strand is assembled in many pieces.

Gaps are sealed by DNA ligase.

The base sequence of each new strand of DNA is complementary to its parent because DNA polymerase follows base-pairing rules ❸. As the enzyme moves along a strand of DNA, it uses the sequence of bases as a template, or guide, to assemble a new strand of DNA from free nucleotides. The polymerase adds a T to the end of the new DNA strand when it reaches an A in the parent DNA sequence; it adds a G when it reaches a C; and so on. Phosphate-group transfers from the nucleotides provide energy for their own attachment to the end of a growing strand of DNA. The enzyme **DNA ligase** seals any gaps, so a new, continuous strand of DNA results ❹.

Numbering the carbons in nucleotides allows us to keep track of the DNA strands in a double helix, because each strand has an unbonded 5' carbon at one end and an unbonded 3' carbon at the other:

5' ▬▬▬▬▬▬▬ 3'
3' ▬▬▬▬▬▬▬ 5'

DNA polymerase can attach free nucleotides only to a 3' carbon. Thus, it can replicate only one strand of a DNA molecule continuously (Figure 8.12). Synthesis of the other strand occurs in segments, in the direction opposite that of unwinding. DNA ligase joins those segments into a continuous strand of DNA.

DNA ligase Enzyme that seals gaps in double-stranded DNA.
DNA polymerase DNA replication enzyme. Uses a DNA template to assemble a complementary strand of DNA.
DNA repair mechanism Any of several processes by which enzymes repair damaged DNA.
DNA replication Process by which a cell duplicates its DNA before it divides.
DNA sequence Order of nucleotide bases in a strand of DNA.
mutation Permanent change in the sequence of DNA.

Proofreading

A DNA molecule is not always replicated with perfect fidelity. Sometimes the wrong base is added to a growing DNA strand; at other times, bases get lost, or extra ones are added. Either way, the new DNA strand will no longer match up perfectly with its parent.

Some of these errors occur after the DNA becomes damaged by exposure to radiation or toxic chemicals. DNA polymerases do not copy damaged DNA very well. In most cases, **DNA repair mechanisms** fix DNA by enzymatically excising and replacing any damaged or mismatched bases before replication begins.

Most DNA replication errors occur simply because DNA polymerases catalyze a tremendous number of reactions very quickly—up to 1,000 bases per second. Mistakes are inevitable, and some DNA polymerases make many of them. Luckily, most DNA polymerases proofread their own work. They correct any mismatches by reversing the synthesis reaction to remove a mismatched nucleotide, then resuming synthesis in the forward direction.

When proofreading and repair mechanisms fail, an error becomes a **mutation**, a permanent change in the DNA sequence. An individual or its offspring may not survive a mutation, because mutations can cause cancer in body cells. Mutations in cells that form eggs or sperm may lead to genetic disorders in offspring. However, not all mutations are dangerous. Some give rise to variations in traits that are the raw material of evolution.

Take-Home Message How is DNA copied?

> A cell replicates its DNA before it divides. Each strand of the double helix serves as a template for synthesis of a new, complementary strand of DNA.

> DNA repair mechanisms and proofreading maintain the integrity of a cell's genetic information. Unrepaired errors may become mutations.

❯ Reproductive cloning is a reproductive intervention that results in an exact genetic copy of an adult individual.

The word "cloning" means making an identical copy of something. In biology, cloning can refer to a laboratory method by which researchers copy DNA fragments (a technique we discuss in Chapter 15). It can also refer to interventions in reproduction that result in an exact genetic copy of an organism.

Genetically identical organisms occur all the time in nature. They arise mainly by the process of asexual reproduction, which we discuss in Chapter 12. Embryo splitting, another natural process, results in identical twins. The first few divisions of a fertilized egg form a ball of cells that sometimes splits spontaneously. If both halves continue to develop independently, identical twins result. Artificial embryo splitting has been routine in research and animal husbandry for decades. With this technique, a ball of cells is grown from a fertilized egg in a laboratory. The ball is teased apart into two halves, each of which goes on to develop as a separate embryo. The embryos are implanted in surrogate mothers, who give birth to identical twins. Artificial twinning and any other technology that yields genetically identical individuals is called **reproductive cloning**.

Twins get their DNA from two parents, which typically differ in their DNA sequence. Thus, although identical twins produced by embryo splitting are identical to one another, they are not identical to either parent. When animal breeders want an exact copy of a specific individual, they may turn to a cloning method that starts with a single cell taken from an adult organism. Such procedures present more of a technical challenge than embryo splitting. Unlike a fertilized egg, a body cell from an adult will not automatically start dividing. It must first be tricked into rewinding its developmental clock.

All cells descended from a fertilized egg inherit the same DNA. Thus, the DNA in each living cell of an individual is like a master blueprint that contains enough information to build an entirely new individual. As different cells in a developing embryo start using different subsets of their DNA, they differentiate, or become different in form and function. In animals, differentiation is usually a one-way path. Once a cell specializes, all of its descendant cells will be specialized the same way. By the time a liver cell, muscle cell, or other specialized cell forms, most of its DNA has been turned off, and is no longer used.

To clone an adult, scientists must first transform one of its differentiated cells into an undifferentiated cell by turning its unused DNA back on. In **somatic cell nuclear transfer** (SCNT), a researcher removes the nucleus from an unfertilized egg, then inserts into the egg a nucleus from an adult animal cell (Figure 8.13). A somatic cell is a body cell, as opposed to a reproductive cell (*soma* is a Greek word for body). If all goes well, the egg's cytoplasm reprograms the transplanted DNA to direct the development of an embryo, which is then implanted into a surrogate mother. The animal that is born to the surrogate is genetically identical with the donor of the nucleus.

A A cow egg is held in place by suction through a hollow glass tube called a micropipette. DNA is identified by a *purple* stain.

B Another micropipette punctures the egg and sucks out the DNA. All that remains inside the egg's plasma membrane is cytoplasm.

C A new micropipette prepares to enter the egg at the puncture site. The pipette contains a cell grown from the skin of a donor animal.

D The micropipette enters the egg and delivers the skin cell to a region between the cytoplasm and the plasma membrane.

E After the pipette is withdrawn, the donor's skin cell is visible next to the cytoplasm of the egg. The transfer is now complete.

F The egg is exposed to an electric current. This treatment causes the foreign cell to fuse with and empty its nucleus into the cytoplasm of the egg. The egg begins to divide, and an embryo forms. After a few days, the embryo may be transplanted into a surrogate mother.

Figure 8.13 Animated Somatic cell nuclear transfer, using cattle cells. This series of micrographs was taken by scientists at Cyagra, a company that specializes in cloning livestock.

Figure 8.14 Champion Holstein dairy cow Nelson's Estimate Liz (*right*) and her clone, Nelson's Estimate Liz II (*left*), who was produced by somatic cell nuclear transfer in 2003. Liz II had already begun to win championships by the time she was a yearling.

SCNT is now a common practice among people who breed prized livestock. Among other benefits, many more offspring can be produced in a given time frame by cloning than by traditional breeding methods. Cloned animals have the same championship features as their DNA donors (Figure 8.14). Offspring can also be produced after a donor animal is castrated or even dead.

The controversial issue with adult cloning is not necessarily about livestock. As the techniques become routine, cloning a human is no longer only within the realm of science fiction. Researchers are already using SCNT to produce human embryos for research, a practice called **therapeutic cloning**. The researchers harvest undifferentiated (stem) cells from the cloned human embryos. Cells from these embryos are being used to study, among many other things, how fatal diseases progress. For example, embryos created using cells from people with genetic heart defects will allow researchers to study how the defect causes developing heart cells to malfunction. Such research may ultimately lead to treatments for people who suffer from fatal diseases. Reproductive cloning of humans is not the intent of such research, but if it were, somatic cell nuclear transfer would indeed be the first step toward that end.

---

reproductive cloning Technology that produces genetically identical individuals.
somatic cell nuclear transfer (SCNT) Method of reproductive cloning in which genetic material is transferred from an adult somatic cell into an unfertilized, enucleated egg.
therapeutic cloning The use of SCNT to produce human embryos for research purposes.

A Hero Dog's Golden Clones (revisited)

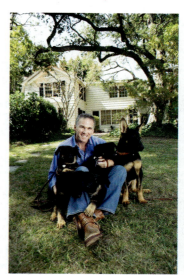

Trakr's clones were produced using SCNT. The ability to clone dogs is a recent development, but the technique is not. Scottish geneticist Ian Wilmut made headlines in 1997 when he announced the first successful demonstration of SCNT. His team had removed the nucleus from an unfertilized sheep egg, then replaced it with the nucleus of a cell taken from the udder of a different sheep. The hybrid egg became an embryo, and then a lamb. The lamb, Dolly, was genetically identical to the sheep that had donated the udder cell.

At first, Dolly looked and acted like a normal sheep. But five years later, she was as fat and arthritic as a twelve-year-old sheep. The following year, Dolly contracted a lung disease that is typical of geriatric sheep, and was euthanized. Dolly's telomeres hinted that she had developed health problems because she was a clone. Telomeres are short, repeated DNA sequences at the ends of chromosomes. They become shorter and shorter as an animal ages. When Dolly was only two years old, her telomeres were as short as those of a six-year-old sheep—the exact age of the adult animal that had been her genetic donor.

Since then, SCNT has also been used to clone mice, rats, rabbits, pigs, cattle, goats, sheep, horses, mules, deer, cats, a camel, a ferret, a monkey, and a wolf. Dolly's telomere problem has not been seen in animals cloned since her, but other problems are common. Many clones are unusually overweight or have enlarged organs. Cloned mice develop lung and liver problems, and almost all die prematurely. Cloned pigs tend to limp and have heart problems. One never did develop a tail or, even worse, an anus.

How Would You Vote? Some view sickly or deformed clones as unfortunate but acceptable casualties of animal cloning research that also yields medical advances for human patients. Should animal cloning be banned? See CengageNow for details, then vote online (cengagenow.com).

Human eggs are difficult to come by. They also come with a hefty set of ethical dilemmas. Thus, researchers have started trying to make hybrid embryos using adult human cells and eggs from other animals, a technique called interspecies nuclear transfer, or iSCNT.

Take-Home Message What is cloning?

> Reproductive cloning technologies produce a clone: an exact genetic copy of an individual.

> The DNA inside a living cell contains all the information necessary to build a new individual.

> Somatic cell nuclear transfer (SCNT) is a reproductive cloning method in which nuclear DNA of an adult donor is transferred to an egg with no nucleus. The hybrid cell develops into an embryo that is genetically identical to the donor.

> Therapeutic cloning uses SCNT to produce human embryos for research.

Summary

Section 8.1 Making **clones**, or exact genetic copies, of adult animals is now a common practice. The techniques, while improving, are still far from perfect; many attempts are required to produce a clone, and clones that survive often have health problems. The practice continues to raise serious ethical questions.

Section 8.2 The DNA of eukaryotes is divided among a characteristic number of **chromosomes** that differ in length and shape. **Histone** proteins organize eukaryotic DNA into **nucleosomes**. When duplicated, a eukaryotic chromosome consists of two **sister chromatids** attached at a **centromere**. **Diploid** cells have two of each type of chromosome.

Chromosome number is the sum of all chromosomes in cells of a given type. A human body cell has twenty-three pairs of chromosomes. Members of a pair of **sex chromosomes** differ among males and females. All others are **autosomes**. Autosomes of a pair have the same length, shape, and centromere location, and they carry the same genes. A **karyotype** can reveal abnormalities in an individual's complement of chromosomes.

Section 8.3 Almost one hundred years of experiments with bacteria and **bacteriophage** offered solid evidence that deoxyribonucleic acid (DNA), not protein, is the hereditary material of life.

Sections 8.4, 8.5 A DNA molecule consists of two strands of DNA coiled into a helix. Nucleotide monomers are joined to form each strand. A DNA nucleotide has a five-carbon sugar (deoxyribose), three phosphate groups, and one of four nitrogen-containing bases after which the nucleotide is named: adenine, thymine, guanine, or cytosine.

Bases of the two DNA strands in a double helix pair in a consistent way: adenine with thymine (A–T), and guanine with cytosine (G–C). The order of bases along the strands varies among species and among individuals.

Section 8.6 The **DNA sequence** of an organism's chromosome(s) constitutes genetic information. A cell passes that information to offspring by copying its DNA before it divides, a process called **DNA replication**. A double-stranded molecule of DNA results in two double-stranded DNA molecules that are identical to the parent. One strand of each molecule is new, and the other is parental. During the replication process, the double helix unwinds. **DNA polymerase** uses each strand as a template to assemble new, complementary strands of DNA from free nucleotides. **DNA ligase** seals any gaps to form a continuous strand.

DNA repair mechanisms fix damaged DNA. Proofreading by DNA polymerases corrects most base-pairing errors. Uncorrected errors may become **mutations**.

Section 8.7 Various **reproductive cloning** technologies produce genetically identical individuals (clones). In **somatic cell nuclear transfer** (SCNT), one cell from an adult is fused with an enucleated egg. The hybrid cell is treated with electric shocks or another stimulus that provokes the cell to divide and begin developing into a new individual. SCNT with human cells, which is called **therapeutic cloning**, produces embryos that are used for stem cell research.

Self-Quiz   Answers in Appendix III

1. Chromosome number _____ .
   a. refers to a particular chromosome pair in a cell
   b. is an identifiable feature of a species
   c. is like a set of books
   d. all of the above

2. Sister chromatids connect at the _____ .

3. The basic unit that structurally organizes a eukaryotic chromosome is the _____ .
   a. higher-order coiling   c. base sequence
   b. double helix   d. nucleosome

4. Which is *not* a nucleotide base in DNA?
   a. adenine   c. uracil   e. cytosine
   b. guanine   d. thymine   f. All are in DNA.

5. What are the base-pairing rules for DNA?
   a. A–G, T–C   c. A–U, C–G
   b. A–C, T–G   d. A–T, G–C

6. One species' DNA differs from others in its _____ .
   a. sugars   c. base sequence
   b. phosphates   d. all of the above

7. When DNA replication begins, _____ .
   a. the two DNA strands unwind from each other
   b. the two DNA strands condense for base transfers
   c. two DNA molecules bond
   d. old strands move to find new strands

8. DNA replication requires _____ .
   a. template DNA   c. DNA polymerase
   b. free nucleotides   d. all of the above

9. Show the complementary strand of DNA that forms on this template DNA fragment during replication:

   5′—GGTTTCTTCAAGAGA—3′

10. _____ is an example of reproductive cloning.
   a. Somatic cell nuclear transfer (SCNT)
   b. Multiple offspring from the same pregnancy
   c. Artificial embryo splitting
   d. a and c
   e. all of the above

Data Analysis Activities

Hershey–Chase Experiments The graph in Figure 8.15 is reproduced from Hershey and Chase's original 1952 publication that showed DNA is the hereditary material of bacteriophage. The data are from the two experiments described in Section 8.3, in which bacteriophage DNA and protein were labeled with radioactive tracers and allowed to infect bacteria. The virus–bacteria mixtures were whirled in a blender to dislodge the viruses, and the tracers were tracked inside and outside of the bacteria.

1. Before blending, what percentage of ^{35}S was outside the bacteria? What percentage was inside? What percentage of ^{32}P was outside the bacteria? What percentage was inside?

2. After 4 minutes in the blender, what percentage of ^{35}S was outside the bacteria? What percentage was inside? What percentage of ^{32}P was outside the bacteria? What percentage was inside?

3. How did the researchers know that the radioisotopes in the fluid came from outside of the bacterial cells (extracellular) and not from bacteria that had been broken apart by whirling in the blender?

4. The extracellular concentration of which isotope, ^{35}S or ^{32}P, increased the most with blending? DNA contains much more phosphorus than do proteins; proteins contain much more sulfur than does DNA. Do these results imply that the viruses inject DNA or protein into bacteria? Why or why not?

Figure 8.15 Detail of Alfred Hershey and Martha Chase's publication describing their experiments with bacteriophage. "Infected bacteria" refers to the percentage of bacteria that survived the blender.

Source: "Independent Functions of Viral Protein and Nucleic Acid in Growth of Bacteriophage." *Journal of General Physiology*, 36(1), Sept. 20, 1952.

---

11. The DNA of each species has unique _____ that set it apart from the DNA of all other species.
    - a. nucleotides
    - b. chromosomes
    - c. sequences
    - d. bases

12. _____ can be used to produce genetically identical organisms (clones).
    - a. SCNT
    - b. Embryo splitting
    - c. Therapeutic cloning
    - d. all of the above

13. A karyotype reveals the _____ of a single cell.
    - a. base sequences
    - b. chromosomes
    - c. hereditary information
    - d. clones

14. SCNT with human cells is called _____ .

15. Match the terms appropriately.
    - ___ bacteriophage
    - ___ clone
    - ___ nucleotide
    - ___ diploid
    - ___ DNA ligase
    - ___ DNA polymerase
    - ___ autosome

    - a. nitrogen-containing base, sugar, phosphate group(s)
    - b. copy of an organism
    - c. does not determine sex
    - d. only DNA and protein
    - e. fills in gaps, seals breaks in a DNA strand
    - f. two chromosomes of each type
    - g. adds nucleotides to a growing DNA strand

Additional questions are available on **CENGAGENOW**.

Critical Thinking

1. Mutations are permanent changes in a cell's DNA base sequence. They typically have negative consequences, but they are also the original source of genetic variation and the raw material of evolution. How can mutations accumulate, given that cells have repair systems that fix changes or breaks in DNA strands?

2. Woolly mammoths have been extinct for about 10,000 years, but we occasionally find one that has been preserved in Siberian permafrost. Resurrecting these huge elephant-like mammals may be possible by cloning DNA isolated from such frozen remains. Researchers are now studying the DNA of a remarkably intact baby mammoth recently discovered frozen in a Siberian swamp. What are some of the pros and cons, both technical and ethical, of cloning an extinct animal?

Animations and Interactions on **CENGAGENOW**:
› Structural organization of eukaryotic chromosomes; Karyotyping; Sex determination in humans; Griffith's experiments; Hershey–Chase experiments; Subunits of DNA; Zooming in on DNA structure; Details of DNA replication; SCNT.

‹ Links to Earlier Concepts

In this chapter, you will see how cells use information in nucleic acids (Section 3.7) to build proteins (3.5, 3.6). You will use what you know about base pairing (8.4) to understand transcription, which is a bit like DNA replication (8.6). You will also revisit ribosomes (4.5) and other organelles (4.6–4.8), energy in metabolism (5.3), and enzymes (5.4). A review of electrons (2.3) and radiant energy (6.2) will be useful as you learn about mutations.

Key Concepts

DNA to RNA to Protein
The sequence of amino acids in a polypeptide chain corresponds to a sequence of nucleotide bases in DNA called a gene. The conversion of information in DNA to protein occurs in two steps: transcription and translation.

DNA to RNA: Transcription
During transcription, one strand of a DNA double helix serves as a template for assembling a single, complementary strand of RNA (a transcript). Each transcript is an RNA copy of a gene.

9 From DNA to Protein

Ricin is a naturally occurring protein that is highly toxic: A dose as small as a few grains of salt can kill an adult. Only botulinum and tetanus toxins are more deadly, and there is no antidote. Ricin effectively deters many animals from eating any part of the castor-oil plant (*Ricinus communis*), which grows wild in tropical regions worldwide and is widely cultivated for its seeds (Figure 9.1). Castor-oil seeds are the source of castor oil, an ingredient in plastics, cosmetics, paints, soaps, polishes, and many other items. After the oil is extracted from the seeds, the ricin is typically discarded along with the leftover seed pulp.

The lethal effects of ricin were known as long ago as 1888, but using ricin as a weapon is now banned by most countries under the Geneva Protocol. However, controlling its production is impossible, because it takes no special skills or equipment to manufacture the toxin from easily obtained raw materials. Thus, ricin appears periodically in the news.

For example, at the height of the Cold War, the Bulgarian writer Georgi Markov had defected to England and was working as a journalist for the BBC. As he made his way to a bus stop on a London street, an assassin used the tip of a modified umbrella to jam a small, ricin-laced ball into Markov's leg. Markov died in agony three days later.

More recently, police in 2003 acted on an intelligence tip and stormed a London apartment, where they found laboratory glassware and castor-oil beans. Traces of ricin were found in a United States Senate mailroom and State Department building, and also in an envelope addressed to the White House in 2004. In 2005, the FBI arrested a man who had castor-oil beans and an assault rifle stashed in his Florida home. Jars of banana baby food laced with ground castor-oil beans also made the news in 2005. In 2006, police found pipe bombs and a baby food jar full of ricin in a Tennessee man's shed. In 2008, castor beans, firearms, and several vials of ricin were found in a Las Vegas motel room after its occupant was hospitalized for ricin exposure.

Figure 9.1 Ricin. Above, one of ricin's polypeptide chains (*green*) helps the molecule cross cell membranes. The other chain (*tan*) destroys a cell's capacity for protein synthesis. Sugars attached to the protein are shown. *Opposite*, seeds of the castor-oil plant, the source of ricin.

One of ricin's two polypeptide chains is an enzyme that inactivates ribosomes. This enzyme also occurs in wheat, barley, and other food plants, but it is not particularly toxic because it cannot cross cell membranes very well. However, ricin's second chain binds to plasma membranes, and the binding causes the cell to take up the ricin molecule by endocytosis. Once inside the cell, ricin's two polypeptide chains separate, and the enzyme begins to inactivate ribosomes. Ribosomes assemble amino acids into proteins. Proteins are critical to all life processes, so cells that cannot make them die very quickly. Someone who inhales ricin can die from low blood pressure and respiratory failure within a few days of exposure.

This chapter details how cells convert information encoded in their DNA to an RNA or a protein product. Even though it is extremely unlikely that your ribosomes will ever encounter ricin, protein synthesis is nevertheless worth appreciating for how it keeps you and all other organisms alive.

RNA
Messenger RNA (mRNA) carries DNA's protein-building instructions. Its nucleotide sequence is read three bases at a time. Sixty-four mRNA base triplets—codons—represent the genetic code. Two other types of RNA interact with mRNA during translation of that code.

RNA to Protein: Translation
Translation is an energy-intensive process by which a sequence of codons in mRNA is converted to a sequence of amino acids in a polypeptide chain. Transfer RNAs deliver amino acids to ribosomes, which catalyze the formation of peptide bonds between the amino acids.

Mutations
Small-scale, permanent changes in the nucleotide sequence of DNA may result from replication errors, the activity of transposable elements, or exposure to environmental hazards. Such mutations can change a gene's product.

> Transcription converts information in a gene to RNA;
> translation converts information in an mRNA to protein.
< Links to Proteins 3.5, Nucleic acids 3.7, Ribosomes 4.5,
Enzymes 5.4, DNA sequence and replication 8.6

You learned in Chapter 8 that genetic information consists of the nucleotide base sequence of DNA. How does a cell convert that information into structural and functional components? Let's start with the nature of the information itself.

DNA is like a book, an encyclopedia that contains all of the instructions for building a new individual. You already know the alphabet used to write the book: the four letters A, T, G, and C, for the four nucleotide bases adenine, thymine, guanine, and cytosine. Each strand of DNA consists of a chain of those four kinds of nucleotides. The linear order, or sequence, of the four bases in the strand is the genetic information. That information occurs in subsets called **genes**.

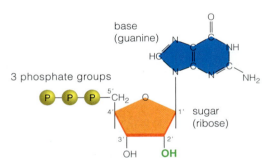

An RNA nucleotide: guanine (G),
or guanosine triphosphate (GTP)

A Guanine, one of the four nucleotides in RNA. The others (adenine, uracil, and cytosine) differ only in their component bases (*blue*). Three of the four bases in RNA nucleotides are identical to the bases in DNA nucleotides.

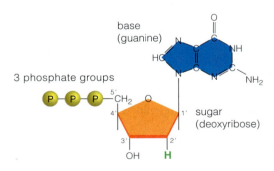

A DNA nucleotide: guanine (G),
or deoxyguanosine triphosphate (dGTP)

B Compare the DNA nucleotide guanine. The only difference between the RNA and DNA versions of guanine (or adenine, or cytosine) is the hydrogen atom or hydroxyl group at the 2′ carbon of the sugar (shown in *green*).

Figure 9.2 Comparison between **A** an RNA nucleotide and **B** a DNA nucleotide.

Converting a Gene to an RNA

Converting the information encoded by a gene into a product starts with RNA synthesis, or **transcription**. By this process, enzymes use the nucleotide sequence of a gene as a template to synthesize a strand of RNA (ribonucleic acid):

$$DNA \xrightarrow{transcription} RNA$$

Except for the double-stranded RNA that is the genetic material of some types of viruses, RNA usually occurs in single-stranded form. A strand of RNA is structurally similar to a single strand of DNA. For example, both are chains of four kinds of nucleotides. Like a DNA nucleotide, an RNA nucleotide has three phosphate groups, a sugar, and one of four bases. However, DNA and RNA nucleotides are slightly different (Figure 9.2). The two nucleic acids are named after their component sugars, ribose and deoxyribose, which differ in one functional group. Three of the bases (adenine, cytosine, and guanine) are the same in RNA and DNA nucleotides, but the fourth base in RNA is uracil, not thymine as it is in DNA.

Despite these small differences in structure, DNA and RNA have very different functions (Figure 9.3). DNA's only role is to store a cell's heritable information. By contrast, a cell transcribes several kinds of RNAs, and each kind has a different function. MicroRNAs are important in gene control, which is the subject of the next chapter. Three types of RNA have roles in protein synthesis. **Ribosomal RNA (rRNA)** is the main component of ribosomes, structures upon which polypeptide chains are built (Section 4.5). **Transfer RNA (tRNA)** delivers amino acids to ribosomes, one by one, in the order specified by a **messenger RNA (mRNA)**.

Converting mRNA to Protein

Messenger RNA is the only kind of RNA that carries a protein-building message. That message is encoded within the sequence of the mRNA by sets of three nucleotide bases, "genetic words" that follow one another along the length of the mRNA. Like the words of a sentence, a series

---

gene Part of a DNA base sequence; specifies an RNA or protein product.
gene expression Process by which the information in a gene becomes converted to an RNA or protein product.
messenger RNA (mRNA) A type of RNA that carries a protein-building message.
ribosomal RNA (rRNA) A type of RNA that becomes part of ribosomes.
transcription Process by which an RNA is assembled from nucleotides using the base sequence of a gene as a template.
transfer RNA (tRNA) A type of RNA that delivers amino acids to a ribosome during translation.
translation Process by which a polypeptide chain is assembled from amino acids in the order specified by an mRNA.

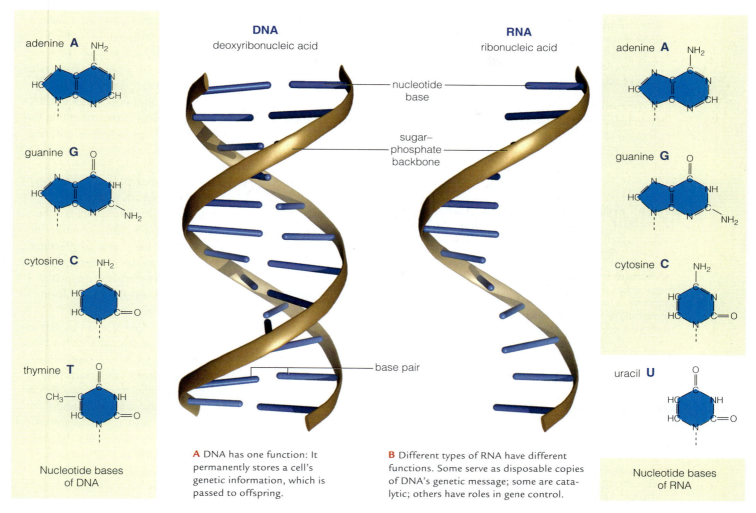

DNA
deoxyribonucleic acid

RNA
ribonucleic acid

adenine **A**

guanine **G**

cytosine **C**

thymine **T**

Nucleotide bases
of DNA

adenine **A**

guanine **G**

cytosine **C**

uracil **U**

Nucleotide bases
of RNA

nucleotide
base

sugar–
phosphate
backbone

base pair

A DNA has one function: It permanently stores a cell's genetic information, which is passed to offspring.

B Different types of RNA have different functions. Some serve as disposable copies of DNA's genetic message; some are catalytic; others have roles in gene control.

Figure 9.3 DNA and RNA compared.

of these genetic words can form a meaningful parcel of information—in this case, the sequence of amino acids of a protein.

By the process of **translation**, the protein-building information in an mRNA is decoded (translated) into a sequence of amino acids. The result is a polypeptide chain that twists and folds into a protein:

$$\text{mRNA} \xrightarrow{\textit{translation}} \text{PROTEIN}$$

Sections 9.4 and 9.5 describe how rRNA and tRNA interact to translate the sequence of base triplets in an mRNA into the sequence of amino acids in a protein.

The processes of transcription and translation are part of **gene expression**, a multistep process by which genetic information encoded by a gene is converted into a structural or functional part of a cell or a body:

$$\text{DNA} \xrightarrow{\textit{transcription}} \text{mRNA} \xrightarrow{\textit{translation}} \text{PROTEIN}$$

A cell's DNA sequence contains all of the information it needs to make the molecules of life. Each gene encodes an RNA, and different types of RNAs interact to assemble proteins from amino acids (Section 3.5). Some of those proteins are enzymes that assemble lipids and complex carbohydrates from simple building blocks (Section 3.2), replicate DNA (Section 8.6), and make RNA, as you will see in the next section.

Take-Home Message **What is the nature of genetic information carried by DNA?**

❯ The nucleotide sequence of a gene encodes instructions for building an RNA or protein product.

❯ A cell transcribes the nucleotide sequence of a gene into RNA.

❯ Although RNA is structurally similar to a single strand of DNA, the two types of molecules differ functionally.

❯ A messenger RNA (mRNA) carries a protein-building code in its nucleotide sequence. rRNAs and tRNAs interact to translate that sequence into a protein.

> RNA polymerase links RNA nucleotides into a chain, in the order dictated by the base sequence of a gene.
> A new RNA strand is complementary in sequence to the DNA strand from which it was transcribed.
< Links to Base pairing 8.4, DNA replication 8.6

Remember that DNA replication begins with one DNA double helix and ends with two DNA double helices (Section 8.6). The two double helices are identical to the

1 RNA polymerase binds to a promoter in the DNA. The binding positions the polymerase near a gene. In most cases, the base sequence of the gene occurs on only one of the two DNA strands. Only the DNA strand complementary to the gene sequence will be translated into RNA.

2 The polymerase begins to move along the DNA and unwind it. As it does, it links RNA nucleotides into a strand of RNA in the order specified by the base sequence of the DNA. The DNA winds up again after the polymerase passes. The structure of the "opened" DNA at the transcription site is called a transcription bubble, after its appearance.

parent molecule because base-pairing rules are followed during DNA replication. A nucleotide can be added to a growing strand of DNA only if it base-pairs with the corresponding nucleotide of the parent strand: G pairs with C, and A pairs with T (Section 8.4). Base-pairing rules also govern RNA synthesis during transcription. An RNA strand is structurally so similar to a DNA strand that the two can base-pair if their nucleotide sequences are complementary. In such hybrid molecules, G pairs with C; A pairs with U (uracil).

During transcription, a strand of DNA acts as a template upon which a strand of RNA—a transcript—is assembled from RNA nucleotides. A nucleotide can be added to a growing RNA only if it is complementary to the corresponding nucleotide of the parent DNA strand. Thus, each new RNA is complementary in sequence to the DNA strand that served as its template. As in DNA replication, each nucleotide provides the energy for its own attachment to the end of a growing strand.

Transcription is similar to DNA replication in that one strand of a nucleic acid serves as a template for synthesis of another. However, in contrast with DNA replication, only part of one DNA strand, not the whole molecule, is used as a template for transcription. The enzyme **RNA polymerase**, not DNA polymerase, adds nucleotides to the end of a growing transcript. Also, transcription results in a single strand of RNA, not two DNA double helices.

Transcription begins with a gene on a chromosome (Figure 9.4). The process gets under way when an RNA polymerase and several regulatory proteins attach to a specific binding site in the DNA called a **promoter** 1. The binding positions the polymerase at a transcription

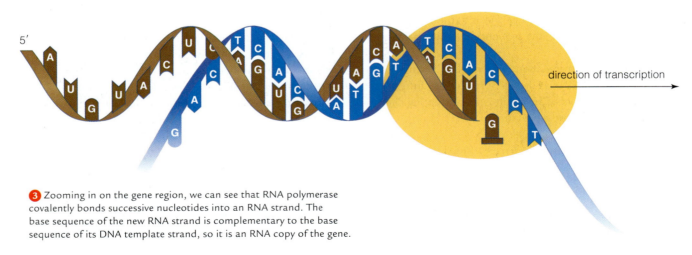

3 Zooming in on the gene region, we can see that RNA polymerase covalently bonds successive nucleotides into an RNA strand. The base sequence of the new RNA strand is complementary to the base sequence of its DNA template strand, so it is an RNA copy of the gene.

Figure 9.4 Animated Transcription. By this process, a strand of RNA is assembled from nucleotides according to a template: a gene region in DNA.

>> **Figure It Out** After the guanine, what is the next nucleotide that will be added to this growing strand of RNA?   Answer: Another guanine (G)

RNA transcripts · DNA molecule

Figure 9.5 Typically, many RNA polymerases simultaneously transcribe the same gene, producing a structure often called a "Christmas tree" after its shape. Here, three genes next to one another on the same chromosome are being transcribed.

›› Figure It Out Are the polymerases transcribing this DNA molecule moving from left to right or from right to left?

Answer: Left to right

start site close to a gene. The polymerase starts moving along the DNA, in the 3′ to 5′ direction over the gene ❷. As it moves, the polymerase unwinds the double helix just a bit so it can "read" the base sequence of the noncoding DNA strand. The polymerase joins free RNA nucleotides into a chain, in the order dictated by that DNA sequence. As in DNA replication, the synthesis is directional: An RNA polymerase adds nucleotides only to the 3′ end of the growing strand of RNA.

When the polymerase reaches the end of the gene, the DNA and the new RNA strand are released. RNA poly-

merase follows base-pairing rules, so the new RNA strand is complementary in base sequence to the DNA strand from which it was transcribed ❸. It is an RNA copy of a gene, the same way that a paper transcript of a conversation carries the same information in a different format. Typically, many polymerases transcribe a particular gene region at the same time, so many new RNA strands can be produced very quickly (Figure 9.5).

Post–Transcriptional Modifications

In eukaryotes, transcription takes place in the nucleus. Eukaryotes also modify their RNA inside the nucleus, then ship it to the cytoplasm. Just as a dressmaker may snip off loose threads or add bows to a dress before it leaves the shop, so do eukaryotic cells tailor their RNA before it leaves the nucleus.

For example, most eukaryotic genes contain **introns**, nucleotide sequences that are removed from a new RNA. Introns intervene between **exons**, sequences that stay in the RNA (Figure 9.6). Introns are transcribed along with exons, but they are removed before the RNA leaves the nucleus. Either all exons remain in the mature RNA, or some are removed and the remaining exons are spliced together. By such **alternative splicing**, one gene can encode different proteins.

New transcripts that will become mRNAs are further tailored after splicing. A modified guanine "cap" gets attached to the 5′ end of each. Later, the cap will help the mRNA bind to a ribosome. A tail of 50 to 300 adenines is also added to the 3′ end of a new mRNA; hence the name, poly-A tail.

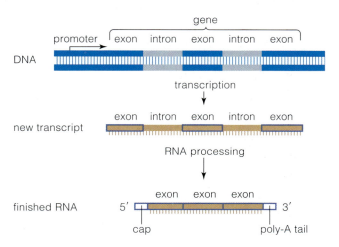

Figure 9.6 Animated Post-transcriptional modification of RNA. Introns are removed and exons spliced together. Messenger RNAs also get a poly-A tail and modified guanine "cap."

alternative splicing RNA processing event in which some exons are removed or joined in various combinations.
exon Nucleotide sequence that is not spliced out of RNA during processing.
intron Nucleotide sequence that intervenes between exons and is excised during RNA processing.
promoter In DNA, a sequence to which RNA polymerase binds.
RNA polymerase Enzyme that carries out transcription.

Take-Home Message How is RNA assembled?

> In transcription, RNA polymerase uses the nucleotide sequence of a gene region in a chromosome as a template to assemble a strand of RNA.

> The new strand of RNA is a copy of the gene from which it was transcribed.

> Base triplets in an mRNA encode a protein-building message. Ribosomal RNA and transfer RNA translate that message into a polypeptide chain.

< Links to Polypeptides 3.5, Ribosomes 4.5, Catalysis 5.4

DNA stores heritable information about proteins, but making those proteins requires messenger RNA (mRNA), transfer RNA (tRNA), and ribosomal RNA (rRNA). The three types of RNA interact to translate DNA's information into a protein.

An mRNA is essentially a disposable copy of a gene. Its job is to carry DNA's protein-building information to the other two types of RNA for translation. That protein-building information consists of a linear sequence of genetic "words" spelled with an alphabet of the four bases A, C, G, and U. Each of the genetic "words" carried by an

mRNA is three bases long, and each is a code—a **codon**—for a particular amino acid. There are four possible bases in each of the three positions of a codon, so there are a total of sixty-four (or 4^3) mRNA codons. Collectively, the sixty-four codons constitute the **genetic code** (Figure 9.7). Which of the four nucleotides is in the first, second, and third position of a base triplet determines which amino acid the codon specifies. For instance, the codon AUG codes for the amino acid methionine (met), and UGG codes for tryptophan (trp).

One codon follows the next along the length of an mRNA, so the order of codons in an mRNA determines the order of amino acids in the polypeptide that will be translated from it. Thus, the base sequence of a gene is transcribed into the base sequence of an mRNA, which is in turn translated into an amino acid sequence (Figure 9.8).

There are only twenty kinds of amino acids found in proteins. Sixty-four codons are more than are needed to specify twenty amino acids, so some amino acids are specified by more than one codon. For instance, GAA and GAG both code for glutamic acid. Other codons signal the beginning and end of a protein-coding sequence. For example, the first AUG in an mRNA is the signal to start translation in most species. AUG also happens to be the codon for methionine, so methionine is always the first amino acid in new polypeptides of such organisms. UAA, UAG, and UGA do not specify an amino acid. They are signals that stop translation, so they are called stop

ala	alanine (A)	leu	leucine (L)
arg	arginine (R)	lys	lysine (K)
asn	asparagine (N)	met	methionine (M)
asp	aspartic acid (D)	phe	phenylalanine (F)
cys	cysteine (C)	pro	proline (P)
glu	glutamic acid (E)	ser	serine (S)
gln	glutamine (Q)	thr	threonine (T)
gly	glycine (G)	trp	tryptophan (W)
his	histidine (H)	tyr	tyrosine (Y)
ile	isoleucine (I)	val	valine (V)

Figure 9.7 Animated The genetic code. Each codon in mRNA is a set of three nucleotide bases. In the large chart, the *left* column lists a codon's first base, the *top* row lists the second, and the *right* column lists the third. Sixty-one of the triplets encode amino acids; the remaining three are signals that stop translation. The amino acid names that correspond to abbreviations in the chart are listed *above*.

>> **Figure It Out** Which codons specify the amino acid lysine (lys)?

Answer: AAA and AAG

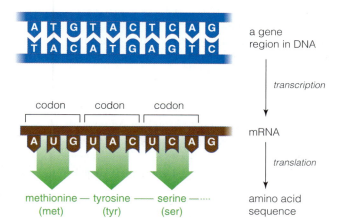

Figure 9.8 Example of the correspondence between DNA, RNA, and proteins. A DNA strand is transcribed into mRNA, and the codons of the mRNA specify a chain of amino acids.

Figure 9.9 **Animated** Ribosome structure. An intact ribosome consists of a large and a small subunit. Structural protein components of the subunits are shown in *green*; catalytic rRNA components, in *tan*. Notice the tunnel through the interior of the large subunit. A polypeptide chain threads through this tunnel as it is being assembled by the ribosome. We show an mRNA (in *red*) attached to the small subunit.

Figure 9.10 **Animated** tRNA structure. **A** Models of the tRNA that carries the amino acid tryptophan. Each tRNA's anticodon is complementary to an mRNA codon. Each also carries the amino acid specified by that codon. **B** During translation, tRNAs dock at an intact ribosome. Here, three tRNAs (*brown*) are docked at the small ribosomal subunit (the large subunit is not shown, for clarity). The anticodons of the tRNAs line up with complementary codons in an mRNA (shown in *red*).

codons. A stop codon marks the end of the protein-coding sequence in an mRNA.

The genetic code is highly conserved, which means that many organisms use the same code and probably always have. Bacteria, archaeans, and some protists have a few codons that differ from the typical code, as do mitochondria and chloroplasts. The variation was a clue that led to a theory of how organelles evolved (we will return to the topic of endosymbiosis in Section 18.6).

rRNA and tRNA—The Translators

A ribosome has one large and one small subunit. Each subunit consists mainly of rRNA, with some associated structural proteins. During translation, a large and a small ribosomal subunit converge as an intact ribosome on an mRNA (Figure 9.9).

Ribosomes and tRNAs interact to translate an mRNA into a polypeptide. Transfer RNAs deliver amino acids to ribosomes in the order specified by the mRNA. Each tRNA has two attachment sites. The first is an anticodon, a triplet of nucleotides that base-pairs with an mRNA codon (Figure 9.10). The other attachment site on a tRNA binds to an amino acid—the one specified by the codon. Transfer RNAs with different anticodons carry different amino acids. You will see in the next section how those tRNAs deliver amino acids, one after the next, to a ribosome during translation of an mRNA.

Ribosomal RNA is one of the few examples of RNA with enzymatic activity: The rRNA of a ribosome, not the protein, catalyzes the formation of a peptide bond

between amino acids. As the amino acids are delivered, the ribosome joins them via peptide bonds into a new polypeptide (Section 3.5). Thus, the order of codons in an mRNA—DNA's protein-building message—becomes translated into a new protein.

Take-Home Message What roles do mRNA, tRNA, and rRNA play during translation?

> mRNA carries protein-building information. The bases in mRNA are "read" in sets of three during protein synthesis. Most of these base triplets (codons) code for amino acids. The genetic code consists of all sixty-four codons.

> Ribosomes, which consist of two subunits of rRNA and proteins, assemble amino acids into polypeptide chains.

> A tRNA has an anticodon complementary to an mRNA codon, and it has a binding site for the amino acid specified by that codon. Transfer RNAs deliver amino acids to ribosomes.

codon In mRNA, a nucleotide base triplet that codes for an amino acid or stop signal during translation.
genetic code Complete set of sixty-four mRNA codons.

start codon (AUG)

initiator tRNA

first amino acid of polypeptide

1 Ribosome subunits and an initiator tRNA converge on an mRNA. A second tRNA binds to the second codon.

peptide bond

2 A peptide bond forms between the first two amino acids.

3 The first tRNA is released and the ribosome moves to the next codon. A third tRNA binds to the third codon.

4 A peptide bond forms between the second and third amino acids.

5 The second tRNA is released and the ribosome moves to the next codon. A fourth tRNA binds the fourth codon.

6 A peptide bond forms between the third and fourth amino acids.

The process repeats until the ribosome encounters a stop codon in the mRNA.

❯ Translation converts the information carried by an mRNA into a polypeptide.

❯ The order of the codons in an mRNA determines the order of amino acids in the translated polypeptide.

❮ Links to Peptide bonds 3.5, Endomembrane system 4.8, Nucleus 4.7, Energy in metabolism 5.3

Translation, the second part of protein synthesis, occurs in the cytoplasm, which has many free amino acids, tRNAs, and ribosomal subunits (Figures 9.11 and 9.12A). Translation proceeds in three stages: initiation, elongation, and termination. In eukaryotes, the initiation stage begins when an mRNA leaves the nucleus and a small ribosomal subunit binds to it. Next, the anticodon of a special tRNA called an initiator base-pairs with the first AUG codon of the mRNA. Then, a large ribosomal subunit joins the small subunit **1**.

In the elongation stage, the ribosome assembles a polypeptide chain as it moves along the mRNA. The initiator tRNA carries the amino acid methionine, so the first amino acid of the new polypeptide chain is methionine. Another tRNA brings the second amino acid to the complex as its anticodon base-pairs with the second codon in the mRNA. The ribosome catalyzes formation of a peptide bond between the first two amino acids **2**.

The first tRNA is released and the ribosome moves to the next codon. Another tRNA brings the third amino acid to the complex as its anticodon base-pairs with the third codon of the mRNA **3**. A peptide bond forms between the second and third amino acids **4**.

The second tRNA is released and the ribosome moves to the next codon. Another tRNA brings the fourth amino acid to the complex as its anticodon base-pairs with the fourth codon of the mRNA **5**. A peptide bond forms between the third and fourth amino acids **6**. The new polypeptide chain continues to elongate as the ribosome catalyzes peptide bonds between amino acids delivered by successive tRNAs.

Termination occurs when the ribosome reaches a stop codon in the mRNA. The mRNA and the polypeptide detach from the ribosome, and the ribosomal subunits separate from each other. Translation is now complete. The new polypeptide will either join the pool of proteins in the cytoplasm, or enter rough ER of the endomembrane system (Section 4.8).

In cells that are making a lot of protein, many ribosomes may simultaneously translate the same mRNA, in which case they are called polysomes (Figure 9.12B).

Figure 9.11 Animated Translation. Translation initiates when ribosomal subunits and an initiator tRNA converge on an mRNA. tRNAs deliver amino acids in the order dictated by successive codons in the mRNA. The ribosome links the amino acids together as it moves along the mRNA, so a polypeptide forms and elongates. Translation terminates when the ribosome reaches a stop codon.

Figure 9.12 **Animated** Overview of translation. **A** In eukaryotes, RNA transcribed in the nucleus moves into the cytoplasm through nuclear pores. Translation occurs in the cytoplasm. Ribosomes simultanously translating the same mRNA are called polysomes. **B** This micrograph shows polypeptides elongating from polysomes on an mRNA.

Labels in figure A: Transcription; RNA transport; ribosome subunits; tRNA; Convergence of RNAs; mRNA; polysomes; Translation; polypeptide; A

Labels in figure B: B; mRNA; polysomes; newly forming polypeptide

Transcription and translation both occur in the cytoplasm of bacteria, and these processes are closely linked in time and in space. Translation begins before transcription is done, so in these cells, a DNA transcription tree may be decorated with polysome "balls."

Given that many polypeptides are translated from one mRNA, why would a cell also make many copies of an mRNA? Compared with DNA, RNA is not very stable. An mRNA may last only a few minutes before it gets disassembled by enzymes in the cytoplasm. The fast turnover allows cells to adjust their protein synthesis quickly in response to changing needs.

Translation is energy intensive. That energy is provided mainly in the form of phosphate-group transfers

from the RNA nucleotide GTP (shown in Figure 9.2A) to molecules involved in the process.

Take-Home Message **How is mRNA translated into protein?**

> Translation is an energy-requiring process that converts the protein-building information carried by an mRNA into a polypeptide.

> During initiation, an mRNA joins with an initiator tRNA and two ribosomal subunits.

> During elongation, amino acids are delivered to the complex by tRNAs in the order dictated by successive mRNA codons. As they arrive, the ribosome joins each to the end of the polypeptide chain.

> Termination occurs when the ribosome encounters a stop codon in the mRNA. The mRNA and the polypeptide are released, and the ribosome disassembles.

A Hemoglobin, an oxygen-binding protein in red blood cells. This protein consists of four polypeptides: two alpha globins (*blue*) and two beta globins (*green*). Each globin has a pocket that cradles a heme (*red*). Oxygen molecules bind to the iron atom at the center of each heme.

····· threonine ——— proline ——— glutamic acid—glutamic acid —— lysine ·······

B Part of the DNA (*blue*), mRNA (*brown*), and amino acid sequence (*green*) of human beta globin.

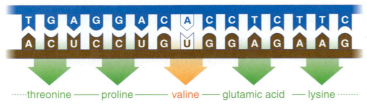

····· threonine ——— proline ——— glycine ——— arginine ——— serine ·······

C A base-pair deletion causes the reading frame for the rest of the mRNA to shift, so a completely different protein product forms. The mutation shown results in a defective beta globin. The outcome is beta thalassemia, a genetic disorder in which a person has an abnormally low amount of hemoglobin.

····· threonine ——— proline ——— valine —— glutamic acid —— lysine ·······

D A base-pair substitution replaces a thymine with an adenine. When the altered mRNA is translated, valine replaces glutamic acid as the sixth amino acid of the polypeptide. Hemoglobin with this form of beta globin is called HbS, or sickle hemoglobin.

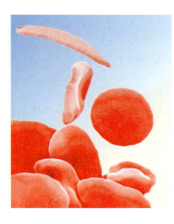

E The substitution of valine for glutamic acid causes the HbS protein to aggregate into rod-like clumps at low oxygen levels. The clumps distort normally round red blood cells into sickle shapes (a sickle is a farm tool with a crescent-shaped blade).

Figure 9.13 Animated Examples of mutations in the human beta globin gene.

> If the nucleotide sequence of a gene changes, it may result in an altered gene product, with harmful effects.
< Links to Electron energy levels 2.3, Protein structure 3.6, Free radicals and hemes as cofactors 5.4, Radiant energy 6.2, DNA replication and mutations 8.6

Mutations are permanent changes in the sequence of a cell's DNA (Section 8.6). If a mutation changes the genetic instructions encoded in the DNA, an altered gene product may result. Remember, more than one codon can specify the same amino acid, so cells have some margin of safety. For example, a mutation that changes a UCU codon to UCC in an mRNA may not have further effects, because both codons specify serine. However, other mutations have drastic consequences.

The oxygen-binding properties of hemoglobin provide an example of how mutations can change the structure (and function) of a protein. As red blood cells circulate through the lungs, the hemoglobin proteins inside of them bind to oxygen molecules. The cells then travel to regions of the body where the oxygen level is low. There, the hemoglobin releases its oxygen cargo. When the red blood cells return to the lungs, the hemoglobin binds to more oxygen.

Hemoglobin's structure allows it to bind and release oxygen. The protein consists of four polypeptides called globins (Figure 9.13A). Each globin folds around a heme, a cofactor with an iron atom at its center (Section 5.4). Oxygen molecules bind to hemoglobin at those iron atoms.

In adult humans, two alpha globins and two beta globins make up each hemoglobin molecule. Defects in either polypeptide can cause a condition called anemia, in which a person's blood is deficient in red blood cells or in hemoglobin. Both outcomes limit the blood's ability to carry oxygen.

For example, the loss of a particular nucleotide—a **deletion**—from the DNA of the beta globin gene causes a type of anemia called beta thalassemia. Like many other deletions, this one causes the reading frame of mRNA codons to shift. The shift garbles the genetic message, just as incorrectly grouping a series of letters garbles the meaning of a sentence:

> The cat ate the rat.
> T hec ata tet her at.
> Th eca tat eth era t.

Expression of a beta globin gene with a beta thalassemia mutation results in a polypeptide that differs drastically from normal beta globin (Figure 9.13B,C). Hemoglobin molecules do not assemble correctly with the altered polypeptide, an outcome that is the source of the anemia.

Frameshifts may also be caused by **insertion** mutations, in which extra bases get inserted into a gene. Other types of mutations do not cause frameshifts. With

Figure 9.14 Two types of DNA damage that can lead to mutations. **A** Chromosomes from a human cell after exposure to gamma rays (a type of ionizing radiation). The broken pieces (*red* arrows) may get lost during DNA replication. **B** A thymine dimer.

A

B

thymine dimer

a **base–pair substitution**, a nucleotide and its partner are replaced by a different base pair. A substitution may result in an amino acid change in a gene's protein product, or a premature stop codon that shortens it. Sickle-cell anemia, a type of anemia that is most common in people of African ancestry, arises because of a base-pair substitution. The substitution causes valine to be the sixth amino acid of beta globin instead of glutamic acid (Figure 9.13D). Hemoglobin with this mutation in its beta globin is called sickle hemoglobin, or HbS.

Unlike glutamic acid, which carries a negative charge, valine carries no charge. As a result of that one substitution, a tiny patch of the beta globin polypeptide changes from hydrophilic to hydrophobic, which in turn causes the hemoglobin's behavior to change slightly. HbS molecules stick together and form large, rodlike clumps under certain conditions. Red blood cells that contain the clumps become distorted into a crescent (sickle) shape (Figure 9.13E). Sickled cells clog tiny blood vessels, thus disrupting blood circulation throughout the body. Over time, repeated episodes of sickling can damage organs and cause death.

What Causes Mutations?

Insertion mutations are often caused by activity of **transposable elements**, segments of DNA that can move spontaneously within or between chromosomes. Transposable elements can be hundreds or thousands of base pairs long, so when one interrupts a gene it becomes a major insertion that changes the gene's product. Transposable elements are common in the DNA of all species; about

45 percent of human DNA consists of transposable elements or their remnants.

Many mutations occur spontaneously during DNA replication (Section 8.6). DNA polymerases make mistakes at predictable rates, but most types fix errors as they occur. Errors that remain uncorrected are mutations.

Environmental agents also cause mutations. For example, x-rays and other forms of ionizing radiation can knock electrons out of atoms. Ionizing radiation breaks chromosomes into pieces that get lost during DNA replication (Figure 9.14A). It also damages DNA indirectly when it penetrates living tissue, because it leaves a trail of destructive free radicals in its wake (Section 5.4).

Nonionizing radiation boosts electrons to a higher energy level, but not enough to knock them out of an atom. DNA absorbs one kind, ultraviolet (UV) light. Exposure to UV light can cause two adjacent thymine bases to bond covalently to one another (Figure 9.14B). The resulting thymine dimer kinks the DNA. DNA polymerase may copy the kinked part incorrectly during replication, so a mutation becomes introduced into the DNA. Mutations that cause certain kinds of cancers begin with thymine dimers. Exposing unprotected skin to sunlight increases the risk of skin cancer because it causes thymine dimers to form in the DNA of skin cells.

Some natural or synthetic chemicals can also cause mutations. For instance, chemicals in cigarette smoke transfer small hydrocarbon groups to the bases in DNA. The altered bases mispair during replication, or stop replication entirely. Both events increase the chance of mutation.

---

base–pair substitution Type of mutation in which a single base-pair changes.
deletion Mutation in which one or more base pairs are lost.
insertion Mutation in which one or more base pairs become inserted into DNA.
transposable element Segment of DNA that can spontaneously move to a new location in a chromosome.

Take-Home Message What is a mutation?

❯ A base-pair substitution, insertion, or deletion are mutations that may alter a gene product.

❯ Most mutations arise during DNA replication as a result of unrepaired DNA polymerase errors. Some mutations occur as a result of transposable elements, or after exposure to harmful radiation or chemicals.

Ricin and Your Ribosomes (revisited)

The enzyme part of ricin inactivates ribosomes by removing a particular adenine base from one of the rRNAs in the heavy subunit. That adenine is part of an RNA binding site for proteins that help with elongation. After the base has been removed, a ribosome can no longer bind to those proteins, and elongation stops.

One molecule of ricin enzyme can inactivate more than 1,000 ribosomes per minute. If enough ribosomes are affected, protein synthesis grinds to a halt, and the cell quickly dies. A modified form of ricin is currently being tested as a treatment for some kinds of cancer. The ricin is attached to an antibody that can find cancer cells in a person's body. Researchers hope that the attached ricin will kill the cancer cells without harming normal ones.

How Would You Vote? Accidental exposure to ricin is unlikely, but terrorists may try to poison food or water supplies with it. Researchers have developed a vaccine against ricin. Do you want to be vaccinated? See CengageNow for details, then vote online (cengage.com).

genetic code. Each tRNA has an anticodon that can base-pair with a codon, and it binds to the amino acid specified by that codon. rRNA and proteins make up the two subunits of ribosomes.

Section 9.5 Genetic information carried by an mRNA directs the synthesis of a polypeptide during translation. First, an mRNA, an initiator tRNA, and two ribosomal subunits converge. The intact ribosome then joins successive amino acids, which are delivered by tRNAs in the order specified by the codons in the mRNA. Translation ends when the polymerase encounters a stop codon.

Section 9.6 **Insertions**, **deletions**, and **base-pair substitutions** are mutations that can arise by replication error, the activity of **transposable elements**, or exposure to environmental hazards such as radiation or chemicals. A mutation that changes a gene's product may have harmful effects. Sickle-cell anemia, which is caused by a base-pair substitution in the hemoglobin beta chain gene, is one example.

Summary

Section 9.1 The ability to make proteins is critical to all life processes. Ricin is toxic because it inactivates ribosomes.

Section 9.2 DNA's genetic information is encoded within its base sequence. **Genes** are subunits of that sequence. A cell uses the information in a gene to make an RNA or protein product. The process of **gene expression** involves **transcription** of a DNA sequence to an RNA, and **translation** of the information in an **mRNA**, or **messenger RNA**, to a protein product. Translation requires the participation of **tRNA** (**transfer RNA**) and **rRNA** (**ribosomal RNA**).

Section 9.3 During transcription, **RNA polymerase** binds to a **promoter** near a gene region of a chromosome. The polymerase assembles a strand of RNA by linking RNA nucleotides in the order dictated by the base sequence of the gene. Thus, the new RNA is complementary to the gene from which it was transcribed.

The RNA of eukaryotes is modified before it leaves the nucleus. **Introns** are removed. With **alternative splicing**, some **exons** may be removed also, and the remaining ones spliced in different combinations. A cap and a poly-A tail are also added to a new mRNA.

Section 9.4 mRNA carries DNA's protein-building information. The information consists of a series of **codons**, sets of three nucleotides. Sixty-four codons, most of which specify amino acids, constitute the

Self-Quiz   Answers in Appendix III

1. A chromosome contains many different gene regions that are transcribed into different _____ .
   a. proteins   c. RNAs
   b. polypeptides   d. a and b

2. A binding site for RNA polymerase is called a _____ .
   a. gene   c. codon
   b. promoter   d. protein

3. Energy that drives transcription is provided mainly by _____ .
   a. ATP   c. GTP
   b. RNA nucleotides   d. all are correct

4. An RNA molecule is typically _____ ; a DNA molecule is typically _____ .
   a. single-stranded; double-stranded
   b. double-stranded; single-stranded
   c. both are single-stranded
   d. both are double-stranded

5. RNAs form by _____ ; proteins form by _____ .
   a. replication; translation
   b. translation; transcription
   c. transcription; translation
   d. replication; transcription

6. Most codons specify a(n) _____ .
   a. protein   c. amino acid
   b. polypeptide   d. mRNA

7. Anticodons pair with _____ .
   a. mRNA codons   c. RNA anticodons
   b. DNA codons   d. amino acids

Data Analysis Activities

Herbicides and Chromosome Damage in Foresters Forestry workers in the U.S. routinely apply herbicides and pesticides as part of their work. A 2001 study by Vincent F. Garry looked at chromosome abnormalities inside white blood cells of forestry employees who worked with the herbicide 2,4-D (dichlorophenoxyacetic acid) more than five days in one year. Most of the workers in the study distributed the herbicide over large areas from a helicopter or airplane. The results, shown in Figure 9.15, were categorized by type of aberration.

Type of Chromosome Aberration	Total Volume of Herbicide Applied			
	None	Low (1–100 gal)	Midrange (100–1,000 gal)	Heavy (>1,000 gal)
Rearrangements	0.65	1.20	1.00	2.22
Missing pieces	0.93	1.17	1.33	2.89
Breaks	1.29	1.03	1.33	3.00
Other	0.29	0.29	0.50	0.33
No. of subjects	14	7	6	9

Figure 9.15 Chromosome abnormalities and use of the herbicide 2,4-D. The table compares chromosome abnormalities in white blood cells of forestry workers who routinely apply herbicide as part of their job. The numbers indicate the average number of aberrations per 100 cells.

Results are categorized by the total volume of herbicide applied, and by type of chromosome damage.

1. Which group of workers, those who sprayed a low amount of 2,4-D or those who sprayed a midrange amount of the herbicide, had the lower frequency of chromosome breaks?
2. What group had the highest frequency of missing chromosome pieces? Breaks?
3. How many herbicide-wielding foresters were tested?

---

8. What is the maximum length of a polypeptide encoded by an mRNA that is 45 nucleotides long?

9. _____ are removed from new mRNA transcripts.
   - a. Introns
   - b. Exons
   - c. Telomeres
   - d. Amino acids

10. Where does transcription take place in a typical eukaryotic cell?
    - a. the nucleus
    - b. ribosomes
    - c. the cytoplasm
    - d. b and c are correct

11. Where does translation take place in a typical eukaryotic cell?
    - a. the nucleus
    - b. ribosomes
    - c. the cytoplasm
    - d. b and c are correct

12. Each amino acid is specified by a set of _____ bases in an mRNA transcript.
    - a. 3
    - b. 20
    - c. 64
    - d. 120

13. _____ different codons constitute the genetic code.
    - a. 3
    - b. 20
    - c. 64
    - d. 120

14. _____ can cause mutations.
    - a. Replication errors
    - b. Transposons
    - c. Ionizing radiation
    - d. Nonionizing radiation
    - e. b and c are correct
    - f. all of the above

15. Match the terms with the best description.
    - ___genetic message
    - ___promoter
    - ___polysome
    - ___exon
    - ___genetic code
    - ___intron
    - ___transposable element
      - a. protein-coding segment
      - b. gets around
      - c. read as base triplets
      - d. removed before translation
      - e. occurs only in groups
      - f. complete set of 64 codons
      - g. binding site for RNA polymerase

Additional questions are available on **CENGAGENOW**.

Critical Thinking

1. Antisense drugs help us fight some types of cancer and viral diseases. The drugs consist of short mRNA strands that are complementary in base sequence to mRNAs linked to the diseases. Speculate on how antisense drugs work.

2. An anticodon has the sequence GCG. What amino acid does this tRNA carry? What would be the effect of a mutation that changed the C of the anticodon to a G?

3. Each position of a codon can be occupied by one of four nucleotides. What is the minimum number of nucleotides per codon necessary to specify all 20 of the amino acids that are typical of eukaryotic proteins?

4. Using Figure 9.7, translate this nucleotide sequence into an amino acid sequence, starting at the first base:

   5'—GGUUUCUUGAAGAGA—3'

5. Translate the sequence of bases in the previous question, starting at the second base.

6. Cigarette smoke contains at least fifty-five different chemicals identified as carcinogenic (cancer-causing) by the International Agency for Research on Cancer (IARC). When these carcinogens enter the bloodstream, enzymes convert them to a series of chemical intermediates that are easier to excrete. Some of the intermediates bind irreversibly to DNA. Propose a hypothesis about why cigarette smoke causes cancer.

Animations and Interactions on **CENGAGENOW**:
> Transcription; RNA processing; The genetic code; Ribosome structure; tRNA structure; Translation; Differences between prokaryotic and eukaryotic protein synthesis; Substitutions; Frameshifts.

‹ Links to Earlier Concepts

A review of what you know about development (Section 1.3) and metabolic control (5.5) will be helpful as we revisit the concept of gene expression (9.2) in more depth. You will be applying what you know about the organization of chromosomal DNA (8.2) and mutations (9.6) as you learn about controls over transcription (9.3), translation (9.5), and other processes that affect gene expression. You will revisit carbohydrates (3.3) and fermentation (7.6) in context of gene control in bacteria.

Key Concepts

Gene Control in Eukaryotes
A variety of molecules and processes alter gene expression in response to changing conditions both inside and outside the cell. Selective gene expression also results in differentiation, by which cell lineages become specialized.

Mechanisms of Control
All cells in an embryo inherit the same genes, but they start using different subsets of those genes during development. The orderly, localized expression of master genes gives rise to the body plan of complex multicelled organisms.

10 Controls Over Genes

10.1 Between You and Eternity

You are in college, your whole life ahead of you. Your risk of developing cancer is as remote as old age, an abstract statistic that is easy to forget. "There is a moment when everything changes—when the width of two fingers can suddenly be the total distance between you and eternity." Robin Shoulla wrote those words after being diagnosed with breast cancer. She was seventeen. At an age when most young women are thinking about school, friends, parties, and potential careers, Robin was dealing with radical mastectomy: the removal of a breast, all lymph nodes under the arm, and skeletal muscles in the chest wall under the breast. She was pleading with her oncologist not to use her jugular vein for chemotherapy and wondering if she would survive to see the next year (Figure 10.1).

Robin's ordeal became part of a statistic, one of more than 200,000 new cases of breast cancer diagnosed in the United States each year. About 5,700 of those cases occur in women and men under thirty-four years of age.

Every second, millions of cells in your skin, bone marrow, gut lining, liver, and elsewhere are dividing and replacing their worn-out, dead, and dying predecessors. They do not divide at random. Many gene expression controls regulate cell growth and division. When those controls fail, cancer is the outcome. **Cancer** is a multistep process in which abnormally growing and dividing cells disrupt body tissues. Mechanisms that normally keep cells from getting overcrowded in tissues are lost, so cancer cell populations may reach extremely high densities. Unless chemotherapy, surgery, or another procedure eradicates them, cancer cells can put an individual on a painful road to death. Each year, cancers cause 15 to 20 percent of all human deaths in developed countries alone.

Cancer typically begins with a mutation in a gene whose product is part of a system of stringent controls over cell growth and division. Such controls govern when and how fast specific genes are transcribed and translated. The mutation may be

normal cells in organized clusters

cancer cells in disorganized clusters

Figure 10.1 A case of breast cancer. *Above*, this light micrograph shows irregular clusters of cancer cells in the milk ducts of human breast tissue. *Opposite*, Robin Shoulla. Diagnostic tests revealed abnormal cells such as these in her body.

inherited, or it may be a new one, as when DNA becomes damaged by environmental agents. If the mutation alters the gene's protein product so that it no longer works properly, one level of control over the cell's growth and division has been lost. You will be considering the impact of gene controls in chapters throughout the book, and in some chapters of your life.

Robin Shoulla survived. Although radical mastectomy is rarely performed today (a modified procedure is less disfiguring), it is the only option when cancer cells invade muscles under the breast. It was Robin's only option. Now, sixteen years later, she has what she calls a normal life: career, husband, children. Her goal as a cancer survivor: "To grow very old with gray hair and spreading hips, smiling."

cancer Disease that occurs when the uncontrolled growth of body cells physically and metabolically disrupts tissues.

Examples in Eukaryotes
One of the two X chromosomes is inactivated in every cell of female mammals. The Y chromosome carries a master gene that causes male traits to develop in the human fetus. Flower development is orchestrated by a set of homeotic genes.

Gene Control in Bacteria
Bacterial gene controls govern responses to short-term changes in nutrient availability and other aspects of the environment. The main gene controls bring about fast adjustments in the rate of transcription.

> Gene controls govern the kinds and amounts of substances that are present in a cell at any given interval.
< Links to Phosphorylation 5.3, Glycolysis 7.3, Histones 8.2, Gene expression 9.2, Promoters and transcription 9.3, Translation 9.5, Globin 9.6

All of the cells in your body are descended from the same fertilized egg, so they all contain the same DNA with the same genes. Some of the genes are transcribed by all cells; such genes affect structural features and metabolic pathways common to all cells.

In other ways, however, nearly all of your body cells are specialized. **Differentiation**, the process by which cells

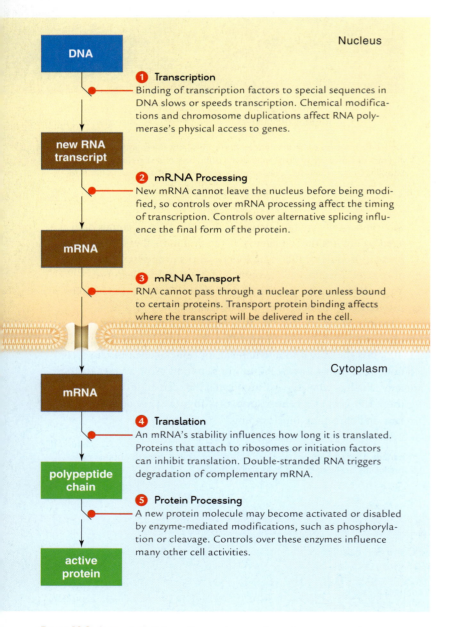

1 Transcription
Binding of transcription factors to special sequences in DNA slows or speeds transcription. Chemical modifications and chromosome duplications affect RNA polymerase's physical access to genes.

2 mRNA Processing
New mRNA cannot leave the nucleus before being modified, so controls over mRNA processing affect the timing of transcription. Controls over alternative splicing influence the final form of the protein.

3 mRNA Transport
RNA cannot pass through a nuclear pore unless bound to certain proteins. Transport protein binding affects where the transcript will be delivered in the cell.

4 Translation
An mRNA's stability influences how long it is translated. Proteins that attach to ribosomes or initiation factors can inhibit translation. Double-stranded RNA triggers degradation of complementary mRNA.

5 Protein Processing
A new protein molecule may become activated or disabled by enzyme-mediated modifications, such as phosphorylation or cleavage. Controls over these enzymes influence many other cell activities.

Figure 10.2 Animated Points of control over eukaryotic gene expression.

Figure 10.3 Hypothetical part of a chromosome that contains a gene. Molecules that affect the rate of transcription of the gene bind at promoter (*yellow*) or enhancer (*green*) sequences.

become specialized, occurs as different cell lineages begin to express different subsets of their genes. Which genes a cell uses determines the molecules it will produce, which in turn determines what kind of cell it will be.

For example, most of your body cells express the genes that encode the enzymes of glycolysis (Section 7.3), but only immature red blood cells use the genes that code for globin (Section 9.6). Only your liver cells express genes for enzymes that neutralize certain toxins.

A cell rarely uses more than 10 percent of its genes at once. Which genes are expressed at any given time depends on many factors, such as conditions in the cytoplasm and extracellular fluid, and the type of cell. These factors affect controls governing all steps of gene expression, starting with transcription and ending with delivery of an RNA or protein product to its final destination (Figure 10.2). Such controls consist of processes that start, enhance, slow, or stop gene expression.

Control of Transcription Many controls affect whether and how fast certain genes are transcribed into RNA **1**. Those that prevent an RNA polymerase from attaching to a promoter near a gene also prevent transcription of the gene. Controls that help RNA polymerase bind to DNA also speed up transcription.

Some types of proteins affect the rate of transcription by binding to special nucleotide sequences in the DNA. For example, an **activator** speeds up transcription when it binds to a promoter. Activators also speed transcription by binding to DNA sequences called **enhancers**. An enhancer is not necessarily close to the gene it affects, and may even be on a different chromosome (Figure 10.3). As another example, a **repressor** slows or stops transcription when it binds to certain sites in DNA.

Regulatory proteins such as activators and repressors are called **transcription factors**. Whether and how fast a gene is transcribed depends on which transcription factors are bound to the DNA.

Interactions between DNA and the histone proteins it wraps around also affect transcription. RNA polymerase can only attach to DNA that is unwound from histones (Section 8.2). Attachment of methyl groups ($-CH_3$) causes DNA to wind tightly around histones; thus, molecules that methylate DNA prevent its transcription.

The number of copies of a gene also affects how fast its product is made. For example, in some cells, DNA is copied repeatedly with no cytoplasmic division between

→ transcription start site   ← transcription end

replications. The result is a cell full of polytene chromosomes, each of which consists of hundreds or thousands of side-by-side copies of the same DNA molecule. All of the DNA strands carry the same genes. Translation of one gene, which occurs simultaneously on all of the identical DNA strands, produces a lot of mRNA, which is translated quickly into a lot of protein. Polytene chromosomes are common in immature amphibian eggs, the storage tissues of some plants, and the saliva gland cells of some insect larvae (Figure 10.4).

mRNA Processing As you know, before eukaryotic mRNAs leave the nucleus, they are modified—spliced, capped, and finished with a poly-A tail (Section 9.3). Controls over these modifications can affect the form of a protein product and when it will appear in the cell ❷. For example, controls that determine which exons are spliced out of an mRNA affect which form of a protein will be translated from it.

mRNA Transport mRNA transport is another point of control ❸. For example, in eukaryotes, transcription occurs in the nucleus, and translation in the cytoplasm. A new RNA can pass through pores of the nuclear envelope only after it has been processed appropriately. Controls that delay the processing also delay an mRNA's appearance in the cytoplasm, and thereby delay its translation.

Controls also govern mRNA localization. A short base sequence near an mRNA's poly-A tail is like a zip code. Certain proteins that attach to the zip code drag the mRNA along cytoskeletal elements and deliver it to the organelle or area of the cytoplasm specified by the code. Other proteins that attach to the zip code region prevent the mRNA from being translated before it reaches its destination. mRNA localization allows cells to grow or move in specific directions. It is also crucial for proper embryonic development.

Translational Control Most controls over eukaryotic gene expression affect translation ❹. Many govern the

production or function of the various molecules that carry out translation. Others affect mRNA stability: The longer an mRNA lasts, the more protein can be made from it. Enzymes begin to disassemble a new mRNA as soon as it arrives in the cytoplasm. The fast turnover allows cells to adjust their protein synthesis quickly in response to changing needs. How long an mRNA persists depends on its base sequence, the length of its poly-A tail, and which proteins are attached to it.

As a different example, microRNAs inhibit translation of other RNA. Part of a microRNA folds back on itself and forms a small double-stranded region. By a process called RNA interference, any double-stranded RNA (including a microRNA) is cut up into small bits that are taken up by special enzyme complexes. These complexes destroy every mRNA in a cell that can base-pair with the bits. So, expression of a microRNA complementary in sequence to a gene inhibits expression of that gene.

Post–Translational Modification
Many newly synthesized polypeptide chains must be modified before they become functional ❺. For example, some enzymes become active only after they have been phosphorylated (Section 5.3). Such post-translational modifications inhibit, activate, or stabilize many molecules, including the enzymes that participate in transcription and translation.

Figure 10.4 Polytene chromosomes in the salivary gland cells of fruit flies. These giant chromosomes form by repeated DNA replication without cell division.

Each of these chromosomes consists of hundreds or thousands of copies of the same DNA strand, aligned side by side. Transcription is visible as puffs (*white* arrows) where the DNA has loosened.

---

activator Regulatory protein that increases the rate of transcription when it binds to a promoter or enhancer.
differentiation The process by which cells become specialized.
enhancer Binding site in DNA for proteins that enhance the rate of transcription.
repressor Regulatory protein that blocks transcription.
transcription factor Regulatory protein that influences transcription; e.g., an activator or repressor.

Take-Home Message What is gene control?

❯ Most cells of multicelled organisms differentiate when they start expressing a unique subset of their genes. Which genes a cell expresses depends on the type of organism, its stage of development, and environmental conditions.

❯ Various control processes regulate all steps between gene and gene product.

> Research with fruit flies yielded the insight that body plans are a result of patterns of gene expression in embryos.
< Links to Development 1.3, Taxa 1.5, Genes 9.2

For about a hundred years, *Drosophila melanogaster* has been the subject of choice for many research experiments on eukaryotic gene expression. Why? It costs almost nothing to feed this fruit fly, which is only about 3 millimeters long and can live its entire life in a bottle. *D. melanogaster* also reproduces quickly and has a short life cycle. As well, experimenting on insects generally viewed as nuisance pests presents few ethical dilemmas.

Many important discoveries about how gene controls guide development have come from *Drosophila* research. The discoveries are clues to understanding similar processes in humans and other organisms, which have a shared evolutionary history.

Homeotic Genes

Homeotic genes control the formation of specific body parts (eyes, legs, segments, and so on) during the development of embryos. All homeotic genes encode transcription factors with a homeodomain, a region of about sixty amino acids that can bind to a promoter or some other DNA sequence in a chromosome.

Homeotic genes are a type of master gene. The products of **master genes** affect the expression of many other genes. Expression of a master gene causes other genes to be expressed, with the final outcome being the comple-

tion of an intricate task such as the formation of an eye during embryonic development.

Such processes begin long before body parts develop, as various master genes are expressed in local areas of the early embryo. The master gene products form in concentration gradients that span the entire embryo. Depending on where they are located within the gradients, embryonic cells begin to transcribe different homeotic genes. Products of the homeotic genes form in specific areas of the embryo. The different products cause cells to differentiate into tissues that form specific structures such as wings or a head.

The function of many homeotic genes has been discovered by manipulating their expression, one at a time. Researchers inactivate a homeotic gene by introducing a mutation or deleting it entirely, an experiment called a **knockout**. An organism that carries the knocked-out gene may differ from normal individuals, and the differences are clues to the function of the missing gene product.

Researchers often name homeotic genes based on what happens in their absence. For instance, fruit flies with a mutated *eyeless* gene develop with no eyes (Figure 10.5A,B). *Dunce* is required for learning and memory. *Wingless*, *wrinkled*, and *minibrain* are self-explanatory. *Tinman* is necessary for development of a heart. Flies with a mutated *groucho* gene have too many bristles above their eyes. One gene was named *toll*, after what its German discoverer exclaimed upon seeing the disastrous effects of the mutation (*toll* is a German slang word that means "cool!").

Figure 10.5 Eyes and *eyeless*. **A** A normal fruit fly has large, round eyes. **B** A fruit fly with a mutation in its *eyeless* gene develops with no eyes. **C** Eyes form wherever the *eyeless* gene is expressed in fly embryos—here, on the head and wing.

Humans, mice, squids, and other animals have a gene called *PAX6*. In humans, *PAX6* mutations result in missing irises, a condition called aniridia **D**. Compare a normal iris **E**. *PAX6* is so similar to *eyeless* that it triggers eye development when expressed in fly embryos.

Figure 10.6 How gene expression control makes a fly, as illuminated by segmentation. The expression of different master genes is shown by different colors in fluorescence microscopy images of whole *Drosophila* embryos at successive stages of development. The bright dots are individual cell nuclei.

A, B The master gene *even-skipped* is expressed (in *red*) only where two maternal gene products (*blue* and *green*) overlap.

C–E The products of several master genes, including the two shown here in *green* and *blue*, confine the expression of *even-skipped* (*red*) to seven stripes.

F One day later, seven segments develop that correspond to the position of the stripes.

Homeotic genes control development by the same mechanisms in all eukaryotes, and many are interchangeable among different species. Thus, we can infer that they evolved in the most ancient eukaryotic cells. Homeodomains often differ among species only in conservative substitutions—one amino acid has replaced another with similar chemical properties.

Consider the *eyeless* gene. Eyes form in embryonic fruit flies wherever this gene is expressed, which is typically in tissues of the head. If the *eyeless* gene is expressed in another part of the developing embryo, eyes form there too (Figure 10.5C). Humans, squids, mice, fish, and many other animals have a gene called *PAX6*, which is very similar to the *eyeless* gene. In humans, mutations in *PAX6* cause eye disorders such as aniridia, in which a person's irises are underdeveloped or missing (Figure 10.5D,E). *PAX6* works across different species. For example, if the *PAX6* gene from a human or mouse is inserted into an *eyeless* mutant fly, it has the same effect as the *eyeless* gene: An eye forms wherever it is expressed. Such studies are evidence of a shared ancestor among these evolutionarily distant animals.

Filling in Details of Body Plans

As an embryo develops, its differentiating cells form tissues, organs, and body parts. Some cells that alternately migrate and stick to other cells develop into nerves, blood vessels, and other structures that weave through the tissues. Events like these fill in the body's details, and all are driven by cascades of master gene expression.

homeotic gene Type of master gene; its expression controls formation of specific body parts during development.
knockout An experiment in which a gene is deliberately inactivated in a living organism.
master gene Gene encoding a product that affects the expression of many other genes.
pattern formation Process by which a complex body forms from local processes during embryonic development.

Pattern formation is the process by which a complex body forms from local processes in an embryo. Pattern formation begins as maternal mRNAs are delivered to opposite ends of an unfertilized egg as it forms. The localized maternal mRNAs get translated right after the egg is fertilized, and their protein products diffuse away in gradients that span the entire embryo. Cells of the developing embryo translate different master genes, depending on where they fall within those gradients. The products of the master genes also form in overlapping gradients. Cells of the embryo translate still other master genes depending on where they fall within these gradients, and so on.

Such regional gene expression during development results in a three-dimensional map that consists of overlapping concentration gradients of master gene products. Which master genes are active at any given time changes, and so does the map. Some master gene products cause undifferentiated cells to differentiate, and specialized tissues are the outcome. The formation of body segments in a fruit fly embryo is an example of how pattern formation works (Figure 10.6).

❯ Selective gene expression gives rise to many traits.
❮ Links to Chromosomes and sex determination 8.2, Mutations 9.6

Many of the traits that are characteristic of humans and other eukaryotic organisms arise as an outcome of gene expression controls, as the following examples illustrate.

X Chromosome Inactivation

In humans and other mammals, a female's cells each contain two X chromosomes, one inherited from her mother, the other one from her father. One X chromosome is always tightly condensed (Figure 10.7). We call the condensed X chromosomes "Barr bodies," after Murray Barr, who discovered them. RNA polymerase cannot access most of the genes on the condensed chromosome. This **X chromosome inactivation** ensures that only one of the two X chromosomes in a female's cells is active. According to a theory called **dosage compensation**, the inactivation equalizes expression of X chromosome genes between the sexes. The body cells of male mammals (XY) have one set of X chromosome genes. The body cells of female mammals (XX) have two sets, but only one is expressed. Normal development of female embryos depends on this control.

X chromosome inactivation occurs when an embryo is a ball of about 200 cells. In humans and many other mammals, it occurs independently in every cell of a female embryo. The maternal X chromosome may get inactivated in one cell, and the paternal or maternal X chromosome may get inactivated in a cell next to it. Once the selection is made in a cell, all of that cell's descen-

Figure 10.7 X chromosome inactivation. Barr bodies are visible as *red* spots in the nucleus of the four XX cells on the *left*. Compare the nucleus of two XY cells to the *right*.

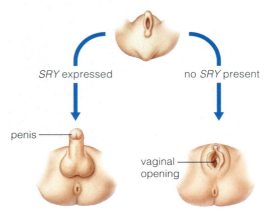

Structures that will give rise to external genitalia appear at seven weeks

SRY expressed   no *SRY* present

penis

vaginal opening

birth approaching

Figure 10.8 Development of reproductive organs in human embryos. An early human embryo appears neither male nor female. Gene expression determines what reproductive organs will form.

In an XY embryo, the *SRY* gene product triggers the formation of testes, male gonads that secrete testosterone. This hormone initiates development of other male traits. In an XX embryo, ovaries form in the absence of the Y chromosome and its *SRY* gene.

dants make the same selection as they continue dividing and forming tissues. As a result of the inactivation, an adult female mammal is a "mosaic" for the expression of genes on the X chromosome. She has patches of tissue in which genes of the maternal X chromosome are expressed, and patches in which genes of the paternal X chromosome are expressed.

How does just one of two X chromosomes get inactivated? An X chromosome gene called *XIST* does the trick. This gene is transcribed on only one of the two X chromosomes. The gene's product, a large RNA, sticks to the chromosome that expresses the gene. The RNA coats the chromosome and causes it to condense into a Barr body. Thus, transcription of the *XIST* gene keeps the chromosome from transcribing other genes. The other chromosome does not express *XIST*, so it does not get coated with RNA; its genes remain available for transcription. It is still unknown how the cell chooses which chromosome will express *XIST*.

Male Sex Determination in Humans

The human X chromosome carries 1,336 genes. Some of those genes are associated with sexual traits, such as the distribution of body fat and hair. However, most of the genes on the X chromosome govern nonsexual traits such as blood clotting and color perception. Such genes are expressed in both males and females. Males, remember, also inherit one X chromosome.

A The pattern in which the floral identity genes *A*, *B*, and *C* are expressed affects differentiation of cells growing in whorls in the plant's tips. Their gene products guide expression of other genes in cells of each whorl; a flower results.

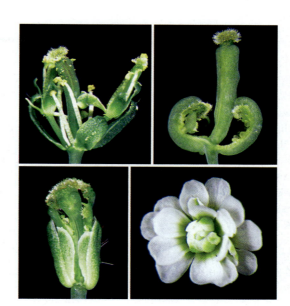

B Mutations in *Arabidopsis* *ABC* genes result in malformed flowers.

Top left, right: A gene mutations lead to petal-less flowers with no structures in place of missing petals.

Bottom left: B gene mutations lead to flowers with sepals instead of petals.

Bottom right: C gene mutations lead to flowers with petals instead of sepals and carpels. Compare the normal flower in **A**.

Figure 10.9 Animated Control of flower formation, as revealed by mutations in *Arabidopsis thaliana*.

The human Y chromosome carries only 307 genes, but one of them is the *SRY* gene—the master gene for male sex determination in mammals. Its expression in XY embryos triggers the formation of testes, which are male gonads (Figure 10.8). Some of the cells in these primary male reproductive organs make testosterone, a sex hormone that controls the emergence of male secondary sexual traits such as facial hair, increased musculature, and a deep voice. We know that *SRY* is the master gene that controls emergence of male sexual traits because mutations in this gene cause XY individuals to develop external genitalia that appear female. An XX embryo has no Y chromosome, no *SRY* gene, and much less testosterone, so primary female reproductive organs (ovaries) form instead of testes. Ovaries make estrogens and other sex hormones that will govern the development of female secondary sexual traits, such as enlarged, functional breasts, and fat deposits around the hips and thighs.

Flower Formation

When it is time for a plant to flower, populations of cells that would otherwise give rise to leaves instead differentiate into floral parts—sepals, petals, stamens, and carpels. How does the switch happen? Studies of mutations in the common wall cress plant, *Arabidopsis thaliana*, elucidated how the specialized parts of a flower develop. Three sets of master genes called *A*, *B*, and *C* guide the process of flower formation. These genes are switched on by environmental cues such as seasonal changes in the length of night.

At the tip of a shoot, cells form whorls of tissue, one over the other like layers of an onion. Cells in each whorl give rise to different tissues depending on which of their *ABC* genes are activated (Figure 10.9A). In the outer whorl, only the *A* genes are switched on, and their products trigger events that cause sepals to form. Cells in the next whorl express both *A* and *B* genes; they give rise to petals. Cells farther in express *B* and *C* genes; they give rise to stamens, the structures that produce male reproductive cells. The cells of the innermost whorl express only the *C* genes; they give rise to carpels, the structures that produce female reproductive cells. The phenotypic effects of mutations in *ABC* genes support this model (Figure 10.9B).

---

dosage compensation Theory that X chromosome inactivation equalizes gene expression between males and females.
X chromosome inactivation Shutdown of one of the two X chromosomes in the cells of female mammals.

Take-Home Message **What are some examples of gene control in eukaryotes?**

❯ X chromosome inactivation balances expression of X chromosome genes between female (XX) and male (XY) mammals. The balance is vital for development of female embryos.

❯ *SRY* gene expression triggers the development of male traits in mammals.

❯ Gene control also guides flower formation. *ABC* master genes that are expressed differently in shoot tissues govern development of flower parts.

> Bacteria control gene expression mainly by adjusting the rate of transcription.

‹ Links to Carbohydrates 3.3, Controls over metabolism 5.5, Lactate fermentation 7.6

Bacteria and archaeans do not undergo development and become multicelled organisms, so these cells do not use master genes. However, they do use gene controls. By adjusting gene expression, they can respond to environmental conditions. For example, when a certain nutrient becomes available, a bacterial cell will begin transcribing genes whose products allow the cell to use that nutrient. When the nutrient is not available, transcription of those genes stops. Thus, the cell does not waste energy and resources producing gene products that are not needed at a particular moment.

Bacteria control their gene expression mainly by adjusting the rate of transcription. Genes that are used together often occur together on the chromosome, one after the other. All of them are transcribed together into a single RNA strand, so their transcription is controllable in a single step.

The Lactose Operon

Escherichia coli lives in the gut of mammals, where it dines on nutrients traveling past. Its carbohydrate of choice is glucose, but it can make use of other sugars, such as the lactose in milk. *E. coli* cells use a set of three enzymes in order to harvest the glucose subunit of lactose molecules. However, unless there is lactose in the gut, *E. coli* cells keep the three genes for those enzymes turned off. There is one promoter for all three genes. Flanking the promoter are two **operators**, regions of DNA that serve as binding sites for a repressor. (Repressors, remember, stop transcription.) A promoter and one or more operators that together control the transcription of multiple genes are collectively called an **operon**. Operons occur in bacteria, archaeans, and eukaryotes. The one that controls lactose metabolism in *E. coli* is called the lac operon (Figure 10.10 ❶).

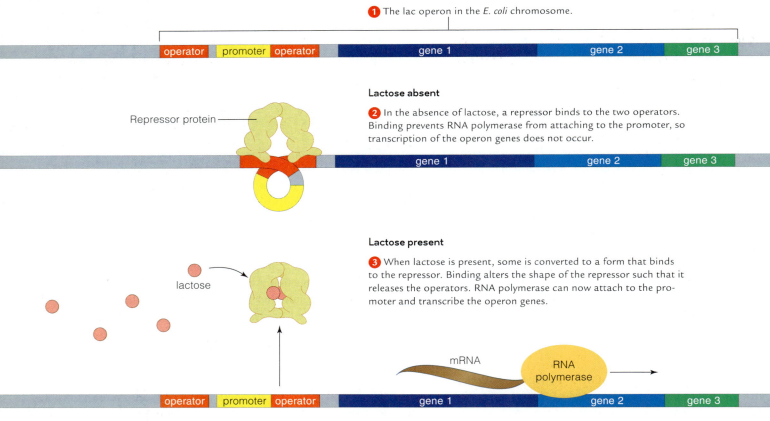

❶ The lac operon in the *E. coli* chromosome.

Lactose absent

❷ In the absence of lactose, a repressor binds to the two operators. Binding prevents RNA polymerase from attaching to the promoter, so transcription of the operon genes does not occur.

Repressor protein

Lactose present

❸ When lactose is present, some is converted to a form that binds to the repressor. Binding alters the shape of the repressor such that it releases the operators. RNA polymerase can now attach to the promoter and transcribe the operon genes.

lactose

mRNA RNA polymerase

Figure 10.10 Animated Example of gene control in bacteria: the lactose operon on a bacterial chromosome. The operon consists of a promoter flanked by two operators, and three genes for lactose-metabolizing enzymes.

›› **Figure It Out** What portion of the operon binds RNA polymerase when lactose is present?

Answer: The promoter

When lactose is not present, the lac operon repressors bind to the *E. coli* DNA, and lactose-metabolizing genes stay switched off. One repressor molecule binds to both operators, so that the region of DNA with the promoter twists into a loop ❷. RNA polymerase cannot bind to the twisted promoter, so it cannot transcribe the operon genes.

When lactose is in the gut, some of it is converted to another sugar, allolactose. Allolactose binds to the repressor and changes its shape. The altered repressor can no longer bind to the operators. The looped DNA unwinds, the promoter is now accessible to RNA polymerase, and transcription begins ❸.

E. coli cells use extra enzymes to metabolize lactose compared with glucose, so it is more efficient for them to use glucose. Accordingly, when both sugars are present, the cells will use up all of the available glucose before switching to lactose metabolism.

Lactose Intolerance

Like infants of other mammals, human infants drink milk. Cells in the lining of the small intestine secrete lactase, an enzyme that cleaves the lactose in milk into its subunit monosaccharides. In most people, lactase production starts to decline at the age of five.

After that, it becomes more difficult to digest lactose in food—a condition called lactose intolerance. Lactose is not absorbed directly by the intestine. Thus, any that is not broken down in the small intestine ends up in the large intestine, which hosts *E. coli* and a variety of other bacteria. These resident organisms respond to the abundant sugar supply by switching on their lac operons. Carbon dioxide, methane, hydrogen, and other gaseous products of their various fermentation reactions accumulate quickly in the large intestine, distending its wall and causing pain. The other products of their metabolism (undigested carbohydrates) disrupt the solute–water balance inside the large intestine, and diarrhea results.

Not everybody is lactose intolerant. Many people carry a mutation in one of the genes responsible for

---

operator Part of an operon; a DNA binding site for a repressor.
operon Group of genes together with a promoter–operator DNA sequence that controls their transcription.

Between You and Eternity (revisited)

Mutations in some genes predispose individuals to develop certain kinds of cancer. Tumor suppressor genes are named because tumors are more likely to occur when these genes mutate. Two examples are *BRCA1* and *BRCA2:* A mutated version of one or both of these genes is often found in breast and ovarian cancer cells. Because mutations in genes such as *BRCA* can be inherited, cancer is not only a disease of the elderly, as Robin Shoulla's story illustrates. Robin is one of the unlucky people who carry mutations in both *BRCA1* and *BRCA2*.

If a *BRCA* gene mutates in one of three especially dangerous ways, a woman has an 80 percent chance of developing breast cancer before the age of seventy. *BRCA* genes are master genes whose protein products help maintain the structure and number of chromosomes in a dividing cell. The multiple functions of these proteins are still being unraveled. We do know they participate directly in DNA repair (Section 8.6), so any mutations that alter this function also alter the cell's capacity to repair damaged DNA. Other mutations are likely to accumulate, and that sets the stage for cancer.

The products of *BRCA* genes also bind to receptors for the hormones estrogen and progesterone, which are abundant on cells of breast and ovarian tissues. Binding suppresses transcription of growth factor genes in these cells. Among other things, growth factors stimulate cells to divide during normal, cyclic renewals of breast and ovarian tissues. When a mutation alters a *BRCA* gene so that its product cannot bind to hormone receptors, the cells overproduce growth factors. Cell division goes out of control, and tissue growth becomes disorganized. In other words, cancer develops.

Two groups of researchers, one at the Dana-Farber Cancer Institute at Harvard, the other at the University of Milan, recently found that the RNA product of the *XIST* gene localizes abnormally in breast cancer cells. In those cells, both X chromosomes are active. It makes sense that two active X chromosomes would have something to do with abnormal gene expression, but why the RNA product of an unmutated *XIST* gene does not localize properly in cancer cells remains a mystery.

Mutations in the *BRCA1* gene may be part of the answer. The Harvard researchers found that the protein product of the *BRCA1* gene physically associates with the RNA product of the *XIST* gene. They were able to restore proper *XIST* RNA localization—and proper X chromosome inactivation—by restoring the function of the *BRCA1* gene product in breast cancer cells.

How Would You Vote? Some women at high risk of developing breast cancer opt for preventive breast removal. Many of them never would have developed cancer. Should the surgery be restricted to cancer treatment? See CengageNow for details, then vote online (cengagenow.com).

the programmed lactase shutdown. These people make enough lactase to continue drinking milk without problems into adulthood.

Take-Home Message Do bacteria control gene expression?

❯ In bacteria, the main gene expression controls regulate transcription in response to shifts in nutrient availability and other outside conditions.

Summary

 Section 10.1 Controls over gene expression are a critical part of the embryonic development and normal functioning of a multi-celled body. When gene controls fail, as occurs as a consequence of some mutations, **cancer** may be the outcome.

 Section 10.2 Which genes a cell uses depends on the type of organism, the type of cell, factors inside and outside the cell, and, in complex multicelled species, the organism's stage of development.

Controls over gene expression are part of homeostasis in all organisms. They also drive development in multi-celled eukaryotes. All cells of an embryo share the same genes. As different cell lineages use different subsets of genes during development, they become specialized, a process called **differentiation**. Specialized cells form tissues and organs in the adult.

Different molecules and processes govern every step between transcription of a gene and delivery of the gene's product to its final destination. Most controls operate at transcription; **transcription factors** such as **activators** and **repressors** influence transcription by binding to promoters, **enhancers**, or other sequences in chromosomal DNA.

 Section 10.3 Knockouts of **homeotic genes** in fruit flies (*Drosophila melanogaster*) revealed that local controls over gene expression govern the embryonic development of all complex, multicelled bodies, a process called **pattern formation**. Various **master genes** are expressed locally in different parts of an embryo as it develops. Their products diffuse through the embryo and affect expression of other master genes, which affect the expression of others, and so on. These cascades of master gene products form a dynamic spatial map of overlapping gradients that spans the entire embryo body. Cells differentiate according to their location on the map.

 Section 10.4 In female mammals, most genes on one of the two X chromosomes are permanently inaccessible. **X chromosome inactivation** balances gene expression between the sexes. Such **dosage compensation** arises because the *XIST* gene gets transcribed on only one of the two X chromosomes. The gene's RNA product shuts down the chromosome that transcribes it.

Studies of mutations in *Arabidopsis thaliana* showed that three sets of master genes (*A*, *B*, and *C*) guide cell differentiation in the whorls of a floral shoot; sepals, petals, stamens, and carpels form.

 Section 10.5 Bacterial cells do not have great structural complexity and do not undergo development. Most of their gene controls reversibly adjust transcription rates in response to environmental conditions, especially nutrient availability. **Operons** are examples of bacterial gene controls. The lactose operon governs expression of three genes, the three products of which allow the bacterial cell to metabolize lactose. Two **operators** that flank the promoter are binding sites for a repressor that blocks transcription.

Self-Quiz   Answers in Appendix III

1. The expression of a gene may depend on _____ .
   a. the type of organism   c. the type of cell
   b. environmental conditions   d. all of the above

2. Gene expression in multicelled eukaryotic cells changes in response to _____ .
   a. conditions outside the cell   c. operons
   b. master gene products   d. a and b

3. Binding of _____ to _____ in DNA can increase the rate of transcription of specific genes.
   a. activators; promoters   c. repressors; operators
   b. activators; enhancers   d. both a and b

4. Proteins that influence gene expression by binding to DNA are called _____ .

5. Polytene chromosomes form in some types of cells that _____ .
   a. have a lot of chromosomes   c. are differentiating
   b. are making a lot of protein   d. b and c are correct

6. Eukaryotic gene controls govern _____ .
   a. transcription   e. translation
   b. RNA processing   f. protein modification
   c. RNA transport   g. a through e
   d. mRNA degradation   h. all of the above

7. Controls over eukaryotic gene expression guide _____ .
   a. natural selection   c. development
   b. nutrient availability   d. all of the above

8. The expression of *ABC* genes _____ .
   a. occurs in layers
   b. controls flower formation
   c. causes mutations in flowers
   d. both a and b

9. Cell differentiation _____ .
   a. occurs in all complex multicelled organisms
   b. requires unique genes in different cells
   c. involves selective gene expression
   d. both a and c
   e. all of the above

10. During X chromosome inactivation _____ .
    a. female cells shut down   c. pigments form
    b. RNA coats chromosomes   d. both a and b

11. A cell with a Barr body is _____ .
    a. a bacterium   c. from a female mammal
    b. a sex cell   d. infected by Barr virus

BRCA Mutations in Women Diagnosed With Breast Cancer

Investigating a correlation between specific cancer-causing mutations and the risk of mortality in humans is challenging, in part because each cancer patient is given the best treatment available at the time. There are no "untreated control" cancer patients, and ideas about which treatments are best change quickly as new drugs become available and new discoveries are made.

Figure 10.11 shows results from a 2007 study by Pal Moller and his colleagues. The researchers looked for BRCA mutations in 442 women who had been diagnosed with breast cancer, and followed their treatments and progress over several years.

All of the women in the study had at least two affected close relatives, so their risk of developing breast cancer due to an inherited factor (such as a BRCA mutation) was estimated to be greater than that of the general population.

1. According to this study, what is a woman's risk of dying of cancer if two of her close relatives have breast cancer?
2. What is her risk of dying of cancer if she carries a mutated BRCA1 gene?
3. According to these results, is a BRCA1 or BRCA2 mutation more dangerous in breast cancer cases?
4. What other data would you have to see in order to make a conclusion about the effectiveness of preventive mastectomy or oophorectomy?

BRCA Mutations in Women Diagnosed With Breast Cancer				
	BRCA1	BRCA2	No BRCA Mutation	Total
Total number of patients	89	35	318	442
Avg. age at diagnosis	43.9	46.2	50.4	
Preventive mastectomy	6	3	14	23
Preventive oophorectomy	38	7	22	67
Number of deaths	16	1	21	38
Percent died	18.0	2.8	6.9	8.6

Figure 10.11 Results from a 2007 study investigating BRCA mutations in women diagnosed with breast cancer. All women in the study had a family history of the disease.

Some of the women underwent preventive mastectomy (removal of the noncancerous breast) during their course of treatment. Others had preventive oophorectomy (surgical removal of the ovaries) to prevent the possibility of getting ovarian cancer.

Top, model of the unmutated BRCA1 protein.

---

12. Homeotic gene products _____ .
    a. flank a bacterial operon
    b. map out the overall body plan in embryos
    c. control the formation of specific body parts

13. A gene that is knocked out is _____ .
    a. deleted   c. expressed
    b. inactivated   d. either a or b

14. A promoter and a set of operators that control access to two or more genes is a(n) _____ .
    a. lactose molecule   c. dosage compensator
    b. operon   d. knockout

15. Match the terms with the most suitable description.
    ____ ABC genes   a. a big RNA is its product
    ____ XIST gene   b. binding site for repressor
    ____ operator   c. cells become specialized
    ____ Barr body   d. —CH₃ additions to DNA
    ____ differentiation e. inactivated X chromosome
    ____ methylation   f. guide flower development

Additional questions are available on **CENGAGENOW**.

Animations and Interactions on **CENGAGENOW**:
❯ Points of control over gene expression; ABC model for flowering; X chromosome inactivation; Structure and function of the lac operon.

Critical Thinking

1. Why does a cell regulate its gene expression?

2. Do the same gene controls operate in bacterial cells and eukaryotic cells? Why or why not?

3. Unlike most rodents, guinea pigs are well developed at the time of birth. Within a few days, they can eat grass, vegetables, and other plant material.

   Suppose a breeder decides to separate baby guinea pigs from their mothers three weeks after they were born. He wants to raise the males and the females in different cages. However, he has trouble identifying the sex of young guinea pigs. Suggest how a quick look through a microscope can help him identify the females.

4. Geraldo isolated an *E. coli* strain in which a mutation has hampered the capacity of the cAMP activator to bind the promoter of the lactose operon. How will this mutation affect transcription of the lactose operon when the *E. coli* cells are exposed to the following conditions?

   a. Lactose and glucose are both available.

   b. Lactose is available but glucose is not.

   c. Both lactose and glucose are absent.

Be sure you understand chromosome structure and chromosome number (Section 8.2) before reading this chapter. You will revisit eukaryotic cell structure (4.2, 4.10, 4.11), particularly the nucleus (4.7). What you know about free radicals (5.4) and mutations (9.6), receptors and recognition proteins (4.4), phosphorylation (5.3), fermentation (7.6), and gene control in eukaryotes (10.2) will help you understand cancer and how it develops.

Key Concepts

The Cell Cycle

A cell cycle starts when a new cell forms by division of a parent cell, and ends when the cell completes its own division. A typical cell cycle proceeds through intervals of interphase, mitosis, and cytoplasmic division.

Mitosis

Mitosis divides the nucleus and maintains the chromosome number. It has four sequential stages: prophase, metaphase, anaphase, and telophase. A spindle parcels the cell's duplicated chromosomes into two nuclei.

11 How Cells Reproduce

Each human starts out as a fertilized egg. By the time of birth, the human body consists of about a trillion cells, all descended from that single cell. Even in an adult, billions of cells divide every day as new cells replace worn-out ones. However, human cells tend to divide a few times and die within weeks when grown in the laboratory.

Researchers started trying to coax human cells to become immortal—to keep dividing outside of the body—in the mid-1800s. Why? Many human diseases occur only in human cells. Immortal cell lineages, or cell lines, would allow researchers to study human diseases (and potential cures for them) without experimenting on people.

At Johns Hopkins University, George and Margaret Gey were among the researchers trying to culture human cells. They had been working on the problem for almost thirty years when, in 1951, their assistant Mary Kubicek prepared a sample of human cancer cells. Mary named the cells HeLa, after the first and last names of the patient from whom the cells had been taken.

The HeLa cells began to divide, again and again. The cells were astonishingly vigorous, quickly coating the inside of their test tube and consuming the nutrient broth in which they were bathed. Four days later, there were so many cells that the researchers had to transfer them to more tubes. The cell populations increased at a phenomenal rate. The cells were dividing every twenty-four hours and coating the inside of the tubes within days.

Sadly, cancer cells in the patient were dividing just as fast. Just six months after she had been diagnosed with cervical cancer, malignant cells had invaded tissues throughout her body. Two months after that, Henrietta Lacks, a young African American woman from Baltimore, was dead.

Although Henrietta passed away, her cells lived on in the Geys' laboratory (Figure 11.1). The Geys were able to grow polio-virus in HeLa cells, a practice that enabled them to find out which strains of the virus cause polio. That work was a critical step in the development of polio vaccines, which have since saved millions of lives.

Figure 11.1 HeLa cells, a legacy of cancer victim Henrietta Lacks (*right*). *Opposite*, fluorescence micrograph of two HeLa cells in the process of dividing. *Blue* and *green* show two proteins that help microtubules (*red*) attach to chromosomes (*white*). Defects in these and other proteins that orchestrate cell division result in descendant cells with too many or too few chromosomes, an outcome that is a hallmark of cancer.

Henrietta Lacks's cells, frozen away in tiny tubes and packed in Styrofoam boxes, continue to be shipped among laboratories all over the world. Researchers use those cells to investigate cancer, viral growth, protein synthesis, and the effects of radiation. They helped several researchers win Nobel Prizes for research in medicine and chemistry. Some HeLa cells even traveled into space for experiments on the Discoverer XVII satellite.

Henrietta Lacks was just thirty-one, a wife and mother of five, when runaway cell divisions killed her. Her legacy continues to help people, through her cells that are still dividing, again and again, more than fifty years after she died. Understanding why cancer cells are immortal—and why we are not—begins with understanding the structures and mechanisms that cells use to divide.

Cytoplasmic Division
After nuclear division, the cytoplasm divides. Typically, one nucleus ends up in each of two new cells. The cytoplasm of an animal cell simply pinches in two. In plant cells, a cross-wall forms in the cytoplasm and divides it.

The Cell Cycle Gone Awry
Built-in mechanisms monitor and control the timing and rate of cell division. On rare occasions, the surveillance mechanisms fail, and cell division becomes uncontrollable. Tumor formation and cancer are outcomes.

> Cells reproduce by dividing.
> Division of a eukaryotic cell typically occurs in two steps: nuclear division followed by cytoplasmic division.
> The sequence of stages through which a cell passes during its lifetime is called the cell cycle.
‹ Links to Cell structure 4.2, Nucleus 4.7, Chromosomes 8.2, DNA replication 8.6, Gene expression controls 10.2

The Life of a Cell

Just as the series of events in an animal's life is called a life cycle, the events that occur from the time a cell forms until the time it divides is called a **cell cycle** (Figure 11.2). A typical cell spends most of its life in **interphase**. During this phase, the cell increases its mass, roughly doubles the number of its cytoplasmic components, and replicates its DNA in preparation for division. Interphase consists of three stages:

❶ G1, the first interval (or gap) of cell growth, before DNA replication

❷ S, the time of synthesis (DNA replication)

❸ G2, the second interval (or gap), when the cell prepares to divide

Gap intervals were named because outwardly they seem to be periods of inactivity. Actually, most cells going about their metabolic business are in G1. Cells preparing to divide enter S, when they copy their DNA. During G2, they make the proteins that will drive cell division. Once the S phase begins, DNA replication usually proceeds at a predictable rate and ends before the cell divides ❹.

The remainder of the cycle consists of the division process itself. When a cell divides, both of its two cellu-lar offspring end up with a blob of cytoplasm and some DNA. Each of the offspring of a eukaryotic cell inherits its DNA packaged inside a nucleus. Thus, a eukaryotic cell's nucleus has to divide ❺ before its cytoplasm does ❻.

There are two processes by which cell nuclei divide. As you will discover in this chapter and the next, these two processes—mitosis and meiosis—have much in common, but their outcomes differ. **Mitosis** is a nuclear division mechanism that maintains the chromosome number. Remember from Section 8.2 that diploid cells have two sets of chromosomes. For example, human body cells have 46 chromosomes, two of each type. Except for a pairing of sex chromosomes (XY) in males, the chromosomes of each pair are homologous. **Homologous chromosomes** have the same length, shape, and genes (*hom*– means alike). Typically, each member of a pair was inherited from one of two parents.

With mitosis followed by cytoplasmic division, a diploid parent cell produces two diploid offspring. Both offspring have the same chromosome number as the parent. However, it is not just the number of chromosomes that matters. If only the total mattered, then one of the cell's offspring might get, say, two pairs of chromosome 22 and no pairs whatsoever of chromosome 9. A cell cannot function properly without a full complement of DNA, which means it needs to have one copy of each type of chromosome. Thus, each of a cell's descendants receives one copy of each chromosome.

When a cell is in G1, each of its chromosomes consists of one double-stranded DNA molecule (Figure 11.3A). The cell replicates its DNA in S, so by G2, each of its chromosomes consists of two double-stranded DNA molecules

❶ G1 is the interval of growth before DNA replication. The cell's chromosomes are unduplicated during this stage.

❷ S is the time of synthesis. The name refers to DNA synthesis, because the cell copies its DNA during this stage.

❻ At the end of mitosis, the cytoplasm typically divides, and the cycle begins anew in interphase for each descendant cell.

❸ G2 is the interval after DNA replication and before mitosis. The cell prepares to divide during this stage.

❺ The nucleus divides during mitosis.

❹ Interphase ends.

Figure 11.2 Animated The eukaryotic cell cycle. The length of the intervals differs among cells. G1, S, and G2 are part of interphase.

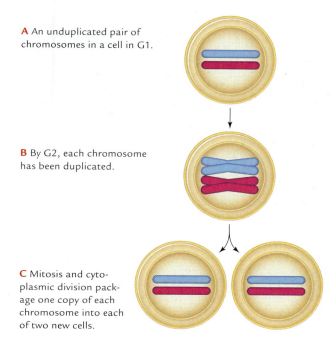

A An unduplicated pair of chromosomes in a cell in G1.

B By G2, each chromosome has been duplicated.

C Mitosis and cytoplasmic division package one copy of each chromosome into each of two new cells.

Figure 11.3 How mitosis maintains the chromosome number.

Figure 11.4 A multicelled eukaryote develops by repeated cell divisions. This photo shows early frog embryos, each a product of three mitotic divisions of one fertilized egg.

›› Figure It Out Each of these embryos consists of how many cells?   Answer: Eight

(Figure 11.3B). These molecules stay attached to one another at the centromere (as sister chromatids) until mitosis is almost over. The next section shows how mitosis parcels sister chromatids into separate nuclei. When the cytoplasm divides, these new nuclei are packaged into separate cells (Figure 11.3C). Each new cell has a full complement of unduplicated chromosomes, and each starts the cell cycle over again in G1 of interphase.

A Bigger Picture of Cell Division

Cell division is complicated business; it requires the coordinated participation of thousands of molecules. A host of gene expression controls (Section 10.2) orchestrates the process. As you will see in Section 11.5, many of these controls function as built-in brakes on the cell cycle. Apply the brakes that work in G1, and the cycle stalls in G1. Lift the brakes, and the cycle runs again. Sometimes the brakes

---

asexual reproduction Reproductive mode by which offspring arise from a single parent only.
cell cycle A series of events from the time a cell forms until its cytoplasm divides.
homologous chromosomes Chromosomes with the same length, shape, and set of genes.
interphase In a eukaryotic cell cycle, the interval between mitotic divisions when a cell grows, roughly doubles the number of its cytoplasmic components, and replicates its DNA.
mitosis Nuclear division mechanism that maintains the chromosome number. Basis of body growth and tissue repair in multicelled eukaryotes; also asexual reproduction in some plants, animals, fungi, and protists.

are not lifted. For example, the nerve cells of adult humans normally stay in G1 of interphase. Because the cell cycle of these cells cannot proceed, they do not divide. Thus, damaged nerve cells cannot be replaced. Other cells divide at rates that depend on cell type. The stem cells in your red bone marrow divide every 12 hours. Their descendants become red blood cells that replace 2 to 3 million worn-out ones in your blood each second. Cells in the tips of a bean plant root divide every 19 hours. The cells in a fruit fly embryo divide every 10 minutes.

Mitosis and cytoplasmic division are the basis of increases in body size during development (Figure 11.4), and ongoing replacements of damaged or dead cells. Individuals of many species of plants, animals, fungi, and protists reproduce by mitosis and cytoplasmic division, a process called **asexual reproduction**. Bacteria and archaeans also reproduce asexually, but they do it by binary fission, a separate mechanism that we will consider in Section 19.6.

> When a nucleus divides by mitosis, each new nucleus has the same chromosome number as the parent cell.
> The four main stages of mitosis are prophase, metaphase, anaphase, and telophase.
< Links to Cytoskeletal elements 4.10, Chromosome condensation 8.2, Transcriptional control 10.2

During interphase, a cell's chromosomes are loosened to allow transcription and DNA replication (Figure 11.5). In preparation for nuclear division, they begin to pack tightly **1**. Transcription and DNA replication stop as the chromosomes condense into their most compact "X" forms (Section 8.2). A cell reaches **prophase**, the first stage of mitosis, when its chromosomes have condensed so much that they are visible under a light microscope **2**. "Mitosis" is from the Greek word for thread, *mitos*, after the threadlike appearance of chromosomes during nuclear division.

Most animal cells have a centrosome, a region near the nucleus that organizes microtubules while they are forming. A centrosome usually includes two centrioles, barrel-shaped organelles that help microtubules assemble (Section 4.10). The centrosome gets duplicated just before prophase. Then, during prophase, one of the two centrosomes moves to the opposite side of the cell. Microtubules that begin to extend from both centrosomes form a **spindle**, a dynamic network of microtubules that moves chromosomes during nuclear division. Motor proteins traveling along the microtubules help the spindle extend in the appropriate directions. Plant cells have no centrosomes, but they do have spindles and structures that organize them.

In prophase, the spindle penetrates the nuclear region as the nuclear envelope breaks up. Some microtubules of the spindle stop lengthening when they reach the middle of the cell. Others lengthen until they reach a chromosome and attach to it at the centromere. By the end of prophase, one sister chromatid of each chromosome has become attached to microtubules extending from one spindle pole, and the other sister has become attached to microtubules extending from the other spindle pole **3**.

The opposing sets of microtubules then begin a tug-of-war by adding and losing tubulin subunits. As the microtubules extend and shrink, they push and pull the chromosomes. When all the microtubules are the same length, the chromosomes are aligned midway between

Onion root cell   **Whitefish embryo cell**

Figure 11.5 Animated Mitosis. Micrographs *here* and *opposite* show plant cells (onion root, *left*), and animal cells (whitefish embryo, *right*). *This page*, interphase cells are shown for comparison, but interphase is not part of mitosis.

Opposite page, the stages of mitosis. The drawings show a diploid (*2n*) animal cell. For clarity, only two pairs of chromosomes are illustrated, but nearly all eukaryotic cells have more than two. The two chromosomes of the pair inherited from one parent are *pink*; the two chromosomes from the other parent are *blue*.

the spindle poles **4**. The alignment marks **metaphase** (from *meta*, the ancient Greek word for between).

During **anaphase**, microtubules of the spindle separate the sister chromatids of each duplicated chromosome, and move them toward opposite spindle poles **5**. Each DNA molecule has now become a separate chromosome.

Telophase begins when the two clusters of chromosomes reach the spindle poles **6**. Each cluster has the same number and kinds of chromosomes as the parent cell nucleus had—two of each type of chromosome, if the parent cell was diploid. A new nucleus forms around each cluster as the chromosomes loosen up again. Once the two nuclei have formed, telophase is over, and so is mitosis.

anaphase Stage of mitosis during which sister chromatids separate and move to opposite spindle poles.
metaphase Stage of mitosis at which the cell's chromosomes are aligned midway between poles of the spindle.
prophase Stage of mitosis during which chromosomes condense and become attached to a newly forming spindle.
spindle Dynamically assembled and disassembled network of microtubules that moves chromosomes during nuclear division.
telophase Stage of mitosis during which chromosomes arrive at the spindle poles and decondense, and new nuclei form.

Take-Home Message **What is the sequence of events that take place during mitosis?**

> Each chromosome in a cell's nucleus was duplicated before mitosis begins, so each consists of two DNA molecules (sister chromatids).

> In prophase, the chromosomes condense and a spindle forms. The spindle microtubules attach to the chromosomes as the nuclear envelope breaks up.

> At metaphase, the (still duplicated) chromosomes are aligned midway between the spindle poles.

> In anaphase, microtubules separate the sister chromatids of each chromosome, and pull them toward opposite spindle poles. Each DNA molecule is now a separate chromosome.

> In telophase, two clusters of chromosomes reach the spindle poles. A new nuclear envelope forms around each cluster.

> Two new nuclei form at the end of mitosis. Each one has the same chromosome number as the parent cell's nucleus.

centrosome

1 Early Prophase
Mitosis begins. In the nucleus, the DNA begins to appear grainy as it starts to condense. The centrosome gets duplicated.

2 Prophase
The duplicated chromosomes become visible as they condense. One of the two centrosomes moves to the opposite side of the nucleus. The nuclear envelope breaks up.

3 Transition to Metaphase
The nuclear envelope is gone, and the chromosomes are at their most condensed. Spindle microtubules assemble and bind to chromosomes at the centromere. Sister chromatids are attached to opposite spindle poles.

microtubule of spindle

4 Metaphase
All of the chromosomes are aligned midway between the spindle poles.

5 Anaphase
Spindle microtubules separate the sister chromatids and move them toward opposite spindle poles. Each sister chromatid has now become an individual, unduplicated chromosome.

6 Telophase
The chromosomes reach the spindle poles and decondense. A nuclear envelope forms around each cluster, and mitosis ends.

› In most eukaryotes, the cell cytoplasm divides between late anaphase and the end of telophase. The mechanism of division differs between plants and animals.
‹ Links to Cytoskeleton 4.10, Primary wall 4.11

A cell's cytoplasm usually divides after mitosis, so two cells form. The process of cytoplasmic division, which is called **cytokinesis**, differs among eukaryotes.

Typical animal cells pinch themselves in two after nuclear division (Figure 11.6). How? The spindle begins to disassemble during telophase ❶. The cell cortex, which is the mesh of cytoskeletal elements just under the plasma membrane, includes a band of actin and myosin filaments that wraps around the cell's midsection. The band is called a contractile ring because it contracts when its component proteins are energized by ATP. When the ring contracts, it shrinks, dragging the plasma membrane inward as it does ❷. The sinking plasma membrane becomes visible on the outside of the cell as an indentation between the former spindle poles ❸. The indentation is called a **cleavage furrow**. The cleavage furrow advances around the cell, deepening as it does, until the cytoplasm (and the cell) is pinched in two ❹. Each of the two cells formed by this division has its own nucleus and some of the parent cell's cytoplasm; each is enclosed by a plasma membrane.

Dividing plant cells face a particular challenge because they have stiff cell walls on the outside of their plasma membrane (Section 4.11). Accordingly, plant cells do not pinch themselves in two; they have a completely different mechanism of cytoplasmic division (Figure 11.7). By the end of anaphase in a plant cell, a set of short micro-

❺ The future plane of division was established before mitosis began. Vesicles cluster here when mitosis ends.

❻ As the vesicles fuse with each other, they form a cell plate along the plane of division.

❼ The cell plate expands outward along the plane of division. When it reaches the plasma membrane, it attaches to the membrane and partitions the cytoplasm.

❽ The cell plate matures as two new cell walls. These walls join with the parent cell wall, so each descendant cell becomes enclosed by its own cell wall.

Figure 11.7 **Animated** Cytoplasmic division of a plant cell.

tubules has formed on either side of the future plane of division. These microtubules now guide vesicles from Golgi bodies and the cell surface to the future plane of division ❺. There, the vesicles and their wall-building contents start to fuse into a disk-shaped **cell plate** ❻. The plate expands at its edges until it reaches the plasma membrane ❼. When the cell plate attaches to the membrane, it partitions the cytoplasm. In time, the cell plate will develop into a primary cell wall that merges with the parent cell's wall. Thus, by the end of division, each of the descendant cells will be enclosed by its own plasma membrane and its own cell wall ❽.

cell plate After nuclear division in a plant cell, a disk-shaped structure that forms a cross-wall between the two new nuclei.
cleavage furrow In a dividing animal cell, the indentation where cytoplasmic division will occur.
cytokinesis Cytoplasmic division.

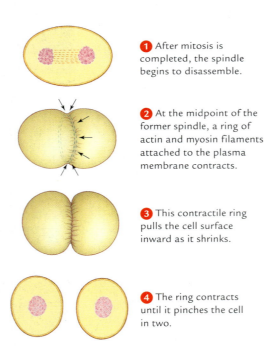

❶ After mitosis is completed, the spindle begins to disassemble.

❷ At the midpoint of the former spindle, a ring of actin and myosin filaments attached to the plasma membrane contracts.

❸ This contractile ring pulls the cell surface inward as it shrinks.

❹ The ring contracts until it pinches the cell in two.

Figure 11.6 **Animated** Cytoplasmic division of an animal cell.

Take-Home Message How do cells divide?

› After mitosis, the cytoplasm of the parent cell typically is partitioned into two descendant cells, each with its own nucleus.

› The process of cytoplasmic division differs between plants and animals.

› In animal cells, a contractile ring pinches the cytoplasm in two. In plant cells, a cell plate that forms midway between the spindle poles partitions the cytoplasm when it reaches and connects to the parent cell wall.

Controls Over Cell Division

> On rare occasions, controls over cell division are lost.
< Links to Receptor proteins 4.4, Phosphorylation 5.3, Free radicals 5.4, UV light and mutations 9.6, Eukaryotic gene control 10.2

Every second, millions of cells in your skin, bone marrow, gut lining, liver, and elsewhere are dividing and replacing their worn-out, dead, and dying predecessors. They do not divide at random. Many gene controls govern DNA replication and cell division in eukaryotic cells.

What happens when something goes wrong? Suppose sister chromatids do not separate as they should during mitosis. As a result, one descendant cell ends up with too many chromosomes and the other with too few. Or suppose DNA gets damaged when a chromosome is being duplicated. A cell's DNA can also be damaged by free radicals (Section 5.4), or environmental assaults such as chemicals or ultraviolet radiation (Section 9.6). Such problems are frequent and inevitable, but a cell may not function properly unless they are countered quickly.

The cell cycle has built-in checkpoints that allow problems to be corrected before the cycle advances. Certain proteins, the products of "checkpoint" genes, can monitor whether a cell's DNA has been copied completely, whether it is damaged, and even whether enough nutrients to support division are available. Such proteins interact to delay or stop the cell cycle while simultaneously enhancing transcription of genes involved in chromosome repair (Figure 11.8). If the problem stays uncorrected, checkpoint gene products cause the cell to self-destruct.

Sometimes a checkpoint gene mutates so that its protein product no longer works properly. In other cases, the controls that regulate its production fail, and a cell makes too much or too little of its product. When enough checkpoint mechanisms fail, a cell loses control over its cell cycle. The cell may skip interphase, so division occurs over and over with no resting period. Signaling mechanisms that make an abnormal cell die may stop working. The problem is compounded because these checkpoint malfunctions are passed along to the cell's descendants, which form a **neoplasm**, an accumulation of cells that lost control over how they grow and divide.

Consider **growth factors**, which are molecules that stimulate cells to divide and differentiate. One kind, an epidermal growth factor (EGF), stimulates a cell to enter mitosis by binding to a receptor on the cell's plasma membrane. Binding to EGF changes the shape of the receptor so that it becomes enzymatic and phosphorylates itself. Phosphorylation activates the EGF receptor, and it starts a cascade of other intracellular events that

growth factor Molecule that stimulates mitosis.
neoplasm An accumulation of abnormally dividing cells.

Figure 11.8 Checkpoint genes in action. Radiation damaged the DNA inside this nucleus. **A** *Green* dots pinpoint the location of the product of the *53BP1* gene, and **B** *red* dots pinpoint the location of the product of the *BRCA1* gene. Both proteins have clustered around the same chromosome breaks in the same nucleus; both function to recruit DNA repair enzymes. The integrated action of these and other checkpoint gene products blocks mitosis until the DNA breaks are fixed.

Figure 11.9 Neoplasms are associated with mutations in checkpoint genes. In this section of human breast tissue, phosphorylated EGF receptor is stained *brown*. Normal cells are the ones with lighter staining.

The heavily stained cells have formed a neoplasm; the abnormal overabundance of the phosphorylated EGF receptor means that mitosis is being continually stimulated in these cells. The EGF receptor is overproduced or overactive in most neoplasms.

ultimately moves the cell cycle out of interphase and into mitosis. The EGF receptor is the product of a checkpoint gene; cells of most neoplasms carry mutations resulting in its overactivity or overabundance (Figure 11.9).

Take-Home Message How does a cell "know" when to divide?

> Gene expression controls advance, delay, or block the cell cycle in response to internal and external conditions.

> The failure of cell cycle checkpoints results in uncontrolled cell divisions.

› Cancer develops as cells of a neoplasm become malignant.
‹ Links to Adhesion proteins 4.4, Fermentation 7.6

A neoplasm that forms a lump in the body is called a **tumor**, but the two terms are sometimes used interchangeably. Mutations that alter the products of checkpoint genes or the rate at which they are made are associated with an increased risk of tumor formation. Once such a tumor-causing mutation has occurred, the mutated gene is called an oncogene. An **oncogene** is any gene that transforms a normal cell into a tumor cell (from the Greek *onkos*, or bulging mass). Some mutations can be passed to offspring, which is one reason that some types of tumors tend to run in families.

Checkpoint genes encoding proteins that promote mitosis are called **proto-oncogenes** because mutations

❶ benign neoplasm   ❷ malignant neoplasm

Figure 11.10 Animated Benign and malignant neoplasms.

❶ Benign neoplasms grow slowly and stay in their home tissue.

❷ Cells of a malignant neoplasm can break away from their home tissue.

❸ The malignant cells become attached to the wall of a blood vessel or lymph vessel. They release digestive enzymes that create an opening in the wall, then enter the vessel.

❹ The cells creep or tumble along inside blood vessels, then leave the bloodstream the same way they got in. They often start growing in other tissues, a process called metastasis.

can turn them into oncogenes. The gene that encodes the EGF receptor is an example of a proto-oncogene.

Checkpoint gene products that inhibit mitosis are called tumor suppressors because tumors form when they are missing. The products of the *BRCA1* and *BRCA2* genes (Chapter 10) are examples of tumor suppressors. These proteins regulate, among other things, the expression of DNA repair enzymes. Mutations in *BRCA* genes are often found in cells of neoplasms. As another example, viruses such as HPV (human papillomavirus) cause a cell to make proteins that interfere with its own tumor suppressors. Infection with HPV causes noncancerous skin growths called warts, and some kinds are associated with neoplasms that form on the cervix.

Benign neoplasms such as ordinary skin moles are not dangerous (Figure 11.10). They grow very slowly, and their cells retain the plasma membrane adhesion proteins that keep them properly anchored to the other cells in their home tissue ❶.

A malignant neoplasm is one that gets progressively worse, and is dangerous to health. The disease called **cancer** occurs when the abnormally dividing cells of a malignant neoplasm disrupt body tissues, both physically and metabolically. Malignant cells typically display the following three characteristics:

First, like cells of all neoplasms, malignant cells grow and divide abnormally. Controls that usually keep cells from getting overcrowded in tissues are lost, so their populations may reach extremely high densities with cell division occurring very rapidly. The number of small blood vessels, or capillaries, that transport blood to the growing cell mass also increases abnormally.

Second, the cytoplasm and plasma membrane of malignant cells are altered. The cytoskeleton may be shrunken, disorganized, or both. Malignant cells typically have an abnormal chromosome number, with some chromosomes present in multiple copies, and others missing or damaged. The balance of metabolism is often shifted, as in an amplified reliance on ATP formation by fermentation rather than by aerobic respiration.

Altered or missing proteins impair the function of the plasma membrane of malignant cells. For example, these cells do not stay anchored properly in tissues because their plasma membrane adhesion proteins are defective

cancer Disease that occurs when a neoplasm physically and metabolically disrupts body tissues.
metastasis The process in which cancer cells spread from one part of the body to another.
oncogene Gene that has the potential to transform a normal cell into a tumor cell.
proto-oncogene Gene that can become an oncogene.
tumor A neoplasm that forms a lump.

A Basal cell carcinoma is the most common type of skin cancer. This slow-growing, raised lump may be uncolored, reddish-brown, or black.

B The second most common form of skin cancer is a squamous cell carcinoma. This pink growth, firm to the touch, grows under the surface of skin.

C Melanoma spreads fastest. Cells form dark, encrusted lumps that may itch or bleed easily.

Figure 11.11 Skin cancer is one type of cancer that can be detected early with periodic screening.

or missing ❷. Malignant cells can slip easily into and out of vessels of the circulatory and lymphatic systems ❸. By migrating through these vessels, the cells establish neoplasms elsewhere in the body ❹. The process in which malignant cells break loose from their home tissue and invade other parts of the body is called **metastasis**. Metastasis is the third hallmark of malignant cells.

Unless chemotherapy, surgery, or another procedure eliminates malignant cells from the body, they can put an individual on a painful road to death. Each year, cancer causes 15 to 20 percent of all human deaths in developed countries alone. The good news is that mutations in multiple checkpoint genes are required to transform a normal cell into a malignant one, and these mutations may take a lifetime to accumulate. Life-style choices such as not smoking and avoiding exposure of unprotected skin to sunlight can reduce one's risk of acquiring mutations in the first place. Some neoplasms can be detected with periodic screening procedures such as Pap tests or dermatology exams (Figure 11.11). Neoplasms that are detected early enough can often be removed before metastasis occurs.

Henrietta's Immortal Cells (revisited)

Cancer is a multistep process. Researchers already know about many of the mutations that contribute to the disease. They are working to identify drugs that target and destroy malignant cells or stop them from dividing. Such research may yield drugs that put the brakes on cancer.

HeLa cells have proven to be indispensable in cancer research. For example, they were used in early tests of taxol, a drug that keeps microtubules from disassembling and so interferes with mitosis. Frequent divisions of cancer cells make them more vulnerable to this poison than normal cells.

The photo on the *right* shows a more recent example of cancer research that relies on HeLa cells. In these telophase cells, the protein identified by the *blue* stain, INCENP, helps sister chromatids stay attached to one another at the centromere. In normal cells, INCENP associates with the enzyme identified by the *green* stain, Aurora B, only at specific times during mitosis. Aurora B helps attach spindle microtubules to centromeres, so defects in this enzyme or its expression result in unequal segregation of chromosomes into descendant cells. Researchers recently correlated overexpression of Aurora B in cancer cells with shortened patient survival rates. Thus, drugs that inhibit Aurora B function are now being tested as potential cancer therapies.

Despite the invaluable cellular legacy of Henrietta Lacks, her body rests in an unmarked grave in an unmarked cemetery. These days, physicians and researchers are required to obtain a signed consent form before they take tissue samples from a patient. No such requirement existed in the 1950s. It was common at that time for doctors to experiment on patients without their knowledge or consent. Thus, the young resident who was treating

Henrietta Lacks's cancerous cervix probably never even thought about asking permission before he took a sample of it. That sample was the one that the Geys used to establish the HeLa cell line. No one in Henrietta's family knew about the cells until 25 years after she died. HeLa cells are still being sold worldwide, but her family has not received any compensation to date.

How Would You Vote? You can legally donate—but not sell—your own organs and tissues. However, companies can profit from research on donated organs or tissues, and also from cell lines derived from these materials. Companies that do so are not obligated to share their profits with the donors. Should profits derived from donated tissues or cells be shared with the donors or their families? See CengageNow for details, then vote online (cengagenow.com).

Summary

Section 11.1 An immortal line of cancer cells is a legacy of cancer victim Henrietta Lacks. HeLa cells have been an invaluable research tool all over the world and even in space. Researchers trying to unravel the mechanisms of cancer continue to work with these cells.

Section 11.2 A **cell cycle** includes all the stages through which a eukaryotic cell passes during its lifetime: **interphase**, **mitosis**, and cytoplasmic division. Most of a cell's activities, including replication of the cell's **homologous chromosomes**, occur in interphase. A cell reproduces by dividing: nucleus first, then the cytoplasm. Each of the two descendant cells receives a complete set of chromosomes and a blob of cytoplasm.

Nuclear division mechanisms partition the duplicated chromosomes into new nuclei. Mitosis maintains the chromosome number. It is the basis of growth, cell replacements, and tissue repair in multicelled species, and **asexual reproduction** in many species.

Section 11.3 During mitosis, duplicated homologous chromosomes line up in the middle of the cell, then are pulled apart. Nuclear envelopes form around the two clusters of chromosomes, forming two new nuclei with the parental chromosome number. Mitosis proceeds in these four stages:

Prophase. Duplicated chromosomes start to condense. Microtubules assemble and form a **spindle**, and the nuclear envelope breaks up. Some microtubules that extend from one spindle pole attach to one chromatid of each chromosome; some that extend from the opposite spindle pole attach to its sister chromatid. Spindle microtubules drag each chromosome toward the center of the cell.

Metaphase. All chromosomes are aligned at the spindle's midpoint.

Anaphase. The sister chromatids of each chromosome detach from each other, and the spindle microtubules start moving them toward opposite spindle poles. Motor proteins drive the movements.

Telophase. A cluster of chromosomes that consists of a complete set of chromosomes reaches each spindle pole. A nuclear envelope forms around each cluster, forming two new nuclei. Both nuclei have the parental chromosome number. Mitosis is over when these nuclei form.

Section 11.4 In most cases, cells divide in two after their nucleus divides. Mechanisms of **cytokinesis** differ. In animal cells, a contractile ring of microfilaments that is part of the cell cortex pulls the plasma membrane inward, forming a **cleavage furrow**. Contraction continues until the cytoplasm is pinched in two. In plant cells, a band of microtubules and microfilaments forms around the nucleus before mitosis. This band marks the site where the **cell plate** forms. The cell plate expands until it fuses with the parent cell wall, thus becoming a cross-wall that partitions the cytoplasm.

Section 11.5 The products of checkpoint genes, including **growth factor** receptors, are part of a host of gene controls that govern the cell cycle. Such controls advance, pause, or stop the cycle in response to conditions inside or outside of the cell. Molecules that work together to monitor the integrity of the cell's DNA can pause the cycle until breaks or other problems are fixed. When checkpoint mechanisms fail, a cell loses control over its cell cycle, and the cell's descendants form a **neoplasm**.

Section 11.6 Mutations can turn **proto-oncogenes** into **oncogenes**. Such mutations typically disrupt checkpoint gene products or their expression, and can result in neoplasms. Neoplasms may form lumps called **tumors**. Mutations in multiple checkpoint genes can transform benign neoplasms into malignant ones. Cells of malignant neoplasms can break loose from their home tissues and colonize other parts of the body, a process called **metastasis**. **Cancer** occurs when malignant neoplasms physically and metabolically disrupt normal body tissues.

Self-Quiz   Answers in Appendix III

1. Mitosis and cytoplasmic division function in _____ .
   a. asexual reproduction of single-celled eukaryotes
   b. growth and tissue repair in multicelled species
   c. gamete formation in bacteria and archaeans
   d. both a and b

2. A duplicated chromosome has _____ chromatid(s).
   a. one   b. two   c. three   d. four

3. Except for a pairing of sex chromosomes, homologous chromosomes _____ .
   a. carry the same genes   c. are the same length
   b. are the same shape   d. all of the above

4. Most cells spend the majority of their lives in _____ .
   a. prophase   d. telophase
   b. metaphase   e. interphase
   c. anaphase   f. a and c

5. The chromosomes align at the midpoint of the spindle during _____ .
   a. prophase   d. telophase
   b. metaphase   e. interphase
   c. anaphase   f. cytokinesis

6. The spindle attaches to chromosomes at the _____ .
   a. centriole   c. centromere
   b. contractile ring d. centrosome

7. Only _____ is not a stage of mitosis.
   a. prophase b. interphase   c. metaphase d. anaphase

8. In intervals of interphase, G stands for _____ .
   a. gap   b. growth   c. Gey   d. gene

Data Analysis Activities

HeLa Cells Are a Genetic Mess HeLa cells continue to be an extremely useful tool in cancer research. One early finding was that HeLa cells can vary in chromosome number. The panel of chromosomes in Figure 11.12, originally published in 1989 by Nicholas Popescu and Joseph DiPaolo, shows all of the chromosomes in a single metaphase HeLa cell.

1. What is the chromosome number of this HeLa cell?
2. How many extra chromosomes does this cell have, compared to a normal human body cell?
3. Can you tell that this cell came from a female? How?

Figure 11.12 Chromosomes in a HeLa cell.

9. In the diagram of the nucleus *below*, fill in the blanks with the name of each interval.

_____   _____   _____   _____   _____

10. Interphase is the part of the cell cycle when _____ .
    a. a cell ceases to function
    b. the spindle forms
    c. a cell grows and duplicates its DNA
    d. mitosis proceeds

11. After mitosis, the chromosome number of a descendant cell is _____ the parent cell's.
    a. the same as   c. rearranged compared to
    b. one-half of   d. doubled compared to

12. Name any checkpoint gene.

13. Which of the following encompasses the other two?
    a. cancer   b. neoplasm   c. tumor

14. Match each term with its best description.
    ____cell plate   a. lump of cells
    ____spindle   b. made of microfilaments
    ____tumor   c. divides plant cells
    ____cleavage furrow   d. organizes the spindle
    ____contractile ring   e. metastatic cells
    ____cancer   f. made of microtubules
    ____centrosomes   g. indentation

15. Match each stage with the events listed.
    ____metaphase   a. sister chromatids move apart
    ____prophase   b. chromosomes start to condense
    ____telophase   c. new nuclei form
    ____anaphase   d. all duplicated chromosomes are
                                aligned at the spindle equator

Additional questions are available on **CENGAGENOW**.

Critical Thinking

1. When a cell reproduces by mitosis and cytoplasmic division, does its life end?

 2. The eukaryotic cell in the photo on the *left* is in the process of cytoplasmic division. Is this cell from a plant or an animal? How do you know?

3. Exposure to radioisotopes or other sources of radiation can damage DNA. Humans exposed to high levels of radiation face a condition called radiation poisoning. Why do you think that hair loss and damage to the lining of the gut are early symptoms of radiation poisoning? Speculate about why exposure to radiation is used as a therapy to treat some kinds of cancers.

4. Suppose you have a way to measure the amount of DNA in one cell during the cell cycle. You first measure the amount at the G1 phase. At what points in the rest of the cycle will you see a change in the amount of DNA per cell?

Animations and Interactions on **CENGAGENOW**:
❯ The cell cycle; Mitosis; Cytoplasmic division; Neoplasms; The cell cycle and cancer.

❮ Links to Earlier Concepts

Be sure you have a clear picture of the structural organization of chromosomes (Section 8.2) and how genes work (9.2) before you begin this chapter on the cellular basis of sexual reproduction. You will draw on your understanding of DNA replication (8.6), cytoplasmic division (11.4), and cell cycle controls (11.5) as we compare meiosis with mitosis (11.2). You also will be revisiting microtubules (4.10), genetically identical organisms (8.7), and the effects of mutation (9.6).

Key Concepts

Sexual Versus Asexual Reproduction

In asexual reproduction, one parent transmits its genes to offspring. In sexual reproduction, offspring inherit genes from two parents who usually differ in some number of alleles. Differences in alleles are the basis of differences in traits.

Stages of Meiosis

Meiosis is a nuclear division process that occurs only in cells set aside for sexual reproduction. Meiosis reduces the chromosome number by sorting a reproductive cell's chromosomes into four new nuclei.

12 Meiosis and Sexual Reproduction

12.1 Why Sex?

If the function of reproduction is the perpetuation of one's genes, then an asexual reproducer would seem to win the evolutionary race. In asexual reproduction, all of an individual's genes are passed to all of its offspring. **Sexual reproduction** mixes up the genes of two parents (Figure 12.1), so only about half of each parent's genetic information is passed to offspring.

So why sex? Variation in the forms and combinations of heritable traits is typical of sexually reproducing populations. Some traits allow their bearers to thrive in particular environments. All offspring of asexual reproducers are clones of their parent, so all are adapted the same way to an environment—and all are equally vulnerable to changes in it. By contrast, the offspring of sexual reproducers have unique combinations of many traits. As a group, their diversity offers them flexibility: a better chance of surviving environmental change than clones. Some of them may have a particular combination of traits that suits them perfectly to the changed environment.

Other organisms are part of the environment, and they, too, can change. Think of predator and prey, say, foxes and rabbits. If one rabbit is better than others at outrunning the foxes, it has a better chance of escaping, surviving, and passing to its offspring the genes that help it evade foxes. Thus, over many generations, rabbits may get faster. If one fox is better than others at outrunning faster rabbits, it has a better chance of eating, surviving, and passing to its offspring genes that help it catch faster rabbits. Thus, over many generations, the foxes may tend to get faster. As one species changes, so does the other—an idea called the Red Queen hypothesis, after Lewis Carroll's book *Through the Looking Glass*. In the book, the Queen of Hearts tells Alice, "It takes all the running you can do, to keep in the same place."

An adaptive trait tends to spread more quickly through a sexually reproducing population than through an asexually repro-

ducing one. Why? In asexual reproduction, new combinations of traits can arise only by mutation. An adaptive trait is passed from one generation to the next along with the same set of other traits, adaptive or not. By contrast, sexual reproduction mixes up the genes of individuals that often have different forms of traits. It generates new combinations of traits in far fewer generations than does mutation alone.

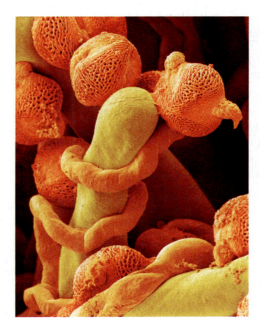

Figure 12.1
Moments in the stages of sexual reproduction of humans (*opposite*) and plants (*left*). Sexual reproduction mixes up the genetic material of two organisms.

In flowering plants, pollen grains (*orange*) germinate on flower carpels (*yellow*). Pollen tubes with male gametes inside grow from the grains down into tissues of the ovary, which house the flower's female gametes.

However, just because sexual reproducers are more genetically diverse does not mean that they win the evolutionary race. In terms of numbers of individuals and how long their lineages have endured, the most successful organisms on Earth—by a long shot—are bacteria, which reproduce by an entirely different mechanism.

---

sexual reproduction Reproductive mode by which offspring arise from two parents and inherit genes from both.

Recombinations and Shufflings
During meiosis, homologous chromosomes come together and swap segments. Then they are randomly sorted into separate nuclei. Both processes lead to novel combinations of alleles among offspring.

Sexual Reproduction in the Context of Life Cycles
Gametes form by different mechanisms in males and females, but meiosis is part of both processes. In most plants, spore formation and other events intervene between meiosis and gamete formation.

Mitosis and Meiosis Compared
Similarities between mitosis and meiosis suggest that meiosis may have originated by evolutionary remodeling of mechanisms that already existed for mitosis and, before that, for repairing damaged DNA.

> Asexual reproduction produces clones.
> Sexual reproduction mixes up alleles from two parents.
> Meiosis, the basis of sexual reproduction, is a nuclear division mechanism that occurs in immature reproductive cells of eukaryotes.
< Links to Clones 8.1, Chromosomes in eukaryotes 8.2, DNA replication 8.6, Genetically identical organisms 8.7, Genes 9.2, Effects of mutations 9.6, Homologous chromosomes and asexual reproduction and mitosis 11.2

Introducing Alleles

An individual's genes collectively contain the information necessary to make a new individual (Section 8.7). All off-spring of an asexual reproducer inherit the same number and kinds of genes; thus, mutations aside, all are clones of the parent (Section 8.1).

Inheritance gets more complicated with sexual repro-duction because two parents contribute genes to offspring. The **somatic** (body) cells of multicelled organisms that reproduce sexually contain pairs of chromosomes. Typi-cally, one chromosome of each pair is maternal and the other is paternal (Figure 12.2). Except for a pairing of nonidentical sex chromosomes, the two chromosomes of every pair carry the same set of genes.

If the DNA sequence of every gene pair were identi-cal, then sexual reproduction would produce clones, just like asexual reproduction does. Just imagine: The entire human population might consist of clones, in which case everybody would look exactly alike.

But the two genes of a pair are often not identical. Why not? Mutations that inevitably accumulate in DNA change its sequence. Thus, the two genes of any pair might differ a bit. If the sequences differ enough, those genes will encode slightly different forms of the gene's product (Section 9.6). Different forms of the same gene are called **alleles**.

Figure 12.2 Homologous chromosomes. Typically, one chromo-some of each pair is inherited from the mother; the other, from the father. Colored patches in this fluorescence micrograph indi-cate corresponding DNA sequences on the chromosomes.

These chromosomes carry the same series of genes, but the DNA sequence of any one of those genes might differ just a bit from that of its partner on the other chromosome. Different forms of a gene are called alleles.

Alleles influence thousands of traits. For example, the beta globin gene you encountered in Section 9.6 has more than 700 alleles: one that causes sickle-cell anemia, one that causes beta thalassemia, and so on. The beta globin gene is only one of about 30,000 human genes, and most genes have multiple alleles. Alleles are one reason that the individuals of a sexually reproducing species do not all look exactly the same. The offspring of sexual reproducers inherit new combinations of alleles, which is the basis of new combinations of traits.

What Meiosis Does

Sexual reproduction involves the fusion of reproductive cells from two parents. It requires **meiosis**, a nuclear divi-

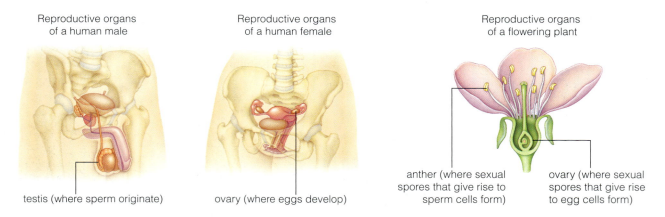

Reproductive organs of a human male

Reproductive organs of a human female

Reproductive organs of a flowering plant

testis (where sperm originate)

ovary (where eggs develop)

anther (where sexual spores that give rise to sperm cells form)

ovary (where sexual spores that give rise to egg cells form)

Figure 12.3 Animated Examples of reproductive organs. Meiosis of germ cells in reproductive organs gives rise to gametes: eggs and sperm in humans, egg cells and sperm cells in flowering plants.

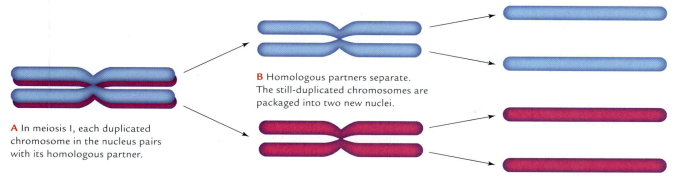

A In meiosis I, each duplicated chromosome in the nucleus pairs with its homologous partner.

B Homologous partners separate. The still-duplicated chromosomes are packaged into two new nuclei.

C Sister chromatids separate in meiosis II. The now unduplicated chromosomes are packaged into four new nuclei.

Figure 12.4 How meiosis halves the chromosome number.

sion mechanism that halves the chromosome number. The process of sexual reproduction begins with meiosis in **germ cells**, which are immature reproductive cells. Meiosis in germ cells produces mature reproductive cells called **gametes**. A sperm is an example of a male gamete. An egg is a female gamete. Gametes usually form inside special reproductive structures or organs (Figure 12.3).

Gametes have a single set of chromosomes, so they are **haploid** (*n*): Their chromosome number is half of the diploid (2*n*) number (Section 8.2). Human body cells are diploid, with 23 pairs of homologous chromosomes. Meiosis of a human germ cell (2*n*) normally produces gametes with 23 chromosomes: one of each pair (*n*). The diploid chromosome number is restored at **fertilization**, when two haploid gametes (one egg and one sperm, for example) fuse to form a **zygote**, the first cell of a new individual.

The first part of meiosis is similar to mitosis. A cell duplicates its DNA before either nuclear division process starts. As in mitosis, the microtubules of a spindle move the duplicated chromosomes to opposite spindle poles. However, meiosis sorts the chromosomes into new nuclei not once, but twice, so it results in the formation of four haploid nuclei. The two consecutive nuclear divisions are called meiosis I and meiosis II:

Interphase	Meiosis I	Meiosis II
DNA is replicated prior to meiosis I	Prophase I Metaphase I Anaphase I Telophase I	Prophase II Metaphase II Anaphase II Telophase II

In some cells, no resting period occurs between these two stages. In others, interphase with no DNA replication separates meiosis I and II.

During meoisis I, every duplicated chromosome aligns with its homologous partner (Figure 12.4A). Then the homologous chromosomes are pulled away from one another (Figure 12.4B). After homologous chromosomes separate, each ends up in one of two new nuclei. At this stage, the chromosomes are still duplicated (the sister chromatids are still attached to one another).

During meiosis II, the sister chromatids of each chromosome are pulled apart, so each becomes an individual, unduplicated chromosome (Figure 12.4C). The chromosomes are sorted into four new nuclei. With one unduplicated version of each chromosome, the new nuclei are all haploid (*n*).

Thus, meiosis partitions the chromosomes of one diploid nucleus (2*n*) into four haploid (*n*) nuclei. The next section zooms in on the details of this process.

alleles Forms of a gene that encode slightly different versions of the gene's product.
fertilization Fusion of two gametes to form a zygote.
gamete Mature, haploid reproductive cell; e.g., an egg or a sperm.
germ cell Diploid reproductive cell that gives rise to haploid gametes by meiosis.
haploid Having one of each type of chromosome characteristic of the species.
meiosis Nuclear division process that halves the chromosome number. Basis of sexual reproduction.
somatic Relating to the body.
zygote Cell formed by fusion of two gametes; the first cell of a new individual.

Take-Home Message **Why do populations that reproduce sexually tend to have the most variation in heritable traits?**

> Paired genes on homologous chromosomes may vary in sequence as alleles.

> Alleles are the basis of traits. Sexual reproduction mixes up alleles from two parents.

> The nuclear division process of meiosis is the basis of sexual reproduction in eukaryotes. It precedes the formation of gametes or spores.

> Meiosis halves the diploid (2*n*) chromosome number, to the haploid number (*n*). When two gametes fuse at fertilization, the chromosome number is restored. The resulting zygote has two sets of chromosomes, one from each parent.

> Meiosis halves the chromosome number.
> During meiosis, chromosomes of a diploid nucleus become distributed into four haploid nuclei.
< Links to Diploid chromosome number 8.2, DNA replication 8.6, Mitosis 11.2

DNA replication occurs prior to meiosis, so a cell's chromosomes are duplicated by the time meiosis I begins: Each chromosome consists of two sister chromatids. The nucleus is diploid (2n): It contains two sets of chromosomes, one from each parent. Let's now turn to the cellular events that occur during meiosis itself.

Figure 12.5 Animated Meiosis. Two pairs of chromosomes are illustrated in a diploid (2n) animal cell. Homologous chromosomes are indicated in *blue* and *pink*. Micrographs show meiosis in a lily plant cell (*Lilium regale*).

>> Figure It Out During which phase of meiosis does the chromosome number become reduced?

Answer: Anaphase I

Meiosis I The first stage of meiosis I is prophase I (Figure 12.5). During this phase, the chromosomes condense, and homologous chromosomes align tightly and swap segments (more about segment-swapping in the next section). The centrosome gets duplicated along with its two centrioles. One centriole pair moves to the opposite side of the cell as the nuclear envelope breaks up. Spindle microtubules begin to extend from the centrosomes **1**.

By the end of prophase I, microtubules of the spindle connect the chromosomes to the spindle poles. Each chromosome is now attached to one spindle pole, and its homologous partner is attached to the other. The microtubules lengthen and shorten, pushing and pulling the chromosomes as they do. At metaphase I, all of the microtubules are the same length, and the chromosomes are aligned midway between the poles of the spindle **2**.

In anaphase I, the spindle microtubules separate the homologous chromosomes and pull them toward opposite

Meiosis I   One diploid nucleus to two haploid nuclei

1 Prophase I. Homologous chromosomes condense, pair up, and swap segments. Spindle microtubules attach to them as the nuclear envelope breaks up.

2 Metaphase I. The homologous chromosome pairs are aligned midway between spindle poles.

3 Anaphase I. The homologous chromosomes separate and begin heading toward the spindle poles.

4 Telophase I. Two clusters of chromosomes reach the spindle poles. A new nuclear envelope forms around each cluster, so two haploid (n) nuclei form.

plasma membrane   spindle

nuclear envelope breaking up   centrosome   one pair of homologous chromosomes spindle poles **❸**. During telophase I, the chromosomes reach the spindle poles **❹**. New nuclear envelopes form around the two clusters of chromosomes as the DNA loosens up. Each of the two haploid (n) nuclei that form contains one set of (duplicated) chromosomes. The cytoplasm may divide at this point to form two haploid cells. Interphase occurs in some cells at the end of meiosis I, but the DNA is not replicated before meiosis II begins.

Meiosis II During prophase II, the chromosomes condense as a new spindle forms. One centriole moves to the opposite side of each new nucleus, and the nuclear envelopes break up. By the end of prophase II, microtubules connect the chromosomes to the spindle poles. Each chromatid is now attached to one spindle pole, and its sister is attached to the other **❺**. The microtubules lengthen and shorten, pushing and pulling the chromosomes as they do. At metaphase II, all of the microtubules are the same

length, and the chromosomes are aligned midway between the spindle poles **❻**.

In anaphase II, the spindle microtubules pull the sister chromatids apart **❼**. Each chromosome now consists of one molecule of DNA. During telophase II, the chromosomes (now unduplicated) reach the spindle poles **❽**. New nuclear envelopes form around the four clusters of chromosomes as the DNA loosens up. Each of the four haploid (n) nuclei that form contains one set of unduplicated chromosomes. The cytoplasm may divide, so four haploid cells form.

Take-Home Message **What happens to a cell during meiosis?**

> During meiosis, the nucleus of a diploid ($2n$) cell divides twice. Four haploid (n) nuclei form, each with a full set of chromosomes—one of each type.

Meiosis II Two haploid nuclei to four haploid nuclei

❺ Prophase II. The chromosomes condense. Spindle microtubules attach to each sister chromatid as the nuclear envelope breaks up.

❻ Metaphase II. The (still duplicated) chromosomes are aligned midway between poles of the spindle.

❼ Anaphase II. All sister chromatids separate. The now unduplicated chromosomes head to the spindle poles.

❽ Telophase II. A cluster of chromosomes reaches each spindle pole. A new nuclear envelope encloses each cluster, so four haploid (n) nuclei form.

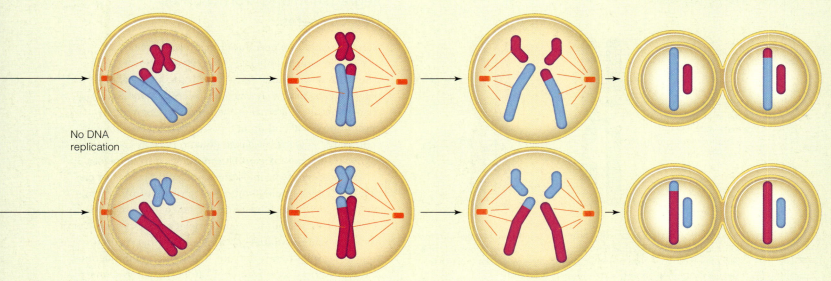

No DNA replication

> Crossovers and the random sorting of chromosomes into gametes result in new combinations of traits among offspring of sexual reproducers.

‹ Link to Chromosome structure 8.2

The previous section mentioned briefly that duplicated chromosomes swap segments with their homologous partners during prophase I. It also showed how each chromosome aligns with and then separates from its homologous partner during anaphase I. Both events introduce novel combinations of alleles into gametes. Along with fertilization, these events contribute to the variation in

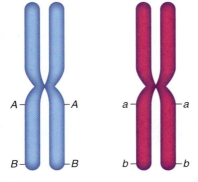

A Here, we focus on only two of the many genes on a chromosome. In this example, one gene has alleles *A* and *a*; the other has alleles *B* and *b*.

B Close contact between homologous chromosomes promotes crossing over between nonsister chromatids. Paternal and maternal chromatids exchange corresponding pieces.

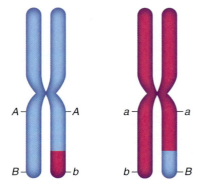

C Crossing over mixes up paternal and maternal alleles on homologous chromosomes.

Figure 12.6 **Animated** Crossing over. *Blue* signifies a paternal chromosome, and *pink*, its maternal homologue. For clarity, we show only one pair of homologous chromosomes and one crossover, but more than one crossover may occur in each chromosome pair.

combinations of traits among the offspring of sexually reproducing species.

Crossing Over in Prophase I

Early in prophase I of meiosis, all chromosomes in a germ cell condense. When they do, each is drawn close to its homologue. The chromatids of one homologous chromosome become tightly aligned with the chromatids of the other along their length:

This tight, parallel orientation favors **crossing over**, a process in which a chromosome and its homologous partner exchange corresponding pieces of DNA (Figure 12.6). Homologous chromosomes may swap any segment or segments of DNA along their length, although crossovers tend to occur more frequently in certain regions.

Swapping segments of DNA shuffles alleles between homologous chromosomes. It breaks up the particular combinations of alleles that occurred on the parental chromosomes, and makes new ones on the chromosomes that end up in gametes. Thus, crossing over introduces novel combinations of traits among offspring. It is a normal and frequent process in meiosis, but the rate of crossing over varies among species and among chromosomes. In humans, between 46 and 95 crossovers occur per meiosis, so on average each chromosome crosses over at least once.

Segregation of Chromosomes Into Gametes

Normally, all of the new nuclei that form in meiosis I receive the same number of chromosomes. However, whether a new nucleus ends up with the maternal or paternal version of a chromosome is entirely random. The chance that the maternal or the paternal version of any chromosome will end up in a particular nucleus is 50 percent. Why? The answer has to do with the way the spindle segregates the homologous chromosomes during meiosis I.

The process of chromosome segregation begins in prophase I. Imagine one of your own germ cells undergoing meiosis. Crossovers have already made genetic mosaics of its chromosomes, but for simplicity let's put crossing over aside for a moment. Just call the twenty-three chromosomes you inherited from your mother the maternal ones, and the twenty-three from your father the paternal ones.

1 The four possible alignments of three pairs of chromosomes in a nucleus at metaphase I.

 2 Resulting combinations of maternal and paternal chromosomes in the two nuclei that form at telophase I.

3 Resulting combinations of maternal and paternal chromosomes in the four nuclei that form at telophase II. Eight different combinations are possible.

Figure 12.7 Animated Hypothetical segregation of three pairs of chromosomes in meiosis I. Maternal chromosomes are *pink*; paternal, *blue*. Which chromosome of each pair gets packaged into which of the two new nuclei that form at telophase I is random. For simplicity, no crossing over occurs in this example, so all sister chromatids are identical.

During prophase I, microtubules fasten your cell's chromosomes to the spindle poles. Chances are fairly slim that all of the maternal chromosomes get attached to one pole and all of the paternal chromosomes get attached to the other. Microtubules extending from a spindle pole bind to the centromere of the first chromosome they contact, regardless of whether it is maternal or paternal. Each homologous partner gets attached to the opposite spindle pole. Thus, there is no pattern to the attachment of the maternal or paternal chromosomes to a particular pole.

Now imagine that your germ cell has just three pairs of chromosomes (Figure 12.7). By metaphase I, those three pairs of maternal and paternal chromosomes are divvied up between the two spindle poles in one of four ways **1**. In anaphase I, homologous chromosomes separate and are pulled toward opposite spindle poles. In telophase I, a new nucleus forms around the chromosomes that cluster at each spindle pole. Each nucleus contains one of eight possible combinations of maternal and paternal chromosomes **2**.

In telophase II, each of the two nuclei divides and gives rise to two new haploid nuclei. The two new nuclei are identical because no crossing over occurred in our hypothetical example, so all of the sister chromatids were

identical. Thus, at the end of meiosis in this cell, two (2) spindle poles have divvied up three (3) chromosome pairs. The resulting four nuclei have one of eight (2^3) possible combinations of maternal and paternal chromosomes **3**.

Cells that give rise to human gametes have twenty-three pairs of homologous chromosomes, not three. Each time a human germ cell undergoes meiosis, the four gametes that form end up with one of 8,388,608 (or 2^{23}) possible combinations of homologous chromosomes. That number does not even take into account crossing over, which mixes up the alleles on maternal and paternal chromosomes, or fusion with another gamete at fertilization. Are you getting an idea of why such fascinating combinations of traits show up among the generations of your own family tree?

crossing over Process in which homologous chromosomes exchange corresponding segments during prophase I of meiosis.

Take-Home Message How does meiosis introduce variation in combinations of traits?

❯ Crossing over is recombination between nonsister chromatids of homologous chromosomes during prophase I. It makes new combinations of parental alleles.

❯ Homologous chromosomes can be attached to either spindle pole in prophase I, so each homologue can be packaged into either one of the two new nuclei. Thus, the random assortment of homologous chromosomes increases the number of potential combinations of maternal and paternal alleles in gametes.

> Details of gamete formation differ among plants and animals, but meiosis is always part of the process.
< Links to Flagella 4.10, Cytoplasmic division 11.4

Gametes are the specialized cells that are the basis of sexual reproduction. All are haploid, but they differ in other details. For example, human male gametes—sperm—have one flagellum (Section 4.10). Opossum sperm have two, and roundworm sperm have none. A flowering plant's male gamete consists simply of a nucleus. We leave details of reproduction for later chapters, but you will need to know a few concepts before you get there.

Gamete Formation in Plants Two kinds of multicelled bodies form during the life cycle of a plant: sporophytes and gametophytes. A **sporophyte** is typically diploid, and spores form by meiosis in its specialized parts. Spores consist of one or a few haploid cells. These cells undergo mitosis and give rise to **gametophytes**, multicelled haploid bodies inside which one or more gametes form. A sequoia tree is an example of a sporophyte (Figure 12.8A). Male and female gametophytes develop inside different types of cones that form on each tree. In flowering plants, gametophytes form in flowers, which are specialized reproductive shoots of the sporophyte body.

Gamete Formation in Animals In the life cycle of a typical animal, a zygote matures as a multicelled body that produces gametes (Figure 12.8B). Animal gametes arise by meiosis of diploid germ cells. In male animals (Figure 12.9), the germ cell develops into a primary spermatocyte **1**. Meiosis I in a primary spermatocyte results in two haploid secondary spermatocytes **2**, which undergo meiosis II and become four spermatids **3**. Each spermatid then matures as a **sperm 4**.

In female animals (Figure 12.10), a germ cell becomes a primary oocyte, which is an immature egg **5**. This cell undergoes meiosis and division, as occurs with a primary spermatocyte. Two haploid cells form when the primary oocyte divides after meiosis I. However, the cytoplasm of a primary oocyte divides unequally, so the cells differ in size and function. One of the cells is called a first polar body. The other cell, the secondary oocyte, is much larger because it gets nearly all of the parent cell's cytoplasm **6**. This larger cell undergoes meiosis II and cytoplasmic division, which again is unequal **7**. One of the two cells that forms is a second polar body. The other cell gets most of the cytoplasm and matures into a female gamete, which is called an ovum (plural, ova), or **egg**.

Polar bodies are not nutrient-rich or plump with cytoplasm, and generally do not function as gametes. In time they degenerate. Their formation simply ensures that the egg will have a haploid chromosome number, and also will get enough metabolic machinery to support early divisions of the new individual.

Fertilization Meiosis of a diploid germ cell produces haploid gametes. When two gametes fuse at fertilization, the resulting zygote is diploid. Thus, meiosis halves the chromosome number, and fertilization restores it. If meiosis did not precede fertilization, the chromosome number would double with every generation. An individual's set

A Plant life cycle

mitosis

zygote (2*n*)

multicelled sporophyte (2*n*)

meiosis

diploid

haploid

fertilization

gametes (*n*)

spores (*n*)

multicelled gametophyte (*n*)

Figure 12.8 Comparing the life cycles of animals and plants.

A Generalized life cycle for most plants. A sequoia tree is a sporophyte.

B Generalized life cycle for animals. The zygote is the first cell to form when the nuclei of two gametes, such as a sperm and an egg, fuse at fertilization.

B Animal life cycle

mitosis

zygote (2*n*)

multicelled body (2*n*)

meiosis

diploid

haploid

fertilization

gametes (*n*)

---

egg Mature female gamete, or ovum.
gametophyte A haploid, multicelled body in which gametes form during the life cycle of plants.
sperm Mature male gamete.
sporophyte Diploid, spore-producing stage of a plant life cycle.

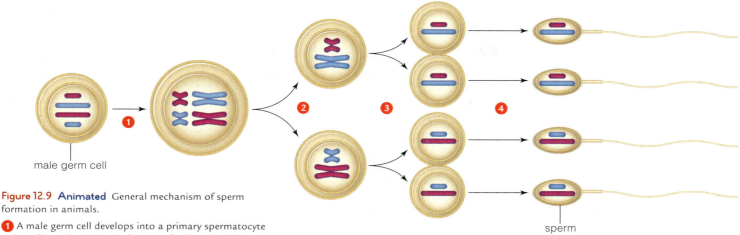

Figure 12.9 Animated General mechanism of sperm formation in animals.

1 A male germ cell develops into a primary spermatocyte as it replicates its DNA. Both types of cell are diploid.

2 Meiosis I in the primary spermatocyte results in two secondary spermatocytes, which are haploid.

3 Four haploid spermatids form when the secondary spermatocytes undergo meiosis II.

4 Spermatids mature as sperm (haploid male gametes).

male germ cell

sperm

female germ cell

egg

Figure 12.10 Animated General mechanism of egg formation in animals. *Left*, an illustration of human sperm surrounding an egg during fertilization.

5 A female germ cell (an oogonium) develops into a primary oocyte as it replicates its DNA. Both types of cell are diploid.

6 Meiosis I in the primary oocyte results in a secondary oocyte and a first polar body. Unequal cytoplasmic division makes the polar body much smaller than the oocyte. Both cells are haploid. Polar bodies typically degenerate.

7 Meiosis II followed by unequal cytoplasmic division in the secondary oocyte results in a polar body and an ovum, or egg. Both cells are haploid.

of chromosomes is like a fine-tuned blueprint that must be followed exactly, page by page, in order to build a body that functions normally. As you will see in the next chapter, chromosome number changes can have drastic consequences, particularly in animals.

Take-Home Message **How does meiosis fit into the life cycle of plants and animals?**

❯ Meiosis and cytoplasmic division precede the development of haploid gametes in animals and spores in plants.

❯ The union of two haploid gametes at fertilization results in a diploid zygote.

> Though they have different results, mitosis and meiosis are fundamentally similar processes.
‹ Links to Mitosis 11.3, Cell cycle controls 11.5

This chapter opened with hypotheses about the evolutionary advantages of asexual and sexual reproduction. It seems like a giant evolutionary step from producing clones to producing genetically varied offspring, but was it really?

By mitosis and cytoplasmic division, one cell becomes two new cells. This process is the basis of growth and tissue repair in all multicelled species. Single-celled eukaryotes (and some multicelled ones) also reproduce asexually by way of mitosis and cytoplasmic division. Mitotic (asexual) reproduction results in clones, which are genetically identical copies of a parent.

Meiosis also begins with a diploid cell, one that is specialized for reproduction. Meiosis in this cell produces haploid gametes. Gametes of two parents fuse to form a zygote, which is a diploid cell of mixed parentage. Meiotic (sexual) reproduction results in offspring that are genetically different from the parent, and from one another.

Though their end results differ, there are striking parallels between the four stages of mitosis and meiosis II (Figure 12.11). As one example, a spindle forms and separates chromosomes during both processes. There are many more similarities at the molecular level.

Long ago, the molecular machinery of mitosis may have been remodeled into meiosis. Evidence for this hypothesis includes a host of molecules, including the products of the *BRCA* genes (Chapters 10 and 11), that are made by all modern eukaryotes. These molecules monitor and repair breaks in DNA, for example during DNA replication prior to mitosis. They actively maintain the integrity of a cell's chromosomes. It turns out that the very same set of molecules monitor and fix the breaks in homologous chromosomes during crossing over in prophase I of meiosis. Some of them function as checkpoint proteins in both mitosis and meiosis, so mutations that affect them or the rate at which they are made can affect the outcomes of both nuclear division processes.

In anaphase of mitosis, sister chromatids are pulled apart. What would happen if the connections between the sisters did not break? Each duplicated chromosome would be pulled to one or the other spindle pole—which is exactly what happens in anaphase I of meiosis.

Sexual reproduction probably originated by mutations that affected processes of mitosis. As you will see in later chapters, the remodeling of existing processes into new ones is a common evolutionary theme.

Take-Home Message Are the processes of mitosis and meiosis related?

> Meiosis may have evolved by the remodeling of existing mechanisms of mitosis.

Meiosis I One diploid nucleus to two haploid nuclei

Prophase I
• Chromosomes condense.
• Homologous chromosomes pair.
• Crossovers occur (not shown).
• Spindle forms and attaches chromosomes to spindle poles.
• Nuclear envelope breaks up.

Metaphase I
• Chromosomes align midway between spindle poles.

Anaphase I
• Homologous chromosomes separate and move toward opposite spindle poles.

Telophase I
• Chromosome clusters arrive at spindle poles.
• New nuclear envelopes form.
• Chromosomes decondense.

Figure 12.11 Comparative summary of key features of mitosis and meiosis, starting with a diploid cell.

Only two paternal and two maternal chromosomes are shown. Both were duplicated in interphase, prior to nuclear division. A spindle of microtubules moves the chromosomes in mitosis as well as meiosis.

Mitosis maintains the parental chromosome number. Meiosis halves it, to the haploid number.

Mitotic cell division is the basis of asexual reproduction among eukaryotes. It also is the basis of growth and tissue repair of multicelled eukaryotic species.

Why Sex? (revisited)

There are a few all-female species of fishes, reptiles, and birds in nature, but not mammals. In 2004, researchers fused two mouse eggs in a test tube and made an embryo using no DNA from a male. The embryo developed into Kaguya, the world's first fatherless mammal (*right*). The mouse grew up healthy, engaged in sex with a male mouse, and gave birth to offspring. The researchers wanted to find out if sperm was required for normal development.

How Would You Vote? Researchers made a "fatherless" mouse from two mouse eggs. Should they be prevented from trying the process with human eggs? See CengageNow for details, then vote online (cengagenow.com).

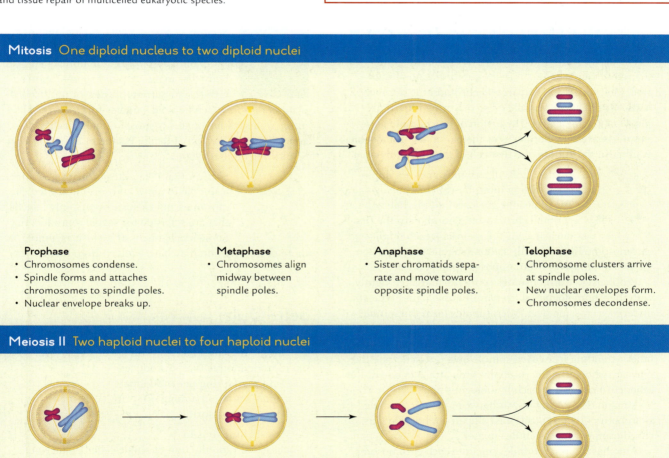

Mitosis One diploid nucleus to two diploid nuclei

Prophase
- Chromosomes condense.
- Spindle forms and attaches chromosomes to spindle poles.
- Nuclear envelope breaks up.

Metaphase
- Chromosomes align midway between spindle poles.

Anaphase
- Sister chromatids separate and move toward opposite spindle poles.

Telophase
- Chromosome clusters arrive at spindle poles.
- New nuclear envelopes form.
- Chromosomes decondense.

Meiosis II Two haploid nuclei to four haploid nuclei

Prophase II
- Chromosomes condense.
- Spindle forms and attaches chromosomes to spindle poles.
- Nuclear envelope breaks up.

Metaphase II
- Chromosomes align midway between spindle poles.

Anaphase II
- Sister chromatids separate and move toward opposite spindle poles.

Telophase II
- Chromosome clusters arrive at spindle poles.
- New nuclear envelopes form.
- Chromosomes decondense.

Summary

Section 12.1 Sexual reproduction mixes up the genetic information of two parents. Under some conditions, the variation in traits among offspring of sexual reproducers provides them with an evolutionary advantage over genetically identical offspring.

Section 12.2 Offspring of asexual reproduction are genetically identical to their one parent: They are clones. The offspring of sexual reproduction differ from parents, and often from one another, in the details of shared traits. Offspring of most sexual reproducers inherit pairs of chromosomes, one of each pair from the mother and the other from the father. Paired genes on homologous chromosomes may vary in DNA sequence, in which case they are called **alleles**.

Meiosis, a nuclear division mechanism that occurs in eukaryotic **germ cells**, precedes the formation of **gametes**. Meiosis halves the parental chromosome number. It also shuffles parental alleles, so offspring inherit new combinations of alleles. The fusion of two **haploid** gametes during **fertilization** restores the parental chromosome number in the **zygote**, the first cell of the new individual.

Section 12.3 All chromosomes are duplicated during interphase, before meiosis. Two divisions, meiosis I and II, halve the parental chromosome number. In the first nuclear division, meiosis I, each duplicated chromosome lines up with its homologous partner; then the two are moved toward opposite spindle poles.

Prophase I. Chromosomes condense and align tightly with their homologues. Each pair of homologues typically undergoes crossing over. Microtubules form the spindle. One of two pairs of centrioles is moved to the other side of the nucleus. The nuclear envelope breaks up, so microtubules extending from each spindle pole can penetrate the nuclear region. The microtubules then attach to one or the other chromosome of each homologous pair.

Metaphase I. A tug-of-war between the microtubules from both poles has positioned all pairs of homologous chromosomes at the spindle equator.

Anaphase I. Microtubules separate each chromosome from its homologue and move one to each spindle pole. Anaphase I ends as a cluster of duplicated chromosomes nears each spindle pole.

Telophase I. Two nuclei form; typically the cytoplasm divides. All of the chromosomes are still duplicated; each still consists of two sister chromatids.

The second nuclear division, meiosis II, occurs in both nuclei that formed in meiosis I. The chromosomes condense in prophase II, and align in metaphase II. Sister chromatids of each chromosome are pulled apart from each other in anaphase II, so each becomes an individual chromosome. By the end of telophase II, four haploid nuclei have formed, each with one set of chromosomes.

Section 12.4 Events in prophase I and metaphase I produce nonparental combinations of alleles. The nonsister chromatids of homologous chromosomes undergo **crossing over** during prophase I: They exchange segments at the same place along their length, so each ends up with new combinations of alleles that were not present in either parental chromosome.

Crossing over during prophase I, and random segregation of maternal and paternal chromosomes into new nuclei, contribute to variation in traits among offspring. Microtubules can attach the maternal or the paternal chromosome of each pair to one or the other spindle pole. Either chromosome may end up in any new nucleus, and in any gamete. Such chromosome shufflings, along with crossovers during prophase I of meiosis, are the basis of variation in traits we see in sexually reproducing species.

Section 12.5 Multicelled diploid bodies are typical in life cycles of plants and animals. A diploid **sporophyte** is a multicelled plant body that makes haploid spores. Spores give rise to **gametophytes**, or multicelled plant bodies in which haploid gametes form. Germ cells in the reproductive organs of most animals give rise to **sperm** or **eggs**. Fusion of haploid gametes at fertilization results in a diploid zygote.

Section 12.6 Like mitosis, meiosis requires a spindle to move and sort duplicated chromosomes, but meiosis occurs only in cells that are set aside for sexual reproduction. Some mechanisms of meiosis resemble those of mitosis, and may have evolved from them.

Self-Quiz   Answers in Appendix III

1. The main evolutionary advantage of sexual over asexual reproduction is that it produces _____ .
   a. more offspring per individual
   b. more variation among offspring
   c. healthier offspring

2. Meiosis functions in _____ .
   a. asexual reproduction of single-celled eukaryotes
   b. growth and tissue repair in multicelled species
   c. sexual reproduction
   d. both a and b

3. Sexual reproduction in animals requires _____ .
   a. meiosis   c. spore formation
   b. fertilization   d. a and b

4. Meiosis _____ the parental chromosome number.
   a. doubles   c. maintains
   b. halves   d. mixes up

5. Crossing over mixes up _____ .
   a. chromosomes   c. zygotes
   b. alleles   d. gametes

Data Analysis Activities

BPA and Abnormal Meiosis In 1998, researchers at Case Western University were studying meiosis in mouse oocytes when they saw an unexpected and dramatic increase of abnormal meiosis events (**Figure 12.12**). Improper segregation of chromosomes during meiosis is one of the main causes of human genetic disorders, which we will discuss in Chapter 14.

The researchers discovered that the spike in meiotic abnormalities began immediately after the mouse facility started washing the animals' plastic cages and water bottles in a new, alkaline detergent. The detergent had damaged the plastic, which began to leach bisphenol A (BPA). BPA is a synthetic chemical that mimics estrogen, the main female sex hormone in animals. BPA is used to manufacture polycarbonate plastic items (including baby bottles and water bottles) and epoxies (including the coating on the inside of metal cans of food).

1. What percentage of mouse oocytes displayed abnormalities of meiosis with no exposure to damaged caging?
2. Which group of mice showed the most meiotic abnormalities in their oocytes?
3. What is abnormal about metaphase I as it is occurring in the oocytes shown in **Figure 12.12B, C,** and **D**?

Caging materials	Total number of oocytes	Abnormalities
Control: New cages with glass bottles	271	5 (1.8%)
Damaged cages with glass bottles		
Mild damage	401	35 (8.7%)
Severe damage	149	30 (20.1%)
Damaged bottles	197	53 (26.9%)
Damaged cages with damaged bottles	58	24 (41.4%)

Figure 12.12 Meiotic abnormalities associated with exposure to damaged plastic caging. Fluorescent micrographs show nuclei of single mouse oocytes in metaphase I. **A** Normal metaphase; **B–D** examples of abnormal metaphase. Chromosomes are *red*; spindle fibers are *green*.

6. Crossing over happens during which phase of meiosis?

7. The stage of meiosis that makes descendant cells haploid is _____ .
   - a. prophase I
   - b. prophase II
   - c. anaphase I
   - d. anaphase II
   - e. metaphase I
   - f. metaphase II

8. Dogs have a diploid chromosome number of 78. How many chromosomes do their gametes have?
   - a. 39
   - b. 78
   - c. 156
   - d. 234

9. _____ contributes to variation in traits among the offspring of sexual reproducers.
   - a. Crossing over
   - b. Random attachment of chromosomes to spindle poles
   - c. Fertilization
   - d. both a and b
   - e. all are factors

10. The cell in the diagram to the *right* is in anaphase I, not anaphase II. I know this because _____ .

11. Which of the following is one of the very important differences between mitosis and meiosis?
   - a. Chromosomes align midway between spindle poles only in meiosis.
   - b. Homologous chromosomes pair up only in meiosis.
   - c. DNA is replicated only in mitosis.
   - d. Sister chromatids separate only in meiosis.
   - e. Interphase occurs only in mitosis.

12. Match each term with its description.
   - ___ interphase
   - ___ metaphase I
   - ___ allele
   - ___ sporophyte
   - ___ gamete
   - a. different molecular form of a gene
   - b. may be none between meiosis I and meiosis II
   - c. all chromosomes are aligned at spindle equator
   - d. haploid
   - e. does not occur in animals

Additional questions are available on **CENGAGENOW**.

Critical Thinking

1. Explain why you can predict that meiosis tends to give rise to greater genetic diversity among offspring in fewer generations than asexual reproduction does.

2. Make a simple sketch of meiosis in a cell with a diploid chromosome number of 4. Now try it when the chromosome number is 3.

3. The diploid chromosome number for the body cells of a frog is 26. What would that number be after three generations if meiosis did not occur before gamete formation?

Animations and Interactions on **CENGAGENOW**:
❯ Reproductive organs; Meiosis step-by-step; Crossing over; Random segregation of chromosomes in meiosis; Egg and sperm formation.

Lindsay, 22

Savannah, 19

Cody, 23

Brandon, 18

Ben, 23

Jeff, 21

< Links to Earlier Concepts

Before starting this chapter, you may want to review what you know about traits (Section 1.5), chromosomes (8.2), DNA (8.3), mutation (8.6), genes (9.2), and alleles (12.2). You will revisit probability and sampling error (1.8), laws of nature (1.9), protein structure (3.5), pigments (6.2), and clones (8.7). As you read, refer back to the the stages of meiosis (12.3). You will consider the roles that crossing over and the segregation of chromosomes (12.4) into gametes (12.5) play in inheritance.

Key Concepts

Where Modern Genetics Started

Gregor Mendel gathered evidence of the genetic basis of inheritance. His meticulous work gave him clues that heritable traits are specified in units. The units, which are distributed into gametes in predictable patterns, were later identified as genes.

Insights From Monohybrid Crosses

During meiosis, pairs of genes on homologous chromosomes separate and end up in different gametes. Inheritance patterns of alleles associated with different forms of a trait can be used as evidence of such gene segregation.

13 Observing Patterns in Inherited Traits

13.1 Menacing Mucus

In 1988, researchers discovered a gene that, when mutated, causes cystic fibrosis (CF). Cystic fibrosis is the most common fatal genetic disorder in the United States. The gene in question, *CFTR*, encodes a protein that moves chloride ions out of epithelial cells. Sheets of these cells line the passageways and ducts of the lungs, liver, pancreas, intestines, reproductive system, and skin. When the CFTR protein pumps chloride ions out of these cells, water follows the ions by osmosis. This two-step process maintains a thin film of water on the surface of the epithelial sheets. Mucus slides easily over the wet sheets of cells.

The most common mutation in CF is a deletion of three base pairs—one codon that specifies the 508th amino acid of the CFTR protein, a phenylalanine. This deletion, which is called Δ*F508*, disrupts membrane trafficking of CFTR so that newly assembled polypeptides are stranded in the endoplasmic reticulum. The protein itself can function properly, but it never reaches the cell surface to do its job.

One outcome is that the transport of chloride ions out of epithelial cells is disrupted. If not enough chloride ions leave the cells, not enough water leaves them either, so the surfaces of epithelial cell sheets are not as wet as they should be. Mucus that normally slips and slides through the body's tubes sticks to them instead. Thick globs of mucus accumulate and clog passageways and ducts throughout the body. Breathing becomes difficult as the mucus obstructs the smaller airways of the lungs.

The CFTR protein also functions as a receptor that alerts the body to the presence of bacteria. Bacteria bind to CFTR. The binding triggers endocytosis, which speeds the immune system's defensive responses. Without the CFTR protein on the surface of epithelial cell linings, disease-causing bacteria that enter the ducts and passageways of the body can persist there. Thus, chronic bacterial infections of the intestine and lungs are hallmarks of cystic fibrosis.

Daily routines of posture changes and thumps on the chest and back help clear the lungs of some of the thick mucus, and

Figure 13.1 Cystic fibrosis. *Opposite*, a few of the many young victims of cystic fibrosis, which occurs most often in people of northern European ancestry. At least one young person dies every day in the United States from complications of this disease.

Above, model of the CFTR protein. The parts shown here are ATP-driven motors that widen or narrow a channel (*gray* arrow) across the plasma membrane. The tiny part of the protein that is deleted in most people with cystic fibrosis is shown on the ribbon in *green*.

antibiotics help control infections, but there is no cure. Even with a lung transplant, most cystic fibrosis patients live no longer than thirty years, at which time their tormented lungs usually fail (Figure 13.1).

More than 10 million people carry the Δ*F508* mutation in one of their two copies of the *CFTR* gene, but most of them do not realize it because they have no symptoms. Cystic fibrosis only occurs when a person inherits two mutated genes, one from each parent. This unlucky event occurs in about 1 of 3,300 births worldwide. Why is it that people with one copy of a mutated *CFTR* gene are healthy, but people with two copies are ill with cystic fibrosis? Why is the mutation that causes cystic fibrosis so common if its effects are so devastating? You will begin to find the answers to such questions in this chapter, in which we introduce principles of inheritance—how new individuals are put together in the image of their parents.

Insights From Dihybrid Crosses Pairs of genes on different chromosomes are typically distributed into gametes independently of how other gene pairs are distributed. Breeding experiments with alternative forms of two unrelated traits can be used as evidence of such independent assortment.

Variations on Mendel's Theme Not all traits appear in Mendelian inheritance patterns. An allele may be partly dominant over a nonidentical partner, or codominant with it. Multiple genes may influence a trait; some genes influence many traits. The environment also influences gene expression.

❯ Some traits are inherited in predictable patterns. Such patterns offer information about the alleles that influence those traits.

❮ Links to Traits 1.5, Chromosomes 8.2, Discovery of DNA's function 8.3, Mutations 8.6, Genes and gene expression 9.2, Mutations 9.6, Alleles 12.2, Gamete formation and fertilization 12.5

By the nineteenth century, most people had an idea that two parents contribute hereditary material to their offspring, but no one knew what that material was. Some thought that hereditary material must be some type of fluid, with fluids from both parents blending at fertilization like milk into coffee. However, the idea of "blending inheritance" failed to explain what people could see with their own eyes. Children sometimes had traits such as freckles that did not appear in either parent. A cross between a black horse and a white one did not produce only gray offspring.

The naturalist Charles Darwin did not accept the idea of blending inheritance, but he could not come up with an alternative even though inheritance was central to his theory of natural selection. Darwin saw that forms of traits often vary among individuals in a population. He realized that if some forms of traits help individuals survive and reproduce, then those forms would tend to appear more frequently in a population over generations (we return to Darwin and his theory of natural selection in Chapter 16). Despite these insights, however, neither Darwin nor anyone else at the time knew that hereditary information (DNA) is divided into discrete units (genes), an insight that is critical to understanding how heredity really works.

Even before Darwin presented his theory of natural selection, someone had been gathering evidence that would support it. Gregor Mendel, an Austrian monk (*left*), had been carefully breeding thousands of pea plants. By meticulously documenting how certain traits are passed from plant to plant, generation after generation, Mendel had been collecting evidence of how inheritance works.

Mendel's Experimental Approach

Mendel studied variation in the traits of the garden pea, *Pisum sativum*. This plant is naturally self-fertilizing, which means that its flowers produce male gametes (in pollen) and female gametes (in carpels) that form viable embryos when they meet up.

In order to study inheritance, Mendel had to breed particular individuals together, then observe and document the traits of their offspring. Control over the reproduction of an individual pea plant begins with preventing it from self-fertilizing. Mendel did this by removing a flower's pollen-bearing anthers, then brushing its carpel with pollen from another plant. When the plant set seed, Mendel collected the seeds, planted them, and recorded the traits of the new plants that grew. Figure 13.2 shows an example of this process.

Many of Mendel's experiments involved plants that "bred true" for a particular trait. Breeding true for a trait means that, mutations aside, all offspring have the same form of the trait as the parent(s), generation after generation. For example, all offspring of pea plants that breed true for white flowers also have white flowers.

Breeders cross-fertilize plants when they transfer pollen among individuals that have different traits. As you will see in the next section, Mendel discovered that the traits of the offspring of such cross-fertilized pea plants often appear in predictable patterns. Mendel's meticulous work tracking pea plant traits led him to conclude (correctly) that hereditary information is passed from one generation to the next in discrete units.

carpel   anther

A Garden pea flower, cut in half. Male gametes form in pollen grains produced by the anthers, and female gametes form in carpels. Experimenters can control the transfer of hereditary material from one flower to another by snipping off a flower's anthers (to prevent the flower from self-fertilizing), and then brushing pollen from another flower onto its carpel.

B In this example, pollen from a plant with purple flowers is brushed onto the carpel of a white-flowered plant.

C Later, seeds develop inside pods of the cross-fertilized plant. An embryo in each seed develops into a mature pea plant.

D Every plant that arises from this cross has purple flowers. Predictable patterns such as this offer evidence of how inheritance works.

Figure 13.2 Animated Breeding garden pea plants (*Pisum sativum*), which can self-fertilize or cross-fertilize.

Figure 13.3 Loci of a few human genes. Genetic diseases that result from mutations in the genes are shown in parenthesis. The number or letter below each chromosome is its name; the characteristic banding patterns appear after staining. A map of all 23 human chromosomes is in Appendix VII.

Inheritance in Modern Terms

DNA was not proven to be hereditary material until the 1950s (Section 8.3), but Mendel discovered its units, which we now call genes, almost a century before then. Today, we know that individuals of a species share certain traits because their chromosomes carry the same genes. Offspring tend to look like their parents because they inherited their parents' genes.

The DNA sequence of each gene occurs at a specific location, or **locus** (plural, loci), on a particular chromosome (Figure 13.3). The somatic cells of humans and other animals are diploid, so they have pairs of genes, on pairs of homologous chromosomes (Figure 13.4). In most cases, both genes of a pair are expressed (Section 9.2).

Genes occur in pairs on homologous chromosomes.

The members of each pair of genes may be identical, or they may differ slightly, as alleles.

Figure 13.4 Animated Genes on chromosomes. Any pair of genes on homologous chromosomes may vary as alleles. Different alleles may result in different versions of a trait.

The two genes of a pair may be identical, or they may be slightly different. Alternative forms of a gene are called alleles (Section 12.2). Organisms that breed true for a specific trait probably have identical alleles governing that trait. An individual with identical alleles of a gene is said to be **homozygous** for the allele. The particular set of alleles that an individual carries is called **genotype**.

Alleles are the major source of variation in a trait. New alleles arise by mutation (Section 8.6). A mutation may cause a trait to change, as when a gene that causes flowers to be purple mutates so the resulting flowers are white. Flower color is an example of **phenotype**, which refers to an individual's observable traits. Any mutated gene is an allele, whether or not it affects phenotype.

The offspring of a cross, or mating, between individuals that breed true for different forms of a trait are **hybrids**. Hybrids carry different alleles of a gene, so they are said to be **heterozygous** for the alleles (*hetero–*, mixed). In many cases, the effect of one allele influences the effect of the other, and the outcome of this interaction is visible in the hybrid phenotype. An allele is **dominant** when its effect masks that of a **recessive** allele paired with it. Usually italic capital letters such as *A* signify dominant alleles, and lowercase italic letters such as *a* signify recessive ones. Thus, a homozygous dominant individual carries a pair of dominant alleles (*AA*). A homozygous recessive individual carries a pair of recessive alleles (*aa*). A heterozygous, or hybrid, individual carries a pair of nonidentical alleles (*Aa*).

---

dominant Refers to an allele that masks the effect of a recessive allele paired with it.
genotype The particular set of alleles carried by an individual.
heterozygous Having two different alleles of a gene.
homozygous Having identical alleles of a gene.
hybrid The offspring of a cross between two individuals that breed true for different forms of a trait; a heterozygous individual.
locus Location of a gene on a chromosome.
phenotype An individual's observable traits.
recessive Refers to an allele with an effect that is masked by a dominant allele on the homologous chromosome.

> Pairs of genes on homologous chromosomes separate during meiosis, so they end up in different gametes.
‹ Links to Probability and sampling error 1.8, Laws of nature 1.9, Meiosis 12.3, Chromosome segregation 12.4

Mendel crossed plants that bred true for purple flowers with plants that bred true for white flowers. All of the offspring of these crosses had purple flowers, but Mendel did not know why this pattern occurred. We now understand that one gene governs purple flower color in pea plants. The allele that specifies purple (let's call it *P*) is dominant over the allele that specifies white (*p*). Thus, a pea plant with two *P* alleles (*PP*) has purple flowers, and one with two *p* alleles (*pp*) has white flowers.

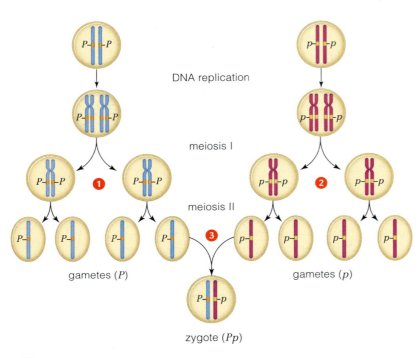

DNA replication

meiosis I

meiosis II

gametes (*P*)   gametes (*p*)

zygote (*Pp*)

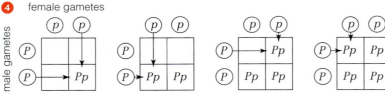

female gametes

male gametes

Figure 13.5 Gene segregation. Homologous chromosomes separate during meiosis, so the pairs of genes they carry separate too. Each of the resulting gametes carries only one of the two members of each gene pair. For clarity, we show only one set of homologous chromosomes.

❶ All gametes made by a parent homozygous for a dominant allele carry that allele.

❷ All gametes made by a parent homozygous for a recessive allele carry that allele.

❸ If these two parents are crossed, the union of any of their gametes at fertilization produces a zygote with both alleles. All offspring of this cross will be heterozygous.

❹ This outcome is easy to see with a Punnett square. Parental gametes are listed in circles on the top and left sides of a grid. Each square is filled in with the combination of alleles that would result if the gametes in the corresponding row and column met up.

When homologous chromosomes separate during meiosis, the gene pairs on those chromosomes separate too. Each gamete that forms carries only one of the two genes of a pair (Figure 13.5). Thus, plants homozygous for the dominant allele (*PP*) can only make gametes that carry the dominant allele *P* ❶. Plants homozygous for the recessive allele (*pp*) can only make gametes that carry the recessive allele *p* ❷. If these homozygous plants are crossed (*PP* × *pp*), only one outcome is possible: A gamete carrying a *P* allele meets up with a gamete carrying a *p* allele ❸. All of the offspring of this cross have one of each allele, so their genotype is *Pp*. A grid called a **Punnett square** makes it easier to predict the genetic outcomes of crosses ❹. Because all of the offspring of this cross carry the dominant allele *P*, all have purple flowers.

This pattern is so predictable that it can be used as evidence of a dominance relationship between alleles. Breeding experiments use such patterns to reveal genotype. In a **testcross**, an individual that has a dominant trait (but an unknown genotype) is crossed with an individual known to be homozygous recessive. The pattern of traits among the offspring of the cross can reveal whether the tested individual is heterozygous or homozygous.

For example, we may do a testcross between a purple-flowered pea plant (which could have a genotype of either *PP* or *Pp*) and a white-flowered pea plant (*pp*). If all of the offspring of this cross had purple flowers, we would know that the genotype of the purple-flowered parent was *PP*.

A **monohybrid cross** is a breeding experiment that checks for a dominance relationship between the alleles of a single gene. Individuals that are identically heterozygous for one gene—(*Pp*) for example—are bred together or self-fertilized. The frequency at which the two traits appear among the offspring of this cross may show that one of the alleles is dominant over the other.

To produce identically heterozygous individuals for a monohybrid cross, we would start with two individuals that breed true for two different forms of a trait (Figure 13.6A). In pea plants, purple or white flowers is one example of a trait with two distinct forms, but there are many others. Mendel investigated seven of them: stem length (tall and short), seed color (yellow and green), pod texture

law of segregation The two members of each pair of genes on homologous chromosomes end up in different gametes during meiosis.
monohybrid cross Breeding experiment in which individuals identically heterozygous for one gene are crossed. The frequency of traits among the offspring offers information about the dominance relationship between the alleles.
Punnett square Diagram used to predict the genetic and phenotypic outcome of a cross.
testcross Method of determining genotype in which an individual of unknown genotype is crossed with one that is known to be homozygous recessive.

parent plant homozygous for purple flowers

parent plant homozygous for white flowers

PP × pp

Pp hybrid

P p

two types of gametes

	P	p
P	PP	Pp
p	Pp	pp

B A cross between the F_1 offspring is a mono-hybrid cross. The phenotype ratio in F_2 offspring in this example is 3:1 (3 purple to 1 white).

A All of the F_1 offspring of a cross between two plants that breed true for different forms of a trait are identically heterozygous. These offspring make two types of gametes: P and p.

Figure 13.6 Animated Example of a monohybrid cross. ›› **Figure It Out** In this example, how many possible genotypes are there in the F_2 generation?   Answer: Three: PP, Pp, and pp

Table 13.1 Mendel's Seven Pea Plant Traits

Trait	Dominant Form	Recessive Form
Seed Shape	Round	Wrinkled
Seed Color	Yellow	Green
Pod Texture	Smooth	Wrinkled
Pod Color	Green	Yellow
Flower Color	Purple	White
Flower Position	Along Stem	At Tip
Stem Length	Tall	Short

(smooth and wrinkled), and so on (Table 13.1). A cross between the two true-breeding individuals yields hybrid offspring: ones that are identically heterozygous for the alleles that govern the trait. When these F_1 (first generation) hybrids are crossed, the frequency at which the two traits appear in the F_2 (second generation) offspring offers information about dominance relationships. F is an abbreviation for filial, which means offspring.

A cross between two purple-flowered heterozygous individuals (Pp) offers an example. Each of these plants can make two types of gametes: ones that carry a P allele, and ones that carry a p allele. So, in a monohybrid cross between two Pp plants ($Pp \times Pp$), the two types of gametes can meet up in four possible ways at fertilization:

Possible Event	Probable Outcome
Sperm P meets egg P ⟶	zygote genotype is PP
Sperm P meets egg p ⟶	zygote genotype is Pp
Sperm p meets egg P ⟶	zygote genotype is Pp
Sperm p meets egg p ⟶	zygote genotype is pp

Three out of four possible outcomes of this cross include at least one copy of the dominant allele P. Each time fertilization occurs, there are 3 chances in 4 that the resulting offspring will inherit a P allele, and have purple flowers. There is 1 chance in 4 that it will inherit two recessive p alleles, and have white flowers. Thus, the prob-

ability that a particular offspring of this cross will have purple or white flowers is 3 purple to 1 white, which we represent as a ratio of 3:1 (Figure 13.6B).

If the probability of one individual inheriting a particular genotype is difficult to imagine, think about probability in terms of the phenotypes of many offspring. In this example, there will be roughly three purple-flowered plants for every white-flowered one. The 3:1 pattern is an indication that purple and white flower color are specified by alleles with a clear dominant–recessive relationship: purple is dominant, and white is recessive.

The phenotype ratios in the F_2 offspring of Mendel's monohybrid crosses were all close to 3:1. These results became the basis of his **law of segregation**, which we state here in modern terms: Diploid cells carry pairs of genes, on pairs of homologous chromosomes. The two genes of each pair are separated from each other during meiosis, so they end up in different gametes.

Take-Home Message What is Mendel's law of segregation?

› Diploid cells carry pairs of genes, on pairs of homologous chromosomes. The two genes of each pair are separated from each other during meiosis, so they end up in different gametes.

› Mendel discovered patterns of inheritance in pea plants by tracking the results of many monohybrid crosses.

❯ Many gene pairs tend to sort into gametes independently of one another.

❮ Links to Meiosis 12.3, Crossing over and chromosome segregation 12.4

A monohybrid cross allows us to track alleles of one gene pair. What about alleles of two gene pairs? How two gene pairs get sorted into gametes depends partly on whether the two genes are on the same chromosome. When homologous chromosomes separate during meiosis, either one of the pair can end up in a particular nucleus. Thus, gene pairs on one chromosome get sorted into gametes independently of gene pairs on other chromosomes (Figure 13.7).

Punnett squares are particularly useful when predicting inheritance patterns of two or more genes simultaneously, such as with a dihybrid cross. A **dihybrid cross** tests for dominance relationships between alleles of two genes. In a typical dihybrid cross, individuals identically heterozygous for alleles of two genes (dihybrids) are crossed, and the traits of the offspring are observed.

To make a dihybrid cross, we would start with individuals that breed true for two different traits. Let's use genes for flower color (*P*, purple; *p*, white) and height (*T*, tall; *t*, short), and assume that these genes occur on separate chromosomes. Figure 13.8 shows a dihybrid cross starting with one parent plant that breeds true for purple flowers and tall stems (*PPTT*), and another that breeds true for white flowers and short stems (*pptt*). Each homo-

zygous plant makes only one type of gamete ❶. So, all of the offspring from a cross between these parent plants (*PPTT* × *pptt*), will be dihybrids (*PpTt*) and have purple flowers and tall stems ❷.

Four combinations of alleles are possible in the gametes of *PpTt* dihybrids ❸. If two *PpTt* plants are crossed (a dihybrid cross, *PpTt* × *PpTt*), the four types of gametes can combine in sixteen possible ways at fertilization ❹. Those sixteen genotypes would result in four different phenotypes. Nine would be tall with purple flowers, three would be short with purple flowers, three would be tall with white flowers, and one would be short with white flowers. Thus, the ratio of these phenotypes is 9:3:3:1.

Mendel discovered the 9:3:3:1 ratio of phenotypes among the offspring of his dihybrid crosses, although he had no idea what it meant. He could only say that "units" specifying one trait (such as flower color) are inherited independently of "units" specifying other traits (such as plant height). In time, Mendel's hypothesis became known as the **law of independent assortment**, which we state here in modern terms: Gene pairs are sorted into gametes independently of other gene pairs.

Mendel published his results in 1866, but apparently his work was read by few and understood by no one. In 1871 he became abbot of his monastery, and his pioneering experiments ended. He died in 1884, never to know that they would be the starting point for modern genetics.

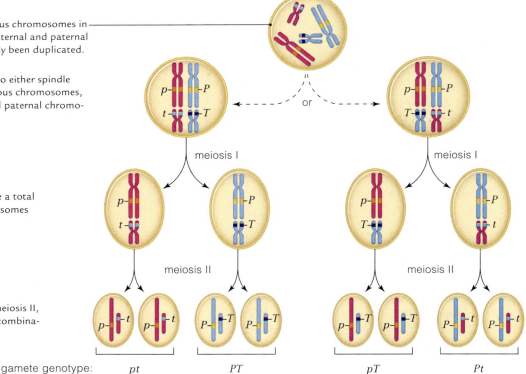

A This example shows just two pairs of homologous chromosomes in the nucleus of a diploid (2*n*) reproductive cell. Maternal and paternal chromosomes, shown in *pink* and *blue*, have already been duplicated.

B Either chromosome of a pair may get attached to either spindle pole during meiosis I. With two pairs of homologous chromosomes, there are two different ways that the maternal and paternal chromosomes can get attached to opposite spindle poles.

C Two nuclei form with each scenario, so there are a total of four possible combinations of parental chromosomes in the nuclei that form after meiosis I.

D Thus, when sister chromatids separate during meiosis II, the gametes that result have one of four possible combinations of maternal and paternal chromosomes.

meiosis I

meiosis II

gamete genotype: *pt*   *PT*   *pT*   *Pt*

Figure 13.7 Independent assortment.

parent plant
homozygous
for purple flowers
and long stems
PPTT

parent plant
homozygous
for white flowers
and short stems
pptt

1 PT × pt

2 PpTt
dihybrid

3 PT Pt pT pt
four types of gametes

4

	PT	Pt	pT	pt
PT	PPTT	PPTt	PpTT	PpTt
Pt	PPTt	PPtt	PpTt	Pptt
pT	PpTT	PpTt	ppTT	ppTt
pt	PpTt	Pptt	ppTt	pptt

Figure 13.8 Animated A dihybrid cross between plants that differ in flower color and plant height. *P* and *p* stand for dominant and recessive alleles for flower color. *T* and *t* stand for dominant and recessive alleles for height.

1 Meiosis in a homozygous individual results in one type of gamete.

2 A cross between two homozygous individuals yields offspring with one possible genotype. All offspring that form in this example are dihybrids (heterozygous for two genes) with purple flowers and tall stems.

3 Meiosis in dihybrid individuals results in four kinds of gametes.

4 If two dihybrid individuals are crossed (a dihybrid cross), the four types of gametes can meet up in 16 possible ways. Out of 16 possible genotypes of the offspring, 9 will result in plants that are purple-flowered and tall; 3, purple-flowered and short; 3, white-flowered and tall; and 1, white-flowered and short. Thus, the ratio of phenotypes in a dihybrid cross is 9:3:3:1.

≫ Figure It Out What do the flowers inside the boxes represent?

Answer: Phenotypes of the F_2 offspring

The Contribution of Crossovers

It makes sense that gene pairs on different chromosomes would assort independently into gametes, but what about gene pairs on the same chromosome? Mendel studied seven genes in pea plants, which have seven chromosomes. Was he lucky enough to choose one gene on each of those chromosomes? As it turns out, some of the genes Mendel studied *are* on the same chromosome. The genes are far enough apart that crossing over occurs between them very frequently—so frequently that they tend to assort into gametes independently, just as if they were on different chromosomes.

By contrast, genes that are very close together on a chromosome do not assort independently, because crossing over does not happen very often between them. Such genes are said to be linked. Alleles of some linked genes stay together during meiosis more frequently than others, an effect due to the relative distance between the genes. Genes that are closer together get separated less frequently by crossovers. Thus, the closer together any two genes are on a chromosome, the more likely gametes

will be to receive parental combinations of alleles of those genes. Genes are said to be tightly linked if the distance between them is relatively small.

All of the genes on a chromosome are called a **linkage group**. Peas have 7 different chromosomes, so they have 7 linkage groups. Humans have 23 different chromosomes, so they have 23 linkage groups.

dihybrid cross Breeding experiment in which individuals identically heterozygous for two genes are crossed. The frequency of traits among the offspring offers information about the dominance relationships between the paired alleles.
law of independent assortment During meiosis, members of a pair of genes on homologous chromosomes get distributed into gametes independently of other gene pairs.
linkage group All genes on a chromosome.

Take-Home Message **What is Mendel's law of independent assortment?**

❯ Each member of a pair of genes on homologous chromosomes tends to be distributed into gametes independently of how other genes are distributed during meiosis.

Beyond Simple Dominance

> Mendel focused on traits that are based on clearly domi-
nant and recessive alleles. However, the expression patterns
of genes for some traits are not as straightforward.
< Links to Fibrous proteins 3.5, Pigments 6.2

The inheritance patterns in the last two sections offer
examples of simple dominance, in which a dominant
allele fully masks the expression of a recessive one. In
many other cases, both genes of a pair are expressed at
the same time. The "one gene equals one trait" equation
does not always apply, either. Several gene products may
influence the same trait, or the expression of a single gene
may influence multiple traits.

Codominance With **codominance**, two nonidentical
alleles of a gene are both fully expressed in heterozygotes,
so neither is dominant or recessive. Codominance may
occur in **multiple allele systems**, in which three or more
alleles of a gene persist among individuals of a population. The three alleles of the *ABO* gene are an example.
An enzyme encoded by the *ABO* gene modifies a carbohydrate on the surface of human red blood cell membranes.
The *A* and *B* alleles encode slightly different versions of

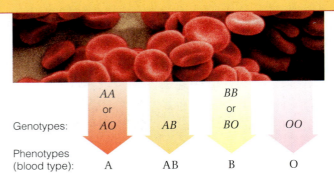

Genotypes:	*AA* or *AO*	*AB*	*BB* or *BO*	*OO*
Phenotypes (blood type):	A	AB	B	O

Figure 13.9 Animated Combinations of alleles that are the basis
of human blood type.

the enzyme, which in turn modify the carbohydrate differently. The *O* allele has a mutation that prevents its
enzyme product from becoming active at all.

The two alleles you carry for the *ABO* gene determine
the form of the carbohydrate on your blood cells, and that
carbohydrate is the basis of your blood type (Figure 13.9).
The *A* and the *B* allele are codominant when paired. If
your genotype is *AB*, then you have both versions of the
enzyme, and your blood is type AB. The *O* allele is recessive when paired with either the *A* or *B* allele. If your genotype is *AA* or *AO*, your blood is type A. If your genotype
is *BB* or *BO*, it is type B. If you are *OO*, it is type O.

Receiving incompatible blood cells in a transfusion
is very dangerous, because the immune system usually
attacks red blood cells bearing molecules that do not occur
in one's own body. The attack can cause the blood cells to
clump or burst, a transfusion reaction with potentially fatal
consequences. People with type O blood can donate blood
to anyone else, so they are called universal donors. However, they can receive transfusions of type O blood only.
People with type AB blood can receive a transfusion of any
blood type, so they are called universal recipients.

homozygous (*RR*) × homozygous (*rr*) ⟶ heterozygous (*Rr*)

A Cross a red-flowered with a white-flowered plant,
and all of the offspring will be pink heterozygotes.

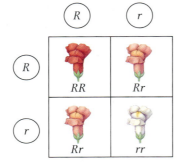

B If two of the pink heterozygotes are crossed, the phenotypes of the resulting offspring
will occur in a 1:2:1 ratio.

	R	*r*
R	*RR*	*Rr*
r	*Rr*	*rr*

Figure 13.10 Animated Incomplete dominance in heterozygous (*pink*) snap-
dragons. An allele that affects red pigment is paired with a "white" allele.

>> **Figure It Out** Is the experiment in **B** a monohybrid or dihybrid cross?

Answer: A monohybrid cross

Incomplete Dominance With **incomplete dominance**,
one allele of a gene pair is not fully dominant over the
other, so the heterozygous phenotype is between the two
homozygous phenotypes. A gene that influences flower
color in snapdragon plants is an example. A cell has one
copy of this gene on each homologous chromosome,
and both copies are expressed. One allele (*R*) encodes an
enzyme that makes a red pigment. The enzyme encoded
by a mutated allele (*r*) cannot make any pigment. Plants
homozygous for the *R* allele (*RR*) make a lot of red pigment, so they have red flowers. Plants homozygous for
the *r* allele (*rr*) do not make any pigment at all, so their
flowers are white. Heterozygous plants (*Rr*) make only
enough red pigment to color their flowers pink (Figure
13.10A). A cross between two pink-flowered heterozygous
plants yields red-, pink-, and white-flowered offspring in a
1:2:1 ratio (Figure 13.10B).

	EB	Eb	eB	eb
EB	EEBB	EEBb	EeBB	EeBb
Eb	EEBb	EEbb	EeBb	Eebb
eB	EeBB	EeBb	eeBB	eeBb
eb	EeBb	Eebb	eeBb	eebb

Figure 13.11 Animated Epistasis in dogs. Epistatic interactions among products of two gene pairs affect coat color in Laborador retrievers. *Left*, all dogs with an *E* and *B* allele have black fur. Those with an *E* and two recessive *b* alleles have brown fur. All dogs homozygous for the recessive *e* allele have yellow fur. *Right*: black, chocolate, and yellow Laborador retrievers.

Epistasis Some traits are affected by multiple gene products, an effect called polygenic inheritance or **epistasis**. Human skin color, which is a result of interactions among several gene products, is an example. Similar genes affect Labrador retriever coat color, which can be black, yellow, or brown (Figure 13.11). One gene is involved in the synthesis of the pigment melanin. A dominant allele of the gene (*B*) specifies black fur, and its recessive partner (*b*) specifies brown fur. A dominant allele of a different gene (*E*) causes melanin to be deposited in fur, and its recessive partner (*e*) reduces melanin deposition. Thus, a dog that carries an *E* and a *B* allele has black fur. One that carries an *E* allele and is homozygous for the *b* allele has brown fur. A dog homozygous for the *e* allele has yellow fur regardless of its *B* or *b* alleles.

Figure 13.12 Animated Marfan syndrome. Rising basketball star Haris Charalambous died suddenly in 2006 when his aorta burst during warm-up exercises. He was 21.

Charalambous was very tall and lanky, with long arms and legs—traits that are valued in professional athletes such as basketball players. These traits are also associated with Marfan syndrome.

About 1 in 5,000 people are affected by Marfan syndrome worldwide. Like many of them, Charalambous did not realize he had the syndrome.

codominant Refers to two alleles that are both fully expressed in heterozygous individuals.
epistasis Effect in which a trait is influenced by the products of multiple genes.
incomplete dominance Condition in which one allele is not fully dominant over another, so the heterozygous phenotype is between the two homozygous phenotypes.
multiple allele system Gene for which three or more alleles persist in a population.
pleiotropic Refers to a gene whose product influences multiple traits.

Pleiotropy A **pleiotropic** gene is one that influences multiple traits. Mutations in such genes are associated with complex genetic disorders such as sickle-cell anemia (Section 9.6) and cystic fibrosis. For example, thickened mucus in cystic fibrosis patients affects the entire body, not just the respiratory tract. The mucus clogs ducts that lead to the gut, which results in digestive problems. Male CF patients are typically infertile because their sperm flow is hampered by the thickened secretions.

Marfan syndrome is another example of a genetic disorder caused by mutation in a pleiotropic gene. In this case, the gene encodes fibrillin. Long fibers of this protein impart elasticity to tissues of the heart, skin, blood vessels, tendons, and other body parts. Mutations in the fibrillin gene result in tissues that form with defective fibrillin or none at all. The largest blood vessel leading from the heart, the aorta, is particularly affected. Muscle cells in the aorta's thick wall do not work very well, and the wall itself is not as elastic as it should be. The aorta expands under pressure, so the lack of elasticity eventually makes it thin and leaky. Calcium deposits accumulate inside. Inflamed, thinned, and weakened, the aorta can rupture abruptly during exercise.

Marfan syndrome is particularly difficult to diagnose. Affected people are often tall, thin, and loose-jointed, but there are plenty of tall, thin, loose-jointed people that do not have the syndrome. Symptoms may not be apparent, so many people die suddenly and early without ever knowing they had the disorder (Figure 13.12).

Take-Home Message Are all alleles clearly dominant or recessive?

❯ An allele may be fully dominant, incompletely dominant, or codominant with its partner on a homologous chromosome.

❯ In epistasis, two or more gene products influence a trait.

❯ The product of a pleiotropic gene influences two or more traits.

> Individuals of most species vary in some of their shared traits. Many traits show a continuous range of variation.
‹ Link to Genetically identical organisms 8.7

Most organic molecules are made in metabolic pathways that involve many enzymes. Genes encoding those enzymes can mutate in any number of ways, so their products may function within a spectrum of activity that ranges from excessive to not at all. Thus, the end product of a metabolic pathway can be produced within a range of concentration and activity. Environmental factors often add further variations on top of that. In the end, phenotype often results from complex interactions among gene products and the environment.

Continuous Variation The individuals of a species typically vary in many of their shared traits. Some of those traits appear in two or three distinct forms. Others occur in a range of small differences that is called **continuous variation**. The more genes and environmental factors that influence a trait, the more continuous is its variation.

How do we determine whether a trait varies continuously? First, we divide the total range of phenotypes into measurable categories, such as inches of height (Figure 13.13). The number of individuals that fall into each category gives the relative frequencies of phenotypes across

Figure 13.14 Example of environmental effects on animal phenotype. The color of the snowshoe hare's fur varies by season. In summer, the fur is brown (*left*); in winter, white (*right*). Both forms offer seasonally appropriate camouflage from predators.

our range of measurable values. Finally, we plot the data as a bar chart. A graph line around the top of the bars shows the distribution of values for the trait. If the line is a bell-shaped curve, or **bell curve**, the trait varies continuously.

Human eye color is another example of a trait that varies continuously. The colored part of the eye is the iris, a doughnut-shaped, pigmented structure. Iris color, like skin color, is the result of epistasis among gene products that make and distribute melanins. The more melanin deposited in the iris, the less light is reflected from it. Dark irises have dense melanin deposits that absorb almost all light, and reflect almost none. Melanin deposits are not as extensive in brown eyes, which reflect some light. Green and blue eyes have the least amount of melanin, so they reflect the greatest amount of light.

Environmental Effects on Phenotype Variations in traits are not always the result of differences in alleles. Environmental factors often affect gene expression, which in turn affects phenotype. For example, seasonal changes

Figure 13.13 Animated Continuous variation in height among male biology students at the University of Florida. The students were divided into categories of one-inch increments in height and counted (*bottom*). A graph of the resulting data produces a bell-shaped curve (*top*), an indication that height varies continuously.

Figure 13.15 Environmental effects on the body form of *Daphnia*. In this animal, the form of the individual on the *left* develops in environments with few predators. The *Daphnia* on the *right* has a longer tail spine and a pointy head—traits that develop in response to chemicals emitted by predatory insects.

in temperature and the length of day affect the production of melanin and other pigments that color the skin and fur of many animals. These animals have different color phases in different seasons (Figure 13.14).

Factors such as the presence of predators can also influence phenotype. Consider *Daphnia pulex*, a microscopic freshwater relative of shrimp. Aquatic insects prey on these invertebrates. *Daphnia* living in ponds with few predators have rounded heads, but those in ponds with many predators have pointy heads (Figure 13.15). The predators emit chemicals that trigger the development of the pointier phenotype.

Yarrow plants offer another example of how environment influences phenotype. Yarrow is useful for genetics

A Plant grown at high elevation (3,060 meters above sea level)

B Plant grown at mid-elevation (1,400 meters above sea level)

C Plant grown at low elevation (30 meters above sea level)

Figure 13.16 Environmental effects on plant phenotype. Cuttings from the same yarrow plant (*Achillea millefolium*) grow to different heights at three different elevations.

bell curve Bell-shaped curve; typically results from graphing frequency versus distribution for a trait that varies continuously.
continuous variation In a population, a range of small differences in a shared trait.

Menacing Mucus (revisited)

The allele most commonly associated with cystic fibrosis, $\Delta F508$, is eventually lethal in homozygous individuals, but not in those who are heterozygous. This allele is codominant with the normal one, so both copies of the gene are expressed in heterozygous individuals. Such individuals make enough of the normal CFTR protein to have normal chloride ion transport.

The $\Delta F508$ allele is at least 50,000 years old and very common: up to 1 in 25 people carry it in some populations. Why has this allele persisted for so long and at such high frequency if it is dangerous? The $\Delta F508$ allele may be the lesser of two evils because it offers heterozygous individuals a survival advantage against certain deadly infectious diseases. The unmutated CFTR protein triggers endocytosis when it binds to bacteria. This process is an essential part of the body's immune response to bacteria in the respiratory tract. However, the same function of CFTR allows bacteria to enter cells of the gastrointestinal tract, where they can be deadly. For example, endocytosis of *Salmonella typhi* (shown at *left*) into epithelial cells lining the gut results in a dangerous infection called typhoid fever. The $\Delta F508$ mutation alters the CFTR protein so that bacteria can no longer be taken up by intestinal cells. People that carry it may have a decreased susceptibility to typhoid fever and other bacterial diseases that begin in the intestinal tract.

How Would You Vote? Tests for predisposition to genetic disorders are now available. Do you support legislation preventing discrimination based on the results of such tests? See CengageNow for details, then vote online (cengagenow.com).

experiments because it grows easily from cuttings. All cuttings of a plant have the same genotype, so experimenters know that genes are not the basis for any phenotypic differences among them. In one study, genetically identical yarrow plants had different phenotypes when grown at different altitudes (Figure 13.16).

The environment also affects human genes. One of our genes encodes a protein that transports serotonin across the membrane of brain cells. Serotonin lowers anxiety and depression during traumatic times. Some mutations in the serotonin transporter gene can reduce the ability to cope with stress. It is as if some of us are bicycling through life without an emotional helmet. Only when we crash does the mutation's phenotypic effect—depression—appear. Other human genes affect emotional state, but mutations in this one reduce our capacity to recover from emotional setbacks.

Take-Home Message How does phenotype vary?

> Some traits have a range of small differences, or continuous variation. The more genes and other factors that influence a trait, the more continuous the distribution of phenotype.

> Enzymes and other gene products control steps of most metabolic pathways. Environmental conditions can affect one or more steps, and thus contribute to variation in phenotypes.

Summary

Section 13.1 Cystic fibrosis, the most common fatal genetic disorder in the United States, is usually caused by a particular deletion in the *CFTR* gene. The allele that carries the mutation persists at high frequency despite its devastating effects. Only those homozygous for the allele have the disorder.

Section 13.2 Genetic information is passed to offspring in discrete units (genes). Each gene occurs at a particular **locus**, or location, on a chromosome. The somatic cells of diploid individuals have paired genes on their homologous chromosomes. Mutations give rise to alleles. Individuals with identical alleles are **homozygous** for the allele. Individuals that breed true for a trait are homozygous for alleles that affect the trait. **Heterozygous**, or **hybrid**, individuals carry two nonidentical alleles. A **dominant** allele masks the effect of a **recessive** allele on the homologous chromosome. **Genotype** (an individual's particular set of alleles) gives rise to **phenotype**, which refers to an individual's particular set of traits.

Section 13.3 Crossing individuals that breed true for two forms of a trait yields identically heterozygous F_1 offspring. A cross between such offspring is a **monohybrid cross**. The frequency at which the traits appear in the F_2 offspring can reveal dominance relationships among the alleles associated with those traits. Predictable patterns of inheritance can be used as evidence of genotype, as in **testcrosses**. **Punnett squares** are useful in determining the probability of the genotype and phenotype of the offspring of crosses. Mendel's monohybrid cross results led to his **law of segregation** (stated here in modern terms): During meiosis, the genes of each pair separate, so each gamete gets one or the other gene.

Section 13.4 Crossing individuals that breed true for two forms of two traits yields F_1 offspring identically heterozygous for alleles governing those traits. A cross between such offspring is a **dihybrid cross**. The frequency at which the traits appear in the F_2 offspring can reveal dominance relationships among the alleles associated with those traits. Mendel's dihybrid cross results led to his **law of independent assortment** (stated here in modern terms): During meiosis, gene pairs on homologous chromosomes tend to sort into gametes independently of other gene pairs. Crossovers can break up **linkage groups**.

Section 13.5 Not all alleles are clearly dominant or recessive. With **incomplete dominance**, an allele is not fully dominant over its partner on a homologous chromosome, so both are expressed. The combination of alleles gives rise to an intermediate phenotype.

Codominant alleles are both expressed at the same time in heterozygotes, as in **multiple allele systems** such as the one underlying ABO blood typing. In **epistasis**, interacting products of one or more genes often affect the same trait. A **pleiotropic** gene affects two or more traits.

Section 13.6 A trait that is influenced by the products of multiple genes often occurs in a range of small increments of phenotype called **continuous variation**. Continuous variation typically occurs as a **bell curve** in the range of values. An individual's phenotype may be influenced by environmental factors.

Self-Quiz   Answers in Appendix III

1. A heterozygous individual has a _____ for a trait being studied.
   a. pair of identical alleles
   b. pair of nonidentical alleles
   c. haploid condition, in genetic terms

2. An organism's observable traits constitute its _____ .
   a. phenotype   c. genotype
   b. variation   d. pedigree

3. Filial means _____ .

4. The second-generation offspring of a cross between individuals who are homozygous for different alleles of a gene are called the _____ .
   a. F_1 generation   c. hybrid generation
   b. F_2 generation   d. none of the above

5. F_1 offspring of the cross $AA \times aa$ are _____ .
   a. all AA c. all Aa
   b. all aa d. 1/2 AA and 1/2 aa

6. Refer to question 4. Assuming complete dominance, the F_2 generation will show a phenotypic ratio of _____ .
   a. 3:1   b. 9:1   c. 1:2:1   d. 9:3:3:1

7. A testcross is a way to determine _____ .
   a. phenotype   b. genotype   c. both a and b

8. Assuming complete dominance, crosses between two dihybrid F_1 pea plants, which are offspring from a cross $AABB \times aabb$, result in F_2 phenotype ratios of _____ .
   a. 1:2:1   b. 3:1   c. 1:1:1:1   d. 9:3:3:1

9. The probability of a crossover occurring between two genes on the same chromosome _____ .
   a. is unrelated to the distance between them
   b. decreases with the distance between them
   c. increases with the distance between them

10. A gene that affects three traits is _____ .
    a. epistatic   c. pleiotropic
    b. a multiple allele system   d. dominant

11. _____ alleles are both expressed.
    a. Dominant   c. Pleiotropic
    b. Codominant   d. Hybrid

12. A bell curve indicates _____ in a trait.

The Cystic Fibrosis Mutation and Typhoid Fever The *ΔF508* mutation disables the receptor function of the CFTR protein, so it inhibits endocytosis of bacteria into epithelial cells. Endocytosis is an important part of the respiratory tract's immune defenses against common *Pseudomonas* bacteria, which is why *Pseudomonas* infections are a chronic problem in cystic fibrosis patients.

The *ΔF508* mutation also inhibits endocytosis of *Salmonella typhi* into cells of the gastrointestinal tract, where internalization of this bacteria can cause typhoid fever. Typhoid fever is a common worldwide disease. Its symptoms include extreme fever and diarrhea, and the resulting dehydration causes delirium that may last several weeks. If untreated, it kills up to 30 percent of those infected. Around 600,000 people die annually from typhoid fever. Most of them are children.

In 1998, Gerald Pier and his colleagues compared the uptake of *S. typhi* by different types of epithelial cells: those homozygous for the normal allele, and those heterozygous for the *ΔF508* mutation. (Cells homozygous for the mutation do not take up any *S. typhi* bacteria.) Some of their results are shown in Figure 13.17.

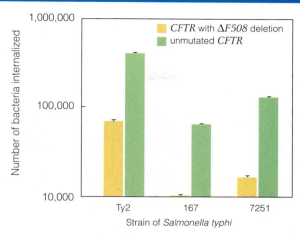

Figure 13.17 In epithelial cells, effect of the *CF* mutation on the uptake of three different strains of *Salmonella typhi* bacteria.

1. Regarding the Ty2 strain of *S. typhi*, about how many more bacteria were able to enter normal cells (those expressing unmutated *CFTR*) than cells expressing the gene with the *ΔF508* deletion?
2. Which strain of bacteria entered normal epithelial cells most easily?
3. The *ΔF508* deletion inhibited the entry of all three *S. typhi* strains into epithelial cells. Can you tell which strain was most inhibited?

---

13. Match the terms with the best description.
    ___ dihybrid cross   a. *bb*
    ___ monohybrid cross   b. *AABB × aabb*
    ___ homozygous condition   c. *Aa*
    ___ heterozygous condition   d. *Aa × Aa*

Additional questions are available on **CENGAGENOW**.

Genetics Problems   Answers in Appendix III

1. Mendel crossed a true-breeding pea plant with green pods and a true-breeding pea plant with yellow pods. All offspring had green pods. Which color is recessive?

2. Assuming that independent assortment occurs during meiosis, what type(s) of gametes will form in individuals with the following genotypes?
    a. *AABB*   b. *AaBB*   c. *Aabb*   d. *AaBb*

3. Refer to problem 2. Determine the frequencies of each genotype among offspring from the following matings:
    a. *AABB × aaBB*   c. *AaBb × aabb*
    b. *AaBB × AABb*   d. *AaBb × AaBb*

4. Suppose you identify a new gene in mice. One of its alleles specifies white fur, another specifies brown. You want to see if the two alleles interact in simple or incomplete dominance. What test would give you the answer?

5. Many genes are so vital for development that mutations in them are lethal. Even so, heterozygous individuals can perpetuate alleles that are lethal, such as the Manx (*M*)

allele in cats. Homozygous cats (*MM*) die before birth. In heterozygous cats (*Mm*), the spine develops abnormally, and these animals end up with no tail (*left*). What is the dominance relationship between the Manx allele and the normal allele? If two *Mm* cats mate, what is the probability that any one of their surviving kittens will be heterozygous (*Mm*)?

6. One gene encodes the second enzyme in a melanin-synthesizing pathway. An individual who is homozygous for a recessive allele of this gene cannot make or deposit melanin in body tissues. *Albinism*, the absence of melanin, is the result. Humans and many other organisms can have this phenotype. In the following situations, what are the probable genotypes of the father, the mother, and their children?
    a. Both parents have normal phenotypes; some but not all of their children have the albino phenotype.
    b. Both parents and all of their children have the albino phenotype.
    c. The mother is unaffected, the father has the albino phenotype, three children are unaffected and one child has the albino phenotype.

Additional genetics problems are available on **CENGAGENOW**.

Animations and Interactions on **CENGAGENOW**:
❯ Crossing garden pea plants; Genes on chromosomes; Monohybrid cross; Dihybrid cross; Codominance and ABO blood group; Incomplete dominance; Pleiotropy in Marfan syndrome; Epistasis in dogs; Continuous variation.

‹ **Links to Earlier Concepts**

Be sure you understand dominance relationships (Sections 13.2, 13.5, and 13.6), gene expression (9.2, 9.3), and mutations (9.6) before you start this chapter. You will use your knowledge of chromosomes (8.2), meiosis (12.3, 12.4), gametes (12.5), DNA (8.6) and sex determination (10.4). Sampling error (1.8), amino acids (3.5), lysosomes (4.8), the cell cortex (4.10), metabolic pathways (5.5), tyrosinase (5.4), pigments (6.2), and oncogenes (11.6) will turn up in context of genetic disorders.

Key Concepts

Tracking Traits in Humans
Inheritance patterns in humans are determined by following traits through generations of family trees. The types of traits followed in such studies include genetic abnormalities or syndromes associated with a genetic disorder.

Autosomal Inheritance
Many human traits can be traced to dominant or recessive alleles on autosomes. These alleles are inherited in characteristic patterns: dominant alleles tend to appear in every generation; recessive ones can skip generations.

14 Human Inheritance

The color of human skin begins with melanosomes. These skin cell organelles make two types of melanin pigments: one brownish-black; the other, reddish. Most people have about the same number of melanosomes in their skin cells. Variations in skin color occur because the kinds and amounts of melanins vary among people, as does the formation, transport, and distribution of the melanosomes.

Variations in skin color may have evolved as a balance between vitamin production and protection against harmful UV radiation. Dark skin would have been beneficial under the intense sunlight of the African savannas where humans first evolved. Melanin acts as a natural sunscreen because it prevents UV radiation in sunlight from breaking down folate, a vitamin essential for normal sperm formation and embryonic development. Children born to light-skinned women exposed to high levels of sunlight have a heightened risk of birth defects.

Early human groups that migrated to regions with colder climates were exposed to less sunlight. In these regions, lighter skin color would have been beneficial. Why? UV radiation stimulates skin cells to make a molecule the body converts to essential vitamin D. Where sunlight exposure is minimal, UV radiation damage is less of a risk than vitamin D deficiency, which has serious health consequences for developing fetuses and children. People with dark, UV-shielding skin have a high risk of this deficiency in regions where sunlight exposure is minimal.

Skin color, like most other human traits, has a genetic basis. More than 100 gene products are involved in the synthesis of melanin, and the formation and deposition of melanosomes. Mutations in at least some of these genes may have contributed to regional variations of human skin color. Consider a gene on chromosome 15, *SLC24A5*, that encodes a transport protein in melanosome membranes. Nearly all people of African, Native American, or east Asian descent carry the same allele of this gene. Between 6,000 and 10,000 years ago, a mutation gave rise to a different allele. The mutation, a single base–pair substitution, changed the 111th amino acid of the transport protein from alanine to threonine. The change results in less melanin—and lighter skin color— than the original African allele does. Today, nearly all people of European descent carry this mutated allele.

A person of mixed ethnicity may make gametes that contain different combinations of alleles for dark and light skin. It is fairly rare that one of those gametes contains all of the alleles for dark skin, or all of the alleles for light skin, but it happens (Figure 14.1).

Skin color is only one of many human traits that vary as a result of single nucleotide mutations. The small scale of such changes offers a reminder that all of us share the genetic legacy of common ancestry.

Figure 14.1 Variation in human skin color (*right*) begins with differences in alleles inherited from parents. *Opposite*, fraternal twin girls Kian and Remee, born in 2006. Both of the children's grandmothers are of European descent, and have pale skin. Both of their grandfathers are of African descent, and have dark skin. The twins inherited different alleles of some of the genes that affect skin color from their mixed-race parents, who, given the appearance of their children, must be heterozygous for those alleles.

Sex–Linked Inheritance
The X chromosome holds about 10 percent of all human genes, so many traits are affected by alleles on this chromosome. Inheritance patterns of such X-linked alleles tend to differ between males and females.

Changes in Chromosome Structure and Number
A chromosome may undergo a large-scale, permanent change in its structure, or the number of autosomes or sex chromosomes may change. In humans, such changes usually result in a genetic disorder.

Genetic Testing
Genetic testing provides information about the risk of passing a harmful allele to one's offspring. After conception, various methods of prenatal testing can reveal a genetic abnormality or disorder in a fetus or embryo.

> Geneticists study inheritance patterns in humans by tracking genetic disorders and abnormalities through families.
> Charting genetic connections with pedigrees reveals inheritance patterns of certain traits.
< Links to Sampling error 1.8, Chromosomes 8.2, Dominance 13.2, Complex inheritance patterns 13.6

Some organisms, including pea plants and fruit flies, are ideal for genetic analysis. They have relatively few chromosomes, they reproduce quickly under controlled conditions, and breeding them poses few ethical problems. It does not take long to track a trait through many generations. Humans, however, are a different story. Unlike flies grown in laboratories, we humans live under variable

conditions, in different places, and we live as long as the geneticists who study us. Most of us select our own mates and reproduce if and when we want to. Our families tend to be on the small side, so sampling error (Section 1.8) is a major factor in human genetics studies.

Thus, inheritance patterns in humans are typically studied by tracking observable traits that crop up in families over many generations. Researchers graph such data as standardized charts of genetic connections called **pedigrees** (Figure 14.2). Pedigree analyses can reveal whether a trait is associated with a dominant or recessive allele, and whether the allele is on an autosome or a sex chromosome. Pedigree analysis also allows geneticists to determine the probability that a trait will recur in future generations of a family or a population.

Types of Genetic Variation

Some easily observed human traits follow Mendelian inheritance patterns. Like the flower color of pea plants, these traits are controlled by a single gene with two alleles, one dominant and the other recessive. For example, some people have earlobes that attach at their base, and others have earlobes that dangle free. The allele for unattached earlobes is dominant and the allele for attached earlobes is recessive. Similarly, the allele that specifies a cleft chin is dominant over the allele for a smooth chin, and the allele for dimples is dominant over that for no dimples. Someone who is homozygous for two recessive alleles of the *MC1R* gene makes the reddish kind of melanin but not the brownish-black kind, so this person has red hair.

Single genes on autosomes or sex chromosomes also govern more than 6,000 genetic abnormalities and disorders. Table 14.1 lists a few examples. A genetic abnormality

---

pedigree Chart showing the pattern of inheritance of a trait through generations in a family.

A Standard symbols used in pedigrees.

□	male
○	female
□—○	marriage/mating
○ □ □ ○	offspring
■ ●	individual showing trait being studied
◇	sex not specified
I, II, III, IV...	generation

B A pedigree for polydactyly, which is characterized by extra fingers, toes, or both. The *black* numbers signify the number of fingers on each hand; the *red* numbers signify the number of toes on each foot. Though it occurs on its own, polydactyly is also one of several symptoms of Ellis–van Creveld syndrome.

* Gene not expressed in this carrier.

C *Right*, pedigree for Huntington's disease, a progressive degeneration of the nervous system. Researcher Nancy Wexler and her team constructed this extended family tree for nearly 10,000 Venezuelans. Their analysis of unaffected and affected individuals revealed that a dominant allele on human chromosome 4 is the culprit. Wexler has a special interest in the disorder: It runs in her family.

Figure 14.2 Animated Pedigrees.

is a rare or uncommon version of a trait, such as having six fingers on a hand or having a web between two toes. Genetic abnormalities are not inherently life-threatening, and how you view them is a matter of opinion. By contrast, a genetic disorder sooner or later causes medical problems that may be severe. A genetic disorder is often characterized by a specific set of symptoms (a syndrome). In general, much more research focuses on genetic disorders than on other human traits, because what we learn helps us develop treatments for affected people.

The next two sections of this chapter focus on inheritance patterns of human single-gene disorders, which affect about 1 in 200 people. Keep in mind that these inheritance patterns are the least common kind. Most human traits, including skin color, are polygenic (influenced by multiple genes) and often have environmental factors too. Many genetic disorders are like this, including diabetes, asthma, obesity, cancers, heart disease, and multiple sclerosis. The inheritance patterns of these disorders are extremely complex, and despite intense research our understanding of the alleles associated with them remains incomplete. For example, abnormalities on almost every chromosome have been found in people with autism, a developmental disorder, but a person who carries one or more of these abnormalities does not necessarily have autism. Mutations in specific regions of chromosomes 1, 2, 6, 7, 13, 15, and 22 increase an individual's chance of developing schizophrenia, a neurobiological disorder, but not everyone with those mutations develops schizophrenia. Appendix VII shows a map of human chromosomes with the locations of some alleles that are known to play a role in genetic disorders and other human traits.

Alleles that give rise to severe genetic disorders are generally rare in populations, because they compromise the health and reproductive ability of their bearers. Why do they persist? Mutations periodically reintroduce them. In some cases, a normal allele in heterozygotes masks the effects of a harmful one. In others, a codominant allele offers a survival advantage in a particularly hazardous environment. You will see examples of how this works in later chapters.

Take-Home Message How do we study inheritance patterns in humans?

> Inheritance patterns in humans are often studied by tracking traits through generations of families.

> A genetic abnormality is a rare version of an inherited trait. A genetic disorder is an inherited condition that causes medical problems.

> Some human genetic traits are governed by a single gene and inherited in a Mendelian fashion. Many others are influenced by multiple genes, as well as the environment.

Table 14.1 Patterns of Inheritance for Some Genetic Abnormalities and Disorders

Disorder or Abnormality	Main Symptoms
Autosomal dominant inheritance pattern	
Achondroplasia	One form of dwarfism
Aniridia	Defects of the eyes
Camptodactyly	Rigid, bent fingers
Familial hypercholesterolemia	High cholesterol level; clogged arteries
Huntington's disease	Degeneration of the nervous system
Marfan syndrome	Abnormal or missing connective tissue
Polydactyly	Extra fingers, toes, or both
Progeria	Drastic premature aging
Neurofibromatosis	Tumors of nervous system, skin
Autosomal recessive inheritance pattern	
Albinism	Absence of pigmentation
Hereditary methemoglobinemia	Blue skin coloration
Cystic fibrosis	Abnormal glandular secretions leading to tissue and organ damage
Ellis–van Creveld syndrome	Dwarfism, heart defects, polydactyly
Fanconi anemia	Physical abnormalities, bone marrow failure
Galactosemia	Brain, liver, eye damage
Hereditary hemochromatosis	Iron overload damages joints and organs
Phenylketonuria (PKU)	Mental impairment
Sickle-cell anemia	Adverse pleiotropic effects on entire body
Tay–Sachs disease	Deterioration of mental and physical abilities; early death
X-linked recessive inheritance pattern	
Androgen insensitivity syndrome	XY individual but having some female traits; sterility
Red–green color blindness	Inability to distinguish red from green
Hemophilia	Impaired blood clotting ability
Muscular dystrophies	Progressive loss of muscle function
X-linked anhidrotic dysplasia	Mosaic skin (patches with or without sweat glands); other effects
X-linked dominant inheritance pattern	
Fragile X syndrome	Intellectual, emotional disability
Changes in chromosome number	
Down syndrome	Mental impairment; heart defects
Turner syndrome (XO)	Sterility; abnormal ovaries and sexual traits
Klinefelter syndrome	Sterility; mild mental impairment
XXX syndrome	Minimal abnormalities
XYY condition	Mild mental impairment or no effect
Changes in chromosome structure	
Chronic myelogenous leukemia (CML)	Overproduction of white blood cells; organ malfunctions
Cri-du-chat syndrome	Mental impairment; abnormal larynx

> An allele is inherited in an autosomal dominant pattern if the trait it specifies appears in homozygous and heterozygous people.
> An allele is inherited in an autosomal recessive pattern if the trait it specifies appears only in homozygous people.
‹ Links to Lysosomes 4.8, Cytoskeletal elements 4.10, Tyrosinase 5.4, Autosomes 8.2, DNA replication and repair 8.6, Gene expression 9.2, RNA processing 9.3, Inheritance 13.2, Codominance and pleiotropy 13.5

Figure 14.4 Examples of autosomal dominant disorders. **A** Achondroplasia affects Ivy Broadhead (*left*), as well as her brother, father, and grandfather. **B** Five-year-old Megan is already showing symptoms of Hutchinson–Gilford progeria.

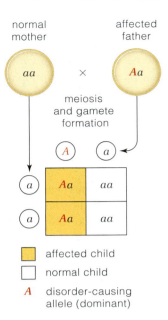

Figure 14.3 Animated Autosomal dominant inheritance, in which a dominant allele (*red*) is fully expressed in heterozygous people.

The Autosomal Dominant Pattern

An allele on an autosome is inherited in a dominant pattern if it is expressed in both homozygotes and heterozygotes. A trait it specifies tends to appear in every generation. When one parent is heterozygous, and the other is homozygous for the recessive allele, each of their children has a 50 percent chance of inheriting the dominant allele and displaying the trait associated with it (**Figure 14.3**).

Achondroplasia is an example of an autosomal dominant disorder (a disorder caused by a dominant allele on an autosome). The allele responsible for achondroplasia interferes with formation of the embryonic skeleton. About 1 out of 10,000 people are heterozygous for this allele. As adults, they average about four feet, four inches tall, and have abnormally short arms and legs relative to other body parts (**Figure 14.4A**). Most homozygotes die before or shortly after birth.

Huntingon's disease is also caused by an autosomal dominant allele. With this genetic disorder, involuntary muscle movements increase as the nervous system slowly deteriorates. Typically, symptoms do not start until after the age of thirty, and people affected by the syndrome die during their forties or fifties. The mutation that causes Huntington's alters a protein necessary for brain cell development. It is an expansion mutation, in which the same three nucleotides have been inserted into DNA many, many times. Hundreds of thousands of other expansion repeats occur harmlessly in and between other genes on human chromosomes. This one alters the function of a critical gene product.

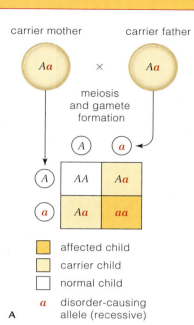

carrier mother   carrier father

Aa   ×   *Aa*

meiosis
and gamete
formation

(A)   (a)

	A	a
A	*AA*	*Aa*
a	*Aa*	*aa*

▢ affected child
▢ carrier child
▢ normal child

a disorder-causing allele (recessive)

A

B

C

Figure 14.5 Animated Autosomal recessive inheritance. Only homozygous people show the trait associated with a recessive allele on an autosome. **A** In this example, both parents are carriers of a recessive autosomal allele (*red*). Each of their children has a 25 percent chance of being homozygous for it. **B** The albino phenotype is associated with recessive alleles that cause a deficiency in a melanin-producing enzyme. **C** Conner Hopf was diagnosed with Tay–Sachs disease, an autosomal recessive disorder, at age 7–1/2 months. He died before his second birthday.

Hutchinson–Gilford progeria is an autosomal dominant disorder characterized by drastically accelerated aging. It usually results from a base-pair substitution in the gene for lamin A, a protein subunit of intermediate filaments that help organize chromosomes. The mutation adds a signal for an intron/exon splice. The resulting mis-spliced mRNA encodes a protein with a large deletion that cannot assemble into intermediate filaments, and instead accumulates on the nuclear membrane. The pleiotropic effects of the mutation include defects in transcription, mitosis, and division. Symptoms begin to appear before age two. Skin that should be plump and resilient starts to thin, muscles weaken, and bones that should lengthen and grow stronger soften. Premature baldness is inevitable. Most people with the disorder die in their early teens as a result of a stroke or heart attack brought on by hardened arteries, a condition typical of advanced age (Figure 14.4B).

Progeria does not run in families because affected people do not usually live long enough to reproduce. Other dominant alleles that cause severe problems can persist if their expression does not interfere with reproduction. The allele that causes achondroplasia is an example. With Huntington's disease and other late-onset disorders, people tend to reproduce before symptoms appear, so the allele may be passed unknowingly to children.

The Autosomal Recessive Pattern

An allele on an autosome is inherited in a recessive pattern if it is expressed only in homozygous people, so traits associated with the allele may skip generations. People heterozygous for the allele are carriers, which means that they have the allele but not the trait. Any child of two carriers has a 25 percent chance of inheriting the allele from both parents (Figure 14.5A). Being homozygous for the allele, such children would have the trait. All children of homozygous parents are also homozygous.

Albinism, a lack of melanin, is inherited in an autosomal recessive pattern. The albino phenotype occurs in people homozygous for an allele that encodes a defective form of the enzyme tyrosinase. Melanocytes produce no melanin in the absence of this enzyme, so the hair, skin, and irises lack typical coloration (Figure 14.5B).

Tay–Sachs disease is an example of an autosomal recessive disorder. In the general population, about 1 in 300 people is a carrier for a Tay–Sachs allele, but the incidence is ten times higher in some groups, such as Jews of eastern European descent. Mutations associated with this disorder cause a deficiency or malfunction of a lysosomal enzyme that breaks down gangliosides, a lipid component of plasma membranes. These lipids can accumulate to toxic levels in nerve cells if they are not recycled properly by lysosomes. Affected infants typically seem normal for the first few months. Symptoms begin to appear as the gangliosides accumulate to higher and higher levels inside their nerve cells. Within three to six months the child becomes irritable, listless, and may have seizures. Blindness, deafness, and paralysis follow. Affected children usually die by the age of five (Figure 14.5C).

Take-Home Message How do we know when a trait is associated with an allele on an autosomal chromosome?

❯ With an autosomal dominant inheritance pattern, persons heterozygous for an allele have the associated trait. Thus, the trait appears in every generation.

❯ With an autosomal recessive inheritance pattern, only persons who are homozygous for an allele have the associated trait, which can skip generations.

> Traits associated with recessive alleles on the X chromosome appear more frequently in men than in women.
> A man cannot pass an X chromosome allele to a son.
< Links to Cell cortex 4.10, Pigments 6.2

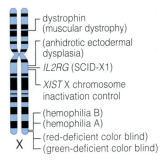

dystrophin (muscular dystrophy)
(anhidrotic ectodermal dysplasia)
IL2RG (SCID-X1)
XIST X chromosome inactivation control
(hemophilia B)
(hemophilia A)
(red-deficient color blind)
(green-deficient color blind)

X

The X chromosome (shown on the *left*) carries about 2,000 human genes, which is almost 10 percent of the total number. A recessive allele on this chromosome (an X-linked recessive allele) leaves certain clues when it causes a genetic disorder. First, more males than females are affected by the disorder. This is because heterozygous males are affected, but heterozygous females are not. Heterozygous females have a dominant, normal allele on one of their X chromosomes that masks the effects of the recessive allele on the other. Heterozygous males have only one X chromosome, so they are not similarly protected if they inherit an X-linked recessive allele (Figure 14.6).

Second, an affected father cannot pass his X-linked recessive allele to a son because all children who inherit their father's X chromosome are female. Thus, a heterozygous female must be the bridge between an affected male and his affected grandson.

X-linked dominant alleles that cause disorders are rarer than X-linked recessive ones, probably because they tend to be lethal in male embryos.

Red–Green Color Blindness

The pattern of X-linked recessive inheritance shows up among individuals who have some degree of color blind-

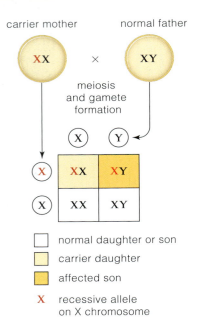

carrier mother   normal father

XX   ×   XY

meiosis and gamete formation

X   Y

normal daughter or son
carrier daughter
affected son
X   recessive allele on X chromosome

Figure 14.6 Animated X-linked recessive inheritance. In this case, the mother carries a recessive allele on one of her two X chromosomes (*red*).

ness (Figure 14.7). The term refers to a range of conditions in which an individual cannot distinguish among some or all colors in the spectrum of visible light. Color vision depends on the proper function of pigment-containing receptors in the eyes. Most of the genes involved in color vision are on the X chromosome, and mutations in those genes often result in altered or missing receptors. Normally, humans can sense the differences among 150 colors. A person who has red–green color blindness sees

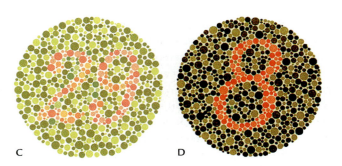

C   D

Figure 14.7 Color blindness. **A** View with red–green color blindness. The perception of blues and yellows is normal, but red and green appear similar. **B** Compare what a person with normal vision sees.

Above, two Ishihara plates, which are standardized tests for color blindness. **C** You may have one form of red–green color blindness if you see the number 7 instead of 29 in this circle. **D** You may have another form if you see a 3 instead of an 8.

Figure 14.8 A classic case of X-linked recessive inheritance: a partial pedigree of the descendants of Queen Victoria of England. At one time, the recessive X-linked allele that resulted in hemophilia was present in eighteen of Victoria's sixty-nine descendants, who sometimes intermarried. Of the Russian royal family members shown, the mother (Alexandra Czarina Nicolas II) was a carrier.

›› Figure It Out How many of Alexis's siblings were affected by hemophilia A?   Answer: None

fewer than 25 colors because receptors that respond to red and green wavelengths are weakened or absent. Some people who have color blindness confuse red and green colors. Others see green as shades of gray, but perceive blues and yellows quite well.

Hemophilia A

Hemophilia A is an X-linked recessive disorder that interferes with blood clotting. Most of us have a blood clotting mechanism that quickly stops bleeding from minor injuries. That mechanism involves factor VIII, a protein product of a gene on the X chromosome. Bleeding can be prolonged in males who carry a mutation in this gene, or in females who are homozygous for one. Affected people tend to bruise very easily, but internal bleeding is their most serious problem. Repeated bleeding inside the joints disfigures them and causes chronic arthritis. Heterozygous females make enough factor VIII to have a clotting time that is close to normal.

In the nineteenth century, the incidence of hemophilia A was relatively high in royal families of Europe and Russia, probably because the common practice of inbreeding kept the allele in their family trees (Figure 14.8). Today, about 1 in 7,500 people is affected, but that number may be rising because the disorder is now a treatable one. More affected people are living long enough to transmit the mutated allele to children.

Duchenne Muscular Dystrophy

Duchenne muscular dystrophy (DMD) is one of several X-linked recessive disorders characterized by muscle degeneration. A gene on the X chromosome encodes dystrophin, a protein in muscle and nerve cells. Dystrophin is part of a complex of proteins that structurally and functionally links the cytoskeleton to extracellular matrix across the plasma membrane of these cells. By a process that is still unknown, abnormal or absent dystrophin causes muscle cells to die. The dead cells are eventually replaced by fat cells and connective tissue.

DMD affects about 1 in 3,500 people, almost all of them boys. Symptoms begin between age three and seven. Anti-inflammatory medication can sometimes slow the progression of this disorder, but there is no cure. When an affected boy is about twelve, he will begin to use a wheelchair and his heart muscle will start to fail. Even with the best care, he will probably die before he is thirty, from a heart disorder or respiratory failure (suffocation).

Take-Home Message **How do we know when a trait is associated with an allele on an X chromosome?**

› Men heterozygous for an X-linked recessive allele have the trait associated with the allele. Heterozygous women do not, because they have a normal allele on their second X chromosome. Thus, the trait appears more often in men.

› Men transmit an X-linked allele to their daughters, but not to their sons.

> Major changes in chromosome structure have been evolutionarily important. More frequently, such changes tend to result in genetic disorders.

‹ Links to Karyotyping 8.2, DNA sequence 8.6, Mutations and hemoglobin 9.6, *SRY* gene 10.4, Oncogenes 11.6, Meiosis 12.3, Crossing over 12.4

Large-scale changes in chromosome structure usually have drastic effects on health; about half of all miscarriages are due to chromosome abnormalities of the developing embryo. These changes are rare, but they do occur spontaneously in nature. They can also be induced by exposure to certain chemicals or radiation. Either way, the scale of such changes often allows them to be detected by karyotyping (Section 8.2).

Duplication Even normal chromosomes have DNA sequences that are repeated two or more times. These repetitions are called **duplications** (Figure 14.9A). Duplications happen during prophase I of meiosis, when crossing over occurs unequally between homologous chromosomes. When homologous chromosomes align side by

A With a duplication, a section of a chromosome gets repeated.

B With a deletion, a section of a chromosome gets lost.

C With an inversion, a section of a chromosome gets flipped so it runs in the opposite orientation.

D With a translocation, a broken piece of a chromosome gets reattached in the wrong place. This example shows a reciprocal translocation, in which two chromosomes exchange chunks.

Figure 14.9 Large-scale changes in chromosome structure.

side, their DNA sequences may misalign at some point along their length. In this case, the crossover deletes a stretch of DNA from one chromosome and splices it into the homologous partner. The probability of misalignment is greater in regions where the same sequence of nucleotides is repeated many times. Some duplications, such as the expansion mutations that cause Huntington's, cause genetic abnormalities or disorders. Others have been evolutionarily important.

Deletion A **deletion** is the loss of some portion of a chromosome (Figure 14.9B). In mammals, deletions usually cause serious disorders and are often lethal. The loss of genes results in the disruption of growth, development, and metabolism. For instance, a small deletion in chromosome 5 causes mental impairment and an abnormally shaped larynx. Affected infants tend to make a sound like the meow of a cat, hence the name of the disorder, cri-du-chat, which is French for "cat's cry."

Inversion With an **inversion**, part of the sequence of DNA within the chromosome becomes oriented in the reverse direction, with no molecular loss (Figure 14.9C). An inversion may not affect a carrier's health if it does not disrupt a gene region, because the individual's cells have a full complement of genes. However, it may affect fertility, because inverted chromosomes tend to mispair during meiosis. Crossovers between mispaired chromosomes can produce large deletions or duplications that reduce the viability of forthcoming embryos. Some carriers do not know that they have an inversion until they are diagnosed with infertility and their karyotype is tested.

Translocation If a chromosome breaks, the broken part may get attached to a different chromosome, or to a different part of the same one. This structural change is called a **translocation**. Most translocations are reciprocal, or balanced, which means that two chromosomes exchange broken parts (Figure 14.9D). A reciprocal translocation between chromosomes 8 and 14 is the usual cause of Burkitt's lymphoma, an aggressive cancer of the immune system. This translocation moves a proto-oncogene to a region that is vigorously transcribed in immune cells, with disastrous results. Many other reciprocal translocations have no adverse effects on health, but, like inversions, they can affect fertility. Translocated chromosomes pair abnormally during meiosis. They segregate improperly about half of the time, so about half of the resulting gametes carry major duplications or deletions. If one of these gametes unites with a normal gamete at fertilization, the resulting embryo almost always dies. As with inversions, many people do not realize they carry a translocation until they have difficulty with fertility.

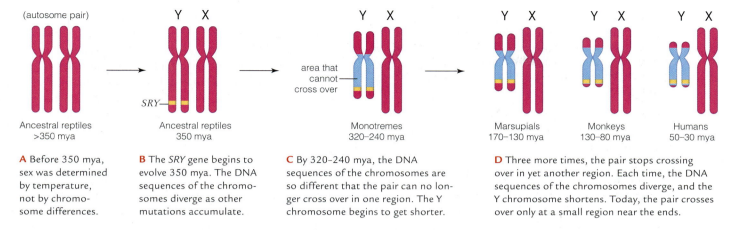

(autosome pair)	Y X	Y X	Y X	Y X	Y X
Ancestral reptiles >350 mya	Ancestral reptiles 350 mya	Monotremes 320–240 mya	Marsupials 170–130 mya	Monkeys 130–80 mya	Humans 50–30 mya

SRY

area that cannot cross over

A Before 350 mya, sex was determined by temperature, not by chromosome differences.

B The *SRY* gene begins to evolve 350 mya. The DNA sequences of the chromosomes diverge as other mutations accumulate.

C By 320–240 mya, the DNA sequences of the chromosomes are so different that the pair can no longer cross over in one region. The Y chromosome begins to get shorter.

D Three more times, the pair stops crossing over in yet another region. Each time, the DNA sequences of the chromosomes diverge, and the Y chromosome shortens. Today, the pair crosses over only at a small region near the ends.

Figure 14.10 Evolution of the Y chromosome. Today, the *SRY* gene determines male sex. Homologous regions of the chromosomes are shown in *pink*; mya, million years ago.

Chromosome Changes in Evolution

As you can see, large-scale alterations in chromosome structure may reduce an individual's fertility. Individuals who are heterozygous for such changes may not be able to produce offspring at all. However, individuals homozygous for an inversion sometimes become the founders of new species. It may seem as if this outcome would be exceedingly rare, but it is not. Speciation can and does occur by large-scale changes in chromosomes.

Karyotyping and DNA sequence comparisons show that the chromosomes of all species contain evidence of major structural alterations. For example, duplications have often allowed a copy of a gene to mutate while the original carried out its unaltered function. The multiple and strikingly similar globin chain genes of humans and other primates apparently evolved by this process. Four globin chains associate in each hemoglobin molecule (Section 9.6). Different alleles specify different versions of the chains. Which versions of the chains get assembled into a hemoglobin molecule determine the oxygen-binding characteristics of the resulting protein.

As another example, X and Y chromosomes were once homologous autosomes in reptilelike ancestors of mammals (Figure 14.10). Ambient temperature probably determined the gender of those organisms, as it still does in turtles and some other modern reptiles. Then, about 350 million years ago (mya), a gene on one of the two homologous chromosomes mutated. The change, which was the beginning of the male sex determination gene *SRY*,

interfered with crossing over during meiosis. A reduced frequency of crossing over allowed the chromosomes to diverge around the changed region. Mutations began to accumulate separately in the two chromosomes. Over evolutionary time, the chromosomes became so different that they no longer crossed over at all in the changed region, so they diverged even more. Today, the Y chromosome is much smaller than the X, and only retains about 5 percent homology with it. The Y crosses over mainly with itself—by translocating duplicated regions of its own DNA.

Some chromosome structure changes contributed to differences among closely related organisms, such as apes and humans. Human somatic cells have twenty-three pairs of chromosomes, but those of chimpanzees, gorillas, and orangutans have twenty-four. Thirteen human chromosomes are almost identical with chimpanzee chromosomes. Nine more are similar, except for some inversions. One human chromosome matches up with two in chimpanzees and the other great apes (Figure 14.11). During human evolution, two chromosomes evidently fused end to end and formed our chromosome 2. How do we know? The region where the fusion occurred contains the remnants of a telomere, which is a special DNA sequence that caps the ends of chromosomes.

telomere sequence

human   chimpanzee

Figure 14.11
Human chromosome 2 compared with chimpanzee chromosomes 2A and 2B.

deletion Loss of part of a chromosome.
duplication Repeated section of a chromosome.
inversion Structural rearrangement of a chromosome in which part of it becomes oriented in the reverse direction.
translocation Structural change of a chromosome in which a broken piece gets reattached in the wrong location.

Take-Home Message Does chromosome structure change?

> A segment of a chromosome may be duplicated, deleted, inverted, or translocated. Such a change is usually harmful or lethal, but may be conserved in the rare circumstance that it has a neutral or beneficial effect.

› Occasionally, abnormal events occur before or during meiosis, and new individuals end up with the wrong chromosome number. Consequences range from minor to lethal changes in form and function.

‹ Links to Sampling error and bias 1.8, Meiosis 12.3, Gamete formation 12.5

Less than 1 percent of children are born with a chromosome number that differs from their parents. In humans, such major changes to the genetic blueprint can have serious effects on an individual's health.

Chromosome number changes often arise through **nondisjunction**, in which a cell's chromosomes do not separate properly during nuclear division. Nondisjunction during meiosis (Figure 14.12) can affect the chromosome number at fertilization. For example, suppose that a normal gamete fuses with an $n+1$ gamete (one with an extra chromosome). The new individual will be trisomic ($2n+1$), having three of one type of chromosome and two of every other type. As another example, if an $n-1$ gamete fuses with a normal n gamete, the new individual will be $2n-1$, or monosomic.

Trisomy and monosomy are types of **aneuploidy**, a condition in which cells have too many or too few copies of a chromosome. Autosomal aneuploidy is usually fatal in humans. However, about 70 percent of flowering plant species, and some insects, fishes, and other animals, are **polyploid**, which means that their cells have three or more of each type of chromosome.

Autosomal Change and Down Syndrome A few trisomic humans are born alive, but only those that have trisomy 21 will survive infancy. A newborn with three chromosomes 21 will develop Down syndrome. This autosomal disorder occurs once in 800 to 1,000 births, and it affects more than 350,000 people in the United States alone (Figure 14.13). Individuals with Down syndrome have upward-slanting eyes, a fold of skin that starts at the inner corner of each eye, a deep crease across the sole of each palm and foot, one (instead of two) horizontal furrows on their fifth fingers, slightly flattened facial features, and other symptoms.

Not all of the outward symptoms develop in every individual. That said, trisomic 21 individuals tend to have moderate to severe mental impairment and heart problems. Their skeleton grows and develops abnormally, so older children have short body parts, loose joints, and misaligned bones of the fingers, toes, and hips. The muscles and reflexes are weak, and motor skills such as speech develop slowly. With medical care, trisomy 21 individuals live about fifty-five years. Early training can help affected individuals learn to care for themselves and to take part in normal activities.

Change in the Sex Chromosome Number

Nondisjunction also causes alterations in the number of X and Y chromosomes, with a frequency of about 1 in 400 live births. Most often, such alterations lead to difficulties in learning and impaired motor skills such as a speech delay, but problems may be so subtle that the underlying cause is never diagnosed.

Female Sex Chromosome Abnormalities Individuals with Turner syndrome have an X chromosome and no corresponding X or Y chromosome (XO). The syndrome probably arises most frequently as an outcome of inheriting an unstable Y chromosome from the father: The

| Metaphase I | Anaphase I | Telophase I | Metaphase II | Anaphase II | Telophase II |

Figure 14.12 An example of nondisjunction during meiosis. Of the two pairs of homologous chromosomes shown here, one fails to separate during anaphase I. The chromosome number is altered in the resulting gametes.

Figure 14.13 Down syndrome, genotype and phenotype.

>> Figure It Out Is the karyotype of an individual who is male or female?

Answer: Male (XY)

zygote was genetically male, but the Y chromosome broke up and got lost early in development.

The incidence of Turner syndrome does not rise with maternal age, and there are fewer people affected by it than other chromosome abnormalities: Only about 1 in 2,500 newborn girls has it. XO individuals grow up well proportioned but short (with an average height of four feet, eight inches). Their ovaries do not develop properly, so they do not make enough sex hormones to become sexually mature. The development of secondary sexual traits such as breasts is also inhibited.

A female may inherit multiple X chromosomes. XXX syndrome occurs in about 1 of 1,000 births. Only one X chromosome is typically active in female cells (Section 10.4), so having extra X chromosomes usually does not result in physical or medical problems.

Male Sex Chromosome Abnormalities About 1 out of every 500 males has an extra X chromosome (XXY). Most cases are an outcome of nondisjunction during meiosis. The resulting disorder, Klinefelter syndrome, develops at puberty. XXY males tend to be overweight, tall, and within a normal range of intelligence. They make more estrogen and less testosterone than normal males, and this hormone imbalance has feminizing effects. Affected men tend to have small testes and prostate glands, low sperm counts, sparse facial and body hair, high-pitched voices, and enlarged breasts. Testosterone injections during puberty can reverse these traits.

About 1 in 1,000 males is born with an extra Y chromosome (XYY). Adults tend to be taller than average and have mild mental impairment, but most are otherwise normal. XYY men were once thought to be predisposed to a life of crime. This misguided view was based on sampling error (too few cases in narrowly chosen groups such as prison inmates) and bias (the researchers who gathered the karyotypes also took the personal histories of the participants). That view was disproven in 1976, when a geneticist reported results from his study of 4,139 tall males, all twenty-six years old, who had registered for military service. Besides their data from physical examinations and intelligence tests, the records offered clues to social and economic status, education, and any criminal convictions. Only twelve of the males studied were XYY, which meant that the "control group" had more than 4,000 males. The only findings? Mentally impaired, tall males who engage in criminal deeds are more likely to get caught, irrespective of karyotype.

aneuploidy A chromosome abnormality in which an individual's cells carry too many or too few copies of a particular chromosome.
nondisjunction Failure of sister chromatids or homologous chromosomes to separate during nuclear division.
polyploid Having three or more of each type of chromosome characteristic of the species.

Take-Home Message **What are the effects of chromosome number changes?**

❯ Nondisjunction can change the number of autosomes or sex chromosomes in gametes. Such changes usually cause genetic disorders in offspring.

❯ Sex chromosome abnormalities are usually associated with some degree of learning difficulty and motor skill impairment.

> Our understanding of human inheritance can be used to provide prospective parents with information about the health of their future children.
‹ Links to Probability 1.8, Amino acids 3.5, Metabolic pathways 5.5

Studying human inheritance patterns has given us many insights into how genetic disorders arise and progress, and how to treat them. Surgery, prescription drugs, hormone replacement therapy, and dietary controls can minimize and in some cases eliminate the symptoms of a genetic disorder. Some disorders can be detected early enough to start countermeasures before symptoms develop. Most hospitals in the United States now screen newborns for mutations in the gene for phenylalanine hydroxylase, an enzyme that catalyzes the conversion of the amino acid phenylalanine to tyrosine. Without a functional form of this enzyme, the body becomes deficient in tyrosine, and phenylalanine accumulates to high levels. The imbalance inhibits protein synthesis in the brain, which in turn results in the severe neurological symptoms characteristic of *phenylketonuria*, or PKU. Restricting all intake of phenylalanine can slow the progression of PKU, so routine early screening has resulted in fewer individuals developing the disorder.

Prospective parents worried about the possibility that a future child of theirs might have a genetic disorder also benefit from human genetics studies. The probability that a child will inherit a genetic disorder can be estimated by checking parental karyotypes and pedigrees, and testing the parents for alleles known to be associated with genetic disorders. This type of genetic screening is typically done before pregnancy, to help the prospective parents make decisions about family planning. Most couples would choose to know if their future children face a high risk of inheriting a severe genetic disorder, but the information may come at a heavy price. Learning about a life-threatening allele in your DNA can be devastating.

Prenatal Diagnosis

Genetic screening can also be done post-conception, in which case it is called prenatal diagnosis. Prenatal means before birth. *Embryo* is a term that applies until eight weeks after fertilization, after which *fetus* is appropriate. Prenatal diagnosis checks for physical and genetic abnormalities in an embryo or fetus. It can often reveal the presence of a genetic disorder in an unborn child. More than 30 conditions, including aneuploidy, hemophilia, Tay–Sachs disease, sickle-cell anemia, muscular dystrophy, and cystic fibrosis, are detectable prenatally. If the disorder is treatable, early detection allows the newborn to receive prompt and appropriate treatment. A few defects are even surgically correctable before birth. Prenatal diagnosis also gives parents time to prepare for the birth of an

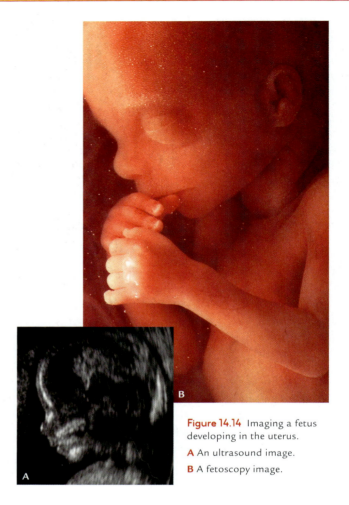

Figure 14.14 Imaging a fetus developing in the uterus.
A An ultrasound image.
B A fetoscopy image.

affected child, and an opportunity to decide whether to continue with the pregnancy or terminate it.

As an example of how prenatal diagnosis works, consider a thirty-five-year-old woman who becomes pregnant. Her doctor will probably use a noninvasive procedure, *obstetric sonography*, in which ultrasound waves directed across the woman's abdomen form images of the fetus's developing limbs and internal organs (Figure 14.14A). An ultrasound image is not very detailed, but there is no detectable risk to the pregnancy. The images may reveal physical defects associated with a genetic disorder, in which case a more invasive technique would be recommended for further diagnosis. *Fetoscopy* yields images of the fetus that are much higher in resolution than ultrasound images (Figure 14.14B). With this procedure, sound waves are pulsed from inside the mother's uterus. A sample of fetal blood is often drawn at the same time.

Human genetics studies show that our thirty-five-year-old woman has about a 1 in 80 chance that her baby will be born with a chromosomal abnormality—a risk more than six times higher than when she was twenty years old. Thus, even if no abnormalities were detected by

amniotic sac

placenta

Figure 14.15 An 8-week-old fetus. With amniocentesis, fetal cells shed into the fluid inside the amniotic sac are tested for genetic disorders. Chorionic villus sampling tests cells of the chorion, which is part of the placenta.

ultrasound, she probably will be offered a more thorough diagnostic procedure, *amniocentesis*, in which a small sample of fluid is drawn from the amniotic sac enclosing the fetus (Figure 14.15). The fluid contains cells shed by the fetus, and those cells can be tested for genetic disorders. *Chorionic villus sampling* (CVS) can be done earlier than amniocentesis. With this technique, a few cells from the chorion are removed and tested for genetic disorders. The chorion is a membrane that surrounds the amniotic sac and helps form the placenta, an organ that allows substances to be exchanged between mother and embryo.

An invasive procedure often carries a risk to the fetus. For example, if a punctured amniotic sac does not reseal itself quickly, too much fluid may leak out of it, resulting in miscarriage. The risks vary by the procedure. Amniocentesis has improved so much that, in the hands of a skilled physician, the procedure no longer increases the risk of miscarriage. CVS occasionally disrupts the placenta's development and thus causes underdeveloped or missing fingers and toes in 0.3 percent of newborns. Fetoscopy raises the miscarriage risk by 2 to 10 percent.

Preimplantation Diagnosis

Reproductive interventions such as preimplantation diagnosis offer an alternative to couples who discover they are at high risk of having a child with a genetic disorder.

Chinese and Europeans do not share any skin pigmentation allele that does not also occur in other populations. However, most people of Chinese descent carry a particular allele of the *DCT* gene, the product of which helps convert tyrosine to melanin. Few people of European or African descent have this allele. Taken together, the distribution of the *SLC24A5* and *DCT* genes suggests that (1) an African population was ancestral to both the Chinese and Europeans, and (2) Chinese and European populations separated before their pigmentation genes mutated and their skin color changed.

How Would You Vote? Physical attributes such as skin color, which have a genetic basis, are often used to define race. Do twins such as Kian and Remee belong to different races? See CengageNow for details, then vote online (cengagenow.com).

Preimplantation diagnosis is a procedure that relies on *in vitro* fertilization, in which sperm and eggs taken from prospective parents are mixed in a test tube. If an egg becomes fertilized, the resulting zygote will begin to divide. In about forty-eight hours, it will have become an embryo that consists of a ball of eight cells (Figure 14.16). All of the cells in this ball have the same genes, but none has yet committed to being specialized one way or another. Doctors can remove one of these undifferentiated cells and analyze its genes. The withdrawn cell will not be missed. If the embryo has no detectable genetic defects, it is inserted into the woman's uterus to develop. Many of the resulting "test-tube babies" are born in good health.

Figure 14.16 Clump of cells formed by three mitotic divisions after *in vitro* fertilization. All eight of the cells are identical and one can be removed for genetic analysis to determine whether the embryo carries any genetic defects.

Take-Home Message How do we use what we know about human inheritance?

> Genetic testing can provide prospective parents with information about the health of their future children.

Summary

Section 14.1 Like most other human traits, skin color has a genetic basis. Minor differences in the alleles that govern melanin synthesis and the deposition of melanosomes affect skin color. The differences probably evolved as a balance between vitamin production and protection against harmful UV radiation.

Section 14.2 Geneticists use **pedigrees** to track certain traits through generations of a family. Such studies can reveal inheritance patterns for alleles that can be predictably associated with specific phenotypes, including genetic abnormalities or disorders. A genetic abnormality is an uncommon version of a heritable trait that does not result in medical problems. A genetic disorder is a heritable condition that sooner or later results in mild or severe medical problems.

Section 14.3 An allele on an autosome is inherited in an autosomal dominant pattern if the trait associated with the allele appears in heterozygous individuals. Such traits tend to appear in every generation of families that carry the allele. An allele on an autosome is inherited in an autosomal recessive pattern if the trait associated with the allele only appears in homozygous individuals. Such traits can skip generations.

Section 14.4 An allele is inherited in an X-linked pattern when it occurs on the X chromosome. Most X-linked inheritance disorders are recessive, because X-linked dominant alleles tend to be lethal in male embryos. X-linked recessive disorders tend to appear in men more often than in women. This is because women have two X chromosomes, so they can be heterozygous for the recessive allele. Men can transmit an X-linked allele to their daughters, but not to their sons. Only a woman can pass an X-linked allele to a son.

Section 14.5 Major changes in chromosome structure include **duplications**, **deletions**, **inversions**, and **translocations**. Most major alterations are harmful or lethal in humans. Even so, many major structural changes have accumulated in the chromosomes of all species over evolutionary time.

Section 14.6 Changes in chromosome number are usually an outcome of **nondisjunction**, in which chromosomes fail to separate properly during meiosis. Such changes tend to cause genetic disorders among the resulting offspring. In **aneuploidy**, an individual's cells have too many or too few copies of a chromosome.

The most common aneuploidy, trisomy 21, causes Down syndrome. Most other cases of autosomal aneuploidy are lethal in embryos. **Polyploid** individuals have three or more of each type of chromosome. Polyploidy is lethal in humans, but many flowering plants, and some insects, fishes, and other animals, are polyploid.

A change in the number of sex chromosomes usually results in some degree of impairment in learning and motor skills. These problems can be so subtle that the underlying cause may not ever be diagnosed, as among XXY, XXX, and XYY children.

Section 14.7 Prospective parents can estimate their risk of transmitting a harmful allele to offspring with genetic screening, in which their pedigrees and genotype are analyzed by a genetic counselor. Prenatal genetic testing of an embryo or fetus can reveal genetic abnormalities or disorders before birth.

Self-Quiz   Answers in Appendix III

1. Constructing a pedigree is useful when studying inheritance patterns in organisms that _____ .
   a. produce many offspring per generation
   b. produce few offspring per generation
   c. have a very large chromosome number
   d. reproduce sexually
   e. have a fast life cycle

2. Pedigree analysis is necessary when studying human inheritance patterns because _____ .
   a. humans have more than 20,000 genes
   b. of ethical problems with experimenting on humans
   c. inheritance in humans is more complicated than in other organisms
   d. genetic disorders occur in humans
   e. all of the above

3. A recognized set of symptoms that characterize a genetic disorder is a(n) _____ .
   a. syndrome   b. disease   c. abnormality

4. If one parent is heterozygous for a dominant allele on an autosome and the other parent does not carry the allele, any child of theirs has a _____ chance of being heterozygous.
   a. 25 percent   c. 75 percent
   b. 50 percent   d. no chance; it will die

5. Is this statement true or false? A son can inherit an X-linked recessive allele from his father.

6. A trait that is present in a male child but not in either of his parents is characteristic of _____ inheritance.
   a. autosomal dominant   d. It is not possible to
   b. autosomal recessive   answer this question
   c. X-linked recessive   without more information.

7. Color blindness is a case of _____ inheritance.
   a. autosomal dominant   c. X-linked dominant
   b. autosomal recessive   d. X-linked recessive

8. What do you think the pattern of inheritance of the human *SRY* gene is called?

Data Analysis Activities

Skin Color Survey of Native Peoples A 2000 study measured average skin color of people native to more than fifty regions, and correlated them to the amount of UV radiation received in those regions. Some of their results are shown in Figure 14.17.

1. Which country receives the most UV radiation? The least?
2. The people native to which country have the darkest skin? The lightest?
3. According to these data, how does the skin color of indigenous peoples correlate with the amount of UV radiation incident in their native regions?

Country	Skin Reflectance	UVMED
Australia	19.30	335.55
Kenya	32.40	354.21
India	44.60	219.65
Cambodia	54.00	310.28
Japan	55.42	130.87
Afghanistan	55.70	249.98
China	59.17	204.57
Ireland	65.00	52.92
Germany	66.90	69.29
Netherlands	67.37	62.58

Figure 14.17
Skin color of indigenous peoples and regional incident UV radiation. Skin reflectance measures how much light of 685 nanometers wavelength is reflected from skin; UVMED is the annual average UV radiation received at Earth's surface.

---

9. A female child inherits one X chromosome from her mother and one from her father. What sex chromosome does a male child inherit from each of his parents?

10. Nondisjunction may occur during _____ .
    a. mitosis
    c. fertilization
    b. meiosis
    d. both a and b

11. Nondisjunction can result in _____ .
    a. polyploidy
    c. crossing over
    b. aneuploidy
    d. pleiotropy

12. Nondisjunction can occur during _____ of meiosis.
    a. anaphase I
    c. anaphase II
    b. telophase I
    d. a or c

13. Is this statement true or false? Body cells may inherit three or more of each type of chromosome characteristic of the species, a condition called polyploidy.

14. Klinefelter syndrome (XXY) can be easily diagnosed by _____ .
    a. pedigree analysis
    c. karyotyping
    b. aneuploidy
    d. phenotypic treatment

15. Match the chromosome terms appropriately.
    ___ polyploidy
    ___ deletion
    ___ aneuploidy
    ___ translocation
    ___ syndrome
    ___ nondisjunction during meiosis
    a. symptoms of a genetic disorder
    b. segment of a chromosome moves to a nonhomologous chromosome
    c. extra sets of chromosomes
    d. results in gametes with the wrong chromosome number
    e. a chromosome segment lost
    f. one extra chromosome

Additional questions are available on **CENGAGE**NOW.

Animations and Interactions on **CENGAGE**NOW:
❯ A human pedigree; Autosomal dominant inheritance; Autosomal recessive inheritance; Duplications, deletions, inversions, and translocations; X-linked inheritance.

Genetics Problems   Answers in Appendix III

1. Does the phenotype indicated by the red circles and squares in this pedigree show an inheritance pattern that is autosomal dominant, autosomal recessive, or X-linked?

2. Human females are XX and males are XY.

   a. With respect to X-linked alleles, how many different types of gametes can a male produce?

   b. If a female is homozygous for an X-linked allele, how many types of gametes can she produce with respect to that allele?

   c. If a female is heterozygous for an X-linked allele, how many types of gametes can she produce with respect to that allele?

3. People homozygous for the *HbS* allele develop sickle-cell anemia (Section 9.6). Heterozygotes have fewer symptoms. A couple who are both heterozygous for the *HbS* allele plan to have children. For each of the pregnancies, state the probability that they will have a child who is:

   a. homozygous for the *HbS* allele
   b. homozygous for the normal allele
   c. heterozygous (having the normal and the *HbS* allele)

4. A few individuals with Down syndrome have forty-six chromosomes: two normal-appearing chromosomes 21, and a longer-than-normal chromosome 14. Speculate on how this chromosome abnormality arises.

5. An allele responsible for Marfan syndrome (Section 13.5) is inherited in an autosomal dominant pattern. What is the chance that any child will inherit it if one parent does not carry the allele and the other is heterozygous?

❮ Links to Earlier Concepts

This chapter builds on your understanding of DNA's structure (Sections 8.3, 8.4, and 13.2) and replication (8.6). Clones (8.1), gene expression (9.2, 9.3), and knockouts (10.3) are important in genetic engineering, particularly as they apply to research on human traits (13.6) and genetic disorders (14.2). You will revisit tracers (2.2), triglycerides (3.4), denaturation (3.6), β-carotene (6.2), the *lac* operon (10.5), and cancer (11.6).

Key Concepts

DNA Cloning

Researchers routinely make recombinant DNA by cutting and pasting together DNA from different species. Plasmids and other vectors can carry foreign DNA into host cells.

Finding Needles in Haystacks

DNA libraries, hybridization, and PCR are techniques that allow researchers to isolate and make many copies of a fragment of DNA they want to study.

15 Biotechnology

15.1 Personal DNA Testing

About 99 percent of your DNA is exactly the same as everyone else's. If you compared your DNA with your neighbor's, about 29.7 billion nucleotides of the two sequences would be identical. The remaining 30 million or so are sprinkled throughout your chromosomes, mainly as single nucleotide differences.

The sprinkling is not entirely random; some regions of DNA vary less than others. Such conserved regions are of particular interest to researchers because they are the ones most likely to have an essential function. When a conserved sequence does vary among people, the variation tends to be in particular nucleotides. A nucleotide difference carried by a measurable percentage of a population, usually above 1 percent, is called a single-nucleotide polymorphism, or SNP (pronounced "snip").

Alleles of most genes differ by single nucleotides, and differences in alleles are the basis of the variation in human traits that makes each individual unique (Section 12.2). Thus, SNPs account for many of the differences in the way humans look, and they also have a lot to do with differences in the way our bodies work—how we age, respond to drugs, weather assaults by pathogens and toxins, and so on.

Consider a gene, *APOE*, that specifies apolipoprotein E, a protein component of lipoprotein particles (Section 3.5). One allele of this gene, *ε4*, is carried by about 25 percent of people. Nucleotide 4,874 of this allele is a cytosine instead of the normal thymine, a SNP that results in a single amino acid change in the protein product of the gene. How this change affects the function of apolipoprotein E is unclear, but we do know that having the ε4 allele increases one's risk of developing Alzheimer's disease later in life, particularly in people homozygous for it.

About 4.5 million SNPs in human DNA have been identified, and that number is growing every day. A few companies are now offering to determine some of the SNPs you carry (Figure 15.1). The companies extract your DNA from the cells in a few drops of spit, then analyze it for SNPs.

Personal genetic testing may soon revolutionize medicine by allowing physicians to customize treatments on the basis of an individual's genetic makeup. For example, an allele associated with a heightened risk of a particular medical condition could be identified long before symptoms actually appear. People with that allele could then be encouraged to make life-style changes

Figure 15.1 Personal genetic testing. *Right*, a SNP-chip. Personal DNA testing companies use chips like this one to analyze their customers' chromosomes for SNPs. This chip, shown actual size, reveals which versions of 906,600 SNPs occur in the individual's DNA.

known to delay the onset of the condition. For some conditions, treatment that begins early enough may prevent symptoms from developing at all. Physicians could design treatments to fit the way a condition is likely to progress in the individual, and also to prescribe only those drugs that will work in the person's body.

You are now at a time when geneticists hold molecular keys to the kingdom of inheritance. As you will see, what they are unlocking is already having an impact on all of us.

DNA Sequencing
Sequencing reveals the linear order of nucleotides in DNA. Comparing genomes offers insights into human genes and evolution. An individual can be identified by unique parts of their DNA.

Genetic Engineering
Genetic engineering, the directed modification of an organism's genes, is now a routine part of research and development. Genetically modified organisms are now quite common.

Gene Therapy
Genetic engineering continues to be tested in medical applications. It also continues to raise ethical questions.

15.2 Cloning DNA

> Researchers cut up DNA from different sources, then paste the resulting fragments together.
> Cloning vectors can carry foreign DNA into host cells.
< Links to Clones 8.1, Discovery of DNA structure 8.3, Base pairing and directionality of DNA strands 8.4, DNA ligase 8.6, mRNA 9.2, Introns 9.3, The *lac* operon 10.5

In the 1950s, excitement over the discovery of DNA's structure gave way to frustration: No one could determine the order of nucleotides in a molecule of DNA. Identifying a single base among thousands or millions of others turned out to be a huge technical challenge. A seemingly unrelated discovery offered a solution. Viruses called bacteriophages infect bacteria by injecting DNA into them (Section 8.3). Some bacteria are resistant to infection, and Werner Arber, Hamilton Smith, and their coworkers discovered why: Special enzymes inside these bacteria chop up any injected viral DNA before it has a chance to integrate into the bacterial chromosome. The enzymes restrict viral growth; hence their name, **restriction enzymes**. A restriction enzyme cuts DNA wherever a specific nucleotide sequence occurs (Figure 15.2). For example, the enzyme *Eco*RI (named after *E. coli,* the bacteria from which it was isolated) cuts DNA at the sequence GAATTC ❶. Other restriction enzymes cut different sequences.

The discovery of restriction enzymes allowed researchers to cut gigantic molecules of chromosomal DNA into manageable and predictable chunks. It also allowed them to combine DNA fragments from different organisms. How? Many restriction enzymes, including *Eco*RI, leave single-stranded tails on DNA fragments ❷. Researchers realized that complementary tails will base-pair ❸. Thus, the tails are called "sticky ends," because two fragments of DNA stick together when their matching tails base-pair. Regardless of the source of DNA, any two fragments will stick together, as long as their tails are complementary.

Figure 15.3 Plasmid cloning vectors. **A** Micrograph of a plasmid. **B** A commercial plasmid cloning vector. Restriction enzyme recognition sequences are indicated on the *right* by the name of the enzyme that cuts them. Researchers insert foreign DNA into the vector at these sites. Bacterial genes (*gold*) help researchers identify host cells that take up a vector with inserted DNA. This vector carries two antibiotic resistance genes and the *lac* operon (Section 10.5).

Base-paired sticky ends can be covalently bonded together with the enzyme DNA ligase ❹. Thus, using appropriate restriction enzymes and DNA ligase, researchers can cut and paste DNA from different organisms. The result, a hybrid molecule composed of DNA from two or more organisms, is called **recombinant DNA**.

Why make recombinant DNA? It is the first step in **DNA cloning**, a set of laboratory methods that uses living cells to mass-produce specific DNA fragments. For example, researchers often insert DNA fragments into **plasmids**, small circles of DNA independent of the chromosome. Before a bacterium divides, it copies any plasmids it carries along with its chromosome, so both descendant cells get one of each. If a plasmid carries a fragment of foreign DNA, that fragment gets copied and distributed to descendant cells along with the plasmid.

❶ A restriction enzyme recognizes a specific base sequence (*orange* boxes) in DNA from any source.

❷ The enzyme cuts DNA from two sources into fragments. This enzyme leaves sticky ends.

❸ When the DNA fragments from the two sources are mixed together, matching sticky ends base-pair with each other.

❹ DNA ligase joins the base-paired DNA fragments. Molecules of recombinant DNA are the result.

Figure 15.2 Animated Using restriction enzymes to make recombinant DNA.

>> **Figure It Out** Why did the enzyme cut both strands of DNA?

Answer: Because the recognition sequence occurs on both strands.

A A restriction enzyme cuts a specific base sequence in chromosomal DNA and in a plasmid cloning vector.

B A fragment of chromosomal DNA and the plasmid base-pair at their sticky ends. DNA ligase joins the two pieces of DNA.

C The recombinant plasmid is inserted into a host cell. When the cell multiplies, it makes multiple copies of the plasmids.

Figure 15.4 Animated An example of cloning. Here, a fragment of chromosomal DNA is inserted into a plasmid.

Thus, plasmids can be used as **cloning vectors**, which are molecules that carry foreign DNA into host cells (Figure 15.3). A host cell into which a cloning vector has been inserted can be grown in the laboratory (cultured) to yield a huge population of genetically identical cells, or clones (Section 8.1). Each clone contains a copy of the vector and the fragment of foreign DNA it carries (Figure 15.4). Researchers can harvest the DNA fragment from the clones in large quantities.

cDNA Cloning

Cloning eukaryotic genes can be tricky, because eukaryotic DNA contains introns (Section 9.3). Unless you are a eukaryotic cell, it is not very easy to find the parts of the DNA that encode gene products. Thus, researchers who study eukaryotic genes and their expression work with mRNA, because the introns have already been snipped out of it. However, restriction enzymes and DNA ligase only work on double-stranded DNA, not single-stranded RNA. In order to study mRNA, researchers first make a DNA copy of it, then clone the DNA.

---

cDNA DNA synthesized from an RNA template by the enzyme reverse transcriptase.
cloning vector A DNA molecule that can accept foreign DNA, be transferred to a host cell, and get replicated in it.
DNA cloning Set of procedures that uses living cells to make many identical copies of a DNA fragment.
plasmid Of many bacteria and archaeans, a small ring of nonchromosomal DNA replicated independently of the chromosome.
recombinant DNA A DNA molecule that contains genetic material from more than one organism.
restriction enzyme Type of enzyme that cuts specific nucleotide sequences in DNA.
reverse transcriptase A viral enzyme that uses mRNA as a template to make a strand of cDNA.

An mRNA can be transcribed into a molecule of double-stranded DNA by a process that is essentially the reverse of RNA transcription. In this process, researchers use **reverse transcriptase**, a replication enzyme made by certain types of viruses, to assemble a strand of complementary DNA, or **cDNA**, on an mRNA template:

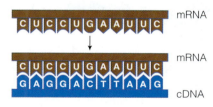

DNA polymerase added to the mixture strips the RNA from the hybrid molecule as it copies the cDNA into a second strand of DNA. The outcome is a double-stranded DNA copy of the original mRNA:

*Eco*RI recognition site

Like any other DNA, double-stranded cDNA may be cut with restriction enzymes and pasted into a cloning vector using DNA ligase.

Take-Home Message What is DNA cloning?

❯ DNA cloning uses living cells to mass-produce particular DNA fragments. Restriction enzymes cut DNA into fragments, then DNA ligase seals the fragments into cloning vectors. Recombinant DNA molecules result.

❯ A cloning vector that holds foreign DNA can be introduced into a living cell. When the host cell divides, it gives rise to huge populations of genetically identical cells (clones), each of which contains a copy of the foreign DNA.

❯ DNA libraries and the polymerase chain reaction (PCR) help researchers isolate particular DNA fragments.
❮ Links to Tracers 2.2, Denaturation 3.6, Base pairing 8.4, DNA replication 8.6

A Individual bacterial cells from a DNA library are spread over the surface of a solid growth medium. The cells divide repeatedly and form colonies—clusters of millions of genetically identical descendant cells.

B A piece of special paper pressed onto the surface of the growth medium will bind some cells from each colony.

C The paper is soaked in a solution that ruptures the cells and releases their DNA. The DNA clings to the paper in spots mirroring the distribution of colonies.

D A probe is added to the liquid bathing the paper. The probe hybridizes (base-pairs) with the spots of DNA that contain complementary base sequences.

E The bound probe makes a spot. Here, one radioactive spot darkens x-ray film. The position of the spot is compared to the positions of the original bacterial colonies. Cells from the colony that made the spot are cultured, and the DNA they contain is harvested.

Figure 15.5 Animated Nucleic acid hybridization. In this example, a radioactive probe helps identify a bacterial colony that contains a targeted sequence of DNA.

Isolating Genes

The entire set of genetic material—the **genome**—of most organisms consists of thousands of genes. To study or manipulate a single gene, researchers must first separate the gene from all of the others. To do that, researchers often begin by cutting an organism's DNA into pieces, and then cloning all the pieces. The result is a genomic library, a set of clones that collectively contain all of the DNA in a genome. Researchers can also harvest mRNA, make cDNA copies of it, and then clone the cDNA to make a cDNA library. A cDNA library represents only those genes being expressed at the time the mRNA was harvested.

Genomic and cDNA libraries are **DNA libraries**—sets of cells that host various cloned DNA fragments. In such libraries, a cell that contains a particular DNA fragment of interest is mixed up with thousands or millions of others that do not. All the cells look the same, so researchers have to get tricky to find that one clone among all of the others—the needle in the haystack.

Using a probe is one trick. A **probe** is a fragment of DNA or RNA labeled with a tracer (Section 2.2). Researchers design probes to match a targeted DNA sequence. For example, they may synthesize an oligomer (a short chain of nucleotides) based on a known DNA sequence, then attach a radioactive phosphate group to it. The nucleotide sequence of a probe is complementary to that of the targeted gene, so the probe can base-pair with the gene. Base pairing between DNA (or DNA and RNA) from more than one source is called **nucleic acid hybridization**.

A probe mixed with DNA from a library base-pairs with (hybridizes to) the targeted gene (Figure 15.5). Researchers pinpoint a clone that hosts the gene by detecting the label on the probe. The clone is then cultured, and the DNA fragment of interest is extracted in bulk from the cultured cells.

PCR

The **polymerase chain reaction** (**PCR**) is a technique used to mass-produce copies of a particular section of DNA without having to clone it in living cells. The reaction can transform a needle in a haystack—that one-in-a-million fragment of DNA—into a huge stack of needles with a little hay in it (Figure 15.6).

DNA library Collection of cells that host different fragments of foreign DNA, often representing an organism's entire genome.
genome An organism's complete set of genetic material.
nucleic acid hybridization Base-pairing between DNA or RNA from different sources.
polymerase chain reaction (**PCR**) Method that rapidly generates many copies of a specific section of DNA.
primer Short, single strand of DNA designed to hybridize with a DNA fragment.
probe Short fragment of DNA labeled with a tracer; designed to hybridize with a nucleotide sequence of interest.

Figure 15.6 Animated Two rounds of PCR. Each cycle of this reaction can double the number of copies of a targeted section of DNA. Thirty cycles can make a billion copies.

targeted section

❶ DNA template (*blue*) is mixed with primers (*pink*), nucleotides, and heat-tolerant *Taq* DNA polymerase.

❷ When the mixture is heated, the double-stranded DNA template separates into single strands. When it is cooled, some of the primers base-pair with the template DNA.

❸ *Taq* polymerase begins DNA synthesis at the primers, so complementary strands of DNA form on the single-stranded templates.

❹ The mixture is heated again, and the double-stranded DNA separates into single strands. When it is cooled, some of the primers base-pair with the template DNA. The copied DNA also serves as a template.

❺ Each round of PCR reactions can double the number of copies of the targeted DNA section.

The starting material for PCR is any sample of DNA with at least one molecule of a target sequence. It might be DNA from a mixture of 10 million different clones, a sperm, a hair left at a crime scene, or a mummy. Essentially any sample that has DNA in it can be used for PCR.

The PCR reaction is based on DNA replication (Section 8.6). First, the starting material is mixed with DNA polymerase, nucleotides, and primers. **Primers** are short single strands of DNA that base-pair with a certain DNA sequence. In PCR, two primers are made. Each base-pairs with one end of the section of DNA to be amplified, or mass-produced **❶**. Researchers expose the reaction mixture to repeated cycles of high and low temperature. High temperature disrupts the hydrogen bonds that hold the two strands of a DNA double helix together (Section 8.4). Thus, during a high-temperature cycle, every molecule of double-stranded DNA unwinds and becomes single-stranded **❷**. During a low-temperature cycle, the single DNA strands hybridize with complementary partner strands, and double-stranded DNA forms again.

The DNA polymerases of most organisms denature at the high temperatures required to separate DNA strands. The kind that is used in PCR reactions, *Taq* polymerase, is from *Thermus aquaticus*. This bacterial species lives in hot springs and hydrothermal vents, so its DNA polymerase is necessarily heat-tolerant.

Taq polymerase recognizes hybridized primers as places to start DNA synthesis. During a low-temperature cycle, the enzyme starts replicating DNA where primers have hybridized with template **❸**. Synthesis proceeds along the template strand until the temperature rises and the DNA separates into single strands **❹**. The newly synthesized DNA is a copy of the targeted section.

When the mixture cools, the primers rehybridize, and DNA synthesis begins again. The number of copies of the targeted section of DNA can double with each cycle of heating and cooling **❺**. Thirty PCR cycles may amplify that number a billionfold.

Take-Home Message How do researchers study one gene in the context of many others?

> Researchers isolate one gene from the many other genes in a genome by making DNA libraries or using PCR.

> Probes are used to identify one clone that hosts a DNA fragment of interest among many other clones in a DNA library.

> PCR, the polymerase chain reaction, quickly mass-produces copies of a particular section of DNA.

❯ DNA sequencing reveals the order of nucleotide bases in a section of DNA.

❮ Links to Tracers 2.2, Nucleotides 8.4, DNA replication 8.6

Researchers determine the order of the nucleotide bases in DNA with **DNA sequencing** (Figure 15.7). The most commonly used method is similar to DNA replication, in that DNA polymerase synthesizes a strand of DNA based on the nucleotide sequence of a template molecule. Researchers mix the DNA to be sequenced (the template) with nucleotides, DNA polymerase, and a primer that hybridizes to the DNA. Starting at the primer, the polymerase joins free nucleotides into a new strand of DNA, in the order dictated by the sequence of the template.

DNA polymerase joins a nucleotide to a DNA strand only at the hydroxyl group on the strand's 3′ carbon (Section 8.6). The DNA sequencing reaction mixture includes four kinds of dideoxynucleotides, which have no hydroxyl group on their 3′ carbon ❶. During the reaction, a polymerase randomly adds either a regular nucleotide or a dideoxynucleotide to the end of a growing DNA strand. If it adds a dideoxynucleotide, the 3′ carbon of the strand will not have a hydroxyl group, so synthesis of the strand ends there ❷. After about 10 minutes, the reaction has produced millions of DNA fragments of all different lengths; most are incomplete copies of the starting DNA. All of the copies end with one of the four dideoxynucleotides ❸. For example, there will be many 10-base-pair-long copies of the template in the mixture. If the tenth base in the original DNA molecule was adenine, every one of those fragments will end with a dideoxyadenine.

The fragments are then separated by **electrophoresis**. With this technique, an electric field pulls the DNA fragments through a semisolid gel. DNA fragments of different sizes move through the gel at different rates. The shorter the fragment, the faster it moves, because shorter fragments slip through the tangled molecules of the gel faster than longer fragments do.

All fragments of the same length move through the gel at the same speed, so they gather into bands. All of the fragments in a given band have the same dideoxynucleotide at their ends. Each of the four types of dideoxynucleotides (A, C, G, or T) was labeled with a different colored pigment tracer, and those tracers now impart distinct colors to the bands ❹. Each color designates one of the four dideoxynucleotides, so the order of colored bands in the gel represents the DNA sequence ❺.

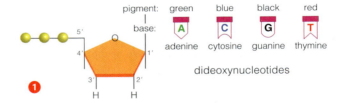

Figure 15.7 Animated DNA sequencing, in which DNA polymerase is used to incompletely replicate a section of DNA.

❶ Sequencing depends on dideoxynucleotides to terminate DNA replication. Each is labeled with a colored pigment. Compare Figure 8.7.

❷ DNA polymerase uses a section of DNA as a template to synthesize new strands of DNA. Synthesis of each new strand stops when a dideoxynucleotide is added.

❸ At the end of the reaction, there are many incomplete copies of the original DNA in the mixture.

❹ Electrophoresis separates the copied DNA fragments into bands according to their length. All of the DNA strands in each band end with the same dideoxynucleotide; thus, each band is the color of that dideoxynucleotide's tracer pigment.

❺ A computer detects and records the color of successive bands on the gel (see Figure 15.8 for an example). The order of colors of the bands represents the sequence of the template DNA.

Figure 15.8 Human genome sequencing. *Left*, some of the supercomputers used to sequence the human genome at Venter's Celera Genomics in Maryland. Information in Celera's SNP database is the basis of many new genetic tests. *Right*, a human DNA sequence, raw data.

The Human Genome Project

The technique we just described was invented in 1975. Ten years later, DNA sequencing had become so routine that people were thinking about sequencing the entire human genome: a daunting proposition, given that it consists of about 3 billion bases. At the time, the task would have required at least 6 million sequencing reactions, all done by hand. Proponents insisted that sequencing the genome would have enormous payoffs for medicine and research. Opponents said it would divert funds from work that was more urgent—and had a better chance of succeeding. But sequencing techniques kept getting better, so every year more bases could be sequenced in less time. Automated (robotic) DNA sequencing and PCR had just been invented. Both of these techniques were still cumbersome and expensive, but many researchers sensed their potential. Waiting for faster technologies seemed the most efficient way to sequence 3 billion bases, but how fast did they need to be before the project could begin?

A few private companies decided not to wait, and started to sequence the human genome. One of them intended to patent the sequence after it was determined. This development provoked widespread outrage, but it also spurred commitments in the public sector. In 1988, the National Institutes of Health (NIH) effectively annexed the project by hiring James Watson (of DNA structure fame) to head the official Human Genome Project, and providing $200 million per year to fund it. A consortium formed between the NIH and international institutions that were sequencing different parts of the genome. Watson set aside 3 percent of the funding for studies of ethical and social issues arising from the research. He later resigned over a patent disagreement, and geneticist Francis Collins took his place.

Amid the ongoing squabbles over patent issues, Celera Genomics formed in 1998. With biologist Craig Venter at its helm, the company intended to commercialize genetic information. Celera started to invent faster techniques for sequencing genomic DNA (Figure 15.8), because the first to have the complete sequence had a legal basis for patenting it. The competition motivated the public consortium to move its efforts into high gear.

Then, in 2000, U.S. President Bill Clinton and British Prime Minister Tony Blair jointly declared that the sequence of the human genome could not be patented. Celera kept on sequencing anyway. Celera and the public consortium separately published about 90 percent of the sequence in 2001. By 2003, fifty years after the discovery of the structure of DNA, the sequence of the human genome was officially completed. At this writing, about 99 percent of its coding regions have been identified. Researchers have not discovered what all of the genes encode, only where they are in the genome. What do we do with this vast amount of data? The next step is to find out what the sequence means.

DNA sequencing Method of determining the order of nucleotides in DNA.
electrophoresis Technique that separates DNA fragments by size.

Take-Home Message How is the order of nucleotides in DNA determined?

> With DNA sequencing, a strand of DNA is partially replicated. Electrophoresis is used to separate the resulting fragments of DNA, which are tagged with tracers, by length.

> Improved sequencing techniques and worldwide efforts allowed the human genome sequence to be determined.

> Comparing the sequence of the human genome with that of other species is helping us understand how the human body works.
> Unique sequences of genomic DNA can be used to distinguish an individual from all others.
< Links to Lipoproteins 3.5, DNA replication 8.6, Knockouts 10.3, Locus 13.2, Complex variation in traits 13.6

It took 15 years to sequence the human genome for the first time, but the techniques have improved so much that sequencing an entire genome now takes less than a month. Full genome sequencing is already available to the general public. However, even though we are able to determine the sequence of an individual's genome, it will be a long time before we understand all the information coded within that sequence.

The human genome contains a massive amount of seemingly cryptic data. Currently, the best way to decipher it is by comparing it to genomes of other organisms, the premise being that all organisms are descended from shared ancestors, so all genomes are related to some extent. We see evidence of such genetic relationships simply by comparing the raw sequence data, which, in some regions, is extremely similar across many species (Figure 15.9).

Comparing genomes is part of **genomics**, the study of genomes. Genomics is a broad field that encompasses whole-genome comparisons, structural analysis of gene products, and the study of small-scale variation. It is also providing powerful insights into evolution. For example, comparing primate genomes revealed how speciation can occur by structural changes in chromosomes (Section 14.5). Comparing genomes also revealed that changes in chromosome structure do not occur randomly. Rather, if a chromosome breaks, it tends to do it in a particular spot. Human, mouse, rat, cow, pig, dog, cat, and horse chromosomes have undergone several translocations at these breakage hot spots during evolution. In humans, chromosome abnormalities that contribute to the progression of cancer also occur at the very same hot spots.

Comparisons between coding regions of a genome are offering medical benefits. We have learned about the function of many human genes by studying their coun-

Figure 15.10 SNP–chip analysis. **A** Each spot is a region where the individual's genomic DNA has hybridized with one SNP. **B** The entire chip tests for 550,000 SNPs. The small *white* box indicates the magnified portion shown in **A**.

terpart genes in other species. For example, researchers comparing the human and mouse genomes discovered a human version of a mouse gene, *APOA5*, that encodes a lipoprotein (Section 3.5). Mice with an *APOA5* knockout have four times the normal level of triglycerides in their blood. The researchers then looked for—and found—a correlation between *APOA5* mutations and high triglyceride levels in humans. High triglycerides are a risk factor for coronary artery disease.

DNA Profiling

As you learned in Section 15.1, only about 1 percent of your DNA is unique. The shared part is what makes you human; the differences make you a unique member of the species. In fact, those differences are so unique that they can be used to identify you. Identifying an individual by his or her DNA is called **DNA profiling**.

```
758  GATAATCCTGTTTTGAACAAAAGGTCAAATTGCTGAATAGAAA-GTCTTGATTAACTAAAAGATGTACAAAGTGGAATTA 836   Human
752  GATAATCCTGTTTTGAACAAAAGGTCAAATTGCTGAATAGAAA-GTCTTGATTAACTAAAAGATGTACAAAGTGGAATTA 830   Mouse
751  GATAATCCTGTTTTGAACAAAAGGTCAAATTGCTGAATAGAAA-GTCTTGATTAACTAAAAGATGTACAAAGTGGAATTA 829   Rat
754  GATAATCCTGTTTTGAACAAAAGGTCAAATTGCTGAATAGAAA-GTCTTGATTAACTAAAAGATGTACAAAGTGGAATTA 832   Dog
782  GATAATCCTGTTTTGAACAAAAGGTCAAATTGCTGAATAGAAA-GTCTTGATTAACTAAAAGATGTACAAAGTGGAATTA 860   Chicken
758  GATAATCCTGTTTTGAACAAAAGGTCAAATTGCTGAATAGAAA-GTCTTGATTAAGTAAAAGATGTACAAAGTGGAATTA 836   Frog
823  GATAATCCTGTTTTGAACAAAAGGTCAGATTGCTGAATAGAAAAGGCTTGATTAAAGCAGAGATGTACAAAGTGGACGCA 902   Zebrafish
763  GATAATCCTGTTTTGAACAAAAGGTCAAATTGTTGAATAGAGACGCTTTGATAAAAGCGGAGGAGGTACAAAGTGGGACC- 841   Fugu
```

Figure 15.9 Genomic DNA alignment. This is a region of the gene for a DNA polymerase. Differences are highlighted. The chance that any two of these sequences would randomly match is about 1 in 10^{46}.

Figure 15.11 An individual's short tandem repeat profile. Each peak on the chart represents a test for one short tandem repeat. The size of a peak indicates the number of repeats at that locus.

For example, this individual has 14 repeats at the D3S1358 locus on chromosome 3, and 17 repeats on the other chromosome 3 (remember that human body cells are diploid, with two sets of chromosomes).

SNP analysis is one example of DNA profiling. SNP-chips are microscopic arrays (microarrays) of DNA samples that have been stamped in separate spots on small glass plates. Each sample is an oligomer with one SNP. In an SNP analysis, an individual's genomic DNA is washed over a SNP-chip. The DNA hybridizes only with oligomers that have a matching SNP sequence. Probes reveal where the genomic DNA has hybridized to an oligomer—and which SNPs are carried by the individual (Figure 15.10).

Another example of DNA profiling involves **short tandem repeats**, sections of DNA in which a series of 4 or 5 nucleotides is repeated several times in a row. Short tandem repeats tend to occur in predictable spots, but the number of repeats in each spot differs among individuals. For example, one person's DNA may have fifteen repeats of the bases TTTTC at a certain locus. Another person's DNA may have this sequence repeated only twice in the same locus. Such repeats slip spontaneously into DNA during replication, and their numbers grow or shrink over generations. Unless two people are identical twins, the chance that they have identical short tandem repeats in even three regions of DNA is 1 in 10^{18}, which is far more than the number of people on Earth. Thus, an individual's array of short tandem repeats is, for all practical purposes, unique.

Analyzing a person's short tandem repeats begins with PCR, which is used to copy ten to thirteen particular regions of chromosomal DNA known to have repeats. The sizes of the copied DNA fragments differ among most individuals, because the number of tandem repeats in those regions also differs. Thus, electrophoresis can be used to reveal an individual's unique array of short tandem repeats (Figure 15.11).

Analysis of short tandem repeats will soon be replaced by full genome sequencing, but for now it continues to be a common DNA profiling method. Geneticists compare short tandem repeats on Y chromosomes to determine relationships among male relatives, and to trace an individual's ethnic heritage. They also track mutations that accumulate in populations over time by comparing DNA profiles of living organisms with those of ancient ones. Such studies are allowing us to reconstruct population dispersals that happened long ago.

Short tandem repeat profiles are routinely used to resolve kinship disputes, and as evidence in criminal cases. Within the context of a criminal or forensic investigation, DNA profiling is called DNA fingerprinting. The Federal Bureau of Investigation maintains a database of DNA fingerprints. As of November 2009, the database contained the short tandem repeat profiles of 7.6 million offenders, and had been used in over 100,000 criminal investigations. DNA fingerprints have also been used to identify the remains of almost 300,000 people, including the individuals who died in the World Trade Center on September 11, 2001.

DNA profiling Identifying an individual by analyzing the unique parts of his or her DNA.
genomics The study of genomes.
short tandem repeats In chromosomal DNA, sequences of 4 or 5 bases repeated multiple times in a row.

Take-Home Message **What do we do with information about genomes?**

❯ Analysis of the human genome sequence is yielding new information about human genes and how they work.

❯ DNA profiling identifies individuals by the unique parts of their DNA.

> Bacteria and yeast are the organisms most commonly subjected to genetic engineering.
< Link to Gene expression 9.2

Traditional cross-breeding methods can alter genomes, but only if individuals with the desired traits will interbreed. Genetic engineering takes gene-swapping to an entirely different level. **Genetic engineering** is a laboratory process by which an individual's genome is deliberately modified. A gene may be altered and reinserted into an individual of the same species, or a gene from one species may be transferred to another to produce an organism that is **transgenic**. Both methods result in a **genetically modified organism**, or **GMO**.

The most common GMOs are bacteria and yeast. These cells have the metabolic machinery to make complex organic molecules, and they are easily modified. For example, the *E. coli* on the *left* have been modified to produce a fluorescent protein from jellyfish. The cells are genetically identical, so the visible variation in fluorescence among them reveals differences in gene expression. Such differences may help us discover why some bacteria of a population become dangerously resistant to antibiotics, and others do not.

Genetically modified bacteria expressing a jellyfish gene emit green light.

Bacteria and yeast have been modified to produce medically important proteins. People with diabetes were among the first beneficiaries of such organisms. Insulin for their injections was once extracted from animals, but it provoked an allergic reaction in some people. Human insulin, which does not provoke allergic reactions, has been produced by transgenic *E. coli* since 1982. Slight modifications of the gene have also yielded fast-acting and slow-release forms of human insulin.

Engineered microorganisms also produce proteins used in food manufacturing. For example, cheese is traditionally made with an extract of calf stomachs, which contain the enzyme chymotrypsin. Most cheese manufacturers now use chymotrypsin made by genetically engineered bacteria. Other examples are GMO-produced enzymes that improve the taste and clarity of beer and fruit juice, slow bread staling, or modify fats.

> Genetically engineered crop plants are widespread in the United States.
< Links to β-carotene 6.2, Promoters 9.3

Agrobacterium tumefaciens is a species of bacteria that infects many plants, including peas, beans, potatoes, and other important crops. It carries a plasmid with genes that cause tumors to form on infected plants; hence the name Ti plasmid (for Tumor-inducing). Researchers use the Ti plasmid as a vector to transfer foreign or modified genes into plants. They remove the tumor-inducing genes from the plasmid, then insert desired genes. Whole plants can be grown from plant cells that integrate the modified plasmid into their chromosomes (Figure 15.12).

Genetically modified *A. tumefaciens* bacteria are used to deliver genes into some food crop plants, including soybeans, squash, and potatoes. Researchers also transfer genes into plants by way of electric or chemical shocks, or by blasting them with DNA-coated pellets.

As crop production expands to keep pace with human population growth, it places unavoidable pressure on ecosystems everywhere. Irrigation leaves mineral and salt residues in soils. Tilled soil erodes, taking topsoil with it. Runoff clogs rivers, and fertilizer in it causes algae to grow so fast that fish suffocate. Pesticides can be harmful to humans, other animals, and beneficial insects.

Pressured to produce more food at lower cost and with less damage to the environment, many farmers have begun to rely on genetically modified crop plants. Some of these modified plants carry genes that impart resistance to devastating plant diseases. Others offer improved yields, such as a strain of transgenic wheat that has twice the yield of unmodified wheat. GMO crops such as Bt corn and soy help farmers use smaller amounts of toxic pesticides. Organic farmers often spray their crops with spores of Bt (*Bacillus thuringiensis*), a bacterial species that makes a protein toxic only to insect larvae. Researchers transferred the gene encoding the Bt protein into plants. The engineered plants produce the Bt protein, but otherwise they are identical to unmodified plants. Insect larvae die shortly after eating their first and only GMO meal. Farmers can use much less pesticide on crops that make their own (Figure 15.13).

Transgenic crop plants are also being developed for Africa and other drought-stricken, impoverished regions of the world. Genes that confer drought tolerance and

---

Take-Home Message **What is genetic engineering?**

> Genetic engineering is the deliberate alteration of an individual's genome, and it results in a genetically modified organism (GMO).

> A transgenic organism carries a gene from a different species. Transgenic bacteria and yeast are used in research, medicine, and industry.

---

genetic engineering Process by which deliberate changes are introduced into an individual's genome.
genetically modified organism (GMO) Organism whose genome has been modified by genetic engineering.
transgenic Refers to a genetically modified organism that carries a gene from a different species.

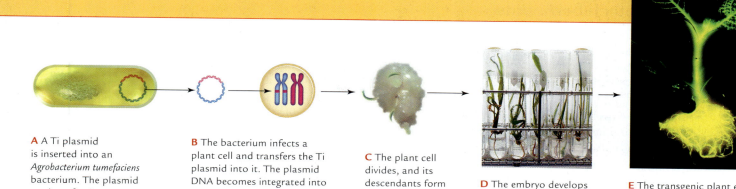

A A Ti plasmid is inserted into an *Agrobacterium tumefaciens* bacterium. The plasmid carries a foreign gene.

B The bacterium infects a plant cell and transfers the Ti plasmid into it. The plasmid DNA becomes integrated into one of the cell's chromosomes.

C The plant cell divides, and its descendants form an embryo.

D The embryo develops into a transgenic plant.

E The transgenic plant expresses the foreign gene. This tobacco plant is expressing a firefly gene.

Figure 15.12 Animated Using the Ti plasmid to make a transgenic plant.

insect resistance are being introduced into plants such as corn, beans, sugarcane, cassava, cowpeas, banana, and wheat. The resulting GMO crops may help people in those regions who rely on agriculture for food and income.

Genetic modifications can make food plants more nutritious. For example, rice plants have been engineered to make β-carotene in their seeds. β-carotene is an orange photosynthetic pigment (Section 6.2) that is remodeled by cells of the small intestine into vitamin A. Two genes in the β-carotene synthesis pathway were transferred into rice plants. One gene was from corn; the other, from bacteria. Both are under the control of a promoter that works in seeds. One cup of the engineered rice seeds—grains of Golden Rice—has enough β-carotene to satisfy a child's daily recommended amount of vitamin A.

The USDA Animal and Plant Health Inspection Service (APHIS) regulates the introduction of GMOs into the environment. At this writing, APHIS has deregulated seventy-four genetically modified crop plants, which means the plants are approved for unregulated use in the United States. The most widely planted GMO crops include corn, sorghum, cotton, soy, canola, and alfalfa engineered for resistance to glyphosate, an herbicide. Rather than tilling the soil to control weeds, farmers can spray their fields with glyphosate, which kills the weeds but not the engineered crops.

After long-term, widespread use of glyphosate, weeds resistant to the herbicide are becoming more common. The engineered gene is also appearing in wild plants and in nonengineered crops, which means that transgenes can (and do) escape into the environment. The genes are probably being transferred from transgenic plants to nontransgenic ones via pollen carried by wind or insects.

Many people are opposed to any GMO. Some worry that our ability to tinker with genetics has surpassed our ability to understand the impact of the tinkering. Controversy raised by such GMO use invites you to read the research and form your own opinions. The alternative is to be swayed by media hype (the term "Frankenfood,"

Figure 15.13 Genetically modified crops can help farmers use less pesticide. *Top*, the Bt gene conferred insect resistance to the genetically modified plants that produced this corn. *Bottom*, corn produced by unmodified plants is more vulnerable to insect pests.

for instance), or by reports from possibly biased sources (such as herbicide manufacturers).

Take-Home Message **Are there genetically modified plants?**

❯ Plants with modified or foreign genes are now common farm crops.

> Genetically engineered animals are invaluable in medical research and in other applications.
< Links to Knockout experiments 10.3, Human genetic disorders Chapter 14

Traditional cross-breeding has produced animals so unusual that transgenic animals may seem a bit mundane by comparison (Figure 15.14A). Cross-breeding is also a form of genetic manipulation, but many transgenic animals would probably never have occurred without laboratory intervention (Figure 15.14B,C).

The first genetically modified animals were mice. Today, such mice are commonplace, and they are invaluable in research (Figure 15.15). For example, we have discovered the function of human genes (including the *APOA5* gene discussed in Section 15.5) by inactivating their counterparts in mice. Genetically modified mice are also used as models of human diseases. For example, researchers inactivated the molecules involved in the control of glucose metabolism, one by one, in mice. Studying the effects of the knockouts has resulted in much of our current understanding of how diabetes works in humans. Genetically modified animals also make proteins that have medical and industrial applications. Various transgenic goats produce proteins used to treat cystic fibrosis, heart attacks, blood clotting disorders, and even nerve gas

exposure. Milk from goats transgenic for lysozyme, an antibacterial protein in human milk, may protect infants and children in developing countries from acute diarrheal disease. Goats transgenic for a spider silk gene produce the silk protein in their milk; researchers can spin this protein into nanofibers that are useful in medical and electronics applications. Rabbits make human interleukin-2, a protein that triggers divisions of immune cells. Genetic engineering has also given us dairy goats with heart-healthy milk, pigs with heart-healthy fat and environmentally friendly low-phosphate feces, extra-large sheep, and cows that are resistant to mad cow disease. Many people think that genetically engineering livestock is unconscionable. Others see it as an extension of thousands of years of acceptable animal husbandry practices. The techniques have changed, but not the intent: We humans continue to have a vested interest in improving our livestock.

Knockouts and Organ Factories

Millions of people suffer with organs or tissues that are damaged beyond repair. In any given year, more than 80,000 of them are on waiting lists for an organ transplant in the United States alone. Human donors are in such short supply that illegal organ trafficking is now a common problem.

Pigs are a potential source of organs for transplantation, because pig and human organs are about the same in both size and function. However, the human immune system battles anything it recognizes as nonself. It rejects a pig organ at once, because it recognizes proteins and

Figure 15.14 Genetically modified animals. **A** Featherless chicken developed by traditional cross-breeding methods in Israel. Such chickens survive in hot deserts where cooling systems are not an option. **B** The pig on the *left* is transgenic for a yellow fluorescent protein; its nontransgenic littermate is on the *right*. **C** Mira the transgenic goat produces a human anticlotting factor in her milk.

Figure 15.15 Example of how genetically engineered animals are useful in research. Mice transgenic for multiple pigments ("brainbow mice") are allowing researchers to map the complex neural circuitry of the brain. Individual nerve cells in the brainstem of a brainbow mouse are visible in this fluorescence micrograph.

carbohydrates on the plasma membrane of pig cells. Within a few hours, blood coagulates inside the organ's vessels and dooms the transplant. Drugs can suppress the immune response, but they also render organ recipients particularly vulnerable to infection. Researchers have produced genetically modified pigs that lack the offending molecules on their cells. The human immune system may not reject tissues or organs transplanted from these pigs.

Transferring an organ from one species into another is called **xenotransplantation**. Critics of xenotransplantation are concerned that, among other things, pig-to-human transplants would invite pig viruses to cross the species barrier and infect humans, perhaps catastrophically. Their concerns are not unfounded. Evidence suggests that some of the worst pandemics arose when animal viruses adapted to new hosts: humans.

Tinkering with the genes of animals raises a host of ethical dilemmas. For example, mice, monkeys, and other animals have been genetically modified to carry mutations associated with certain human diseases. These animals are allowing researchers to study—and test treatments for—conditions such as multiple sclerosis, cystic fibrosis, diabetes, cancer, and Huntington's disease without experimenting on humans. However, the engineered animals often suffer the same terrible symptoms of these conditions as humans do.

xenotransplantation Transplantation of an organ from one species into another.

15.9 Safety Issues

> The first transfer of foreign DNA into bacteria ignited an ongoing debate about potential dangers of transgenic organisms that enter the environment.

When James Watson and Francis Crick presented their model of the DNA double helix in 1953, they ignited a global blaze of optimism about genetic research. The very book of life seemed to be open for scrutiny. In reality, no one could read it. Scientific breakthroughs are not very often accompanied by the simultaneous discovery of the tools needed to study them. New techniques would have to be invented before that book would become readable.

Twenty years later, Paul Berg and his coworkers discovered how to make recombinant organisms by fusing DNA from two species of bacteria. By isolating DNA in manageable subsets, researchers now had the tools to be able to study its sequence in detail. They began to clone and analyze DNA from many different organisms. The technique of genetic engineering was born, and suddenly everyone was worried about it.

Researchers knew that DNA itself was not toxic, but they could not predict with certainty what would happen every time they fused genetic material from different organisms. Would they accidentally make a superpathogen? Could they make a new, dangerous form of life by fusing DNA of two normally harmless organisms? What if that new form escaped from the laboratory and transformed other organisms?

In a remarkably quick and responsible display of self-regulation, scientists reached a consensus on new safety guidelines for DNA research. Adopted at once by the NIH, these guidelines included precautions for laboratory procedures. They covered the design and use of host organisms that could survive only under the narrow range of conditions inside the laboratory. Researchers stopped using DNA from pathogenic or toxic organisms for recombinant DNA experiments until proper containment facilities were developed.

Now, all genetic engineering should be done under these laboratory guidelines, but the rules are not a guarantee of safety. We are still learning about escaped GMOs and their effects, and enforcement is a problem. For example, the expense of deregulating a GMO for release and importation is prohibitive for endeavors in the public sector. Thus, most commercial GMOs were produced by large, private companies—the same ones that typically wield tremendous political influence over the very government agencies charged with regulating them.

Take-Home Message Why do we genetically engineer animals?

> Animals that would be impossible to produce by traditional breeding methods are being created by genetic engineering.

Take-Home Message Is genetic engineering safe?

> Guidelines for DNA research have been in place for decades in the United States and other countries. Researchers are expected to comply, but the guidelines are not a guarantee of safety.

> We as a society continue to work our way through the ethical implications of applying new DNA technologies.
> The manipulation of individual genomes continues even as we are weighing the risks and benefits of this research.
< Links to Proto-oncogenes and cancer 11.6, Locus 13.2, Human genetic disorders 14.2

Getting Better We know of more than 15,000 serious genetic disorders. Collectively, they cause 20 to 30 percent of infant deaths each year, and account for half of all mentally impaired patients and a fourth of all hospital admissions. They also contribute to many age-related disorders, including cancer, Parkinson's disease, and diabetes. Drugs and other treatments can minimize the symptoms of some genetic disorders, but gene therapy is the only cure. **Gene therapy** is the transfer of recombinant DNA into an individual's body cells, with the intent to correct a genetic defect or treat a disease. The transfer, which occurs by way of lipid clusters or genetically engineered viruses, inserts an unmutated gene into an individual's chromosomes.

Human gene therapy is a compelling reason to embrace genetic engineering research. It is now being tested as a treatment for heart attack, sickle-cell anemia, cystic fibrosis, hemophilia A, Parkinson's disease, Alzheimer's disease, several types of cancer, and inherited diseases of the eye, the ear, and the immune system. The results are encouraging. For example, little Rhys Evans (Figure 15.16) was born with SCID-X1, a severe X-linked genetic disorder that stems from a mutated allele of the *IL2RG* gene. The gene encodes a receptor for an immune signaling molecule. Children affected by this disorder can survive only in germ-free isolation tents, because they cannot fight infections. In the late 1990s, researchers used a genetically engineered virus to insert unmutated copies of *IL2RG* into cells taken from the bone marrow of twenty boys with SCID-X1. Each child's modified cells were infused back into his bone marrow. Within months of their treatment eighteen of the boys left their isolation tents for good. Rhys was one of them. Gene therapy had permanently repaired their immune systems.

Getting Worse Manipulating a gene within the context of a living individual is unpredictable even when we know its sequence and locus. No one, for example, can predict where a virus-injected gene will become integrated into a chromosome. Its insertion might disrupt other genes. If it interrupts a gene that is part of the controls over cell division, then cancer might be the outcome. Five of the twenty boys treated with gene therapy for SCID-X1 have since developed a type of bone marrow cancer called leukemia, and one of them has died. The researchers had wrongly predicted that cancer related to the gene therapy would be rare. Research now implicates the very gene targeted for repair, especially when combined with the virus

Figure 15.16 Rhys Evans, who was born with SCID–X1. His immune system has been permanently repaired by gene therapy.

that delivered it. Apparently, integration of the modified viral DNA activated nearby proto-oncogenes (Section 11.6) in the children's chromosomes.

Other unanticipated problems sometimes occur with gene therapy. Jesse Gelsinger had a rare genetic deficiency of a liver enzyme that helps the body rid itself of ammonia, a toxic by-product of protein breakdown. Jesse's health was fairly stable while he was on a low-protein diet, but he had to take a lot of medication. In 1999, Jesse volunteered to be in a clinical trial testing a gene therapy for his condition. He had a severe allergic reaction to the viral vector, and four days after receiving the treatment, his organs shut down and he died. He was 18.

Getting Perfect The idea of selecting the most desirable human traits, **eugenics**, is an old one. It has been used as a justification for some of the most horrific episodes in human history, including the genocide of 6 million Jews during World War II. Thus, it continues to be a hotly debated social issue. For example, using gene therapy to cure human genetic disorders seems like a socially acceptable goal to most people. However, imagine taking this idea a bit further. Would it also be acceptable to engineer the genome of an individual who is within a normal range of phenotype in order to modify a particular trait? Researchers have already produced mice that have improved memory, enhanced learning ability, bigger muscles, and longer lives. Why not people?

eugenics Idea of deliberately improving the genetic qualities of the human race.
gene therapy The transfer of a normal or modified gene into an individual with the goal of treating a genetic defect or disorder.

Given the pace of genetics research, the eugenics debate is no longer about how we would engineer desirable traits, but how we would choose the traits that are desirable. Realistically, cures for many severe but rare genetic disorders will not be found, because the financial return will not even cover the cost of the research. Eugenics, however, might just turn a profit. How much would potential parents pay to be sure that their child will be tall or blue-eyed? Would it be okay to engineer "superhumans" with breathtaking strength or intelligence? How about a treatment that can help you lose that extra weight, and keep it off permanently? The gray area between interesting and abhorrent can be very different depending on who is asked. In a survey conducted in the United States, more than 40 percent of those interviewed said it would be fine to use gene therapy to make smarter and cuter babies. In one poll of British parents, 18 percent would be willing to use it to keep a child from being aggressive, and 10 percent would use it to keep a child from growing up to be homosexual.

Getting There Some people are adamant that we must never alter the DNA of anything. The concern is that gene therapy puts us on a slippery slope that may result in irreversible damage to ourselves and to the biosphere. We as a society may not have the wisdom to know how to stop once we set foot on that slope. One is reminded of our peculiar human tendency to leap before we look. And yet, something about the human experience allows us to dream of such things as wings of our own making, a capacity that carried us into space. In this brave new world, the questions before you are these: What do we stand to lose if serious risks are not taken? And, do we have the right to impose the consequences of taking such risks on those who would choose not to take them?

Take-Home Message Can people be genetically modified?

> Genes can be transferred into a person's cells to correct a genetic defect or treat a disease. However, the outcome of altering a person's genome remains unpredictable given our current understanding of how the genome works.

Personal DNA Testing (revisited)

The results of SNP analysis by a personal DNA testing company also include estimated risks of developing conditions associated with your particular set of SNPs. For example, the test will probably determine whether you are homozygous for one allele of the *MC1R* gene. If you are, the company's report will tell you that you have red hair. Very few SNPs have such a clear cause-and-effect relationship as the *MC1R* allele for red hair, however. Most human traits are polygenic, and many are also influenced by environmental factors such as life-style (Section 13.6). Thus, although a DNA test can reliably determine the SNPs in an individual's genome, it cannot reliably predict the effect of those SNPs on the individual.

For example, if you carry one ε4 allele of the *APOE* gene, a DNA testing company cannot tell you whether you will develop Alzheimer's disease later in life. Instead, the company will report your lifetime risk of developing the disease, which is about 29 percent, as compared with about 9 percent for someone who has no ε4 allele.

What does a 29 percent lifetime risk of developing Alzheimer's disease mean? The number is a probability statistic; it means that, on average, 29 of every 100 people who have the ε4 allele eventually get the disease. Having a high risk does not mean you are certain to end up with Alzheimer's, however. Not everyone who develops the disease has the ε4 allele, and not everyone with the ε4 allele develops Alzheimer's disease. Other as yet unknown alleles—some protective, some not—contribute to the disease.

How Would You Vote? The plunging cost of genetic testing has spurred an explosion of companies offering personal DNA sequencing and SNP profiling. The results of such testing may in some cases be of clinical use, for example in diagnosis of early-onset genetic disorders, or in predicting how an individual will respond to certain medications. However, we are still at an extremely early stage in our understanding of how genes contribute to most conditions, particularly age-related disorders such as Alzheimer's disease. Geneticists believe that it will be five to ten more years before we can use genotype to accurately predict an individual's risk of these conditions. Until then, should genetic testing companies be prohibited from informing clients of their estimated risk of developing such disorders based on SNPs? See CengageNow for details, then vote online (cengagenow.com).

Summary

Section 15.1 Personal DNA testing companies identify a person's unique array of single-nucleotide polymorphisms. Personal genetic testing may soon revolutionize the way medicine is practiced.

Section 15.2 The discovery of **restriction enzymes** allowed researchers to cut huge molecules of chromosomal DNA into manageable, predictable chunks. It also allowed them to combine DNA fragments from different organisms to make **recombinant DNA**. With **DNA cloning**, restriction enzymes cut DNA into pieces,

then DNA ligase splices the pieces into **plasmids** or other **cloning vectors**. The resulting hybrid molecules are inserted into host cells such as bacteria. When a host cell divides, it forms huge populations of genetically identical descendant cells. Each of these clones has a copy of the foreign DNA.

RNA cannot be cloned directly. **Reverse transcriptase**, a viral enzyme, is used to transcribe single-stranded RNA into **cDNA** for cloning.

Section 15.3 A **DNA library** is a collection of cells that host different fragments of DNA, often representing an organism's entire **genome**. Researchers can use **probes** to identify cells that host a specific fragment of DNA. Base pairing between nucleic acids from different sources is called **nucleic acid hybridization**.

The **polymerase chain reaction (PCR)** uses **primers** and a heat-resistant DNA polymerase to rapidly increase the number of copies of a section of DNA.

Section 15.4 DNA sequencing reveals the order of bases in a section of DNA. DNA polymerase is used to partially replicate a DNA template. The reaction produces a mixture of DNA fragments of all different lengths. **Electrophoresis** separates the fragments by length into bands. The entire genomes of several organisms have now been sequenced.

Section 15.5 Genomics, or the study of genomes, is providing insights into the function of the human genome. Similarities between genomes of different organisms are evidence of evolutionary relationships, and can be used as a predictive tool in research.

DNA profiling identifies a person by the unique parts of his or her DNA. Examples include methods of determining an individual's array of SNPs or **short tandem repeats**. Within the context of a criminal investigation, a DNA profile is called a DNA fingerprint.

Sections 15.6–15.9 Recombinant DNA technology is the basis of **genetic engineering**, the directed modification of an organism's genetic makeup with the intent to modify its phenotype. A gene is modified and reinserted into an individual of the same species, or a gene from one species is inserted into an individual of a different species to make a **transgenic** organism. The result of either process is a **genetically modified organism (GMO)**.

Transgenic bacteria and yeast produce medically valuable proteins. Transgenic crop plants are helping farmers produce food more efficiently. Genetically modified animals produce human proteins, and may one day provide a source of organs and tissues for **xenotransplantation** into humans.

Safety guidelines minimize potential risks to researchers in genetic engineering labs. Although these and other government regulations limit the release of genetically

modified organisms into the environment, such laws are not guarantees against accidental releases or unforeseen environmental effects.

Section 15.10 With **gene therapy**, a gene is transferred into body cells to correct a genetic defect or treat a disease. As with any new technology, the potential benefits of genetically modifying humans must be weighed against the potential risks. The practice raises ethical issues such as whether **eugenics** is desirable.

Self-Quiz   Answers in Appendix III

1. _____ cut(s) DNA molecules at specific sites.
   a. DNA polymerase   c. Restriction enzymes
   b. DNA probes   d. Reverse transcriptase

2. A _____ is a small circle of bacterial DNA that contains a few genes and is separate from the chromosome.
   a. plasmid   c. nucleus
   b. chromosome   d. double helix

3. By reverse transcription, _____ is assembled on a(n) _____ template.
   a. mRNA; DNA   c. DNA; ribosome
   b. cDNA; mRNA   d. protein; mRNA

4. For each species, all _____ in the complete set of chromosomes is the _____ .
   a. genomes; phenotype   c. mRNA; start of cDNA
   b. DNA; genome   d. cDNA; start of mRNA

5. A set of cells that host various DNA fragments collectively representing an organism's entire set of genetic information is a _____ .
   a. genome   c. genomic library
   b. clone   d. GMO

6. _____ is a technique to determine the order of nucleotide bases in a fragment of DNA.

7. Fragments of DNA can be separated by electrophoresis according to _____ .
   a. sequence   c. species
   b. length   d. composition

8. PCR can be used _____ .
   a. to increase the number of specific DNA fragments
   b. in DNA fingerprinting
   c. to modify a human genome
   d. a and b are correct

9. An individual's set of unique _____ can be used as a DNA profile.
   a. DNA sequences   c. SNPs
   b. short tandem repeats   d. all of the above

10. Which of the following can be used to carry foreign DNA into host cells? Choose all correct answers.
    a. RNA   e. lipid clusters
    b. viruses   f. blasts of pellets
    c. PCR   g. xenotransplantation
    d. plasmids   h. sequencing

Data Analysis Activities

Enhanced Spatial Learning in Mice with Autism Mutation

Autism is a neurobiological disorder with a range of symptoms that include impaired social interactions, stereotyped patterns of behavior such as hand-flapping or rocking, and, occasionally, greatly enhanced intellectual abilities. Some autistic people have a mutation in neuroligin 3, a type of cell adhesion protein (Section 4.4) that connects brain cells to one another. One mutation changes amino acid 451 from arginine to cysteine.

Mouse and human neuroligin 3 are very similar. In 2007, Katsuhiko Tabuchi and his colleagues genetically modified mice to carry the same arginine-to-cysteine substitution in their neuroligin 3. The mutation caused an increase in transmission of some types of signals between brain cells. Mice with the mutation had impaired social behavior, and, unexpectedly, enhanced spatial learning ability (Figure 15.17).

1. In the first test, how many days did unmodified mice need to learn to find the location of a hidden platform within 10 seconds?
2. Did the modified or the unmodified mice learn the location of the platform faster in the first test?
3. Which mice learned faster the second time around?
4. Which mice showed the greatest improvement in memory between the first and the second test?

Figure 15.17 Enhanced spatial learning ability in mice with a mutation in neuroligin 3 (R451C), compared with unmodified (wild-type) mice.

A The mice were tested in a water maze, in which a platform is submerged a few millimeters below the surface of a deep pool of warm water. The platform is not visible to swimming mice. Mice do not particularly enjoy swimming, so they locate a hidden platform as fast as they can. When tested again, they can remember its location by checking visual cues around the edge of the pool.

B How quickly they remember the platform's location is a measure of spatial learning ability. The platform was moved and the experiment was repeated for the second test.

---

11. A transgenic organism _____ .
    a. carries a gene from another species
    b. has been genetically modified
    c. both a and b

12. _____ can be used to correct a genetic defect.
    a. Cloning vectors   d. Xenotransplantation
    b. Gene therapy   e. a and b
    c. Cloning   f. all of the above

13. Match the recombinant DNA method with the appropriate enzyme.
    ____ PCR   a. *Taq* polymerase
    ____ cutting DNA   b. DNA ligase
    ____ cDNA synthesis   c. reverse transcriptase
    ____ DNA sequencing   d. restriction enzyme
    ____ pasting DNA   e. DNA polymerase

14. Match the terms with the most suitable description.
    ____ DNA fingerprint   a. having a foreign gene
    ____ Ti plasmid   b. alleles have them
    ____ nucleic acid   c. a person's unique collection
        hybridization   of short tandem repeats
    ____ eugenics   d. base pairing of DNA or
    ____ SNP   DNA and RNA from
    ____ transgenic   different sources
    ____ GMO   e. selecting "desirable" traits
                             f. genetically modified
                             g. used in some gene transfers

Additional questions are available on **CENGAGENOW**.

Critical Thinking

1. Restriction enzymes in bacterial cytoplasm cut injected bacteriophage DNA wherever certain sequences occur. Why do you think the bacterial chromosome does not get chopped up too?

2. The *FOXP2* gene encodes a transcription factor associated with vocal learning in mice, bats, birds, and humans. The chimpanzee, gorilla, and rhesus FOXP2 proteins are identical; the human version differs in only 2 of 715 amino acids, a change thought to have contributed to the development of spoken language. In humans, loss-of-function mutations in *FOXP2* result in severe speech and language disorders. In mice, they hamper brain function and impair vocalizations. Mice genetically engineered to carry the human version of *FOXP2* show changes in their vocal patterns, and more growth and greater adaptability of neurons involved in memory and learning. Biologists do not anticipate that a similar experiment in chimpanzees would confer the ability to speak, because spoken language is a complex, epistatic trait (Section 13.5). What do you think might happen if their prediction is incorrect?

Animations and Interactions on **CENGAGENOW**:
❯ Recombinant DNA; Cloning; Using a radioactive probe; DNA sequencing; PCR; DNA fingerprinting; Genetically engineering plants.

❮ Links to Earlier Concepts

This chapter explores a clash between traditional beliefs and science. You may wish to review critical thinking (Section 1.6) before you begin. What you know about inheritance of alleles (13.2) will help you understand natural selection. Finding the age of ancient rocks and fossils depends on the properties of radioisotopes (2.2). You will see how master genes (10.3, 10.4) are evidence of shared ancestry, and revisit evolution by gene duplication (14.5).

Key Concepts

Emergence of Evolutionary Thought

Nineteenth-century naturalists started to think about the global distribution of species. They discovered similarities and differences among major groups, including those represented as fossils.

A Theory Takes Form

Evidence of evolution, or change in lines of descent, led Charles Darwin and Alfred Wallace to independently develop a theory of natural selection. The theory explains how traits that define each species change over time.

16 Evidence of Evolution

16.1 Reflections of a Distant Past

How do you think about time? Perhaps you can conceive of a few hundred years of human events, but how about a few million? Envisioning the distant past requires an intellectual leap from the familiar to the unknown.

One way to make that leap involves, surprisingly, asteroids. Asteroids are small planets hurtling through space. They range in size from 1 to 1,500 kilometers (roughly 0.5 to 1,000 miles) wide. Millions of them orbit the sun between Mars and Jupiter —cold, stony leftovers from the formation of our solar system. Asteroids are difficult to see even with the best telescopes, because they do not emit light. Many cross Earth's orbit, but most of those pass us by before we know about them. Some have not passed us at all.

The mile-wide Barringer Crater in Arizona is difficult to miss (Figure 16.1). A 300,000-ton asteroid made this impressive pockmark in the desert sandstone when it slammed into Earth 50,000 years ago. The impact was 150 times more powerful than the bomb that leveled Hiroshima.

No humans were in North America at the time of the impact. If there were no witnesses, how do we know what happened? We often reconstruct history by studying physical evidence of events that took place long ago. Geologists were able to infer the most probable cause of the Barringer Crater by analyzing tons of meteorites, melted sand, and other rocky clues at the site.

Similar evidence points to even larger asteroid impacts in the more distant past. For example, a **mass extinction**, or permanent loss of major groups of organisms, occurred 65.5 million years ago. The event is marked by an unusual, worldwide layer of rock called the K–T boundary layer. There are plenty of dinosaur fossils below this layer. Above it, in rock layers that were deposited more recently, there are no dinosaur fossils, anywhere. An impact crater off the coast of what is now the Yucatán Peninsula dates to about 65.5 million years ago. Coincidence? Many

mass extinction Simultaneous loss of many lineages from Earth.

Figure 16.1 From evidence to inference. What made the Barringer Crater (*opposite*)? Rocky evidence points to a 300,000-ton asteroid that collided with Earth 50,000 years ago. *Above*, bands that are part of a unique layer of rock that formed 65.5 million years ago, worldwide. The layer marks an abrupt transition in the fossil record that implies a mass extinction. The *red* pocketknife gives an idea of scale.

scientists say no. They have inferred from the evidence that the impact of an asteroid about 20 km (12 miles) wide caused a global catastrophe that wiped out the dinosaurs.

You are about to make an intellectual leap through time, to places that were not even known a few centuries ago. We invite you to launch yourself from this premise: Natural phenomena that occurred in the past can be explained by the very same physical, chemical, and biological processes that operate today. That premise is the foundation for scientific research into the history of life. The research represents a shift from experience to inference—from the known to what can only be surmised— and it gives us astonishing glimpses into the distant past.

Evidence From Fossils
The fossil record provides physical evidence of past changes in many lines of descent. We use the property of radioisotope decay to determine the age of rocks and fossils.

Evidence From Biogeography
Geologic events have influenced evolution. Correlating geologic and evolutionary events helps explain the distribution of species, past and present.

Evidence in Form and Function
Different lineages may have similar body parts that reflect descent from a shared ancestor. Lineages with common ancestry often develop in similar ways.

❯ Belief systems are influenced by the extent of our understanding of the natural world. Those that are inconsistent with systematic observations tend to change over time.

The seeds of biological inquiry were taking hold in the Western world more than 2,000 years ago. Aristotle, the Greek philosopher, was making connections between observations in an attempt to explain the order of the natural world. Like few others of his time, Aristotle viewed nature as a continuum of organization, from lifeless matter through complex plants and animals. Aristotle was one of the first **naturalists**, people who observe life from a scientific perspective.

By the fourteenth century, Aristotle's earlier ideas about nature had been transformed into a rigid view of life, in which a "great chain of being" extended from the lowest form (snakes), through humans, to spiritual beings. Each link in the chain was a species, and each was said to have been forged at the same time in a perfect state. The chain itself was complete and continuous. Because everything that needed to exist already did, there was no room for change. Once every species had been discovered, the meaning of life would be revealed.

European naturalists that embarked on globe-spanning survey expeditions brought back tens of thousands of plants and animals from Asia, Africa, North and South America, and the Pacific Islands. Each newly discovered species was carefully catalogued as another link in the chain of being.

By the late 1800s, naturalists such as Alfred Wallace were seeing patterns in where species live and how they might be related, and had started to think about the natural forces that shape life. These naturalists were pioneers in **biogeography**, the study of patterns in the geographic distribution of species. Some of the patterns raised questions that could not be answered within the framework of prevailing belief systems. For example, globe-trotting explorers had discovered plants and animals living in extremely isolated places. The isolated species looked suspiciously similar to species living across vast expanses of open ocean, or on the other side of impassable mountain ranges. Could different species be related? If so, how could the related species end up geographically isolated?

For example, the three birds in Figure 16.2 live on different continents, but they share a set of unusual features.

Figure 16.3 Similar-looking, unrelated species that are native to distant geographic realms: *above*, an American spiny cactus; and *left*, an African spiny spurge.

A Emu, native to Australia

B Rhea, native to South America

C Ostrich, native to Africa

Figure 16.2 Similar-looking, related species that are native to distant geographic realms. The three types of ratite birds are unlike most other birds in several traits, including their long, muscular legs and their inability to fly.

These flightless birds sprint about on long, muscular legs in flat, open grasslands about the same distance from the equator. All raise their long necks to watch for predators. Wallace thought that the shared set of unusual traits might mean that these three birds descended from a common ancestor (and he was right), but he had no idea how they could have ended up on different continents.

Naturalists of the time also had trouble classifying organisms that are very similar in some features, but different in others. For example, the plants in Figure 16.3 are native to different continents. Both live in hot deserts where water is seasonally scarce. Both have rows of sharp spines that deter herbivores, and both store water in their thick, fleshy stems. However, their reproductive parts are very different, so these plants cannot be as closely related as their outward appearance might suggest.

Observations like these are part of **comparative morphology**, the study of body plans and structures among groups of organisms. Organisms that are outwardly very similar may be quite different internally; think of fishes and porpoises. Others that differ greatly in outward appearance may be very similar in underlying structure. For example, a human arm, a porpoise flipper, an elephant leg, and a bat wing have comparable internal bones, as Section 16.8 will explain.

Comparative morphology in the nineteenth century revealed body parts that have no apparent function, an idea that added to the naturalists' confusion. According to prevailing beliefs, every species had been created in a perfect state. If that were so, then why were there useless parts such as leg bones in snakes (which do not walk), or the vestiges of a tail in humans (Figure 16.4)?

Fossils were puzzling too. A **fossil** is the remains or traces of an organism that lived in the ancient past—physical evidence of ancient life. Geologists mapping rock formations exposed by erosion or quarrying had discovered identical sequences of rock layers in different parts of the world. Deeper layers held fossils of simple marine life. Layers above them held similar but more intricate fossils. In higher layers, fossils that were similar but even more intricate looked like they belonged to modern species. The photos on the *right* show one such series, ten fossils of shelled protists, each from a successive layer of rock in a stack. What did these fossil sequences mean?

Fossils of many animals that had no living representatives were also being unearthed. If the animals had been perfect at the time of creation, then why had they become extinct?

Taken as a whole, the accumulating findings from biogeography, comparative morphology, and geology did not fit with prevailing beliefs of the nineteenth century. If species had not been created in a perfect state (and extinct species, fossil sequences and "useless" body parts implied that they had not), then perhaps species had indeed changed over time.

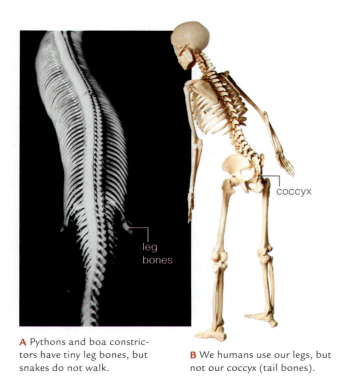

A Pythons and boa constrictors have tiny leg bones, but snakes do not walk.

B We humans use our legs, but not our coccyx (tail bones).

Figure 16.4 Animated Vestigial body parts.

biogeography Study of patterns in the geographic distribution of species and communities.
comparative morphology Study of body plans and structures among groups of organisms.
fossil Physical evidence of an organism that lived in the ancient past.
naturalist Person who observes life from a scientific perspective.

> In the 1800s, many scholars realized that life on Earth had changed over time, and began to think about what could have caused the changes.
‹ Link to Critical thinking and how science works 1.6

Squeezing New Evidence Into Old Beliefs

In the nineteenth century, naturalists were faced with increasing evidence that life on Earth, and even Earth itself, had changed over time. Around 1800, Georges Cuvier, an expert in zoology and paleontology, was trying to make sense of the new information. He had observed abrupt changes in the fossil record, and knew that many fossil species seemed to have no living counterparts. Given this evidence, he proposed a startling idea: Many species that had once existed were now extinct. Cuvier also knew about evidence that Earth's surface had changed. For example, he had seen fossilized seashells on mountainsides far from modern seas. Like most others of his time, he assumed Earth's age to be in the thousands, not millions, of years. He reasoned that geologic forces unlike those known today would have been necessary to raise sea floors to mountaintops in this short time span. Catastrophic geological events would have caused

extinctions, after which surviving species repopulated the planet. Cuvier's idea came to be known as **catastrophism**. We now know it is incorrect; geologic processes have not changed over time.

Another scholar, Jean-Baptiste Lamarck, was thinking about processes that drive **evolution**, or change in a line of descent. A line of descent is also called a **lineage**. Lamarck thought that a species gradually improved over generations because of an inherent drive toward perfection, up the chain of being. The drive directed an unknown "fluida" into body parts needing change. By Lamarck's hypothesis, environmental pressures cause an internal need for change in an individual's body, and the resulting change is inherited by offspring. Try using Lamarck's hypothesis to explain why a giraffe's neck is very long. We might predict that some short-necked ancestor of the modern giraffe stretched its neck to browse on leaves beyond the reach of other animals. The stretches may have even made its neck a bit longer. By Lamarck's hypothesis, that animal's offspring would inherit a longer neck, and after many generations strained to reach ever loftier leaves, the modern giraffe would have been the result. Lamarck was correct in thinking that environmental factors affect a species' traits, but his understanding of how inheritance works was incomplete.

Darwin and the HMS *Beagle*

In 1831, the twenty-two-year-old Charles Darwin was wondering what to do with his life. Ever since he was eight, he had wanted to hunt, fish, collect shells, or watch insects and birds—anything but sit in school. After an attempt to study medicine in college, he earned a degree in theology from Cambridge. All through school, however, Darwin spent most of his time with faculty members and other students who embraced natural history. Botanist John Henslow arranged for Darwin to become a naturalist aboard the *Beagle*, a ship about to embark on a survey expedition to South America.

The *Beagle* set sail for South America in December, 1831 (Figure 16.5). The young man who had hated school and had no formal training in science quickly became an enthusiastic naturalist. During the *Beagle*'s five-year voyage, Darwin found many unusual fossils. He saw diverse species living in environments that ranged from the sandy shores of remote islands to the plains high in the Andes. Along the way, he read the first volume of a new and popular book, Charles Lyell's *Principles of Geology*.

Figure 16.5 Voyage of the HMS *Beagle*. With Darwin aboard as ship's naturalist, the vessel (*top*) originally set sail to map the coast of South America, but ended up circumnavigating the globe over a period of five years (*bottom*). Darwin's detailed observations of the geology, fossils, plants, and animals he encountered on this expedition changed the way he thought about evolution.

Figure 16.6 Charles Darwin, *left. Right*, this page from Darwin's 1836 notes on the "Transmutation of Species" reads, "Let a pair be introduced and increase slowly, from many enemies, so as often to intermarry who will dare say what result / According to this view animals on separate islands ought to become different if kept long enough apart with slightly differing circumstances. — Now Galapagos Tortoises, Mocking birds, Falkland Fox, Chiloe fox, — Inglish and Irish Hare."

Lyell was a proponent of what became known as the **theory of uniformity**, the idea that gradual, repetitive change had shaped Earth. For many years, geologists had been chipping away at the sandstones, limestones, and other types of rocks that form from accumulated sediments in lakebeds, river bottoms, and ocean floors. These rocks held evidence that gradual processes of geologic change operating in the present were the same ones that operated in the distant past.

The theory of uniformity held that strange catastrophes were not necessary to explain Earth's surface. Over great spans of time, gradual, everyday geologic processes such as erosion could have sculpted Earth's current landscape.

The theory challenged the prevailing belief that Earth was 6,000 years old. According to traditional scholars, people had recorded everything that happened in those 6,000 years—and in all that time, no one had mentioned seeing a species evolve. However, by Lyell's calculations, it must have taken millions of years to sculpt Earth's surface. Darwin's exposure to Lyell's ideas gave him insights into the geologic history of the regions he would encounter on his journey. Was millions of years enough time for species to evolve? Darwin thought that it was (Figure 16.6).

---

catastrophism Now-abandoned hypothesis that catastrophic geologic forces unlike those of the present day shaped Earth's surface.
evolution Change in a line of descent.
lineage Line of descent.
theory of uniformity Idea that gradual repetitive processes occurring over long time spans shaped Earth's surface.

Take-Home Message How did new evidence change the way people in the nineteenth century thought about the history of life?

> In the 1800s, fossils and other evidence led some naturalists to propose that Earth and the species on it had changed over time. The naturalists also began to reconsider the age of Earth.

> Darwin's detailed observations of nature during a five-year voyage around the world changed his ideas about how evolution occurs.

❯ Darwin's observations of species in different parts of the world helped him understand a driving force of evolution.
❮ Link to Alleles and traits 13.2

Old Bones and Armadillos

Darwin sent to England the thousands of specimens he had collected on his voyage. Among them were fossil glyptodons from Argentina. These armored mammals are extinct, but they have many traits in common with modern armadillos (Figure 16.7). For example, armadillos live only in places where glyptodons once lived. Like glyptodons, armadillos have helmets and protective shells that consist of unusual bony plates. Could the odd shared traits mean that glyptodons were ancient relatives of armadillos? If so, perhaps traits of their common ancestor had changed in the line of descent that led to armadillos. But why would such changes occur?

A Key Insight—Variation in Traits

Back in England, Darwin pondered his notes and fossils. He also read an essay by one of his contemporaries, economist Thomas Malthus. Malthus had correlated increases in human population size with famine, disease,

and war. He proposed that humans run out of food, living space, and other resources because they tend to reproduce beyond the capacity of their environment to sustain them. When that happens, the individuals of a population must either compete with one another for the limited resources, or develop technology to increase their productivity. Darwin realized that Malthus's ideas had wider application: All populations, not just human ones, must have the capacity to produce more individuals than their environment can support.

Darwin also knew that individuals of a species are not always identical; they have many traits in common, but they also vary in size, color, or other features. He saw such variation among many of the finch species that live on isolated islands of the Galápagos archipelago. This island chain is separated from South America by 900 kilometers (550 miles) of open ocean, so Darwin realized that most of the species living on the islands had been isolated there for a long time.

Darwin also knew about **artificial selection**, the process whereby humans choose traits that they favor in a domestic species. For example, he was familiar with dramatic variations in traits that breeders of dogs and horses had produced through selective breeding. He recognized that an environment could similarly select traits that make individuals of a population suited to it.

Finches living on individual islands of the Galápagos resembled species Darwin saw living in South America, but many of them had unique traits that suited them to their particular island habitat. It dawned on Darwin that having a particular version of a variable trait might give an individual an advantage over competing members of its species. The trait might enhance the individual's ability to survive and reproduce in its particular environment. Darwin realized that in any population, some individuals have traits that make them better suited to their environment than others. In other words, individuals of a natural population vary in fitness. We define **fitness** as the degree of adaptation to a specific environment, and measure it as relative genetic contribution to future generations. A trait that enhances an individual's fitness is called an evolutionary **adaptation**, or **adaptive trait**.

Figure 16.7 Ancient relatives. **A** A modern armadillo, about a foot long. **B** Fossil of a glyptodon, an automobile-sized mammal that lived between 2 million and 15,000 years ago.

Glyptodons and armadillos are widely separated in time, but they share a restricted distribution and unusual traits, including a shell and helmet of keratin-covered bony plates—a material similar to crocodile and lizard skin. (The fossil in **B** is missing its helmet.) Their unique shared traits were a clue that helped Darwin develop a theory of evolution by natural selection.

Figure 16.8 Alfred Wallace, codiscoverer of natural selection.

Table 16.1 Principles of Natural Selection
Observations About Populations
❯ Natural populations have an inherent reproductive capacity to increase in size over time.
❯ As a population expands, resources that are used by its individuals (such as food and living space) eventually become limited.
❯ When resources are limited, the individuals of a population compete for them.
Observations About Genetics
❯ Individuals of a species share certain traits.
❯ Individuals of a natural population vary in the details of those shared traits.
❯ Shared traits have a heritable basis, in genes. Alleles (slightly different forms of a gene) arise by mutation.
Inferences
❯ A certain form of a shared trait may make its bearer better able to survive.
❯ The individuals of a population that are better able to survive tend to leave more offspring.
❯ Thus, an allele associated with an adaptive trait tends to become more common in a population over time.

Over many generations, individuals with the most adaptive traits tend to survive longer and reproduce more than their less fit rivals. Darwin understood that this process, which he called **natural selection**, could be a process by which evolution occurs. If an individual has a form of a trait that makes it better suited to an environment, then it is better able to survive. If an individual is better able to survive, then it has a better chance of living long enough to produce offspring. If individuals that bear an adaptive, heritable trait produce more offspring than those that do not, then the frequency of that trait will tend to increase in the population over successive generations. Table 16.1 summarizes this reasoning in modern terms.

Great Minds Think Alike

Darwin wrote out his ideas about natural selection, but let ten years pass without publishing them. In the meantime, Alfred Wallace, who had been studying wildlife in the Amazon basin and the Malay Archipelago, wrote an essay and sent it to Darwin for advice. Wallace's essay outlined evolution by natural selection—the very same theory as Darwin's. Wallace had written earlier letters to Darwin and Lyell about patterns in the geographic distribution of species; he too had connected the dots. Wallace is now called the father of biogeography (Figure 16.8).

adaptation (**adaptive trait**) A heritable trait that enhances an individual's fitness.
artificial selection Selective breeding of animals by humans.
fitness Degree of adaptation to an environment, as measured by an individual's relative genetic contribution to future generations.
natural selection A process in which environmental pressures result in the differential survival and reproduction of individuals of a population who vary in the details of shared, heritable traits.

In 1858, just weeks after Darwin received Wallace's essay, the theory of evolution by natural selection was presented at a scientific meeting. Both Darwin and Wallace were credited as authors. Wallace was still in the field and knew nothing about the meeting, which Darwin did not attend. The next year, Darwin published *On the Origin of Species*, which laid out detailed evidence in support of the theory. Many scholars had already accepted the idea of descent with modification, or evolution. However, there was a fierce debate over the idea that evolution occurs by natural selection. Decades would pass before experimental evidence from the field of genetics led to its widespread acceptance in the scientific community.

As you will see in the remainder of this chapter, the theory of evolution by natural selection is supported by and helps explain the fossil record as well as similarities in the form, function, and biochemistry of living things.

Take-Home Message What is natural selection?

❯ Natural selection is a process in which individuals of a population who vary in the details of shared, heritable traits survive and reproduce with differing success as a result of environmental pressures.

❯ Traits favored by natural selection are said to be adaptive. An adaptive trait increases the chances that an individual bearing it will survive and reproduce.

> Fossils are remnants or traces of organisms that lived in the past. The fossil record holds clues to life's evolution.
‹ Link to Radioisotopes 2.2

Even before Darwin's time, fossils were recognized as stone-hard evidence of earlier forms of life. Most fos-

sils are mineralized bones, teeth, shells, seeds, spores, or other hard body parts. Trace fossils such as footprints and other impressions, nests, burrows, trails, eggshells, or feces (*left*) are evidence of an organism's activities.

A 30-million-year-old *Elomeryx*. This small terrestrial mammal was a member of the same artiodactyl group that gave rise to hippopotamuses, pigs, deer, sheep, cows, and whales.

B *Rodhocetus*, an ancient whale, lived about 47 million years ago. Its distinctive ankle bones point to a close evolutionary connection to artiodactyls.

C *Dorudon atrox*, an ancient whale that lived about 37 million years ago. Its artiodactyl-like ankle bones were much too small to have supported the weight of its huge body on land, so this mammal had to be fully aquatic.

D Modern cetaceans such as the sperm whale have remnants of a pelvis and leg, but no ankle bones.

Figure 16.9 Links in the ancient lineage of whales. The ancestors of whales probably walked on land. The skull and lower jaw of cetaceans—which include whales, dolphins, and porpoises—have distinctive features that are also characteristic of ancient carnivorous land animals. DNA sequence comparisons suggested that those animals were probably artiodactyls, hooved mammals with two or four toes on each foot. With their artiodactyl-like ankle bones, *Rodhocetus* and *Dorudon* were probably offshoots of the ancient artiodactyl-to-modern-whale lineage as it transitioned back to life in water. The photo compares the ankle bones of a *Rodhocetus* (*left*) with those of a modern artiodactyl, a pronghorn antelope (*right*).

The process of fossilization begins when an organism or its traces become covered by sediments or volcanic ash. Water seeps into the remains, and metal ions and other inorganic compounds dissolved in the water gradually replace the minerals in the bones and other hard tissues. Sediments that accumulate on top of the remains exert increasing pressure on them. After a very long time, the pressure and mineralization process transform the remains into rock.

Most fossils are found in layers of sedimentary rock such as mudstone, sandstone, and shale. These rocks form as rivers wash silt, sand, volcanic ash, and other materials from land to sea. Mineral particles in the materials settle on sea floors in horizontal layers that vary in thickness and composition. After hundreds of millions of years, the layers of sediments become compacted into layers of rock.

We study layers of sedimentary rock in order to understand the historical context of fossils we find in them. Usually, the deeper layers in a stack were the first to form, and those closest to the surface formed most recently. Thus, the deeper the layer of sedimentary rock, the older the fossils it contains. A layer's composition and thickness relative to other layers is also a clue about local and global events that were occurring as it formed. For instance, layers of sedimentary rock deposited during ice ages are thinner than other layers. Why? Tremendous volumes of water froze and became locked in glaciers during the ice ages. Rivers dried up, and sedimentation slowed. When the glaciers melted, sedimentation resumed and the layers became thicker.

The Fossil Record

We have fossils for more than 250,000 known species. Considering the current range of biodiversity, there must have been many millions more, but we will never know all of them. Why not?

The odds are against finding evidence of an extinct species, because fossils are relatively rare. Most of the time, an organism's remains are quickly obliterated by scavengers or decay. Organic materials decompose in the presence of oxygen, so remains can endure only if they are encased in an air-excluding material such as sap, tar, ice, or mud. Remains that do become fossilized are often deformed, crushed, or scattered by erosion and other geologic assaults. For us to find a fossil of an extinct species, at least one specimen had to be buried before it decomposed or something ate it. The burial site had to escape destructive geologic events, and it had to be a place accessible enough for us to find.

Despite these challenges, the fossil record is substantial enough to help us reconstruct patterns in the history of life. We have been able to find fossil evidence of the evolutionary history of many species (Figure 16.9).

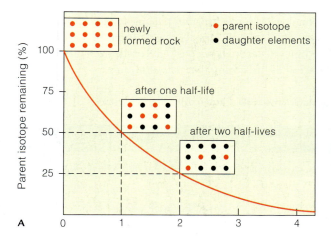

newly formed rock
● parent isotope
● daughter elements

after one half-life

after two half-lives

A

(y-axis) Parent isotope remaining (%) — 100, 75, 50, 25
(x-axis) 0, 1, 2, 3, 4

B Long ago, trace amounts of ^{14}C and a lot more ^{12}C were incorporated into the tissues of a nautilus. The carbon atoms were part of organic molecules in the nautilus's food. ^{12}C is stable and ^{14}C decays, but the proportion of the two isotopes in the nautilus's tissues remained the same. Why? As long as it was alive, the nautilus continued to gain both types of carbon atoms in the same proportions from its food.

C When the nautilus died, it stopped eating, so its body stopped gaining carbon. The ^{12}C atoms already in its tissues were stable, but the ^{14}C atoms (represented as *red* dots) were decaying into nitrogen atoms. Thus, over time, the amount of ^{14}C decreased relative to the amount of ^{12}C. After 5,370 years, half of the ^{14}C had decayed; after another 5,370 years, half of what was left had decayed, and so on.

Figure 16.10 Animated Radiometric dating. **A** Half-life, the time it takes for half of the atoms in a sample of radioisotope to decay.

B–D Using radiometric dating to find the age of a fossil. Carbon 14 (^{14}C) is a radioisotope of carbon that decays into nitrogen. It forms in the atmosphere and combines with oxygen to become CO_2, which enters food chains by way of photosynthesis.

❯❯ **Figure It Out** How much of any radioisotope remains after two of its half–lives have passed?

Answer: 25 percent.

D Fossil hunters discover the fossil and measure its ^{14}C and ^{12}C content—the number of atoms of each isotope. The ratio of those numbers can be used to calculate how many half-lives passed since the organism died. For example, if the ^{14}C to ^{12}C ratio is one-eighth of the ratio in living organisms, then three half-lives $(½)^3$ must have passed since the nautilus died. Three half-lives of ^{14}C is 16,110 years.

Radiometric Dating

Remember from Section 2.2 that a radioisotope is a form of an element with an unstable nucleus. Atoms of a radioisotope become atoms of other elements as their nucleus disintegrates. The predictable products of this process, radioactive decay, are called daughter elements.

Radioactive decay is not influenced by temperature, pressure, chemical bonding state, or moisture; it is influenced only by time. Thus, like the ticking of a perfect clock, each type of radioisotope decays at a constant rate. The time it takes for half of a radioisotope's atoms to decay is a characteristic of the radioisotope called **half–life** (Figure 16.10A). For example, radioactive uranium 238 decays into thorium 234, which decays into something else, and so on until it becomes lead 206. The half-life of the decay of uranium 238 to lead 206 is 4.5 billion years.

The predictability of radioactive decay can be used to find the age of a volcanic rock—the date it cooled. Rock deep inside Earth is hot and molten, so atoms swirl and mix in it. Rock that reaches the surface cools and hardens. As the rock cools, minerals crystallize in it. Each kind of mineral has a characteristic structure and composition. For example, the mineral zircon consists primarily of

ordered arrays of zircon silicate molecules ($ZrSiO_4$). Some of the molecules in a zircon crystal have uranium atoms substituted for zirconium atoms, but never lead atoms. Thus, new zircon crystals that form as molten rock cools contain no lead. However, uranium decays into lead at a predictable rate. Thus, over time, uranium atoms disappear from a zircon crystal, and lead atoms accumulate in it. The ratio of uranium atoms to lead atoms in a zircon crystal can be measured precisely. That ratio can be used to calculate how long ago the crystal formed (its age).

zircon

We have just described **radiometric dating**, a method that can reveal the age of a material by measuring its content of a radioisotope and daughter elements. The oldest known terrestrial rock, a tiny zircon crystal from Australia, is 4.404 billion years old.

Recent fossils that still contain carbon can be dated by measuring their carbon 14 content (Figure 16.10B–D). Most of the ^{14}C in a fossil will have decayed after about 60,000 years. The age of fossils older than that can be estimated only by dating volcanic rocks in lava flows above and below the fossil-containing rock.

half–life Characteristic time it takes for half of a quantity of a radioisotope to decay.
radiometric dating Method of estimating the age of a rock or fossil by measuring the content and proportions of a radioisotope and its daughter elements.

Take-Home Message What are fossils?

❯ Fossils are evidence of organisms that lived in the remote past, a stone-hard historical record of life.

❯ Researchers use the predictability of radioisotope decay to estimate the age of rocks and fossils.

> Transitions in the fossil record are boundaries for great intervals of the geologic time scale.

Radiometric dating and fossils allow us to recognize similar sequences of sedimentary rock layers around the world. Transitions between layers mark boundaries between great intervals of time in the **geologic time scale**, which is a chronology of Earth's history (Figure 16.11). Each layer's composition offers clues about conditions on Earth during the time the layer was deposited. Fossils in the layers are a record of life during that period of time.

geologic time scale Chronology of Earth's history.

Eon	Era	Period	Epoch	mya	Major Geologic and Biological Events
Phanerozoic	Cenozoic	Quaternary	Recent	0.01	Modern humans evolve. Major extinction event is now under way.
			Pleistocene	1.8	
		Tertiary	Pliocene	5.3	Tropics, subtropics extend poleward. Climate cools; dry woodlands and grasslands emerge. Adaptive radiations of mammals, insects, birds.
			Miocene	23.0	
			Oligocene	33.9	
			Eocene	55.8	
			Paleocene	65.5 ◄	Major extinction event, perhaps precipitated by asteroid impact. Mass extinction of all dinosaurs and many marine organisms.
	Mesozoic	Cretaceous	Late		
				99.6	
			Early		Climate very warm. Dinosaurs continue to dominate. Important modern insect groups appear (bees, butterflies, termites, ants, and herbivorous insects including aphids and grasshoppers). Flowering plants originate and become dominant land plants.
				145.5	
		Jurassic			Age of dinosaurs. Lush vegetation; abundant gymnosperms and ferns. Birds appear. Pangea breaks up.
				199.6 ◄	Major extinction event
		Triassic			Recovery from the major extinction at end of Permian. Many new groups appear, including turtles, dinosaurs, pterosaurs, and mammals.
	Paleozoic			251 ◄	Major extinction event
		Permian			Supercontinent Pangea and world ocean form. Adaptive radiation of conifers. Cycads and ginkgos appear. Relatively dry climate leads to drought-adapted gymnosperms and insects such as beetles and flies.
				299	
		Carboniferous			High atmospheric oxygen level fosters giant arthropods. Spore-releasing plants dominate. Age of great lycophyte trees; vast coal forests form. Ears evolve in amphibians; penises evolve in early reptiles (vaginas evolve later, in mammals only).
				359 ◄	Major extinction event
		Devonian			Land tetrapods appear. Explosion of plant diversity leads to tree forms, forests, and many new plant groups including lycophytes, ferns with complex leaves, seed plants.
				416	
		Silurian			Radiations of marine invertebrates. First appearances of land fungi, vascular plants, bony fishes, and perhaps terrestrial animals (millipedes, spiders).
				443 ◄	Major extinction event
		Ordovician			Major period for first appearances. The first land plants, fishes, and reef-forming corals appear. Gondwana moves toward the South Pole and becomes frigid.
				488	
		Cambrian			Earth thaws. Explosion of animal diversity. Most major groups of animals appear (in the oceans). Trilobites and shelled organisms evolve.
				542	
Proterozoic					Oxygen accumulates in atmosphere. Origin of aerobic metabolism. Origin of eukaryotic cells, then protists, fungi, plants, animals. Evidence that Earth mostly freezes over in a series of global ice ages between 750 and 600 mya.
				2,500	
Archaean and earlier					3,800–2,500 mya. Origin of bacteria and archaeans.
					4,600–3,800 mya. Origin of Earth's crust, first atmosphere, first seas. Chemical, molecular evolution leads to origin of life (from protocells to anaerobic single cells).

Figure 16.11 Animated The geologic time scale correlated with sedimentary rock exposed by erosion in the Grand Canyon. **A** Transitions between layers of sedimentary rock mark great time spans in Earth's history (not to the same scale). mya: millions of years ago. Dates are from the International Commission on Stratigraphy, 2007.

B We can reconstruct some of the events in the history of life by studying rocky clues in the layers. Here, the *red* triangles mark times of great mass extinctions. "First appearance" refers to appearance in the fossil record, not necessarily the first appearance on Earth; we often discover fossils that are significantly older than previously discovered specimens.

Permian

Kaibab Limestone

Toroweap Formation

Coconino Sandstone

Hermit Shale

Carboniferous

Esplanade Sandstone

Wescogame Formation

Manakacha Formation

Watahomigi Formation

Redwall Limestone

Temple Butte Formation

Cambrian

Muav Limestone

Bright Angel Shale

Tapeats Sandstone

Proterozoic

Chuar Group*

Nankoweap Formation*

Unkar Group*

Vishnu Basement Rocks

*Layers not visible in this view of the Grand Canyon

C Each rock layer has a composition and set of fossils that reflect events during its deposition. For example, Coconino Sandstone, which stretches from California to Montana, is mainly weathered sand. Ripple marks and reptile tracks are the only fossils in it. Many think it is the remains of a vast sand desert, like the Sahara is today.

Take-Home Message What is the geologic time scale?

> The geologic time scale is a chronology of Earth's history that correlates geologic and evolutionary events of the ancient past.

> Over billions of years, movements of Earth's outer layer have changed the land, atmosphere, and oceans, with profound effects on the evolution of life.

Wind, water, and other natural forces continuously sculpt the surface of Earth, but they are only part of a bigger picture of geologic change. Earth itself also changes dramatically. For instance, the Atlantic coasts of South America and Africa seem to "fit" like jigsaw puzzle pieces. By one theory, all continents that exist today were once part of a bigger supercontinent—**Pangea**—that had split into fragments and drifted apart. The idea explained why the same types of fossils occur in sedimentary rock on both sides of the Atlantic Ocean.

At first, most scientists did not accept this theory, which was called continental drift. Continents drifting about Earth seemed to be an outrageous idea, and no one knew what would drive such movement. However, evidence that supported the model kept turning up. For instance, molten rock deep inside Earth wells up and solidifies on the surface. Some iron-rich minerals become magnetic as they solidify, and their magnetic poles align with Earth's poles when they do. If continents never moved, then all of these ancient rocky magnets would be aligned north-to-south, like compass needles. Indeed, the magnetic poles of the rock formations are aligned—but not north-to-south. The poles of rock formations on different continents point in all different directions. Either Earth's magnetic poles veer dramatically from their north–south axis, or the continents wander.

Deep-sea explorers also discovered that ocean floors are not as static and featureless as had been assumed.

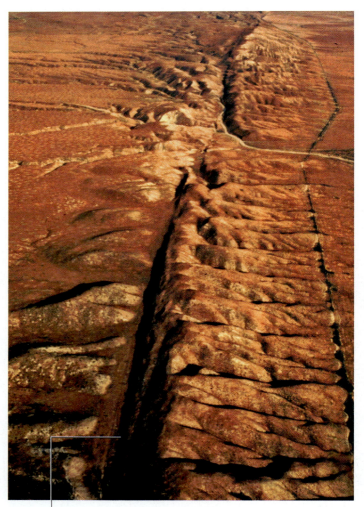

Figure 16.12 Animated Plate tectonics. Huge pieces of Earth's outer rock layer slowly drift apart and collide. As the plates move, they convey continents around the globe. The current configuration of the plates is shown in Appendix VIII.

1 At oceanic ridges, huge plumes of molten rock welling up from Earth's interior drive the movement of tectonic plates. New crust spreads outward as it forms on the surface, forcing adjacent tectonic plates away from the ridge and into trenches elsewhere.

2 At trenches, the advancing edge of one plate plows under an adjacent plate and buckles it.

3 Faults are ruptures in Earth's crust where plates meet. The diagram shows a type of fault called a rift, in which plates move apart. The aerial photo on the *left* shows about 4.2 kilometers (2.6 miles) of the San Andreas Fault, which extends 1,300 km (800 miles) through California. This fault is a boundary between two tectonic plates slipping past one another.

4 Plumes of molten rock rupture a tectonic plate at what are called "hot spots." The Hawaiian Islands have been forming from molten rock that continues to erupt from a hot spot under the Pacific (tectonic) Plate.

3 fault   **2** trench   **1** ridge   **4** hot spot   trench

| **A** 420 mya | **B** 237 mya | **C** 152 mya | **D** 65.5 mya | **E** 14 mya |

Figure 16.13 A series of reconstructions of the drifting continents. **A** The supercontinent Gondwana (*yellow*) had begun to break up by the Silurian. **B** The supercontinent Pangea formed during the Triassic, then **C** began to break up in the Jurassic. **D** K–T boundary. **E** The continents reached their modern configuration in the Miocene.

Immense ridges stretch thousands of kilometers across the sea floor (Figure 16.12). Molten rock spewing from the ridges pushes old sea floor outward in both directions **1**, then cools and hardens into new sea floor. Elsewhere, older sea floor plunges into deep trenches **2**.

Such discoveries swayed the skeptics. Finally, there was a plausible mechanism for continental drift, which is now called **plate tectonics**. By this theory, Earth's relatively thin outer layer of rock is cracked into immense plates, a bit like a gigantic cracked eggshell. Molten rock emanating from an undersea ridge or continental rift at one edge of a plate pushes old rock at the opposite edge into a trench. The movement is like that of a colossal conveyer belt that transports continents on top of it to new locations. Each plate moves no more than 10 centimeters (4 inches) a year—about half as fast as your toenails grow—but that is enough to carry a continent all the way around the world after 40 million years or so.

Evidence of tectonic movement is all around us, in faults **3** and other various geologic features of our landscapes. For example, volcanic island chains (archipelagos) form as a plate moves across an undersea hot spot. Hot spots are places where a narrow plume of molten rock wells up from deep inside Earth and ruptures a plate **4**.

Plate tectonics solved some long-standing puzzles. Consider an unusual geologic formation that occurs in a belt across Africa. The sequence of rock layers in this formation is so complex that it is quite unlikely to have formed more than once, but identical sequences also occur in huge belts that span India, South America, Africa, Madagascar, Australia, and Antarctica. The most likely explanation for the wide distribution is that the formations were deposited together on a single continent that later broke up. This explanation is supported by fossils in the rock layers: the remains of a type of fern (*Glossopteris*) whose seeds were too heavy to float or to be wind-blown over an ocean, and of an early reptile (*Lystrosaurus*) whose body was not built for swimming between continents.

Here is the puzzle: *Glossopteris* disappeared in the Permian–Triassic mass extinction event (251 million years ago), and *Lystrosaurus* disappeared 6 million years after that. Both organisms were extinct millions of years before Pangea formed. Could they have evolved together on a different supercontinent, one that predated Pangea? Evidence suggests that they did. The older supercontinent, which we now call **Gondwana**, included most of the land masses that currently exist in the Southern Hemisphere as well as India and Arabia (Figure 16.13). Many modern species, including the ratite birds pictured in Figure 16.2, live only in places that were once part of Gondwana.

After Gondwana formed, it drifted south, across the South Pole, then north until it merged with other continents to form Pangea. We now know that at least five times since Earth's outer layer of rock solidified 4.55 billion years ago, a single supercontinent with one ocean lapping at its coastline formed and then split up again. The resulting changes in Earth's surface, atmosphere, and waters have had a profound impact on the course of life's evolution. A continent's climate changes—often dramatically so—along with its position on Earth. Colliding continents physically separate organisms living in oceans, and bring together those that had been living apart on land. As continents break up, they separate organisms living on land, and bring together ones that had been living in separate oceans. Such changes are a major driving force of evolution, as you will see in Chapter 17. Lineages that cannot adapt to the changes die out, and new evolutionary opportunities open up for the survivors.

Gondwana Supercontinent that existed before Pangea, more than 500 million years ago.
Pangea Supercontinent that formed about 237 million years ago and broke up about 152 million years ago.
plate tectonics Theory that Earth's outer layer of rock is cracked into plates, the slow movement of which rafts continents to new locations over geologic time.

Take-Home Message How has Earth changed over geologic time spans?

❯ Over geologic time, movements of Earth's crust have caused dramatic changes in the continents, atmosphere, and oceans.

❯ The course of life's evolution has been influenced by these changes.

❯ Physical similarities may be evidence of shared ancestry.

How do we know about evolution that occurred in the ancient past? Like asteroid impacts, evolution leaves evidence. Fossils are one example, but organisms that are alive today provide others. To a biologist, remember, evolution means change in a line of descent. Clues about the history of a lineage may be encoded in body form, function, or biochemistry. For example, similarities in the structure of body parts often reflect shared ancestry. Comparative morphology can be used to unravel evolutionary relationships in such cases. Similar body parts that evolved in a common ancestor are called **homologous structures** (*hom–* means "the same"). Homologous structures may be used for different purposes in different groups, but the very same genes direct their development.

Morphological Divergence

A body part that appears very different in different lineages may be similar in some underlying aspect of form. For example, even though vertebrate forelimbs are not the same in size, shape, or function from one group to the next, they clearly are alike in the structure and positioning of bony elements. They also are alike in the patterns of nerves, blood vessels, and muscles that develop inside of them. Such similarities are evidence of shared ancestry.

As you will see in the next chapter, populations that are not interbreeding tend to diverge genetically, and in time they diverge morphologically. Change from the body form of a common ancestor is an evolutionary pattern called **morphological divergence**. We have evidence from fossilized limb bones that all modern land vertebrates are descended from a family of ancient "stem reptiles" that crouched low to the ground on four legs. Descendants of this ancestral family diversified into many new habitats on land, and gave rise to the groups we call reptiles, birds, and mammals. A few lineages that had become adapted to walking on land even returned to life in the seas.

Over millions of years, the stem reptile's five-toed limbs become adapted for very different purposes across many lineages (**Figure 16.14**). In extinct reptiles called pterosaurs, most birds, and bats, they were modified for flight. In penguins and porpoises, the limbs are now flippers useful for swimming. In humans, five-toed forelimbs became arms and hands, in which the thumb evolved in opposition to the fingers. An opposable thumb was the basis of more precise motions and a firmer grip. Among elephants, the limbs are now strong and pillarlike, capable of supporting a great deal of weight. Limbs degenerated to nubs in pythons and boa constrictors, and to nothing at all in other snakes.

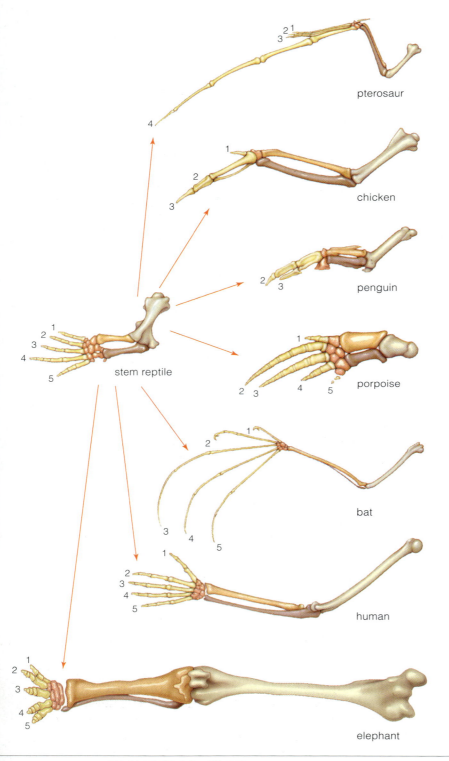

Figure 16.14 Morphological divergence among vertebrate forelimbs, starting with the bones of a stem reptile. The number and position of many skeletal elements were preserved when these diverse forms evolved; notice the bones of the forearms. Certain bones were lost over time in some of the lineages (compare the digits numbered 1 through 5). The drawings are not to the same scale.

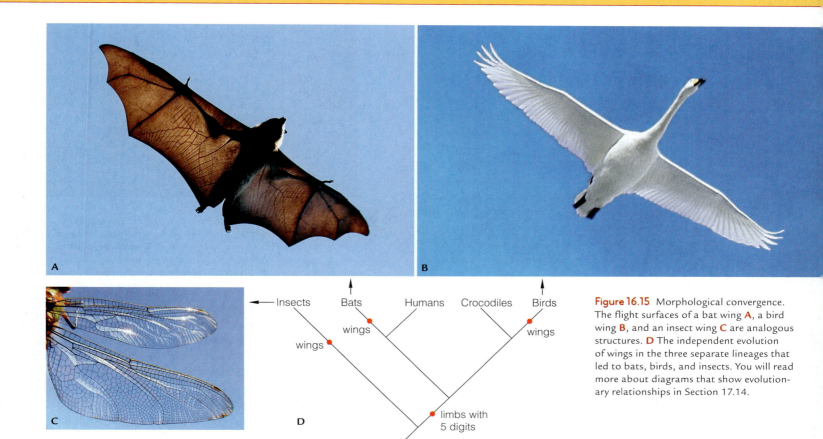

Insects　Bats　Humans　Crocodiles　Birds

wings

wings

wings

limbs with
5 digits

A

B

C

D

Figure 16.15 Morphological convergence. The flight surfaces of a bat wing **A**, a bird wing **B**, and an insect wing **C** are analogous structures. **D** The independent evolution of wings in the three separate lineages that led to bats, birds, and insects. You will read more about diagrams that show evolutionary relationships in Section 17.14.

Morphological Convergence

Similar body parts are not always homologous. Similar structures may evolve independently in separate lineages as adaptations to the same environmental pressures. In this case they are called **analogous structures**. Analogous structures look alike in different lineages but did not evolve in a shared ancestor; they evolved independently after the lineages diverged. Evolution of similar body parts in different lineages is **morphological convergence**.

We can sometimes identify analogous structures by studying their underlying form. For example, bird, bat, and insect wings all perform the same function: flight. However, several clues tell us that the flight surfaces of these wings are not homologous. The wing surfaces are adapted to the same physical constraints that govern flight, but the adaptations are different. In the case of

birds and bats, the limbs themselves are homologous, but the adaptations that make those limbs useful for flight differ. The surface of a bat wing is a thin, membranous extension of the animal's skin. By contrast, the surface of a bird wing is a sweep of feathers, which are specialized structures derived from skin. Insect wings differ even more. An insect wing forms as a saclike extension of the body wall. Except at forked veins, the sac flattens and fuses into a thin membrane. The veins are reinforced with chitin, which structurally support the wing. The unique adaptations for flight are evidence that wing surfaces of birds, bats, and insects are analogous structures—they evolved after the ancestors of these modern groups diverged (Figure 16.15).

analogous structures Similar body structures that evolved separately in different lineages.
homologous structures Similar body parts that evolved in a common ancestor.
morphological convergence Evolutionary pattern in which similar body parts evolve separately in different lineages.
morphological divergence Evolutionary pattern in which a body part of an ancestor changes in its descendants.

Take-Home Message Are similar body parts indicative of an evolutionary relationship?

> In morphological divergence, a body part inherited from a common ancestor becomes modified differently in different lines of descent. Such parts are called homologous structures.

> In morphological convergence, body parts that appear alike evolved independently in different lineages, not in a common ancestor. Such parts are called analogous structures.

› Similar patterns of embryonic development may be evidence of evolutionary relationships.

‹ Links to Master genes in development 10.3, Floral identity gene mutations 10.4, Evolution by gene duplications 14.5

The development of an embryo into the body of a plant or animal is orchestrated by layer after layer of master gene expression. The failure of any single master gene to participate in this symphony of expression can result in a drastically altered body plan, typically with devastating consequences. Because a mutation in a master gene typically unravels development, these genes tend to be highly conserved, which means they have changed very little or not at all over evolutionary time. Thus, a master gene with a similar sequence and function across different lineages is strong evidence that those lineages are related.

Similar Genes in Plants

The master genes called homeotic genes guide formation of specific body parts during development. A mutation in one homeotic gene can disrupt details of the body's form. For example, any mutation that inactivates a floral identity gene, *Apetala1*, in wild cabbage plants (*Brassica oleracea*) results in mutated flowers. Such flowers form with male reproductive structures (stamens) where petals are supposed to be. At least in the laboratory, these abundantly stamened flowers are exceptionally fertile, but such alterations usually are selected against in nature. *Apetala1* mutations in common wall cress plants (*Arabidopsis thaliana*) also results in flowers that have no petals (Section 10.4). The *Apetala1* gene affects the formation of petals across many different lineages, so it is very likely that this gene evolved in a shared ancestor.

Developmental Comparisons in Animals

The embryos of many vertebrate species develop in similar ways. Their tissues form the same way, as embryonic cells divide, differentiate, and interact. For example, all vertebrates go through a stage in which they have four limb buds, a tail, and a series of somites—divisions of the body that give rise to a backbone (Figure 16.16). Given that

Figure 16.17 Expression of the *antennapedia* gene in the embryonic tissues of the insect thorax causes legs to form. Normally, the gene is never expressed in cells of any other tissue. A mutation that causes *antennapedia* to be expressed in the embryonic tissues of a *Drosophila*'s head (*left*) causes legs to form there too (*right*).

the same genes direct development in these lineages, how do the adult forms end up so different? Part of the answer is that there are differences in the onset, rate, or completion of early steps in development. These differences are brought about by variations in the expression patterns of master genes that govern development. The variation has apparently arisen mainly as a result of gene duplications followed by mutation, the same way that multiple globin genes evolved in primates (Section 14.5).

For example, master genes called *Hox* sculpt details of the body's form during embryonic development. The pattern of expression of these genes determines the identity of particular zones along the body axis. Insects and other arthropods have ten *Hox* genes. One of them, *antennapedia*, determines the identity of the thorax (the part with legs). Legs develop wherever *antennapedia* is expressed in an embryo (Figure 16.17). Vertebrates have four sets of the same ten *Hox* genes that occur in insects. A vertebrate version of *antennapedia*, the *Hoxc6* gene, determines the identity of the back (as opposed to the neck or tail). Expression of this gene causes ribs to develop on a vertebra (Figure 16.18). Vertebrae of the neck and tail normally develop with no *Hoxc6* expression, and no ribs.

Figure 16.16 Visual comparison of vertebrate embryos. All vertebrates go through an embryonic stage in which they have four limb buds, a tail, and divisions called somites along their back. Embryos *left* to *right:* human, mouse, bat, chicken, alligator.

Figure 16.18 An example of comparative embryology. Expression of the *Hoxc6* gene is indicated by *purple* stain in two vertebrate embryos, chick (*left*) and garter snake (*right*). Expression of this gene causes a vertebra to develop ribs as part of the back. Chickens have 7 vertebrae in their back and 14 to 17 vertebrae in their neck; snakes have upwards of 450 back vertebrae and essentially no neck.

Hox genes also regulate limb formation. Body appendages as diverse as crab legs, beetle legs, sea star arms, butterfly wings, fish fins, and mouse feet start out as clusters of cells that bud from the surface of the embryo. The buds form wherever the homeotic gene *Dlx* is expressed. *Dlx* encodes a transcription factor that signals clusters of embryonic cells to "stick out from the body" and give rise to an appendage. *Hox* genes suppress *Dlx* expression in all parts of an embryo that will not have appendages.

Forever Young At an early stage, a chimpanzee skull and a human skull appear quite similar. As development continues, both skulls change shape as different parts grow at different rates (Figure 16.19). However, the human skull undergoes less pronounced differential growth than the chimpanzee skull does. As a result, a human adult has a rounder braincase, a flatter face, and a less protruding jaw compared with an adult chimpanzee.

In its proportions, a human adult skull is more like the skull of an infant chimpanzee than the skull of an adult chimpanzee. The similarity suggests that human evolution involved changes that slowed the rate of development, causing traits that were previously typical of juvenile stages to persist into adulthood.

Juvenile features also persist in other adult animals, notably salamanders called axolotls. The larvae of most species of salamander live in water and use external gills to breathe. Lungs that replace the gills as development continues allow the adult to breathe air and live on land. By contrast, axolotls never give up their aquatic life-style; their external gills and other larval traits persist into adulthood.

The closest relatives of axolotls are tiger salamanders. As you might expect, tiger salamander larvae resemble axolotls, although they are smaller.

Reflections of a Distant Past (revisited)

The K–T boundary layer (*left*) consists of an unusual clay that formed 65 million years ago, worldwide (the *red* pocketknife is shown for scale). The clay is rich in iridium, an element rare on Earth's surface but common in asteroids. After finding the iridium, researchers looked for evidence of an asteroid big enough to cover the entire Earth with its debris. They found a crater that is about 65 million years old, buried under sediments off the coast of Mexico's Yucatán Peninsula. It is so big—273.6 kilometers (170 miles) across and 1 kilometer (3,000 feet) deep—that no one had realized it was a crater. This crater is evidence of an asteroid impact 40 million times more powerful than the one that made the Barringer Crater, certainly big enough to have influenced life on Earth in a big way.

How Would You Vote? Many theories and hypotheses about events in the ancient past are necessarily based on traces left by those events, not on data collected by direct observations. Is indirect evidence ever enough to prove a theory about a past event? See CengageNow for details, then vote online (cengagenow.com).

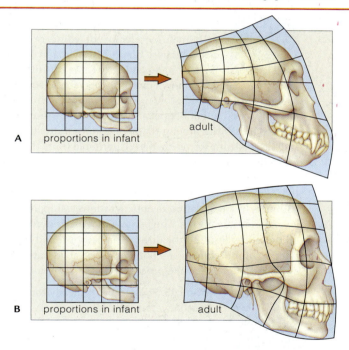

Figure 16.19 Animated Morphological differences between two primates. These skulls are depicted as paintings on a rubber sheet divided into a grid. Stretching the sheets deforms the grid. Differences in how they are stretched are analogous to different growth patterns. Shown here, proportional changes during skull development in **A** the chimpanzee and **B** the human. Chimpanzee skulls change more than human skulls, so the relative proportions in bones of adult and infant humans are more similar than those of adult and infant chimpanzees.

Take-Home Message Are similarities in development indicative of shared ancestry?

> Similarities in patterns of development are the result of master genes that have been conserved over evolutionary time.

> Some differences between closely related species arose as a result of changes in the rate of development.

Summary

Section 16.1 Events of the ancient past can be explained by the same physical, chemical, and biological processes that operate today. An asteroid impact may have caused a **mass extinction** 65.5 million years ago.

Section 16.2 Expeditions by nineteenth-century **naturalists** yielded increasingly detailed observations of nature. Geology, **biogeography**, and **comparative morphology** of organisms and their **fossils** led to new ways of thinking about the natural world.

Section 16.3 Prevailing belief systems may influence interpretation of the underlying cause of a natural event. The nineteenth-century naturalists proposed **catastrophism** and the **theory of uniformity** in their attempts to reconcile traditional beliefs with physical evidence of **evolution**, or change in a **lineage** over time.

Section 16.4 Humans select desirable traits in animals by selective breeding, or **artificial selection**. Charles Darwin and Alfred Wallace independently came up with a theory of how environments also select traits: A population tends to grow until it exhausts environmental resources. As that happens, competition for those resources intensifies among the population's individuals. Individuals with forms of shared, heritable traits that make them more competitive for the resources tend to produce more offspring. Thus, **adaptive traits (adaptations)** that impart greater **fitness** to an individual become more common in a population over generations, compared with less competitive forms. The process in which environmental pressures result in the differential survival and reproduction of individuals of a population is called **natural selection**. It is one of the processes that drives evolution.

Section 16.5 Fossils are typically found in stacked layers of sedimentary rock. Younger fossils usually occur in layers deposited more recently, on top of older fossils in older layers. Fossils are relatively scarce, so the fossil record will always be incomplete. The characteristic **half-life** of a radioisotope allows us to determine the age of rocks and fossils using **radiometric dating**.

Section 16.6 Transitions in the fossil record are the boundaries of great intervals of the **geologic time scale**, a chronology of Earth's history that correlates geologic and evolutionary events.

Section 16.7 By the theory of **plate tectonics**, the movements of Earth's tectonic plates carry land masses to new positions. Such movements had profound impacts on the directions of life's evolution. Several times in Earth's history, all land masses have converged as supercontinents. **Gondwana** and **Pangea** are examples.

Section 16.8 Comparative morphology can reveal evolutionary connections among lineages. **Homologous structures** are similar body parts that, by **morphological divergence**, became modified differently in different lineages. Such parts are evidence of a common ancestor. **Analogous structures** are body parts that look alike in different lineages but did not evolve in a common ancestor. By the process of **morphological convergence**, they evolved separately after the lineages diverged.

Section 16.9 Similarities among patterns of embryonic development reflect shared ancestry. Genes that affect development tend to be conserved. Mutations that alter the rate of development may allow juvenile traits to persist into adulthood.

Self-Quiz   Answers in Appendix III

1. The number of species on an island depends on the size of the island and its distance from a mainland. This statement would most likely be made by _____ .
   - a. an explorer
   - b. a biogeographer
   - c. a geologist
   - d. a philosopher

2. Evolution _____ .
   - a. is natural selection
   - b. is heritable change in a line of descent
   - c. can occur by natural selection
   - d. b and c are correct

3. Which of the following is a fossil?
   - a. An insect encased in 10-million-year-old tree sap
   - b. A woolly mammoth frozen in Arctic permafrost for the last 50,000 years
   - c. Mineral-hardened remains of a whale-like animal found in an Egyptian desert
   - d. An impression of a plant leaf in a rock
   - e. All of the above could be considered fossils

4. Did Pangea or Gondwana form first?

5. The bones of a bird's wing are similar to the bones in a bat's wing. This observation is an example of _____ .
   - a. uniformity
   - b. evolution
   - c. comparative morphology
   - d. a lineage

6. If the half-life of a radioisotope is 20,000 years, then a sample in which three-quarters of that radioisotope has decayed is _____ years old.
   - a. 15,000   b. 26,667   c. 30,000   d. 40,000

7. Forces of geologic change include _____ (select all that are correct).
   - a. erosion
   - b. fossilization
   - c. volcanic activity
   - d. evolution
   - e. tectonic plate movement
   - f. wind
   - g. asteroid impacts
   - h. hot spots

8. The Cretaceous ended _____ million years ago.

Data Analysis Activities

Abundance of Iridium in the K–T Boundary Layer In the late 1970s, geologist Walter Alvarez was investigating the composition of the 1-centimeter-thick layer of clay that marks the Cretaceous–Tertiary (K–T) boundary all over the world. He asked his father, Nobel Prize–winning physicist Luis Alvarez, to help him analyze the elemental composition of the layer. The photo shows Luis and Walter Alvarez with a section of the K–T boundary layer.

The Alvarezes and their colleagues tested the layer in Italy and in Denmark. The researchers discovered that the K–T boundary layer had a much higher iridium content than the surrounding rock layers. Some of their results are shown in the table in Figure 16.20.

Iridium belongs to a group of elements (Appendix IV) that are much more abundant in asteroids and other solar system materials than they are in Earth's crust. The Alvarez group concluded that the K–T boundary layer must have originated with extraterrestrial material. They calculated that an asteroid 14 kilometers (8.7 miles) in diameter would contain enough iridium to account for the iridium in the K–T boundary layer.

Sample Depth	Average Abundance of Iridium (ppb)
+ 2.7 m	< 0.3
+ 1.2 m	< 0.3
+ 0.7 m	0.36
boundary layer	41.6
− 0.5 m	0.25
− 5.4 m	0.30

Figure 16.20 Abundance of iridium in and near the K–T boundary layer in Stevns Klint, Denmark. Many rock samples taken from above, below, and at the boundary layer were tested for iridium content. Depths are given as meters above or below the boundary layer.

The iridium content of an average Earth rock is 0.4 parts per billion (ppb) of iridium. An average meteorite contains about 550 parts per billion of iridium.

1. What was the iridium content of the K–T boundary layer?
2. How much higher was the iridium content of the boundary layer than the sample taken 0.7 meter above the layer?

---

9. Life originated in the _____ .

10. Through _____ , a body part of an ancestor is modified differently in different lines of descent.
   a. morphological divergence
   b. adaptive divergence
   c. morphological convergence
   d. homologous evolution

11. Homologous structures among major groups of organisms may differ in _____ .
   a. size   c. function
   b. shape   d. all of the above

12. Match the terms with the most suitable description.
   ___ fitness
   ___ fossils
   ___ homeotic genes
   ___ half-life
   ___ homologous structures
   ___ uniformity
   ___ analogous structures
   ___ natural selection

   a. evidence of life in distant past
   b. geologic change occurs continuously
   c. human arm and bird wing
   d. big role in development
   e. measured by reproductive success
   f. insect wing and bird wing
   g. survival of the fittest
   h. characteristic of radioisotope

Additional questions are available on **CENGAGENOW**.

Animations and Interactions on **CENGAGENOW**:
› Vestigial body parts; Geologic time scale; Half-life; Radiometric dating; Plate tectonics; Proportional changes in embryonic development.

Critical Thinking

1. Radiometric dating does not measure the age of an individual atom. It is a measure of the age of a quantity of atoms—a statistic. As with any statistical measure, its values may deviate around an average (see sampling error, Section 1.8). Imagine that one sample of rock is dated ten different ways. Nine of the tests yield an age close to 225,000 years. One test yields an age of 3.2 million years. Do the nine consistent results imply that the one that deviates is incorrect, or does the one odd result invalidate the nine that are consistent?

2. If you think of geologic time spans as minutes, life's history might be plotted on a clock such as the one shown on the *right*. According to this clock, the most recent epoch started in the last 0.1 second before noon. Where does that put you?

‹ Links to Earlier Concepts

This chapter builds on the theory of natural selection (Section 16.4). You may wish to review life's organization (1.2), the geologic time scale (16.6), and plate tectonics (16.7). You will revisit experiments (1.7), sampling error (1.8), alleles (12.2), cellular reproduction (11.3, 12.2, 12.3), the genetic basis of traits (10.3, 13.2, 13.5, 13.6), effects of genetic changes (9.6, 14.2, 14.6), bacteria (4.1, 4.5), melanin (13.5), and transgenic plants (15.7).

Key Concepts

Microevolution

Individuals of a population inherit different alleles, and so they differ in phenotype. Over generations, any allele may increase or decrease in frequency in a population. Such change is called microevolution.

Processes of Microevolution

Natural selection may maintain or shift the range of variation of a shared, heritable trait in a population. Gene flow counters the evolutionary effects of mutation, natural selection, and genetic drift.

17 Processes of Evolution

17.1 Rise of the Super Rats

Slipping in and out of the pages of human history are rats—*Rattus*—the most notorious of mammalian pests. Rats thrive in urban centers, where garbage is plentiful and natural predators are not (Figure 17.1). The average city in the United States sustains about one rat for every ten people. Part of their success stems from an ability to reproduce very quickly. Rat populations can expand within weeks to match the amount of garbage available for them to eat. Unfortunately for us, rats carry pathogens and parasites associated with infectious diseases such as bubonic plague and typhus. They chew their way through walls and wires, and eat or foul 20 to 30 percent of our total food production. Rats cost us about $19 billion per year.

For years, people have been fighting back with poisons, including arsenic and cyanide. Baits laced with warfarin, an organic compound that interferes with blood clotting, were popular in the 1950s. Rats that ate the poisoned baits died within days after bleeding internally or losing blood through cuts or scrapes. Warfarin was very effective, and compared to other rat poisons, it had much less impact on harmless species. It quickly became the rodenticide of choice.

In 1958, however, a Scottish researcher reported that warfarin was not working against some rats. Similar reports from other European countries followed. About twenty years later, about 10 percent of rats caught in urban areas of the United States were resistant to warfarin. What happened?

To find out, researchers compared rats that were resistant to warfarin with those who were not. The difference was traced to a gene on one of the rat chromosomes. Certain mutations in the gene were common among warfarin-resistant rat populations but rare among vulnerable ones. Warfarin inhibits the gene's product, an enzyme that recycles vitamin K after it has been used to activate blood clotting factors. The mutations made the enzyme less active, but also insensitive to warfarin.

"What happened" was evolution by natural selection. As warfarin exerted pressure on rat populations, the populations changed. The previously rare alleles became adaptive. Rats that had an unmutated gene died after eating warfarin. The lucky ones that had one of the warfarin-resistance alleles survived and passed it to their offspring. The rat populations recovered quickly, and a higher proportion of rats in the next generation carried the alleles. With each onslaught of warfarin, the frequency of the alleles in rat populations increased.

Selection pressures can and often do change. When warfarin resistance increased in rat populations, people stopped using warfarin. The frequency of warfarin-resistance alleles in rat populations declined, probably because rats with the alleles are not as healthy as normal rats. Now, savvy exterminators in urban areas know that the best way to control a rat infestation is to exert another kind of selection pressure: Remove their source of food, which is usually garbage. Then the rats will eat each other.

Figure 17.1 Rats as pests. *Above*, rats infesting rice fields in the Philippine Islands ruin more than 20 percent of the crop. *Opposite*, rats thrive wherever people do. Dousing buildings and soil with poisons does not usually exterminate rat populations, which recover quickly. Rather, the practice selects for rats that are resistant to the poisons.

How Species Arise
Speciation varies in its details, but it always involves the end of gene flow between populations. Microevolutionary events that occur independently lead to genetic divergences, which are reinforced by reproductive isolation.

Macroevolution
Patterns of genetic change that involve more than one species are called macroevolution. Recurring patterns of macroevolution include the origin of major groups, one species giving rise to many, and mass extinction.

Cladistics
Evolutionary tree diagrams are based on the premise that all species interconnect through shared ancestors. Grouping species by shared ancestry better reflects evolutionary history than do traditional ranking systems.

> Evolution starts with mutations in individuals.
> Mutation is the source of new alleles.
> Sexual reproduction can quickly spread a mutation through a population.
‹ Links to Life's organization 1.2, Mutation 9.6, Alleles 12.2, Mendelian inheritance 13.2, Complex traits 13.5 and 13.6

Variation in Populations

A **population** is a group of interbreeding individuals of the same species in some specified area (Section 1.2). The individuals of a species—and a population—share certain features. For example, giraffes normally have very long necks, brown spots on white coats, and so on. These are examples of morphological traits (*morpho*– means form). Individuals of a species also share physiological traits, such as metabolic activities. They also respond the same way to certain stimuli, as when hungry giraffes feed on tree leaves. These are behavioral traits.

Individuals of a population have the same traits because they have the same genes. However, almost every shared trait varies a bit among individuals of a population (Figure 17.2). Alleles of the shared genes are the main source of this variation. Many traits have two or more distinct forms, or morphs. A trait with only two forms is dimorphic (*di*– means two). Purple and white flower color in the pea plants that Gregor Mendel studied is an example of a dimorphic trait (Section 13.3). Dimorphic flower color occurs in this case because the interaction of two alleles with a clear dominance relationship gives rise to the trait. Traits with more than two distinct forms are polymorphic (*poly*–, many). Human blood type, which is determined by the codominant *ABO* alleles, is an example (Section 13.5). Traits that vary continuously among the individuals of a population often arise by interactions among alleles of several genes, and may be influenced by environmental factors (Sections 13.5 and 13.6).

In earlier chapters, you learned about the processes that introduce and maintain variation in traits among individuals of a species. Table 17.1 summarizes the key events involved. Mutation is the original source of new alleles. Other events shuffle alleles into different combinations, and what a shuffle that is! There are $10^{116,446,000}$ possible combinations of human alleles. Not even 10^{10} people are living today. Unless you have an identical twin, it is unlikely that another person with your precise genetic makeup has ever lived, or ever will.

An Evolutionary View of Mutations

Being the original source of new alleles, mutations are worth another look—this time within the context of their impact on populations. We cannot predict when or in which individual a particular gene will mutate. We can, however, predict the average mutation rate of a species, which is the probability that a mutation will occur in a given interval. In humans, that rate is about 2.2×10^{-9} mutations per base pair per year. That means almost 70 mutations accumulate in the human genome per decade.

Many mutations give rise to structural, functional, or behavioral alterations that reduce an individual's chances of surviving and reproducing. Even one biochemical change may be devastating. For instance, the skin, bones, tendons, lungs, blood vessels, and other vertebrate organs incorporate the protein collagen. If one of the genes for collagen mutates in a way that changes the protein's function, the entire body may be affected. A mutation such as this can change phenotype so drastically that it results in death, in which case it is a **lethal mutation**.

A **neutral mutation** changes the base sequence in DNA, but the alteration has no effect on survival or reproduction. It neither helps nor hurts the individual. For instance, if you carry a mutation that keeps your earlobes attached to your head instead of swinging freely, attached earlobes should not in itself stop you from surviving and

Table 17.1	Sources of Variation in Traits Among Individuals of a Species
Genetic Event	**Effect**
Mutation	Source of new alleles
Crossing over at meiosis I	Introduces new combinations of alleles into chromosomes
Independent assortment at meiosis I	Mixes maternal and paternal chromosomes
Fertilization	Combines alleles from two parents
Changes in chromosome number or structure	Transposition, duplication, or loss of chromosomes

Figure 17.2 Sampling phenotypic variation in **A** (*opposite*) a type of snail found on islands in the Caribbean, and **B** humans. The variation in shared traits among individuals is mainly an outcome of variations in alleles that influence those traits.

reproducing as well as anybody else. So, natural selection does not affect the frequency of this particular mutation in a population.

Occasionally, a change in the environment favors a mutation that had previously been neutral or even somewhat harmful. The warfarin resistance gene in rats is an example. Even if a beneficial mutation bestows only a slight advantage, its frequency tends to increase in a population over time. This is because natural selection operates on traits with a genetic basis. With natural selection, remember, environmental pressures result in an increase in the frequency of a beneficial trait in a population over generations (Section 16.4).

Mutations have been altering genomes for billions of years, and they are still at it. Cumulatively, they have given rise to Earth's staggering biodiversity. Think about it: The reason you do not look like an avocado or an earthworm or even your next-door neighbor began with mutations that occurred in different lines of descent.

Allele Frequencies

Together, all the alleles of all the genes of a population comprise a pool of genetic resources called a **gene pool**. Members of a population breed with one another more often than they breed with members of other populations, so their gene pool is more or less isolated. **Allele frequency** refers to the abundance of a particular allele among the individuals of a population. Change in an allele's frequency in a population is the same thing as change in a line of descent—evolution. Evolution within a population or species is called **microevolution**.

A theoretical reference point, **genetic equilibrium**, occurs when the allele frequencies of a population do not change (in other words, the population is not evolving). Genetic equilibrium can only occur if every one of the following five conditions are met: (1) Mutations never occur; (2) the population is infinitely large; (3) the population is isolated from all other populations of the species; (4) mating is random; and (5) all individuals survive to produce the same number of offspring. As you can imagine, all five conditions are never met in nature, so natural populations are never in genetic equilibrium.

Microevolution is always occurring in natural populations because the processes that drive it are always operating. The remaining sections of this chapter explore microevolutionary processes—mutation, natural selection, genetic drift, and gene flow—and their effects. Remember, even though we can recognize patterns of evolution, none of them are purposeful. Evolution simply fills the nooks and crannies of opportunity.

allele frequency Abundance of a particular allele among members of a population.
gene pool All of the alleles of all of the genes in a population; a pool of genetic resources.
genetic equilibrium Theoretical state in which a population is not evolving.
lethal mutation Mutation that drastically alters phenotype; causes death.
microevolution Change in allele frequencies in a population or species.
neutral mutation A mutation that has no effect on survival or reproduction.
population A group of organisms of the same species who live in a specific location and breed with one another more often than they breed with members of other populations.

> Researchers know whether a population is evolving by tracking deviations from a baseline of genetic equilibrium.
< Link to Codominant alleles 13.5

The Hardy–Weinberg Formula Early in the twentieth century, Godfrey Hardy (a mathematician) and Wilhelm Weinberg (a physician) independently applied the rules of probability to sexually reproducing populations. They realized that gene pools can remain stable only when five conditions are being met:

1. Mutations do not occur.

2. The population is infinitely large.

3. The population is isolated from all other populations of the species (no gene flow).

4. Mating is random.

5. All individuals survive and produce the same number of offspring.

These conditions never occur all at once in nature. Thus, allele frequencies for any gene in the shared pool always change. However, we can think about a hypothetical situation in which the five conditions are being met and a population is not evolving.

Hardy and Weinberg developed a simple formula that can be used to track whether a population of any sexually reproducing species is in a state of genetic equilibrium. Consider a hypothetical gene that encodes a blue pigment in butterflies. Two alleles of this gene, *B* and *b*, are codominant. A butterfly homozygous for the *B* allele (*BB*) has dark-blue wings. A butterfly homozygous for the *b* allele (*bb*) has white wings. A heterozygous butterfly (*Bb*) has medium-blue wings (Figure 17.3).

At genetic equilibrium, the proportions of the wing-color genotypes are

$$p^2(BB) + 2pq(Bb) + q^2(bb) = 1.0$$

where *p* and *q* are the frequencies of alleles *B* and *b*. This equation became known as the Hardy–Weinberg equilibrium equation. It defines the frequency of a dominant allele (*B*) and a recessive allele (*b*) for a gene that controls a particular trait in a population.

The frequencies of *B* and *b* must add up to 1.0. To give a specific example, if *B* occupies 90 percent of the loci, then *b* must occupy the remaining 10 percent (0.9 + 0.1 = 1.0). No matter what the proportions,

$$p + q = 1.0$$

At meiosis, remember, paired alleles are assorted into different gametes. The proportion of gametes with the *B* allele is *p*, and the proportion with the *b* allele is *q*. The Punnett square *below* shows the genotypes possible in the next generation (*BB*, *Bb*, and *bb*). Note that the frequencies of the three genotypes add up to 1.0:

$$p^2 + 2pq + q^2 = 1.0$$

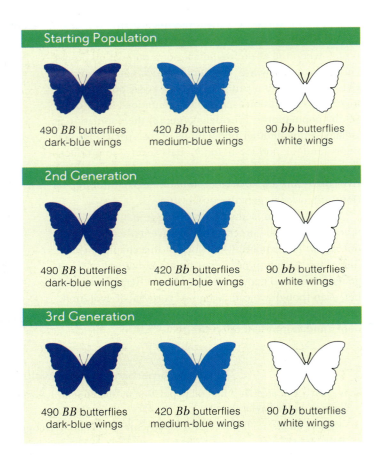

Starting Population

490 *BB* butterflies — dark-blue wings
420 *Bb* butterflies — medium-blue wings
90 *bb* butterflies — white wings

2nd Generation

490 *BB* butterflies — dark-blue wings
420 *Bb* butterflies — medium-blue wings
90 *bb* butterflies — white wings

3rd Generation

490 *BB* butterflies — dark-blue wings
420 *Bb* butterflies — medium-blue wings
90 *bb* butterflies — white wings

	p *B*	*q* *b*
p *B*	*BB* (p^2)	*Bb* (*pq*)
q *b*	*Bb* (*pq*)	*bb* (q^2)

Figure 17.3 Animated Finding out whether a population is evolving. The frequencies of wing-color alleles among all of the individuals in this hypothetical population of butterflies are not changing; thus, the population is not evolving.

Suppose that the population has 1,000 individuals and that each one produces two gametes:

490 *BB* individuals make 980 *B* gametes
420 *Bb* individuals make 420 *B* and 420 *b* gametes
90 *bb* individuals make 180 *b* gametes

The frequency of alleles B and b among 2,000 gametes is:

$$B = \frac{980 + 420}{2{,}000 \text{ alleles}} = \frac{1{,}400}{2{,}000} = 0.7 = p$$

$$b = \frac{180 + 420}{2{,}000 \text{ alleles}} = \frac{600}{2{,}000} = 0.3 = q$$

At fertilization, gametes combine at random and start a new generation. If the population size stays constant at 1,000, there will be 490 BB, 420 Bb, and 90 bb individuals. The frequencies of the alleles for dark-blue, medium-blue, and white wings are the same as they were in the original gametes. Thus, dark-blue, medium-blue, and white wings occur at the same frequencies in the new generation.

As long as the assumptions that Hardy and Weinberg identified continue to hold, the pattern persists. If traits show up in different proportions from one generation to the next, though, one or more of the five assumptions is not being met. The hunt can begin for the evolutionary forces driving the change.

Applying the Rule How does the Hardy–Weinberg formula work in the real world? Researchers can use it to estimate the frequency of carriers of alleles that cause genetic traits and disorders.

As an example, hereditary hemochromatosis (HH) is the most common genetic disorder among people of Irish ancestry. Affected individuals absorb too much iron from food. The symptoms of this autosomal recessive disorder include liver problems, fatigue, and arthritis. A study in Ireland found the frequency for one allele that causes HH to be 0.14. If $q = 0.14$, then p is 0.86. Based on this study, the carrier frequency ($2pq$) can be calculated to be about 0.24. Such information is useful to doctors and to public health professionals.

Another example: A mutation in the *BRCA2* gene has been linked to breast cancer in adults. A deviation from the birth frequencies predicted by the Hardy–Weinberg formula suggests that this mutation can also have effects even before birth. In one study, researchers looked at the mutation's frequency among newborn girls. They found fewer homozygotes than expected, based on the number of heterozygotes and the Hardy–Weinberg formula. Thus, it seems that in homozygous form the mutation impairs the survival of female embryos.

17.4 Patterns of Natural Selection

17.4 Patterns of Natural Selection

❯ Natural selection occurs in different patterns depending on the organisms involved and their environment.
❮ Link to Natural selection theory and fitness 16.4

The remainder of this chapter explores the mechanisms and effects of processes that drive evolution, including natural selection. **Natural selection** is a process in which environmental pressures result in the differential survival and reproduction of individuals of a population. It influences the frequency of alleles in a population by operating on phenotypes that have a genetic basis.

We observe different patterns of natural selection, depending on the selection pressures and the organisms involved. Sometimes, individuals with a trait at one extreme of a range of variation are selected against, and those at the other extreme are favored. We call this pattern directional selection. With stabilizing selection, mid-range forms are favored, and the extremes are selected against. With disruptive selection, forms at the extremes of the range of variation are favored, and the intermediate forms are selected against. We will discuss these three modes of natural selection, which **Figure 17.4** summarizes, in the following two sections.

Section 17.7 explores sexual selection, a mode of natural selection that operates on a population by influencing mating success. This section also discusses balanced polymorphism, a particular case of natural selection in which the fitness (Section 16.4) of heterozygous individuals is greater than that of homozygous individuals in a particular environment.

Natural selection and other processes of evolution can alter a population or species so much that it becomes a new species. We discuss mechanisms of speciation in later sections of this chapter.

Figure 17.4
Overview of three modes of natural selection.

natural selection A process in which environmental pressures result in the differential survival and reproduction of individuals of a population who vary in the details of shared, heritable traits.

> Directional selection favors a phenotype at one end of a range of variation.
< Links to Experimental design 1.7, Bacteria 4.1 and 4.5, Melanin deposition in fur 13.5, Continuous variation 13.6

Directional selection shifts an allele's frequency in a consistent direction, so forms at one end of a range of phenotypic variation become more common over time (Figure 17.5). The following examples show how field observations provide evidence of directional selection.

The Peppered Moth Peppered moths feed and mate at night, and rest motionless on trees during the day. Their behavior and coloration offer camouflage from day-flying,

Figure 17.6 Animated Directional selection in the peppered moth. **A** Light peppered moths on a nonsooty tree trunk are hidden from predators. **B** Dark ones stand out. In places where soot darkens tree trunks, the dark color **C** is more adaptive than **D** the light color.

Time 1

Time 2

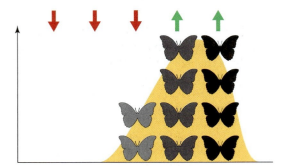

Time 3

Figure 17.5 Animated Directional selection. The bell-shaped curves indicate continuous variation in a butterfly wing-color trait. *Red* arrows show which forms are being selected against; *green*, forms that are being favored.

moth-eating birds. Light-colored moths were the most common form in preindustrial England. A dominant allele that resulted in the dark color was rare. The air was clean, and light-gray lichens grew on the trunks and branches of most trees. Light moths were camouflaged when they rested on the lichens, but dark moths were not (Figure 17.6A,B). By the 1850s, the dark moths had become much more common. Why? The industrial revolution had begun, and smoke from coal-burning factories was beginning to change the environment. Air pollution was killing the lichens. Dark moths were better camouflaged on soot-darkened trees (Figure 17.6C,D). Researchers hypothesized that birds were selectively eliminating light moths from populations of peppered moths living in industrialized areas. In the 1950s, H. B. Kettlewell bred both moth forms in captivity and marked hundreds so that they could be easily identified. He released them near highly industrialized areas around Birmingham and near an unpolluted part of Dorset. His team recaptured more of the dark moths in the polluted area and more light ones near Dorset. They also observed predatory birds eating more light moths in Birmingham, and more dark moths in Dorset.

Pollution controls went into effect in 1952. As a result, tree trunks gradually became free of soot, and lichens made a comeback. Kettlewell observed that moth phenotypes shifted too: Wherever pollution decreased, the frequency of dark moths decreased as well. Many other researchers since Kettlewell have confirmed the rise and fall of the dark-colored form of the peppered moth.

Rock Pocket Mice Directional selection also affects the color of rock pocket mice in Arizona's Sonoran Desert. Rock pocket mice are small mammals that spend the day sleeping in underground burrows, emerging at night to forage for seeds.

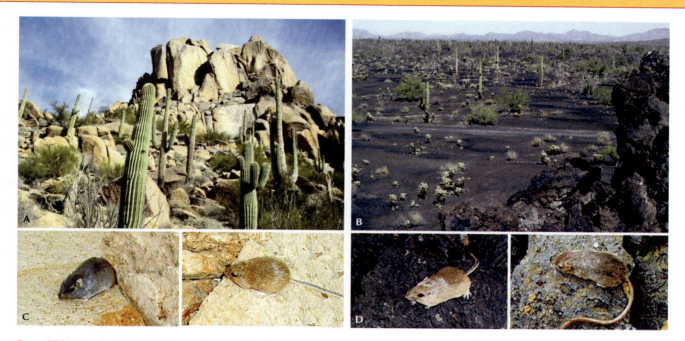

Figure 17.7 Directional selection in populations of rock pocket mice. **A** Mice with light fur are more common in areas with light-colored granite. **B** Mice with dark fur are more common in areas with dark basalt. **C,D** Mice with coat colors that do not match their surroundings are more easily seen by predators, so they are preferentially eliminated from the populations.

The Sonoran Desert is dominated by outcroppings of light brown granite (Figure 17.7A). There are also patches of dark basalt rock, the remains of ancient lava flows (Figure 17.7B). Most of the mice in populations that inhabit the dark rock have dark gray coats. Most of the mice in populations that inhabit the light brown rock have light brown coats. The difference arises because mice that match the rock color in each habitat are camouflaged from their natural predators. Night-flying owls more easily see mice that do not match the rocks, and they preferentially eliminate easily seen mice from each population (Figure 17.7C,D). The preferential predation results in a directional shift in the frequency of alleles that affect coat color. Compared to granite-dwelling populations, populations living on the dark basalt have a much higher frequency of four alleles that cause increased melanin deposition in fur.

Antibiotic Resistance Human attempts to control the environment can result in directional selection, as is the case with the warfarin-resistant rats. The use of antibiotics is another example. Prior to the 1940s, scarlet fever, tuberculosis, and pneumonia caused one-fourth of the annual deaths in the United States alone. Since the 1940s, we have been relying on antibiotics such as penicillin to fight these and other dangerous bacterial diseases. We also use them in other, less dire circumstances. Antibiotics are used preventively, both in humans and in livestock. They are part of the daily rations of millions of cattle, pigs, chickens, fish, and other animals that are raised on factory farms.

Bacteria evolve at an accelerated rate compared with humans, in part because they reproduce very quickly. For example, the common intestinal bacteria *E. coli* can divide every 17 minutes. Each new generation is an opportunity for mutation, so the gene pool of a bacterial population varies greatly. Thus, in any population of bacteria, some cells are likely to carry alleles that allow them to survive an antibiotic treatment. When the survivors reproduce, the frequency of antibiotic-resistance alleles increases in the population. A typical two-week course of antibiotics can potentially exert selection pressure on over a thousand generations of bacteria, and antibiotic-resistant strains may be the outcome.

Antibiotic-resistant bacteria have plagued hospitals for many years, and now they are becoming similarly common in schools. Even as researchers scramble to find new antibiotics, this trend is bad news for the millions of people each year who contract cholera, tuberculosis, or another dangerous bacterial disease.

directional selection Mode of natural selection in which phenotypes at one end of a range of variation are favored.

Take-Home Message What is the effect of directional selection?

❯ Directional selection causes allele frequencies underlying a range of variation to shift in a consistent direction.

> Stabilizing selection is a form of natural selection that maintains an intermediate phenotype.
> Disruptive selection favors forms of a trait at both ends of a range of variation.

Natural selection can bring about a directional shift in a population's range of phenotypes. Depending on the environment and the organisms involved, the process may also favor a midrange form of a trait, or it may eliminate the midrange form and favor extremes.

Stabilizing Selection

With **stabilizing selection**, an intermediate form of a trait is favored, and extreme forms are not. This mode of natural selection is also called balancing selection because it tends to preserve the midrange phenotypes in a population (Figure 17.8). For example, the body weight of sociable weavers (*Philetairus socius*) is subject to stabilizing selection (Figure 17.9). Weaver birds build large communal nests in areas of the African savanna. Between 1993 and 2000, Rita Covas and her colleagues captured, tagged, weighed, and released birds living in communal nests before the breeding season began. The researchers then recaptured and weighed the surviving birds after the breeding season was over.

Covas's field studies indicated that body weight in sociable weavers is a trade-off between the risks of starvation and predation. Foraging is not easy in the sparse habitat of an African savanna, and leaner birds do not store enough fat to avoid starvation. A meager food supply selects against birds with low body weight. Fatter birds may be more attractive to predators, and not as agile when escaping. Predators select against birds of high body weight. Thus, birds of intermediate weight have the selective advantage, and make up the bulk of sociable weaver populations.

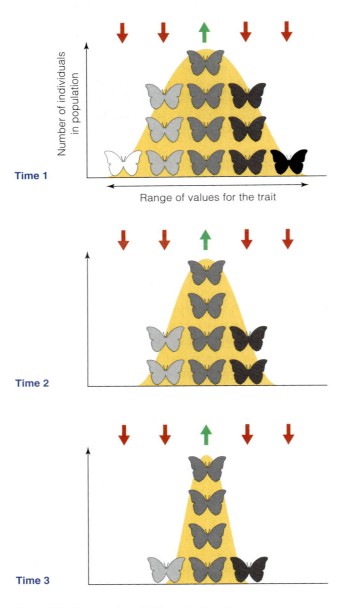

Time 1

Range of values for the trait

Time 2

Time 3

Figure 17.8 Animated Stabilizing selection eliminates extreme forms of a trait, and maintains the predominance of an intermediate phenotype in a population. *Red* arrows indicate which forms are being selected against; *green*, forms that are being favored. Compare the data set from a field experiment shown in Figure 17.9.

Figure 17.9 Stabilizing selection in sociable weavers. Graph shows the number of birds (out of 977) that survived a breeding season.

>> **Figure It Out** What is the optimal weight of a sociable weaver?

Answer: About 29 grams

Disruptive Selection

Conditions that favor forms of a trait at both ends of a range of variation drive **disruptive selection**. With this mode of natural selection, intermediate forms are selected against (Figure 17.10). Consider the black-bellied seedcracker (*Pyrenestes ostrinus*), a colorful finch species native to Cameroon, Africa. In these birds, there is a genetic basis for bill size. The bill of a typical black-bellied seedcracker, male or female, is either 12 millimeters wide, or wider than fifteen millimeters (Figure 17.11). Birds that have a bill between 12 and 15 millimeters wide are uncommon. Seedcrackers with the large and small bill forms inhabit the same geographic range, and they breed randomly with respect to bill size. It is as if every human adult were four feet or six feet tall, with no one of intermediate height.

Environmental factors that affect seedcracker feeding performance maintain the dimorphism in bill size. The finches feed mainly on the seeds of two types of sedge, which is a grasslike plant. One sedge produces hard seeds; the other, soft seeds. Small-billed birds are better at opening the soft seeds, but large-billed birds are better at cracking the hard ones. All seeds are abundant during Cameroon's wet seasons, and all seedcrackers feed on both types of seeds. However, during the region's dry seasons, sedge seeds are scarce, and each bird focuses on eating the seeds that it opens most efficiently. Small-billed birds feed mainly on soft seeds, and large-billed birds feed mainly on hard seeds. Birds with intermediate-sized bills cannot open either type of seed as efficiently as the other birds, so they are less likely to survive the dry seasons.

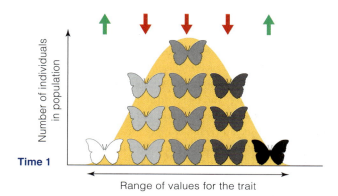

Time 1

Range of values for the trait

Time 2

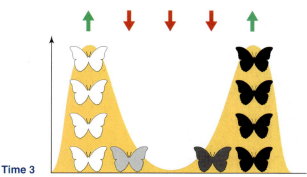

Time 3

Figure 17.10 Animated Disruptive selection eliminates midrange forms of a trait, and maintains extreme forms. *Red* arrows indicate which forms are being selected against; *green*, forms that are being favored.

Figure 17.11 Animated Disruptive selection in African seedcracker populations. In these birds, a distinct dimorphism in bill size is a result of competition for scarce food during dry seasons. These conditions favor birds with bills that are **A** twelve millimeters wide or **B** fifteen to twenty millimeters wide. Birds with bills of intermediate size are selected against.

---

disruptive selection Mode of natural selection that favors forms of a trait at the extremes of a range of variation; intermediate forms are selected against.
stabilizing selection Mode of natural selection in which intermediate forms of a trait are favored over extremes.

Take-Home Message What types of natural selection favor intermediate or extreme forms of traits?

> With stabilizing selection, an intermediate phenotype is favored, and extreme forms are selected against.

> With disruptive selection, an intermediate form of a trait is selected against, and extreme phenotypes are favored.

> Individuals may be selective agents for their own species.
> Any mode of natural selection may maintain two or more alleles in a population.
< Links to Sickle-cell anemia 9.6, Codominance 13.5

Selection pressures that operate on natural populations are often not as clear-cut as the examples in the previous sections might suggest. An allele may be adaptive in one circumstance but harmful in another, as the story about warfarin-resistance in rats illustrates. Even individuals of the same species can play a role.

Nonrandom Mating Not all natural selection occurs as a result of interactions between a species and its environment. Competition within a species also drives evolution. Consider how the individuals of many sexually reproducing species have a distinct male or female phenotype. Individuals of one sex (often males) tend to be more colorful, larger, or more aggressive than individuals of the other sex. These traits seem puzzling because they take energy and time away from an individual's survival activities. Some are probably maladaptive because they attract predators. Why do they persist?

The answer is **sexual selection**, in which the genetic winners outreproduce others of a population because they are better at securing mates. With this mode of natural selection, the most adaptive forms of a trait are those that help individuals defeat same-sex rivals for mates, or are the ones most attractive to the opposite sex.

For example, the females of some species cluster in defensible groups when they are sexually receptive.

Males of these species typically compete for access to the clusters of females. Competition for ready-made harems favors males that are combative (Figure 17.12A).

As another example, males or or females that are choosy about mates act as selective agents on their own species. The females of some species shop for a mate among males that display species-specific cues such as a specialized appearance or courtship behavior (Figure 17.12B). The cues often include flashy body parts or behaviors, traits that can be a physical hindrance and they may attract predators. However, a flashy male's survival despite his obvious handicap implies health and vigor, two traits that are likely to improve a female's chances of bearing healthy, vigorous offspring. The selected males pass alleles for their attractive traits to the next generation of males, and females pass alleles that influence mate preference to the next generation of females.

Sexual selection can give rise to highly exaggerated traits. For example, the eyes of the Malaysian stalk-eyed fly are on the tips of long, horizontal eyestalks that provide no obvious survival advantage to their bearers. Their adaptive value is sexual: Female flies prefer to mate with males sporting the longest eyestalks (Figure 17.12C).

---

balanced polymorphism Maintenance of two or more alleles for a trait at high frequency in a population as a result of natural selection against homozygotes.
sexual selection Mode of natural selection in which some individuals outreproduce others of a population because they are better at securing mates.

Figure 17.12
Sexual selection in action.

A Male elephant seals fight for sexual access to a cluster of females.

B A male bird of paradise engaged in a flashy courtship display has caught the eye (and, perhaps, the sexual interest) of a female. Female birds of paradise are choosy; a male mates with any female that accepts him.

C Stalk-eyed flies cluster on aerial roots to mate. Females prefer males with the longest eyestalks. This photo, taken in Malaysia, shows a male with very long eyestalks (*top*) that has captured the interest of the three females below him.

Balanced Polymorphism Any mode of natural selection may result in a **balanced polymorphism**. In this state, two or more alleles are maintained in a population at relatively high frequency by an environment that favors heterozygotes (individuals with nonidentical alleles).

Consider the gene that encodes the beta globin chain of hemoglobin, which is the oxygen-transporting protein in blood. *HbA* is the normal allele of this gene. The codominant *HbS* allele carries a mutation that causes sickle-cell anemia (Section 9.6).

Individuals homozygous for the *HbS* allele often die in their teens or early twenties. Despite being so harmful, the *HbS* allele persists at very high frequency among the human populations in tropical and subtropical regions of Asia and Africa. Why? Populations with the highest frequency of the *HbS* allele also have the highest incidence of malaria (Figure 17.13). Mosquitoes transmit the parasitic protist that causes malaria, *Plasmodium*, to human hosts. *Plasmodium* multiplies in the liver and then in red blood cells. The cells rupture and release new parasites during recurring bouts of severe illness.

It turns out that people who make both normal and sickle hemoglobin are more likely to survive malaria than people who make only normal hemoglobin. Several mechanisms are possible. For example, in *HbA/HbS* heterozygotes, *Plasmodium*-infected red blood cells sometimes take on a sickle shape. The abnormal shape brings the cells to the attention of the immune system, which destroys them—along with the parasites they harbor. By contrast, *Plasmodium*-infected red blood cells cells of *HbA/HbA* homozygotes do not sickle, so the parasite may remain hidden from the immune system.

In areas where malaria is common, the persistence of the *HbS* allele is a matter of relative evils. Malaria and sickle-cell anemia are both potentially deadly. *HbA/HbS* heterozygotes are more likely to survive malaria than *HbA/HbA* homozygotes. Heterozygotes are not completely healthy, but they do make enough normal hemoglobin to survive. With or without malaria, heterozygotes are more likely to live long enough to reproduce than *HbS/HbS* homozygotes. The result is that nearly one-third of the people living in the most malaria-ridden regions of the world are heterozygous for the *HbS* allele.

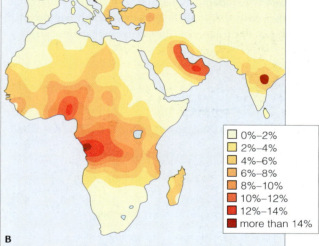

0%–2%
2%–4%
4%–6%
6%–8%
8%–10%
10%–12%
12%–14%
more than 14%

Figure 17.13 Malaria and sickle-cell anemia.

A Distribution of malaria cases (*orange*) reported in Africa, Asia, and the Middle East in the 1920s, before the start of programs to control mosquitoes, which transmit the parasitic protist that causes the disease.

B Distribution (by percentage) of people that carry the sickle-cell allele. Notice the correlation between the maps.

C Physician searching for mosquito larvae in Southeast Asia.

Take-Home Message How does natural selection maintain diversity?

> With sexual selection, a trait is adaptive if it gives an individual an advantage in securing mates. Sexual selection reinforces phenotypical differences between males and females, and sometimes gives rise to exaggerated traits.

> Environmental pressures that favor heterozygotes can lead to a balanced polymorphism.

> Especially in small populations, random changes in allele frequencies can lead to a loss of genetic diversity.
< Links to Probability and sampling error 1.8, Locus 13.2, Ellis–van Creveld syndrome 14.2

Genetic drift is a random change in allele frequencies over time, brought about by chance alone. We explain genetic drift in terms of probability—the chance that some event will occur. Remember, sample size is important in probability (Section 1.8). For example, every time you flip a coin, there is a 50 percent chance it will land heads up. With 10 flips, the proportion of times heads actually land up may be very far from 50 percent. With 1,000 flips, that proportion is more likely to be near 50 percent. We can apply the same rule to populations: the larger the population, the smaller the impact of random changes in allele frequencies.

Imagine two populations, one with 10 individuals, and the other with 100. If allele X occurs in both populations at a 10 percent frequency, then only one person carries

A The size of these populations of beetles was maintained at 10 breeding individuals. Allele b^+ was lost in one population (one graph line ends at 0).

B The size of these populations was maintained at 100 individuals. Drift in these populations was less than the small populations in **A**.

Figure 17.14 Animated Genetic drift in flour beetles (*Tribolium castaneum*, shown *below left* on a flake of cereal). Randomly selected beetles heterozygous for alleles b^+ and b were maintained in populations of **A** 10 individuals or **B** 100 individuals for 20 generations.

Graph lines in **B** are smoother than in **A**, indicating that drift was greatest in the sets of 10 beetles and least in the sets of 100. Notice that the average frequency of allele b^+ rose at the same rate in both groups, an indication that natural selection was at work too: Allele b^+ was weakly favored.

>> **Figure It Out** In how many populations did allele b^+ become fixed?   Answer: Six

the allele in the small population. If that person dies without reproducing, allele X will be lost from that population. However, ten people in the large population carry the allele. All ten would have to die without reproducing for the allele to be lost. Thus, the chance that the small population will lose allele X is greater than that for the large population. Steven Rich and his colleagues demonstrated this effect in populations of flour beetles (Figure 17.14).

Genetic drift can lead to the loss of genetic diversity in a population. This outcome is possible in all populations, but it is more likely to occur in small ones. When all individuals of a population are homozygous for an allele, we say that the allele is **fixed**. The frequency of an allele that is fixed will not change unless mutation or another process introduces a different allele into the population.

Bottlenecks Genetic drift can be dramatic when a few individuals rebuild a population or start a new one, such as occurs after a **bottleneck**. A bottleneck is a drastic reduction in population size brought about by severe selection pressure. For example, excessive hunting had reduced the population of northern elephant seals (shown in Figure 17.12A) to a mere twenty individuals by the late 1890s. The population has recovered to about 170,000 individuals since hunting restrictions were implemented, but every seal is homozygous at every gene locus analyzed to date. Genetic drift after the bottleneck has fixed all of the alleles in this population.

Bottlenecking can also occur when a small group of individuals founds a new population. If the group is not representative of the original population in terms of allele frequencies, then the new population will not be representative of it either. This outcome is called the **founder effect**. If a founding group is very small, the new population's genetic diversity may be quite reduced. Imagine that a seabird lands in the middle of a population of plants on a mainland. In this population's gene pool, half of the alleles governing flower color specify white flowers; the other half specify yellow flowers. A few seeds stick to the bird's feathers. The bird flies to a remote island and drops the seeds, which later sprout and form a new population on the island. If most of the seeds happened to be homozygous for the yellow flower allele, then the frequency of that allele in the new population will be much greater than 50 percent.

bottleneck Reduction in population size so severe that it reduces genetic diversity.
fixed Refers to an allele for which all members of a population are homozygous.
founder effect Change in allele frequencies that occurs when a small number of individuals establish a new population.
genetic drift Change in allele frequencies in a population due to chance alone.
inbreeding Mating among close relatives.

Figure 17.15 An Amish child with Ellis–van Creveld syndrome. The syndrome is characterized by dwarfism, polydactyly, and heart defects, among other symptoms. The recessive allele that causes it is common in the Old Order Amish of Lancaster County, an outcome of the founder effect and moderate inbreeding.

Genetic drift affects allele frequencies in inbred populations. **Inbreeding** is nonrandom breeding or mating between close relatives, which share more alleles than nonrelatives do. Inbreeding lowers a population's genetic diversity. Loss of diversity tends to be a bad thing, because it means more individuals in the population are homozygous for recessive alleles with harmful effects. This is why most societies discourage or forbid incest, or mating between parents and children or between siblings.

The Old Order Amish in Lancaster County, Pennsylvania, offer an example of the effects of inbreeding. Amish people marry only within their community. Intermarriage with other groups is not permitted, and no "outsiders" are allowed to join the community. As a result, Amish populations are moderately inbred, and many of their individuals are homozygous for harmful alleles. The Lancaster population has an unusually high frequency of a recessive allele that causes Ellis–van Creveld syndrome (Figure 17.15). This allele has been traced to a man and his wife, two of a group of 400 Amish who immigrated to the United States in the mid-1700s. As a result of the founder effect and inbreeding since then, about 1 of 8 people in the Lancaster population is now heterozygous for the allele, and 1 in 200 is homozygous for it.

17.9 Gene Flow

> Individuals, along with their alleles, move into and out of populations. This flow of alleles counters genetic change that tends to occur within a population.
‹ Link to Transgenic plants 15.7

Individuals tend to mate or breed most frequently with other members of their own population. However, most populations of a species are not completely isolated from one another, so there may be intermating among nearby populations. Also, individuals sometimes leave one population and join another. **Gene flow**, the movement of alleles among populations, occurs in both cases. Gene flow stabilizes allele frequencies, so it counters the effects of mutation, natural selection, and genetic drift.

Gene flow is typical among populations of animals, which tend to be more mobile, but it also occurs in plant populations. Consider the acorns that blue jays disperse when they gather nuts for the winter. Every fall, jays visit acorn-bearing oak trees repeatedly, then bury acorns in the soil of home territories that may be as much as a mile away (Figure 17.16). The jays transfer acorns—and the alleles inside them—among populations of oak trees that would otherwise be genetically isolated.

Figure 17.16 Blue jay, a mover of acorns that helps keep genes flowing between separate oak populations.

Gene flow in plants also occurs when pollen is transferred from one individual to another, often over great distances. Many opponents of genetic engineering cite gene flow from transgenic organisms into wild populations via pollen transfer. Herbicide-resistance genes and the Bt gene (Section 15.7) have been found in weeds and unmodified crop plants that are growing near fields of transgenic plants. The long-term effects of this gene flow are currently unknown.

gene flow The movement of alleles into and out of a population.

> Speciation differs in its details, but reproductive isolating mechanisms are always part of the process.
‹ Links to Zygote 12.2, Meiosis 12.3

Mutation, natural selection, and genetic drift operate on all natural populations, and they do so independently in populations that are not interbreeding. When gene flow does not keep populations alike, different genetic changes accumulate in each one. Sooner or later, the populations become so different that we call them different species. The evolutionary process by which new species arise is called **speciation**.

Evolution is a dynamic, extravagant, messy, and ongoing process that can be challenging for people who like their categories neat. Speciation offers a perfect example, because it rarely occurs at a precise moment in time. Individuals often continue to interbreed even as populations are diverging, and populations that have already diverged may come together and interbreed again.

Every time speciation happens, it happens in a unique way, because each species is a product of its own unique evolutionary history. However, we can identify some recurring patterns. For example, reproductive isolation is always part of speciation. **Reproductive isolation** refers to the end of gene flow between populations. It is part of the process by which sexually reproducing species attain and maintain their separate identities. By preventing successful interbreeding, reproductive isolation reinforces differences between diverging populations. We say the isolation is prezygotic if pollination or mating cannot occur, or if zygotes cannot form. If hybrids form but are unfit or infertile, the isolation is postzygotic (Figure 17.17).

Mechanisms of Reproductive Isolation

Temporal Isolation Some populations cannot interbreed because the timing of their reproduction differs. The periodical cicada (*inset*) offers an example. Cicadas feed on roots as they mature underground, then emerge to reproduce. Three species of cicada reproduce every 17 years. Each has a sibling species with nearly identical form and behavior, except that the siblings emerge on a 13-year cycle instead of a 17-year cycle. Sibling species have the potential to interbreed, but they can only get together once every 221 years!

Mechanical Isolation In some cases, the size or shape of an individual's reproductive parts prevent it from mating with members of another population. For example, black sage (*Salvia mellifera*) and white sage (*S. apiana*) grow in the same areas, but hybrids rarely form because the flowers of the two species have become specialized for different pollinators (Figure 17.18). Carpenter bees, hawkmoths, and other large insects pollinate white sage when they force open the petals to access nectar hidden inside the flowers. Honeybees seeking nectar are too small to touch the reproductive parts of a white sage flower, but they are just the right size to pollinate flowers of black sage. The weight of larger bees perching on the tiny flowers of black sage pulls the delicate petals closed. Large bees access the nectar of this species by piercing the petals, so they usually avoid touching the flower's reproductive parts.

Ecological Isolation Populations adapted to different microenvironments in the same region may be ecologically isolated. For example, two species of manzanita (a plant) native to the Sierra Nevada mountain range rarely hybridize. One species that is better adapted for conserving water inhabits dry, rocky hillsides high in the foothills. The other lives on lower slopes where water stress is not as intense. The physical separation makes cross-pollination unlikely.

Prezygotic reproductive isolation

Different species form and . . .

- Individuals reproduce at different times (temporal isolation).
- Physical incompatibilities prevent individuals from interbreeding (mechanical isolation).
- Individuals live in different places so they never meet up for sex (ecological isolation).
- Individuals ignore or do not get the required cues for sex (behavioral isolation).

Mating occurs and . . .

- No fertilization occurs (gamete incompatibility).

Zygotes form and . . .

Postzygotic reproductive isolation

- Hybrid embryos die early, or new individuals die before they can reproduce (hybrid inviability).
- Hybrid individuals or their offspring do not make functional gametes (hybrid sterility).

Interbreeding is successful

Figure 17.17 Animated How reproductive isolation prevents interbreeding.

anthers

stigma

A Black sage is pollinated mainly by honeybees and other small insects.

Figure 17.18
Mechanical isolation in sage.

B The flowers of black sage are too delicate to support larger insects. Big insects access the nectar of small sage flowers only by piercing from the outside, as this carpenter bee is doing. When they do so, they avoid touching the flower's reproductive parts.

C The reproductive parts (anthers and stigma) of white sage flowers are too far away from the petals to be brushed by honeybees, so honeybees cannot pollinate this species. White sage is pollinated mainly by larger bees and hawkmoths, which brush the flower's stigma and anthers as they pry apart the petals to access nectar.

Behavioral Isolation In animals, behavioral differences can stop gene flow between related species. For instance, males and females of some bird species engage in courtship displays before sex (Figure 17.19). A female recognizes the vocalizations and movements of a male of her species as an overture to sex, but females of different species usually do not.

Gamete Incompatibility Even if gametes of different species do meet up, they often have molecular incompatibilities that prevent them from fusing. Gamete incompatibility may be the primary speciation route of animals that fertilize their eggs by releasing free-swimming sperm in water.

Figure 17.19
Behavioral isolation. Species-specific courtship displays precede sex among many birds, including these albatrosses.

Hybrid Inviability As populations begin to diverge, so do their genes. Even chromosomes of species that diverged recently may have major differences. Thus, a hybrid zygote may have extra or missing genes, or genes with incompatible products. If genetic incompatibilities disrupt development, a hybrid embryo may die. Hybrid offspring that survive may have reduced fitness. For example, ligers and tigons (offspring of lions and tigers) have more health problems and a shorter life expectancy than individuals of either parent species.

Hybrid Sterility Some interspecies crosses produce robust but sterile offspring. For example, mating a female horse (64 chromosomes) with a male donkey (62 chromosomes) produces a mule. The mule's 63 chromosomes cannot pair up evenly during meiosis, so this animal makes few viable gametes.

If hybrids are fertile, their offspring usually have lower and lower fitness with each successive generation. A mismatch between nuclear and mitochondrial DNA may be the cause (mitochondrial DNA is inherited from the mother only).

reproductive isolation Absence of gene flow between populations; always part of speciation.
speciation One of several processes by which new species arise.

> **Take-Home Message** **How do species attain and maintain separate identities?**
>
> ❯ Speciation is an evolutionary process by which new species form. It varies in its details and duration.
>
> ❯ Reproductive isolation, which occurs by one of several mechanisms, is always a part of speciation.

› In allopatric speciation, a physical barrier arises and ends gene flow between populations.
‹ Links to Galápagos archipelago 16.4, Plate tectonics 16.7

Genetic changes that lead to a new species can begin with physical separation between populations, so **allopatric speciation** is one way that new species form (*allo*– means different; *patria*, fatherland). By this speciation mode, a physical barrier separates two population and ends gene flow between them. Then, reproductive isolating mechanisms arise, so even if the populations meet up again later, their individuals could not interbreed.

Populations of most species are separated by some distance, and gene flow between them is usually intermittent. Whether a geographic barrier can block that gene flow depends on whether and how an organism travels (such as by swimming, walking, or flying), and how it reproduces (for example, by internal fertilization or by pollen dispersal). A geographic barrier can arise in an instant, or over an eon. The Great Wall of China is an example of a barrier that arose abruptly. As it was being built, the wall cut off gene flow among nearby populations of insect-pollinated plants. DNA sequencing comparisons show that trees, shrubs, and herbs on either side of the wall are now diverging genetically.

Geographic isolation usually occurs much more slowly. For example, it took millions of years of tectonic plate movements (Section 16.7) to bring the two continents of North and South America close enough to collide. The land bridge where the two continents now connect is called the Isthmus of Panama. When this isthmus formed about 4 million years ago, it cut off the flow of water—and gene flow among populations of aquatic organisms—as it separated one large ocean into what are now the Pacific and the Atlantic oceans (Figure 17.20).

Speciation in Archipelagos

The Florida Keys and some other island chains are so close to a mainland that gene flow is more or less unimpeded, so they foster little if any speciation. The Hawaiian Islands, the Galápagos Islands, and some other island chains are archipelagos. These remote, isolated islands were born of hot spots on the ocean floor. They are the tops of volcanoes, so we can assume that their fiery surfaces were initially barren and inhospitable to life.

Winds or currents sometimes carry a few individuals of a mainland species to an island in an archipelago. If the individuals reproduce, their descendants may establish a population on the island. The vast expanse of ocean that isolates the island from the mainland functions as a geographic barrier to gene flow. Thus, over generations, the island population diverges from the mainland species.

Individuals of the diverging population may in turn colonize other islands in the archipelago. Habitats and selection pressures that differ within and between the islands can foster even more divergences from the ances-

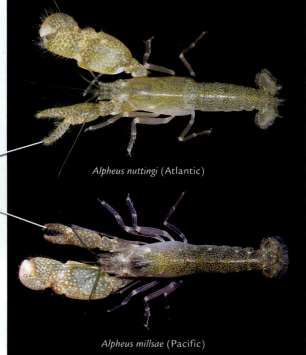

Figure 17.20 Allopatric speciation in snapping shrimp. The Isthmus of Panama (*above*) cut off gene flow among populations of these aquatic shrimp when it formed 4 million years ago. Today, individuals from opposite sides of the isthmus are so similar that they might interbreed, but they are behaviorally isolated: Instead of mating when they are brought together, they snap their claws at one another aggressively. The photos on the *right* show two of the many closely related species that live on opposite sides of the isthmus.

Alpheus nuttingi (Atlantic)

Alpheus millsae (Pacific)

Akepa
(*Loxops coccineus*)

Insects, spiders, nectar; high mountain rain forest

Akekee
(*Loxops caeruleirostris*)

Insects, spiders, nectar; high mountain rain forest

Nihoa finch
(*Telespiza ultima*)

Insects, buds, seeds, flowers, seabird eggs; rocky or shrubby slopes

Palila
(*Loxioides bailleui*)

Mamane seeds, buds, flowers, berries, insects; high mountain dry forests

Maui parrotbill
(*Pseudonestor xanthophrys*)

Insect larvae, pupae, caterpillars; mountain forests, dense underbrush

Apapane
(*Himatione sanguinea*)

Nectar, caterpillars and other insects, spiders; high mountain forests

Poouli
(*Melamprosops phaeosoma*)

Tree snails, insects in understory; last one died in 2004

Maui Alauahio
(*Paroreomyza montana*)

Bark or leaf insects, some nectar; high mountain rain forest

Kauai Amakihi
(*Hemignathus kauaiensis*)

Bark-picker; insects, spiders, nectar; high mountain rain forest

Akiapolaau
(*Hemignathus munroi*)

Probes, digs insects from big trees; high mountain rain forest

Akohekohe
(*Palmeria dolei*)

Mostly nectar from flowering trees, some insects, pollen; high mountain rain forest

Iiwi
(*Vestiaria coccinea*)

Mostly nectar (ohia flowers, lobelias, mints), some insects; high mountain rain forest

Figure 17.21 Animated Allopatric speciation on an archipelago.

Archipelagos such as the Hawaiian Islands (*right*) are separated from mainland continents by thousands of miles of open ocean—a geographic barrier that prevents gene flow between any island colonizers and mainland populations. Further divergences occur as colonizers spread to the other islands in the chain.

Above, a few of the known species of Hawaiian honeycreepers, with some of their dietary and habitat preferences. Specialized bills and behaviors adapt the honeycreepers to feed on certain insects, seeds, fruits, nectar, or other foods. DNA sequence comparisons suggest that the ancestor of all Hawaiian honeycreepers resembled the housefinch (*Carpodacus*) at *left*.

tral species. Even as the island populations become different species, some individuals may return to an island colonized by their ancestors.

The Hawaiian archipelago includes 19 islands and more than 100 atolls that stretch 1,500 miles in the Pacific Ocean. The habitats on the land masses of this archipelago range from lava beds, rain forests, and grasslands to dry woodlands and snow-capped peaks. The first birds to colonize it found a buffet of fruits, seeds, nectars, and tasty insects. The near absence of competitors and preda-

tors in an abundance of rich and vacant habitats spurred rapid speciation. Figure 17.21 hints at the variation among Hawaiian honeycreepers, descendants of one mainland finch species that arrived on the archipelago about 3.5 million years ago. Hawaiian honeycreepers and thousands of other species are unique to the Hawaiian archipelago.

Take-Home Message What happens after a physical barrier arises and prevents populations from interbreeding?

❯ A physical barrier that intervenes between populations or subpopulations of a species prevents gene flow among them. As gene flow ends, genetic divergences give rise to new species. This process is called allopatric speciation.

allopatric speciation Speciation pattern in which a physical barrier that separates members of a population ends gene flow between them.

> Populations sometimes speciate even without a physical barrier that bars gene flow between them.
< Links to Polyploidy 14.6, Fitness 16.4

Sympatric Speciation

In **sympatric speciation**, populations inhabiting the same geographic region speciate in the absence of a physical barrier between them (*sym*– means together).

Sympatric speciation can occur in an instant with a change in chromosome number. About 95 percent of ferns and 70 percent of flowering plant species are polyploid, as well as a few conifers, insects and other arthropods, mollusks, fishes, amphibians, and reptiles. Remember, chromosome number can change as a result of nondisjunction (Section 14.6). Common bread wheat originated after related species hybridized, and then the chromosome number multiplied in the offspring (Figure 17.22). In flowering plants, a chromosome number change can also originate in one parent. If a somatic cell fails to divide during mitosis, the resulting polyploid cells may proliferate to form shoots and flowers. If the flowers self-fertilize, a new species may result.

Sympatric speciation can also occur with no change in chromosome number. The mechanically isolated sage plants you learned about in Section 17.10 speciated with no physical barrier to gene flow. As another example, more than 500 species of cichlid, a freshwater fish, speciated in the shallow waters of Lake Victoria. This large freshwater lake sits isolated from river inflow on an elevated plain in Africa's Great Rift Valley. Since Lake Victoria formed about 400,000 years ago, it has dried up three times. DNA sequence comparisons indicate that almost all of the cichlid species in this lake arose since the last dry

spell, which was 12,400 years ago. How could hundreds of species arise so quickly? The answer begins with differences in the color of ambient light and water clarity in different parts of the lake. Water absorbs blue light, so the deeper it is, the less blue light penetrates it. The light in shallower, clear water is mainly blue; the light that penetrates deeper, muddier water is mainly red.

Lake Victoria cichlids vary in color and in patterning (Figure 17.23). Outside of captivity, female cichlids rarely mate with males of other species. Given a choice, they prefer brightly colored males of their own species. Their preference has a molecular basis in genes that encode light-sensitive pigments of the eye. The pigments made by species that live mainly in shallower, clear water are more sensitive to blue light. The males of these species are also the bluest. The pigments made by species that live mainly in deeper, murkier water are more sensitive to red light. Males of these species are redder. In other words, the colors that a female cichlid sees best are the same colors displayed by males of her species. Thus, mutations in genes that affect color perception are likely to affect the choice of mates and of habitats. Such mutations are probably the way sympatric speciation occurs in these fish.

Sympatric speciation has also occurred in greenish warblers of central Asia (*Phylloscopus trochiloides*). A chain of populations of this bird encircles the Tibetan plateau. Adjacent populations of greenish warblers interbreed easily, except for two populations in northern Siberia. Individuals of these two populations overlap in range, but they do not interbreed because they do not recognize one another's songs. They have become behaviorally isolated. Small genetic differences between the other populations have added up to major differences between

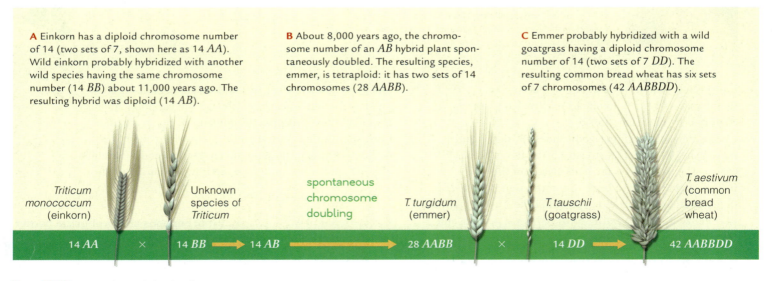

A Einkorn has a diploid chromosome number of 14 (two sets of 7, shown here as 14 *AA*). Wild einkorn probably hybridized with another wild species having the same chromosome number (14 *BB*) about 11,000 years ago. The resulting hybrid was diploid (14 *AB*).

B About 8,000 years ago, the chromosome number of an *AB* hybrid plant spontaneously doubled. The resulting species, emmer, is tetraploid: it has two sets of 14 chromosomes (28 *AABB*).

C Emmer probably hybridized with a wild goatgrass having a diploid chromosome number of 14 (two sets of 7 *DD*). The resulting common bread wheat has six sets of 7 chromosomes (42 *AABBDD*).

| *Triticum monococcum* (einkorn) | | Unknown species of *Triticum* | | spontaneous chromosome doubling | *T. turgidum* (emmer) | | *T. tauschii* (goatgrass) | | *T. aestivum* (common bread wheat) |

14 *AA* × 14 *BB* → 14 *AB* → 28 *AABB* × 14 *DD* → 42 *AABBDD*

Figure 17.22 Sympatric speciation in wheat.

Figure 17.23 Red fish, blue fish: Males of four closely related species of cichlid native to Lake Victoria, Africa. Hundreds of cichlids speciated in sympatry in this lake. Mutations in genes that affect females' perception of the color of ambient light in deeper or shallower regions of the lake also affect their choice of mates. Female cichlids prefer to mate with brightly colored males of their own species.

›› Figure It Out What form of natural selection has been driving sympatric speciation in Lake Victoria cichlids?

Answer: Sexual selection

the two populations at the ends of the chain. The chain of greenish warbler populations are collectively called a ring species. Such species present one of those paradoxes for people who like neat categories: Gene flow occurs continuously all around the chain, but the two populations at the ends of the chain are different species. Where should we draw the line that divides those two species?

Parapatric Speciation

Parapatric speciation may occur when one population extends across a broad region encompassing diverse habitats. The different habitats exert distinct selection pressures on parts of the population, and the result may be divergences that lead to speciation. Hybrids that form in a contact zone between habitats are less fit than individuals on either side of it.

Consider velvet walking worms, which resemble caterpillars but may be more related to spiders. These worms are predatory; they shoot streams of glue from their head at insects. Once entangled in the sticky glue, the insects are easy prey for the worms. Two rare species of velvet walking worm are native to the island of Tasmania. The giant velvet walking worm (*Tasmanipatus barretti*) and the blind velvet walking worm (*T. anophthalmus*) can interbreed, but they only do so in a tiny area where their habitats overlap. Hybrid offspring are sterile, which

parapatric speciation Speciation model in which different selection pressures lead to divergences within a single population.
sympatric speciation Pattern in which speciation occurs in the absence of a physical barrier.

Table 17.2	Comparison of Speciation Models		
Speciation Model:	Allopatric	Sympatric	Parapatric
Original population			
Initiating event:	barrier arises	genetic change	new niche entered
Reproductive isolation occurs			
New species arises:	in isolation	within population	in new niche

may be the main reason the two species are maintaining separate identities in the absence of a physical barrier between their populations. Table 17.2 compares parapatric speciation with other speciation models.

Take-Home Message Can speciation occur without a physical barrier to gene flow?

> By a sympatric speciation model, new species arise from a population even in the absence of a physical barrier.

> By a parapatric speciation model, populations maintaining contact along a common border evolve into distinct species.

❯ Macroevolution includes patterns of change such as one species giving rise to multiple species, the origin of major groups, and major extinction events.

❮ Links to Homeotic genes 10.3, Geologic time scale 16.6, Plate tectonics 16.7

Patterns of Macroevolution

Microevolution is change in allele frequencies within a single species or population. Macroevolution is our name for evolutionary patterns on a larger scale. Flowering plants evolved from seed plants, animals with four legs (tetrapods) evolved from fish, birds evolved from dinosaurs—all of these are examples of macroevolution that occurred over millions of years.

A central theme of macroevolution is that major evolutionary novelties have often stemmed from the adaptation of an existing structure for a completely different purpose. This theme is called preadaptation or **exaptation**. Some traits serve a very different purpose in modern species than they did when they first evolved. For example, the feathers that allow modern birds to fly are derived from feathers that first evolved in some dinosaurs. Those dinosaurs could not have used their feathers

Figure 17.24 An example of stasis. *Top*, 320-million-year-old coelacanth fossil found in Montana. *Bottom*, a live coelacanth (*Latimeria chalumnae*) caught off the waters of Sulawesi in 1998. The coelacanth lineage has changed very little over evolutionary time.

for flight, but they probably did use them for insulation. Thus, we say that flight feathers in birds are an exaptation of insulating feathers in dinosaurs.

Stasis With the simplest macroevolutionary pattern, **stasis**, lineages persist for millions of years with little or no change. Consider coelacanths, an order of ancient lobe-finned fish that had been assumed extinct for at least 70 million years until a fisherman caught one in 1938. The modern coelacanth species are very similar to fossil specimens hundreds of millions of years old (Figure 17.24).

Mass Extinctions By current estimates, more than 99 percent of all species that ever lived are now **extinct**, or irrevocably lost from Earth. In addition to continuing small-scale extinctions, the fossil record indicates that there have been more than twenty mass extinctions, which are simultaneous losses of many lineages. These include five catastrophic events in which the majority of species on Earth disappeared (Section 16.6).

Adaptive Radiation In an evolutionary pattern called **adaptive radiation**, a lineage rapidly diversifies into several new species. Adaptive radiation can occur after individuals colonize a new environment that has a variety of different habitats with few or no competitors. The adaptation of populations to different regions of the new environment produces new species. The Hawaiian honeycreepers that you read about in Section 17.11 arose this way (Figure 17.25).

Adaptive radiation may occur after a key innovation evolves. A **key innovation** is a new trait that allows its bearer to exploit a habitat more efficiently or in a novel way. The evolution of lungs offers an example, because

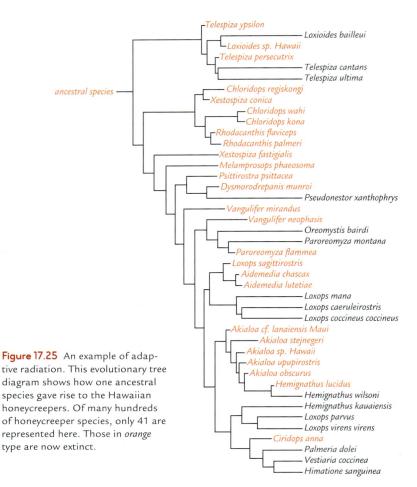

Figure 17.25 An example of adaptive radiation. This evolutionary tree diagram shows how one ancestral species gave rise to the Hawaiian honeycreepers. Of many hundreds of honeycreeper species, only 41 are represented here. Those in *orange* type are now extinct.

ancestral species

- Telespiza ypsilon
- Loxioides bailleui
- Loxioides sp. Hawaii
- Telespiza persecutrix
- Telespiza cantans
- Telespiza ultima
- Chloridops regiskongi
- Xestospiza conica
- Chloridops wahi
- Chloridops kona
- Rhodacanthis flaviceps
- Rhodacanthis palmeri
- Xestospiza fastigialis
- Melamprosops phaeosoma
- Psittirostra psittacea
- Dysmorodrepanis munroi
- Pseudonestor xanthophrys
- Vangulifer mirandus
- Vangulifer neophasis
- Oreomystis bairdi
- Paroreomyza montana
- Paroreomyza flammea
- Loxops sagittirostris
- Aidemedia chascax
- Aidemedia lutetiae
- Loxops mana
- Loxops caeruleirostris
- Loxops coccineus coccineus
- Akialoa cf. lanaiensis Maui
- Akialoa stejnegeri
- Akialoa sp. Hawaii
- Akialoa upupirostris
- Akialoa obscurus
- Hemignathus lucidus
- Hemignathus wilsoni
- Hemignathus kauaiensis
- Loxops parvus
- Loxops virens virens
- Ciridops anna
- Palmeria dolei
- Vestiaria coccinea
- Himatione sanguinea

Figure 17.26 An example of coevolved species.

A The large blue butterfly (*Maculinea arion*) parasitizes a species of red ant, *Myrmica sabuleti*.

B To an ant, a honey-exuding, hunched-up *Maculinea arion* caterpillar appears to be an ant larva. This deceived ant is preparing to carry the caterpillar back to its nest, where the caterpillar will eat ant larvae for the next 10 months until it pupates.

lungs were a key innovation that opened the way for an adaptive radiation of vertebrates on land.

Adaptive radiations also occur after geologic or climatic events eliminate some species from a habitat. The surviving species then can exploit resources from which they had been previously excluded. This is the way mammals were able to undergo an adaptive radiation after the dinosaurs disappeared.

Coevolution The process by which close ecological interactions between two species cause them to evolve jointly is called **coevolution**. One species acts as an agent of selection on the other; each adapts to changes in the other. Over evolutionary time, the two species may become so interdependent that they can no longer survive without one another.

Relationships between coevolved species can be quite intricate. For example, the large blue butterfly (*Maculinea arion*) parasitizes red ants (**Figure 17.26A**). After hatching, the larva (caterpillar) of a large blue butterfly feeds on wild thyme flowers and then drops to the ground. A red ant that finds the caterpillar strokes it, whereupon the caterpillar exudes honey. The ant eats the honey and continues to stroke the caterpillar, which secretes more honey. This interaction continues for hours, until the caterpillar suddenly hunches itself up into a shape that appears, to

an ant anyway, very much like an ant larva (**Figure 17.26B**). The beguiled ant then picks up the caterpillar and carries it back to the ant nest, where, in most cases, other ants kill it—except, however, if the ants are of one particular species, *Myrmica sabuleti*. Secretions of a *M. arion* caterpillar fool these ants into treating it just like a larva of their own. For the next 10 months, the caterpillar lives in the nest and grows to gigantic proportions by feeding on ant larvae. After it metamorphoses into a butterfly, it lays its eggs on wild thyme near another *M. sabuleti* nest, and the cycle starts anew. This coevolved relationship between ant and butterfly is extremely specific. Any increase in the ants' ability to identify a caterpillar in their nest selects for caterpillars that better deceive the ants, which in turn select for ants that can better identify the caterpillars. Each species exerts directional selection on the other.

Evolutionary Theory

Biologists do not doubt that macroevolution occurs, but many disagree about how it occurs. However we choose to categorize evolutionary processes, the very same genetic change may be at the root of all evolution—fast or slow, large-scale or small-scale. Dramatic jumps in morphology, if they are not artifacts of gaps in the fossil record, may be the result of mutations in homeotic or other regulatory genes. Macroevolution may include more processes than microevolution, or it may not. It may be an accumulation of many microevolutionary events, or it may be an entirely different process. Evolutionary biologists may disagree about these and other hypotheses, but all of them are trying to explain the same thing: how all species are related by descent from common ancestors.

---

adaptive radiation A burst of genetic divergences from a lineage gives rise to many new species.
coevolution The joint evolution of two closely interacting species; each species is a selective agent for traits of the other.
exaptation Adaptation of an existing structure for a completely different purpose; a major evolutionary novelty.
extinct Refers to a species that has been permanently lost.
key innovation An evolutionary adaptation that gives its bearer the opportunity to exploit a particular environment more efficiently or in a new way.
stasis Evolutionary pattern in which a lineage persists with little or no change over evolutionary time.

Take-Home Message **What is macroevolution?**

> Macroevolution comprises large-scale patterns of evolutionary change such as adaptive radiations, coevolution, and mass extinction.

> Cladistics allows us to organize our knowledge about how species are related.
< Link to Taxonomy 1.5

Ranking Versus Grouping

Linnaeus devised his system of taxonomy before anyone knew about evolution, so taxonomy rankings do not necessarily reflect evolutionary relationships. Our increasing understanding of evolution is prompting a major, ongoing overhaul of the way biologists view life's diversity. Instead of trying to divide that tremendous diversity into a series of ranks, most biologists are now focusing on evolutionary connections. Each species is viewed not as a member or representative of a rank in a hierarchy, but rather as part of a bigger picture of evolution.

Phylogeny is the evolutionary history of a species or a group of them, a kind of genealogy that follows a lineage's evolutionary relationships through time. The central question of phylogeny is, "Who is related to whom?" Methods of finding the answer to that question are an important part of phylogeny. One method, **cladistics**, groups species on the basis of their shared characters. A **character** is a quantifiable, heritable characteristic—any physical, behavioral, physiological, or molecular trait of a species (Table 17.3).

The result of a cladistic analysis is a **cladogram**, a diagram that shows a network of evolutionary relationships (Figure 17.27). Each line in a cladogram represents a lineage, which may branch into two lineages at a node. The node represents a common ancestor of the two lineages. Every branch ends with a **clade** (from *klados*, a Greek word for twig or branch), a species or group of species that share a set of characters.

Ideally, each clade is a **monophyletic group** that comprises an ancestor and all of its descendants. However, as you learned in Section 17.10, evolution can be challenging for those who like neat categories. Each species has many characters, and researchers discover more species and more characters all the time. Because we do not yet know all species, or all characters, cladistic groupings are necessarily hypotheses. They may differ depending on which characters are used for the analysis, so clades often change when new discoveries are made.

Cladograms and other types of **evolutionary trees** summarize our best data-supported hypotheses about how a group of species evolved. We use these diagrams to visualize evolutionary trends and patterns. For instance, the two lineages that emerge from a node on a cladogram are called **sister groups**. Sister groups are, by default, the same age. We may not know what that age is, but we can compare sister groups on a cladogram and say something about their relative rates of evolution.

Like other hypotheses, evolutionary tree diagrams get revised. However, the diagrams are based on the solid premise that all species are interconnected by shared ancestry. Every living thing is related if you just go back far enough in time. An evolutionary biologist's job is to figure out where the connections are.

How We Use Evolutionary Biology

The first Polynesians arrived on the Hawaiian Islands sometime before 1000 A.D., and Europeans followed in 1778. Hawaii's rich ecosystem was hospitable to all newcomers, including the settlers' dogs, cats, pigs, cows, goats, deer, and sheep. Escaped livestock began to eat and trample rain forest plants that had provided Hawaiian honeycreepers with food and shelter. Entire forests were cleared to grow imported crops, and plants that escaped

Table 17.3	Examples of Characters			
	Multicellular	Backbone	Legs	Hair or Fur
Earthworm	✔	–	–	–
Tuna	✔	✔	–	–
Lizard	✔	✔	✔	–
Mouse	✔	✔	✔	✔
Human	✔	✔	✔	✔

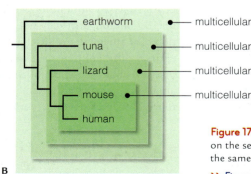

Figure 17.27 Animated Cladograms. **A** This example is based on the set of characters chosen in Table 17.3. **B** We can visualize the same cladogram as "sets within sets" of characters.

>> **Figure It Out** In this cladogram, which is the sister group of the mouse?

Answer: Human

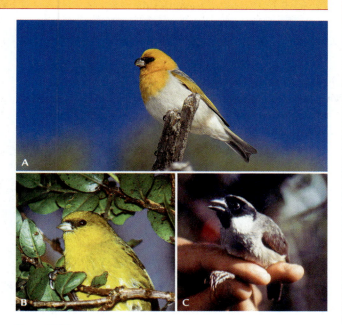

Figure 17.28 Three honeycreeper species: going, going, and gone.

A The palila has an adaptation that allows it to feed on mamane seeds, which are toxic to most other birds. The one remaining palila population is declining because mamane plants are being trampled by cows and gnawed to death by goats and sheep. Only about 2,640 palila remained in 2009.

B Avian malaria carried by mosquitoes to higher altitudes is decimating the last population of the akekee. Between 2000 and 2007, the number of akekee plummeted from 7,839 birds to 3,536.

C This male poouli—rare, old, and missing an eye—died in 2004 from avian malaria. There were two other poouli alive at the time, but neither has been seen since then.

cultivation began to crowd out native plants. Mosquitoes introduced in 1826 spread diseases from imported chickens to native bird species. Stowaway rats and snakes ate their way through populations of native birds and their eggs. Mongooses deliberately imported to eat the rats and snakes preferred to eat birds and eggs.

Ironically, the very isolation that spurred adaptive radiations made the honeycreepers vulnerable to extinction. The birds had no built-in defenses against predators or diseases of the mainland. Specializations such as extravagantly elongated beaks became hindrances when the birds' habitats suddenly changed or disappeared.

---

character Quantifiable, heritable characteristic or trait.
clade A species or group of species that share a set of characters.
cladistics Method of determining evolutionary relationships by grouping species into clades based on shared characters.
cladogram Evolutionary tree that shows a network of evolutionary relationships among clades.
evolutionary tree Type of diagram that summarizes evolutionary relationships among a group of species.
monophyletic group An ancestor and all of its descendants.
phylogeny Evolutionary history of a species or group of species.
sister groups The two lineages that emerge from a node on a cladogram.

Rise of the Super Rats (revisited)

The gene most commonly involved in warfarin resistance in rats is inherited in a simple dominance pattern. The gene's normal allele is recessive when paired with the mutated allele that confers resistance. In the presence of warfarin, the dominant allele is clearly adaptive. Rats with this allele require a lot of extra vitamin K, but being vitamin K–deficient is not so bad when compared with being dead from rat poison. However, in the absence of warfarin, individuals that have the warfarin resistance allele are at a serious disadvantage compared with those who do not. Rats with the allele cannot easily obtain enough vitamin K from their diet to sustain normal blood clotting and bone formation. Thus, the allele is adaptive when warfarin is present, and maladaptive when it is not, so periodic exposure to warfarin maintains a balanced polymorphism of the resistance gene in rat populations.

How Would You Vote? Antibiotic–resistant strains of bacteria are now widespread. One standard animal husbandry practice includes continually dosing healthy livestock with the same antibiotics prescribed for people. Should this practice stop? See CengageNow for details, then vote online (cengagenow.com).

Thus, at least 43 species of honeycreeper that had thrived on the islands before the arrival of humans were extinct by 1778. Today, 32 of the remaining 71 species are endangered, and 26 are extinct despite tremendous conservation efforts since the 1960s (Figure 17.28). Invasive, non-native species of plants and animals are now established, and the rise in global temperatures is allowing disease-bearing mosquitoes to invade high-altitude habitats that had previously been too cold for them.

The story of the Hawaiian honeycreepers is a dramatic illustration of how evolution works. It also shows how finding ancestral connections can help species that are still living. As more and more honeycreeper species become extinct, the group's reservoir of genetic diversity dwindles. The lowered diversity means the group as a whole is less resilient to change, and more likely to suffer catastrophic species losses. Deciphering their phylogeny can tell us which honeycreeper species are most different from the others—and those are the ones most valuable in terms of preserving genetic diversity. Such research allows us to concentrate our resources and conservation efforts on those species that hold the best hope for the survival of the entire group. For example, we now know the poouli (Figure 17.28C) to be the most distantly related member of the genus. Unfortunately, that knowledge came too late; the species is probably extinct.

Take-Home Message **How do evolutionary biologists study life's diversity?**

> Evolutionary biologists study phylogeny in order to understand how all species are connected by shared ancestry.

Summary

Section 17.1 Our efforts to control pests have resulted in directional selection for resistant populations. Populations tend to change along with the selection pressures that are acting on them.

Sections 17.2, 17.3 Individuals of a **population** share traits with a heritable basis. All of their alleles form a **gene pool**. New alleles that arise by mutation may be **neutral**, **lethal**, or advantageous. **Microevolution**, or changes in the **allele frequencies** of a population, occurs constantly by processes of mutation, natural selection, genetic drift, and gene flow. We use deviations from **genetic equilibrium** to study how a population is evolving.

Sections 17.4–17.6 **Natural selection** occurs in different patterns. **Directional selection** shifts the range of variation in traits in one direction. **Stabilizing selection** favors intermediate forms of a trait. **Disruptive selection** favors forms at the extremes of a range of variation.

Section 17.7 Traits that differ between males and females are an outcome of **sexual selection**, in which adaptive traits are ones that make their bearers better at securing mates. With **balanced polymorphism**, an environment that favors heterozygotes maintains two or more alleles at high frequency.

Sections 17.8, 17.9 Random change in allele frequencies, or **genetic drift**, can lead to the loss of genetic diversity by causing alleles to become **fixed**. Genetic drift is most pronounced in small or **inbreeding** populations. The **founder effect** may occur after an evolutionary **bottleneck**. **Gene flow** counters the effects of mutation, natural selection, and genetic drift.

Section 17.10 The details of **speciation** differ every time it occurs, but **reproductive isolation**, the absence of gene flow between populations, is always a part of the process. The exact moment at which two populations become separate species is often impossible to pinpoint.

Section 17.11 In **allopatric speciation**, a geographic barrier arises and interrupts gene flow between populations. After gene flow ends, genetic divergences then give rise to new species.

Section 17.12 Speciation can occur with no physical barrier to gene flow. In **sympatric speciation**, populations in physical contact speciate. Polyploid species of many plants (and a few animals) often originate in sympatry. With **parapatric speciation**, populations in contact along a common border speciate.

Section 17.13 Macroevolution refers to patterns of evolution above the species level: stasis, adaptive radiation, coevolution, and extinction. Major evolutionary novelty typically arises by **exaptation**, in which a lineage uses a structure for a different purpose than its ancestor did. With **stasis**, a lineage changes very little over evolutionary time. A **key innovation** can result in an **adaptive radiation**, or rapid diversification into new species. **Coevolution** occurs when two species act as agents of selection upon one another. A lineage that is permanently lost from Earth is **extinct**.

Section 17.14 **Cladistics** allows us to reconstruct evolutionary history (**phylogeny**). Species are grouped into **clades** based on shared **characters**. Ideally, a clade is a **monophyletic group**, but clades often change as a result of new information. A **cladogram** is a type of **evolutionary tree** in which each line represents one lineage. A lineage can branch into two **sister groups** at a node, which represents a shared ancestor.

Self-Quiz   Answers in Appendix III

1. Individuals don't evolve, _____ do.

2. Biologists define evolution as _____ .
   a. purposeful change in a lineage
   b. heritable change in a line of descent
   c. acquiring traits during the individual's lifetime

3. _____ is the original source of new alleles.
   a. Mutation   d. Gene flow
   b. Natural selection   e. All are original sources of
   c. Genetic drift   new alleles

4. Evolution can only occur in a population when _____ .
   a. mating is random
   b. there is selection pressure
   c. neither is necessary

5. A fire devastates all trees in a wide swath of forest. Populations of a species of tree-dwelling frog on either side of the burned area diverge to become separate species. This is an example of _____ .

6. Stabilizing selection tends to _____ (select all that apply).
   a. eliminate extreme forms of a trait
   b. favor extreme forms of a trait
   c. eliminate intermediate forms of a trait
   d. favor intermediate forms of a trait
   e. shift allele frequencies in one direction

7. Disruptive selection tends to _____ (select all that apply).
   a. eliminate extreme forms of a trait
   b. favor extreme forms of a trait
   c. eliminate intermediate forms of a trait
   d. favor intermediate forms of a trait
   e. shift allele frequencies in one direction

Data Analysis Activities

Resistance to Rodenticides in Wild Rat Populations Beginning in 1990, rat infestations in northwestern Germany started to intensify despite continuing use of rat poisons. In 2000, Michael H. Kohn and his colleagues analyzed the genetics of wild rat populations around Munich. For part of their research, they trapped wild rats in five towns, and tested those rats for resistance to warfarin and the more recently developed poison bromadiolone. The results are shown in Figure 17.29.

1. In which of the five towns were most of the rats susceptible to warfarin?
2. Which town had the highest percentage of poison-resistant wild rats?
3. What percentage of rats in Olfen were resistant to warfarin?
4. In which town do you think the application of bromadiolone was most intensive?

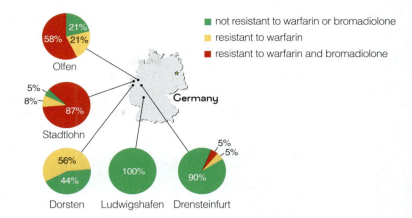

- not resistant to warfarin or bromadiolone
- resistant to warfarin
- resistant to warfarin and bromadiolone

Figure 17.29 Resistance to rat poisons in wild populations of rats in Germany, 2000.

8. Directional selection tends to _____ (select all that apply).
   a. eliminate extreme forms of a trait
   b. favor an extreme form of a trait
   c. eliminate intermediate forms of a trait
   d. favor intermediate forms of a trait
   e. shift allele frequencies in one direction

9. Sexual selection, such as occurs when males compete for access to fertile females, frequently influences aspects of body form and can lead to _____ .
   a. male/female differences
   b. male aggression
   c. exaggerated traits
   d. all of the above

10. The persistence of malaria and sickle-cell anemia in a population is a case of _____ .
    a. bottlenecking
    b. balanced polymorphism
    c. natural selection
    d. artificial selection
    e. both b and c

11. _____ tends to keep populations of a species similar to one another.
    a. Genetic drift
    b. Gene flow
    c. Mutation
    d. Natural selection

12. In evolutionary trees, each node represents a(n) _____ .
    a. single lineage
    b. extinction
    c. point of divergence
    d. adaptive radiation

13. In cladograms, sister groups are _____ .
    a. inbred
    b. the same age
    c. represented by nodes
    d. in the same family

14. Match the evolution concepts.
    ___ gene flow
    ___ natural selection
    ___ mutation
    ___ genetic drift
    ___ adaptive radiation
    ___ coevolution
    ___ phylogeny
    ___ cladogram

    a. can lead to interdependent species
    b. changes in a population's allele frequencies due to chance alone
    c. alleles enter or leave a population
    d. evolutionary history
    e. occurs in different patterns
    f. burst of divergences from one lineage into many
    g. source of new alleles
    h. diagram of sets within sets

Additional questions are available on **CENGAGENOW**.

Critical Thinking

1. Rama the cama, a llama–camel hybrid, was born in 1997. The idea was to breed an animal that has the camel's strength and endurance and the llama's gentle disposition. However, instead of being large, strong, and sweet, Rama is smaller than expected and has a camel's short temper. The breeders plan to mate him with Kamilah, a female cama, but they wonder if offspring from such a match would be fertile. What does Rama's story tell you about the genetic changes required for reproductive isolation in nature? Explain why a biologist might not view Rama as evidence that llamas and camels are the same species.

2. Some theorists have hypothesized that many of our uniquely human traits arose by sexual selection. Over many thousands of years, women attracted to charming, witty men perhaps prompted the development of human intellect far beyond what was necessary for mere survival. Men attracted to women with juvenile features may have shifted the species as a whole to be less hairy and softer featured than any of our simian relatives. Can you think of a way to test these hypotheses?

Animations and Interactions on **CENGAGENOW**:
❯ Hardy–Weinberg analysis; Directional selection; Directional selection in the peppered moth; Stabilizing selection; Disruptive selection; Disruptive selection in African finches; Genetic drift; Reproductive isolation; Allopatric speciation on an archipelago; Cladograms.

❮ Links to Earlier Concepts

This chapter starts your survey of biodiversity, introduced in Section 1.4. Here, we will place the origin organic compounds (3.2), bacteria and archaea (4.5), and eukaryotes (4.6) on the time line of Earth history (16.6).

We return to connections between photosynthesis and aerobic respiration (Chapters 6 and 7) and consider how the nucleus, ER, mitochondria, and chloroplasts (4.7–4.9) might have originated.

Key Concepts

Setting the Stage for Life
Earth formed about 4 billion years ago from matter distributed in space by the big bang (the origin of the universe). The early Earth was an inhospitable place, where meteorite impacts and volcanic eruptions were common and the atmosphere held little or no oxygen.

Building Blocks of Life
All life is composed of the same organic subunits. Simulations of conditions on the early Earth show that these molecules could have formed by reactions in the atmosphere or sea. Organic subunits also form in space and could have been delivered to Earth by meteorites.

18 Life's Origin and Early Evolution

18.1 Looking for Life

The photo at the *left* shows the Eagle Nebula, an unimaginably huge cloud of gas and dust 7,000 light years away. We live in a vast universe that we have only begun to explore. So far, we know of only one planet that has life—Earth. In addition, biochemical, genetic, and metabolic similarities among Earth's species imply that all evolved from a common ancestor that lived billions of years ago. What properties of the ancient Earth allowed life to arise, survive, and diversify? Could similar processes occur on other planets? These are some of the questions posed by **astrobiology**, the study of life's origins and distribution in the universe.

Astrobiologists study Earth's extreme habitats to determine the range of conditions that living things can tolerate. One group found bacteria living about 30 centimeters (1 foot) below the soil surface of Chile's Atacama Desert, a place said to be the driest on Earth (Figure 18.1). Another group drilled 3 kilometers (almost 2 miles) beneath the soil surface in Virginia, where they found bacteria thriving at high pressure and temperature. They named their find *Bacillus infernus*, or "bacterium from hell."

Knowledge gained from studies of life on Earth inform the search for extraterrestrial life. On Earth, all metabolic reactions involve interactions among molecules in aqueous solution (dissolved in water). We assume that the same physical and chemical laws operate throughout the universe, so liquid water is considered an essential requirement for life. Thus, scientists were excited when a robotic lander discovered water frozen in the soil of Mars, our closest planetary neighbor. If there is life on Mars, it is likely to be underground. Mars has no ozone layer, so ultraviolet radiation would fry organisms at the planet's surface. However, Martian life may exist in deep rock layers just as it does on Earth.

Suppose scientists do find evidence of microbial life on Mars or another planet. Why would it matter? Such a discovery would

Figure 18.1 The Mars-like landscape of Chile's Atacama Desert. Scientist Jay Quade, visible in the distance at the *right*, was a member of a team that found bacteria living beneath this desert soil.

support the hypothesis that life on Earth arose as a consequence of physical and chemical processes that occur throughout the universe. The discovery of extraterrestrial microbes would also make the possibility of nonhuman intelligent life in the universe more likely. The more places life exists, the more likely it is that complex, intelligent life evolved on other planets in the same manner that it did on Earth.

---

astrobiology The scientific study of life's origin and distribution in the universe.

The First Cells Form
All cells have enzymes that carry out reactions, a plasma membrane, and a genome of DNA. Experiments provide insight into how cells arose through physical and chemical processes, such as the tendency of lipids to form membrane-like structures when mixed with water.

Life's Early Evolution
The first cells were probably anaerobic. An early divergence separated bacteria from archaeans and ancestors of eukaryotic cells. Evolution of oxygen-producing photosynthesis in bacteria altered Earth's atmosphere, creating conditions that favored aerobic organisms.

Eukaryotic Organelles
A nucleus, ER, and other membrane-enclosed organelles are defining features of eukaryotic cells. Some organelles may have evolved from infoldings of the plasma membrane. Mitochondria and chloroplasts probably descended from bacteria that lived inside other cells.

> Physical and geological forces produced Earth, its seas, and its atmosphere.
> The early atmosphere had little oxygen.
❮ Link to Elements 2.2

From the Big Bang to the Early Earth

No one was around to witness the birth of the universe, so our ideas about what happened will forever remain in the realm of conjecture. However, the ancient events that led to our universe and our planet left their signature in the energy and matter that exist today. Scientists who study stars and space continue to discover clues about how our universe originated.

The widely accepted **big bang theory** states that the universe began in a single instant, about 13 to 15 billion years ago. In that instant all existing matter and energy suddenly appeared and exploded outward from a single point. Simple elements such as hydrogen and helium formed within minutes. Then, over millions of years, gravity drew the gases together and they condensed to form giant stars (Figure 18.2).

Explosions of these early stars scattered the heavier elements from which today's galaxies formed. Our own galaxy, the Milky Way, probably began as a cloud of stellar debris trillions of kilometers wide. Some of that debris condensed and formed the galaxy's stars.

About 5 billion years ago, the star we call our sun was orbited by a cloud of dust and rocks (asteroids), but no planets. The asteroids collided and merged into bigger asteroids. The heavier these pre-planetary rocks became, the more gravitational pull they exerted, and the more material they gathered. By about 4.6 billion years ago, Earth and the other planets of our solar system had formed.

Figure 18.3 Artist's depiction of early Earth, at a time when volcanic activity and meteor strikes were still common events.

Conditions on the Early Earth

Planet formation did not clear out all of the debris orbiting the sun, so the early Earth received a constant hail of meteorites. Earth's surface was molten, and more molten rock and gases spewed continually from volcanoes. Gases released by volcanoes and meteorite impacts were the main components of the early atmosphere.

What was Earth's early atmosphere like? Studies of volcanic eruptions, meteorites, ancient rocks, and other planets suggest that the air contained water vapor, carbon dioxide, and gaseous hydrogen and nitrogen. We know that there was little or no oxygen, because the oldest existing rocks show no evidence of iron oxidation (rusting). If oxygen had been present in Earth's early atmosphere, it would have caused rust to form. More important, it would have interfered with assembly of the organic compounds necessary for life. Oxygen would have reacted with and destroyed the compounds as fast as they formed.

At first, any water falling on Earth's molten surface evaporated immediately. As the surface cooled, rocks formed. Later, rains washed mineral salts out of these rocks and the salty runoff pooled in early seas (Figure 18.3). It was in these seas that life began.

---

big bang theory Model describing formation of the universe as a nearly instant distribution of matter through space.

Figure 18.2 What the cloud of dust, gases, rocks, and ice around the early sun may have looked like.

Take-Home Message What were conditions like on the early Earth?

> Meteor impacts were common.
> The atmosphere had little or no oxygen.
> The seas contained mineral salts leached from the rocks.

❯ All living things are made from organic subunits: simple sugars, amino acids, fatty acids, and nucleotides. Where did the subunits that made up the first life come from? Here we look at three well-researched possibilities.
❮ Link to Molecules of life 3.2

Lightning–Fueled Atmospheric Reactions

In 1953, Stanley Miller and Harold Urey proposed that reactions in Earth's early atmosphere could have produced building blocks for the first life. At that time, many scientists thought Earth's early atmosphere consisted of methane, ammonia, and hydrogen gas. Miller and Urey placed water and these gases into a reaction chamber (Figure 18.4). As the mix circulated, sparks from electrodes simulated lightning. Within a week, a variety of amino acids and other small molecules formed.

The Miller–Urey experiment was initially hailed as a breakthrough that demonstrated the first step on the road to life. Then, the idea that Earth's early atmosphere consisted mainly of carbon dioxide and nitrogen dioxide gained favor. When Miller redid his experiment using these gases in his apparatus, he was unable to detect any amino acid formation.

Miller died in 2007, but experiments by other scientists have given new credence to his ideas. One experiment showed that amino acids do form in a simulated carbon dioxide and nitrogen atmosphere. Miller failed to detect amino acids when he used these gases because his experiment also formed compounds that break down amino acids. On the early Earth, rains would have washed amino acids formed by atmospheric reactions into the sea, where the breakdown reactions might not occur.

Reactions at Hydrothermal Vents

Reactions near deep-sea hydrothermal vents also produce organic building blocks. A **hydrothermal vent** is like an underwater geyser, a place where hot, mineral-rich water streams out through a rocky opening (Figure 18.5). The water is heated by geothermal energy. Günter Wächtershäuser and Claudia Huber simulated conditions near a hydrothermal vent by combining hot water with carbon monoxide (CO) and potassium cyanide (KCN) and metal ions like those in rocks near the vents. Their results showed amino acids formed within a week.

Delivery From Space

The presence of amino acids, sugars, and nucleotide bases in meteorites that fell to Earth suggests another possible origin for life's building blocks. These molecules may have formed in interstellar clouds of ice, dust, and gases and been delivered to Earth by meteorites. During Earth's early years, meteorites fell to Earth thousands of times more frequently than they do today.

Figure 18.4 Animated Diagram of an apparatus designed by Stanley Miller and Harold Urey to test whether organic compounds could have formed by chemical interactions in Earth's early atmosphere. Water, hydrogen gas (H_2), methane (CH_4), and ammonia (NH_3) were kept circulating through the apparatus. Sparks from an electrode simulated lightning.

❯❯ **Figure It Out** Which gas in this mixture provided the nitrogen for the amino group in the amino acids?   Answer: Ammonia

Figure 18.5 A hydrothermal vent on the sea floor. Mineral-rich water heated by geothermal energy streams out of the vent, into the cold ocean water. The drop in temperature causes dissolved minerals to come out of solution and form a chimney-like structure around the vent.

hydrothermal vent Rocky, underwater opening where mineral-rich water heated by geothermal energy streams out.

Take-Home Message Where did the simple organic building blocks of the first life come from?

❯ Simulation experiments support the hypothesis that simple organic compounds could have formed by chemical reactions in Earth's early atmosphere or in the sea near a hydrothermal vent.

❯ Observations and experiments also support the hypothesis that such compounds could have formed in space and been carried to Earth on meteorites.

> Experiments demonstrate how traits and processes seen in all living cells could have begun with physical and chemical reactions among nonliving collections of molecules.

❮ Link to Cell structure 4.2

Steps on the Road to Life

In addition to sharing the same molecular components, all cells have a plasma membrane with a lipid bilayer. They have a genome of DNA that enzymes transcribe into RNA, and ribosomes that translate RNA into proteins. All cells replicate, and pass on copies of their genetic material to their descendants. The many similarities in structure, metabolism, and replication processes among all life are evidence of descent from a common cellular ancestor.

Time has erased all evidence of the earliest cells, but scientists can still investigate this first chapter in life's history. They use their knowledge of chemistry to design experiments that test whether a particular hypothesis about how life began is plausible. Such studies support the hypothesis that cells arose as a result of a stepwise process that began with inorganic materials (Figure 18.6). Each step on the road to life can be explained by familiar chemical and physical mechanisms that still occur today.

Origin of Metabolism

Modern cells take up organic subunits, concentrate them, and assemble them into organic polymers. Before there were cells, a nonbiological process that concentrated organic subunits in one place would have increased the chance that the subunits would combine.

By one hypothesis, this process occurred on clay-rich tidal flats. Clay particles have a slight negative charge, so positively charged molecules in seawater stick to them. At low tide, evaporation would have concentrated the subunits even more, and energy from the sun might have caused them to bond together as polymers. In simulations of tidal flat conditions, amino acids form short chains.

By another hypothesis, metabolic reactions began in the high-temperature, high-pressure environment near a hydrothermal vent. Rocks around the vents contain iron sulfide (pyrite) and are porous, with many tiny chambers

Inorganic molecules
↓ self-assemble on Earth and in space
Organic monomers
↓ self-assemble in aquatic environments on Earth
Organic polymers
↓ interact in early metabolism
↓ self-assemble as vesicles
↓ become the first genome
Protocells in an RNA world
↓ are subject to selection that favors a DNA genome
DNA-based cells

Figure 18.6 Proposed sequence for the evolution of cells. Scientists investigate this process by carrying out experiments and simulations that test hypotheses about feasibility of individual steps.

about the size of cells (Figure 18.7). Metabolism may have begun when iron sulfide in the rocks donated electrons to dissolved carbon monoxide, setting in motion reactions that formed larger organic compounds. Researchers who tested this hypothesis by simulating vent conditions found that organic compounds such as pyruvate, do form and accumulate in the chambers. In addition, all modern organisms require iron-sulfide cofactors to carry out some reactions. The universal requirement for these cofactors may be a legacy of life's rocky beginnings.

Origin of the Cell Membrane

Molecules formed by early synthetic reactions would have simply floated away from one another unless something enclosed them. In modern cells, a plasma membrane serves this function. If the first reactions took place in tiny rock chambers, the rock would have acted as a boundary. Over time, lipids produced by reactions inside a chamber could have accumulated and lined the chamber wall. Such lipid-enclosed collections of interacting molecules could have been the first **protocells**. A protocell is a membrane-enclosed collection of molecules that takes up material and replicates itself.

Experiments by Jack Szostak and others have shown that rock chambers are not necessary for protocell formation. Figure 18.8A illustrates one type of protocell that Szostak investigates. Figure 18.8B is a photo of a protocell that formed in his laboratory. A membrane of lipid bilayer encloses strands of RNA. The protocell "grows" by adding fatty acids to its membrane and nucleotides to its RNA. Mechanical force causes protocell division.

Figure 18.7 Cell-sized chambers in iron-sulfide-rich rocks formed by simulations of conditions near hydrothermal vents. Similar chambers could have served as protected environments in which the first metabolic reactions took place.

20 µm

Figure 18.8 Protocells. **A** Illustration of a laboratory-produced protocell. It has a bilayer membrane of fatty acids and holds strands of RNA. Ribonucleotides that diffuse into the protocell become incorporated into complementary strands of RNA. The vesicle can enlarge by incorporating additional fatty acids. **B** Laboratory-formed protocell consisting of RNA-coated clay (*red*) surrounded by fatty acids and alcohols.

C Field-testing a hypothesis about protocell formation. David Deamer pours a mix of small organic molecules and phosphates into a hot acidic pool in Russia.

David Deamer studies protocell formation in both the laboratory and the field. In the lab, he has shown that the small organic molecules carried to Earth on meteorites can react with minerals and seawater to form vesicles with a bilayer membrane. However, Deamer has yet to locate a natural environment that facilitates the same process. In one experiment, he added a mix of organic subunits to the acidic waters of a clay-rich volcanic pool in Russia (Figure 18.8C). The organic subunits bound tightly to the clay, but no vesicle-like structures formed. Deamer concluded that hot acidic waters of volcanic springs do not provide the right conditions for protocell formation. He continues to carry out experiments to determine what naturally occurring conditions do favor this process.

Origin of the Genome

All modern cells have a genome of DNA. They pass copies of their DNA to descendant cells, which use instructions encoded in the DNA to build proteins. Some of these proteins are enzymes that synthesize new DNA, which is passed along to descendant cells, and so on. Thus, protein synthesis depends on DNA, which is built by proteins. How did this cycle begin?

In the 1960s, Francis Crick and Leslie Orgel addressed this dilemma by suggesting that RNA may have been the first molecule to encode genetic information. Since then, evidence for an early **RNA world**—a time when RNA both stored genetic information and functioned like an enzyme in protein synthesis—has accumulated. **Ribozymes**, or RNAs that function as enzymes, have been discovered in living cells. The rRNA in ribosomes speeds formation of peptide bonds during protein synthesis. Other ribozymes cut noncoding bits (introns) out of newly formed RNAs. Researchers have also produced self-replicating ribozymes that copy themselves by assembling free nucleotides.

If the earliest self-replicating genetic systems were RNA-based, then why do all organisms have a genome of DNA? The structure of DNA may hold the answer. Compared to a double-stranded DNA molecule, single-stranded RNA breaks apart more easily and mutates more often. Thus, a switch from RNA to DNA would make larger, more stable genomes possible.

---

protocell Membranous sac that contains interacting organic molecules; hypothesized to have formed prior to the earliest life forms.
ribozyme RNA that functions as an enzyme.
RNA world Hypothetical early interval when RNA served as the genetic information.

Take-Home Message What have experiments revealed about the steps that led to the first cells?

> All living cells carry out metabolic reactions, are enclosed within a plasma membrane, and can replicate themselves.

> Concentration of molecules on clay particles or in tiny rock chambers near hydrothermal vents may have helped start metabolic reactions.

> Vesicle-like structures with outer membranes can form spontaneously.

> An RNA-based system of inheritance may have preceded DNA-based systems.

> ❯ Fossils and molecular comparisons among living species inform us about the history of life on Earth.
> ❮ Links to Bacteria and archaeans 4.5, Eukaryotic cells 4.6, Photosynthesis 6.4, Evolution of aerobic respiration 7.2

Origin of Bacteria and Archaea

How old is life on Earth? Different analyses provide slightly different answers. Given the genetic differences among living species and current mutation rates, scientists estimate that the common ancestor of all cells lived about 4.3 billion years ago. Some microscopic filaments from Australia that date back 3.5 billion years may be fossil cells (Figure 18.9A). Microfossils from another Australian location are widely accepted as evidence that cells were living around hydrothermal vents on the sea floor by 3.2 billion years ago.

The small size and simple structure of early fossil cells suggests that they were not eukaryotes. This finding is consistent with evidence from gene comparisons among living organisms that places bacteria and archaea near the base of the tree of life. Because Earth's early air and seas held little oxygen, the first cells were probably anaerobic.

Genetic analysis tells us that a divergence early in the history of life separated the domains Bacteria and Archaea. After the split, light-capturing pigments evolved in some members of both groups. The oxygen-releasing noncyclic pathway of photosynthesis evolved only in one bacterial lineage, the cyanobacteria (Figure 18.9B,C). Cyanobacteria are a relatively recent branch on the bacterial family tree, so noncyclic photosynthesis presumably arose through mutations that modified the cyclic pathway.

Cyanobacteria and other photosynthetic bacteria grew as dense mats in shallow sunlit water. The mats trapped minerals and sediments. Over many years, continual cell growth and deposition of minerals formed large dome-shaped, layered structures called **stromatolites** (Figure 18.9D,E). Such structures still form in some seas today.

Effects of Increasing Oxygen

By about 2.4 billion years ago, the oxygen produced by cyanobacteria had began to accumulate in Earth's waters and atmosphere. Here we pick up the story that we began in Section 7.2.

The rise in Earth's oxygen levels had three important consequences for life:

1. Oxygen interferes with the self-assembly of complex organic compounds, so life could no longer arise from nonliving materials.

2. The presence of oxygen put organisms that thrived in aerobic conditions at an advantage. Species that could not adapt to higher oxygen levels became extinct, or became restricted to the remaining low-oxygen environments, such as deep ocean sediments. Aerobic respiration evolved and became widespread. This pathway uses oxygen, and it is far more efficient at releasing energy than other reactions. Aerobic respiration would later allow the evolution of multicelled eukaryotes with high energy requirements.

Figure 18.9 Fossils of early life. **A** Strand of what may be some type of bacterial cells dates back 3.5 billion years. **B,C** Fossils of two types of cyanobacteria that lived approximately 850 million years ago in what is now Bitter Springs, Australia. **D** Artist's depiction of stromatolites in an ancient sea. **E** Cross-section through a fossilized stromatolite. Each layer formed when a mat of living cells trapped sediments. Descendant cells grew over the sediment layer, then trapped more sediment, forming the next layer.

Figure 18.10 Fossil history of eukaryotes. **A** *Grypania spiralis*, dates to 2.1 billion years ago. It may be the oldest known eukaryote, but some scientists think the coils are colonial bacteria. **B** *Tawuia*, probably an early alga. **C** Fossils of a red alga, *Bangiomorpha pubescens*. This multicelled species lived 1.2 billion years ago. Some cells formed a holdfast that anchored the body. Other cells produced sexual spores.

3. As oxygen enriched the atmosphere, some oxygen molecules broke apart, then recombined as ozone (O_3). Formation of an ozone layer in the upper atmosphere reduced the amount of solar ultraviolet (UV) radiation that reached Earth's surface. UV radiation can damage DNA and other biological molecules. It does not penetrate deep into water, but without the protective effect of the ozone layer, life could not have moved onto land.

The Rise of Eukaryotes

The third domain of life arose when eukaryotic cells branched off from the archaean lineage. Trace amounts of lipids in 2.7-billion-year-old rocks give us hints about when this second great branching took place. The lipids are biomarkers for eukaryotes. A **biomarker** is a compound made only by a particular type of cell; it is like a molecular signature.

Fossilized coils so large that they can be seen with the naked eye may also be evidence of early eukaryotes (Figure 18.10A,B). The fossil in Figure 18.10C is certainly a eukaryote. It is a red alga that lived about 1.2 billion years ago. This alga also has the distinction of being the oldest species known to reproduce sexually. (Only eukaryotes reproduce sexually.) The alga grew as hairlike strands, with cells at one end forming a holdfast that held it in

place. Cells at the strand's other end were specialized to produce sexual spores by meiosis.

The evolution of sexual reproduction and multicellularity were milestones in the history of life. Sex gave some eukaryotic organisms a new way to exchange genes. Multicellularity coupled with cellular differentiation opened the way to evolution of larger bodies that have specialized parts adapted to specific functions.

Trace fossils and biomarkers indicate that sponge-like animals may have evolved by 870 million years ago. By 570 million years ago, animals with more complex bodies shared the oceans with bacteria, archaeans, fungi, and protists, including the lineage of green algae that would later give rise to land plants.

Animal diversity increased greatly during a great adaptive radiation in the Cambrian, 543 million years ago. When that period finally ended, all of the major animal lineages, including the vertebrates (animals with backbones), were represented in the seas.

biomarker Molecule produced only by a specific type of cell.
stromatolite Dome-shaped structures composed of layers of bacterial cells and sediments.

Take-Home Message **What do we know about events that occurred early in the history of life?**

❯ The first cells evolved by 3.5 billion years ago. They did not have a nucleus and were probably anaerobic.

❯ An early diverge separated the bacteria and archaeans.

❯ After the noncyclic pathway of photosynthesis evolved, oxygen accumulated in the atmosphere and ended the further spontaneous chemical origin of life. The stage was set for the evolution of eukaryotic cells.

> Eukaryotic cells have a composite ancestry, with different components derived from archaea and bacteria.
> Scientists study modern cells to test hypotheses about how organelles evolved in the past.
< Links to Nucleus 4.7, Chloroplasts and mitochondria 4.9

Origin of the Nucleus

In all eukaryotes, the DNA resides in a nucleus. The outer layer of the nucleus, the nuclear envelope, consists of a double layer of membrane with protein-lined pores that control flow of material into and out of the nucleus. By contrast, the DNA of archaeans, the ancestors of eukaryotes, typically lies unenclosed in the cytoplasm.

The nucleus and endomembrane system probably evolved when the plasma membrane of an ancestral cell

Figure 18.11 **Animated** One model for the origin of the nuclear envelope and the endoplasmic reticulum. These organelles may have formed when portions of the plasma membrane folded inward.

A Marine bacterium (*Nitrosococcus oceani*) with highly folded internal membranes visible across its midline.

B Freshwater bacterium (*Gemmata obscuriglobus*) with DNA enclosed by a two-layer membrane (indicated by the arrow).

Figure 18.12 Bacteria with internal membranes. Both photos are electron micrographs.

folded inward (Figure 18.11). In support of this hypothesis, a few modern bacteria do have some internal membrane-enclosed compartments. Membrane infoldings can be selectively advantageous because the folds increase the surface area available for membrane-associated reactions. For example, the marine bacterium *Nitrosococcus oceani* has a system of highly folded internal membranes (Figure 18.12A). Enzymes embedded in the membranes allow the cell to meet its energy needs by breaking down ammonia.

An infolded membrane that cordons off a cell's genetic material can help protect the genome from physical or biological threats. For example, *Gemmata obscuriglobus* is one of the few bacteria that has a membrane around its DNA (Figure 18.12B). Compared to typical bacteria, it can withstand much higher levels of mutation-causing radiation. Researchers attribute this cell's radiation resistance to the tight packing of its DNA within the membrane-enclosed compartment. In other bacteria, the DNA spreads out through a broader area of the cytoplasm. Enclosing the genetic material within a membrane could also help protect it from viruses that inject their genetic material into bacteria or from interference caused by bits of DNA absorbed from the environment.

Mitochondria and Chloroplasts

Mitochondria and chloroplasts are eukaryotic organelles that resemble bacteria in their size and structure. Like bacteria, these organelles have a genome arranged as a circle of DNA. The organelles also behave somewhat independently, duplicating their DNA and dividing at a different time than the cell that holds them. Taken together, these observations prompted the hypothesis that mitochondria and chloroplasts evolved as a result of **endosymbiosis**, a relationship in which one type of cell (the symbiont) lives and replicates inside another cell (the host). The host in an endosymbiotic relationship passes some symbionts along to its descendants when it divides.

Genetic similarities between mitochondria and modern aerobic bacteria called rickettsias (Figure 18.13A) suggests that the two groups share a common ancestor. Presumably, a rickettsia-like cell infected an early eukaryote. The host began to use ATP produced by its aerobic symbiont while the symbiont began to rely on the host for raw materials. Over time, genes that occurred in both the host and symbiont were free to mutate. If a gene lost its function in one partner, a gene from the other could take up the slack. Eventually, the host and symbiont both became incapable of living independently.

Similarly, chloroplasts are structurally and genetically similar to a group of modern oxygen-producing photosynthetic bacteria called cyanobacteria. These similarities cause biologists to infer that chloroplasts evolved from an ancient relative of these cells.

photosynthetic organelle with a bacteria-like cell wall

mitochondrion

nucleus

A *Rickettsia prowazekii*, an aerobic bacterium that infects human cells and causes the disease typhus. Of all bacterial genomes sequenced so far, that of *R. prowazekii* is most similar to the mitochondrial genome. Like mitochondria, these bacteria take up pyruvate from the cytoplasm and break it down by aerobic respiration.

B *Cyanophora paradoxa*, one of the flagellated protists called glaucophytes. Its photosynthetic structures resemble cyanobacteria. They even have a wall similar in composition to the wall around a cyanobacterial cell.

Figure 18.13 Some modern cells that provide evidence in support of the endosymbiotic hypothesis for the origin of mitochondria and chloroplasts.

Additional Evidence of Endosymbiosis

A chance discovery made by microbiologist Kwang Jeon supports the hypothesis that bacteria can evolve into organelles. In 1966, Jeon was studying *Amoeba proteus*, a species of single-celled protist. By accident, one of his cultures became infected by a rod-shaped bacterium. Some infected amoebas died right away. Others kept growing, but only slowly. Intrigued, Jeon maintained those infected cultures to see what would happen. Five years later, the descendant amoebas were host to many bacterial cells, yet they seemed healthy. In fact, when these amoebas were treated with bacteria-killing drugs that usually do not harm amoebas, they died.

Experiments confirmed that the amoebas had come to rely on the bacteria. When Jeon swapped the nucleus from a bacteria-tolerant amoeba for the nucleus in a typical amoeba, the recipient cell died. Yet, when bacteria were included with the nucleus transplant, most cells survived. It seemed that the amoebas had come to require the bacteria for some life-sustaining function. Additional studies showed that the amoebas had lost the ability to make an essential enzyme. They now depended on their bacterial endosymbionts to make that enzyme for them.

We also have evidence to support the hypothesis that cyanobacteria can become organelles. The interior of the single-celled protists called glaucophytes is taken up largely by green photosynthetic organelles that resemble cyanobacteria (Figure 18.13B). The organelle even has a cell wall that contains peptidoglycan, a material made by some bacteria, but no eukaryotes. Many other aquatic protists have cyanobacteria living inside them. However, in most cases these bacteria are endosymbionts, not organelles; the bacteria can still live on their own if removed from their host. However, the photosynthetic organelles of glaucophytes, like chloroplasts, have evolved a dependence on their host. They cannot survive on their own.

However they arose, early eukaryotic cells had a nucleus, endomembrane system, mitochondria, and—in certain lineages—chloroplasts. These cells were the first protists. Over time, their many descendants came to include the modern protist lineages, as well as the plants, fungi, and animals. The next section provides a time frame for these pivotal evolutionary events.

Take-Home Message How did eukaryotic organelles evolve?

❯ The nucleus and endomembrane system may have evolved from infoldings of the plasma membrane.

❯ Mitochondria and chloroplasts may have evolved when bacterial endosymbionts and their hosts became mutually dependent.

endosymbiosis One species lives and reproduces inside another.

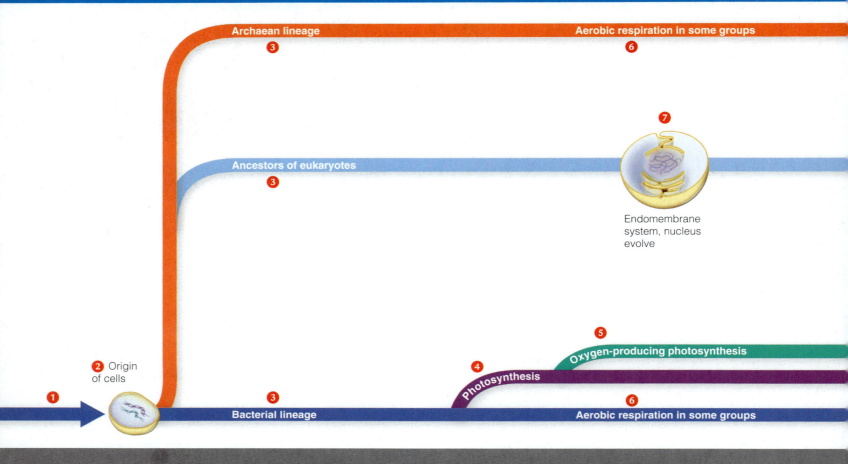

Hydrogen-rich, oxygen-poor atmosphere

Atmospheric oxygen level begins to increase

Archaean lineage 3

Aerobic respiration in some groups 6

7

Ancestors of eukaryotes 3

Endomembrane system, nucleus evolve

5

Oxygen-producing photosynthesis

4 Photosynthesis

2 Origin of cells

1

3 Bacterial lineage

6 Aerobic respiration in some groups

3.8 billion years ago

3.2 billion years ago

2.7 billion years ago

Steps Preceding Cells

1 Between 5 billion and 3.8 billion years ago, as an outcome of chemical and molecular evolution, complex carbohydrates, lipids, proteins, and nucleic acids formed from the simple organic compounds present on early Earth.

Origin of Cells

2 The first living cells evolved by 3.8 billion years ago. They did not have a nucleus or other organelles. Atmospheric oxygen was low, so early cells probably made ATP by anaerobic pathways.

Three Domains of Life

3 The first major divergence gave rise to bacteria and to the common ancestor of the archaeans and all eukaryotic cells.

Not long after, the ancestors of archaeans and eukaryotic cells diverged.

Photosynthesis, Aerobic Respiration Evolve

4 A cyclic pathway of photosynthesis evolved in some bacterial groups.

5 An oxygen-releasing noncyclic pathway evolved later in the cyanobacteria and, over time, changed the atmosphere.

6 Aerobic respiration evolved independently in many bacterial groups.

Origin of Endomembrane System, Nucleus

7 Cell sizes and the amount of genetic information continued to expand in ancestors of what would become the eukaryotic cells. The endomembrane system, including the nuclear envelope, arose through the modification of cell membranes between 3 and 2 billion years ago.

Atmospheric oxygen reaches current levels; ozone layer gradually forms

11 **Archaea**

11 **Eukarya**

8

Origin of animals

Origin of fungi

Animals

Fungi

Heterotrophic protists

9

Endosymbiotic origin of mitochondria

Protists with chloroplasts that evolved from algae

Protists with chloroplasts that evolved from bacteria

Origin of lineage leading to plants

Plants

Endosymbiotic origin of chloroplasts

10

11 **Bacteria**

Oxygen-producing photosynthetic bacteria

Other photosynthetic bacteria

Heterotrophic bacteria

| 1.2 billion years ago | 900 million years ago | 435 million years ago |

Endosymbiotic Origin of Mitochondria

8 An aerobic bacterium entered an anaerobic eukaryotic cell. Over many generations, the two species established a symbiotic relationship. Descendants of the bacterial cell became mitochondria.

Endosymbiotic Origin of Chloroplasts

9 A heterotrophic protist took in oxygen-producing bacteria (cyanobacteria). Descendants of the bacteria evolved into chloroplasts. Later, some photosynthetic protists would evolve into chloroplasts inside other protist hosts.

Plants, Fungi, and Animals Evolve

10 All major lineages— including fungi, animals, and the algae that would give rise to plants evolved in the seas.

Lineages That Have Endured to the Present

11 Today, organisms live in nearly all regions of Earth's waters, crust, and atmosphere. They are related by descent and share certain traits. However, each lineage encountered different selective pressures, and unique traits evolved in each one.

Figure 18.14 Animated Milestones in the history of life, based on the most widely accepted hypotheses. This figure also shows the evolutionary connections among all groups of organisms. The time line is not to scale.

>> **Figure It Out** Which organelle evolved first, mitochondria or chloroplasts?

Answer: Mitochondria

Looking For Life (revisited)

When it comes to sustaining life, Earth is just the right size. If the planet were much smaller, it would not exert enough gravitational pull to keep atmospheric gases from drifting off into space. The photo at the *right* shows the relative sizes of Earth

and Mars. As you can see, Mars is only about half the size of Earth. As a result, it has a much thinner atmosphere. What atmosphere there is consists mainly of carbon dioxide, some nitrogen, and only traces of oxygen. Thus, if life exists on Mars it is almost certainly anaerobic.

How would you vote? Martian soil may contain microbes that could provide new information about the origin and evolution of life. Should we bring samples of Martian soil to Earth for analysis? See CengageNow for details, then vote online (cengagenow.com).

Summary

Section 18.1 Astrobiology is the study of life's origin and distribution in the universe. The presence of cells in deserts and deep below Earth's surface suggests life may exist in similar settings on other planets.

Section 18.2 According to the **big bang theory**, the universe formed in an instant 13 to 15 billion years ago. Earth and other planets formed more than 4 billion years ago. Early in Earth's history, there was little oxygen in the air, volcanic eruptions were common, and there was a constant hail of meteorites.

Section 18.3 Laboratory simulations offer indirect evidence that organic compounds self-assemble spontaneously under conditions like those in Earth's early atmosphere or in the hot, mineral-rich water around **hydrothermal vents**. Examination of meteorites shows that such compounds might have formed in deep space and reached Earth in meteorites.

Section 18.4 Proteins that speed metabolic reactions might have first formed when amino acids stuck to clay, then bonded under the heat of the sun. Or, reactants could have begun interacting in rocks near deep-sea hydrothermal vents. Membrane-like structures and vesicles form when proteins or lipids are mixed with

water. They serve as a model for **protocells**, which may have preceded cells. An **RNA world**, a time in which RNA was the genetic material, may have preceded DNA-based systems. RNA still is a part of ribosomes that carry out protein synthesis in all organisms. Discovery of **ribozymes**, RNAs that act as enzymes, lends support to the RNA world hypothesis. A later switch from RNA to DNA would have made the genome more stable.

Section 18.5 The first cells evolved when oxygen levels in the atmosphere and seas were low, so they probably were anaerobic. An early divergence separated bacteria from the common ancestor of archaeans and eukaryotes. An oxygen-releasing, non-cyclic pathway of photosynthesis evolved in one bacterial lineage (cyanobacteria). These bacteria grew in mats that collected sediment and, over countless generations, formed dome-shaped structures called **stromatolites**.

Over time, oxygen released by cyanobacteria changed Earth's atmosphere. The increased oxygen level prevented evolution of new life from nonliving molecules, created a protective ozone layer, and favored cells that carried out aerobic respiration. This ATP-forming metabolic pathway was a key innovation in the evolution of eukaryotic cells.

Protists were the first eukaryotes. Their **biomarkers** and fossils date back more than 2 billion years. Diversification of protists gave rise to plants, fungi, and animals.

Section 18.6 By one hypothesis, the internal membranes that are typical of eukaryotic cells may have evolved through infoldings of the plasma membrane of prokaryotic ancestors. Existence of some bacteria with internal membranes supports this hypothesis.

Mitochondria and chloroplasts resemble bacteria, and these organelles most likely evolved by **endosymbiosis**. By this evolutionary process, one cell enters and survives inside another. Then, over generations, host and guest cells come to depend on one another for essential metabolic processes. Some modern protists have bacterial symbionts inside them.

Section 18.7 Evidence from many sources allows scientists to reconstruct the order of events and make a hypothetical time line for the history of life.

Self–Quiz   Answers in Appendix III

1. An abundance of _____ in the atmosphere would have prevented the spontaneous assembly of organic compounds on early Earth.
   a. hydrogen   b. methane   c. oxygen   d. nitrogen

2. The prevalance of iron-sulfide cofactors in organisms supports the hypothesis that life arose _____ .
   a. in outer space   c. near deep-sea vents
   b. on tidal flats   d. in the upper atmosphere

Data Analysis Activities

A Changing Earth Modern conditions on Earth are unlike those when life first evolved. Figure 18.15 shows how the frequency of asteroid impacts and composition of the atmosphere have changed over time. Use this figure and information in the chapter to answer the following questions.

1. Which occurred first, a decline in asteroid impacts, or a rise in the atmospheric level of oxygen?
2. How do modern levels of carbon dioxide and oxygen compare to those at the time when the first cells arose?
3. Which is now more abundant, oxygen or carbon dioxide?

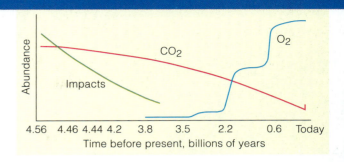

Figure 18.15 How asteroid impacts (*green*), atmospheric carbon dioxide concentration (*pink*), and oxygen concentration (*blue*) changed over geologic time.

---

3. Ribosomes can catalyze formation of peptide bonds. This supports the hypothesis that _____ .
   a. an RNA world preceded DNA-based genomes
   b. RNA can hold more information than DNA
   c. the first protists had RNA as their genetic material
   d. all of the above

4. By one hypothesis, clay _____ .
   a. facilitated assembly of early polypeptides
   b. was present at hydrothermal vents
   c. provided energy for early metabolism
   d. all of the above

5. The evolution of _____ resulted in an increase in the levels of atmospheric oxygen.
   a. sexual reproduction
   b. aerobic respiration
   c. the noncyclic pathway of photosynthesis
   d. the cyclic pathway of photosynthesis

6. Mitochondria most resemble _____ .
   a. archaeans   c. cyanobacteria
   b. aerobic bacteria   d. early eukaryotes

7. What was the energy source in the Miller–Urey simulation of conditions on the early Earth?

8. The first sexual reproducers were _____ .
   a. archaeans   c. cyanobacteria
   b. aerobic bacteria   d. eukaryotes

9. Oxygen released by _____ accumulated in the atmosphere and produced the ozone layer.
   a. archaeans   c. cyanobacteria
   b. aerobic bacteria   d. early eukaryotes

10. What is a ribozyme made of?

11. A rise in oxygen in Earth's air and seas put organisms that engaged in _____ at a selective advantage.
    a. aerobic respiration   c. photosynthesis
    b. fermentation   d. sexual reproduction

12. Which of the following was not present on Earth when mitochondria first evolved?
    a. archaeans   c. protists
    b. bacteria   d. animals

13. Chloroplasts most resemble _____ .
    a. archaeans   c. cyanobacteria
    b. aerobic bacteria   d. early eukaryotes

14. Which provides a more stable genome, DNA or RNA?

15. Arrange these events in order of occurence, with 1 being the earliest and 6 the most recent.
    ___1   a. emergence of the noncyclic
    ___2   pathway of photosynthesis
    ___3   b. origin of mitochondria
    ___4   c. origin of protocells
    ___5   d. emergence of the cyclic
    ___6   pathway of photosynthesis
                  e. origin of chloroplasts
                  f. the big bang

Additional questions are available on **CENGAGENOW**.

Critical Thinking

1. Researchers looking for fossils of the earliest life forms face many hurdles. For example, few sedimentary rocks date back more than 3 billion years. Review what you learned about plate tectonics (Section 16.7). Explain why so few remaining samples of these early rocks remain.

2. Craig Venter and Claire Fraser are working to create a "minimal organism." They are starting with *Mycoplasma genitalium*, a bacterium that has 517 genes. By disabling its genes one at a time, they discovered that 265–350 of them code for essential proteins. The scientists are synthesizing the essential genes and inserting them, one by one, into an engineered cell consisting only of a plasma membrane and cytoplasm. They want to see how few genes it takes to build a new life form. What properties would such a cell have to exhibit for you to conclude that it was alive?

Animations and Interactions on **CENGAGENOW**:
› Miller–Urey experiment; Milestones in history of life.

‹ **Links to Earlier Concepts**

This introduction to viruses and viroids touches on reverse transcription (Section 15.2), ribozymes (18.4), DNA repair (8.6), and cancer (10.1). The chapter also covers bacteria and archaeans (4.5). You will learn more about the three domain classification system (1.5), bacteria that gave rise to organelles (18.6), and antibiotic resistance (17.5). You will also draw on your knowledge of coevolution (17.13) and cladistics (17.14).

Key Concepts

Viruses and Viroids

Viruses are noncellular, with a protein coat and a genome of nucleic acid, but no metabolic machinery. Viruses must infect cells to replicate. Some infect humans and cause disease. Viroids are RNA bits that do not encode proteins. Even so, they can infect plant cells and replicate inside them.

Structure and Function of Bacteria

Bacteria are small cells with DNA and ribosomes, but no nucleus or typical eukaryotic organelles. They are also the most abundant and metabolically diverse organisms, with autotrophs (self-feeders) and heterotrophs (feeders on others) among them.

19 Viruses, Bacteria, and Archaeans

In this chapter, we explore the diversity of two of Earth's oldest lineages. Billions of years before there were plants or animals, Earth's seas were home to bacteria and archaeans. These small cells do not have a nucleus or other typical eukaryotic organelles. Viruses are simpler still, with no chromosomes, ribosomes, or metabolic machinery. By many definitions, viruses are not even alive. Despite their simplicity, viruses can evolve because they have genes that mutate.

For example, scientists have learned quite a bit about the origin and evolution of HIV (human immunodeficiency virus). This virus causes AIDS (acquired immunodeficiency syndrome). Researchers first isolated HIV in the early 1980s, and have since determined that there are two strains (subtypes), HIV-1 and the less prevalant HIV-2. By sequencing the HIV-1 genome and comparing it to the genomes of primate viruses, researchers found that the human virus evolved from simian immunodeficiency virus (SIV). SIV infects wild chimpanzees.

By one hypothesis, the first human infected was someone who butchered or ate meat from an SIV-infected chimp. To test the plausibility of this hypothesis, researchers looked for evidence of simian foamy virus (SFV), another primate virus, among people in an African village who commonly hunt and eat monkeys and apes. The researchers found that one percent of the villagers showed sign of prior infection by SFV—evidence that ape-to-human transmission of a virus is possible.

To find out when HIV jumped to humans, researchers have looked for the virus in old tissue samples stored from routine hospital tests. Samples from two people who lived in Africa's Democratic Republic of the Congo are the earliest evidence of HIV in humans. One is a sample of blood stored since 1959. The other, a woman's lymph node, was removed in 1960. Gene sequences of the two viral samples differ a bit, which implies that HIV had already been around and mutating by the time these people became infected. Given what we know about the mutation rates for viruses, the common ancestor of the two genotypes (the earliest HIV-1) must have first infected humans in the early 1900s.

Comparing genes of HIV in stored and modern blood samples has allowed researchers to trace the movement of the virus out of Africa. This data shows that HIV-1 was carried from Africa to Haiti in about 1966. The virus diversified in Haiti. Then, in about 1969, one person infected by HIV with mutations that arose in Haiti brought the virus to the United States. Once there, it spread quietly for 12 years until AIDS was identified as a threat in 1981.

Today more than 20 million people have died from AIDS and about 30 million are infected with HIV. The virus infects and replicates inside white blood cells essential to immune responses (Figure 19.1). Eventually, the infected white blood cells die. Death of such cells destroys the body's ability to defend itself. As a result, many disease-causing organisms run rampant, causing symptoms of AIDS and health problems that can be fatal.

Figure 19.1 Micrographs of a new HIV particle budding from an infected white blood cell. The photo on the *opposite page* shows many viral particles (*blue* dots) on an infected cell.

Replication and Gene Exchange
Bacteria have a single chromosome and some also have one or more plasmids. They reproduce by fission, a type of asexual reproduction. Cells can exchange genes by swapping plasmids, and by other processes.

Bacterial Diversity
Bacteria are well studied and highly diverse. They put oxygen into the air, supply nutrients to plants, and break down wastes and remains. Some live in or on our bodies and have beneficial effects. Others are pathogens that cause human disease.

Archaean Diversity
Archaeans were discovered relatively recently. Many are adapted to life in very hot or very salty places. Others live in low-oxygen environments and make methane. Still others live beside bacteria in soils and seas. None cause human disease.

❯ Viruses hover near the border between living and nonliving things. They have genes, but none of the cellular machinery required to express those genes or replicate them.
❮ Link to Reverse transcription 15.2

Viral Traits and Diversity

A **virus** is a noncellular infectious agent composed of a protein coat wrapped around genetic material (RNA or DNA) and a few viral enzymes. A virus is far smaller than any cell and has no ribosomes or other metabolic machinery. To replicate, the virus must infect a cell of a specific organism, which we call its host.

Each type of virus has structural adaptations that allow it to infect and replicate in hosts of a particular type. **Bacteriophages** infect bacteria. One well-studied group has a complex coat (Figure 19.2A). DNA is encased in a protein "head." Attached to the head is a rodlike "tail" with fibers that attach the virus to its host.

The tobacco mosaic virus infects plants. It has a helical structure, with coat proteins arranged around a strand of RNA to form a rod (Figure 19.2B).

Many animal viruses have a 20-sided protein coat. In adenoviruses, this coat has a protein spike at each corner (Figure 19.2C). Adenoviruses are "naked," but most animal viruses are enveloped; a bit of membrane from a prior host encloses the virus. Herpesviruses are enveloped DNA viruses (Figure 19.2D). HIV is an enveloped RNA virus.

Viral Replication

Viral replication begins when a virus attaches to proteins in the host's plasma membrane. A virus cannot seek out a host, but rather relies on a chance encounter. The virus, or just its genetic material, enters the cell and hijacks the cell's metabolic machinery. Viral genes direct the replication of viral genetic material and the production of viral proteins. These components then self-assemble to form new viral particles. New virus buds from the infected host cell or is released when the host bursts.

Bacteriophage Replication Two replication pathways are common among bacteriophages (Figure 19.3). In the **lytic pathway**, viral genes enter a host and immediately direct it to make new viral particles. Soon the cell dies by lysis (breaks open), allowing new viral particles to escape.

In the **lysogenic pathway**, viral DNA becomes integrated into the host chromosome and a latent period precedes formation of new viruses. The viral DNA is copied along with host DNA, and is passed along to all descendants of the host cell. Like tiny time bombs, viral DNA inside these descendant cells awaits a signal to enter the lytic pathway.

HIV Replication HIV replicates inside a human white blood cell (Figure 19.4). Spikes of viral protein that extend beyond the envelope attach to proteins in the host cell's membrane ❶. The viral envelope fuses with this membrane, and viral enzymes and genetic material (RNA) enter the cell ❷. A viral enzyme (reverse transcriptase) converts viral RNA into double-stranded DNA that the cell's genetic machinery can read ❸. Viral DNA is moved to the nucleus, where a viral enzyme integrates it into a host chromosome ❹. Viral DNA is transcribed along with host genes ❺. Some of the resulting RNA is translated into viral proteins ❻, and some becomes the genetic material of new HIV particles ❼. The viral particles self-assemble at the plasma membrane ❽. As the virus buds

A T4 bacteriophage — DNA inside protein coat, sheath, tail fiber

B Tobacco mosaic virus — RNA, protein subunits of coat

C Adenovirus — 20-sided protein coat that encloses DNA

D Herpesvirus — viral DNA and enzymes, lipid envelope with protein components, 20-sided protein coat beneath the envelope

Figure 19.2 Models illustrating viral structure.　❯❯ **Figure It Out** Which virus acquired membrane from its host?　Answer: The herpesvirus

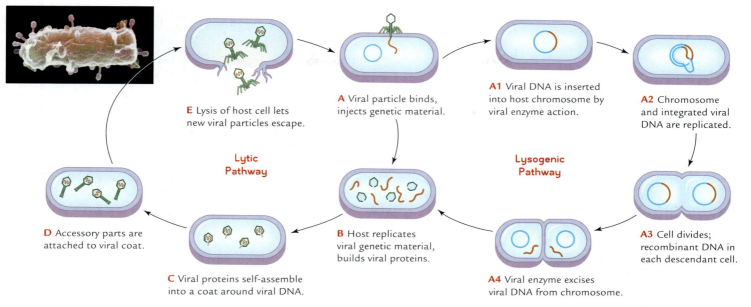

E Lysis of host cell lets new viral particles escape.

A Viral particle binds, injects genetic material.

A1 Viral DNA is inserted into host chromosome by viral enzyme action.

A2 Chromosome and integrated viral DNA are replicated.

Lytic Pathway

Lysogenic Pathway

D Accessory parts are attached to viral coat.

B Host replicates viral genetic material, builds viral proteins.

A3 Cell divides; recombinant DNA in each descendant cell.

C Viral proteins self-assemble into a coat around viral DNA.

A4 Viral enzyme excises viral DNA from chromosome.

Figure 19.3 Animated Bacteriophage replication pathways.

›› Figure It Out What does the blue circle in A represent?

Answer: The bacterial chromosome

Figure 19.4 Animated Replication cycle of HIV, an eveloped RNA virus.

1 Virus binds to a host cell.

2 Viral RNA and enzymes enter cell.

3 Viral reverse transcriptase uses viral RNA to make double-stranded viral DNA.

4 Viral DNA integrates into host genome.

5 Transcription produces viral RNA.

6 Some viral RNA is translated to produce viral proteins.

7 Other viral RNA forms the new viral genome.

8 Viral proteins and viral RNA self-assemble at the host membrane.

9 New virus buds from the host cell, with an envelope of host plasma membrane.

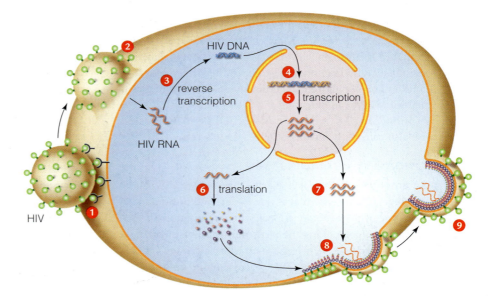

from the host cell, some of the host's plasma membrane becomes the viral envelope **9**. Each viral particle can now infect other white blood cells.

Drugs that fight HIV interfere with viral binding to the host, reverse transcription, integration of DNA, or processing of viral polypeptides to form viral proteins.

bacteriophage Virus that infects bacteria.
lysogenic pathway Bacteriophage replication path in which viral DNA becomes integrated into the host's chromosome and is passed to the host's descendants.
lytic pathway Bacteriophage replication pathway in which a virus immediately replicates in its host and kills it.
virus Noncellular, infectious particle of protein and nucleic acid; replicates only in a host cell.

Take-Home Message **What is a virus and how does it replicate?**

❯ A virus is a noncellular infectious particle that consists of nucleic acid enclosed in a protein coat and sometimes an outer envelope.

❯ A virus replicates by binding to a specific type of host cell, taking over the host's metabolic machinery, and using that machinery to produce viral components. These components self-assemble to form new viral particles.

> Viral particles far outnumber cells, and viruses have wide-ranging effects on all forms of life, including humans.
< Links to DNA repair 8.6, Cancer 10.1

Some viruses have a positive effect on human health, as when bacteriophages kill bacteria that could cause food poisoning. However, other viruses are **pathogens**, meaning they cause disease.

Common Viral Diseases

A variety of nonenveloped viruses, including adenoviruses, infect membranes of the upper respiratory system and cause common colds. Other nonenveloped viruses infect the lining of the intestine and cause viral gastroenteritis, commonly referred to as a stomach flu. Nonenveloped viruses also cause warts. Human papillomavirus (HPV), the cause of genital warts, is the most common sexually transmitted virus. Some strains of HPV can also cause cervical cancer.

Cold sores are caused by one herpesvirus (an enveloped virus). Another herpesvirus causes genital herpes. Still others cause infectious mononucleosis and chicken pox. Once a person has been infected by a herpesvirus, the virus persists in the body for life. It may enter a latent state, similar to that in the lysogenic cycle of a bacteriophage, only to resume activity later on. For example, the virus that causes chicken pox in children sometimes reemerge as shingles (a painful rash) at an older age.

Influenza (flu), mumps, measles, and German measles are also caused by enveloped viruses.

Emerging Viral Diseases

An **emerging disease** is a disease that suddenly expands its range, or a disease that is newly detected in humans. A new type of disease may appear when a mutation alters a viral genome, making the virus more easily spread or more deadly. RNA viruses have an especially high mutation rate because a host's DNA proofreading and repair mechanisms (Section 8.6) do not work on RNA. Viruses also evolve by picking up new genes from their host, or from another virus that infects a cell at the same time.

AIDS is one example of an emerging disease. Here we consider a few others.

West Nile Fever West Nile virus is an enveloped RNA virus that replicates in birds. Mosquitoes carry the virus from host to host, so we say they are the **vector** for this virus. Sometimes a bite from a virus-carrying mosquito causes a human infection. This is a dead end for the virus. Although it can replicate in human cells, not enough virus gets into the blood for the infection to be passed on.

Most people infected with West Nile virus are not sickened, but some develop West Nile fever, which causes

Figure 19.5 A health care worker putting on protective gear during the SARS epidemic.

flulike symptoms. In about 1 percent of West Nile fever cases, the virus attacks the nervous system, with results that can be fatal.

West Nile virus had long been present in Africa, the Middle East, and parts of Europe, but it was unknown in the Western Hemisphere until 1999, when it began killing people and birds in New York City. Over the next few years, infected migratory birds spread the virus across the country. Today, West Nile fever is an **endemic disease** throughout the continental United States, meaning the disease remains present, but at a low level.

SARS Sudden acute respiratory syndrome (SARS) first appeared in late 2002 in China (Figure 19.5). The disease became an **epidemic**, a disease that is widespread in one region. Then, with the help of air travelers, SARS became a **pandemic**, an outbreak of disease that encompasses many regions and poses a threat to human health. Over the course of 9 months, SARS sickened about 8,000 people in 37 countries and killed 774.

A previously unknown type of coronavirus (*right*) causes SARS. Coronaviruses are named for the "corona" or crown of protein spikes that extends through their envelope. Where did this new virus come from? Researchers found a SARS-like virus in Chinese horseshoe bats, but it does not infect humans. They hypothesize that bats with the SARS-like virus were captured and brought to wildlife markets, where they gave the virus to other animals. Inside those animals, the virus evolved the ability to infect humans.

Influenza H5N1 and H1N1 Influenzaviruses commonly cause flu outbreaks during the winter in temperate regions. Because of mutations, each year's flu virus is a bit different than that of previous years. However, a typical seasonal flu has a low mortality rate and causes deaths mainly among older people. Doctors recommend that people over 50 get a yearly flu shot that will protect against the anticipated version of the seasonal flu. Such shots do not protect people from dramatically different strains of influenzavirus, such as those described below.

Avian influenza H5N1, commonly called bird flu, occasionally infects people who have direct contact with birds. When the virus does infect people, the death rate is disturbingly high. From 2003 to 2009, the World Health Organization received reports of 417 human cases of influenza H5N1, mainly in Asia. Of these, 257 (about 60 percent) were fatal. Fortunately, person-to-person transmission of the H5N1 virus is exceedingly rare.

By contrast, influenza H1N1, shown in the micrograph at the *left*, is easily transmitted by a cough or a sneeze. It is commonly called swine flu. The H1N1 virus first appeared in Mexico in early 2009, where it caused severe respiratory symptoms. From Mexico, the virus spread to the United States, and then throughout the world. Fortunately, initial fears of a high death rate proved unfounded, antiviral drugs were released to treat the virus, and a vaccine was created.

Health officials continue to monitor H5N1 and H1N1 influenza. Either virus could mutate and their coexistence raises the possibility of a potentially disastrous gene exchange. The influenza H1N1 virus already has a composite genome, with material from a human flu virus, bird flu virus, and two different swine flu viruses. If it picked up genes from avian H5N1, the result could be an influenzavirus that is easily transmissible and deadly.

---

emerging disease A disease that was previously unknown or has recently begun spreading to a new region.
endemic disease Disease that persists at a low level in a region or population.
epidemic Disease outbreak limited to one region.
pandemic Outbreak of disease that affects many separate regions and poses a serious threat to human health.
pathogen Disease-causing agent.
vector Animal that carries a pathogen from one host to the next.

> Tiny bits of RNA, just a few hundred nucleotides long, can use enzymes in a plant cell to replicate themselves.
< Links to Plasmodesmata 4.12, Ribozyme 18.4

In 1971, plant pathologist Theodor Diener announced the discovery of a new type of pathogen, a small RNA without a protein coat. He named it a **viroid**, because it seemed like a stripped-down version of a virus.

Diener had been investigating potato spindle tuber disease, an illness that stunts potato plants and causes them to produce only a few small, deformed potatoes. Diener expected to find a viral pathogen, but was forced to consider other options after discovering that the infectious agent passed through filters too fine to allow passage of even the smallest virus.

To identify components of the apparently minuscule pathogen, Diener treated extracts from infected plants with enzymes to see what would inactivate the pathogen. Extracts treated with enzymes that destroyed DNA, lipids, or protein still infected plants. Only RNA-digesting enzymes made the extracts harmless. Diener concluded that the pathogen must consist of RNA.

Plant pathologists have now described about thirty viroids that cause disease in commercially valuable plants, including citrus, apples, coconuts, avocados, and chrysanthemums. All known viroids are circular, single-stranded RNAs. Base pairing between different parts of a viroid usually causes it to fold up into a rodlike shape (*right*). The viroid is remarkably small, with fewer than 400 nucleotides. By comparison, even the smallest viral genome consists of thousands of nucleotides.

Unlike the genetic material of a virus, viroid RNA does not encode proteins. However, the viroid itself has enzymatic activity. In other words, it is a ribozyme (Section 18.4). Typically, viroid replication occurs when a host enzyme (RNA polymerase) moves along the circular RNA repeatedly. The result is a long RNA strand, with many copies of the viroid attached end to end. The strand then cuts itself up in appropriate places, forming new viroids. The viroids spread through the plant via plasmodesmata that connect cells, and phloem (food-carrying vessels).

---

viroid Small noncoding RNA that can infect plants.

> Bacteria are small, structurally simple, widely dispersed, and highly abundant cells.
< Links to Bacteria 4.5, Evolution of organelles 18.6

Cell Size, Structure, and Motility

The typical bacterial cell cannot be seen without a light microscope. It is far smaller than a eukaryotic cell, about the size of a mitochondrion. In fact, there is evidence that certain bacteria were the ancestors of mitochondria (Section 18.6).

Biologists describe bacteria by their shapes (Figure 19.6A). A spherical cell is called a coccus; a rod-shaped cell a bacillus; and a spiral cell a spirillum.

Nearly all bacteria have a semirigid, porous cell wall around the plasma membrane (Figure 19.6B). A secreted slime layer or a capsule may enclose the wall. Slime helps a cell stick to surfaces. A capsule is tougher and helps some bacterial pathogens evade the immune defenses of their vertebrate hosts.

Inside the cell, the **bacterial chromosome** is a circle of double-stranded DNA. It attaches to the plasma membrane and resides in a cytoplasmic region called the **nucleoid**. Ribosomes scattered through the cytoplasm make proteins. There is no endomembrane system like that of eukaryotes, although a few kinds of bacteria do have internal membranes of some sort (Section 18.6).

coccus

bacillus

A spirillum

Many bacterial cells have one or more flagella. Unlike eukaryotic flagella, bacterial flagella do not contain microtubules and do not bend side to side. Instead, they rotate like a propeller.

Hairlike filaments called **pili** (singular, pilus) often extend from the cell surface. Some cells use pili to stick to surfaces. Others glide along by using their pili as grappling hooks. A pilus extends out to a surface, sticks to it, then shortens, drawing the cell forward. Another type of retractable pilus draws cells together for gene exchanges as described in the next section.

Abundance and Metabolic Diversity

In terms of sheer numbers the bacteria are unparalleled among cells. Biologists at the University of Georgia have estimated that 5 million trillion trillion bacterial cells live on Earth.

Metabolic diversity contributes to bacterial success. There are four known modes of nutrition and, as a group, bacteria use them all (Table 19.1).

Photoautotrophs are photosynthetic; they use light energy to build organic compounds from carbon dioxide and water. This group includes nearly all plants, and some protists, as well as many bacteria.

Chemoautotrophs get energy by removing electrons from inorganic molecules such as sulfides. They use this energy to build organic compounds from carbon dioxide and water. All are bacteria or archaeans.

Photoheterotrophs use light energy and get carbon by breaking down organic compounds in their environment. All are bacteria or archaeans.

Chemoheterotrophs get carbon and energy by breaking down organic compounds assembled by other organisms. Many bacteria are in this group, as are some archaeans, protists, and all animals and fungi. Some bacterial chemoheterotrophs feed on living organisms. Others are decomposers that break down organic wastes or remains.

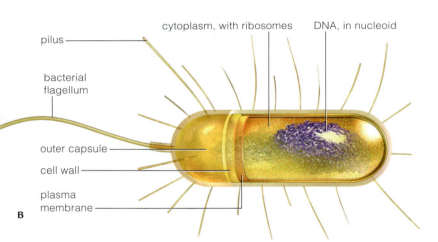

pilus

bacterial flagellum

cytoplasm, with ribosomes   DNA, in nucleoid

outer capsule

cell wall

plasma membrane

B

Figure 19.6 Animated Bacterial cell shapes **A** and body plan **B**.

bacterial chromosome Circle of double-stranded DNA that resides in the bacterial cytoplasm.
nucleoid Cytoplasmic region where prokaryotic chromosome lies.
pilus Hairlike extension from the cell wall of some bacteria.

Table 19.1 Nutritional Modes		
Mode of Nutrition	Energy Source	Carbon Source
Photoautotrophic	Sunlight	Carbon dioxide
Chemoautotrophic	Inorganic substances	Carbon dioxide
Photoheterotrophic	Sunlight	Organic compounds
Chemoheterotrophic	Organic compounds	Organic compounds

Take-Home Message What are are features of bacterial cells?

> Bacteria are small, typically walled cells with no nucleus. Their single chromosome lies in the cytoplasm and there is no endomembrane system.

> As a group, bacteria are the most numerous and the most metabolically diverse organisms.

❯ Bacteria can only reproduce asexually, but new gene combinations arise when they exchange genetic material.
❮ Link to Asexual reproduction 11.2

Bacteria have staggering reproductive potential. Most reproduce by **binary fission**, a type of asexual reproduction (Figure 19.7). The process begins when the cell replicates its single chromosome, which is attached to the inside of the plasma membrane ❶. The DNA replica attaches to the plasma membrane adjacent to the parent molecule. Addition of new membrane and wall material elongates the cell and moves the two DNA molecules apart ❷. Then, membrane and cell wall material are deposited across the cell's midsection ❸. Addition of this material partitions the cell, producing two identical cells ❹.

❶ A bacterium has one circular chromosome that attaches to the inside of the plasma membrane.

❷ The cell duplicates its chromosome, attaches the copy beside the original, and adds membrane and wall material between them.

❸ When the cell has almost doubled in size, new membrane and wall are deposited across its midsection.

❹ Two genetically identical cells result.

Figure 19.7 Animated Binary fission. The micrograph *above* shows step 3 of this process in the bacteria *Bacillus cereus*.

Figure 19.8 Prokaryotic conjugation. One cell extends a sex pilus out to another, draws it close, and gives it a copy of a plasmid.

❯❯ **Figure It Out** Does conjugation increase the number of cells?

Answer: No. It is not a mode of reproduction.

Horizontal Gene Transfers

Besides inheriting DNA "vertically" from a parent cell, bacteria engage in **horizontal gene transfer**: the transfer of genetic material between existing individuals. In the process of **conjugation**, one cell gives a plasmid to the other. A **plasmid** is a small circular DNA molecule that is separate from the bacterial chromosome and has only a few genes. Cells get together for conjugation when one cell extends a sex pilus out to a prospective partner and reels it in (Figure 19.8). Once the cells are close together, the cell that made the sex pilus passes a copy of its plasmid to its partner. The cells then separate. Afterwards, each cell will pass the plasmid on to its descendants. Each cell can also donate the plasmid to other cells during another gene transfer.

Two other processes can also introduce new genes. First, a cell can take up DNA from its environment, a process called transformation. Second, viruses that infect bacteria sometimes move genes between their hosts.

The ability of bacteria to acquire new genes has important implications. Suppose a gene for antibiotic resistance arises by mutation in a bacterial cell. This gene can not only be passed on to that cell's descendants, but also be transferred to other existing cells. Gene transfers speed the rate at which a gene spreads through a population, thus accelerating the response to selective pressure.

---

binary fission Method of asexual reproduction that divides one bacterial or archaean cell into two identical descendant cells.
conjugation Mechanism of gene exchange in which one bacterial or archaean cell passes a plasmid to another.
horizontal gene transfer Transfer of genetic material.
plasmid Of many bacteria and archaeans, a small ring of nonchromosomal DNA replicated independently of the chromosome.

Take-Home Message **How do bacteria reproduce and exchange genes among cells?**

❯ Bacteria reproduce by binary fission, a type of asexual reproduction.

❯ Gene exchange occurs by conjugation. Bacteria also obtain new genes from viruses and directly from their environment.

> Bacteria serve as decomposers, cycle nutrients, and form partnerships with many other species.
< Links to PCR 15.3, Antibiotics 17.5, Chloroplast origin 18.6

Bacteria that cause human disease often get the spotlight, but most bacteria are either harmless or beneficial. As you will see, they live in many habitats and show an amazing degree of ecological diversity.

Heat-Loving Bacteria

If life emerged in thermal pools or near hydrothermal vents, the modern heat-loving bacteria may resemble those early cells. Biochemical comparisons put them near the base of the bacterial family tree. One species, *Thermus aquaticus*, was discovered in a volcanic spring in Yellowstone National Park. Biochemist Kary Mullis isolated a heat-stable DNA polymerase from *T. aquaticus* and put the enzyme to work in the first PCR reactions (Section 15.3). He won a Nobel Prize for inventing this process, which is now widely used in biotechnology.

Oxygen-Producing Cyanobacteria

Photosynthesis evolved in many bacterial lineages, but only the **cyanobacteria** release free oxygen by a noncyclic pathway, as plants do (Figure 19.9A). If, as evidence suggests, chloroplasts evolved from ancient cyanobacteria, we have cyanobacteria and their chloroplast descendants to thank for the oxygen in Earth's atmosphere (Section 18.6).

Some cyanobacteria partner with fungi and form lichens (Chapter 22), and some live on the surface of the soil, but most are aquatic. One of these, *Spirulina*,

is grown commercially and sold as a health food. Other aquatic cyanobacteria carry out the ecologically important task of **nitrogen fixation**: They incorporate gaseous nitrogen ($N\equiv N$) into ammonia (NH_3). Plants and algae need nitrogen but cannot use nitrogen gas because they cannot break its triple bond. They can, however, take up the ammonia produced and released by cyanobacteria.

Highly Diverse Proteobacteria

Proteobacteria, the most diverse bacterial lineage, also includes some nitrogen fixers. *Rhizobium* lives in roots of legumes, a group of plants that includes peas and beans. Nitrogen-fixing by *Rhizobium* benefits host plants and also enriches the soil.

Myxobacteria, another group of soil proteobacteria, show remarkable cooperative behavior. They glide about as a cohesive group, feeding on other bacteria. When food runs out, thousands of cells join together to form a multicelled fruiting body, a structure with spores (dormant cells) atop a stalk. Wind disperses spores to new habitats, where each germinates and releases a single cell.

The largest known bacterium is a marine proteobacterium, *Thiomargarita namibiensis* (Figure 19.9B). This chemoautotroph stores nitrogen and sulfur in a huge vacuole, making it big enough to be visible to the naked eye.

Escherichia coli is a chemoheterotroph that lives in the mammalian gut. It is part of the **normal flora**, a collection of microorganisms that typically live in and on a body. Most *E. coli* benefit their human host by producing vitamin K. However, the strain *E. coli* O157:H7 is among the top three causes of food poisoning. The other two, *Salmonella* and *Campylobacter*, are also proteobaceria.

Other proteobacterial pathogens that affect the gut include *Helicobactor pylori* (the main cause of stomach ulcers) and *Vibrio cholerae* (the agent of cholera).

Rickettsias are a proteobacterial subgroup of tiny cells that live as intracellular parasites and are transmitted by ticks or insects. Tick-borne rickettsias cause Rocky Mountain spotted fever. Rickettsias are also notable as the closest living relatives of the ancient cells that evolved into mitochondria.

The Thick-Walled Gram Positives

Gram-positive bacteria are a lineage characterized by thick cell walls that are tinted purple when prepared for microscopy by **Gram staining**. Thinner-walled bacteria such as cyanobacteria and proteobacteria are stained pink by this staining process, and are described as Gram-negative.

Most Gram-positive bacteria are chemoheterotrophs. For example, *Lactobacillus* is a lactate fermenter (Section 7.6). It is a common decomposer and sometimes spoils milk. We use it to produce yogurt, cheese, sauerkraut, and sour foods (Figure 19.9C). *L. acidophilus* is part of

nitrogen-fixing cell

A

B

C

6 μm

0.2 mm

Figure 19.9 Ecologically important bacteria. **A** Chain of cyanobacteria, with many photosynthetic, oxygen-producing cells and one nitrogen-fixing cell.

B *Thiomargarita namibiensis*, a proteobacterium and the largest known bacterim. It has an enormous vacuole that holds sulfur and nitrate.

C Lactate-fermenting bacteria (*Lactobacillus*) in yogurt. Related cells are decomposers or part of the human normal flora.

Figure 19.10 A Staphylococci, common skin bacteria.
B An abscess caused by an antibiotic-resistant staph infection.

Figure 19.11 A Spirochete that causes Lyme disease. **B** Bull's-eye rash at the site of a tick bite is often the first sign of infection.

the normal flora on skin and in the gut and vagina. The lactate that the bacteria produce lowers the pH of the surroundings, and helps keep pathogenic bacteria in check. *L. acidophilus* spores are available in capsules to be taken to promote gut health.

Clostridium and *Bacillus* are Gram-positive soil bacteria that can form an **endospore**. Unlike a typical bacterial spore, an endospore can survive heating, freezing, radiation, and disinfectants.

Toxins made by some endospore-forming bacteria can be deadly. Inhale *Bacillus anthracis* endospores and you may get anthrax, a disorder in which the bacterial toxin interferes with breathing. *Clostridium tetani* endospores that germinate in wounds cause tetanus, in which toxins lock muscles in ongoing contraction. *C. botulinum*, an anaerobe that grows in improperly canned foods, makes a toxin that causes a paralyzing food poisoning known as botulism. The same toxin, prepared as Botox, can be injected to temporarily paralyze facial muscles that tug on the skin and cause wrinkling.

Worldwide, about one-third of the population is infected by *Mycobacterium tuberculosis*, the cause of tuberculosis. Droplets from coughs spread the disease, which kills about 1.6 million people each year. Bacteria-laden droplets also spread *Streptococcus*, the cause of strep throat. If these bacteria get into a wound they can become what the media calls "flesh-eating bacteria." The result is fast-spreading infection that kills surface tissue and can be fatal. *Staphylococcus*, a common skin bacteria, can have the same effect if it enters a cut. More commonly, a staph infection will cause a boil or an abscess (Figure 19.10).

Most staph infections can be cured with methicillin, a type of antibiotic. However, directional selection has favored antibiotic resistance (Section 17.5). Antibiotic-resistant staph infections previously occurred mainly in hospitals and nursing homes. Now such infections are breaking out in schools and prisons. The bacteria are transmitted by contact with an infected person or something an infected person has touched, as by sharing towels and razors.

The sexually transmitted disease gonorrhea is caused by a diplococcus, a Gram-positive spherical bacterium that usually occurs as paired cells.

Spring–Shaped Spirochetes

Spirochetes look like a stretched-out spring (Figure 19.11). Some live in the cattle gut and help their host by breaking down cellulose. Others are aquatic decomposers and some fix nitrogen. A pathogenic spirochete causes the sexually-transmitted disease syphilis. Another spirochete, transmitted by ticks, causes Lyme disease.

Parasitic Chlamydias

Chlamydias are tiny cocci. Like the rickettsias, they can only live and replicate in eukaryotic host cells. Chlamydia infection is the most common sexually transmitted bacterial disease in the United States.

chlamydias Tiny round bacteria that are intracellular parasites of eukaryotic cells.
cyanobacteria Photosynthetic, oxygen-producing bacteria.
endospore Resistant resting stage of some soil bacteria.
Gram staining Process used to prepare bacterial cells for microscopy, and to distinguish groups based on cell wall structure.
nitrogen fixation Incorporation of nitrogen gas into ammonia.
normal flora Normally harmless or beneficial microorganisms that typically live in or on a body.
proteobacteria Largest bacterial lineage.
spirochetes Bacteria that resemble a stretched-out spring.

Take-Home Message How do bacteria affect other organisms?

> Most bacteria play beneficial roles in nutrient cycles by adding oxygen to the air, making nitrogen available to plants, or serving as decomposers.

> Some bacteria are human pathogens.

› Archaeans are more similar to eukaryotes than they are to bacteria. Many survive in extreme environments, but new species are turning up almost everywhere.
‹ Links to Three-domain system 1.5, Cladistics 17.14

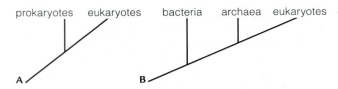

Figure 19.12 Comparison of **A** two-domain and **B** three-domain trees of life. The two-domain model was widely accepted until new evidence revealed previously unknown differences between bacteria and archaeans. The three-domain model is now in wide use.

Discovery of the Third Domain

Science is a self-correcting process in which new discoveries can overturn even long-held ways of thinking. For example, biologists historically divided all life into two groups, prokaryotes and eukaryotes. These groups were very different in size and structure, so scientists thought they represented distinct lineages (clades) that parted ways early in the history of life (Figure 19.12A).

Then, in the late 1970s Carl Woese began investigating evolutionary relationships among the prokaryotes. At that time, all were considered bacteria. By comparing ribosomal RNA gene sequences, Woese found that some methane-making cells were as similar to eukaryotes as they were to typical bacteria. Woese concluded that the methane makers were not bacteria, but rather a previously unrecognized branch on the tree of life. To accommodate this branch, he proposed a new classification system with three domains of life: bacteria, archaea, eukaryotes (Figure 19.12B).

Woese's ideas were greeted with skepticism. He had been trained as a physicist, and many biologists could not accept that he had found something they had missed. They were sure his methods were flawed. But as years went by, evidence in support of Woese's conclusions mounted. Archaeans and bacteria have different cell wall and membrane components. Archaeans and eukaryotes organize their DNA around histone proteins, which bacteria do not have. Sequencing the genome of *Methanococcus jannaschii* (*left*) provided the definitive evidence. Most of this archaean's genes have no counterpart in bacteria.

Today, the hypothesis that bacteria and archaeans constitute a single lineage has been discarded, and the three-domain classification system is in widespread use. Woese compares the discovery of archaeans to the discovery of a new continent, which he and others are now exploring.

0.5 µm

Archaean Diversity

On the basis of their physiology, many archaeans fall into one of three groups: methanogens, extreme halophiles, and extreme thermophiles.

Methanogens are organisms that produce methane (CH_4), commonly known as natural gas. They are adapted to anaerobic conditions, and exposure to oxygen inhibits their growth or kills them. Methane-making archaeans live near deep-sea hydrothermal vents, in soils, and in ocean sediments (Figure 19.13A). They also live in the gut of humans and grazers such as cows and sheep. Methane produced in the gut escapes as belches or flatulence.

By their metabolic activity, methanogens produce 2 billion tons of methane annually. Release of methane into the air has important environmental effects, because methane is a greenhouse gas (an atmospheric gas that traps heat near Earth, thus causing global warming).

Extreme halophiles are adapted to life in a highly salty environment. Salt-loving archaeans live in the Dead Sea, the Great Salt Lake, saltwater evaporation ponds, and other highly salty habitats (Figure 19.13B). Some have a purple pigment (bacteriorhodopsin) that allows them to use light energy to produce ATP.

Extreme thermophiles are adapted to life at a very high temperature. They live in hot springs (Figure 19.13C) and near deep-sea hydrothermal vents, where temperatures can exceed 110°C (230°F). Their existence is cited as evidence that life could have originated on the sea floor.

As biologists continue to explore archaean diversity, they are finding that archaeans are not restricted to extreme environments. Archaeans live alongside bacteria nearly everywhere and even exchange genes with them by conjugation. So far, scientists have not found any archaeans that pose a major threat to human health, although some that live in the mouth may encourage periodontal (gum) disease.

---

extreme halophile Organism adapted to life in a highly salty environment.
extreme thermophile Organism adapted to life in a very high-temperature environment.
methanogen Organism that produces methane gas as a metabolic by-product.

Take-Home Message **What are archaeans?**

› Archaeans belong to a lineage that is structurally and genetically distinct from bacteria.

› Many archaeans are adapted to extremely hot or salty conditions. Some produce methane as a metabolic by-product.

› Some archaeans live in human bodies, but none are known to be important as pathogens.

A Deep-sea sediments. Bubbles of methane rising from the floor of the Black Sea are evidence of methanogens in sediments below.

B Highly salty waters. Pigmented extreme halophiles color the brine in this California lake.

C Thermally heated waters. Pigmented archaeans color the rocks in waters of this hot spring in Nevada.

Figure 19.13 Examples of archaean habitats.

Evolution of a Disease (revisited)

❮ Links to Directional selection 17.5, Coevolution 17.13

As noted earlier, viruses have genes and can evolve. Just as bacteria evolve resistance to antibiotics, HIV has adapted to some antiviral drugs. For example, AZT, the first drug approved to fight AIDS, inhibits HIV replication by interfering with reverse transcription of viral RNA. When random mutations made some HIV particles resistant to AZT, directional selection favored those mutations. As a result of AZT treatment, evolution occurred: AZT-resistant strains of HIV became increasingly common.

Pathogens also coevolve with their hosts. For example, HIV seems to be adapting to human immune defenses. Our white blood cells have recognition proteins at their surface that allows them to detect HIV and fight it. There are different alleles for these recognition proteins and they vary among regions, with some alleles being common in Asia, others in Africa, and others in Europe. A recent study found a corresponding difference in the frequency of "escape mutations" in populations of HIV. Escape mutations help the virus evade detection by the recognition proteins of white blood cells. If escape mutations arose randomly and were not under selection, all types would be similarly prevalent in all HIV populations. Instead, escape mutations that evade common Asian recognition proteins prevail in Asia, while those that evade common African proteins predominate in Africa. Directional selection has apparently favored HIV mutations that evade the most common white blood cell defense in each region.

Selection also acts on hosts, favoring those that can fight off or evade a pathogen. For example, about 10 percent of people of European ancestry have a mutation that lessens the likelihood of infection by most HIV strains. The frequency of this mutation is highest among northern Europeans and declines with latitude. It is absent in American Indian, east Asian, and African populations. The protective mutation now enjoys a selective advantage as a result of the AIDS epidemic. However, the mutation did not arise as a result of AIDS. Mutation is a random process and studies of ancient remains tell us that this mutation has been in the northern European gene pool for thousands of years.

By one hypothesis, the mutation's current frequency and distribution reflect previous positive selection during epidemics of other diseases. In one test of this hypothesis, researchers compared the frequency of the mutation among inhabitants of small islands off the coast of Croatia. The researchers chose these islands because they knew from historical records that during the mid-1400s repeated outbreaks of an unknown disease occurred on some islands, but not others. Where epidemics did occur, the population declined by an average of 70 percent. Results supported the hypothesis. Inhabitants of islands affected by epidemics in the 1400s are significantly more likely to have the protective allele than inhabitants of islands that were unaffected.

How would you vote? Antiviral drugs help keep people with HIV healthy and lessen the likelihood of viral transmission. However, an estimated 25 percent of HIV–infected Americans do not know they are infected. Annual, voluntary HIV tests with drug treatment for those infected could help curtail the AIDS pandemic. Do you favor an expanded, voluntary testing program? See CengageNow for details, then vote online (cengagenow.com).

Summary

Section 19.1 Scientists use their knowledge of evolution to investigate how a new disease such as AIDS can arise and spread in the human population.

Section 19.2 A **virus** is a noncellular infectious agent that consists of a protein coat around a core of DNA or RNA. In some viruses, the coat is enveloped in a bit of plasma membrane derived from a previous host.

Because a virus lacks ribosomes and other metabolic machinery, it must replicate inside a host cell. Viruses attach to a host cell, then enter it or insert viral genetic material into it. Viral genes and enzymes direct the host to replicate viral genetic material and make viral proteins. New viral particles self-assemble and are released.

Bacteriophages, viruses that infect bacteria, have two types of replication pathways. In a **lytic pathway**, multiplication is rapid, and the new viral particles are released by lysis. In a **lysogenic pathway**, the virus enters a latent state that extends the cycle.

HIV is an enveloped RNA virus that replicates in human white blood cells. Viral RNA that enters the cell must be reverse transcibed to DNA to begin the process of replication. The virus acquires its envelope as it buds from the cell membrane.

Section 19.3 Some viruses are human **pathogens**, agents that cause disease. A **vector** is an animal that carries a pathogen between hosts. An **emerging disease** is new to humans or spreading to a new region. An **endemic disease** is one that is present but not spreading. An outbreak of disease in one region is an **epidemic**. If the outbreak affects many regions and threatens human health it is a **pandemic**. Changes to viral genomes as a result of mutation or gene exchanges can alter the properties of a viral disease.

Section 19.4 Viroids are bits of RNA that do not encode proteins. They infect plants and replicate inside them.

Section 19.5 Bacteria are small, structurally simple cells. They do not have a nucleus or cytoplasmic organelles typical of eukaryotes. The single **bacterial chromosome**, a circle of double-stranded DNA, resides in a cytoplasmic region called the **nucleoid**. Cell shapes vary. Typical surface structures include a cell wall, a protective capsule or slime layer, one or more flagella, and hairlike extensions called **pili**.

As a group, bacteria are metabolically diverse, including both autotrophs and heterotrophs.

Section 19.6 Bacteria reproduce by **binary fission**: replication of a single, circular chromosome and division of a parent cell into two genetically equivalent descendants. **Horizontal gene transfers** move genes between existing cells, as when **conjugation** moves a **plasmid** with a few genes from one cell into another.

Section 19.7 Bacteria are widespread, abundant, and diverse. Many have essential ecological roles. **Cyano-bacteria** produce oxygen during photosynthesis. Some also carry out **nitrogen fixation**, producing ammonia that algae and plants need as a nutrient. **Proteobacteria**, the largest bacterial lineage, also includes nitrogen-fixers. In addition, it includes soil bacteria that show cooperative behavior, the closest relatives of mitochondria, cells that are part of our **normal flora**, and some pathogens. **Gram staining** is a method used to prepare bacteria for examination under a microscope. Gram-positive bacteria have thick walls. Some Gram-positive soil bacteria produce **endospores** that allow them to survive boiling and disinfectants. **Chlamydias** are tiny bacteria that live inside vertebrate cells. **Spirochetes** resemble a stretched-out spring and some are pathogens.

Section 19.8 Archaeans superficially resemble bacteria. However, comparisons of structure, function, and genetic sequences position archaeans in a separate domain, closer to eukaryotes than to bacteria. Ongoing research is showing that archaeans are more diverse and widely distributed than was previously thought. In their physiology, many archaeans are **methanogens** (methane makers), **extreme halophiles** (salt lovers), and **extreme thermophiles** (heat lovers). Archaeans coexist with bacteria in many habitats and can exchange genes with them.

Self-Quiz   Answers in Appendix III

1. A(n) _____ may have a genome of RNA or DNA.
   a. bacterium   b. viroid   c. virus   d. archaean

2. Which is smallest?
   a. bacterium   b. viroid   c. virus   d. archaean

3. In _____ , viral DNA is integrated into a bacterial chromosome and passed to descendant cells.
   a. prokaryotic fission   c. the lysogenic pathway
   b. the lytic pathway   d. both b and c

4. The genetic material of HIV is _____ .

5. Viral genomes can be altered by _____ .
   a. mutation   b. gene exchanges   c. both a and b

6. True or false? Prokaryotic conjugation is a type of asexual reproduction.

Data Analysis Activities

Maternal Transmission of HIV Since the AIDS pandemic began, there have been more than 8,000 cases of mother-to-child HIV transmission in the United States. In 1993, American physicians began giving antiretroviral drugs to HIV-positive women during pregnancy and treating both mother and infant in the months after birth. Only about 10 percent of mothers were treated in 1993, but by 1999 more than 80 percent got antiviral drugs.

Figure 19.14 shows the number of AIDS diagnoses among children in the United States. Use the information in this graph to answer the following questions.

1. How did the number of children diagnosed with AIDS change during the late 1980s?
2. What year did new AIDS diagnoses in children peak, and how many children were diagnosed that year?
3. How did the number of AIDS diagnoses change as the use of antiretrovirals in mothers and infants increased?

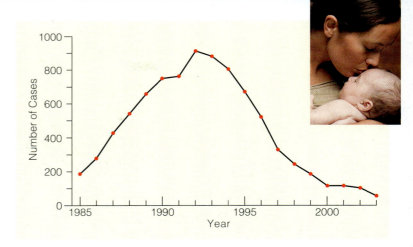

Figure 19.14 Number of new AIDS diagnoses in the United States per year among children exposed to HIV during pregnancy, birth, or by breast-feeding.

7. How many chromosomes does an archaean have?

8. All _____ are oxygen-releasing photoautotrophs.
   a. spirochetes
   b. archaeans
   c. cyanobacteria
   d. bacteria

9. Nitrogen-fixing bacteria produce _____ .
   a. methane   b. ammonia   c. nitrogen gas

10. Vitamin-producing *E. coli* cells in your gut are _____ .
    a. normal flora
    b. chemoheterotrophs
    c. bacteria
    d. all of the above

11. Some soil bacteria such as *Bacillus anthracis* survive harsh conditions by forming a(n) _____ .
    a. endospore   b. heterocyst   c. plasmid

12. A Gram-positive coccus is _____ .
    a. spherical   b. rod-shaped   c. spiral-shaped

13. The vector for Lyme disease is a(n) _____ .
    a. spirochete   b. archaean   c. tick

14. Production of _____ by archaeans may contribute to global warming.
    a. methane   b. nitrogen   c. oxygen

15. Match the terms with their most suitable description.
    ___ archaean
    ___ cyanobacteria
    ___ virus
    ___ viroid
    ___ plasmid
    ___ extreme halophile
    ___ nucleoid
    ___ extreme thermophile

    a. plant-infecting RNA
    b. noncellular infectious particle; nucleic acid core, protein coat
    c. likes it hot
    d. site of bacterial chromosome
    e. sister group to the eukaryotes
    f. evolved into chloroplasts
    g. small circle of bacterial DNA
    h. salt lover

Additional questions are available on **CENGAGENOW**.

Critical Thinking

1. Viruses that do not have a lipid envelope tend to remain infectious outside the body longer than enveloped viruses. "Naked" viruses are also less likely to be rendered harmless by soap and water. Can you explain why?

2. Methanogens have been found in the human gut and deep-sea sediments, but not in the human mouth or the surface waters of the ocean. What physiological trait of methanogens could explain this distribution?

3. Review the description of Fred Griffith's experiments with *Streptococcus pneumoniae* in Section 18.3. Using your knowledge of bacterial biology, explain the process by which the harmless bacteria became dangerous.

4. The antibiotic penicillin acts by interfering with the production of new bacterial cell wall. Cells treated with penicillin do not die immediately, but they cannot reproduce. Explain how penicillin halts binary fission. Explain also why the cancer drug taxol, which stops eukaryotic division by interfering with spindle formation, has no effect on bacterial cells.

5. About 1 percent of Europeans are homozygous for an allele that provides protection against infection by HIV. Would you expect more or fewer heterozygous for this allele? Explain your reasoning.

Animations and Interactions on **CENGAGENOW**:
❯ Bacteriophage replication; Bacterial structure; Binary fission, Conjugation.

‹ Links to Earlier Concepts

This chapter covers the protists, a diverse collection of lineages introduced in Section 1.4. You will learn about how protists reproduce (11.2) and how they are classified (17.14). We reexamine eukaryotic cell structures such as chloroplasts (6.4), pseudopods, and flagella (4.10); reconsider the effects of osmosis on cells (5.6); and return to photosynthetic pigments (6.2). We also delve again into evolution of organelles by endosymbiosis (18.6).

Key Concepts

A Collection of Lineages
Protists include many lineages of eukaryotic organisms, some autotrophs and others heterotrophs. The protists are not a clade; some groups are more closely related to plants, or to fungi and animals, than to other protists.

Single-Celled Lineages
Most protist lineages are entirely single-celled. These groups include flagellated protozoans, shelled cells called foraminiferans and radiolarians, and the alveolates (ciliates, dinoflagellates, and apicomplexans).

20 The Protists

If you sample water from just about any aquatic habitat, you will find a variety of single-celled protists. A **protist** is a eukaryotic organism that is not a fungus, plant, or animal. Aquatic protists include single-celled and multicellular autotrophs and heterotrophs. Photosynthetic protists play an important ecological role by taking up carbon dioxide and releasing oxygen. They also serve as food for aquatic animals.

However, some of these single-celled producers occasionally become a threat. When conditions are unusually favorable for growth, as occurs when extra nutrients are present, the cells multiply fast. The result is an **algal bloom**, a higher than normal concentration of aquatic microorganisms. The photo at the *left*, taken in 2007, shows an enormous algal bloom in coastal waters near Pensacola, Florida. Algal blooms are commonly known as "red tides" because the protists involved often have a reddish pigment. However, the term is misleading: Not all algal blooms color the water red, and the event is not related to tidal changes.

Algal blooms in the Gulf of Mexico frequently involve the protist *Karenia brevis*, a dinoflagellate (Figure 20.1). This species makes brevetoxin. A **toxin** is a substance that is produced by one organism and is harmful to others. Brevetoxin interferes with animal nerve cells by binding to a protein in their cell membrane. It sickens and even kills marine invertebrates, fish, sea turtles, sea birds, dolphins, and manatees.

Human nerve cells have the same kinds of membrane proteins as those of other vertebrates and are harmed in the same way. Eating shellfish tainted by brevetoxin causes intestinal problems and nervous system symptoms such as headache, vertigo, loss of coordination, and temporary paralysis. People also are exposed to brevetoxin in spray from onshore winds during an algal bloom. When inhaled, brevetoxin irritates nasal membranes and constricts airways, making breathing difficult. Inside the lungs, metabolic breakdown of brevetoxins creates chemicals

Figure 20.1 Colorized, scanning electron micrograph of the photosynthetic protist *Karenia brevis*. This dinoflagellate benefits us by taking up carbon dioxide and releasing oxygen. However, it also produces a toxin that can sicken people who inhale or ingest it.

that damage DNA. Thus, repeated inhalation of the toxin might increase the risk of lung cancer.

The effects of brevetoxin on humans are interesting, and may be a threat to public health. However, these effects are an evolutionary accident; *K. brevis* does not gain any advantage by sickening humans. Most likely, brevetoxin benefits *K. brevis* by providing protection against potential predators such as heterotrophic protists and tiny animals. Keep this point in mind as you read the chapter: Although we will often mention the ways that protists affect humans, the lineages and traits we describe here had their origins long before humans evolved.

---

algal bloom Population explosion of tiny aquatic producers.
protist Eukaryote that is not a fungus, animal, or plant.
toxin Chemical that is made by one organism and harms another.

Brown Algae and Relatives
Brown algae are an entirely multicellular group of protists. They are members of the same lineage as single-celled photosynthetic cells called diatoms, and filamentous heterotrophs called water molds.

Red Algae, Green Algae
Red algae and green algae are single-celled and multicelled aquatic producers. Red algae have pigments that allow them to live in deep waters. Green algae are the closest relatives of land plants and have the same pigments as them.

Amoebozoans
A lineage of unwalled heterotrophic protists that includes the amoebas and slime molds is the protist group most closely related to the fungi and animals. Members of this lineage live in aquatic habitats or on the forest floor.

> Protists are members of many separate eukaryotic lineages, some only distantly related to one another.
< Links to Asexual Reproduction 11.2, Cladistics 17.14

Figure 20.2 Protist diversity. The single-celled protists include **A** amoebas, **B** euglenoids, and **C** diatoms. Most red algae **D** and all brown algae **E** are multicelled. **F** One proposed eukaryotic family tree with traditional protist groups indicated by *tan* boxes. The protists are not a single lineage.

>> Figure It Out Are land plants more closely related to the red algae or the brown algae?

Answer: Red algae

Protists are collection of lineages, rather than a clade, or monophyletic group (Section 17.14). In fact, some protists are more closely related to plants, fungi, or animals than to other protists.

Figure 20.2 shows a few protists and illustrates where the groups we cover in this chapter fit in the eukaryotic family tree. There are many additional protist lineages, but this sampling of major groups will suffice to demonstrate protist diversity, ecological importance, and the ways that protists can affect human health.

Most protist lineages include only unicellular species (Figure 20.2A–C). However, some lineages include colonial cells, and true multicellularity evolved independently in several groups (Figure 20.2D,E). A multicellular organism consists of cells that cannot survive and reproduce on their own. Cells of colonial organisms live together, but retain an ability to survive and reproduce independently.

Protists have diverse life cycles, but most reproduce both sexually and asexually. Depending on the group, asexually reproducing individuals may be haploid or diploid. For example, the parasite that causes malaria has a haploid stage that divides in human blood and liver cells. The only diploid stage in the parasite's life cycle is the zygote, which undergoes meiosis within a few hours of fertilization. By contrast, the silica-shelled cells called diatoms have a diploid-dominant life cycle. Most of the time, diploid cells divide by mitosis and only the gametes are haploid.

Like plants, some multicelled algae have a life cycle with both haploid and diploid multicelled bodies. Protists are often described as "simple," but the multicelled bodies of some algae can be large and complex.

Protists may be heterotrophs, autotrophs, or switch between nutritional modes. Heterotrophic protists serve as decomposers, prey on smaller organisms such as bacteria, or live inside larger organisms. Protistan autotrophs have chloroplasts. In the lineage that includes red algae and green algae, the chloroplasts evolved from bacteria as described in Section 18.6. We call this event *primary* endosymbiosis. All other protists have chloroplasts that evolved through *secondary* endosymbiosis: A protist that had chloroplasts was engulfed by a heterotrophic one and evolved into an organelle.

Take-Home Message What are protists?

> Protists are a collection of eukaryotic lineages. Most are single-celled, but there are some multicelled species.

> Some protists are autotrophs; other are heterotrophs. A few can switch between modes.

> Most protists reproduce both sexually and asexually.

> Flagellated protozoans swim through lakes, seas, and the body fluids of animals.
< Links to Flagella 4.10, Tonicity 5.6

Flagellated protozoans are unwalled cells with one or more flagella. All groups are entirely or largely heterotrophic. A **pellicle**, a layer of elastic proteins beneath the plasma membrane, helps the cells retain their shape. Haploid cells dominate the life cycle of these groups.

Diplomonads and parabasalids have multiple flagella and are among the few protists that can live in places where oxygen is scarce. Instead of mitochondria, they have organelles that produce ATP by an anaerobic pathway. These organelles evolved from mitochondria and are an adaptation to anaerobic habitats. Free-living diplomonads and parabasalids thrive deep in seas and lakes. Others live inside the bodies of animals.

Diplomonads are unusual in that they have two more or less identical nuclei. The diplomonad *Giardia lamblia* (*left*) causes giardiasis, a waterborne human disease. The protist attaches to the intestinal lining and sucks out nutrients. Symptoms of giardiasis include cramps, nausea, and severe diarrhea. Infected people and animals excrete cysts (a hardy resting stage) of *G. lamblia* in their feces. Drinking water contaminated with the cysts spreads the infection.

The parabasalid *Trichomonas vaginalis* (Figure 20.3A) infects human reproductive tracts and causes the disease trichomoniasis. *T. vaginalis* does not make cysts, so it cannot survive very long outside the human body. Fortunately for the parasite, sexual intercourse puts it directly

---

contractile vacuole In freshwater protists, an organelle that collects and expels excess water.
euglenoid Flagellated protozoan with multiple mitochondria; may be heterophic or have chloroplasts decended from algae.
flagellated protozoan Protist belonging to an entirely or mostly heterotrophic lineage with no cell wall and one or more flagella.
pellicle Layer of proteins that gives shape to many unwalled, single-celled protists.
trypanosome Parasitic flagellate with a single mitochondrion and a membrane-encased flagellum.

Take-Home Message What are flagellated protozoans?

> Parabasalids and diplomonads are heterotrophs that lack mitochondria. Some are important human pathogens.

> Trypanosomes are parasites with a large mitochondrion. Biting insects transmit some that cause human disease.

> Euglenoids include heterotrophs and photoautotrophs with chloroplasts that evolved from green algae.

A The parabasilid *Trichomonas vaginalis* causes a sexually transmitted disease.

B The trypanosome *Trypanosoma brucei*, causes African sleeping sickness.

Figure 20.3 Two parasitic flagellates.

into hosts. In the United States, about 6 million people are infected.

Trypanosomes are long tapered cells, with a single large mitochondrion. A flagellum encased in a membrane runs the length of the cell (Figure 20.3B). Action of the flagellum causes a wavelike motion in the membrane around it and moves the cell.

All trypanosomes are parasites and insects serve as vectors for some that cause human disease. Tsetse flies in sub-Saharan Africa carry the trypanosome that causes African sleeping sickness, a nervous system disease that can be fatal. In Central and South America, bloodsucking bugs transmit a trypanosome that causes Chagas disease, which can damage the heart. Desert sandflies are the vector for leishmaniasis, another trypanosome disease. Untreated leishmaniasis can produce disfiguring scars and harm the liver.

Euglenoids are flagellated cells closely related to the trypanosomes. Unlike trypanosomes, euglenoids have many mitochondria. Some also have chloroplasts that evolved from a green alga (Figure 20.4). Photosynthetic euglenoids have an eyespot near the base of a long flagellum that helps the cell detect light. The pellicle consists of translucent strips of protein that spiral around the cell.

The interior of a euglenoid is saltier than its freshwater habitat, so water tends to diffuse into the cell. Like many other freshwater protists, euglenoids have one or more **contractile vacuoles**, organelles that collect excess water, then contract and expel it to the outside through a pore.

Figure 20.4 Animated Body plan of a euglenoid (*Euglena*), a freshwater species with chloroplasts that evolved from a green alga. See also Figure 20.2B.

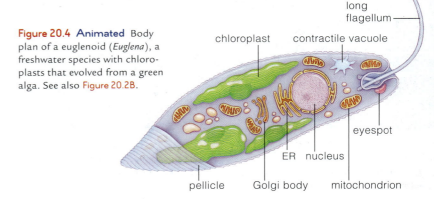

long flagellum
chloroplast
contractile vacuole
eyespot
ER   nucleus
mitochondrion
pellicle   Golgi body

> Heterotrophic single cells with chalky or glassy shells live in great numbers in the world's oceans.
< Link to Microtubules 4.10

Foraminiferans and radiolarians are single-celled marine protists with sieve-like shells. They feed by capturing food with microtubule reinforced cytoplasmic extensions that protrude through the shell's many openings.

Foraminiferans have a calcium carbonate shell. They typically prey on bacteria and smaller protists in ocean sediments. Others are **plankton**; a general term for micro-organisms that drift or swim in an aquatic habitat. Plank-tonic foraminiferans often have small photosynthetic protists inside them (Figure 20.5A). Chalk and limestone deposits (Figure 20.5B) include foraminiferan shells that accumulated on the sea floor over billions of years.

Radiolarians secrete a glassy silica shell (Figure 20.5C). They are a major component of the marine plankton in tropical waters. They capture food with their cytoplasmic extensions and, like planktonic foraminiferans, some have photosynthetic protists living inside them.

---

foraminiferan Heterotrophic single-celled protist with a porous calcium carbonate shell and long cytoplasmic extensions.
plankton Community of tiny drifting or swimming organisms.
radiolarian Heterotrophic single-celled protist with a porous shell of silica and long cytoplasmic extensions.

Take-Home Message What are foraminiferans and radiolarians?

> Two related lineages of heterotrophic marine cells have porous secreted shells. The chalky-shelled foraminiferans live in sediments or drift as part of the plankton. The silica-shelled radiolarians are planktonic.

> Dinoflagellates, ciliates, and apicomplexans are single cells that belong to the alveolate lineage.
> Most dinoflagellates and ciliates are aquatic and free-living, but all apicomplexans are parasites.
< Link to Endocytosis 5.8

Dinoflagellates, ciliates, and apicomplexans belong to a lineage known as the **alveolates**. "Alveolus" means sac, and the characteristic trait of alveolates is a layer of sacs beneath the plasma membrane.

Dinoflagellates

The name **dinoflagellate** means "whirling flagellate." These single-celled protists typically have two flagella, one at the cell's tip and the other running in a groove around its middle like a belt. Combined action of the two flagella causes the cell to rotate as it moves forward. Most dinofla-gellates deposit cellulose in the sacs beneath their plasma membrane, and the deposits form thick protective plates. The vast majority of dinoflagellates are marine plank-ton. They are especially abundant in tropical waters. Some prey on bacteria, and others have chloroplasts that evolved from algae. As Section 20.1 explained, runaway growth of some dinoflagellates such as *Karenia brevis* sometimes results in an algal bloom.

Photosynthetic dinoflagellates live inside reef-building coral (a type of invertebrate animal). The protists supply the coral with sugars and oxygen, and receive shelter and carbon dioxide in return. A coral depends on its protist partners. If the coral loses them it will die.

A few marine dinoflagellates are bioluminescent. Like fireflies, they convert ATP energy into light (Figure 20.6). Emitting light may protect a cell by startling a predator

Figure 20.5 Protists with secreted shells. **A** Live foraminiferan with algal cells inside it (*yellow* dots). **B** Chalk cliffs of Dover, England are remains of calcium carbonate–rich shells of marine protists that accumulate on the sea floor. **C** Silica shell of a radiolarian.

Figure 20.6 Dinoflagellate bioluminescence. A tropical dinoflagel-late (*inset*) emits light when disturbed, as by the motion of an oar. The flash of light may benefit the cell by warding off predators.

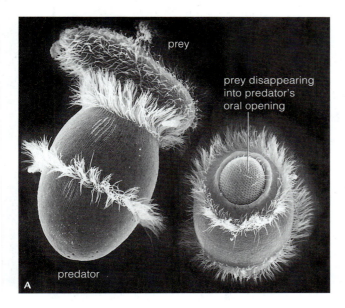

prey

prey disappearing into predator's oral opening

predator

A

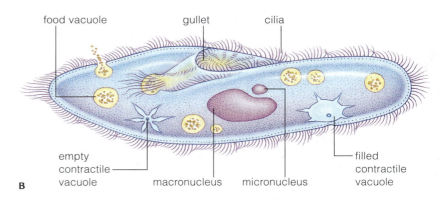

food vacuole   gullet   cilia

empty contractile vacuole   macronucleus   micronucleus   filled contractile vacuole

B

Figure 20.7 Animated Freshwater ciliates. **A** *Didinium*, a barrel-shaped ciliate with tufts of cilia, catching and engulfing a *Paramecium* covered with cilia. **B** Body plan of *Paramecium*. Mitochondria are present but not shown.

that was about to eat it. Alternatively the flash of light may function like a car alarm, attracting the attention of other organisms, including predators that pursue the potential dinoflagellate eaters.

Ciliates

Ciliates, or ciliated protozoans, are unwalled, heterotrophic cells that use their many cilia in locomotion and feeding. Most ciliates are aquatic predators that feed on bacteria, algae, and one another (Figure 20.7A).

Figure 20.7B shows the body plan of *Paramecium*, a free-living ciliate common in lakes and ponds. The cell feeds by using its cilia to sweep water laden with bacteria, algae, and other food particles into a gullet. Food particles enter the cell by endocytosis and are digested inside food vacuoles. Digestive waste is then expelled by exocytosis. Like euglenoids, *Paramecium* has contractile vacuoles that collect and squirt out excess water.

There are two types of nuclei, a macronucleus that controls daily activities and one or more smaller micronuclei that function in sexual reproduction. Like most single-celled protists, *Paramecium* reproduces asexually: A cell duplicates its DNA and organelles, then divides in half, producing two identical cells. Sexual reproduction

occurs by a process that involves meiosis and a swap of micronuclei between two cells.

Although most ciliates are free-living and aquatic, some have adapted to life in the animal gut. Gut-dwelling ciliates help cattle, sheep, and related grazing animals break down the cellulose in plant material. Similarly, ciliates that live in the termite gut help these insects break down wood.

Only one ciliate is known to be a human pathogen. It usually lives in the gut of pigs, but can also survive in the human gut, where it causes nausea and diarrhea. Human infections occur when pig feces get into drinking water.

Apicomplexans

Apicomplexans are parasitic protists that spend part of their life inside cells of their hosts. Their name refers to a complex of microtubules at their apical (top) end that allows them to pierce and enter a host cell. They are also sometimes called sporozoans. Apicomplexans infect a variety of animals, from worms and insects to humans. Their life cycle often involves more than one host species. The next section looks in detail at the disease malaria, and the biology of the apicomplexan that causes it.

---

alveolate Member of a protist lineage having small sacs beneath the plasma membrane; dinoflagellate, ciliate, or apicomplexan.
apicomplexan Single-celled protist that lives as a parasite inside animal cells.
ciliate Single-celled, heterotrophic protist with many cilia.
dinoflagellate Single-celled, aquatic protist with cellulose plates and two flagella; may be heterotrophic or photosynthetic.

Take-Home Message What are alveolates?

> Alveolates are single cells with an array of membrane-bound sacs (alveoli) beneath the plasma membrane.

> Dinoflagellates are common in plankton. These flagellated heterotrophs or photoautotrophs have cellulose plates and move with a whirling motion.

> The ciliates are heterotrophs. Cilia cover all or part of the cell surface and function in locomotion and feeding.

> Apicomplexans are intracellular parasites with a special host-piercing device.

> Malaria, caused by an apicomplexan, has the highest death toll of any protist disease.
< Link to Balanced polymorphism 17.7

Malaria is a leading cause of human death, killing more than 1.3 million people every year. *Plasmodium*, a single-celled apicomplexan, causes malaria. Mosquitoes carry the protist from one human host to another.

Figure 20.8 shows the *Plasmodium* life cycle. A bite from a female mosquito transmits the infective stage of *Plasmodium*, a haploid sporozoite, to a human host **1**. The sporozoite travels through blood vessels to liver cells, where it reproduces asexually **2**. Offspring, called merozoites, enter red blood cells and produce more merozoites **3**. Merozoites also can enter red blood cells and develop into immature gametes, or gametocytes **4**. When a mosquito bites an infected person, gametocytes are taken up with blood and mature in the mosquito gut. Gametes fuse and form zygotes **5**. Meiosis produces cells that develop into new sporozoites that migrate to the insect's salivary glands and await transfer to a new host **6**.

Malaria symptoms usually start a week or two after a bite, when the infected liver cells rupture and release merozoites, metabolic wastes, and cellular debris into blood. Shaking, chills, a burning fever, and sweats follow. After the initial episode, symptoms may subside for a few weeks or even months. Infected people often feel healthy. However, ongoing infection damages the liver, spleen, and kidneys, clogs blood vessels, and cuts blood flow to the brain. The result is convulsions, coma, and eventual death.

Plasmodium cannot survive at low temperatures, so malaria is mainly a tropical disease. It remains common in Mexico, South and Central America, as well as Asia and the Pacific Islands, but the greatest toll is in Africa. One African child dies of malaria every 30 seconds.

Malaria has been a potent selective force on humans in Africa. As Section 17.7 explained, the allele that produces sickle-cell anemia was favored in African populations because it also provides protection against malaria.

Natural selection also acts on *Plasmodium*. The protist has recently become resistant to several antimalarial drugs. Over a longer time frame, it has evolved an amazing capacity to alter the behavior of its hosts. *Plasmodium* makes the mosquitoes that carry it more likely to feed several times a night, and thus more likely to bite several people. It also makes infected humans especially appetizing to a hungry mosquito when gametocytes are present. By manipulating its insect and human hosts, the protist maximizes chances that its offspring will reach a new host.

Take-Home Message *What is malaria?*

> Malaria is a deadly tropical disease caused by an apicomplexan, a parasite that is transmitted by mosquitoes.

> The parasite reproduces asexually in human liver cells and blood cells. Death of these cells causes the symptoms that characterize the disease.

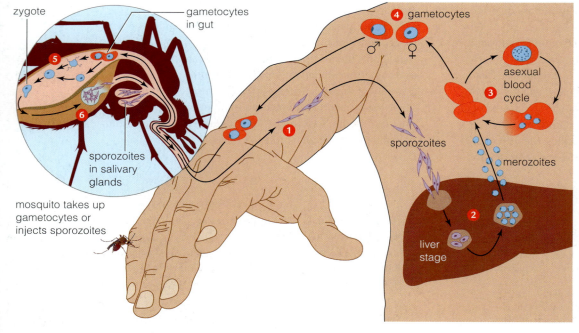

zygote — gametocytes in gut

sporozoites in salivary glands

mosquito takes up gametocytes or injects sporozoites

gametocytes **4**

asexual blood cycle

sporozoites

merozoites

liver stage

Figure 20.8 Animated Life cycle of one of the *Plasmodium* species that cause malaria.

1 Infected mosquito bites a human. Sporozoites enter blood, which carries them to the liver.

2 Sporozoites reproduce asexually in liver cells, mature into merozoites. Merozoites leave the liver and infect red blood cells.

3 Merozoites reproduce asexually in some red blood cells.

4 In other red blood cells, merozoites differentiate into gametocytes.

5 A female mosquito bites and sucks blood from the infected person. Gametocytes in blood enter her gut and mature into gametes, which fuse to form zygotes.

6 Meiosis of zygotes produces cells that develop into sporozoites. The sporozoites migrate to the mosquito's salivary glands.

20.7 Stramenopiles

> The stramenopile lineage includes heterotrophic water molds as well as single-celled diatoms and multicelled brown algae (both photosynthetic).

< Link to Accessory pigments 6.2

Stramenopile means "straw-haired" and refers to a shaggy flagellum that occurs during the life cycle of some members of this group. However, stramenopiles are defined mainly by genetic similarities, rather than visible traits.

Water molds are filamentous heterotrophs that were once mistakenly classified as fungi. Like fungi, they form a mesh of nutrient-absorbing filaments, but the two groups differ in many structural and genetic traits.

Most water molds decompose organic debris in aquatic habitats, but a few are parasites that have significant economic effects. Some grow as fuzzy white patches on fish in fish farms and aquariums (*left*). Others infect land plants, destroying crops and forests. Members of the genus *Phytophthora* are especially notorious. Their name means "plant destroyer." In the mid-1800s, one species destroyed Irish potato crops, causing a famine that killed and displaced millions of people. Today, another species is causing an epidemic of sudden oak death in Oregon, Washington, and California.

The closest relatives of water molds are two photosynthetic groups: diatoms and brown algae. Both have chloroplasts with a brown accessory pigment (fucoxanthin) that

Figure 20.10 Kelp "forest." These brown algae are the largest protists.

tints them olive, golden, or dark brown (Figure 20.9A). The chloroplasts evolved from a red alga.

Diatoms have a two-part silica shell, with upper and lower parts that fit together like a shoe box. Some cells live individually and others form chains. Diatoms can be found in damp soil, lakes, and seas. They are particularly abundant in cool waters. When marine diatoms die, their shells fall to the sea floor. Ancient sea floor that has been uplifted and is now land is the source of silica-rich diatomaceous earth. This substance is an inert powder of glassy bits. It is used in filters, abrasive cleaners, and as an insecticide that does not harm vertebrates (Figure 20.9B).

Brown algae are multicelled inhabitants of temperate or cool seas. In size, they range from microscopic filaments to giant kelps 30 meters (100 feet) tall. Giant kelps form forestlike stands in coastal waters of the Pacific Northwest (Figure 20.10). Like trees in a forest, kelps shelter a wide variety of other organisms.

Brown algae have commercial uses. Alginic acid from their cell walls is used to produce algins that serve as thickeners, emulsifiers, and suspension agents. You will find algins in ice cream, pudding, jelly beans, toothpaste, cosmetics, and many other products.

Figure 20.9 **A** Living diatom. Chloroplasts are visible through the glassy silica shell. See also Figure 20.2C. **B** Diatomaceous earth is shells of ancient diatoms. It kills crawling insects by scratching their surface so they lose water, dry out, and die.

---

brown alga Multicelled marine protist with a brown accessory pigment in its chloroplasts.
diatom Single-celled photosynthetic protist with brown accessory pigments in its choroplasts and a two-part silica shell.
water mold Protist that grows as nutrient-absorbing filaments.

Take-Home Message What are stramenopiles?

> Stramenopiles include diverse lineages that are united on the basis of their genetic similarity, rather than any visible traits.

> Water molds are filamentous decomposers and parasites.

> Diatoms are silica-shelled cells. Brown algae are multicellular. Both groups have a brown accessory pigment in chloroplasts that evolved from a red alga.

> Red algae and green algae belong to the same lineage as the land plants.
< Links to Plant cell walls 4.11, Photosynthetic pigments 6.2 and 6.3, Plant cell division 11.4

Red Algae Do It Deeper

Red algae are photosynthetic protists that are typically multicellular and tropical, although there are some single-celled species. Coralline algae (red algae with cell walls that contain calcium carbonate) are a component of some tropical coral reefs. Compared to brown algae or green algae, red algae can live at greater depths. In addition to chlorophyll *a*, they have phycobilins, pigments that absorb green light. This light penetrates deepest into water. Shallow-water red algae tend to have little phycobilin and appear green. Deep dwellers are almost black.

Figure 20.11 Animated Life cycle of a multicellular red alga (*Porphyra*).

❶ The haploid gametophyte is sheetlike.

❷ Gametes form at its edges.

❸ Fertilization produces a diploid zygote.

❹ The zygote develops into a diploid sporophyte.

❺ Haploid spores form by meiosis on the sporophyte body, and are released.

❻ Spores germinate and develop into a new gametophyte.

>> **Figure It Out** In the sheets of algae used to wrap sushi, are the cells haploid or diploid?   Answer: haploid

Red algae have many commercial uses. Agar, a polysaccharide extracted from cell walls of red algae, is used to keep baked goods and cosmetics moist, to set jellies, and as a vegetarian substitute for gelatin. Carrageenan, another polysaccharide, is added to soy milk and dairy products. Nori, dry sheets of the red alga *Porphyra*, wraps some kinds of sushi. Nori production is big business, with more than 130,000 tons harvested annually.

Like many other multicelled algae, *Porphyra* has an **alternation of generations**, a life cycle that alternates between haploid and diploid multicelled bodies (Figure 20.11). The sheetlike seaweed used as nori is the haploid body, or **gametophyte** ❶. Gametes form along edges of the sheets ❷. Fertilization of gametes produces a diploid zygote ❸. The zygote grows by mitosis into the diploid body, or **sporophyte**. In *Porphyra*, the sporophyte is a microscopic branching filament that grows on the shells of mollusks ❹. Some cells of the sporophyte undergo meiosis and produce haploid spores ❺. Germination of the spore is followed by growth and development of a new gametophyte, thus completing the life cycle ❻.

Green Algae

Green algae are photosynthetic protists with chloroplasts that have chlorophylls *a* and *b*. Most green algae live in fresh water, but some live in the ocean, in the soil, or on surfaces such as tree trunks. Single-celled green algae also partner with fungi to form the composite organisms called lichens.

Chlamydomonas, a flagellated single-celled green alga, lives in standing fresh water (Figure 20.12A). Haploid cells reproduce asexually when conditions favor growth. When nutrients become scarce, gametes form by mitosis. Fusion of two gametes produces a diploid zygote that has a thick protective wall (*right*). When conditions become favorable again, the zygote undergoes meiosis and produces four haploid, flagellated cells.

The colonial green alga *Volvox* also lives in lakes and ponds. A colony consists of flagellated cells joined together by thin cytoplasmic strands to form a whirling sphere (Figure 20.12B). New colonies form inside the parental sphere, which eventually ruptures and releases them.

Sheets of the multicelled species *Ulva* cling to rocks along marine coasts (Figure 20.12C). The sheets grow longer than your arm, but are no more than 40 microns thick. *Ulva* is commonly known as sea lettuce and is a popular food in Scotland. It has an alternation of generations with large, sheetlike bodies in both the haploid and diploid generations.

Studies of green algae helped biologists understand the mechanisms of photosynthesis. Section 6.3 explained how

A *Chlamydomonas* is a single celled species that uses its two flagella to swim in fresh water.

B *Volvox* colonies, each with many flagellated cells joined by thin strands of cytoplasm. New colonies are visible as bright green spheres inside each parent colony.

C Multicellular sheets of sea lettuce (*Ulva*). The long thin sheets are common along coasts in zones where there is little wave action.

Theodor Engelmann used filaments of a green alga to determine the most effective wavelengths of light for photosynthesis. Melvin Calvin used *Chlorella*, a single-celled green alga, to clarify the steps in the reactions we now call the Calvin–Benson cycle.

Evolutionary Connections to Land Plants

Red algae, green algae, and land plants all have a cell wall made of cellulose, store sugars as starch, and have chloroplasts that evolved from a cyanobacterial ancestor by primary endosymbiosis. Thus they are thought to be descended from a common ancestor.

The closest relatives of the land plants belong to a subgroup of freshwater green algae known as the charophyte algae. One modern member of this group, *Chara*, is native to freshwater habitats in Florida (Figure 20.12D). Like plants, and unlike most other green algae, *Chara* cells divide their cytoplasm by cell plate formation (Section 11.4) and have plasmodesmata, cytoplasmic connections between neighboring cells (Section 4.11).

"Green algae" does not refer to a clade because the term excludes land plants, which are on the same evolutionary branch as the charophyte algae.

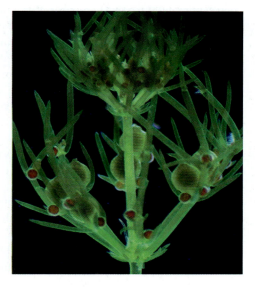

D *Chara*, a charophyte alga known as muskgrass or stinkweed for its strong odor.

Figure 20.12 A sampling of green algal diversity.

---

alternation of generations Of land plants and some protists, a life cycle in which haploid and diploid multicelled bodies form.
gametophyte Gamete-producing haploid body that forms in the life cycle of land plants and some protists.
green alga Photosynthetic protist that deposits cellulose in its cell wall, stores sugars as starch, and has chloroplasts containing chlorophylls *a* and *b*.
red alga Photosynthetic protist that deposits cellulose in its cell wall, stores sugars as starch, and has chloroplasts containing chlorophyll *a* and red pigments called phycobilins.
sporophyte Spore-forming diploid body that forms in the life cycle of land plants and some protists.

Take-Home Message What are red algae and green algae?

> Red algae and green algae are protists that belong to the same clade as the land plants. All members of this clade deposit cellulose in their cell wall and store excess sugars as starch.

> Red algae are mostly multicellular and marine. They have red accessory pigments that allow them to live at greater depths than other algae.

> Green algae include single-celled, colonial, and multicelled species. One subgroup, the charophyte algae, includes the closest living relatives of land plants.

> The amoebas and their relatives are shape-shifting heterotrophs. Many are solitary, but some display communal behavior and cell differentiation that hint at complexities to come in animals.

< Link to Pseudopods 4.10

Amoebozoans are one of the monophyletic groups now being carved out of the former kingdom Protista. Few amoebozoans have a cell wall, shell, or pellicle; nearly all undergo dynamic changes in shape. A compact blob of a cell can quickly send out pseudopods, move about, and capture food (Section 4.10).

Solitary Amoebas

The **amoebas** live as single cells. Figure 20.13 shows *Amoeba proteus*. Like most amoebas, it is a predator in freshwater habitats. Other amoebas can live in the gut of humans and other animals. Some gut-dwelling species do no harm or aid their host's digestive process. Others can cause disease. Each year, about 50 million people are affected by amebic dysentery after drinking water contaminated with cysts of a pathogenic amoeba.

Slime Molds

Slime molds are sometimes described as "social amoebas." There are two types, and both are common on the floor of temperate forests.

Plasmodial slime molds spend most of their life cycle as a plasmodium, a slimy, multinucleated mass. The plasmodium forms when a diploid cell undergoes mitosis many times without cytoplasmic division. It streams out along the forest floor feeding on microbes and organic matter (Figure 20.14). The plasmodium can be as big as a dinner plate. When food runs out, a plasmodium develops into spore-bearing fruiting bodies.

Figure 20.14 Plasmodial slime mold (*Physarum*). **A** These protists feed as a large multinucleated mass—a plasmodium—that oozes along the forest floor and over logs, devouring bacteria. **B** When food runs low, the mass forms spore-bearing structures.

Cellular slime molds such as *Dictyostelium discoideum* spend most of their existence as individual amoeba-like cells (Figure 20.15). Each cell eats bacteria and reproduces by mitosis ❶. If food runs out, thousands of cells stream together, forming a multicelled mass ❷. Environmental gradients in light and moisture induce the mass to crawl along as a cohesive multicelled unit often referred to as a "slug" ❸. Cells of the slug are held together by adhesion proteins and a secreted extracellular maxtrix. When the slug reaches a suitable spot, the cells differentiate and develop into a fruiting body: A stalk forms and a group of cells at its tip become spores ❹. Germination of a spore releases an amoeboid cell that starts the life cycle anew ❺.

Dictyostelium and other amoebozoans provide clues to how signaling pathways of multicelled organisms evolved. Coordinated behavior—an ability to respond to stimuli as a unit—is a hallmark of multicellularity. It requires cell-to-cell communication, which may have originated in

Figure 20.13 An amoeba. The cell has no fixed shape. It feeds or shifts position by extending lobes of cytoplasm (pseudopods).

amoeba Single-celled protist that extends pseudopods to move and to capture prey.
amoebozoan Shape-shifting heterotrophic protist with no pellicle or cell wall; an amoeba or slime mold.
cellular slime mold Amoeba-like protist that feeds as a single predatory cell; joins with others to form a multicellular spore-bearing structure when conditions are unfavorable.
plasmodial slime mold Protist that feeds as a multinucleated mass; forms a spore-bearing structure when enviromental conditions become unfavorable.

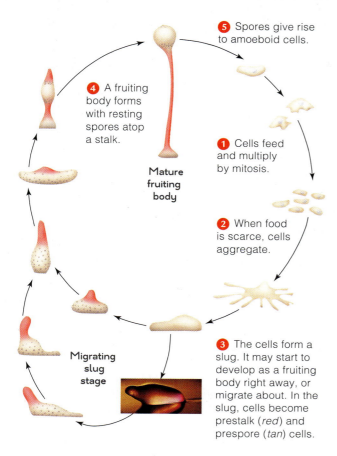

⑤ Spores give rise to amoeboid cells.

④ A fruiting body forms with resting spores atop a stalk.

Mature fruiting body

① Cells feed and multiply by mitosis.

② When food is scarce, cells aggregate.

③ The cells form a slug. It may start to develop as a fruiting body right away, or migrate about. In the slug, cells become prestalk (*red*) and prespore (*tan*) cells.

Migrating slug stage

Figure 20.15 Animated Life cycle of *Dictyostelium discoideum*, a cellular slime mold.

amoeboid ancestors. In *Dictyostelium*, a nucleotide called cyclic AMP is the signal that induces solitary amoeboid cells to stream together. It also triggers changes in gene expression. The changes cause some cells to differentiate into components of a stalk or into spores. Cyclic AMP also functions as the signal among cells in an animal body.

Intriguingly, molecular comparisons suggest that the fungi and animals are a monophyletic group. They also suggest that animals and fungi descended from an ancient amoebozoan-like ancestor.

Take-Home Message What are amoebozoans?

❯ Amoebozoans are heterotrophic protists with cells that lack a cell wall or pellicle, and so can constantly change shape.

❯ Amoebas live as single cells, usually in fresh water.

❯ Slime molds live on forest floors. Plasmodial slime molds feed as a big multinucleated mass. Cellular slime molds feed as single cells, but come together as a multicelled mass when conditions are unfavorable. Both types of slime molds form fruiting bodies that release spores.

Harmful Algal Blooms (revisited)

Harmful algal blooms affect every coastal region of the United States. As described in Section 20.1, blooms of the dinoflagellate *Karenia brevis* are common in the Gulf of Mexico. Other toxin-producing dinoflagellates cause problems along the Atlantic coast. Along the Pacific coast, population explosions of diatoms that produce domoic acid have a similar effect. Like breve-toxin, domoic acid binds to nerve cells and interferes with their function. It kills fish, seabirds, and marine mammals. Humans who eat domoic acid–tainted shellfish or crabs typically suffer from headaches, dizziness, and con-fusion. High doses of domoic acid can kill brain cells, causing permanent loss of short-term memory and, in rare cases, coma or death.

Keeping harmful algal toxins out of the human food supply requires constant vigilance. These toxins have no color or odor, and are unaffected by heating or freezing. Government agencies use laboratory tests to detect harmful algal toxins in water samples and shellfish. When the toxins reach a threatening level, a shore is closed to shellfishing and warning signs are posted (*above right*).

Harmful algal blooms have devastating economic effects. A 2005 bloom of toxin-producing dinoflagellates caused an estimated $18 million loss in shellfish sales in Massachusetts alone. Blooms of *Karenia brevis* cost Florida $19–32 million a year.

Even blooms of nontoxic algae can have negative environmental effects. When huge numbers of these cells die, bacterial decomposers go to work breaking down their remains. Metabolic activity of the bacteria can deplete the water of oxygen, causing fish and other aquatic animals to smother.

What causes algal blooms? The nutrient content of the water plays a major role. Just like land plants, algae require certain nutrients. If you fertil-ize a houseplant or a lawn, a spurt of growth results. Similarly, the addition of nutrients to an aquatic habitat encourages algal cell divisions. We do not deliberately fertilize our waters, but fertilizers that drain off from croplands and lawns, and nutrient-rich wastes from animals at factory farms, get into rivers and are carried to the sea. Add sewage from human communities, and you have the nutrients required for runaway algal growth.

How Would You Vote? Preventing nutrient pollutions from entering coastal water could help reduce the incidence of algal blooms, but taking preventative mea-sures can be costly for farmers, developers, industries, and water treatment plants. Do you favor tightening regulations governing nutrient discharges into coastal waters? See CengageNow for details, then vote online (cengagenow.com).

Summary

Section 20.1 An **algal bloom** is a population explosion of an aquatic **protist**, or another aquatic microorganism. **Toxins** released during some algal blooms can harm wildlife and endanger human health.

Section 20.2 Protists are a collection of mostly single-celled eukaryotes. Many have chloroplasts that evolved from cyanobacteria or from another protist. The dominant stage of the life cycle may be haploid or diploid. The protists are not a natural group, but rather a collection of lineages, some of them only distantly related to one another.

Section 20.3 Flagellated protozoans are single cells with no cell wall. A protein covering, or **pellicle**, helps maintain the cell's shape.

Diplomonads and parabasalids are adapted to oxygen-poor habitats and do not have mitochondria. Members of both groups include species that infect humans.

Trypanosomes are parasites with a single mitochondrion. Insects transmit trypanosomes that cause human diseases. The related **euglenoids** typically live in fresh water. They have a **contractile vacuole** that squirts out excess water. Some have chloroplasts that evolved by secondary endosymbiosis from a green alga.

Section 20.4 Foraminiferans are single cells with a chalky shell. Deposits of their remains are mined for chalk and limestone. **Radiolarians** are single-celled and have a glassy shell. Both groups are marine heterotrophs and may be part of **plankton**. Long cytoplasmic extensions stick out through the porous shell and capture prey.

Sections 20.5, 20.6 Tiny sacs (alveoli) beneath the plasma membrane characterize **alveolates**. All members of this group are single-celled. **Dinoflagellates** are whirling aquatic heterotrophs and autotrophs with cellulose plates. Some are bioluminescent. The **ciliates** are aquatic predators and parasites with many cilia. **Apicomplexans** live as parasites in the cells of animals. Mosquitoes transmit the apicomplexan that causes malaria.

Section 20.7 Water molds are decomposers and parasites that grow as a mesh of absorptive filaments. Some parasitic species are important plant pathogens. Genetic similarities unite the water molds with diatoms and brown algae as stramenopiles.

Diatoms are silica-shelled photosynthetic cells. Deposits of ancient diatom shells are mined as diatomaceous earth. Diatoms contain the pigment fucoxanthin, as do **brown algae**, which include microscopic strands and giant kelps, the largest protists. Brown algae are the source of algins, compounds used as thickeners and emulsifiers.

Section 20.8 Most **red algae** are multicelled and marine. Accessory pigments called phycobilins allow them to capture light even in deep waters. Red algae are commercially important as the source of agar, carrageenan, and as dry sheets (nori) used for wrapping sushi. **Green algae** may be single cells, colonial, or multicelled. They are the closest relatives of land plants.

Like land plants, some of the multicelled algae have an **alternation of generations**. In this type of life cycle, two kinds of multicelled bodies form: a diploid, spore-producing **sporophyte** and a haploid gamete-producing **gametophyte**.

Section 20.9 Amoebozoans are heterotrophic free-living **amoebas** and slime molds. The **plasmodial slime molds** feed as a multinucleated mass. Amoeba-like cells of **cellular slime molds** aggregate when food is scarce and form multicelled fruiting bodies that disperse resting spores. Animal signaling mechanisms may have started in amoebozoan ancestors.

Self-Quiz   Answers in Appendix III

1. All flagellated protozoans _____ .
   a. lack mitochondria   c. live as single cells
   b. are photosynthetic   d. cause disease

2. Deposits of shells from ancient _____ are mined as chalk and limestone.
   a. dinoflagellates   c. radiolarians
   b. diatoms   d. foraminiferans

3. The presence of a contractile vacuole indicates that a single-celled protist _____ .
   a. is marine   c. is photosynthetic
   b. lives in fresh water   d. secretes a toxin

4. Cattle benefit when _____ in their gut help them digest plant material.
   a. trypanosomes   c. ciliates
   b. diatoms   d. foraminiferans

5. An insect bite can transmit a disease-causing _____ to a human host.
   a. trypanosome   c. ciliate
   b. apicomplexan   d. both a and b

6. _____ are the closest protistan relatives of the fungi and animals.
   a. Stramenopiles   c. Apicomplexans
   b. Radiolarians   d. Amoebozoans

7. Accessory pigments of _____ allow them to carry out photosynthesis at greater depths than other algae.
   a. euglenoids   c. brown algae
   b. green algae   d. red algae

8. The protist that causes malaria is _____ .
   a. multicellular   b. a single cell   c. colonial

Data Analysis Activities

Tracking Changes in Algal Blooms Reports of fish kills along Florida's southwest coast date back to the mid-1800s, suggesting that algal blooms are a natural phenomenon in this region. However, University of Miami researchers suspected that a rise in nutrient delivery from land has contributed to an increase in the abundance of the dinoflagellate *K. brevis*. Since the 1950s, the population of coastal cities in southwestern Florida has soared and the amount of agriculture has increased. Did these changes add nutrients that favor *K. brevis* growth to nearshore waters? To find out, the researchers looked at records for coastal waters that have been monitored for more than 50 years. Figure 20.16 shows the average abundance and distribution of *K. brevis* during two time periods: 1954–1963 and 1994–2002.

1. How did the average concentration of *K. brevis* in waters less than 5 kilometers from shore change between the two time periods?
2. How did the average concentration of *K. brevis* in waters more than 25 kilometers from shore change between the two time periods?
3. Does this data support the hypothesis that human activity increased the abundance of *K. brevis* by adding nutrients to coastal waters?
4. Suppose the two graph lines became farther apart as distance from the shore increased. What would that suggest about the nutrient source?

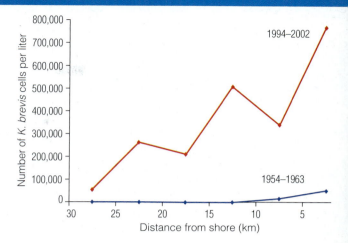

Figure 20.16 Average concentration of *K. brevis* cells detected at various distances from the shore during two time periods: 1954–1963 (*blue* line) and 1994–2002 (*red* line). Samples were collected from offshore waters between Tampa Bay and Sanibel Island.

9. The _____ are important plant pathogens.
   a. dinoflagellates
   b. ciliates
   c. water molds
   d. slime molds

10. Silica-rich shells of ancient diatoms are the source of diatomaceous earth that can be used _____ .
   a. to thicken foods
   b. as a gelatin substitute
   c. as a fertilizer
   d. as an insecticide

11. The sporophyte of a multicellular alga _____ .
   a. is haploid
   b. is a single cell
   c. produces spores
   d. produces gametes

12. Where would you find a cellular slime mold?
   a. on the forest floor
   b. in a tropical sea
   c. in an animal gut
   d. in a mountain lake

13. The organism that causes the sexually transmitted disease trichomoniasis is a _____ .
   a. flagellated protozoan
   b. radiolarian
   c. ciliate
   d. apicomplexan

14. All green algae _____ .
   a. have a cell wall
   b. are marine
   c. are multicellular
   d. all of the above

15. Match each item with its description.
   ___diplomonad   a. cause of leishmaniasis
   ___apicomplexan   b. silica-shelled producer
   ___trypanosome   c. unwalled cell with pseudopods
   ___diatom   d. anaerobic, no mitochondria
   ___brown alga   e. closest relative of land plants
   ___red alga   f. multicelled, with fucoxanthin
   ___green alga   g. cause of malaria
   ___amoeba   h. deep dweller with phycobilins

Additional questions are available on **CENGAGENOW**.

Critical Thinking

1. Suppose you vacation in a developing country where sanitation is poor. Having read about parasitic flagellates in water and damp soil, what would you consider safe to drink? What foods might be best to avoid or which food preparation methods might make them safe to eat?

2. The water in abandoned swimming pools often turns green. If you took a drop of this water and examined it under the microscope, you would see many green flagellated cells. What additional information would you need to determine which protist group the cells belong to?

3. The most common "snow alga," *Chlamydomonas nivalis* (*right*), lives on glaciers. It is a green alga, but it has so many carotenoid pigments that it appears red. Think about the intense sunlight striking its icy habitat during the summer. Besides their role in photosynthesis, what other function might these light-absorbing carotenoids serve?

Animations and Interactions on **CENGAGENOW**:
❯ Body plan of a euglenoid; Action of a contractile vacuole; Body plan of a ciliate; Life cycle of an apicomplexan; Life cycle of a red alga; Life cycle of a cellular slime mold.

‹ Links to Earlier Concepts

Section 20.8 introduced the green algal group that is the closest relative of plants. Section 12.5 introduced you to the general plant life cycle. Here you will see specfic examples. You will learn about the evolution of cell walls with lignin (4.11) and a waxy cuticle perforated by stomata (6.8). You will learn about plant fossils (16.5) and coevolution with pollinators (17.13).

Key Concepts

Adaptive Trends Among Plants

Plants evolved from an aquatic green alga. Over time, new traits evolved that made them increasingly adapted to life in dry climates. The process of adaptation involved changes in plant structure, life cycle, and reproductive processes.

The Bryophytes

Bryophytes are three lineages of low-growing plants. They are the only modern plants in which the gamete-producing body dominates the life cycle, and the spore-producing body is dependent on it. Bryophytes require water for fertilization to occur, and disperse by releasing spores.

21 Plant Evolution

The world's great forests influence life in profound ways. They take up carbon dioxide from the air and release oxygen. At the same time, forests act like giant sponges that absorb water from rain, then release it slowly. By wicking up water and holding soil in place, forests prevent erosion, flooding, and sedimentation that can disrupt rivers, lakes, and reservoirs. Remove trees, and exposed soil washes away or loses nutrients.

All countries rely on their forests for fuel and lumber. The photo at the *left* shows a heavily logged forest in the Canadian province of British Columbia. The developing nations of Asia, Africa, and Latin America have fast-growing populations and high demands for food, fuel, and lumber. To meet their own needs and to create products for sale in the global market, these nations turn to their main resource: local tropical forests. These forests are a biological treasure; they have held 50 percent or more of all land-dwelling species for at least 10,000 years. In less than four decades, humans have destroyed more than half of the world's tropical forests by logging.

Deforestation affects evaporation rates, runoff, and regional patterns of rainfall. In tropical forests, most of the water vapor in the air is released from trees. In heavily logged regions, annual rainfall declines, Rain that does fall swiftly drains away from the exposed, nutrient-poor soil. As the region gets hotter and drier, the fertility and moisture content of the soil decrease.

Like many environmental issues, large-scale deforestation can seem like an overwhelming problem. Wangari Maathai, a Kenyan biologist, suggested a simple solution: Plant new trees (Figure 21.1). Maathai founded the Green Belt Movement, which began by giving poor women in Kenya the resources they need to plant and care for trees. Maathai reasoned that a woman who plants a tree seedling, cares for it, and watches it grow, will value and protect that tree. As many women plant trees, they will observe the improved environment and understand that they can each be a force for positive change.

Figure 21.1 Biologist and Nobel Peace Prize winner Wangari Maathai (*right*) holding one of the tree seedlings her Green Belt Movement raised and planted to restore forests in Kenya.

Maathai began her tree-planting campaign in 1977. By 2007, members of her organization had planted 40 million trees throughout Africa. These trees provide shade, reduce erosion, and have begun to restore the diverse forests that had been disappearing. Just as important, Maathai's practical solution to a seemingly overwhelming problem has inspired others to do what they can to stem the tide of environmental degradation. In 2004, the Nobel Committee honored Maathai's efforts with the Nobel Peace Prize. As Maathai recognized, we are dependent on plants for our well-being. This chapter explains how these green partners of ours evolved and diversified.

Seedless Vascular Plants
Vascular plants have internal pipelines that carry materials through the body and provide structural support. A body that makes spores dominates their life cycle. Seedless vascular plants such as ferns require water for fertilization to occur and they disperse by releasing spores.

Gymnosperms
Seed plants make pollen grains that allow fertilization to occur even in dry times. They also make eggs inside an ovule that develops into a seed. Gymnosperms such as pine trees are seed plants with "naked" seeds, meaning their seeds do not form inside an ovary.

Angiosperms
Angiosperms, or flowering plants, are the most recently evolved seed plants. They alone make flowers, and their seeds form inside a floral ovary that develops into a fruit. Angiosperms are the most widely dispersed and diverse group of plants.

> Plants evolved from a green alga and underwent an adaptive radiation on land. Structural and developmental changes adapted plants to increasingly drier habitats.
‹ Link to Green algae 20.8

The invasion of land got under way about 500 million years ago. Green algae grew at the water's edge, and one lineage, the charophytes, gave rise to the first land plants. The defining trait of land plants is a multicelled embryo that develops within tissues of the parental plant and is nourished by the parent. For this reason, the clade of land plants is called the **embryophytes**.

Structural Adaptations to Life on Land

Plants have unique traits that adapt them to life on land. A multicelled green alga absorbs the water it needs across its body surface. Water also buoys algal parts, helping an alga stand upright. In contrast, land plants face the threat of drying out and must hold themselves upright.

The earliest evolving plant lineages are referred to as bryophytes. Mosses are one example. Some bryophytes, and all other land plants, have a **cuticle**, a secreted waterproof coating that reduces evaporative water loss. Tiny pores called **stomata** (singular stoma) extend across the cuticle and outer leaf surface. Stomata open and close as needed to balance the competing demands for water conservation and gas exchange with air outside the plant.

vascular tissue (a leaf vein)   layer of waxy cuticle

xylem

phloem

cuticle

stoma

Figure 21.2 Diagram of a vascular plant leaf in cross-section, showing some traits that contribute to the success of this group. The micrograph on the *right* shows a stoma, a pore at the surface of a squash plant's leaf. Photosynthetic cells on either side of the stoma control its opening and closing to regulate water loss and gas exchange.

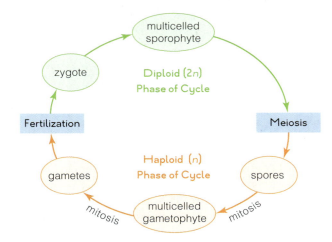

multicelled sporophyte

zygote

Diploid (2n) Phase of Cycle

Fertilization

Meiosis

Haploid (n) Phase of Cycle

gametes

spores

mitosis

multicelled gametophyte

mitosis

Figure 21.3 Generalized plant life cycle.

Bryophytes have structures that hold them in place, but these are not true roots; those evolved later. True roots anchor plants and also contain **vascular tissue**: internal pipelines that transport water and nutrients through a plant body. **Xylem** is the vascular tissue that distributes water and mineral ions. **Phloem** is the vascular tissue that distributes sugars made by photosynthetic cells. Of the 295,000 or so modern plant species, more than 90 percent have xylem and phloem. They are known as the **vascular plants**.

A variety of adaptations contribute to the success of vascular plants. Vascular tissues are reinforced by **lignin**, an organic compound that stiffens cell walls. Vascular tissues with lignin not only distribute materials, they also help plants stand upright and allow them to branch. Vascular tissue extends through stem, roots, and leaves. Evolution of leaves increased the surface area for capturing sunlight and exchanging gases. Figure 21.2 shows a cross-section through a vascular plant leaf.

The Plant Life Cycle

Life cycle changes also adapted vascular plants to life in drier habitats. Like some algae, all land plants have an alternation of generations (Figure 21.3). A diploid body, the sporophyte, makes spores. A plant spore is a walled haploid cell. The spore germinates, then divides by mitosis and develops into a haploid body, or gametophyte. A gametophyte makes gametes (eggs and sperm). When gametes come together at fertilization, they form a zygote that develops into a new sporophyte.

The gametophyte is the largest and longest-lived part of the bryophyte life cycle, but the sporophyte dominates the life cycle of vascular plants. Dry conditions favor an increased emphasis on spore production; spores withstand drying out better than gametes do.

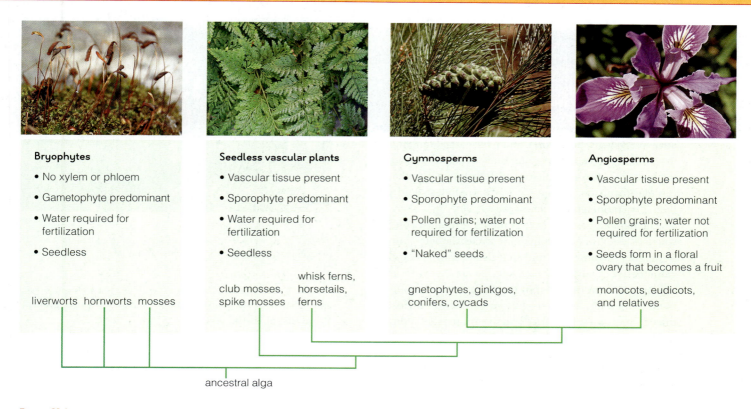

Bryophytes
- No xylem or phloem
- Gametophyte predominant
- Water required for fertilization
- Seedless

liverworts hornworts mosses

Seedless vascular plants
- Vascular tissue present
- Sporophyte predominant
- Water required for fertilization
- Seedless

club mosses, spike mosses

whisk ferns, horsetails, ferns

Gymnosperms
- Vascular tissue present
- Sporophyte predominant
- Pollen grains; water not required for fertilization
- "Naked" seeds

gnetophytes, ginkgos, conifers, cycads

Angiosperms
- Vascular tissue present
- Sporophyte predominant
- Pollen grains; water not required for fertilization
- Seeds form in a floral ovary that becomes a fruit

monocots, eudicots, and relatives

ancestral alga

Figure 21.4 Traits of the four major plant groups and relationships among them. Note that the bryophytes are a collection of lineages, rather than a clade.

Pollen and Seeds

Evolution of new reproductive traits gave the lineage of vascular plants known as the seed plants, a competitive edge in many habitats. Seed plants do not release spores. Instead, their spores give rise to gametophytes inside specialized structures on the sporophyte body.

A **pollen grain** is a walled, immature male gametophyte of a seed plant. It is released into the environment, where it can be transported to another plant by winds or animals even in the driest of times. By contrast, plants that do not make pollen (bryophytes and seedless vascular plants) can only reproduce when a film of water allows their sperm to swim to eggs.

Fertilization of a seed plant also takes place in structures on the sporophyte body. It results in development of a **seed**—an embryo sporophyte with a supply of nutritive tissue inside a waterproof seed coat. Seed plants disperse a new generation by releasing seeds.

There are two lineages of seed plants. The gymnosperms were the first to evolve. A pine tree is a familiar gymnosperm. Angiosperms, the flowering plants, branched off from a gymnosperm lineage and became the most widely distributed and diverse plant group. They alone make flowers and release their seeds inside a fruit. Figure 21.4 summarizes the relationships among the major plant groups and the traits of each group.

cuticle Secreted covering at a body surface.
embryophyte Member of the land plant clade.
lignin Material that stiffens cell walls of vascular plants.
phloem Plant vascular tissue that distributes sugars.
pollen grain Male gametophyte of a seed plant.
seed Of seed plants, embryo sporophyte and nutritive tissue inside a waterproof coat; a mature ovule.
stoma Opening across a plant's cuticle and epidermis; can be opened for gas exchange or closed to prevent water loss.
vascular plant Plant with xylem and phloem.
vascular tissue Internal pipelines of xylem and phloem in the body of a vascular plant.
xylem Plant vascular tissue that distributes water and dissolved mineral ions.

> ### Take-Home Message **What adaptive trends shaped plants?**
>
> ❯ Early plant lineages (bryophytes) have a life cycle dominated by a haploid gametophyte. In most bryophytes, a waxy cuticle and stomata minimize evaporative water loss.
>
> ❯ Most plants have a life cycle dominated by a sporophyte with vascular tissue. Like bryophytes, the oldest vascular plant lineages have flagellated sperm and disperse by releasing spores.
>
> ❯ Seed plants, the most recently evolved plant lineage, can reproduce even in dry times because they make pollen. They disperse by releasing seeds, not spores.

21.3 The Bryophytes

> Bryophytes are low-growing plants that require the presence of water droplets to reproduce sexually.
> A haploid gametophyte dominates the bryophyte life cycle.

Bryophyte Characteristics

The term **bryophyte** refers to members of three separate lineages: mosses, hornworts, and liverworts. Some mosses have tubes that transport water and sugar, but none of the bryophytes have the lignin-stiffened pipelines that the vascular plants do. As a result, few bryophytes stand more than 20 centimeters (8 inches) tall.

The haploid gametophyte is the largest, most conspicuous body in a bryophyte life cycle. Gametes form in a chamber (a gametangium) that develops in or on the gametophyte's surface. Some species are bisexual; eggs and sperm form in different gametangia on the same plant. Other species make eggs and sperm on different plants. Sperm are flagellated and swim to eggs, although insects and mites sometimes aid their journey.

The diploid sporophyte is unbranched and remains attached to the gametophyte even when mature. It makes

Figure 21.6 Cutting blocks of peat for use as fuel in Ireland. Peat is the compressed, carbon-rich remains of peat moss.

spores by meiosis inside a chamber called a sporangium. Wind disperses the spores.

Many of the 24,000 or so bryophytes live in constantly moist places. Others can tolerate periodic drought. They shrivel up and become dormant when water gets scarce,

Figure 21.5 Animated Life cycle of a common moss (*Polytrichum*).

❶ The leafy green part of a moss is the gametophyte.

❷ It supports a sporophyte (stalk and capsule).

❸ Spores form by meiosis in the capsule, are released, and drift with the winds.

❹ Spores develop into gametophytes that produce eggs or sperm in gametangia at their tips.

❺ Sperm released from tips of sperm-producing gametophytes swim through water to eggs at tips of egg-producing gametophytes.

❻ Fertilization produces a zygote.

❼ The zygote grows and develops into a sporophyte while remaining attached to and nourished by its egg-producing parent.

Diploid (2n) Phase

zygote

sporophyte (2n)

Fertilization

Meiosis

gametophyte (n)

spore (n)

male gametophyte

sperm

female gametophyte

egg

Haploid (n) Phase

then resume growth when rains return. Drought tolerance and wind-dispersed spores make bryophytes important pioneer species in rocky places. Decomposition of the bryophytes helps form and improve soil, allowing less hardy plants to take root.

Most bryophytes can reproduce asexually by fragmentation. A piece breaks off and develops into a new plant. Unlike sexual reproduction, fragmentation can take place even during dry times.

Mosses

Mosses are the most familiar bryophytes. Figure 21.5 shows the life cycle of one common group (*Polytrichum*). The gametophyte has leaflike green parts that grow from a central stalk ❶. Threadlike **rhizoids** hold the gametophyte in place. Unlike the roots of vascular plants, rhizoids do not distribute water or nutrients. These resources must be absorbed across the gametophyte's leafy surface.

The moss sporophyte spends its life attached to the gametophyte, which it depends on for nourishment. The sporophyte consists of a capsule (a sporangium) on a stalk ❷. Meiosis of cells inside the capsule yields spores ❸. After dispersal by the wind, a spore germinates and grows into a gametophyte. The moss we are using as our example has separate sexes, with each gametophyte producing eggs or sperm ❹. Other mosses are bisexual. In either case, rain triggers the release of flagellated sperm that swim through a film of water to eggs ❺. Fertilization occurs in the egg chamber, producing a zygote ❻ that develops into a new sporophyte ❼.

Mosses include about 15,000 species. Of these, the 350 or so species of peat moss (*Sphagnum*) are the most economically important. Peat mosses are the dominant plants in peat bogs that cover hundreds of millions of acres in high-latitude regions of Europe, Asia, and North America. Many peat bogs have persisted for thousands of years, and layer upon layer of plant remains have become compressed as a carbon-rich material called **peat**. Blocks of peat are cut, dried, and burned as fuel, especially in Ireland (Figure 21.6).

Freshly harvested peat moss is also an important commercial product. The moss is dried and added to planting mixes to help soil retain moisture. Dead strands of peat moss absorb water, then release it slowly, keeping plant roots moist.

sporophyte

gametophyte

Figure 21.7 A liverwort (*Marchantia*).
A Ribbonlike gametophyte with stalked structures that produce eggs for sexual reproduction.
B Asexual reproduction. Clumps of cells form by mitosis in cups on the gametophyte surface. After raindrops knock the clumps out of the cup, they develop into new plants.

Figure 21.8 Hornwort, with a photosynthetic sporophyte that can survive even after death of the gametophyte.

Liverworts and Hornworts

Liverworts commonly grow in moist places, often alongside mosses. In most of the 6,000 or so species, the gametophyte is flattened and attaches to soil by rhizoids. The widespread genus *Marchantia* forms eggs and sperm on separate plants in gametangia that are elevated on stalks above the main body (Figure 21.7A). In addition to fragmentation, members of this genus also reproduce asexually by producing small clumps of cells in cups on the gametophyte surface (Figure 21.7B). Some *Marchantia* species can be pests in commercial greenhouses. Liverwort infestations are difficult to eradicate because the tiny spores can persist even after all plants are killed.

The 150 species of hornworts are named for a pointy, hornlike sporophyte that can be several centimeters tall (Figure 21.8). The base of the sporophyte is embedded in gametophyte tissues and spores form in a capsule at its tip. Unlike the sporophytes of mosses and liverworts, those of hornworts grow continually from their base, and can survive even after the death of the gametophyte.

---

bryophyte Member of an early plant lineage with a gametophyte-dominant life cycle; for example, a moss.
moss Nonvascular plant with a leafy green gametophyte and an attached, dependent sporophyte consisting of a capsule on a stalk.
peat Carbon-rich moss remains; can be dried for use as fuel.
rhizoid Threadlike structure that anchors a bryophyte.

Take-Home Message What are bryophytes?

❭ Bryophytes include three lineages of plants: mosses, liverworts, and hornworts. All are low-growing plants with no lignin-reinforced vascular tissues. All have flagellated sperm that require a film of water to swim to eggs, and all disperse by releasing spores.

❭ Among land plants, only the bryophytes have a life cycle in which the gametophyte is dominant. The sporophyte remains attached to the gametophyte even when mature.

> Lignified vascular tissue allowed the evolution of larger sporophyte bodies in the ferns and their relatives.
‹ Link to Plant cell walls 4.11

Some mosses have internal pipelines that transport fluid within their body. However, only the vascular plants have lignin-strengthened vascular tissue with xylem and phloem. This innovation in support and plumbing allowed the evolution of the larger, more complex sporophytes that dominate the life cycle of all vascular plants.

Vascular plants evolved from a bryophyte, and the oldest lineages, the **seedless vascular plants**, share traits with this group. They have flagellated sperm that swim to eggs, and they disperse a new generation by releasing spores, not seeds. Club mosses, horsetails, and ferns are examples of seedless vascular plants.

Club Mosses

The 400 or so species of club mosses occur worldwide. Members of the genus *Lycopodium* frequently grow on the floor of temperate forests. The sporophytes resemble miniature pine trees and are known as ground pines (Figure 21.9A). The plant has a horizontal stem, or **rhizome**, that runs along the ground. Roots and upright stems with tiny leaves grow from the rhizome. When a *Lycopodium* plant is several years old, it begins making spores seasonally. The spores form on a strobilus, a soft cone-shaped structure composed of modified leaves.

Lycopodium is gathered from the wild for a variety of uses. Stems are used in wreaths and bouquets. Spores have a waxy coating and are sold as "flash powder" for creating special effects. When spores are sprayed out as a

Figure 21.10 Horsetail (*Equisetum*). **A** Photosynthetic stems with leaflike branches. **B** Spore-bearing strobilus.

fine mist and ignited, they produce a bright flame (Figure 21.9B). *Lycopodium* may also have medicinal properties. It is used as an herbal medicine, and compounds made by the plant may help fight cancer or memory loss.

Horsetails and Rushes

The 25 or so species of the genus *Equisetum* are commonly known as horsetails or rushes. The sporophyte has a rhizome that gives rise to underground roots and to upright hollow stems with tiny nonphotosynthetic leaves. In some species the stems have few or no branches, while other species have thin, leafless branches that extend out in whorls around the stem (Figure 21.10A). Depending on the species, strobili form at tips of photosynthetic stems or on special chlorophyll-free stems (Figure 21.10B).

Silica deposits inside cell walls support the plant and give stems a sandpapery texture that helps deter grazers. Before scouring powders and pads were widely available, people used some *Equisetum* stems as pot scrubbers; thus the common name "scouring rush."

Ferns—The Most Diverse Seedless Plants

With 12,000 or so species, ferns are the most diverse seedless vascular plants. All but 380 or so species live in the tropics. A typical fern sporophyte has fronds (leaves) and roots that grow from rhizomes (Figure 21.11 ❶). Ferns do not have strobili. Instead, **sori** (singular, sorus), clusters of spore-forming capsules, develop on the lower surface of fronds ❷. When the capsules pop open, wind disperses

Figure 21.9 Club moss. **A** *Lycopodium* sporophyte. Waxy yellow spores are produced on the surface of strobili, conelike structures composed of modified leaves. **B** *Lycopodium* spores in use as a special effect at a concert. The towers of fire are created by a device that blows a puff of spores out and ignites them.

---

epiphyte Plant that grows on another plant but does not harm it.
rhizome Stem that grows horizontally along or under the ground.
seedless vascular plant Plant such as a fern or horsetail that has vascular tissue and disperses by producing spores.
sorus Cluster of spore-producing capsules on a fern leaf.

Figure 21.11 Animated
Life cycle of a fern.

1 The familiar leafy form is the diploid sporophyte.

2 Meiosis in cells on the underside of fronds produces haploid spores.

3 After their release, the spores germinate and grow into tiny gametophytes that produce eggs and sperm.

4 Sperm swim to eggs and fertilize them, forming a zygote.

5 The sporophyte begins its development attached to the gametophyte, but it continues to grow and live independently after the gametophyte dies.

the spores. Each spore grows into a tiny gametophyte a few centimeters wide **3**. Eggs and sperm form in chambers on the gametophyte's underside. When it rains, sperm swim through a film of water to reach and fertilize the eggs **4**. The resulting zygote develops into a new sporophyte, and its parental gametophyte dies **5**.

Fern sporophytes vary greatly in size (Figure 21.12). Some floating ferns have fronds only 1 millimeter long, but tree ferns can be 25 meters (80 feet) high. Many tropical ferns are **epiphytes**, plants that attach to and grow on a trunk or branch of another plant but do not withdraw any nutrients from it.

Ferns are sold as potted plants, and the trunks of some tree ferns are shredded to produce a potting medium used to grow orchids. Fiddleheads, the young, tightly coiled fronds of some ferns, are harvested from the wild as food.

Take-Home Message What are seedless vascular plants?

❯ Sporophytes with vascular tissue in their roots, stems, and leaves dominate the life cycle of seedless vascular plants.

❯ Tiny, independent gametophytes produce sperm that swim to eggs through films or droplets of water.

❯ Ferns are the most diverse seedless vascular plants.

Figure 21.12 A sampling of fern diversity. **A** The floating fern *Azolla pinnata*. The whole plant is not as wide a finger. Chambers in the leaves shelter nitrogen-fixing cyanobacteria. Southeast Asian farmers grow this species in rice fields as a natural alternative to chemical fertilizers. **B** Bird's nest fern (*Asplenium nidus*), one of the epiphytes. **C** Lush forest of tree ferns (*Cyathea*) in Australia's Tarra-Bulga National Park. A tree fern's "trunk" is an enlarged rhizome.

> Seed-bearing trees dominate today's forests, but before they evolved, forests of seedless vascular plants stood tall.
> Evolution of seed plants set off a great adaptive radiation.
❮ Link to Geologic time scale 16.6

From Tiny Branchers to Coal Forests

The oldest fossils of vascular plants are spores that date to about 450 million years ago, during the late Ordovician period (Figure 21.13). Early vascular plants such as *Cooksonia* (*left*) stood only a few centimeters high and had a simple branching pattern, with no leaves or roots. Spores formed at branch tips. By the early Devonian, taller species with a more complex branching pattern were common worldwide (Figure 21.14).

Fossil *Cooksonia*

As the Devonian continued, some seedless vascular plants evolved a taller structure. The oldest forest we know about existed about 385 million years ago in what is now upstate New York. Fossil stumps and fronds discovered at this site indicate that the main plants of this forest stood about 8 meters (26 feet high) and resembled modern tree ferns in their structure.

Later, during the Carboniferous period (350–299 million years ago), ancient relatives of club mosses and horsetails evolved into massively stemmed giants (Figure 21.15). Some stood 40 meters (more than 130 feet) high. After these forests first formed, climates changed, and the sea level rose and fell many times. When the waters receded, the forests flourished. After the sea moved back in, submerged trees became buried in sediments that protected them from decomposers. Layers of sediments accumulated one on top of the other. Their weight squeezed the water out of the saturated, undecayed remains, and the compaction generated heat. Over time, pressure and heat transformed the compacted organic remains into **coal**.

Coal is one of our premier fossil fuels. It took millions of years of photosynthesis, burial, and compaction to form each seam of coal. Often you will hear about annual production rates for coal or some other fossil fuel. In fact, we do not "produce" these materials. No one is making more coal. Coal is a nonrenewable source of energy, and one that we are well on the way to depleting.

Figure 21.14 Painting of a Devonian swamp, with branching seedless vascular plants.

Rise of the Seed Plants

The first gymnosperms evolved from a seedless ancestor late in the Devonian period. Cycads and ginkgos were among the earliest gymnosperm lineages. Conifers such as pine trees evolved a bit later. Angiosperms, or flowering plants, descended from a gymnosperm ancestor by about 120 million years ago, during the reign of the dinosaurs (Figure 21.16).

Reproductive traits of seed-bearing plants put them at an advantage in dry habitats. Gametophytes of seedless vascular plants develop from spores that were released into the environment. By contrast, gametophytes of seed plants form inside reproductive parts on a sporophyte body (Figure 21.17). Sperm-producing gametophytes (pollen grains) develop from spores that form inside **pollen sacs**. These spores are called **microspores**. Egg-producing gametophytes develop from **megaspores** that form inside a protective chamber, or **ovule**.

A sporophyte releases pollen grains, but holds onto its eggs. Wind or animals can deliver pollen from one seed plant to the ovule of another, a process called **pollination**.

Bryophytes evolve, diversify; seedless vascular plants evolve.	Diversification of seedless vascular plants.	First treelike plants (fern relatives), first seed plants.	Giant horsetails, club mosses, and relatives in swamp forests. Conifers arise late in the period.	Ginkgos, cycads evolve. Most horsetails and club mosses die off by end of the period.	Adaptive radiations of ferns, cycads, conifers; by start of Cretaceous, conifers are dominant trees.	Flowering plants appear in the early Cretaceous, undergo adaptive radiation, and become dominant.
Ordovician	Silurian	Devonian	Carboniferous	Permian	Triassic	Jurassic	Cretaceous	Tertiary
488	443	416	359	299	251	200	146	66

Millions of years ago (mya)

Figure 21.13 A time line for major events in plant evolution.

Figure 21.15 Painting of a Carboniferous "coal forest." An understory of ferns is shaded by tree-sized relatives of modern horsetails and club mosses.

Figure 21.16 Early angiosperms such as magnolias (*foreground*) evolved while dinosaurs walked on Earth.

Because the sperm of seed plants do not need to swim through a film of water to reach eggs, these plants can reproduce even during dry times.

After pollination and fertilization, the ovule develops into a seed. Dispersing seeds puts the seed plants at an advantage over spore-bearing plants. A seed contains a multicelled embryo sporophyte and enough food to nourish it until conditions favor growth. By contrast, seedless plants release single-celled spores without stored food. The spores cannot wait out a dry period.

Structural traits also gave seed plants an advantage over seedless lineages. Many seed plants undergo secondary growth (growth in diameter) and produce wood. Wood is tissue produced by adding new lignin-stiffened xylem. The giant nonvascular plants that lived in Carboniferous forests did undergo secondary growth. However, their trunks were softer and more flexible than those of seed plants. Only seed plants produce true wood.

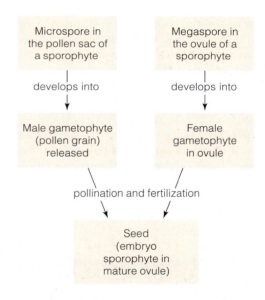

Figure 21.17 How a seed forms.

coal Fossil fuel formed over millions of years by compaction and heating of plant remains.
megaspore Haploid spore formed in ovule of seed plants; develops into an egg-producing gametophyte.
microspore Haploid spore formed in pollen sacs of seed plants; develops into a sperm-producing gametophyte (a pollen grain).
ovule Of seed plants, reproductive structure in which egg-bearing gametophyte develops; after fertilization, matures into a seed.
pollen sac Of seed plants, reproductive structure in which sperm-bearing gametophytes (pollen grains) develop.
pollination Delivery of a pollen grain to the egg-bearing part of a seed plant.

Take-Home Message **What are the major events in the evolution of vascular plants?**

> Seedless vascular plants arose before the seed-bearing lineages. They were widespread during the Carboniferous.

> Seed plants produce sperm-producing gametophytes as pollen grains that wind or animals can carry to eggs. As a result, seed plants can reproduce under drier conditions than other plants.

> The egg-bearing gametophyte of a seed plant forms in an ovule that becomes a seed after fertilization. Dispersing seeds rather than spores increases the reproductive success of seed plants in dry climates.

❯ The gymnosperms are the less diverse of the two modern lineages of seed-bearing plants.

Gymnosperms are vascular seed plants that produce seeds on the surface of ovules. The seeds are said to be "naked" because unlike those of angiosperms they are not inside a fruit. (*Gymnos* means naked and *sperma* is taken to mean seed.) However, many gymnosperms enclose their seeds in a fleshy or papery covering.

The Conifers

The 600 or so species of **conifers** are trees and shrubs with woody cones. Conifers typically have needlelike or scalelike leaves with a thick cuticle. They tend to be more resistant to drought and cold than flowering plants and are the main plants in cool Northern Hemisphere forests. Conifers include the tallest trees in the Northern Hemisphere (redwoods), as well as the most abundant (pines). They also include the long-lived bristlecone pines (Figure 21.18A). One of these trees is now 4,600 years old.

We use fir bark as mulch in our gardens, use oils from cedar in cleaning products, and eat the seeds, or "pine nuts," of some pines. Pines also provide lumber for building homes, and some pines make a sticky resin that deters insects from tunneling into them. We use this resin to make turpentine, a paint solvent.

Lesser Known Gymnosperms

Cycads and ginkgos were most diverse in dinosaur times. They are the only modern seed plants with flagellated sperm. Sperm emerge from pollen grains, then swim in fluid produced by the plant's ovule.

The 130 species of modern **cycads** live mainly in the dry tropics and subtropics (Figure 21.18B). Cycads often resemble palms but the two groups are not close relatives. The "sago palms" commonly used in landscaping and as houseplants are actually cycads.

The only living **ginkgo** species is *Ginkgo biloba*, the maidenhair tree (Figure 21.18C,D). It is a deciduous native of China. Deciduous plants drop all their leaves at once seasonally. The ginkgo's attractive fan-shaped leaves and resistance to insects, disease, and air pollution make it a popular tree along urban streets. Dietary supplements made from ginkgo may slow memory loss in people who have Alzheimer's disease.

Gnetophytes include tropical trees, leathery vines, and desert shrubs. Extracts from the stems of *Ephedra* (Figure 21.18E) are sold as an herbal stimulant and a weight loss aid. Such supplements can be dangerous; a few people have died while using them.

A Representative Life Cycle

A ponderosa pine tree is a sporophyte, and its life cycle is typical of conifers (Figure 21.19). Pollen cones and ovule-bearing cones develop on the same tree. Ovules form on cone scales ❶. Inside the ovule, a megaspore forms by meiosis and develops into a female gametophyte ❷. Male cones hold pollen sacs ❸, where microspores form and develop into pollen grains ❹.

The pollen grains are released and drift with the winds. Pollination occurs when one lands on an ovule ❺. The pollen grain germinates: Some cells develop into a pollen tube that grows through the ovule tissue and delivers sperm to the egg ❻. Pollen tube growth is an astonishingly leisurely process. It typically takes about a year for the tube to grow through the ovule tissue to the egg. When fertilization finally occurs, it produces a zygote ❼. Over about six months, the zygote develops into an

Figure 21.18 Gymnosperm diversity. **A** Bristlecone pine on a mountaintop in the Sierra Nevada. **B** An Australian cycad with its fleshy seeds. **C** Ginkgo's fan-shaped leaves and **D** its fleshy seeds. **E** The gnetophyte *Ephedra*. Yellow structures on stems are pollen-bearing cones.

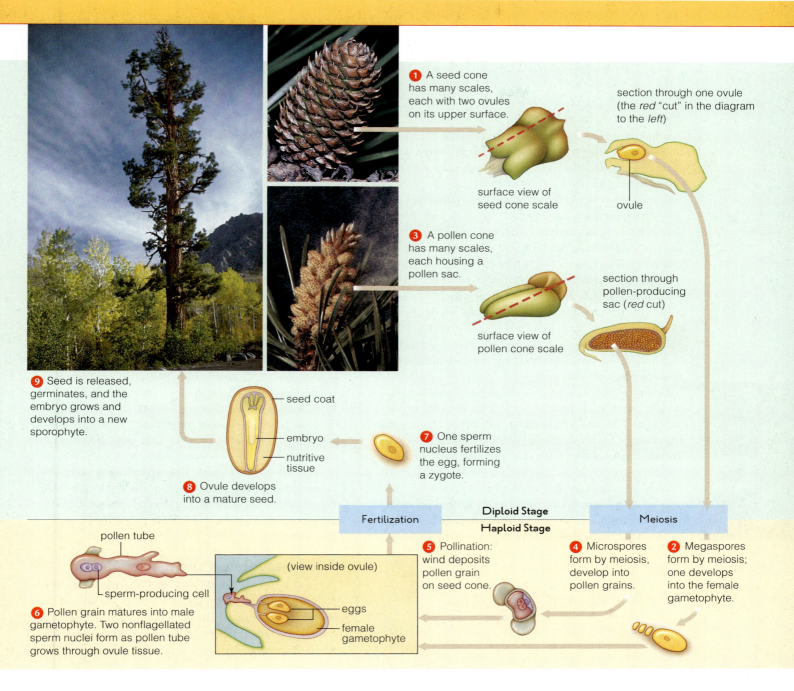

① A seed cone has many scales, each with two ovules on its upper surface.

section through one ovule (the *red* "cut" in the diagram to the *left*)

surface view of seed cone scale

ovule

③ A pollen cone has many scales, each housing a pollen sac.

section through pollen-producing sac (*red* cut)

surface view of pollen cone scale

⑨ Seed is released, germinates, and the embryo grows and develops into a new sporophyte.

seed coat

embryo

nutritive tissue

⑧ Ovule develops into a mature seed.

⑦ One sperm nucleus fertilizes the egg, forming a zygote.

Fertilization	Diploid Stage	Meiosis
	Haploid Stage	

pollen tube

sperm-producing cell

⑥ Pollen grain matures into male gametophyte. Two nonflagellated sperm nuclei form as pollen tube grows through ovule tissue.

(view inside ovule)

eggs

female gametophyte

⑤ Pollination: wind deposits pollen grain on seed cone.

④ Microspores form by meiosis, develop into pollen grains.

② Megaspores form by meiosis; one develops into the female gametophyte.

embryo sporophyte that, along with tissues of the ovule, becomes a seed ⑧. The seed is released, germinates, then grows and develops into a new sporophyte ⑨.

Figure 21.19 Animated Life cycle of a conifer, the ponderosa pine.

➤➤ **Figure It Out** Does the pollen tube grow through tissue of a male cone or a female cone?

Answer: A female cone. It grows after a pollen grain alights on a female cone.

conifer Gymnosperm with nonmotile sperm and woody cones; for example, a pine.
cycad Tropical or subtropical gymnosperm with flagellated sperm, palmlike leaves, and fleshy seeds.
ginkgo Deciduous gymnosperm with flagellated sperm, fan-shaped leaves, and fleshy seeds.
gnetophyte Shrubby or vinelike gymnosperm, with nonmotile sperm; for example *Ephedra*.
gymnosperm Seed plant that does not make flowers or fruits.

Take-Home Message What are gymnosperms?

➤ Gymnosperms are seed-bearing plants. Their eggs and seeds form on the surface of an ovule, not in ovaries as in flowering plants.

➤ All gymnosperms produce pollen. In the cycads and ginkgos, sperm emerge from pollen grains and swim through fluid released by the ovule. In other groups, the sperm are nonmotile.

❯ Angiosperms are seed plants that produce flowers.
Their seeds form inside an ovary that matures into a fruit.
❮ Link to Coevolution 17.13

Angiosperm Traits and Diversity

Angiosperms are vascular seed plants that make flowers and fruits. Nearly 90 percent of modern plant species are flowering plants. What accounts for their great diversity? For one thing, many can grow faster than gymnosperms. Think of how a flowering plant such as a dandelion or grass can grow from a seed and make seeds of its own within a few weeks. In contrast, most gymnosperms grow for years before they produce their first seeds.

Flowers also provide a selective advantage. A **flower** is a specialized reproductive shoot that consists of modified leaves arranged in concentric whorls (Figure 21.20). The outermost whorl is the sepals. Inside that are the petals, then the pollen-bearing parts called **stamens**. The innermost part of the flower is the **carpel**, where eggs form. The base of the carpel is the **ovary**, a chamber that contains one or more ovules.

After pollen-producing plants evolved, some insects began feeding on pollen. In the process, they sometimes served as **pollinators**: animals that facilitate fertilization by moving pollen from male parts of a flower to female parts. The bee in the photo at the *left* is gathering pollen and also functioning as a pollinator. Plants and their pollinators coevolve (Section 17.13). In many flowering plants vibrant colors, strong fragrances, and plentiful nectar that attracted insects evolved. Instead of randomly looking for food, some insects began seeking out particular flowers. These insects made more direct pollen deliveries. Not coincidentally, the great adaptive radiation of flowering plants coincided with an adaptive radiation of insects. Some plants such as grasses continued to rely on wind for pollination. Today, these plants have small, pale, unscented flowers without any fragrance or nectar.

After fertilization, an ovule matures into a seed and the ovary around it becomes a **fruit**. (*Angio*– means enclosed chamber, and *sperma*, seed.) Some gymnosperms have seeds with winglike structures or a fleshy coat, but fruits come in a far greater variety. Many fruits attract animals with sweet taste and bright color. Other fruits float in water, ride the winds, or stick to animal fur. All of these dispersal-related traits help distribute seeds to new habitats where they can thrive, thus increase a parent plant's reproductive success.

Major Lineages

The vast majority of flowering plants belong to one of two lineages. The 80,000 or so **monocots** include orchids, palms, lilies, and grasses. The 170,000 or so **eudicots** include most herbaceous (nonwoody) plants such as tomatoes, cabbages, roses, poppies, most flowering shrubs and trees, and cacti. Monocots and eudicots are named for the number of seed leaves, or cotyledons, their embryos have. Monocots have one cotyledon; eudicots have two. The two groups also differ in some structural details such as the arrangement of their vascular tissues and the number of flower petals. In addition, some eudicots undergo secondary growth and become woody. No monocots produce true wood.

A Representative Life Cycle

Figure 21.21 shows the life cycle of a lily, a monocot. A lily plant is a sporophyte ❶. Pollen forms in stamens of the lily flower ❷. A thin filament holds the upper part of the stamen (the anther) aloft. The anther contains pollen sacs where microspores develop into pollen grains. Eggs form

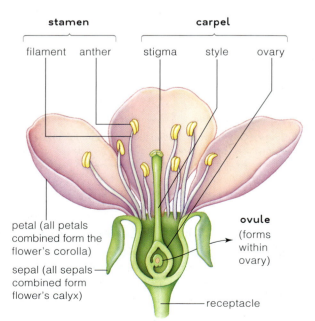

stamen — filament anther

carpel — stigma style ovary

Figure 21.20
Animated
A cherry flower, with pollen-bearing stamens and ovule-bearing carpels. Flowers of some other plants have either carpels or stamens.

petal (all petals combined form the flower's corolla)

sepal (all sepals combined form flower's calyx)

ovule (forms within ovary)

receptacle

angiosperms Largest seed plant lineage. Only group that makes flowers and fruits.
carpel Female reproductive part of a flower.
endosperm Nutritive triploid tissue in angiosperm seeds.
eudicots Largest lineage of angiosperms; includes herbaceous plants, woody trees, and cacti.
flower Specialized reproductive shoot of a flowering plant.
fruit Mature flowering plant ovary; encloses a seed or seeds.
monocots Lineage of angiosperms that includes grasses, orchids, and palms.
ovary Of flowering plants, floral chamber that encloses ovule.
pollinator Animal that moves pollen, thus facilitating pollination.
stamen Male reproductive part of a flower.

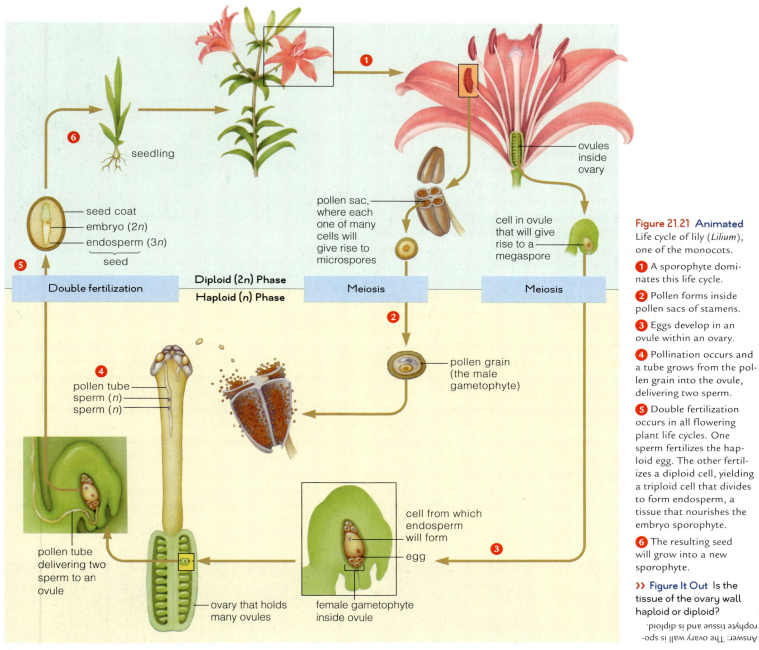

seedling

seed coat
embryo (2n)
endosperm (3n)
seed

Double fertilization

pollen tube
sperm (n)
sperm (n)

pollen tube
delivering two
sperm to an
ovule

ovary that holds
many ovules

Diploid (2n) Phase
Haploid (n) Phase

pollen sac,
where each
one of many
cells will
give rise to
microspores

Meiosis

pollen grain
(the male
gametophyte)

cell from which
endosperm
will form

egg

female gametophyte
inside ovule

ovules
inside
ovary

cell in ovule
that will give
rise to a
megaspore

Meiosis

Figure 21.21 Animated
Life cycle of lily (*Lilium*),
one of the monocots.

❶ A sporophyte domi-
nates this life cycle.

❷ Pollen forms inside
pollen sacs of stamens.

❸ Eggs develop in an
ovule within an ovary.

❹ Pollination occurs and
a tube grows from the pollen grain into the ovule,
delivering two sperm.

❺ Double fertilization
occurs in all flowering
plant life cycles. One
sperm fertilizes the hap-
loid egg. The other fertilizes a diploid cell, yielding
a triploid cell that divides
to form endosperm, a
tissue that nourishes the
embryo sporophyte.

❻ The resulting seed
will grow into a new
sporophyte.

›› **Figure It Out** Is the
tissue of the ovary wall
haploid or diploid?

Answer: The ovary wall is sporophyte tissue and is diploid.

in a flower's carpels. A carpel consists of the stigma (a
slender projection where pollen will land) and the style
(the part where the ovary is located). Egg-bearing gametophytes form from megaspores in the ovary ❸.

Pollination occurs when a pollen grain reaches a receptive stigma. After pollination, a pollen tube forms ❹. The
tube grows down through the style and into the ovary. It
then delivers two sperm to the ovule, where double fertilization occurs ❺. One sperm fertilizes the egg, forming
what will become the embryo. The other sperm fertilizes
a cell that has two nuclei, forming a triploid cell. This cell
will divide and become the **endosperm**, a nutrient-rich

tissue unique to angiosperms. Ovary tissue matures into
a fruit with seeds that can grow into a new sporophyte ❻.

Take-Home Message What are angiosperms?

› Angiosperms are flowering plants. A flower is a special reproductive shoot that
makes pollen in stamens and eggs in carpels. The base of the carpel is an ovary,
a chamber that encloses the ovules and develops into a fruit after fertilization.

› Angiosperms are the most diverse plants. A short life cycle, coevolution with
pollinators, and many fruit dispersal mechanisms contribute to their success.

› Eudicots and monocots are the main lineages of flowering plants.

> Angiosperms feed and shelter animals, and they provide us with food, fabric, oils, medicines, drugs, and more.
< Link to C3 and C4 plants 6.8

It would be nearly impossible to overestimate the importance of the angiosperms. As the most abundant plants in the majority of habitats, they provide food and shelter for a variety of animals. Animals feed on angiosperm roots, shoots, and fruits. They also sip floral nectar (Figure 21.22).

Angiosperms also provide nearly all human food, either directly or by serving as food for livestock. Cereal grains are the most widely planted crops. All are grasses. Like other monocots, cereals store nutrients in their endosperm mainly as starch. In the United States, corn is the top cereal crop, with hundreds of millions of tons harvested every year. Worldwide, rice feeds the greatest number of people. Rye, oats, barley, sorghum, and wheat are other commonly grown cereal crops (Figure 21.23A).

Variation in photosynthetic pathways contributes to the widespread use of cereal crops. Grasses include both C3 plants (rice, wheat, oats, and barley) and C4 plants (corn and sorghum). As Section 6.8 explained, C3 plants grow best in cool conditions, whereas C4 plants do well in hot, dry climates. Thus, people throughout the world can grow a cereal grain that is well suited to their climate.

Soybeans, lentils, peas, and peanuts are among the legumes, the second most important source of human food. Legumes are eudicots and their endosperm stores nutrients as proteins and oils, as well as carbohydrates. Legumes can be paired with grains to provide all the amino acids a human body needs to build proteins.

Figure 21.22 Flowering plants feed animals. This hummingbird is sipping nectar, which is mainly sucrose, from a columbine flower.

All legumes are C3 plants, so they cannot be grown successfully in the hottest regions.

The third major staple of the human diet is plants that store starch in roots or underground stems. Yams, sweet potatoes, potatoes, and cassava are examples. All are eudicots and C3 plants, so their distribution is limited.

In addition to plants that serve as staples, humans enliven their diet with a remarkable variety of plant parts. For example, we dine on the leaves of lettuce and spinach, the stems of asparagus and celery, the immature floral shoots of broccoli and cauliflower, and the fleshy fruits of tomatoes and blueberries. The stamens of crocus flowers provide the spice saffron, and we grate the bark of a tropi-

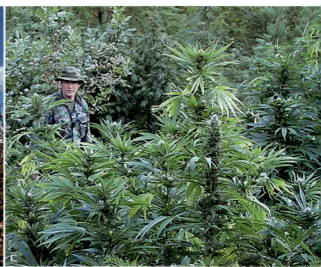

Figure 21.23 Angiosperms as crop plants. **A** Mechanized harvesting of wheat, a monocot.
B Field of cotton, a eudicot. **C** Marijuana (a eudicot) discovered growing illegally in Oregon.

cal tree to make cinnamon. Maple syrup is sap that has been collected from maple tree xylem and boiled until it reaches a syrupy consistency.

Angiosperms are also an important source of fiber. Fabrics made from plant fibers include linen, ramie, hemp, burlap, and cotton (Figure 21.23B). A cotton boll is the fruit of a cotton plant. Cotton fibers are nearly pure cellulose. In addition to fabrics, we use fibers from seed plants to weave rugs. For example, fibers from the leaves of agave plants are used to make sisal rugs.

Oils extracted from the seeds of some eudicots including rapeseed and hemp are used in detergents, skin care products, and as industrial lubricants. Plant oils may also be used as a source of biodiesel fuel.

Angiosperm woods are referred to as "hardwoods," as opposed to gymnosperm "softwoods." Furniture and flooring are often made from oak or other hardwoods. Hardwoods are also preferred as firewood.

Some flowering plants make secondary metabolites that we use as medicines or as mood-altering drugs. A **secondary metabolite** is a compound that has no known metabolic role in the organism that produces it. Many plant secondary metabolites evolved as a defense against grazing animals. Aspirin is derived from a compound discovered in willows. Digitalis from foxglove strengthens a weak heartbeat. The caffeine in coffee beans and nicotine in tobacco leaves are widely used stimulants. Marijuana, a drug illegal in the United States, is one of the country's most valuable cash crops (Figure 21.23C). Worldwide, the cultivation of opium poppies (the source of heroin) and coca (the source of cocaine) have wide-reaching health, economic, and political effects.

Some secondary metabolites of flowering plants are extracted and sold as insecticides that are safe for humans and other vertebrates. For example, chrysanthemums make natural insecticides called pyrethrums. Neem tree oil is another natural insecticide. Compared to synthetic insecticides, plant-derived compounds tend to be less persistent in the environment.

---

secondary metabolite Chemical that has no known role in an organism's normal metabolism; often deters predation.

Take-Home Message **What are the ecological and economic roles of angiosperms?**

> Angiosperms serve as food and shelter for land animals.

> Humans depend on angiosperms as food plants, and as sources of fiber, wood, oils, medicines, and insecticides.

> Cereal grains are the most widely grown food plants. They are monocots with a starchy endosperm. Different members of the group are adapted to different climates.

■ Speaking for the Trees (revisited)

Plants make organic compounds by absorbing energy from the sun, carbon dioxide from the air, and water and dissolved minerals from soil. By the noncyclic pathway of photosynthesis, they split water molecules and release oxygen. Their oxygen output and carbon uptake sustains the atmosphere. Amazingly, every atom of carbon in a tree that stands a hundred meters high (328 feet) and weighs thousands of tons was taken up from the air in the form of carbon dioxide.

A tree is about 20 percent carbon by weight, so enormous amounts of carbon are stored in the living trees of Earth's forests. Additional carbon remains tied up in the leaf litter and decaying trees on the forest floor. The United Nations Food and Agriculture Organization estimates that forest plants and soils hold about one and a half times as much carbon as the air. Trees take up most carbon when they actively growing. Therefore, an acre of tropical forest where trees grow continually stores more carbon than an acre of forest in a region with a limited growing season.

The great carbon-storing capacity of tropical forests means that threats to these forests have global implications. When a tropical forest is converted to cropland, the rate of carbon dioxide storage per acre goes down. In addition, if the forest is cleared by burning, the carbon that was stored in wood and leaf litter enters the atmosphere (Figure 21.24). Rising levels of carbon dioxide in Earth's atmosphere contribute to global warming, by a mechanism we explain in detail in Section 28.8.

Figure 21.24 Burning forests to make way for agriculture or other uses adds carbon dioxide to the air, reduces carbon uptake, and thus contributes to global warming.

How Would You Vote? One proposed way to discourage tropical deforestation is through international carbon offsets. For example, a company in the United States could agree to offset its greenhouse gas production by funding a tropical reforestation or forest preservation project in South America or Africa. Proponents of such agreements say they minimize a company's cost of reducing greenhouse gases and benefit all involved. Some opponents would rather see companies make pollution-reducing changes at home, rather than pay for programs elsewhere. Do you think international carbon offsets are a good idea? See CengageNow for details, then vote online (cengagenow.com).

Summary

Section 21.1 Deforestation has far-reaching environmental effects, including increased erosion and climate change. Individual efforts to reverse deforestation can collectively have a large positive impact.

Section 21.2 Land plants evolved from green algae. These plants are **embryophytes**; they form a multicelled embryo on the parental body. Key adaptations that allowed plants to move into dry habitats include a waterproof cuticle with **stomata**, and internal pipelines of **vascular tissue** (**xylem** and **phloem**) reinforced by **lignin**.

Plant life cycles include two multicelled bodies: the haploid gametophyte and the diploid sporophyte. The gametophyte dominates in early-evolving lineages, but in **vascular plants**, the sporophyte is larger and longer lived. **Seeds** and male gametophytes that can be dispersed without water (**pollen grains**) are important adaptations that contribute to the success of seed plants.

sporophyte

gametophyte

Section 21.3 Bryophytes are non-vascular (no xylem or phloem). Their sperm swim through water droplets to eggs. The sporophyte remains attached to the gametophyte. **Rhizoids** attach a gametophyte to the soil or a surface. **Mosses** are the most diverse bryophytes. The remains of some mosses form **peat**, which is dried and burned as fuel.

Section 21.4 In **seedless vascular plants**, sporophytes have vascular tissues and are the larger, longer-lived phase of the life cycle. Typically the sporophyte's roots and shoots grow from a horizontal stem, or **rhizome**. Tiny free-living gametophytes make flagellated sperm that require water for fertilization. Ferns, the most diverse group of seedless vascular plants, produce spores in **sori**. Many ferns grow as **epiphytes**. Club mosses and horsetails produce spores in conelike structures.

Section 21.5 Forests of giant seedless vascular plants thrived during the Carboniferous period. Heat and pressure transformed the remains of these forests to **coal**.

The first seed plants evolved during the Carboniferous. Seed plant sporophytes have **pollen sacs**, where **microspores** form and develop into male gametophytes (pollen grains). They also have **ovules**, where **megaspores** form and develop into female gametophytes. **Pollination** unites the egg and sperm of a seed plant.

Section 21.6 Gymnosperms include **conifers**, **cycads**, **ginkgos**, and **gnetophytes**. Many are well adapted to dry climates. In conifers, ovules form on the surfaces of woody cones.

Section 21.7 Angiosperms are the most diverse and widespread land plants. They alone make **flowers** and **fruits**. The **stamens** of a flower produce pollen. The female part of a flower is the **carpel**. An **ovary** at the base of the carpel holds one or more ovules.

Many flowering plants coevolved with **pollinators**. After pollination, the flower's ovary becomes a fruit that contains one or more seeds. A flowering plant seed includes an embryo sporophyte and **endosperm**, a nutritious tissue. The two main types, **eudicots** and **monocots**, differ in seed structure and other traits.

Section 21.8 As the dominant plants in most land habitats, flowering plants are ecologically important, as well as essential to human existence. The nutrient-rich endosperm of angiosperm seeds makes them staples of human diets throughout the world. Angiosperms also supply us with fiber, furniture, oils, medicines, and mood-altering drugs. Many useful plant compounds are secondary metabolites, compounds that probably help defend the plant against predation.

Self-Quiz   Answers in Appendix III

1. The first plants were _____ .
   a. ferns
   b. flowering plants
   c. bryophytes
   d. conifers

2. Which of the following statements is false?
   a. Ferns produce seeds inside sori.
   b. Bryophytes do not have xylem or phloem.
   c. Gymnosperms and angiosperms produce seeds.
   d. Only angiosperms produce flowers.

3. In bryophytes, eggs are fertilized in a chamber on the _____ and a zygote develops into a _____ .
   a. sporophyte; gametophytes
   b. gametophyte; sporophyte
   c. sorus; cone

4. Horsetails and ferns are _____ plants.
   a. multicelled aquatic
   b. nonvascular seed
   c. seedless vascular
   d. seed-bearing vascular

5. Coal consists primarily of compressed remains of the _____ that dominated Carboniferous swamp forests.
   a. seedless vascular plants
   b. conifers
   c. flowering plants
   d. a and c

6. The _____ produce flagellated sperm.
   a. mosses
   b. ferns
   c. conifers
   d. monocots
   e. a and b
   f. a through c

7. A seed is a _____ .
   a. female gametophyte
   b. mature ovule
   c. mature pollen grain
   d. modified microspore

Data Analysis Activities

A Global View of Deforestation The U.N. Food and Agriculture Organization (UNFAO) recognizes the importance of forests to human populations and maintains records of forest abundance. Figure 21.25 shows UNFAO data on the amount of forested land throughout the world in 1990, 2000, and 2005.

1. How many hectares of forested land were there in North America in 2005?
2. In which region(s) did the amount of forested land increase between 1990 and 2005?
3. How many hectares of forest did the world lose between 1990 and 2005?
4. In 2002, China embarked on an ambitious campaign to add 76 million hectares of trees over a ten-year period. Do you see any indication that this campaign is successful?

Region	Forested Area (in millions of hectares)		
	1990	**2000**	**2005**
Africa	699	656	635
Asia	574	567	572
Central America	28	24	22
Europe	989	988	1,001
North America	678	678	677
Oceania	233	208	206
South America	891	853	832
World total	4,077	3,988	3,952

Figure 21.25 Changes in forested area by region from 1990 to 2005. One hectare is 2.47 acres. The full report on the world's forests is online at www.fao.org/forestry/en/.

8. True or false? Both spores and sperm of a seedless vascular plant are haploid.

9. Only angiosperms produce _____ .
   a. pollen
   b. seeds
   c. fruits
   d. all of the above

10. Lignin is not found in stems of _____ .
   a. mosses   b. ferns   c. monocots   d. a and b

11. A waxy cuticle helps land plants _____ .
   a. conserve water
   b. take up carbon dioxide
   c. reproduce
   d. stand upright

12. Cereal crops such as rice and corn are _____ .
   a. monocots
   b. vascular plants
   c. seed plants
   d. all of the above

13. Some plant secondary metabolites are used as _____ .
   a. fuel
   b. medicines
   c. textiles
   d. flooring

14. Match the terms appropriately.
   ___ bryophyte
   ___ seedless vascular plant
   ___ gymnosperm
   ___ angiosperm

   a. seeds, but no fruit
   b. flowers and fruits
   c. no xylem or phloem
   d. xylem and phloem, but no ovule

15. Match the terms appropriately.
   ___ ovule
   ___ cuticle
   ___ gametophyte
   ___ sporophyte
   ___ fruit
   ___ endosperm
   ___ rhizome
   ___ sorus
   ___ microspore

   a. gamete-producing body
   b. spore-producing body
   c. becomes seed
   d. horizontal stem
   e. mature ovary
   f. nutritive tissue in seed
   g. where fern spores form
   h. single haploid cell
   i. waterproofing layer

Additional questions are available on **CENGAGENOW**.

Critical Thinking

1. Early botanists admired ferns but found their life cycle perplexing. In the 1700s, they learned to propagate ferns by sowing what appeared to be tiny dustlike "seeds" that they collected from the undersides of fronds. Despite many attempts, the botanists could not locate the pollen source, which they assumed must stimulate these "seeds" to develop. Imagine you could write to these botanists. Compose a note that explains the fern life cycle and clears up their confusion.

2. In most plants the largest, longest-lived body is a diploid sporophyte. By one hypothesis, diploid dominance was favored because it allowed a greater level of genetic diversity. Suppose that a recessive mutation arises. It is mildly disadvantageous now, but it will be useful in some future environment. Explain why such a mutation would be more likely to persist in a fern than in a moss.

3. The photo at the *right* is a micrograph of a longitudinal section through the stem of a squash plant. The stem has been dyed with a substance that tints lignin red. Can you identify the red-ringed structures? Would you expect to find similar structures in the stem of a corn plant? Would you find them in the leafy green part of a moss?

Animations and Interactions on **CENGAGENOW**:
› Moss life cycle; Fern life cycle; Conifer life cycle; Lily life cycle.

‹ Links to Earlier Concepts

Before starting, review Figure 20.2 to get a sense of where fungi fit in the eukaryotic family tree. In this chapter you will learn about fungal interactions with cyanobacteria (19.7), and green algae (20.8). You will also draw on your knowledge of the organic molecules chitin (3.3) and lignin (21.2), and your understanding of the processes of fermentation (7.6) and nutrient cycling (1.3).

Key Concepts

Traits and Classification

Fungi are single-celled and multicelled heterotrophs that feed by secreting digestive enzymes onto organic matter, then absorbing the released nutrients. Multicelled species form a mesh of absorptive filaments. In the most diverse groups, cross-walls reinforce the filaments.

The Oldest Lineages

The oldest fungal lineages have filaments with few or no cross-walls. The mostly aquatic chytrid fungi make flagellated spores. Zygote fungi make thick-walled zygotes that release sexual spores. Glomeromycetes are fungi that branch inside plant cell walls.

22 Fungi

22.1 ## High-Flying Fungi

Fungi are not known for their mobility. You probably do not think of mushrooms and their relatives as world travelers, but some do get around. Fungi produce microscopic spores that drift on air currents. The spores can also lodge in crevices on tiny dust particles. When winds lift these particles aloft, spores go along for the ride. Dust borne fungal spores can disperse great distances, riding winds that swirl high above Earth's surface.

We know, for example, that dust storms in the deserts of North Africa lift fungus-laden particles more than 4.5 kilometers (3 miles) above the desert floor. Winds then transport the dust and spores long distances. Some dust particles travel as far as the Caribbean and the east coast of the Americas.

Most fungi that hitchhike as spores on African dust are harmless. However, fungi that cause plant disease occasionally make the journey. For example, winds of a 1978 cyclone introduced sugar cane rust (a fungal disease) from Cameroon to the Dominican Republic. Similarly, winds probably facilitated the spread of coffee rust fungus from Angola to Brazil in 1980.

Today, an African outbreak of an old fungal foe has agricultural officials around the world on edge. The fungus is wheat stem rust (*upper right*). An infection begins when a spore lands on a wheat plant and fungal filaments invade the plant through its stomata. Within about a week, the fungus grows through the plant's tissues, matures, and produces ten of thousands of rust-colored spores at the stem surface. Each spore can disperse and infect a new plant.

Wheat stem rust was common worldwide until the 1960s, when plant breeders developed new varieties of wheat that resist infection by this fungus. Global distribution of the rust-resistant wheats provided a respite from outbreaks of wheat stem rust for decades. Then, in 1999, a new strain of wheat stem rust (called Ug99) was discovered in Uganda, a country in eastern Africa. Mutations allow Ug99 to infect most wheat varieties that were previously considered resistant to wheat stem rust.

By 2009, windborne spores of Ug99 had dispersed to Kenya, Ethiopia, and Sudan, crossed the Red Sea to Yemen, and from there crossed the Persian Gulf to Iran. India, the world's second largest wheat producer, is expected to be affected next. Most likely, winds will eventually distribute spores of Ug99 throughout the world.

With one of the world's most important food crops under threat, scientists are hard at work screening plants for genes that provide protection against Ug99. They have already bred some resistant wheats. However, creating a resistant wheat suited to each region where Ug99-susceptible wheat currently grows will be a challenge. Thus, many agricultural experts consider Ug99 the single greatest threat to the world food supply at this time.

This chapter discusses the pathogenic fungi that threaten our crops and infect our bodies, but it also introduces many beneficial fungi. Keep in mind that most fungi are decomposers with an important ecological role. Fungi break down organic wastes and remains into inorganic subunits, thereby making essential nutrients available to plants. Fungi also associate with plant roots, helping them to grow better. Other fungi partner with single photosynthetic cells, forming the composite organisms known as lichens. We value fungi as sources of medicine and as food. Some single-celled fungi help us make bread and beer, and countless mushrooms end up atop our pizzas and in our salads and sauces.

Sac Fungi
Sac fungi, the most diverse fungal lineage, include single cells such as the yeasts we use in baking, and multicelled species that have filaments reinforced with cross-walls. When sac fungi reproduce sexually, they produce spores inside a saclike structure.

Club Fungi
Most club fungi are multi-celled. Like the closely related sac fungi, they have filaments reinforced with cross walls. Sexual reproduction involves formation of spores in club-shaped cells. A mushroom is an example of a club fungus fruiting body.

Interactions With Other Organisms
Some fungi partner with photosynthetic cells as lichens. Others live in or on plant roots and benefit the plant by sharing nutrients. A minority of fungi are parasites. Some of these cause diseases in plants or in humans.

> Like animals, fungi are heterotrophs. Unlike animals, fungi have walled cells and digest food outside their body.
< Link to Chitin 3.3

Structure and Function

Fungi are spore-making eukaryotic heterotrophs that have cell walls reinforced with chitin. This polysaccharide also hardens the body covering of animals such as insects or crabs. Like animals, fungi make digestive enzymes, but fungi do not digest food inside their body. Instead, they secrete digestive enzymes onto organic matter and absorb the released breakdown products.

A typical fungus is a free-living **saprobe**, an organism that feeds on organic wastes and remains. Some fungi live inside other organisms. The fungus may benefit its host, have no effect, or be a harmful parasite.

The fungi that live as single cells are informally called yeasts. More often the fungus is multicellular. Molds and mushrooms are familiar examples of multicellular fungi (Figure 22.1). Such fungi grow as a mesh of threadlike filaments collectively called a **mycelium** (plural, mycelia). Each filament is one **hypha** (plural, hyphae), consisting of cells attached end to end (Figure 22.2). The fungus grows by adding cells to the tip of its hyphae.

---

fungus Eukaryotic heterotroph with cell walls of chitin; obtains nutrients by digesting them outside the body and absorbing them.
dikaryotic Having two genetically distinct nuclei.
hypha Component of a fungal mycelium; a filament made up of cells arranged end to end.
mycelium Mass of threadlike filaments (hyphae) that make up the body of a multicelled fungus.
saprobe Organism that feeds on wastes and remains.

one cell of a hypha in the mycelium

Figure 22.2 Close-up of a mycelium, showing the threadlike hyphae, each consisting of cells arranged end to end.

Depending on the fungal group, there may or may not be cross-walls, or septa (singular, septum), between cells of a hypha. When cross-walls do exist, they are porous, so materials can still flow among hyphal cells. Thus nutrients or water taken up in one part of the mycelium can be shared with cells in other regions of the fungal body.

Life Cycles

In fungi, the diploid stage is the least conspicuous part of the life cycle. Only the zygote is diploid. Depending on the fungal group, a haploid stage or a dikaryotic stage dominates the cycle. **Dikaryotic** means that a cell contains two genetically different nuclei ($n+n$).

Most fungi disperse by producing spores. A fungal spore is a cell or cluster of cells, often with a thick wall that allows it to survive harsh conditions. Spores may form by mitosis (asexual spores) or by meiosis (sexual spores). Each fungal group produces sexual spores in a distinctive way, a trait that aids in classification.

Fungal Diversity

More than 70,000 species of fungi have been named, and there may be a million more. Three relatively small groups—chytrids, glomeromycetes, and zygote fungi—have a haploid mycelium that consists of hyphae with few or no cross-walls.

Most fungi belong to two lineages: sac fungi and club fungi. Hyphae with cross-walls at regular intervals evolved in the common ancestor of these lineages and contribute to the success of both groups. Cross-walls strengthen hyphae, allowing development of larger, more complex spore-producing bodies. In addition, cross-walls divide the cytoplasm, so damage to one part of a hypha does not cause the whole hypha to dry out and die.

A Green mold on grapefruit. A mold is a fungus that grows as a mat of microscopic hyphae that reproduce asexually.

B Scarlet hood mushroom in a Virginia forest. A mushroom forms during sexual reproduction of many club fungi.

Figure 22.1 Examples of multicelled fungi.

Take-Home Message What are fungi?

> Fungi are heterotrophs that secrete digestive enzymes onto organic matter and absorb the released nutrients. They live as single cells or as a multicelled mycelium and disperse by producing spores.

❯ Chytrids, zygote fungi, and glomeromycetes are the oldest and least diverse groups of fungi.

Chytrids

Chytrids include the oldest fungal lineages, and are the only living fungi that make flagellated spores. Most of the 1,000 or so species are aquatic decomposers. A few kinds swim about in the gut of sheep, cattle, and other herbivores and help them digest cellulose. Others are parasites.

A chytrid that parasitizes amphibians has been implicated in the global decline in frog populations. The parasite was discovered in the late 1990s when scientists investigated plummeting frog numbers in Australia and South America. Since then, the parasite has been detected in frogs in the Americas, Europe, Africa, and Asia. Chytrid infections may push some frog species to extinction.

Zygote Fungi

Zygote fungi (zygomycetes) include about 1,100 species that produce a thick-walled zygospore during sexual reproduction. Many grow as molds, a mat of asexually reproducing hyphae. Black bread mold (*Rhizopus stolonifer*) has a typical life cycle (Figure 22.3). When food is plentiful, a haploid mycelium grows and produces spores asexually at the tips of specialized hyphae ❶.

There are two mating strains. Sexual reproduction occurs when hyphae of different strains meet ❷. Special cells form on the hyphae and their cytoplasm fuses, producing a young dikaryotic zygospore ❸. Fusion of nuclei produces a mature diploid zygospore, with a thick, protective wall ❹. Meiosis occurs as the zygospore germinates. A hypha emerges, bearing a sac with haploid spores at its tip ❺. After release, the spores germinate and each can give rise to a haploid mycelium.

Most zygote fungi are saprobes, but some are parasites. One species infects flies. Fungal hyphae grow into the fly's brain and alter its behavior. The fly climbs to a high perch, holds on with its proboscis, and dies. Its posture and height facilitate dispersal of the fungal spores. Another zygote fungus infects people who have a weak immune system. Hyphae grow in blood vessels and cause zygomycosis, an often fatal disease. "Mycosis" is a general term for a disease that is caused by a fungus.

Glomeromycetes

The 150 or so species of **glomeromycetes** were previously placed among the zygote fungi, but are now considered a separate group. All form an association with plant

---

chytrid Fungus that makes flagellated spores.
glomeromycete Fungus with hyphae that grow inside the wall of a plant root cell.
zygote fungus Fungus that forms a zygospore during sexual reproduction.

Figure 22.3 Animated Life cycle of a zygote fungus, the black bread mold (*Rhizopus*).

❶ The fungus reproduces asexually as it grows through a slice of bread.

❷ Sexual reproduction begins when hyphae of two different mating strains meet.

❸ Cytoplasmic fusion of hyphal cells yields an immature zygospore that is dikaryotic.

❹ Fusion of the zygospore's haploid nuclei produces the mature diploid zygospore.

❺ The zygospore undergoes meiosis, germinates, and produces an aerial hypha with haploid spores at its tip. Each spore can give rise to a new mycelium.

Figure 22.4 Glomeromycete hypha branching inside a root cell wall.

roots. A hypha grows into a root and branches inside the wall of a root cell (Figure 22.4). The relationship is mutually beneficial. The plant shares sugars with its fungal roommate, which provides it with nutrients taken up from the soil.

Take-Home Message What traits distinguish chytrids, zygote fungi, and glomeromycetes?

❯ The mostly aquatic chytrids are the only fungi with flagellated spores.

❯ Zygote fungi produce a unique walled zygospore during sexual reproduction.

❯ Glomeromycetes have hyphae that branch inside cells of plant roots.

> Sac fungi are the most diverse fungal group. They include single-celled species and multicelled species that have hyphae reinforced by cross-walls at regular intervals.

❮ Link to Fermentation 7.6

Life Cycle

Sac fungi (ascomycetes) are the most diverse group of fungi, with more than 32,000 named species. Sac fungi may be multicelled or single-celled. Hyphae of multicelled sac fungi have cross-walls at regular intervals. The characteristic trait of the group is the ascus (plural, asci), a baglike structure that encloses sexually produced spores.

As in zygote fungi, sexual reproduction of multicelled sac fungi begins when hyphae of different mating strains meet. However, in sac fungi, cytoplasmic fusion is followed by mitosis that produces dikaryotic hyphae. These hyphae intertwine with haploid hyphae to form a fruiting body called an ascocarp. The spore-forming asci develop on the ascocarp (Figure 22.5A).

Multicelled sac fungi also produce spores asexually. In this case, spores form on specialized hyphae that develop from the haploid mycelium.

A Sampling of Diversity

Morels (Figure 22.5B) and truffles (Figure 22.5C) are among the edible ascocarps. A truffle forms underground. When spores mature, the fungus produces an odor like that of a sexually excited male pig. Female pigs detect

part of one hypha that forms a nooselike ring   roundworm

Figure 22.6 Animated A predatory fungus (*Arthrobotrys*) that captures and feeds on roundworms. Rings on the hyphae constrict and entrap the worms, then hyphae grow into the captive and digest it.

this scent and disperse truffle spores as they root through the soil in search of their seemingly subterranean suitor. Dogs can also be trained to snuffle out truffles. The search can be worthwhile. In 2006, a single 1.5-kilogram (about 3-pound) Italian truffle sold for $160,000.

Many sac fungi grow as molds. Some species of the sac fungus *Penicillium* spoil fruits, such as the grapefruit in Figure 22.1. We use other members of this genus to give blue cheese its blue veins and tangy taste. Yet another *Penicillium* species lives in soils and was the original source of the antibiotic drug penicillin.

The most familiar yeasts are sac fungi. These single cells reproduce asexually by budding, as shown in the photo at the *right*. A packet of baking yeast holds spores of a sac fungus. When bread dough is set out to rise in a warm place, the spores germinate and release cells that use sugar in fermentation reactions. Bubbles of carbon dioxide produced by the fermentation cause dough to expand. Fermentation by other sac fungus yeasts helps us produce beer and wine. Still other sac fungi cause human yeast infections.

Figure 22.6 shows an unusual predatory sac fungus. It has special hyphae with loops that can trap roundworms. After feeding on a worm, the fungus makes asexual spores. Spreading these spores in fields can help control populations of roundworms that infect and harm crops.

---

sac fungi Most diverse fungal group; sexual reproduction produces ascospores inside a saclike structure (an ascus).

haploid spore in ascus

Figure 22.5 Sac fungus fruiting bodies (ascocarps). **A** The cup of the scarlet cup fungus is an ascocarp. Asci, each containing eight ascospores (sexual spores) form on the cup's inner surface. **B** Morels and **C** truffles are edible ascocarps. Morels develop aboveground. Truffles form underground.

Take-Home Message What are sac fungi?

> Sac fungi make sexual spores inside sac-shaped asci. They include multicelled species with hyphae divided by internal walls, and yeasts that can reproduce asexually by budding.

❯ Most club fungi are multicelled, with cross-walled hyphae. Their fruiting bodies include the familiar mushrooms.
❮ Link to Lignin 21.2

A **club fungus** (basidiomycete) is typically multicelled. Its sexual spores form inside club-shaped cells on a fruiting body, or basidiocarp, composed of interwoven dikaryotic hyphae. Fruiting bodies of club fungi can be quite large and have diverse shapes (Figure 22.7).

The button mushrooms common in markets are fruiting bodies of a club fungus. Haploid hyphae of this fungus grow underground. When hyphae of two different mating strains meet, they fuse, forming a dikaryotic mycelium (Figure 22.8 ❶). The mycelium grows through the soil and, when conditions favor sexual reproduction, forms mushrooms ❷. Thin tissue sheets (gills) fringed with club-shaped cells line the underside of the mushroom's cap ❸. Fusion of the nuclei in these dikaryotic cells forms a diploid zygote ❹. The zygote undergoes meiosis, forming four haploid spores ❺. These spores are dispersed by wind, germinate, and start a new cycle ❻.

Club fungi play an essential role as decomposers in forests. They are the only fungi capable of breaking down the lignin in wood. Some forest fungi are long-lived giants. In one Oregon forest, mycelium of a honey mushroom fungus extends through more than 2,000 acres of soil. By one estimate, this fungus is 2,400 years old.

Smuts and rusts are important plant pathogens. Unlike most club fungi, these do not produce a big fruiting body. Wheat stem rust (Section 22.1) is one example.

---

club fungi Fungi that produce spores in club-shaped cells.

Puffball · Shelf fungus · Coral fungus · Chanterelles

Figure 22.7 Club fungus fruiting bodies (basidiocarps).

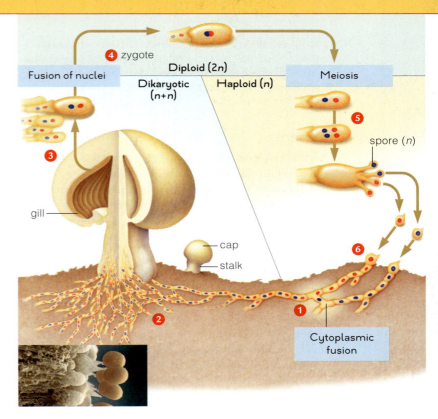

Figure 22.8 **Animated** Club fungus life cycle. Different mating strains are indicated here by *red* or *blue* nuclei. The photo at *lower left* shows spores at the edge of gills.

❶ Haploid hyphal cells meet and their cytoplasm fuses, forming a dikaryotic (*n+n*) cell.
❷ Mitotic cell divisions form a mycelium that produces a mushroom.
❸ Spore-making cells form at the edges of the mushroom's gills.
❹ Inside these dikaryotic cells, nuclei fuse, producing diploid (2*n*) cells.
❺ The diploid cells undergo meiosis, forming haploid (*n*) spores.
❻ Spores are released and give rise to a new haploid mycelium.

❯❯ **Figure It Out** How many nuclei are in each cell of a mushroom's stalk?

Answer: Two. The cells are dikaryotic.

Fungi cannot run away from predators, but toxins protect some species. Each year thousands of people are sickened after eating poisonous mushrooms that they mistook for an edible species. In some cases, mushroom poisoning is fatal. Edible mushrooms often have poisonous look-alikes, so eating mushrooms found in the wild is risky unless you are expert in identifying species.

Some mushroom toxins alter mood and cause hallucinations. The drug LSD is a compound that was initially isolated from such a mushroom. Psilocybin-containing mushrooms are also eaten for their mind-altering effects. Both LSD and psilocybin-containing mushrooms are illegal in the United States.

Take-Home Message What are club fungi?

❯ Club fungi produce their sexual spores at the tips of club-shaped cells. These cells form on a short-lived reproductive body, such as a mushroom.

> Symbiosis refers to two organisms that interact closely throughout their life cycle. Many fungal species are symbiotic with photosynthetic cells in lichens, while others are symbionts that live in or on plants.
❮ Links to Cyanobacteria 19.7, Green algae 20.8

Lichens

A **lichen** is a composite organism composed of a fungus and a single-celled photosynthetic partner (Figure 22.9). Most often it consists of a sac fungus and cyanobacteria or green algae. Some club fungi also form lichens.

The fungus makes up most of the lichen's mass. Its hyphae entwine around the photosynthetic cells that provide it with sugars. Cyanobacteria also provide fixed nitrogen. A lichen may be a **mutualism**, a symbiotic interaction that benefits both partners. However, in some cases the fungus may be parasitically exploiting captive photosynthetic cells that would do better on their own.

Lichens disperse by fragmentation or by releasing small packages with cells of both partners. In addition, the fungus can release spores. The fungus that germinates from a spore can form a new lichen only if it alights near an appropriate photosynthetic cell. This is not as unlikely as it may seem; the required algae and bacteria are common as free-living cells in many environments.

Lichens colonize many places too hostile for most organisms, such as newly exposed bedrock. They help break down the rock by releasing acids and by holding water that freezes and thaws. When soil conditions improve, plants move in and take root. Long ago, lichens may have preceded the invasion of plants onto land.

Figure 22.10 **A** Mycorrhiza formed by a fungus and young hemlock tree root. **B** Experimental demonstration of the effects of mycorrhizae on plant growth in phosphorus-poor, sterilized soil. Seedlings at left were grown without a fungus. Seedlings at right, the experimental group, were grown with a partner fungus.

Mycorrhizae: Fungus + Roots

Nearly all plants have a mutually beneficial relationship with a fungus. A mutualism between a plant root and a fungus is called a **mycorrhiza** (plural, mycorrhizae). In some cases, fungal hyphae form a dense net around roots but do not penetrate them (Figure 22.10A). In other cases, the hyphae enter roots and branch inside root cell walls, as shown earlier in Figure 22.4. About 80 percent of plants form such relationships with fungi.

Hyphae of both kinds of mycorrhizae increase the absorptive surface area of their plant partner. The fungus shares minerals it takes up with the plant, and the plant gives up sugars to the fungus. Many plants do not as grow well when deprived of their fungal partner (Figure 22.10B).

---

lichen Composite organisms, consisting of a fungus and a single-celled alga or a cyanobacterium.
mutualism Mutually beneficial relationship between two species.
mycorrhiza Mutually beneficial partnership between a fungus and a plant root.

Figure 22.9 Lichens. **A** Leaflike lichen on a birch tree. **B** Cross-section of a leaflike lichen. **C** Encrusting lichens on granite.

dispersal fragment (cells of fungus and of photosynthetic species)

outer layer of fungal cells

photosynthetic species

inner layer of loosely woven hyphae

outer layer of fungal cells

Take-Home Message What types of partnerships do fungi form with other organisms?

> In lichens, a fungus shelters one or more photoautotrophs and shares carbon dioxide and mineral ions with them, while receiving some carbohydrates in return. Often the photosymbionts are nitrogen fixers, such as cyanobacteria.

> In mycorrhizae, a fungus living in or on a plant's young roots increases the plant's uptake of water and dissolved mineral ions and helps protect it from pathogens. The fungus withdraws some nutrients from its partner.

> Most fungi feed on organic wastes and remains, but some are pathogens that attack living plants or animals.

Plant Pathogens

Most fungi decompose dead plant material, but some species are parasites that feed on living plants. Such fungi can be ecologically and economically important plant pathogens. As noted earlier, airborne spores spread wheat stem rust, a type of club fungus. A wheat stem rust infection can reduce crop yield by up to 70 percent. As another example, rose growers must contend with powdery mildews. A whitish powder that appears on leaves is the spores of sac fungi that feed on the leaf tissues.

One fungus, accidentally introduced to the United States from Asia in the early 1900s, caused a dramatic change in the country's eastern forests. The fungus infects the American chestnut, causing chestnut blight. Before the fungus arrived, the American chestnut was an important timber species and its nuts nourished forest animals. The chestnut blight first appeared in New York. As winds spread fungal spores westward, most chestnut trees east of the Mississippi became infected and died. Today, few stands of mature trees remain. Shoots occasionally sprout from old root systems, but they cannot reach mature size.

Fungal pathogens of plants have even affected human history. One notorious sac fungus, *Claviceps purpurea*, parasitizes rye and other cereal grains (Figure 22.11). It produces toxic alkaloids that, when eaten, cause ergotism. Symptoms of ergotism include hallucinations, hysteria, and convulsions. Ergotism epidemics were common in Europe in the Middle Ages. They thwarted Peter the Great, a Russian czar who became obsessed with conquering ports along the Black Sea. Soldiers laying siege to those ports ate rye bread and fed it to their horses. The soldiers went into convulsions and the horses into "blind staggers." Ergotism may also have played a role in witch hunts that occurred in Salem, Massachusetts. Descriptions of the behavior of the supposedly bewitched women resemble the symptoms of ergotism.

Human Pathogens

In humans, fungal infections most frequently involve the skin. Typically the fungus feeds on the outermost dead

Figure 22.11 Grains of rye with spore-bearing structures of *Claviceps purpurea*. Eating grain infected by this species causes a type of food poisoning called ergotism.

High–Flying Fungi (revisited)

Species of *Fusarium*, a sac fungus, cause plant and human diseases. In 2006, *Fusarium* spores got into contact lens solution and caused eye infections worldwide. Of 122 infections in the United States, a third were so bad that patients required surgery to replace the cornea, the eye's clear layer. David Schmale of Virginia Tech (*above*) has found spores of many *Fusarium* species in air samples he collected using remote-controlled model aircraft.

How Would You Vote? One *Fusarium* strain kills opium poppies. Should spores of this fungus be sprayed in Afghanistan to reduce supplies of opium and heroin? See CengageNOW for details, then vote online (cengagenow.com).

skin layers by secreting enzymes that dissolve keratin, the main skin protein. Infected areas become raised, red, and itchy. For example, several species of fungus can take up residence in the skin between toes and on the sole of the foot, causing what is commonly called "athlete's foot" (*right*). Fungi also cause the skin infections misleadingly known as "ringworm." No worm is involved. Instead, a ring-shaped lesion appears and expands as fungal hyphae grow outward from the site of the initial infection.

Low numbers of single-celled fungi normally live in the vagina, but sometimes their population explodes, resulting in fungal vaginitis (a vaginal yeast infection). Symptoms usually include itching or burning sensations and a thick, odorless, whitish vaginal discharge. Intercourse is often painful. Disrupting the normal populations of bacteria in the vagina by douching or using antibiotics increases the risk of fungal vaginitis, as does use of oral contraceptives. Nonprescription medication placed into the vagina usually cures the infection. If this is not effective, a woman should consult a doctor.

Inhaled fungal spores also cause disease. Soil in the midwestern and south central United States holds spores that cause histoplasmosis. Most people that inhale the spores have only a brief episode of coughing. However, in immune-suppressed people, the fungus may spread from the lungs, through the blood, and into other organs, with fatal results. Similarly, spores in soil of the American Southwest can cause coccidioidomycosis, or valley fever.

Take-Home Message What are the effects of fungal pathogens?

> Fungal pathogens of plants reduce crop yield and sometimes kill plants.

> In humans, fungi cause annoying skin diseases and vaginitis. Fungal lung diseases can be deadly in people with weakened immune systems.

Summary

Section 22.1 Fungi produce microscopic spores that winds sometimes disperse long distances. As a result, it is difficult to prevent the spread of a fungal pathogen, such as wheat stem rust.

Section 22.2 Fungi are heterotrophs that secrete digestive enzymes on organic matter and absorb released nutrients. Most are **saprobes** that feed on organic remains; they serve as essential decomposers in most ecosystems. Other fungi live in or on organisms. They may be harmless, beneficial, or parasitic.

Fungi include single-celled yeasts and multicelled species. They disperse by releasing spores. In a multicelled species, spores germinate and give rise to filaments called **hyphae**. The filaments typically grow as an extensive mesh called a **mycelium**. Depending on the group, the cells of a hyphae may be haploid (*n*) or **dikaryotic** (*n+n*). In the oldest lineages, the hyphae have few or no cross-walls (septa). More recently evolved lineages have hyphae with porous cross-walls at regular intervals. In all groups, water and nutrients move freely between cells of a hypha.

Section 22.3 The oldest fungal lineage, the **chytrids**, are a mostly aquatic group and the only fungi with flagellated spores. **Zygote fungi** include many familiar molds that grow on fruits, breads, and other foods. Their hyphae are continuous tubes that have no cross-walls. Sexual reproduction produces a thick-walled zygospore. Zygote fungi also produce asexual spores atop specialized hyphae. **Glomeromycetes** are relatives of the zygote fungi that live in soil and extend their hyphae into plant roots. The hyphae branch inside plant cell walls and deliver nutrients to the cell.

Section 22.4 Sac fungi are the most diverse group. They include single-celled yeasts and multicelled species such as cup fungi and morels. Sac fungi produce sexual spores in asci. In multicelled species, these saclike structures form on an ascocarp, a fruiting body composed of haploid and dikaryotic hyphae that have internal cross-walls at regular intervals. Yeasts that are sac fungi help us make wine, breads, and alcoholic beverages. The antibiotic penicillin was originally isolated from a mold that is a sac fungus. Related molds are used to flavor cheeses.

Section 22.5 Like sac fungi, multicelled **club fungi** have hyphae with cross-walls and can produce complex reproductive structures, such as mushrooms. Club fungi are the only fungi that can break down lignin in wood, and are important as decomposers in forest habitats.

Typically, a dikaryotic mycelium dominates the life cycle of multicelled species. It grows by mitosis and, in some species, extends through a vast volume of soil.

When conditions favor reproduction, a reproductive structure, also made up of dikaryotic hyphae, develops. A mushroom is an example. Nuclei fuse in club-shaped cells at the edges of sheets of tissue called gills. Meiosis of the resulting diploid cell produces haploid spores at the tips of the club-shaped cells.

Section 22.6 Many fungi take part in some type of **mutualism**.

A **lichen** is a composite organism composed of a fungus and photosynthetic cells of a green alga or cyanobacterium. The fungus makes up the bulk of the lichen body and obtains a supply of nutrients from its photosynthetic partner. Lichens disperse by releasing fragments that include cells of both partners. The fungus also produces spores. Lichens are important as pioneers in new habitats. They facilitate the breakdown of rock to form soil.

A **mycorrhiza** is an interaction between a fungus and a plant root. Fungal hyphae surround or penetrate the roots and supplement the root's absorptive surface area. The fungus shares absorbed mineral ions with the plant and gets some photosynthetic sugars in return. Most plants have mycorrhizal fungi and grow less well without them.

Section 22.7 Some fungi are pathogens. Fungal pathogens cause powdery mildew of roses and led to the demise of American chestnuts. A fungus that infects cereal grains makes a toxin that causes ergotism, a type of food poisoning. Fungi also infect the human body, most frequently the skin. Yeast infections of the vagina are also common. Inhaled fungal spores can germinate in lungs and cause diseases that can be fatal to people with a weakened immune system.

Self-Quiz   Answers in Appendix III

1. All fungi _____ .
   a. are multicelled   c. are heterotrophs
   b. form flagellated spores   d. all of the above

2. Saprobic fungi obtain nutrients from _____ .
   a. nonliving organic matter   c. living animals
   b. living plants   d. both b and c

3. In _____ , a hypha has few or no cross-walls.
   a. zygomcyetes   c. club fungi
   b. sac fungi   d. all of the above

4. The yeasts whose fermentation reactions produce carbon dioxide that makes bread rise are a type of _____ .
   a. chytrid   c. sac fungus
   b. zygote fungus   d. club fungus

5. In many _____ , an extensive dikaryotic mycelium is the most conspicuous phase of the life cycle.
   a. chytrids   c. sac fungi
   b. zygote fungi   d. club fungi

6. The mycelium of a multicelled fungus is a mesh of filaments, each called a _____ .

Data Analysis Activities

Fighting a Forest Fungus The club fungus *Armillaria ostoyae* infects living trees and acts as a parasite, withdrawing nutrients from them. After the tree dies, the fungus continues to feed on its remains. Fungal hyphae grow out from the roots of infected trees as well as the roots of dead stumps. If these hyphae contact roots of a healthy tree, they can invade and cause a new infection.

Canadian forest pathologists hypothesized that removing fungus-infected stumps after logging could help prevent tree deaths. To test this hypothesis, they carried out an experiment. In half of a forest they removed stumps after logging. In a control area, they left stumps behind. For more than 20 years, they recorded tree deaths and whether *A. ostoyae* caused them. Figure 22.12 shows the results.

1. Which tree species was most often killed by *A. ostoyae* in control forests? Which was least affected by the fungus?
2. For the most affected species, what percentage of deaths did *A. ostoyae* cause in control and in experimental forests?
3. Looking at the overall results, do the data support the hypothesis? Does stump removal reduce tree mortality from *A. ostoyae*?

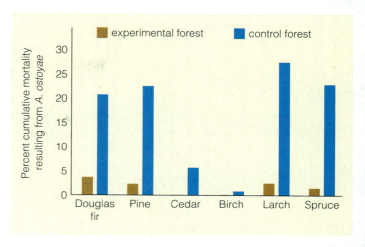

Figure 22.12 Results of a long-term study of how logging practices affect tree deaths caused by the fungus *A. ostoyae*. In the experimental forest, whole trees—including stumps—were removed (*brown* bars). The control half of the forest was logged conventionally, with the stumps left behind (*blue* bars).

---

7. A mushroom is _____ .
   a. the digestive organ of a club fungus
   b. the only part of the fungal body made of hyphae
   c. a reproductive structure that releases sexual spores
   d. the only diploid phase in the club fungus life cycle

8. Spores released from a mushroom's gills are _____ .
   a. diploid   b. haploid   c. dikaryotic

9. _____ are fungi that produce flagellated spores.
   a. Chytrids   c. Zygote fungi
   b. Sac fungi   d. Club fungi

10. Nitrogen-fixing cyanobacteria often interact with a fungus as a _____ .
   a. mycelium   c. mycorrhiza
   b. lichen   d. mycosis

11. _____ are mycorrhizal fungi with hyphae that grow into a root cell and branch inside it.
   a. Glomeromycetes   c. Zygote fungi
   b. Chytrids   d. Club fungi

12. Histoplasmosis is an example of a _____ .
   a. mycelium   c. mycorrhiza
   b. lichen   d. mycosis

13. Chestnut blight _____ .
   a. altered the species composition of eastern forests
   b. was caused by an introduced fungal pathogen
   c. was spread by wind-dispersed spores
   d. all of the above

14. Ergotism _____ .
   a. results from inhalation of fungal spores
   b. produces hallucinations and convulsions
   c. affects only those with weak immune systems
   d. all of the above

15. Match the terms appropriately.
   ____ hypha   a. produces flagellated spores
   ____ chitin   b. component of fungal cell walls
   ____ chytrid   c. partnership between a fungus and
   ____ sac fungus   one or more photoautotrophs
   ____ club fungus   d. filament of a mycelium
   ____ lichen   e. fungus–root partnership
   ____ mycorrhiza   f. forms sexual spores in an ascus
                       g. many form mushrooms

Additional questions are available on **CENGAGENOW**.

Critical Thinking

1. Fungal skin diseases are persistent. Ointments and creams may not reach the deepest infected skin layers. Oral antifungal drugs are far less common than antibacterials and often have bad side effects. Reflect on the evolutionary relationships among bacteria, fungi, and humans. Explain why it is more difficult to create drugs against fungi than bacteria.

2. Certain toxic mushrooms have bright, distinctive colors that mushroom-eating animals learn to recognize. Once sickened, day-feeding animals tend to avoid them. Other toxic mushrooms are as dull-looking as edible ones, but they have a much stronger odor. Some scientists think the strong odors aid in defense against predators that are active at night. Explain their reasoning.

Animations and Interactions on **CENGAGENOW**:
❯ Zygote fungus life cycle; Predatory fungus; Club fungus life cycle.

Key Concepts

Introducing the Animals

Animals are multicelled heterotrophs that digest food inside their body. Most animals can move from place to place during some or all of the life cycle. Animals probably evolved from a colonial protist. The vast majority of modern animals are invertebrates.

Structurally Simple Invertebrates

Sponges filter food from the water and have an asymmetrical body with no true tissues. Cnidarians are predators. Their body is radially symmetrical, with two tissue layers and a gelatinous matrix between the two.

23 Animals I: Major Invertebrate Groups

23.1 Old Genes, New Drugs

When asked to think of an animal, you probably envision one of the **vertebrates**, an animal that has a backbone. Mammals, birds, reptiles, amphibians, and fishes fall into this category. However, the vast majority of animals (more than 97 percent) have no backbone and are commonly called **invertebrates**.

The many invertebrate species represent a vast and largely unexplored reservoir of genetic diversity. Consider for example the cone snails (genus *Conus*), a group most common in warm, nearshore waters of the South Pacific. Humans have long valued the snails for their elaborately patterned shells (*right*). But cone snails interest biologists for additional reasons. The snails are predators and, like some snakes, they produce venom that helps them subdue their prey.

Studies of how cone snail venoms function have led to the development of new drugs. For example, one fish-eating cone snail, *C. magus*, captures prey by harpooning it, pumping it full of venom, then reeling it in. The snail's venom contains a peptide that interferes with the ability to feel pain. When injected into a fish, this peptide keeps the fish from writhing in pain and possibly damaging the snail's harpoon. An injectable pain-killing drug called ziconotide (sold under the brand name Prialt) is a synthetic version of the *C. magus* peptide. Unlike morphine and similar drugs, ziconotide is nonaddictive. Peptides from the venom of another fish-eating cone snail, *C. geographus* (shown at the *left*), may help control convulsions and seizures.

Studies of cone snail peptides also shed light on the evolution of animal genomes. For example, one gene that encodes an enzyme in the venom production pathway of *C. geographus* also exists in fruit flies and humans. Thus, this gene must have existed in the common ancestor of snails, insects, and vertebrates, which lived about 500 million years ago. After these groups diverged, the gene mutated independently in each lineage, and its product diverged in function. In humans, the gene product helps repair blood vessels. We have yet to figure out what role the gene plays in fruit flies.

This example supports an organizing principle in the study of life. Look back through time, and you discover that all organisms are related. At each branch point in the animal family tree, mutations gave rise to changes in biochemistry, body plans, or behavior. The mutations were the source of unique traits that define each lineage.

This chapter focuses on the major groups of invertebrates. The next chapter continues the story, focusing on vertebrates and their closest invertebrate relatives. Even with two chapters, we can scarcely do justice to so many animal species. We can only consider the scope of diversity and the major evolutionary trends. Keep in mind that although invertebrates are structurally simpler than vertebrates, they are not evolutionarily inferior. The diversity and longevity of the invertebrate lineages attests to how well they are adapted to their environment.

invertebrate Animal that does not have a backbone.
vertebrate Animal with a backbone.

Bilateral Invertebrates
Most animals have a body with bilateral symmetry and a concentration of nerve cells in the head. All bilateral animals have organ systems. In flatworms, the organs are hemmed in by tissue. In other bilateral animals, the organs reside inside a fluid-filled body cavity.

The Most Successful Animals
In terms of diversity, arthropods are the most successful animals. Crustaceans abound in the seas and insects are the most diverse animal group on land. Insects play essential roles in ecosystems. They also have economic and health effects.

On the Road to Chordates
Echinoderms are on the same branch of the animal family tree as the vertebrates (animals that have a backbone). Echinoderm adults have a spiny skin and a radially symmetrical body. The larvae show bilateral symmetry, indicating descent from a bilateral ancestor.

> No one trait defines all animals. It takes a list of features to see what sets them apart from all other organisms.
< Links to Animal life cycle 12.5, Tissues 1.2

What Is an Animal?

An **animal** is a multicelled heterotroph with unwalled cells. In contrast to fungi, animals typically ingest food (take it in) and digest it inside their body. Nearly all animals are motile (can move from place to place) during all or some part of their life cycle. Cells of an animal body are diploid. During sexual reproduction, meiosis of specialized germ cells produces haploid gametes. The flagellated sperm swim to eggs, and fertilization produces a zygote. Cells differentiate as the animal develops from an embryo (an early developmental stage) to an adult.

Evolution of Animal Body Plans

Figure 23.1 shows an evolutionary tree for the major animal groups covered in this book. All animals are descended from a common multicelled ancestor ❶ and constitute the clade Metazoa.

The earliest animals were probably aggregations of cells, and sponges still show this level of organization. However, most animals have tissues ❷. Each tissue consists of specific cell types arranged in a way that allows the tissue to carry out a particular task. Tissues develop from embryonic germ layers. In the early animal lineages,

Figure 23.2 How a three-layer animal embryo forms. Most animals have this type of embryo.

embryos had two such layers: an outer **ectoderm** and an inner **endoderm**. In later lineages, cells began to rearrange themselves and form a middle embryonic layer called **mesoderm** (Figure 23.2). The evolution of a three-layer embryo allowed an increase in structural complexity. Most animal groups have organs derived from mesoderm.

The structurally simplest animals such as sponges are asymmetrical. You cannot divide their body into halves that are mirror images. Jellies, sea anemones, and other cnidarians have **radial symmetry**: Body parts are repeated around a central axis, like the spokes of a wheel ❸. Radial animals have no front or back end. They attach to an underwater surface or drift along. Thus, their food can arrive from any direction. Most animals have **bilateral symmetry**. They have many paired structures, so the

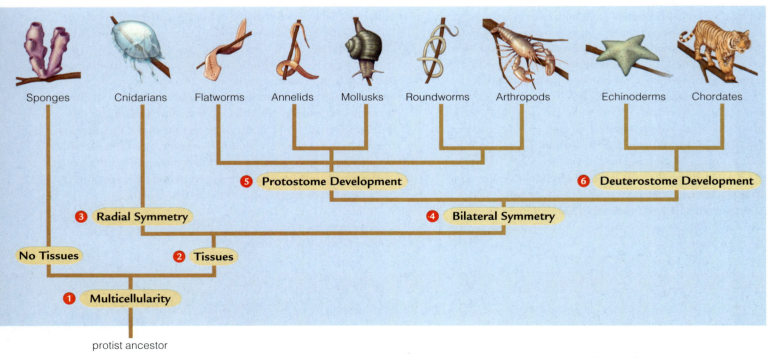

Sponges Cnidarians Flatworms Annelids Mollusks Roundworms Arthropods Echinoderms Chordates

❺ **Protostome Development** ❻ **Deuterostome Development**

❸ **Radial Symmetry** ❹ **Bilateral Symmetry**

No Tissues ❷ **Tissues**

❶ **Multicellularity**

protist ancestor

Figure 23.1 One proposed evolutionary tree for major animal groups based on morphological, developmental, and genetic comparisons. Like all such trees, it is a hypothesis and open to revision in light of new data.

>> Figure It Out What invertebrate group is most closely related to chordates?

Answer: Echinoderms

right and left half of the body are mirror images ❹. Bilateral animals typically have **cephalization**. Many nerve cells and sensory structures have become concentrated at the anterior (front) end of the body. Bilateral animals move through the world head first, and they use the sensory structures on their head to seek out food and detect potential threats.

The head end of most bilateral animals has an opening for taking in food. Flatworms have a saclike digestive cavity. Food and wastes leave through the same body opening. However, most bilateral animals have a tubular gut, with a mouth at one end and an anus at the other. A tubular gut, known as a complete digestive system, has advantages. Parts of the tube can be specialized for taking in food, digesting food, absorbing nutrients, or compacting the wastes. Unlike a saclike cavity, a complete digestive system can carry out all of these tasks simultaneously.

Two lineages of bilateral animals differ in their embryonic development. In **protostomes**, the first opening that appears on the embryo becomes the mouth ❺. *Proto–* means first and *stoma* means opening. In **deuterostomes**, the mouth develops from the second embryonic opening, and the first becomes the anus ❻.

A mass of tissues and organs surrounds the flatworm gut (Figure 23.3A), but a fluid-filled body cavity surrounds the gut of most bilateral animals. A few animals such as roundworms enclose their gut in an unlined cavity called a **pseudocoelom** (Figure 23.3B). More typically, bilateral animals have a **coelom**, a body cavity lined with a tissue derived from mesoderm (Figure 23.3C).

A fluid-filled coelom or pseudocoelom provides three advantages. First, materials can diffuse through the fluid to body cells. Second, muscles can redistribute the fluid to alter the body shape and aid locomotion. Third, organs are not hemmed in by a mass of tissue, so they can grow larger and move more freely.

A No coelom (acoelomate animal)

epidermis — gut cavity

organs packed between gut and body wall

B Pseudocoelom (pseudocoelomate animal)

epidermis — gut cavity

unlined body cavity around gut

C Coelom (coelomate animal)

epidermis — gut cavity

body cavity with a lining (*dark blue*) derived from mesoderm

Figure 23.3 Animated Main body cavity of bilateral animals. **A** Flatworm, with no body cavity. **B** Roundworm, with an unlined, fluid-filled body cavity (a pseudocoelom). **C** Vertebrate (human) with a body cavity lined with tissue derived from mesoderm (*dark blue*).

---

animal Multicelled heterotroph with unwalled cells. Most ingest food and are motile during at least part of the life cycle.
bilateral symmetry Having paired structures so the right and left halves are mirror images.
cephalization Having a concentration of nerve and sensory cells at the head end.
coelom Body cavity lined with tissue derived from mesoderm.
deuterostomes Lineage of bilateral animals in which the second opening on the embryo surface develops into a mouth.
ectoderm Outermost tissue layer of an animal embryo.
endoderm Innermost tissue layer of an animal embryo.
mesoderm Middle tissue layer of a three-layered animal embryo.
protostomes Lineage of bilateral animals in which the first opening on the embryo surface develops into a mouth.
pseudocoelom Unlined body cavity.
radial symmetry Having parts arranged around a central axis, like spokes around a wheel.
segmentation Having a body composed of similar units that repeat along its length.

Most bilaterally symmetrical animals have some degree of **segmentation**, a division of a body into similar units, repeated one after the other along the main axis. We clearly see body segments in annelids such as earthworms. Segmentation opened the way to evolutionary innovations in body form. When many segments have structures that carry out the same function, some segments can become modified without endangering the animal's survival.

Take-Home Message What are animals?

> Animals are multicelled heterotrophs with bodies made of diploid, unwalled cells. Animals develop in stages and most are motile.

> Early animals had no body symmetry. Later, radially symmetrical, and then bilaterally symmetrical bodies evolved. Most animals with bilateral symmetry have a coelom, a fluid-filled body cavity with a lining derived from mesoderm.

❭ Fossils and gene comparisons among modern species provide insights into how animals arose and diversified.
❬ Links to Homeotic gene 10.3, Biomarker 18.5, Speciation 17.10

Colonial Origins

According to the colonial theory of animal origins, the first animals evolved from a colonial protist. At first, all cells in the colony were similar. Each could reproduce and carry out all other essential tasks. Later, mutations produced cells that specialized in some tasks and did not carry out others. Perhaps these cells captured food more efficiently, but did not make gametes. Colonies that had interdependent cells and a division of labor were at a selective advantage, and new specialized cell types evolved. Eventually this process produced the first animal.

What was the ancestral cell like? **Choanoflagellates**, the modern protists most closely related to animals, provide clues. Their name means "collared flagellate." Each choanoflagellate cell has a collar of threadlike structures, called microvilli, surrounding a flagellum (Figure 23.4A). Movement of the flagellum directs food-laden water past microvilli, which filter out food. As you will see, sponges feed the same way. Some choanoflagellates live as single cells, while others form colonies (Figure 23.4B).

The Simplest Living Animal

Studies of the modern representatives of ancient animal lineages can give us an idea of what the earliest animals may have been like. For example, **placozoans** are the oldest group with living representatives. The single species, *Trichoplax adhaerens*, is an asymmetrical marine animal about 2 millimeters in diameter and 2 micrometers thick (Figure 23.4C). It has the simplest body structure of any animal, with only four types of cells. It also has the smallest genome. Ciliated cells on the placozoan's surface allow it to move about. It feeds on bacteria and algae.

Genetic studies of *T. adhaerens* revealed the genetic foundations for traits that evolved in more recent animal lineages. For example, *T. adhaerens* has no nerve cells, but it has genes like those that encode signaling molecules in human nerves. It also has a gene similar to the homeotic genes that regulate development of body parts in more complex animals.

Figure 23.5 Fossil of *Spriggina*, an Ediacaran, that lived about 570 million years ago. It was about 3 centimeters (1 inch) long and may have been a soft-bodied ancestor of arthropods, a major group that diversified during the Cambrian.

Fossil Evidence

Biomarkers characteristic of certain sponges have been found in deposits that date back more than 600 million years. Morphological fossils show that by 570 million years ago, a diverse collection of multicelled organisms, including some animals, lived in the seas (Figure 23.5). These groups are collectively called Ediacarans because their fossils were first discovered in Australia's Ediacara Hills.

Animals underwent a dramatic adaptive radiation during the Cambrian (542–488 million years ago). By the end of this period, all major animal lineages were present in the seas. What caused this Cambrian explosion in diversity? Rising oxygen levels and changes in global climate may have played a role. Also, supercontinents were breaking up. Movement of land masses isolated populations, thus increasing opportunities for allopatric speciation. Biological factors also encouraged speciation. Once the first predators arose, mutations that produced protective hard parts were favored. Evolution of homeotic genes may have sped things along. Mutations in these genes would have allowed adaptive changes to body form in response to predation or altered habitat conditions.

choanoflagellate Member of the protist group most closely related to animals.
placozoan The simplest modern animal, with an asymmetrical flat body, four types of cells, and a small genome.

Figure 23.4 A One free-living choanoflagellate. A collar of microvilli rings its flagellum. **B** Colony of choanoflagellates. Some researchers view it as a model for the origin of animals. **C** The placozoan, *Trichoplax adhaerens*. Its asymmetrical, two-layered body has ciliated surfaces. The red color of this individual results from a diet of red algae.

Take-Home Message How did animals originate and diversify?

❭ The ancestor of all animals was probably a colonial protist. It may have resembled choanoflagellates, the modern protist group most closely related to animals.

❭ The oldest known animal lineage, placozoans, includes one marine species. It is tiny, with only four types of cells.

❭ Animals may have originated more than 600 million years ago. A great adaptive radiation that occurred during the Cambrian period was probably encouraged by a combination of environmental and biological factors.

> Sponges have no tissues or organs. They filter food from the water that flows through their porous body.
‹ Link to Phagocytosis 5.8

General Characteristics

Sponges (phylum Porifera) are aquatic animals with no symmetry, tissues, or organs. They resemble a colony of choanoflagellates but with more kinds of cells and a greater division of labor. Most live in tropical seas.

An adult sponge lives attached to a surface and has a body riddled with pores. Flat, nonflagellated cells form the outer surface, flagellated collar cells line the inner one, and a jellylike matrix lies in between (Figure 23.6A).

Most sponges feed on bacteria filtered from water. As in choanoflagellates, movement of flagella drives food-laden water past a collar of microvilli. As water flows through pores in the body wall, collar cells trap bits of food and engulf them by phagocytosis. All digestion is intracellular. Amoeba-like cells in the matrix receive food from the collar cells and distribute it through the body.

In many species, cells in the matrix secrete fibrous proteins or glassy silica spikes. These materials structurally support the body and discourage predators. Some of the protein-rich sponges are commercially important. These sponges are harvested from the sea, dried, cleaned, and bleached. The rubbery protein remains are then sold for use in bathing and cleaning (Figure 23.6B). About $40 million worth of sponges are harvested each year.

Sponge Reproduction

A typical sponge is a **hermaphrodite**; it produces both eggs and sperm. Sperm are released into the water (Figure 23.7A). Eggs are held in the parental body. Fertilization produces a zygote that develops into a ciliated larva (Figure 23.7B). A **larva** (plural, larvae) is a free-living, sexually immature stage in an animal life cycle. Sponge larvae swim briefly, then settle and become adults.

Many sponges reproduce asexually when small buds or fragments break away and grow into new sponges. Some freshwater species survive oxygen-poor water, drying out, or freezing by producing gemmules: tiny clumps of resting cells encased in a hardened coat. Wind can disperse gemmules. Those that land in a hospitable habitat become active again and grow into new sponges.

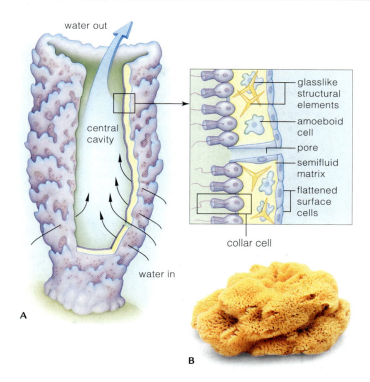

Figure 23.6 Animated The sponge body. **A** Body plan of a simple sponge. Flattened cells cover the outer surface and line pores. Flagellated collar cells line inner canals and chambers. Microvilli of these cells act like a sieve that strains food from the water. Cells in the matrix distribute nutrients and secrete structural elements. **B** Dried protein-rich remains of a bath sponge.

Figure 23.7 Sexual reproduction in sponges. **A** Barrel sponge releasing sperm. **B** Colorized scanning electron micrograph of a sponge larva.

---

hermaphrodite Animal that makes both eggs and sperm.
larva Preadult stage in some animal life cycles.
sponge Aquatic invertebrate that has no tissues or organs and filters food from the water.

Take-Home Message What are sponges?

> Sponges are filter-feeders that have no symmetry, tissues, or organs. Fibers and glassy spikes in the body wall support the body and help deter predators.

> Sponges are hermaphrodites. Their ciliated larvae swim about briefly before settling and developing into adults.

❯ Nearly all cnidarians live in the seas. They have radially symmetrical bodies and a mouth surrounded by tentacles.
❮ Link to Dinoflagellates 20.5

General Characteristics

Cnidarians (phylum Cnidaria) include radially symmetrical animals such as corals, sea anemones, and jellyfish (also called jellies). Nearly all are marine predators. There are two cnidarian body shapes, the medusa and the polyp (Figure 23.8). In both, a tentacle-ringed mouth opens onto a saclike gastrovascular cavity that functions in digestion and gas exchange.

Medusae such as jellyfish are shaped like a bell or umbrella, with a mouth on the lower surface. Most swim or drift about. Polyps such as sea anemones are tubular and one end usually attaches to a surface.

Both medusae and polyps consist of two tissue layers. The outer layer (epidermis) develops from ectoderm, and the inner layer (gastrodermis) from endoderm. Mesoglea, a jellylike secreted matrix, fills the space between the tissue layers. Medusae tend to have a lot of mesoglea. Polyps usually have somewhat less.

The name Cnidaria is from *cnidos*, the Greek word for nettle, a kind of stinging plant. Cnidarian tentacles have stinging cells (cnidocytes) with unique organelles called **nematocysts**. Nematocysts help capture prey and also

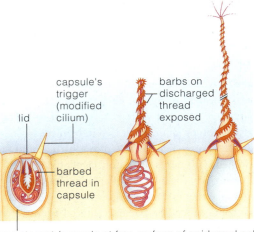

nematocyst (capsule at free surface of epidermal cell)

Figure 23.9 Animated Nematocyst action. In response to a touch of the trigger, the capsule around the nematocyst becomes more permeable. As water diffuses in, pressure builds up, and the thread is forced to turn inside out. The barbed tip pierces prey.

function as a defense. Like a jack-in-the-box, a nematocyst holds a coiled thread beneath a hinged lid (Figure 23.9). When something brushes against the nematocyst's trigger, the lid opens. The thread inside pops out and ejects a barbed thread. Typically, the barbs snag prey. Tentacles then move the prey through the mouth and deposit it in the gastrovascular cavity. Gland cells in the lining of this cavity secrete enzymes that digest the prey.

Occasionally, people trigger nematocyst release by brushing against a coral or jellyfish while swimming. Nematocysts irritate human skin and the stings of one group of jellyfish, known as box jellies, can be deadly.

Coordinated movement of body parts such as tentacles requires interactions among nerves and contractile cells. Cnidarians have a decentralized nervous system called a **nerve net**. This network of nerve cells extends through the body. Tentacles and other body parts move when nerve cells signal contractile cells to shorten. Contraction of these cells redistributes mesoglea, just as a water-filled balloon changes shape when you squeeze it. Mesoglea is a **hydrostatic skeleton**, an enclosed fluid that contractile cells exert force on to bring about movement.

Cnidarian Life Cycles and Diversity

Depending on the group, a cnidarian life cycle may include both polyps and medusae, only polyps, or only medusae. Figure 23.10 shows the life cycle of *Obelia*, a colonial hydroid that produces both polyps and medusae. Sexual reproduction occurs after medusae form on special polyps. Medusae release gametes that fuse, forming a zygote. As in most cnidarians, bilateral ciliated larvae develop. They swim briefly, then settle and develop into

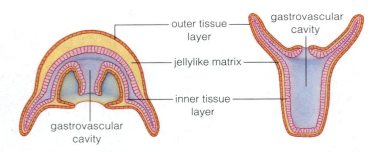

A Diagram showing cross-section of the medusa (*left*) and the polyp (*right*).

B Medusa of the hydroid *Obelia*.

C A sea anemone, a polyp.

Figure 23.8 Animated Cnidarian body plans.

Figure 23.10 **Animated** Life cycle of *Obelia*, a hydroid with medusa and polyp generations. Figure 23.8B is a photo of the medusa.

1 Medusae are the sexual generation. Each medusa makes either eggs or sperm and releases them into the water. The gametes combine to form a zygote.

2 The zygote develops into a ciliated bilateral larva.

3 The larva swims about for a while, then settles and develops into a polyp.

4 The polyp grows and reproduces asexually, eventually producing a branching colony.

5 Some branches of the colony are specialized for capturing and eating prey.

6 Other branches produce and release medusae that begin the sexual phase of the life cycle again.

reproductive polyp — female medusa — male medusa — ovum — sperm — zygote — feeding polyp — one branch of a colony — growth of a polyp — ciliated bilateral larva

polyps. The colony grows asexually when new polyps branch from older ones.

Sea anemones (Figure 23.8C), corals (Figure 23.11A), and hydras (Figure 23.11B) do not have a medusa stage in their life cycle. During sexual reproduction, gametes form on polyps. Polyps also reproduce asexually by budding.

Coral reefs are colonies of polyps enclosed in a skeleton of secreted calcium carbonate. In a mutually beneficial relationship, photosynthetic dinoflagellates live inside each polyp's tissues. The protists receive shelter and carbon dioxide from the coral, which gets sugars and oxygen in return. If a reef-building coral loses its protist symbionts, an event called "coral bleaching," it may die.

Siphonophores are cnidarian colonies that float or drift in the seas. For example, the Portuguese man-of-war (*Physalia*) in Figure 23.11C appears to be a single animal. In fact, it consists of many specialized polyps. The gas-filled float is one specialized polyp. Polyps that catch prey, digest prey, or produce gametes cluster beneath it.

Decline of the Corals, Rise of the Jellies

Increasing sea temperature and marine pollution may be contributing to two opposing trends among cnidarian populations. Reef-building coral populations are in decline, while population explosions of jellyfish have become increasingly common. The reef declines threaten not only corals, but also the many species that live in, on, or around reefs. Increasing jellyfish numbers present problems too. They sting swimmers, force beach closures, and can clog underwater pipes.

Figure 23.11 Variations on polyp structure.

A Polyps of a reef-building coral. Each polyp secretes calcium carbonate, and the colony's accumulated secretions form the skeleton of the reef.

B A hydra, one of the relatively few freshwater cnidarians, shown here with a captured water flea.

C The Portuguese man-of-war (also known as the blue bubble) consists of four kinds of polyp: the gas-filled float, prey-catching polyps, food-digesting polyps, and gamete-producing polyps. Swimmers who brush against the tentacles of this species end up with painful red welts.

cnidarian Radially symmetrical invertebrate with two tissue layers; uses tentacles with stinging cells to capture food.
hydrostatic skeleton Of soft-bodied invertebrates, a fluid-filled chamber that muscles act on, redistributing the fluid.
nematocyst Stinging organelle unique to cnidarians.
nerve net Mesh of nerve cells that allows movement and other behavior in cnidarians.

Take-Home Message What are cnidarians?

❯ Cnidarians are radially symmetrical predators with two layers of tissue.

❯ The two body forms, medusa and polyp, have tentacles with nematocysts around the mouth, which opens into a gastrovascular cavity.

❯ Movements occur when a nerve net signals contractile cells.

> Most flatworms live in the sea, but some live in fresh water or inside the bodies of other animals.
❮ Links to Organs and organ systems 1.2, Osmosis 5.6

With this section, we begin our survey of protostomes, one of the two lineages of bilaterally symmetrical animals. **Flatworms** (phylum Platyhelminthes) are bilateral worms with an array of organ systems, but (as shown in Figure 23.3A) no body cavity. Like other protostomes, flatworms develop from an embryo with three layers.

Turbellarians, flukes, and tapeworms are the main flatworm lineages. The typical turbellarian lives in tropical marine waters (Figure 23.12). However, a lesser number live in fresh water, and a few live in damp places on land. All flukes and tapeworms are parasites of animals.

Structure of a Free–Living Flatworm

Planarians, a type of turbellarian, commonly live in ponds. They glide along, propelled by the movement of cilia that cover the body surface. The planarian's saclike gut is highly branched (Figure 23.13A). Nutrients and oxygen diffuse from the gut's fine branches to all body cells. There is no anus; food enters and wastes leave through the mouth. Surprisingly, the mouth is not at the planarian's head end. Instead, it is at the tip of a muscular tube that extends from the animal's ventral (lower) surface.

There is clear cephalization. The planarian head has chemical receptors and light-detecting eyespots. These sensory structures send messages to a simple brain consisting of paired nerve cell bodies (ganglia). Nerve cords

Figure 23.12 A marine flatworm. Brightly colored flatworms are common inhabitants of coral reef ecosystems.

extend out from the head and run the length of the body (Figure 23.13B).

A planarian is a hermaphrodite, with both female and male sex organs (Figure 23.13C). However, individuals cannot fertilize their own eggs. They must mate. Planarians also reproduce asexually. The body splits in two near the middle, then each piece regrows the missing parts. Regrowth also occurs if a planarian is cut in two.

The body fluid of a planarian holds more solutes than its freshwater environment, so water tends to move into the body by osmosis. A system of tubes regulates internal water and solute levels. Flame cells with a tuft of "flickering" cilia drive any excess water into these tubes, which open to the body surface at a pore (Figure 23.13D–F).

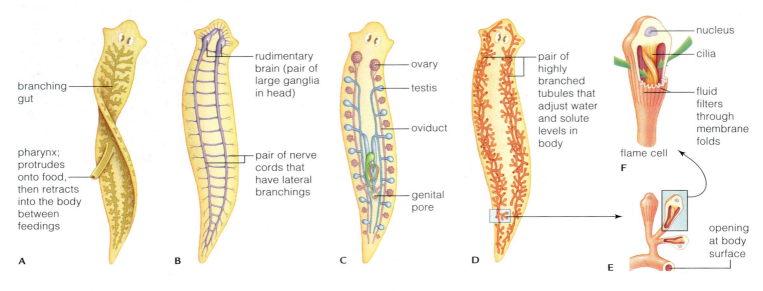

branching gut

pharynx; protrudes onto food, then retracts into the body between feedings

A

rudimentary brain (pair of large ganglia in head)

pair of nerve cords that have lateral branchings

B

ovary

testis

oviduct

genital pore

C

pair of highly branched tubules that adjust water and solute levels in body

D

flame cell

F

nucleus

cilia

fluid filters through membrane folds

opening at body surface

E

Figure 23.13 **Animated** Organ systems of a planarian, one of the flatworms. **A** Digestive system. This branching saclike gut is connected to a pharynx, which protrudes onto food, then retracts into a chamber between feedings. **B** Nervous system. **C** Reproductive system. **D–F** Water-regulating system.

Figure 23.14 Animated Life cycle of the beef tapeworm.

❶ A person eats undercooked beef containing the resting stage of the tapeworm.

❷ In the human intestine, the tapeworm uses its barbed scolex to attach to the intestinal wall. The worm grows by adding new body units called proglottids. Over time, it can become many meters long.

❸ Each proglottid produces both eggs and sperm, which combine. Proglottids with fertilized eggs leave the body in feces.

❹ Cattle eat grass contaminated with proglottids or early larvae.

❺ The larval tapeworm forms a cyst in the cattle muscle.

beef with larval tapeworm

proglottid · scolex

tapeworm in cattle muscle

proglottid with fertilized eggs

larva

Flukes and Tapeworms—The Parasites

Flukes and tapeworms are parasitic flatworms. Immature stages often live in one or more intermediate hosts, then reproduction occurs in the main host. For example, aquatic snails are the intermediate host for the blood fluke (*Schistosoma*). But the fluke can only reproduce inside a mammal, such as a human. Humans become infected when they wash or walk in water where infected snails live. The blood fluke crosses their skin and reproduces in their body, causing schistosomiasis. Worldwide, about 200 million people have this disease. It is most common in Asia.

Like flukes, tapeworms are parasites that live and reproduce in the vertebrate gut. The tapeworm head has a scolex, a structure with hooks or suckers that allow the worm to attach to the intestinal wall. Behind the head are body units called proglottids.

Unlike planarians and flukes, the tapeworm has no gut. Nutrients from the food its host ingests reach tapeworm cells by diffusing across the worm's body wall.

Figure 23.14 shows the life cycle of a beef tapeworm. Larvae enter the body when a person eats undercooked beef that contains larvae ❶. In the intestine, the tapeworm grows as new proglottids form in the region behind the scolex ❷. Each proglottid is hermaphroditic, and sperm from one can fertilize eggs in another. Older proglottids (farthest from the scolex) contain fertilized eggs. The oldest proglottids break off and exit the body in feces ❸. Fertilized eggs can survive for months before being eaten by cattle, the intermediate host ❹. Inside cattle, the tapeworm forms a cyst in muscle ❺.

Eating undercooked pork or fish can also cause a human tapeworm infection. Several outbreaks of human infection by fish tapeworms have been traced to raw salmon. Freezing fish for a week before it is served raw can eliminate the risk of tapeworms.

---

flatworm Bilaterally symmetrical invertebrate with organs but no body cavity; for example, a planarian or tapeworm.
planarian Free-living freshwater flatworm.

Take-Home Message What are flatworms?

❯ Flatworms are bilateral, cephalized animals with a saclike gut, simple nervous system, and a system for regulating the content of internal body fluids. The organs develop from three tissue layers that form in the embryo. Flatworms do not have a body cavity.

❯ Free-living flatworms include many tropical marine species and the freshwater planarians.

❯ Flukes and tapeworms are parasitic flatworms. Both groups include species that infect humans.

> Annelids are segmented worms that live in aquatic environments or burrow through the soil on land.
< Link to Chitin 3.3

Annelids are bilateral worms with a coelom and a body that is distinctly segmented, inside and out. The majority of species are marine worms called polychaetes. Other groups are oligochaetes (including earthworms) and leeches. In both polychaetes and oligochaetes nearly all segments have chitin-reinforced bristles called chaetae. Hence the names polychaete and oligochaete (*poly*–, many; *oligo*–, few).

Marine Polychaetes

The best-known polychaetes are the sandworms (Figure 23.15A). They are often sold as bait for saltwater fishing. Sandworms use their chitin-strengthened jaws to capture other soft-bodied invertebrates. Each body segment has a pair of paddlelike appendages that help the worm burrow in sediments and pursue prey.

Other polychaetes have modifications of this basic body plan. The fan worms and feather duster worms live inside a tube made of secreted mucus and sand grains. The head end protrudes from the tube and its elaborate tentacles capture food that drifts by (Figure 23.15B).

Leeches

Leeches live most commonly in fresh water, but some are marine or dwell in damp habitats on land. The leech body lacks conspicuous bristles and has a sucker at either end. Most leeches are scavengers and predators of small invertebrates. The infamous minority attach to a vertebrate, pierce its skin, and suck blood (Figure 23.16).

The saliva of bloodsucking leeches contains a protein that prevents blood from clotting while the leech feeds. For this reason, doctors who reattach a severed finger or

Figure 23.16 The leech *Hirudo medicinalis* feeding on human blood. A leech sticks to the skin with suckers at either end of the body, then draws blood with chitin-hardened jaws.

ear sometimes apply leeches to the reattached part. The presence of the leeches prevents unwanted clots from forming inside blood vessels of the reattached part.

Earthworms—Oligochaetes That Burrow

Oligochaetes include marine and freshwater species, but the land-dwelling earthworms are most familiar. We consider their body in detail as our example of annelid structure (Figure 23.17).

An earthworm body is segmented inside and out. The outer layer is a cuticle of secreted proteins. Visible grooves on its surface correspond to internal partitions. A fluid-filled coelom runs the length of the body. It is divided into coelomic chambers, one per segment.

Gases are exchanged across the body surface, and a **closed circulatory system** helps distribute oxygen. In such a system, blood flows through a continuous system of vessels. All exchanges between blood and the tissues take place across a vessel wall. Hearts in the anterior of the worm provide the pumping power to move the blood.

The earthworm has a complete digestive system that extends through all coelomic chambers. Earthworms are scavengers that eat their way through the soil and digest organic debris. The worms improve soil by loosening its particles and excreting tiny bits of organic matter that

"jaws"
toothlike structures
pharynx (everted)
antenna
palp (food handling)
tentacle
eyes
chemical-sensing pit

parapod

Figure 23.15 Polychaetes. **A** The sandworm uses its many appendages to burrow into sediment on marine mudflats. It is an active predator with hard jaws. **B** The feather duster worm secretes and lives inside a hard tube. Feathery extensions on the head capture food from the water. Cilia move captured food to the mouth.

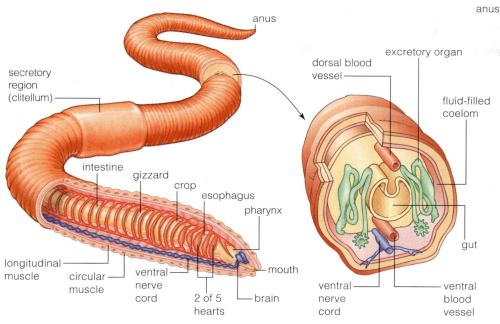

Figure 23.17 **Animated** Earthworm body plan. Each segment contains a coelomic chamber full of organs. A gut, ventral nerve cord, and dorsal and ventral blood vessels run through all coelomic chambers.

>> **Figure It Out** Why do people sometimes describe the body of a coelomate animal such as an earthworm as "a tube inside a tube"?

Answer: One tube, the gut, is inside another tube, the body wall. The fluid-filled coelom lies between the tubes.

decomposers can easily break down. Excreted earthworm "castings" are sold as a fertilizer.

Most segments have paired excretory organs that regulate the solute composition and volume of coelomic fluid. Each organ collects coelomic fluid, adjusts its composition, and expels waste through a pore in the body wall.

Earthworm behavior is coordinated by a simple brain. A pair of nerve cords extend from the brain and run the length of the body. The brain receives information from sensory organs, such as light-sensing organs in the body wall. It also coordinates locomotion.

The earthworm has two sets of muscle. Longitudinal muscles parallel the body's long axis, and circular muscles ring the body. Contraction of these muscles puts pressure on coelomic fluid trapped inside body segments, causing the segments to change shape. When a segment's longitudinal muscles contract, the segment gets shorter and fatter. When circular muscles contract, a segment gets longer and thinner. The worm moves as coordinated waves of contraction run along the body (Figure 23.18).

Figure 23.18 How earthworms move through soil. An earthworm's coelomic fluid serves as a hydrostatic skeleton. Circular and longitudinal muscles that work in opposition exert pressure on this fluid. Coordinated activity of these muscles alters segment shape and produces waves of contraction. The worm's front end is pushed forward, then bristles are extended to anchor the front as rear segments are pulled up behind it.

Earthworms are hermaphrodites, but they cannot fertilize themselves. During mating, a secretory organ (the clitellum) produces mucus that glues two worms together while they swap sperm. Later, that organ secretes a silky case that protects the fertilized eggs.

annelid Segmented worm with a coelom, complete digestive system, and closed circulatory system.
closed circulatory system Circulatory system in which blood flows through a continuous network of vessels; all materials are exchanged across the walls of those vessels.

Take-Home Message What are annelids?

> Annelids are bilateral, coelomate, segmented worms that have a complete digestive system and a closed circulatory system.

> The group includes aquatic species and the land-dwelling earthworms.

> Is there a "typical" mollusk? No. The group has more than 100,000 named species, including tiny snails in treetops, burrowing clams, and giant predators of the open ocean.
< Link to Analogous structures 16.8

Mollusks (phylum Mollusca) are bilaterally symmetrical invertebrates with a reduced coelom. Most dwell in seas, but some live in fresh water or on land. All have a mantle, a skirtlike extension of the upper body wall that covers a mantle cavity (Figure 23.19). In shelled mollusks, the shell consists of a calcium-rich, bonelike material secreted by the mantle. Aquatic mollusks typically have one or more respiratory organs called gills inside their fluid-filled mantle cavity. The mollusk digestive system is always complete. In most mollusks, the mouth contains a radula, a tonguelike organ hardened with chitin.

With more than 100,000 living species, mollusks are second only to arthropods in level of diversity. There are three main groups: gastropods, bivalves, and cephalopods (Figure 23.20).

Gastropods

Gastropods are the most diverse mollusk lineage. Their name means "belly foot," and most species glide about on the broad muscular foot that makes up most of the lower body mass (Figure 23.20A). A gastropod shell, when present, is one-piece and often coiled.

Gastropods have a distinct head that usually has eyes and sensory tentacles. In many aquatic species, a part of the mantle forms an inhalant siphon, a tube through which water is drawn into the mantle cavity. Gastropods and bivalves have an **open circulatory system**. In such a system, blood vessels do not form a continuous loop. Instead, blood leaves vessels and seeps around tissues before returning to the heart. Cells exchange substances with the blood while it is outside of vessels.

Gastropods include the only terrestrial mollusks. In land-dwelling snails and slugs (Figure 23.21A,B), a lung

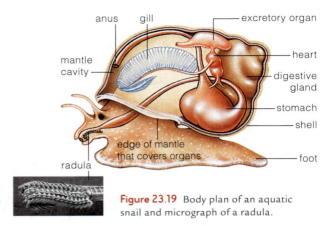

Figure 23.19 Body plan of an aquatic snail and micrograph of a radula.

replaces the gill. Glands on the foot continually secrete mucus that protects the animal as it moves across dry, abrasive surfaces. Most mollusks have separate sexes, but land-dwellers tend to be hermaphrodites. Unlike other mollusks, which produce a swimming larva, the embryos of these groups develop directly into adults.

Lacking a shell, slugs and sea slugs must defend themselves in other ways. Some make and secrete distasteful substances. Certain sea slugs eat cnidarians such as jellyfish and store undischarged nematocysts that serve as a defense. For example, frilly extensions on the back of a Spanish shawl nudibranch function in gas exchange and hold nematocysts (Figure 23.21C).

Bivalves

Bivalves include many mollusks that end up on dinner plates, including mussels, oysters, clams, and scallops (Figure 23.20B). All bivalves have a hinged, two-part shell. Powerful muscles hold the valves together. Contraction of these muscles pulls the two valves shut, enclosing the body and protecting it from predation or drying out. A bivalve has no obvious head. In many, simple eyes arrayed around the edge of the mantle help detect dangers. Most bivalves feed by drawing water into their mantle cavity

A Gastropods (belly footed)   **B** Bivalves (two-part shell)   **C** Cephalopods (jet-propelled)

Figure 23.20 Representatives of three main mollusk groups.

Figure 23.21 Variations on the gastropod body plan. Land snails **A** and slugs **B** are adapted to life in a dry habitat. They have a lung in place of gills. The slime trail left behind after they move over a surface is mucus secreted by their big foot.

C Two Spanish shawl nudibranchs (*Flabellina iodinea*). These sea slugs feed on cnidarians and store undischarged nematocysts inside their bright red respiratory organs.

and trapping food in mucus on the gills. Movement of cilia directs particle-laden mucus to the mouth.

Cephalopods

Cephalopods include squids (Figure 23.20C), nautiluses (Figure 23.22A), octopuses (Figure 23.22B), and cuttlefish. All are predators and most have beaklike, biting mouthparts in addition to a radula. Cephalopods move by jet propulsion. They draw water into the mantle cavity, then force it out through a funnel-shaped siphon. The foot has been modified into arms and/or tentacles that extend from the head. Cephalopod means "head-footed."

Five hundred million years ago, large cephalopods with a long conelike shell were the top predators in the seas. Today, nautiluses have a coiled external shell, but other cephalopods have a highly reduced shell or none at all. Jawed fishes evolved 400 million years ago, and competition with this group may have favored a shift in body form. Cephalopods with the smallest shell could be fastest and most agile. A speedier life-style required other changes as well. Of all mollusks, only cephalopods have a closed circulatory system. Competition with fishes also favored improved eyesight. Like vertebrates, cephalopods have eyes with lenses that focus light. Because mollusks and vertebrates are not closely related, this type of eye is assumed to have evolved independently in these groups.

Cephalopods include the fastest (squids), biggest (giant squid), and smartest (octopuses) invertebrates. Of all invertebrates, octopuses have the largest brain relative to body size, and show the most complex behavior.

Figure 23.22 Cephalopods. **A** Nautilus and **B** an octopus.

bivalve Mollusk with a hinged two-part shell.
cephalopod Predatory mollusk with a closed circulatory system; moves by jet propulsion.
gastropod Mollusk in which the lower body is a broad "foot."
mollusk Invertebrate with a reduced coelom and a mantle.
open circulatory system System in which blood leaves vessels and seeps through tissues before returning to the heart.

Take-Home Message What are mollusks?

❯ Mollusks are bilateral, soft-bodied, coelomate animals, the only ones with a mantle over the body mass.

❯ They include snails and slugs that glide about on a huge foot, bivalves with a hinged two-part shell, and the jet-propelled cephalopods.

> Roundworms are among the most abundant animals. Sediments in shallow water may hold a million per square meter; a cupful of topsoil teems with them.

Roundworms, or nematodes (phylum Nematoda), are bilateral, unsegmented animals that have a pseudocoelom and a cuticle-covered, cylindrical body (Figure 23.23). The collagen-rich cuticle is pliable, and it is repeatedly molted between growth spurts. **Molting** is the shedding of a body covering (or hair, or feathers).

Most roundworms are free-living species less than 5 millimeters long. The soil roundworm *Caenorhabditis elegans* is a favorite for genetic experiments. It has the same tissue types as complex animals. Yet it is transparent, it has only 959 body cells, and it reproduces fast. Its genome is about 1/30 the size of ours. Such traits make it easy for scientists to monitor each cell's fate during development.

Although most roundworms are free-living decomposers in soil or water, some parasitize plants and animals. Some of the animal parasites reach impressive lengths. One that lives inside sperm whales can be 13 meters long (42.65 feet).

Some parasitic roundworms impair human health, especially in developing countries. Eating undercooked pork or wild game can lead to an infection by *Trichinella spiralis*. The roundworm moves from the intestines, through blood, and into muscles. The result is trichinosis, a painful disease that can be fatal.

Wuchereria bancrofti and certain other roundworms cause lymphatic filariasis. Repeated infections injure lymph vessels, so lymph pools inside the legs and feet (Figure 23.24A). Elephantiasis, the common name for this disease, refers to fluid-filled, "elephant-like" legs. Mosquitoes carry the larval roundworms to new hosts.

Ascaris lumbricoides, a large roundworm, currently infects more than 1 billion people, primarily in Asia and in Latin America (Figure 23.24B). People become infected when they eat its eggs, which survive in soil and get on hands and into food. When enough adults occupy a host, they can clog the digestive tract.

---

molting Periodic shedding of an outer body layer or part.
roundworm Unsegmented worm with a pseudocoel and a cuticle that is molted as the animal grows.

Figure 23.24 Parasitic roundworms. **A** The grossly enlarged leg of the man on the *left* is symptomatic of lymphatic filariasis. The damage to lymphatic vessels is permanent. Fluid will continue to pool in the leg even if the roundworms that caused the damage are killed. **B** Live roundworms (*Ascaris lumbricoides*). These intestinal parasites cause stomach pain, vomiting, and appendicitis.

Hookworms, too, infect more than 1 billion people. Juveniles in the soil cut into human skin and migrate through blood vessels to the lungs. They climb up the windpipe, then enter the digestive tract when their host swallows. Once inside the small intestine, they attach to the intestinal wall and suck blood.

Pinworms commonly infect children. Female worms less than a millimeter long leave the rectum at night and lay eggs near the anus. The migration causes itching, and scratching puts eggs under fingernails. From there, they get into food and onto toys. If swallowed, the eggs spread the infection.

pharynx   intestine   eggs in uterus   gonad

false coelom (unlined body cavity)   muscular body wall   anus

Figure 23.23 Animated Body plan of *Caenorhabditis elegans*, one of the free-living roundworms. Sexes are separate, and this is a female.

Take-Home Message What are roundworms?

> Roundworms are bilateral worms with a cuticle that they molt as they grow. Organ systems almost fill their false coelom. Some are agricultural pests or human parasites.

> Arthropods are the most diverse invertebrate group. A variety of features contribute to their success.

< Link to Chitin 3.3

Major Groups

Arthropods (phylum Arthropoda) are bilateral animals with a hardened, jointed external skeleton, a complete gut, an open circulatory system, and a reduced coelom. Modern lineages include chelicerates (spiders and close relatives), crustaceans (such as lobsters, shrimps, and crabs), millipedes and centipedes, and insects. Before we discuss arthropod diversity, consider six key adaptations that contributed to their evolutionary success.

Key Arthropod Adaptations

Hardened Exoskeleton Arthropods have a cuticle (a secreted covering) of chitin, proteins, and waxes. It is an **exoskeleton**, an external skeleton. The exoskeleton protected aquatic species against predators. When some groups invaded land, the exoskeleton also helped them conserve water and supported their body.

A hard exoskeleton does not restrict size increases, because arthropods, like roundworms, molt periodically. Hormones control molting. They encourage the formation of a soft, new cuticle under the old one, which is then shed (Figure 23.25A).

Jointed Appendages If a cuticle were uniformly hard and thick like a plaster cast, it would prevent movement. Arthropod cuticles thin at joints, the regions where hardened body parts meet. Contraction of muscles that span joints makes the cuticle bend, so parts move relative to one another. *Arthropod* means jointed leg (Figure 23.25B).

Modified Segments In early arthropods, body segments were more or less alike. In many of their descendants, segments became fused together or modified for specialized tasks. For example, among insects, thin extensions of the wall of some segments evolved into wings (Figure 23.25C).

Respiratory Structures Aquatic arthropods have a gill or gills for gas exchange. Those that live on land have air-conducting tubes that start at surface pores, branch into finer tubes, and deliver oxygen to all internal tissues.

Figure 23.25 Some features that contribute to arthopod success.

A Hardened exoskeleton. This centipede is molting its old exoskeleton (*gray*).

B Jointed appendages, such as crab legs.

C Modified segments. This wing attached to one segment on the thorax of a fly.

D Specialized developmental stages. This larva of a monarch butterfly feeds on plant leaves. The adult monarch sips nectar.

Sensory Specializations Most arthropods have one or more pairs of eyes: organs that sample the visual world. With the exception of chelicerates, they also have paired **antennae** (singular, antenna) that detect touch and odors.

Specialized Stages of Development Especially among insects, the tasks of surviving and reproducing are divided among different stages of development. Individuals of many species undergo **metamorphosis**: Tissues get remodeled as embryonic stages make the transition to the adult form. A typical immature stage specializes in eating and growing rapidly, whereas the adult specializes in reproduction and dispersal. For instance, caterpillars chew leaves and metamorphose into butterflies, which fly about, sip nectar, mate, then deposit eggs (Figure 23.25D). Differences among stages of development are adaptations to specific conditions in the environment, including seasonal shifts in food, water, and shelter. Having such different bodies also prevents adults and juveniles from competing for the same resources.

antenna Of some arthropods, sensory structure on the head that detects touch and odors.
arthropod Invertebrate with jointed legs and a hard exoskeleton that is periodically molted.
exoskeleton Hard external parts that muscles attach to and move.
metamorphosis Dramatic remodeling of body form during the transition from larva to adult.

Take-Home Message **What traits contribute to the evolutionary success of the arthropods?**

> As a group, arthropods are abundant and widespread. A hardened, jointed exoskeleton, modified body segments, and specialized appendages, respiratory structures, and sensory structures underlie their evolutionary success.

> In many species success also arises from a division of labor among different stages of the life cycle, such as larvae, juveniles, and adults.

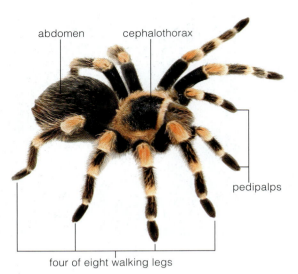

abdomen cephalothorax

pedipalps

four of eight walking legs

A

B

C

D

E

> Chelicerates include the marine horseshoe crabs, but most are arachnids that live on land.
< Link to Tick-borne bacterial diseases 19.7

Chelicerates have a body with a cephalothorax (fused head and thorax) and abdomen (Figure 23.26A). Paired walking legs attach to the cephalothorax. The head has eyes, but no antennae. Paired feeding appendages called chelicerae near the mouth give the group its name. Another pair of appendages, called pedipalps, lie between the chelicerae and the first legs. The appearance and function of the pedipalps varies among chelicerate groups.

Horseshoe crabs are an ancient chelicerate lineage and the only one that is marine. They eat clams and worms. A horseshoe-shaped shield covers the cephalothorax (Figure 23.26B). The last body segment, the telson, has evolved into a long spine that acts as a rudder when the horseshoe crab swims. It also helps the animal right itself if a wave knocks it over. The spine does not produce venom. Each spring, horseshoe crab eggs laid along Atlantic and Gulf coasts serve as food for migratory shorebirds.

Arachnids include spiders, scorpions, ticks, and mites. All have four pairs of walking legs and live on land. Scorpions and spiders produce venom that helps them subdue their prey. Scorpions dispense venom through a stinger on their telson (Figure 23.26C). Their pedipalps have evolved into large claws. Spiders deliver venom with a bite. Their fanglike chelicerae have poison glands. Of 38,000 spider species, about 30 produce venom that can harm humans.

Most spiders indirectly help us by eating insect pests. Some are sit-and-wait predators. They use paired spinners on their abdomen to eject silk and weave prey-catching webs. Others such as tarantulas and jumping spiders are active hunters that sneak up and grab unwary prey.

All ticks are parasites that suck blood from vertebrates (Figure 23.26D). Some that bite humans can transmit the bacteria that cause Lyme disease or other diseases.

Mites are a highly diverse group that include parasites, predators, and scavengers (Figure 23.26E). Most are less than a millimeter long.

arachnids Land-dwelling arthropods with four pairs of walking legs; spiders, scorpions, mites, and ticks.
chelicerates Arthropod subgroup with specialized feeding structures (chelicerae) and no antennae.

Figure 23.26 Examples of chelicerates. **A** Tarantula, a spider. **B** Horseshoe crab (*Limulus*). The long spine helps it steer as it swims. **C** Scorpion. Its last segment has a stinger and the pedipalps are claws that can seize prey. **D** A blood-engorged tick. **E** Dust mite, less than 0.5 millimeter in length.

Take-Home Message What are chelicerates?

> Chelicerates include horseshoe crabs and arachnids. The horseshoe crabs are an ancient marine lineage. Arachnids are land-dwellers with eight walking legs and no antennae.

> Most marine arthropods are crustaceans. Their amazing diversity and abundance has given them the nickname "the insects of the seas."

Crustaceans are a group of mostly marine arthropods with two pairs of antennae. Some live in fresh water. A few such as wood lice live on land.

Small crustaceans reach great numbers in the seas and are an important food source for larger animals. Krill (euphausiids) have a shrimplike body a few centimeters long and swim in upper ocean waters (Figure 23.27A). They are so plentiful and nutritious that a blue whale weighing more than 100 tons can subsist almost entirely on krill it filters from seawater. Most copepods are also marine, but others live in fresh water (Figure 23.27B).

Larval barnacles swim, but adults are enclosed in a calcified shell and live attached to piers, rocks, and even whales. They filter food from the water with feathery legs (Figure 23.27C). As adults, they cannot move about, so you would think that mating might be tricky. But barnacles tend to settle in groups, and most are hermaphrodites. An individual extends a penis, often several times its body length, out to neighbors.

Lobsters, crayfish, crabs, and shrimps all belong to the same crustacean subgroup (the decapods). All are bottom feeders with five pairs of walking legs (Figure 23.28). In some lobsters, crayfish, and crabs, the first pair of legs has become modified as a pair of claws.

Like all arthropods, crabs molt as they grow (Figure 23.29). Some spider crabs do quite a bit of growing. With legs that can reach more than a meter in length, these crabs are the largest living arthropods.

Figure 23.27 Examples of crustaceans. **A** Antarctic krill no more than 6 centimeters long. **B** Free-living female copepod from the Great Lakes, about 1 millimeter long. Note the eggs. **C** Goose barnacles capturing food with their jointed legs.

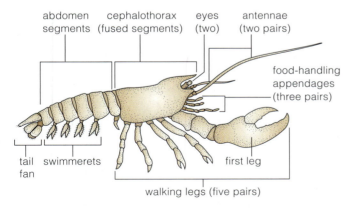

Figure 23.28 Body plan of a clawed lobster (*Homarus americanus*).

Figure 23.29 **Animated** Crab life cycle. Larval and juvenile stages molt repeatedly and grow in size before they are mature adults.

Take-Home Message What are crustaceans?

> Crustaceans are a group of mostly marine arthropods with antennae. They include the largest arthropods. Many species serve as food for other animals, including humans.

crustaceans Mostly marine arthropod group with two pairs of antennae; for example shrimp, crabs, lobsters, and barnacles.

❯ Arthropods are the most diverse animal phylum, and insects are the most diverse arthropods.

Characteristic Features

Insects have a three-part body plan, with a head, thorax, and abdomen (Figure 23.30). The head has one pair of antennae and two compound eyes. Such eyes consist of many individual units, each with a lens. Near the mouth are jawlike mandibles and other feeding appendages.

Three pairs of legs attach to the insect thorax. In some groups, the thorax also has one or two pairs of wings. Insects are the only winged invertebrates.

A few insects spend some time in the water, but the group is overwhelmingly terrestrial. A respiratory system consisting of tracheal tubes carries air from openings at the body surface to tissues deep inside the body. An insect abdomen contains digestive organs and sex organs.

Insect developmental patterns vary. Depending on the group, a fertilized egg either hatches into a small version of the adult, or a juvenile that will later undergo metamorphosis. During metamorphosis, tissues of a juvenile are reorganized. Incomplete metamorphosis means that the alterations in body form take place a bit at a time. Juveniles, called nymphs, change a little with each molt (Figure 23.31A). Complete metamorphosis is more dramatic. The juvenile, called a larva, grows and molts with no change in body plan. Then, the juvenile becomes a pupa. The pupa undergoes tissue remodeling that produces an adult (Figure 23.31B).

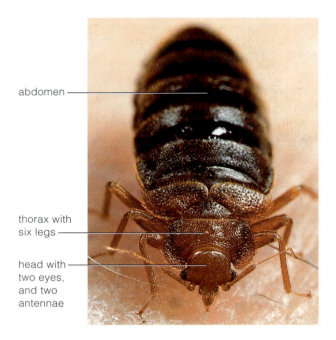

Figure 23.30 A bedbug illustrates the basic insect body plan: a head, thorax, and abdomen. The bug is 7 millimeters long.

Diversity and Abundance

With more than a million species, insects are the most diverse arthropod group. They are also breathtakingly abundant. By some estimates, the ants alone make up about 10 percent of the world's terrestrial animal biomass (the total weight of all living land animals).

Figure 23.32 shows a few of the more than 800,000 insect species. Silverfish are the most ancient lineage of modern insects (Figure 23.32A). They do not undergo metamorphosis. Their young are simply miniature versions of adults. Silverfish are scavengers that require a humid environment.

Insects that undergo incomplete metamorphosis include bugs, earwigs, lice, cicadas, damselflies, termites, and grasshoppers. Earwigs are wingless scavengers that have a flattened body (Figure 23.32B). Males have large pincers on their abdomen. Lice are wingless parasites that suck blood from warm-blooded animals (Figure 23.32C). Fleas are also wingless bloodsucking insects. Cicadas (Figure 23.32D) and the related leafhoppers and aphids suck juices from plants. Damselflies (Figure 23.32E) and the related dragonflies are agile aerial predators of other insects. Termites are social insects. They live in big family groups. Bacterial, archaean, and protistan symbionts in their gut allow them to digest wood (Figure 23.32F). They are unwelcome when they devour buildings or decks, but can be important decomposers. Grasshoppers cannot eat wood, but they can chew through tough, nonwoody plant parts (Figure 23.32G).

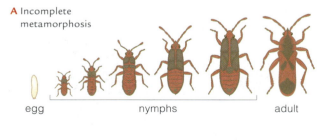

A Incomplete metamorphosis

egg   nymphs   adult

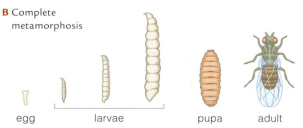

B Complete metamorphosis

egg   larvae   pupa   adult

Figure 23.31 Insect development. **A** Incomplete metamorphosis of a bug. Small changes occur with each molt. **B** Complete metamorphosis of a fruit fly. A larva develops into a pupa that undergoes a dramatic remodeling into the adult form.

The four most successful insect lineages all have wings and undergo complete metamorphosis. There are approximately 150,000 species of flies (Figure 23.32H), and at least as many beetles (Figure 23.32I,J). The wasp in Figure 23.32K is one of about 130,000 hymenopterans. This group also includes the bees and ants. The moths and butterflies (Figure 23.32L) constitute a group of about 120,000 species. As a comparison, consider that there are about 4,500 species of mammals.

---

insects Most diverse arthropod group; members have six legs, two antennae and, in some groups, wings.

Take-Home Message What are insects?

› Insects are arthropods that have six legs, two antennae, and compound eyes. They are diverse and abundant on land.

› The most widespread and diverse insect groups have wings and undergo complete metamorphosis.

Figure 23.32 A sampling of insect diversity.
A Silverfish, an ancient lineage that does not undergo metamorphosis.

Insects that undergo incomplete metamorphosis:
B European earwig, a common household pest.
C Duck louse. It eats bits of feathers and skin.
D Cicada. Male cicadas are among the loudest of all insects.
E Damselfly, one of the insects that has aquatic larvae.
F Termites. These sterile soldiers can shoot a gluelike substance from their head to protect their colony.
G Grasshopper.

Members of the four most diverse insect orders. All are winged and undergo complete metamorphosis:
H A fly, dusted with pollen.
I Ladybird beetles with a distinctive red and black spotted wing cover.
J Staghorn beetle from New Guinea.
K Bald-faced hornet, a type of wasp. This is a fertile female, or queen. She lives in a papery nest with her many offspring.
L Swallowtail butterfly, a lovely lepidopteran, shown acting as pollinator.

›› **Figure It Out** Are insect wings modified legs?

Answer: No. Insect wings are extensions from the body wall. Winged insects have the same number of legs (6) as wingless ones.

> It would be hard to overestimate the importance of insects, for either good or ill.
< Links to Pollinators 21.8, Malaria 20.6

Ecological Services

As you learned in Section 21.8, the flowering plants coevolved with insect pollinators. The vast majority of these plants are pollinated by members of one of the four most diverse insect groups. Other groups contain few or no pollinators. By one hypothesis, the close interactions between pollinator lineages and flowering plants contributed to an increased rate of speciation in both.

Today, declines in populations of insect pollinators concern many biologists. Development of natural areas, use of pesticides, and the spread of newly introduced diseases are reducing the populations of insects that pollinate native plants and agricultural crops.

Insects serve as food for a variety of wildlife. Most songbirds nourish their nestlings on an insect diet. Those that migrate often travel long distances in order to nest and raise their young in areas where insect abundance is seasonally high. Some insects such as dragonflies and mayflies have aquatic larvae that serve as food for trout and other freshwater fish. Most amphibians and reptiles feed mainly on insects. Even humans eat insects. In many cultures, they are considered a tasty source of protein.

Insects dispose of wastes and remains. Flies and beetles quickly discover an animal corpse or a pile of feces (Figure 23.33). They lay their eggs in or on this organic material, and the larvae that hatch devour it. By their actions, the insects keep wastes and remains from piling up, and help distribute nutrients through the ecosystem.

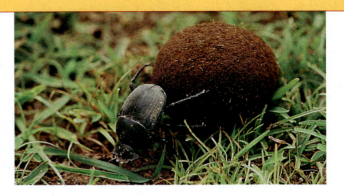

Figure 23.33 Cleanup crew. A dung beetle gathers a ball of dung (feces). The beetle will lay its eggs in the dung and the larval beetles will feed on it as they grow.

Competitors for Crops

Insects are our main competitors for food and other plant products. They devour about a quarter to a third of all crops grown in the United States. Sadly, in an age of global trade and travel, we have more than just home-grown pests to worry about. Consider the Mediterranean fruit fly (Figure 23.34). The Med fly, as it is known, lays eggs in citrus and other fruits, as well as many vegetables. Damage done to plants and fruits by larvae of the Med fly can cut crop yield in half. Med flies are not native to the United States and there is an ongoing inspection program for imported produce. Even so, some Med flies still slip in. So far, all introduced Med fly populations have been successfully eradicated, but these control efforts have cost hundreds of millions of dollars. This amount is probably a bargain. If the Med fly were to become established, crop losses would likely run into the billions.

Vectors for Disease

Some insects spread human pathogens. As you learned earlier, mosquitoes transmit malaria, which kills more than a million people each year (Figure 23.35). Mosquitoes are also vectors for viruses and for roundworms that cause disease. Biting flies transmit African sleeping sickness; biting bugs spread Chagas disease. Fleas that bite rats and then bite humans can transmit bubonic plague. Body lice transmit typhus. So far as we know, bedbugs (Figure 23.30) do not cause disease. However, a heavy bedbug infestation can cause weakness as a result of blood loss, especially in children.

Figure 23.34 Mediterranean fruit fly (Med fly), a major threat to fruit crops.

Figure 23.35 The deadliest animal. When it comes to bites, no animal is more dangerous than the mosquito. Mosquito-borne diseases kill more than 1 million people each year.

Take-Home Message What role do insects play in ecosystems?

> Insects pollinate plants, serve as food for wildlife, and dispose of wastes and remains.

> Insects harm us by eating our crops and spreading diseases.

23.15 Echinoderms

> With this section we begin our survey of deuterostomes. We will continue this survey in the next chapter.

Echinoderm means "spiny skinned." The **echinoderms** (phylum Echinodermata) have interlocking spines and plates of calcium carbonate embedded in their skin. Adults have a radial body plan, with five parts (or multiples of five) around a central axis. However, gene comparisons and the bilateral body plan of the larvae indicate echinoderms evolved from a bilateral ancestor.

Sea stars (also called starfish) are the most familiar echinoderms, and we will use them to illustrate the body plan (Figure 23.36A,B). Sea stars do not have a brain, but do have a decentralized nervous system. Eyespots at the tips of the arms detect light and movement.

A typical sea star is an active predator that moves about on tiny, fluid-filled tube feet. Tube feet are part of a **water–vascular system** unique to echinoderms. Fluid-filled canals extend into each arm, and side canals deliver coelomic fluid into muscular ampullae that function like the bulb on a medicine dropper. Contraction of an ampulla forces fluid into a tube foot, extending the foot.

Sea stars usually prey on bivalve mollusks. They can slide their stomach out through their mouth and into the bivalve's shell. The stomach secretes acid and enzymes that kill the mollusk and begin to digest it. Partially digested food is taken into the stomach and digestion is completed with the aid of digestive glands in the arms.

Gas exchange occurs by diffusion across the tube feet and tiny skin projections at the body surface. There are no specialized excretory organs.

A sea star is either male or female. Gonads (reproductive organs) in the arms make eggs or sperm that are released into the water. Fertilization produces a zygote that develops into a ciliated, bilateral larva. The larva swims about and develops into the adult form.

Sea urchins and sea cucumbers are less familiar echinoderms. In sea urchins, calcium carbonate plates form a stiff, rounded cover from which spines protrude (Figure 23.36C). Sea urchin roe (eggs) are used in some sushi. Overharvesting for markets in Asia threatens species that produce the most highly prized roe. Sea cucumbers are cucumber-shaped, with microscopic plates embedded in a soft body (Figure 23.36D). Some filter food from the seawater. Most feed like earthworms, by burrowing through sediments and ingesting them. If attacked, a sea cucumber can expel organs through the anus to distract the predator. If this defense works and the sea cucumber escapes, its missing parts grow back.

---

echinoderms Invertebrates with a water–vascular system and hardened plates and spines embedded in the skin or body.
water–vascular system Of echinoderms, a system of fluid-filled tubes and tube feet that function in locomotion.

Old Genes, New Drugs (revisited)

Marine invertebrates are important components of ecosystems, a source of food, and a treasure trove of molecules with potential for use in industrial applications or as medicines. Various species of cone snails, sponges, corals, crabs, and sea cucumbers make compounds that show promise as drugs. However, even as we begin to explore this potential, marine biodiversity is on the decline as a result of habitat destruction and overharvesting.

How Would You Vote? Bottom trawling is a fishing technique that helps keep seafood prices low, but can destroy invertebrate habitats. Should it be banned? See CengageNow for details, then vote online (cengagenow.com).

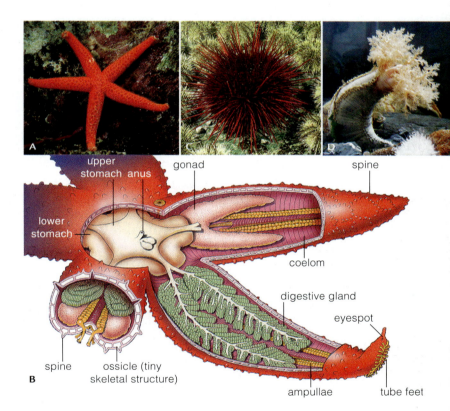

Figure 23.36 Animated Echinoderms. **A,B** Sea star. **C** Sea urchin. **D** Sea cucumber. Other members of this group include brittle stars and sand dollars.

>> **Figure It Out** Does the sea star have a complete digestive system?

Answer: Yes. It has a mouth and an anus.

Take-Home Message **What are echinoderms?**

> Echinoderms are spiny-skinned invertebrates that are radial as adults, but probably had bilateral ancestors. They have a decentralized nervous system and a unique water–vascular system that functions in locomotion.

Summary

Section 23.1 Invertebrates (with no backbone) are the most diverse animal group, and **vertebrates** (with a backbone) evolved from an invertebrate ancestor. As lineages evolved, genes that were present in ancestral lineages took on new functions. Thus, the same gene that functions in the synthesis of venom in a cone snail can play a role in the clotting of human blood.

Section 23.2 Animals are multicelled heterotrophs that digest food inside their body. Most have an embryo with three layers: **ectoderm**, **endoderm**, and **mesoderm**. Early animals had no body symmetry. Cnidarians have **radial symmetry**, but most animals have **bilateral symmetry** and **cephalization**. Bilateral animals typically have a body cavity, either a **pseudocoelom** or (more often) a **coelom**. Some also have **segmentation**, with repeating body units. There are two lineages of bilateral animals, **protostomes** and **deuterostomes**. Table 23.1 summarizes the traits of major animal lineages.

Section 23.3 The first animals evolved from a colonial protist. **Choanoflagellates** are the closest protistan relative of animals. Among living animals, **placozoans** are the oldest and structurally simplest lineage. An adaptive radiation during the Cambrian gave rise to most modern animal lineages.

Section 23.4 Sponges have a porous body with no tissues or organs. They filter food from water, and each is a **hermaphrodite** that produces both eggs and sperm. The ciliated **larva** is the only motile stage.

Section 23.5 Cnidarians such as jellyfishes, corals, and sea anemones are carnivores. They have two tissue layers with a jellylike layer in between. Only cnidarians make **nematocysts**, which they use to capture prey. A gastrovascular cavity functions in both respiration and digestion. Cnidarians have a **hydrostatic skeleton**. A **nerve net** gives commands to contractile cells that redistribute fluid and change the body's shape.

Section 23.6 Flatworms, the simplest animals with organ systems, include marine species and the freshwater **planarians**, as well as parasitic tapeworms and flukes. Some tapeworms and flukes infect humans.

Section 23.7 Annelids are segmented worms. The **closed circulatory system** and digestive, solute-regulating, and nervous systems extend through coelomic chambers. Annelids move when muscle action exerts force on coelomic fluid, altering the shape of all segments in a coordinated manner. Oligochaetes include aquatic species and the familiar earthworms. Polychaetes are predatory marine worms. Leeches are scavengers, predators, or bloodsucking parasites.

Section 23.8 Mollusks have a reduced coelom and a sheetlike mantle that drapes back over itself. They include **gastropods** (such as snails), **bivalves** (such as scallops), and **cephalopods** (such as squids and octopuses). Except in the cephalopods, which are adapted to a speedy, predatory life-style, blood flows through an **open circulatory system**.

Section 23.9 The **roundworms** (nematodes) have an unsegmented, cylindrical body covered by a cuticle that is **molted** as the animal grows. Most are decomposers in soil, but some are parasites.

Sections 23.10–23.14 There are more than 1 million **arthropod** species. Their diversity is attributable to traits such as a hardened **exoskeleton**, jointed appendages, specialized segments, sensory structures such as **antennae**, and **metamorphosis**. The **chelicerates** include horseshoe crabs and **arachnids** (spiders, scorpions, ticks, and mites). They are predators, parasites, or scavengers. The mostly marine **crustaceans** include crabs, lobsters, barnacles, krill, and copepods. **Insects** include the only winged invertebrates. Some serve as decomposers and pollinators; others harm crops and transmit diseases.

Section 23.15 Echinoderms such as sea stars belong to the deuterostome lineage. Spines and other hard parts embedded in the skin support the body. There is no central nervous system. A **water–vascular system** with tube feet functions in locomotion. Adults are radial, but larvae are bilateral, implying a bilateral ancestry.

Table 23.1 Comparative Summary of Animal Body Plans

Group	Body Symmetry	Germ Layers	Digestive System	Main Cavity	Segmented
Placozoans	None	None	None	None	No
Sponges	None	None	None	None	No
Cnidarians	Radial	2	Gut cavity	None	No
Protostomes					
Flatworms	Bilateral	3	Incomplete	None*	No
Annelids	Bilateral	3	Complete	Coelom	Yes
Mollusks	Bilateral	3	Complete	Coelom	No
Roundworms	Bilateral	3	Complete	Pseudocoelom	No
Arthropods	Bilateral	3	Complete	Coelom	Yes
Deuterostomes					
Echinoderms	Radial**	3	Complete	Coelom	No
Chordates	Bilateral	3	Complete	Coelom	Yes

* An ancestor may have had a coelom.
** Radial with some bilateral features; probably had a bilateral ancestor.

Data Analysis Activities

Sustainable Use of Horseshoe Crabs Eggs of Atlantic horseshoe crabs are essential food for many migratory shorebirds. People catch horseshoe crabs for use as bait. In addition, the blood of these crabs is used to test injectable drugs for potentially deadly bacterial toxins. To keep horseshoe crab populations stable, blood is extracted from captured animals, which are then returned to the wild. Concerns about the survival of animals after bleeding led researchers to do an experiment. They compared survival of animals captured and maintained in a tank with that of animals captured, bled, and kept in a similar tank. Figure 23.37 shows the results.

1. In which trial did the most control crabs die? In which did the most bled crabs die?
2. Looking at the overall results, how did the mortality of the two groups differ?
3. Based on these results, would you conclude that bleeding harms horseshoe crabs more than capture alone does?

	Control Animals		Bled Animals	
Trial	Number of crabs	Number that died	Number of crabs	Number that died
1	10	0	10	0
2	10	0	10	3
3	30	0	30	0
4	30	0	30	0
5	30	1	30	6
6	30	0	30	0
7	30	0	30	2
8	30	0	30	5
Total	200	1	200	16

Figure 23.37 Mortality of young male horseshoe crabs kept in tanks during the two weeks after their capture. Blood was taken from half the animals on the day of their capture. Control animals were handled, but not bled. This procedure was repeated eight times with different sets of horseshoe crabs.

1. True or false? Animal cells do not have walls.

2. A coelom is a _____ .
   a. type of bristle   c. sensory organ
   b. resting stage   d. lined body cavity

3. Cnidarians alone have _____ .
   a. nematocysts   c. a hydrostatic skeleton
   b. a mantle   d. a radula

4. Flukes are most closely related to _____ .
   a. tapeworms   c. arthropods
   b. roundworms   d. echinoderms

5. Which group has six legs and two antennae?
   a. crustaceans   c. spiders
   b. insects   d. horseshoe crabs

6. The _____ are mollusks with a hinged shell.
   a. bivalves   c. gastropods
   b. barnacles   d. cephalopods

7. A spider's chelicerae _____ .
   a. detect light   c. produce silk
   b. inject venom   d. eliminate excess water

8. Which of these groups includes the most species?
   a. protostomes   c. arthropods
   b. roundworms   d. mollusks

9. The _____ include the only winged invertebrates.
   a. cnidarians   c. arthropods
   b. echinoderms   d. placozoans

10. The _____ have a coelom and are radial as adults.
    a. cnidarians   c. annelids
    b. echinoderms   d. placozoans

11. Annelids and cephalopods have a(n) _____ circulatory system.

12. Match the organisms with their descriptions.
    ___ choanoflagellates   a. complete gut, pseudocoelom
    ___ placozoan   b. protists closest to animals
    ___ sponges   c. simplest organ systems
    ___ cnidarians   d. body with lots of pores
    ___ flatworms   e. jointed exoskeleton
    ___ roundworms   f. mantle over body mass
    ___ annelids   g. segmented worms
    ___ arthropods   h. tube feet, spiny skin
    ___ mollusks   i. nematocyst producers
    ___ echinoderms   j. simplest known animal

Additional questions are available on **CENGAGENOW**.

Critical Thinking

1. Most animals that are hermaphrodites cannot fertilize their own eggs, but tapeworms can. Explain the advantages and disadvantages of self-fertilization.

2. Both birds and squids have eyes with a lens. Are these structures analogous or homologous?

3. Why are pesticides designed to kill insects more likely to harm lobsters and crabs than fish?

Animations and Interactions on **CENGAGENOW**:
❯ Types of animal body cavity; Sponge body plan; Cnidarian body plan and life cycle; Nematocyst action; Planarian organs; Tapeworm life cycle; Annelid body plan; Roundworm body plan; Crab life cycle; Sea star body plan; Sea star tube feet.

‹ Links to Earlier Concepts

As we continue our survey of animals you may wish to review the discussion of vertebrates and the animal evolutionary tree (Sections 23.1, 23.2). The geologic time scale (16.6) will help put the events discussed here in perspective. You will see more examples of fossils (16.5) and of homologous structures (16.8). You will draw on your understanding of speciation, adaptive radiation, and extinctions (17.10–17.13) and of cladistics (17.14).

Key Concepts

Characteristics of Chordates

Four traits characterize chordate embryos: a supporting rod (notochord), a dorsal nerve cord, a pharynx with gill slits in the wall, and a tail that extends past the anus. Certain invertebrates and all vertebrates belong to this group.

The Fishes

The first vertebrates were jawless fishes. Most modern fishes have jaws. Sharks and their relatives are jawed fishes with a cartilage skeleton. Ray-finned fishes and lobe-finned fishes are jawed bony fishes. Ray-finned fishes are the most diverse vertebrate group.

24 Animals II: The Chordates

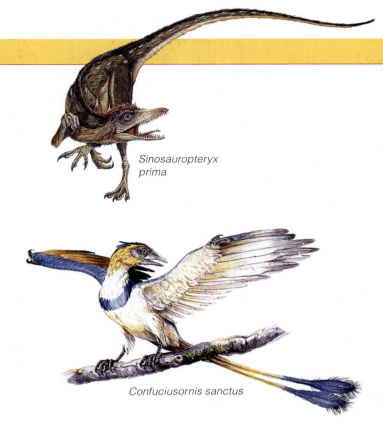

Sinosauropteryx prima

Confuciusornis sanctus

In Darwin's time, the presumed absence of transitional fossils was an obstacle to acceptance of his new theory of evolution. Skeptics wondered, if new species evolve from existing ones, then where are the fossils that document these transitions? Where are the fossils that bridge major groups? In fact, one of these "missing links" was unearthed by workers at a limestone quarry in Germany just one year after Darwin's *On the Origin of Species* was published.

That fossil, about the size of a large crow, looked like a small meat-eating dinosaur. It had a long, bony tail, three clawed fingers on each forelimb, and a heavy jaw with short, spiky teeth. It also had feathers. The new fossil species was named *Archaeopteryx* (ancient winged one). Eventually, eight *Archaeopteryx* fossils were unearthed. One is shown at the *left*. Radiometric dating indicates that *Archaeopteryx* lived 150 million years ago.

Archaeopteryx is the most widely known transitional fossil in the bird lineage, but there are others. In 1994, a farmer in China discovered a fossil dinosaur with small forelimbs and a long tail. Unlike most dinosaurs, this one (depicted in the reconstruction at the *upper right*) had downy feathers like those that cover modern chicks. Researchers named the farmer's fuzzy find *Sinosauropteryx prima*, which means first Chinese feathered dragon. Given its shape and lack of long feathers, *S. prima* was certainly flightless. The feathers probably helped the animal stay warm.

Both *S. prima* and *Archaeopteryx* had a long tail and a mouth full of sharp teeth. In contrast, modern birds have a small tailbone and their toothless jaws are covered with hard layers of protein that form a beak. Another fossil from China, *Confuciusornis sanctus* (sacred Confucius bird), is the earliest known bird with a beak. As shown at the *lower right*, it had a bird's typically short tail with long tail feathers. Even so, its dinosaur ancestry remains apparent. Unlike the wings of modern birds, the wings of *C. sanctus* had digits with claws at their tips.

No human witnessed the major transitions that led to modern animal diversity. However, fossils are physical evidence of changes, and radiometric dating assigns the fossils to places in time. The structure, biochemistry, and genetic makeup of living organisms provide information about branchings. Biological evolution is not a "theory" in the common sense of the word—an idea off the top of someone's head. It is a *scientific* theory. Its predictive power has been tested many times in the natural world, and the theory has withstood these tests (Section 1.9).

Evolutionists often argue among themselves. They argue over how to interpret data and which of the known mechanisms can best explain life's history. At the same time, they eagerly look to new evidence to support or disprove hypotheses. As you will see, fossils and other evidence form the foundation for this chapter's account of vertebrate evolution, including our own origins.

The Transition From Water to Land
Tetrapods, animals that walk on four legs, evolved from a lineage of lobe-finned fishes. Amphibians were the first tetrapod lineage. They can live on land, but their eggs must develop in water and their skin is not waterproof.

The Amniotes
Amniotes have waterproof skin, and their eggs contain membranes that enclose the embryo in fluid. These and other traits allowed the amniotes to expand into dry habitats. Reptiles (including birds) and mammals are the two modern amniote lineages.

Early Primates to Humans
Primates have grasping hands with nails instead of claws. Within the group, there is a trend toward increased brain size and manual dexterity, a flatter face, and upright posture. Fossils provide information about lineages related to our species, which most likely evolved in Africa.

> Chordates are distinguished by their embryonic traits.
> The group includes two lineages of marine invertebrates, as well as the vertebrates.
< Link to Animal classification 23.1

Chordate Characteristics

The previous chapter ended with the echinoderms, a deuterostome lineage. The other major deuterostome lineage, the **chordates**, is defined by four embryonic traits:

1. A **notochord**, a rod of stiff but flexible connective tissue, extends the length of the body and supports it.
2. A dorsal, hollow nerve cord parallels the notochord.
3. Gill slits (narrow openings) extend across the wall of the pharynx (the throat region).
4. A muscular tail extends beyond the anus.

Depending on the group, some, none, or all of these embryonic traits persist in the adult.

Most chordate species are vertebrates. However, the group also includes some marine invertebrates.

Invertebrate Chordates

In **lancelets**, all characteristic distinguishing chordate features are present in the adult (Figure 24.1). A lancelet is about 3 to 7 centimeters long. Its dorsal nerve cord extends into the head, where a group of nerve cells serves as a simple brain. An eyespot at the end of the nerve cord detects light, but there are no paired sensory organs like those of fishes. The lancelet wiggles backward into sediments until it is buried up to its mouth. Movement of cilia lining its pharynx moves water into the pharynx and

Figure 24.2 Animated Tunicates. **A** Larval tunicate. It is free-swimming and has all characteristic chordate features. **B** Diagram and **C** photo of the adult tunicate. It traps food in gill slits of its pharynx.

out through gill slits. Gill slits filter food particles out of the water and also function in gas exchange.

In **tunicates**, larvae have typical chordate traits, but adults retain only the pharynx with gill slits (Figure 24.2). Larvae swim about briefly, then undergo metamorphosis. The tail breaks down and other parts become rearranged into the adult body form. A secreted carbohydrate-rich covering, or "tunic," encloses the adult body and gives the group its common name. The adult feeds by drawing water in through an oral opening, past the gill slits, then expelling it through a second opening. Cilia on the gill slits capture food particles and move them to the gut.

Overview of Chordate Evolution

Until recently, lancelets were considered the closest invertebrate relatives of vertebrates. An adult lancelet looks more like fish than an adult tunicate does, but superficial similarities can be deceiving. Studies of developmental processes and gene sequences have revealed that tunicates are closer to the vertebrates, as illustrated in Figure 24.3.

A braincase, or cranium, evolved in the common ancestor of all fishes, amphibians, reptiles, birds, and mammals. We refer to the members of this clade as **craniates** ❶. The braincase may be entirely cartilage, or, like your skull, it may be a combination of cartilage and bone.

With the exception of one group of fish (hagfishes) all craniates are **vertebrates**. As Section 23.1 explained, the defining trait of this group is a backbone ❷. Biologists refer to the backbone as the vertebral column. It consists of stacked vertebrae. Vertebrae and other skeletal elements are components of the vertebrate **endoskeleton**. This internal skeleton consists of cartilage and, in most groups, bone.

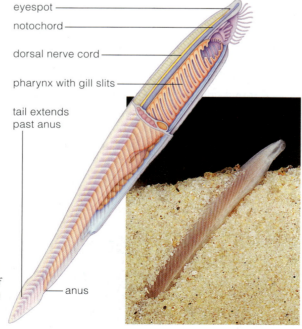

eyespot

notochord

dorsal nerve cord

pharynx with gill slits

tail extends past anus

anus

Figure 24.1 Animated Photo and body plan of a lancelet. Adults retain the set of four traits that define chordates.

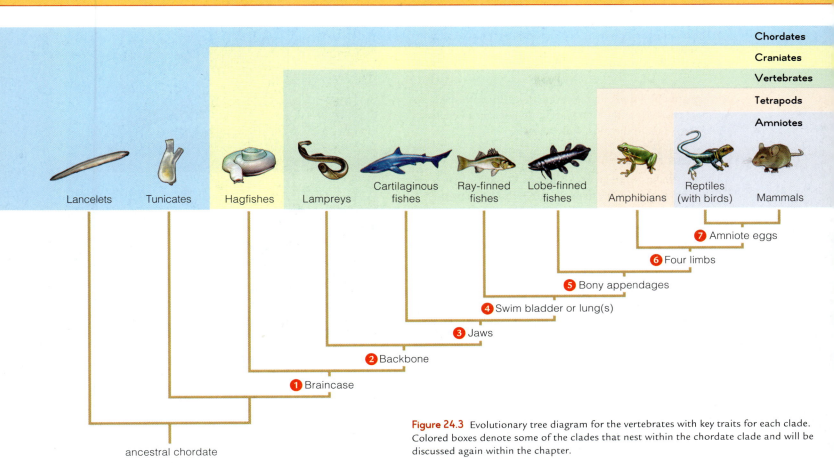

Chordates
Craniates
Vertebrates
Tetrapods
Amniotes

Lancelets · Tunicates · Hagfishes · Lampreys · Cartilaginous fishes · Ray-finned fishes · Lobe-finned fishes · Amphibians · Reptiles (with birds) · Mammals

7 Amniote eggs
6 Four limbs
5 Bony appendages
4 Swim bladder or lung(s)
3 Jaws
2 Backbone
1 Braincase

ancestral chordate

Figure 24.3 Evolutionary tree diagram for the vertebrates with key traits for each clade. Colored boxes denote some of the clades that nest within the chordate clade and will be discussed again within the chapter.

Early fishes sucked up or scraped up food. Then, the evolution of hinged skeletal elements called jaws opened the way to new feeding strategies **3**. Jaws evolved by expansion of the bony parts that structurally supported the gill slits of early jawless fishes. The vast majority of modern fishes have jaws, as do all other vertebrates.

Evolutionary modifications allowed animals to move from water onto the land. In one group of fishes, two small outpouchings on the side of the gut wall evolved into lungs: moist, internal sacs that enhance gas exchange with the air **4**. A subgroup of these fishes also evolved fins with bony supports inside them **5**. These fins would later evolve into the limbs of the first four-legged walkers, or **tetrapods** **6**.

Early tetrapods remained tied to the water. Later, evolution of eggs in which the embryo developed within a series of waterproof membranes allowed the animals known as **amniotes** to disperse widely on land **7**.

amniote Vertebrate in which the embryo develops surrounded by fluid enclosed by membranes inside the egg.
chordate Animal with an embryo that has a notochord, dorsal nerve cord, pharyngeal gill slits, and a tail that extends beyond the anus. For example, a lancelet or a vertebrate.
craniate Chordate with a braincase.
endoskeleton Internal skeleton made up of hardened components such as bones.
lancelet Invertebrate chordate that has a fishlike shape and retains the defining chordate traits into adulthood.
notochord Stiff rod of connective tissue that runs the length of the body in chordate larvae or embryos.
tetrapod Vertebrate with four legs, or a descendant thereof.
tunicate Invertebrate chordate that loses its defining chordate traits during the transition to adulthood.
vertebrate Animal with a backbone.

Take-Home Message **What are chordates and what traits defined the major subgroups?**

› All chordate embryos have a notochord, a dorsal tubular nerve cord, a pharynx with gill slits in its wall, and a tail that extends past the anus. These traits are legacies from a shared ancestor, an early invertebrate chordate.

› A braincase evolved early in chordate evolution and defines the craniates. Most craniates also have a backbone and are vertebrates. Some of the vertebrates evolved four limbs. These tetrapods colonized the land. Amniotes, a tetrapod subgroup with specialized eggs, are now the main land animals.

> The number and diversity of fishes exceed those of all other vertebrate groups combined.

< Link to Adaptive radiation 17.13

We begin our survey of vertebrate diversity with the fishes. These were the first vertebrate lineages to evolve and they remain the most fully aquatic.

Jawless Fishes

The first fishes were jawless, and two lineages of jawless fishes have survived to the present. Both have a cylindrical body with no fins or scales (Figure 24.4). Like a lancelet, they move with a wiggling motion.

Hagfishes have a cranium, but no backbone. Thus they are craniates but not vertebrates. A flexible notochord supports the body. Sensory tentacles near the mouth help a hagfish find food such as soft invertebrates and dead or dying fish. Hagfishes secrete slimy mucus when threatened, a useful defense for a soft-bodied animal. The slime deters most predators, but it has not kept humans from harvesting hagfish. Most products sold as "eelskin" are actually hagfish skin.

Lampreys have a backbone made of cartilage, and so are vertebrates. Some lampreys have a parasite-like feeding style. They attach to another fish using an oral disk with horny teeth made of the protein keratin. Once attached, the lamprey uses a tooth-covered tongue to scrape up bits of the host's flesh.

Figure 24.6 Two cartilaginous fishes. **A** Galápagos sharks are fast-swimming predators. **B** Manta rays glide along slowly. Two fleshy projections on the head funnel small organisms into the mouth. Note the gill slits visible at the body surface of both fishes.

Fishes With Jaws

Jawed fishes likely evolved from jawless ancestors when some of the gill supports became modified. Gill supports are skeletal elements that sit beside the gill slits and hold them open (Figure 24.5). Jawed fishes typically have paired fins that help them maneuver and move quickly through the water. Most also have a covering of scales.

Cartilaginous Fishes The mostly marine **cartilaginous fishes** (Chondrichthyes) such as sharks and rays are jawed fishes with a skeleton of cartilage. Their gills are visible at the body surface. Scales do not grow with the fish; instead new new scales grow between old ones. The scales consist of dentin and enamel, the same hard materials that make up teeth. Shark teeth grow in rows and are continually shed and replaced. Sharks include streamlined

A Hagfish. It feeds on worms and scavenges on the sea floor.

B Parasitic lamprey. It attaches to another fish with its oral disk and scrapes up flesh.

Figure 24.4 Two modern jawless fishes.

Figure 24.5 **Animated** Proposed steps in the evolution of jaws.

predators that chase down prey (Figure 24.6A), placid plankton feeders that strain food from the water, and bottom feeders. Rays are cartilaginous fish with a body that is flattened from top to bottom. Manta rays filter plankton from water and some reach 6 meters (20 feet) wide (Figure 24.6B). Stingrays are bottom feeders. Their barbed tail has a venom gland.

Bony Fishes The **bony fishes** (Osteichthyes) have an adult skeleton consisting largely of bone. Typically, a bony covering also protects the gills. Scales are flattened, and they increase in size as a fish grows. The unique trait of bony fishes is a gas-filled organ derived from an outpouching of the gut. In some bony fishes this organ serves as a lung, and functions in gas exchange. In others, it acts as a swim bladder, an adjustable flotation device. By regulating the volume of gas inside its swim bladder, a bony fish can stay suspended in water at different depths.

Figure 24.7A shows the internal anatomy of a bony fish. Like all vertebrates, it has a closed circulatory system with one heart and a urinary system with paired kidneys that filter blood, adjust its volume and solute composition, and eliminate wastes. The digestive system is complete. In most bony fishes, urinary waste, digestive waste, and gametes exit the body through three separate openings. By contrast, cartilaginous fishes, amphibians, and reptiles have one multipurpose opening called a cloaca.

There are two lineages of bony fishes. **Ray-finned fishes** have flexible fin supports derived from skin. With more than 21,000 freshwater and marine species, they are the most diverse vertebrate group. They include most fish that end up on dinner plates, such as salmon, sardines, bass, swordfish, trout, tuna, halibut, carp, and cod. Some have a highly modified body plan (Figure 24.7B–D).

Lobe-finned fishes include coelacanths and lungfishes (Figure 24.8). Their pelvic and pectoral fins are fleshy body extensions that have supporting bones inside them. As the name suggests, lungfishes have gills and one or two air-filled lungs. A lung consists of an air-filled sac or sacs with an associated network of tiny blood vessels. A lungfish fills its lungs by surfacing and gulping air, then oxygen diffuses from the lungs into the blood.

As the next section explains, bony fins and simple lungs proved selectively advantageous when descendants of some lobe-finned fishes ventured onto land.

---

bony fish Fish with a lung or swim bladder and a skeleton consisting largely of bone.
cartilaginous fish Fish with a skeleton of cartilage.
hagfish Jawless fish with a skull case but no backbone.
lamprey Jawless fish with a backbone of cartilage.
lobe-finned fish Fish with fleshy fins that contain bones.
ray-finned fish Fish with fins supported by thin rays derived from skin; member of most diverse lineage of fishes.

Figure 24.7 Animated Ray-finned bony fishes. **A** Body plan of a perch. **B** Sea horse. **C** Monkfish, a stealthy predator. **D** Long-nose gar, a speedy predator.

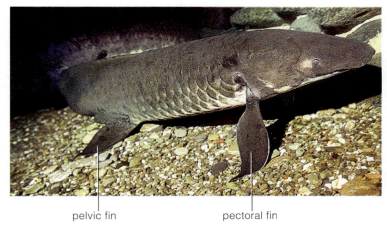

Figure 24.8 Lungfish, a lobe-finned fish. Its pelvic and pectoral fins are supported by bones inside the fins.

Take-Home Message What are the different groups of fishes?

❯ Hagfishes and lampreys are jawless fishes with no fins or scales. Hagfishes do not have a backbone, and so are not considered vertebrates.

❯ Cartilaginous fishes include the sharks and rays. They have a backbone and other skeletal elements made of cartilage.

❯ There are two lineages of bony fishes. The ray-finned fishes are the most diverse group. The lobe-finned fishes are the closest relatives of tetrapods.

❯ Amphibians often spend much of their life out of water, but return to their aquatic roots to reproduce.

❮ Links to Homologous structures 16.8, Chytrid fungi 22.3, Flukes 23.6

The Move Onto Land

Amphibians are land-dwelling predators that require water to breed. The recent discovery of fossil footprints indicates that a four-legged body plan had evolved by 395 million years ago, during the Devonian. Fossilized bones reveal how fishes, adapted to swimming, evolved into four-legged walkers, or tetrapods (Figure 24.9). Bones of a lobe-finned fish's pectoral fins and pelvic fins are homologous with bones of an amphibian's front and hind limbs.

The transition to land was not simply a matter of skeletal changes. Division of the heart into three chambers allowed blood flow in two circuits, one to the body and one to the increasingly important lungs. Changes to the inner ear improved detection of airborne sounds. Eyes became protected from drying out by eyelids.

What was the selective advantage to living on land? An ability to spend some time out of water would have been favored in seasonally dry places. On land, individuals escaped aquatic predators and had a new source of food, insects, which also evolved during the Devonian.

Modern Amphibians

Salamanders and newts resemble the early tetrapods in body form. They have a long tail, and their forelimbs are about the same size as their back limbs (Figure 24.10A). When a salamander walks, the movements of its limbs and bending of its trunk resemble movements made by

Figure 24.10 Animated Salamander gait. **A** Forelimbs and hindlimbs of a salamander are similar sizes and project from the body at right angles. **B** The salamander walks by bending its trunk back and forth, as it moves its limbs in a coordinated fashion. These movements resemble motions made by a swimming fish **C**.

a fish swimming through water (Figure 24.10B,C). Early tetrapods probably used this motion to walk in the water before one lineage ventured onto land.

Larval salamanders look like small versions of adults, except for the presence of gills. In a few species, called axolotls, gills are retained into adulthood. More typically gills disappear as the animal matures and lungs develop.

Figure 24.9 Transition to tetrapods. Skeleton of a Devonian lobe-finned fish **A**, and two early amphibians, *Acanthostega* **B**, and *Ichthyostega* **C**. The painting **D** shows what *Acanthostega* (foreground), and *Ichthyostega* (background) may have looked like.

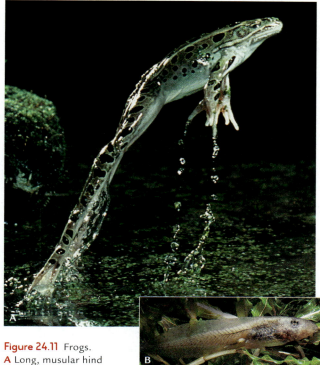

Figure 24.11 Frogs.
A Long, musular hind
limbs allow the adult frog
to leap. B The larval frog, commonly known as a tadpole,
is a swimmer with a long tail and no legs.

Figure 24.12 American toad. The flattened disk visible behind the eye is the eardrum.

Figure 24.13 Frog deformity. Infection by a fluke caused abnormal limb development.

Frogs and toads belong to the most diverse amphibian lineage, with more than 5,000 species. Long, muscular hindlimbs allow the tailless adults to swim, hop, and make spectacular leaps (Figure 24.11A). The much smaller forelimbs help absorb the impact of landings. Toads tend to have somewhat shorter hind legs than frogs and to be better adapted to dry conditions (Figure 24.12).

Larval frogs and toads, commonly called tadpoles or pollywogs, have gills and a tail but no limbs (Figure 24.11B). All frogs and toads have lungs as adults.

Declining Amphibian Diversity

Amphibian populations throughout the world are declining or disappearing. Researchers correlate many declines with shrinking or deteriorating habitats. People commonly fill in low-lying ground where rains once collected and formed pools of standing water. Nearly all amphibians need to deposit their eggs and sperm in water, and their larvae must develop in water.

Other factors that contribute to amphibian declines include introduction of new species in amphibian habitats, long-term shifts in climate, increases in ultraviolet radiation, and the spread of pathogens and parasites. Section 22.3 noted the effects of chytrid fungus infections on amphibians. Figure 24.13 provides an example of the deforming effects of a parasitic fluke (a subgroup of flatworms). Chemical pollution of aquatic habitats also harms amphibians. Their thin skin, unprotected by scales, allows waste carbon dioxide to diffuse out of their body. Unfortunately, it also allows chemical pollutants to enter.

amphibian Tetrapod with a three-chambered heart and scaleless skin; it typically develops in water, then lives on land as a carnivore with lungs.

Take-Home Message How did amphibians evolve and what traits define them?

❯ The tetrapod lineage branched off from the lobe-finned fishes. Amphibians are the oldest tetrapod lineage.

❯ Lungs and a three-chambered heart adapt amphibians to life on land, but nearly all must have access to water to complete their life cycle.

❯ Salamanders, frogs, and toads are the best-known modern amphibians.

> Amniotes are vertebrates that have adapted to a life lived entirely on land.

< Links to Asteroid impact 16.1, Cladistics 17.14

About 300 million years ago, during the Carboniferous, amniotes branched off from an amphibian ancestor. A variety of traits that evolved in this lineage adapted its members to life in dry places. **Amniote eggs** have membranes inside them that keep an embryo moist even away from water (Figure 24.14). In addition, amniote skin is rich in keratin, a protein that makes it waterproof. A pair of well-developed kidneys help conserve water, and fertilization usually takes place inside the female's body.

Shortly after the amniotes arose, the lineage branched in two. One branch gave rise to the mammals. The other gave rise to the members of the clade that biologists call **reptiles** (Figure 24.15). This clade includes turtles, lizards, snakes, crocodilians, and birds.

---

amniote egg Egg with internal membranes that allow the amniote embryo to develop away from water.
dinosaur Reptile lineage abundant in the Jurassic to Cretaceous; now extinct with the exception of birds.
reptile Amniote subgroup that includes lizards, snakes, turtles, crocodilians, and birds.

Figure 24.14 Animated Amniote egg. A bird's egg has a hard shell that encloses the embryo and the characteristic amniote membranes (yolk sac, amnion, chorion, and allantois).

Dinosaurs are extinct members of the reptile clade. Biologists define them by skeletal features, such as the shape of their pelvis and hips. As Section 24.1 explained, evidence suggests that the first birds evolved from a group of feathered dinosaurs. Figure 24.16 shows a Jurassic scene, with the early bird *Archaeopteryx* and an early mammal. With the exception of the birds, all members of the dinosaur lineage disappeared by the end of the Cretaceous. An asteroid impact is considered the most likely cause of their demise (Section 16.1).

Figure 24.15 One proposed evolutionary tree for the living amniote groups. The *blue* box denotes members of the reptile clade. Note that it includes the birds.

Take-Home Message **What are the amniotes?**

> Amniotes are animals that produce eggs in which the young can develop away from water. They are also adapted to life on land by waterproof skin, highly efficient kidneys, and internal fertilization.

> An early divergence separated the ancestors of mammals from the ancestors of reptiles, a group in which biologists include turtles, lizards, snakes, crocodilians, and birds.

Figure 24.16 Painting of a Jurassic scene. In the *middle foreground*, the early bird *Archaeopteryx* flies by. *Behind* the birds, a meat-eating dinosaur sizes up a larger plant-eating dinosaur. At the *far right*, an early mammal surveys the scene from its perch on a tree.

>> **Figure It Out** Which of the animals mentioned above do biologists consider amniotes? Which do they group as reptiles?

Answer: All are amniotes. The dinosaurs and birds are reptiles.

Figure 24.17 Examples of reptiles. **A** Hognose snakes emerging from leathery amniote eggs. **B** Galápagos tortoise. **C** Spectacled caiman, a crocodilian. All crocodilians spend much of their time in water.

❯ All nonbird reptiles—turtles, lizards, snakes, crocodiles, and alligators—have a scale-covered body and acquire heat from their environment.

❮ Link to Metabolic heat 7.5

Modern reptiles include turtles, lizards, snakes, crocodilians, and birds. With the exception of birds, which we consider in the next section, all of these groups are **ectotherms**, meaning "heated from outside." People commonly describe such animals as "cold-blooded," but ectotherms adjust their internal temperature by their behavior. They bask on a warm rock to heat up, or burrow into the soil to cool off. Fish and amphibians are also ectotherms. In contrast, **endotherms** such as birds and mammals produce their own heat through metabolic processes.

Major Groups

Lizards and Snakes

Lizards and snakes constitute the most diverse group of modern reptiles. Overlapping scales cover the body. The smallest lizard can fit on a dime (*left*). The largest, the Komodo dragon, grows up to 3 meters (10 feet) long. Most lizards are predators, but iguanas are herbivores.

The first snakes evolved during the Cretaceous, from short-legged, long-bodied lizards. Some modern snakes have bony remnants of hindlimbs, but most lack limb bones. All are carnivores. Many have flexible jaws that help them swallow prey whole. All snakes have teeth, but not all have fangs. Rattlesnakes and other fanged types bite and subdue prey with venom made in modified salivary glands. On average, only about 2 of the 7,000 snake bites reported annually in the United States are fatal.

Most lizards and snakes lay eggs (Figure 24.17A), but females of some species brood eggs in their body and give birth to well-developed young. The live-bearers do not provide any nourishment to the offspring inside their body, as some mammals do.

Turtles

Turtles and tortoises have a bony shell attached to their skeleton (Figure 24.17B). When threatened, many withdraw into the shell for protection. A fossil recently discovered in China revealed an intermediate stage in turtle evolution. The fossil turtle lived in the sea about 200 million years ago and had a shell only on its ventral (belly) side. Turtles do not have teeth. Instead a horny "beak" made of keratin covers their jaws. Some feed on plants and others are predators.

Crocodilians

Crocodiles, alligators, and caimans are the closest living relatives of birds. Like birds, they have a four-chambered heart. (Lizards, snakes, and turtles have an amphibian-like three-chambered heart.) Like most birds, crocodilians display complex parental behavior. For example, they make and guard a nest, then feed and care for the hatchlings. All crocodilians spend much of their time in water. They have powerful jaws, a long snout, and sharp teeth (Figure 24.17C). They clench prey, drag it under water, tear it apart as they spin around, and gulp the chunks. The feared saltwater crocodile and Nile crocodile weigh up to 1,000 kilograms (2,200 pounds).

---

ectotherm Animal that controls its internal temperature by altering its behavior; for example a lizard.
endotherm Animal that controls its internal temperature by adjusting its metabolism; for example a bird or mammal.

Take-Home Message What are the traits of nonbird reptiles?

❯ All nonbird reptiles are ectotherms, meaning they adjust their temperature by altering their behavior.

❯ Turtles have a bony shell and a horny beak. Lizards and snakes have overlapping scales on an elongated body. Crocodiles and alligators are the closest relatives of birds, and like birds they have a four-chambered heart.

> Birds are the descendants of feathered dinosaurs. They have a body highly modified inside and out for flight.

‹ Links to Aerobic respiration 7.2, Honeycreepers 17.11

Birds are the only living animals with feathers. These lightweight structures derived from scales have roles in insulation, flight, and courtship displays. They also shed water and thus help keep the skin beneath dry.

The bird body has many adaptations related to flight (Figure 24.18). Like humans, birds stand upright on their hindlimbs. The bird forelimbs have evolved into wings. Each wing is covered with long feathers that extend outward and increase its surface area. The feathers give the wing a shape that helps lift the bird as air passes over it. Air cavities inside bones keep body weight low, and make it easier for a bird to become and remain airborne. Flight muscles connect an enlarged breastbone, or sternum, to bones of the upper limb. When the muscles contract, they produce a powerful downstroke.

Flight requires a lot of energy, which is provided by aerobic respiration (Section 7.2). A unique system of air sacs keeps air flowing continually through lungs, ensuring an adequate oxygen supply. A four-chambered heart pumps blood quickly from the lungs to wing muscles and back. A bird also has a rapid heartbeat, and its heart is large relative to its body size.

Archaeopteryx had teeth and a long tail, but modern birds have neither. Layers of the protein keratin cover their jawbones, forming a beak, or bill. Differences in the size and shape of the bill adapt birds to different diets. For example, some honeycreepers have beaks adapted to feed on nectar while other have beaks suited to eating seeds or insects (Section 17.11).

bird An animal with feathers.

Figure 24.19 Mating house sparrows. Lacking a penis, the male must bend his body so his cloaca covers his mate's.

Most nonbird reptiles have a penis, an organ that the male uses to insert sperm into his partner. Male birds do not have a penis. To mate, a male must balance atop a female's back and press his cloaca to hers, a maneuver sometimes referred to as a cloacal kiss (Figure 24.19).

Birds vary in size, proportions, coloration, and capacity for flight. A bee hummingbird weighs 1.6 grams (0.06 ounce). The ostrich, a flightless sprinter, weighs 150 kilograms (330 pounds). More than half of all bird species belong to the subgroup of perching birds. Birds seen at backyard feeders usually belong to this group. Examples include sparrows, jays, starlings, swallows, finches, robins, warblers, orioles, and cardinals.

Take-Home Message **What are birds?**

> Birds are amniotes with feathers. Lightweight bones and highly efficient respiratory and circulatory systems also adapt a bird to flight.

porous, lightweight bone

pectoral girdle

sternum (breastbone)

pelvic girdle

Figure 24.18 Adapted to flight. **A** Among modern vertebrates, only birds and bats fly by flapping a pair of wings. **B** Birds have lightweight bones and a large sternum.

Mammalian Traits

> Mammals scurried about while dinosaurs dominated the land, then diversified after they were gone.

Mammals are hairy or furry animals in which females nourish their offspring with milk secreted from mammary glands. The group name is derived from the Latin *mamma*, meaning breast. Like birds, mammals are endotherms. Their coat of fur or head of hair helps them retain internally produced heat.

Only mammals have four differently shaped kinds of teeth (Figure 24.20). In other vertebrates, an individual's teeth may vary in size, but they are all the same shape. Having teeth of multiple shapes allows mammals to process a wider variety of foods than most other vertebrates.

Three Mammalian Lineages

Monotremes, the oldest mammalian lineage, lay eggs that have a leathery shell similar to that of reptiles. Only three monotreme species survive, the duck-billed platypus (Figure 24.21A) and two kinds of spiny anteater. Monotreme offspring hatch in a relatively undeveloped state, while still tiny, hairless, and blind. They cling to the mother or are held in a skin fold on her belly. Milk oozes from openings on the mother's skin. There are no nipples.

Marsupials are pouched mammals. Young marsupials develop briefly in their mother's body, nourished by

egg yolk and by nutrients that diffuse from the mother's tissues. They are born at an early developmental stage, and crawl along their mother's body to a permanent pouch on her ventral surface. A nipple inside the pouch supplies milk that nourishes them as they continue to develop.

Most of the 240 modern species of marsupials live in Australia and on nearby islands. Kangaroos and koalas are the best known. The opossum (Figure 24.21B) is the only marsupial native to North America.

In **placental mammals**, maternal and embryonic tissues combine as a placenta. This organ allows for a highly efficient transfer of nutrients and gases between mother and embryo. Placental embryos grow faster than those of other mammals, and the offspring are born more fully formed. After birth, the young suckle milk secreted by nipples on their mother's surface (Figure 24.21C).

Placental mammals tend to outcompete other mammalian lineages and are now dominant on all continents except Australia. They are the only mammals that live in the seas. Of the approximately 4,000 placental mammals, nearly half are rodents such as rats and mice.

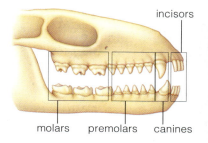

Figure 24.20 Generalized mammalian skull, showing the four differently shaped types of teeth. Compare the teeth of the caiman in Figure 24.17.

Labels: incisors, molars, premolars, canines

---

mammal Animal with hair or fur; females secrete milk from mammary glands.
marsupial Mammal in which young are born at an early stage and complete development in a pouch on the mother's surface.
monotreme Egg-laying mammal.
placental mammal Mammal in which a mother and her embryo exchange materials by means of an organ called the placenta.

Take-Home Message What are mammals?

> Mammals are endothermic amniotes that have hair or fur. The females feed their young milk secreted by mammary glands.

> Egg-laying monotremes, pouched marsupials, and placental mammals are subgroups. Placental mammals are now the dominant lineage in most regions.

A Platypuses are monotremes.

B Opossums are marsupials.

C Humans are placental mammals.

Figure 24.21 Mammalian mothers and their young. All female mammals produce milk to feed offspring.

> Primates include humans and their closest mammalian relatives. As a group, they put their grasping hands to many uses.

‹ Link to Geologic time scale 16.6

Primates, the mammalian subgroup to which humans belong, have five-digit hands and feet capable of grasping (Figure 24.22). The digits are tipped by flattened nails, rather than the sharp claws of related mammals.

Primates first appear in the fossil record about 55 million years ago in Europe, North America, and Asia. Prosimians ("before monkeys") are the most ancient of the surviving lineages. Lemurs and tarsiers are examples. See Table 24.1 and Figure 24.23. An early branching separated the anthropoid lineage to which monkeys, apes, and humans belong, from the prosimians. Further branchings produced hominoids—apes and humans. Our closest living relatives are chimpanzees and bonobos (previously called pygmy chimpanzees). Humans and a variety of extinct humanlike species are referred to as **hominids**.

Figure 24.22 A chimpanzee shows off its grasping five-digit hands and feet. Primates have nails on their digits, rather than claws.

Table 24.1	Primate Classification
Prosimians	Lemurs, tarsiers
Anthropoids	New World monkeys (e.g., spider monkeys, howler monkeys)
	Old World monkeys (e.g., baboons, rhesus monkey)
	Hominoids:
	Hylobatids (gibbons, siamangs)
	Pongids (orangutans, gorillas, chimpanzees, bonobos)
	Hominids (humans, extinct humanlike species)

Key Trends in Primate Evolution

Five trends define the lineage that led to primates, hominids, and then to humans. First, the structure of the face changed (Figure 24.24). Primates are close relatives of tree shrews, and like this group early primates had a long, pointy snout and eyes at the side of the head. Later groups evolved a flattened face with both eyes facing forward. Forward-facing eyes provide better depth perception and probably helped the animals as they made leaps from branch to branch. As the face flattened, the amount of brain area devoted to odor detection declined.

Lemurs   Tarsiers   New World monkeys   Old World monkeys   Gibbons   Orangutans   Gorillas   Chimpanzees   Hominids

Tree shrewlike ancestor

Figure 24.23 Proposed evolutionary tree for the primates.

Tree shrew, a close relative of primates, has a pointy face and its eyes are on either side of its head.

Early primate
(*Plesiadapis*)

Early anthropoid
(*Aegyptopithecus*)

Early hominid
(*Proconsul*)

Figure 24.24 Evolution of the primate skull. One trend is toward a more flattened face, with eyes at the front, rather than the sides of the skull. In hominids, teeth became smaller and more uniform.

A Gorillas are knuckle-walkers. They walk on all fours, with their hands in a partially closed fist. The backbone is straight and the arms are proportionately longer than those of a human.

B Humans are bipedal. They walk upright. Their backbone has a distinct S shape and their skull has a hole at the bottom where it attaches to the spinal cord. In gorillas, this hole is toward the back of the skull.

Figure 24.25 Some skeletal differences associated with differences in posture and locomotion between a human and a gorilla.

Second, teeth became less specialized. Modifications to the jaws and teeth accompanied a shift from eating insects to eating fruits and leaves, then to a mixed diet. Rectangular jaws and lengthy canines evolved in monkeys and apes. A bow-shaped jaw and smaller, more uniform teeth evolved in early hominids.

Third, skeletal changes allowed upright standing and walking, known as **bipedalism**. The backbone developed an S-shaped curve that kept the body aligned over the feet (Figure 24.25). The skull became modified so it perched atop the backbone, rather than connecting to it at the rear.

Fourth, freed from a primary role in locomotion, hands became adapted for novel tasks. Changes to the arrangement of bone and muscle in the hands provided an ability to grasp objects in a fist, or to pick them up between fingertips and the thumb. These abilities allowed some primates to manipulate objects in unique ways, and to use them as tools:

Fifth, the braincase and the brain increased in size and complexity. As brain size increased, so did the length of pregnancy and extent of maternal care. Compared to early primates, later ones had fewer offspring and invested more in each one.

Early primates were solitary. Later, some started to live in small groups. Social interactions and cultural traits began to affect reproductive success. **Culture** is the sum of all learned behavioral patterns transmitted among members of a group and between generations.

---

bipedalism Standing and walking on two legs.
culture Learned behavior patterns transmitted among members of a group and between generations.
hominid Human or extinct humanlike species.
primate Mammal having grasping hands with nails; includes prosimians, monkeys, apes, and hominids such as humans.

power grip   precision grip

Take-Home Message **What traits and trends characterize the primate lineage?**

❯ Primates have grasping hands with nails on their five digits.

❯ Early primates were solitary and shrewlike. Some later lineages evolved a flatter face with more uniformly sized teeth, an upright stance, hands with more dexterity, a larger brain, and a tendency to live in groups.

> Fossils from central, eastern, and southern Africa show that the Miocene through the Pliocene was a "bushy" time of hominid evolution. In other words, many forms were rapidly evolving. We still do not know how they are related.
< Link to Adaptive radiation 17.13

Early Hominids

Genetic comparisons indicate that hominids diverged from apelike ancestors about 6 to 8 million years ago. Fossils that may be hominids are about 6 million years old. *Sahelanthropus tchadensis* had a hominid-like flat face, prominent brow, and small canines, but its brain was relatively small, about the size of a chimpanzee's (Figure 24.26A). *Ardipithecus ramidus* was a hominid that lived 4.4 million years ago. Analysis of a recently discovered fossil female, nicknamed Ardi, suggests that this species may have walked upright when on the ground, but more often used all four limbs to climb in trees.

An indisputably bipedal hominid, *Australopithecus afarensis*, was walking in Africa by about 3.9 million years ago. Remarkably complete skeletons reveal that it habitu-ally walked upright (Figure 24.27A). About 3.7 million years ago, two *A. afarensis* individuals walked across a layer of newly deposited volcanic ash. A light rain fell, and it transformed the powdery ash into a fast-drying cement, which preserved their footprints (Figure 24.27B,C).

A. afarensis was one of the **australopiths** ("southern apes"). This informal group includes *Australopithecus* and *Paranthropus* species. *Australopithecus* species were petite; they had a small face and teeth (Figure 24.26B). One or more species are suspected to be ancestral to modern humans. In contrast, *Paranthropus* species had a stockier build, a wider face, and larger molars. Jaw muscles attached to a pronounced bony crest at the top of their skull (Figure 24.26C). The big molars and strong jaw muscles indicate that fibrous, difficult-to-chew plant parts accounted for a large part of the diet. *Paranthropus* died out about 1.2 million years ago.

Early Humans

What do the fossilized fragments of early hominids tell us about human origins? The record is still too sketchy

A	B	C	D	E
Sahelanthropus tchadensis 6 million years ago	*Australopithecus africanus* 3.2–2.3 million years ago	*Paranthropus boisei* 2.3–1.2 million years ago	*Homo habilis* 1.9–1.6 million years ago	*Homo erectus* 1.9 million to 53,000 years ago

Figure 24.26 A sampling of fossilized hominid skulls from Africa, all to the same scale.

Figure 24.27 A Fossilized bones of Lucy, a female australopith (*Australopithecus afarensis*). This bipedal hominid lived in Africa 3.2 million years ago.

B,C At Laetoli in Tanzania, Mary Leakey discovered footprints made in soft, damp, volcanic ash 3.7 million years ago. The arch, big toe, and heel marks of these footprints are signs of bipedal hominids.

for us to interpret how all the diverse forms were related, let alone which might have been our ancestors. Besides, exactly which traits should we use to define **humans**— members of the genus *Homo*?

Well, what about brains? Our brain is the basis of unsurpassed analytical skills, verbal skills, complex social behavior, and technological innovations. Early hominids did use simple tools, but so do chimpanzees and bonobos. So how did an early hominid make the evolutionary leap to becoming human?

Comparing the brains of modern primates can give us clues. Genes for some brain proteins underwent repeated duplication as the primate lineage evolved. Further studies of how these proteins function may provide additional insight into how our uniquely human traits arose.

Until then, we are left to speculate on the evidence of physical traits among diverse fossils. They include a skeleton that permitted bipedalism, a smaller face, larger cranium, and smaller teeth with more enamel. These traits emerged during the late Miocene and can be observed in *Homo habilis*. The name of this early human means "handy man" (Figure 24.28).

Most of the early known forms of *Homo* are from the East African Rift Valley. Fossil teeth indicate that these early humans ate hard-shelled nuts, dry seeds, soft fruits and leaves, and insects. *H. habilis* may have enriched its diet by scavenging carcasses left behind by carnivores such as saber-tooth cats, but it did not have teeth adapted to a diet rich in meat.

Our close relatives, the chimpanzees and bonobos, use sticks and other natural objects as tools (Section 25.7). Early hominids most likely did the same. They smashed nuts open with rocks and used sticks to dig into termite nests and capture insects. By 2.5 million years ago, some hominids were modifying rocks in ways that made them better tools. Pieces of volcanic rock chipped to a sharp edge were found with animal bones that look like they were scraped by such tools.

The layers of Tanzania's Olduvai Gorge document refinements in toolmaking abilities (Figure 24.29). The layers that date to about 1.8 million years ago hold crudely chipped pebbles. More recent layers contain more complex tools, such as knifelike cleavers.

Olduvai Gorge also held hominid fossils. At the time of their discovery, these fossils were classified as *Homo erectus*. The name means "upright man." Today, some researchers reserve that name for fossils in Asia, and they assign the African fossils to *H. ergaster*. In our discus-

Figure 24.28 Painting of a band of *Homo habilis* in an East African woodland. Two australopiths are shown in the distance at the *left*.

Figure 24.29 A sample of stone tools from Olduvai Gorge in Africa. From *left* to *right*, crude chopper, more refined chopper, hand ax, and cleaver.

sions, we will adopt a traditional approach using "*H. erectus*" in reference to African populations and to descendant populations who, over generations, migrated into Europe and Asia.

H. erectus adults averaged about 5 feet (1.5 meters) tall, and had a larger brain than *H. habilis*. Improved hunting skills may have helped *H. erectus* get the food needed to maintain a large body and brain. Also, *H. erectus* built fires. Cooking probably broadened their diet by softening previously inedible hard foods.

---

australopiths Collection of now-extinct hominid lineages, some of which may be ancestral to humans.
humans Members of the genus *Homo*.

Take-Home Message **What do we know about the australopiths and the early species of the genus *Homo*?**

> Australopiths and certain hominids that preceded them walked upright. Some *Australopithecus* species were close to or on the lineage that led to humans.

> Like some hominids that preceded them, *Homo habilis*, the earliest known human species, walked upright. *Homo erectus* had a larger brain and dispersed out of Africa.

> Judging from the fossil record, the earliest members of the human lineage emerged about 2.5 million years ago, in the great East African Rift Valley.

< Link to Surface-to-volume ratio 4.2

Branchings of the Human Lineage

By 1.7 million years ago *Homo erectus* populations had become established in places as far away from Africa as the island of Java and eastern Europe. At the same time, African populations continued to thrive. Over thousands of generations, genetic divergences led to adaptations to local conditions. Some populations became so different from parental *H. erectus* groups that we call them new species: *H. neanderthalensis* (commonly called Neandertals), *H. floresiensis*, and *H. sapiens*, or modern humans (Figure 24.30).

A fossil from Ethiopia shows that *Homo sapiens* had evolved by 195,000 years ago. Compared to *H. erectus*, *H. sapiens* had smaller teeth, facial bones, and jawbones. Many fossils reveal another new feature: a prominent chin. Compared to earlier hominids, *H. sapiens* had a higher, rounder skull, a larger brain, and the capacity for spoken language.

From 200,000 to 30,000 years ago, Neandertals lived in Africa, the Middle East, Europe, and Asia. They had a stocky body that may have been an adaptation to colder climates. A stocky body has a lower surface-to-volume ratio than a thin one and loses heat to the environment less quickly.

Neandertals had a big brain. Did they believe in an afterlife? Did they have a spoken language? We do not know. They vanished when *H. sapiens* entered the same regions. Did the new arrivals drive them to extinction through direct warfare or indirectly, by outcompeting Neandertals for the same resources? We do not know.

Homo neanderthalensis

Homo sapiens

Homo floresiensis

Figure 24.30 Recent *Homo* species. Whether *H. floresiensis* belongs in this genus is still debated.

Even if a few matings between the species did take place, sequencing data indicates that Neandertal genes did not contribute to the gene pool of modern humans. Even so, because of our common ancestry, modern humans share about 99.5 percent of their genes with Neandertals.

In 2003, scientists discovered human fossils about 18,000 years old on the Indonesian island of Flores. Like *H. erectus*, they had a heavy brow and a relatively small brain for their body size. However, adults would have stood only a meter tall. Scientists who found the fossils assigned them to a new species, *H. floresiensis*. Not everyone is convinced. Some think the fossils belong to *H. sapiens* individuals who had a disease or disorder.

Where Did Modern Humans Originate?

Neandertals evolved from *H. erectus* populations in Europe and western Asia. *H. floresiensis* may have evolved from *H. erectus* in Indonesia. Where did *H. sapiens* originate? Two major models agree that *H. sapiens* evolved from *H. erectus* but differ in the details of where and how fast these events took place. Both attempt to explain the distribution of *H. erectus* and *H. sapiens* fossils, as well as genetic differences among modern humans who live in different regions.

Multiregional Model According to the **multiregional model**, populations of *H. erectus* living in Africa and other regions evolved into populations of *H. sapiens* gradually, over more than a million years. Gene flow among the populations maintained the species through the transition to fully modern humans (Figure 24.31A).

If this model is correct, much of the genetic variation present among modern Africans, Asians, and Europeans began to build up soon after their ancestors branched from an ancestral *H. erectus* population. The multiregional model is based primarily on interpretation of fossils. For example, faces of *H. erectus* fossils from China are said to look more like modern Asians than *H. erectus* that lived in Africa. The idea is that much variation seen among modern *H. sapiens* evolved long ago, in *H. erectus*.

Replacement Model By the more widely accepted **replacement model**, *H. sapiens* arose from a single *H. erectus* population in sub-Saharan Africa within the past 200,000 years. Later, bands of *H. sapiens* entered regions already occupied by *H. erectus* populations, and drove them all to extinction (Figure 24.31B). If this model is correct, then the regional variations observed among modern *H. sapiens* populations arose relatively recently. This model emphasizes the enormous degree of genetic similarity among all living humans.

Fossils support the replacement model. *H. sapiens* fossils date back to 195,000 years ago in East Africa and

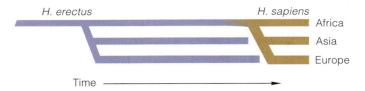

A Multiregional model. *H. sapiens* slowly evolves from *H. erectus* in many regions.

B Replacement model. *H. sapiens* rapidly evolves from one *H. erectus* population in Africa, then disperses and replaces *H. erectus* populations in all regions.

Figure 24.31 Two models for the origin of *H. sapiens*. Arrows represent ongoing gene flow among populations.

100,000 years ago in the Middle East. In Australia, the oldest date to 60,000 years ago and, in Europe, to 40,000 years ago. Global comparisons of markers in mitochondrial DNA, and in the X and Y chromosomes, place the modern Africans closest to the root of the family tree. They also reveal that the most recent common ancestor of all modern humans was alive in Africa 60,000 years ago.

Leaving Home

Fossils and genetic evidence allow scientists to trace human dispersal routes (Figure 24.32). About 120,000 years ago, Africa's interior was becoming cooler and drier. As patterns and amounts of rainfall changed, so did the distribution of herds of grazing animals and the humans who hunted them. A few hunters may have journeyed north from East Africa and into Israel, where fossils 100,000 years old were found in a cave. They have no living descendants. Eruption of Mount Toba in Indonesia may have killed them, along with other ancient travelers.

Later waves of travelers had better luck, as some individuals left established groups and ventured into new territory. Successive generations continued along the coasts of Africa, then Australia and Eurasia. In the Northern Hemisphere, much of Earth's water became locked in vast ice sheets, which lowered the sea level by hundreds of meters. Previously submerged land was drained off between some regions. About 15,000 years ago, one small band of humans crossed such a land bridge from Siberia into North America.

With each step of their journey, humans faced and overcame extraordinary hardships. During this time, they devised cultural means to survive in inhospitable environments. Unrivaled capacities for modifying the habitat

---

multiregional model Model that postulates *H. sapiens* populations in different regions evolved from *H. erectus* in those regions.
replacement model Model for origin of *H. sapiens*; humans evolved in Africa, then migrated to different regions and replaced the other hominids that lived there.

Windows on the Past (revisited)

Interpreting vertebrate fossils can be a challenge. Scientists often disagree over whether a fossil represents a new species, a variant of a known species, or a collection of bones from more than one species. Such disputes are typically sorted out over time, as many scientists examine the fossils and publish their findings.

For example, consider what happened after one group of researchers uncovered hominid bones in a cave in Indonesia and declared their find a new species, *Homo floresiensis*. Some other scientists suggested that the small stature of the fossils could indicate that they were remains of *H. erectus* or *H. sapiens* with a genetic or nutritional disorder. A scientist who examined the fossil wrist disagreed, concluding that the bones are unlike those of *H. erectus* or *H. sapiens*, and so do represent a distinct species. Still other scientists noted that the fossil skull resembles that of some australopiths. They suggest that the fossils could be a new species that descended from australopiths who migrated out of Africa. Further study of existing fossils, a search for more fossils, and possibly DNA studies will help test the competing hypotheses.

How Would You Vote? Collecting fossils is big business, but it encourages poaching on public lands and decreases access to fossils. Should the government regulate fossil sales? See CengageNow for details, then vote online (cengagenow.com).

---

and for language served them well. Cultural evolution is ongoing. Hunters and gatherers persist in a few parts of the world, but others moved on from "stone-age" technology to the age of "high tech." This continued coexistence of such diverse groups is a tribute to the deep behavioral plasticity of the human species.

Take-Home Message **What are modern humans?**

> Fossils and genetic evidence indicate that modern humans, *H. sapiens*, most likely evolved from a *H. erectus* population in Africa.

> As modern humans dispersed out of Africa, geographic and climate factors affected their journey.

Figure 24.32 Some dispersal routes for small bands of *Homo sapiens*. This map shows ice sheets and deserts that prevailed about 60,000 years ago. Fossils provide evidence for the dates when modern humans appeared in these regions:

Africa	by 195,000 years ago
Israel	100,000
Australia	60,000
China	50,000
Europe	40,000
North America	14,000

The oldest evidence of humans in North America is fossil feces from a cave in Oregon.

Summary

Section 24.1 Fossils are an important source of information about evolutionary transitions between major groups. For example, fossils reveal that birds evolved from feathered dinosaurs, and that some early birds had a bony tail and teeth.

Section 24.2 Four features define **chordate** embryos: a **notochord**, a dorsal hollow nerve cord, a pharynx with gill slits, and a tail extending past the anus. These features may or may not persist in adults. Invertebrate chordates include **tunicates** and **lancelets**. The **craniates** are chordates with a braincase of cartilage or bone. Most craniates are **vertebrates**. Their backbone is part of an **endoskeleton**. The **tetrapods** are vertebrates with four limbs. **Amniotes** are a subset of tetrapods that produce eggs that allow embryos to develop away from water.

Section 24.3 **Hagfishes** are jawless fishes with a cranium but no backbone. **Lampreys** are jawless vertebrates. Some are parasites of other fish. **Cartilaginous fishes** such as sharks have jaws with rows of teeth. **Ray-finned** fish are **bony fishes** and the most diverse vertebrate group. **Lobe-finned** fishes, also bony fishes, have fins reinforced with bones. Aspects of the fish body plan adapt the fish to life in water. For example, a swim bladder allows a ray-finned fish to adjust its buoyancy.

Section 24.4 **Amphibians**, the first lineage of tetrapods, branched off from the lobe-finned fishes. Modern amphibians such as frogs, toads, and salamanders are carnivores. Most spend some time on land, but return to the water to reproduce.

Section 24.5 Amniotes, the first vertebrates able to complete their life cycle on dry land, have water-conserving skin and kidneys, and **amniote eggs**. Fertilization usually takes place inside the female's body. There are two lineages. **Reptiles** (including birds and the extinct **dinosaurs**) belong to one lineage, and mammals to the other.

Section 24.6 The most diverse reptile lineage includes lizards and snakes. All are carnivores with flattened scales. Turtles have a bony shell and a horny beak rather than teeth. Crocodilians are carnivores and the closest relatives of birds. Snakes, lizards, turtles, and crocodilians are **ectotherms**: They adjust their temperature by their behavior. By contrast, birds and mammals are **endotherms**, which produce heat by metabolic activity.

Section 24.7 **Birds** are reptiles with feathers. The body plan of most species has been highly modified for flight, with lightweight bones and a highly efficient respiratory system. Male birds do not have a penis.

Section 24.8 **Mammals** have hair or fur, and the females nourish young with milk secreted from mammary glands. They include egg-laying **monotremes**, pouched **marsupials**, and **placental mammals**, which are the most diverse and widespread group.

Section 24.9 **Primates** have hands and feet capable of grasping objects. Prosimians are the oldest lineages. **Hominids** are the most recent. Early primates were shrewlike. Better depth perception, **bipedalism**, increased dexterity, bigger brains, social behavior, extended parental care, and **culture** evolved in some hominid lineages.

Sections 24.10, 24.11 **Australopiths** are early hominids known from Africa. The **human** lineage (*Homo*) arose by 2 million years ago (Figure 24.33), with *H. habilis* as an early toolmaking species. Neandertals and modern humans are relatives, but have distinct gene pools. By the **multiregional model**, *H. erectus* populations in far-flung regions evolved into *H. sapiens*. The **replacement model** has modern humans evolving from *H. erectus* in Africa, then dispersing into regions already occupied by *H. erectus* and driving them to extinction.

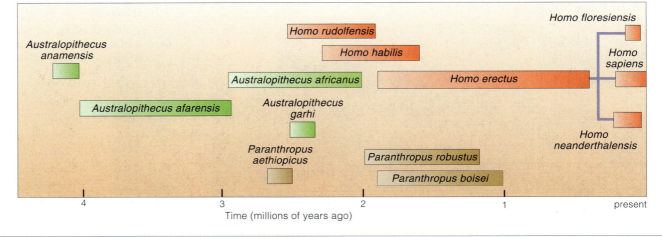

Figure 24.33 Summary of the estimated dates for the origin and extinction of three hominid genera. *Purple* lines are one view of how human species are related. The number of species, which fossils belong to each group, and how the species relate to one another are being debated.

Time (millions of years ago)

Data Analysis Activities

Primate Life History Trends One trend in primate evolution involved changes in life history traits, such as the length of infancy and the time it takes to reach adulthood. Figure 24.34 compares five primate lineages, from most ancient to most recent. It graphs life spans starting with years spent as "infants," when ongoing maternal care is required. It shows time spent as subadults, when individuals are no longer dependent on their mother for care but have not yet begun to breed. It shows the length of the reproductive years, and the length of time lived after reproductive years have passed.

1. What is the average life span for a lemur? A gibbon?
2. Which group reaches adulthood most quickly?
3. Which group has the longest expanse of reproductive years?
4. Which groups survive past their reproductive years?

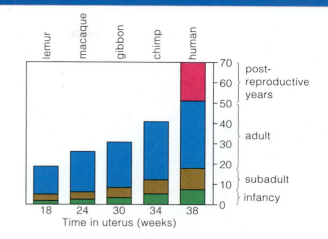

Figure 24.34 Trend toward longer life spans and greater dependency of offspring on adults for five primate lineages.

1. All chordates have (a) _____ .
   a. backbone   b. jaws   c. notochord   d. both b and c

2. The lancelet pharynx functions in _____ .
   a. respiration   c. reproduction
   b. filter-feeding   d. both a and b

3. Vertebrate jaws evolved from _____ .
   a. gill supports   b. ribs   c. scales   d. teeth

4. Lampreys and sharks both have _____ .
   a. jaws   d. a swim bladder
   b. a bony skeleton   e. a four-chambered heart
   c. a cranium   f. lungs

5. A divergence from _____ gave rise to tetrapods.
   a. ray-finned fishes   c. cartilaginous fishes
   b. lizards   d. lobe-finned fishes

6. Reptiles, including birds, belong to one major lineage of amniotes, and _____ belong to another.
   a. sharks   c. mammals
   b. frogs and toads   d. salamanders

7. Reptiles are adapted to life on land by _____ .
   a. tough skin   d. amniote eggs
   b. internal fertilization   e. both a and c
   c. good kidneys   f. all of the above

8. The closest modern relatives of birds are _____ .
   a. crocodilians   b. prosimians   c. turtles   d. lizards

9. Only birds have _____ .
   a. a cloaca   c. feathers
   b. a four-chambered heart   d. amniote eggs

10. An australopith is a(n) _____ .
    a. craniate   d. amniote
    b. vertebrate   e. placental mammal
    c. hominoid   f. all of the above

11. *Homo erectus* _____ .
    a. was the earliest member of the genus *Homo*
    b. was one of the australopiths
    c. evolved in Africa and dispersed to many regions
    d. disappeared as the result of an asteroid impact

12. True or false? Primates are endotherms.

13. Match the organisms with the appropriate description.
    ___ lancelets   a. pouched mammals
    ___ fishes   b. invertebrate chordates
    ___ amphibians   c. feathered amniotes
    ___ primates   d. egg-laying mammals
    ___ birds   e. humans and close relatives
    ___ monotremes   f. have grasping hands with nails
    ___ marsupials   g. first land tetrapods
    ___ placental   h. most diverse mammal
        mammals   lineage
    ___ hominids   i. most diverse vertebrates

Additional questions are available on **CENGAGENOW**.

Critical Thinking

1. Monotremes and marsupials evolved before the super-continent Pangea broke up. Placental mammals evolved after the breakup. How does this explain why Australia has no native placental mammals?

2. Does Alaska have more native birds or reptiles?

3. Describe some mechanisms that could have prevented hybrid births even if humans and Neandertals did occasionally meet and have intercourse.

Animations and Interactions on **CENGAGENOW**:
> Lancelet body plan; Salamander gait; Structure of an amniote egg.

❮ Links to Earlier Concepts

This chapter builds on your understanding of the concepts of sexual selection (17.7) and adaptation (16.4). In our discussion of human behavior, you will be reminded once again of the limits of science (1.9).

Key Concepts

Genetic Foundations
Genes that affect the ability to detect stimuli or to respond to nervous or hormonal stimulation influence behavior. Studying behavioral differences within and between species allows scientists to determine the behavior's proximate and ultimate causes.

Instinct and Learning
Instinctive behavior can be performed without any practice. Most behavior has a learned component. Even instinctive behavior is often modified over time. Some learning can only occur during a certain portion of the lifetime.

25 Animal Behavior

25.1 An Aggressive Defense

Many animals sting or bite when threatened. For example, honeybees sting skunks, bears, and other animals that attack their hive. Honeybees live in large family groups, gather nectar from flowers, and store the nectar as honey. A hive's stored honey, along with the protein-rich bodies of developing bees, makes it a tempting target for hungry animals. Stinging is an adaptive response, an evolved defense against the threat posed by potential hive raiders.

Humans tend to label an animal whose defensive response is easily triggered by humans as "aggressive." Africanized honeybees, known in the popular press as "killer bees," are a prime example. These bees are a hybrid between the European honeybees favored by beekeepers and a bee subspecies native to Africa (Figure 25.1). The European bees raised for commercial use have been selectively bred to have a high threat threshold, but Africanized bees retain their normal defensiveness. Both strains of honeybee can sting only once and make the same kind of venom, but the Africanized bees sting with less provocation, respond to threats in greater numbers, and pursue intruders with more persistence.

Africanized honeybees arose in Brazil in the 1950s. Bee breeders there had imported African bees in the hope of breeding an improved pollinator for this region's tropical orchards. Some of the African imports escaped and mated with European honeybees that had already become established there. Descendants of the resulting hybrids expanded northward and reached Texas in 1990. By 2009, Africanized bees had been detected in New Mexico, Nevada, Utah, southern California, Oklahoma, Louisiana, Alabama, and Florida.

Honeybees live in cavities and Africanized bees can take up residence in spaces between walls and in boxes that hold utility meters. Vibrations from machinery such as lawn mowers can trigger a defensive response. In 2008, a Florida man died after being stung about 100 times. He had inadvertently disturbed an Africanized bee colony while demolishing a trailer.

Fatalities from Africanized bee stings are rare (there have been fewer than 20 since the bees arrived in 1990) but even a single sting can be fatal to someone allergic to honeybee venom. In addition, bee stings are highly painful and people who receive a large number of stings often require hospitalization. Thus, the ongoing spread of Africanized bees through the United States is a matter of concern.

What makes Africanized bees so testy? One factor is a greater response to alarm pheromone. A **pheromone** is a chemical signal that transmits information between two individuals of the same species. When a honeybee guarding the entrance to a hive detects a threat, she releases an alarm pheromone. Bees inside the hive detect this chemical signal and rush out to join the guards in driving off the intruder.

Researchers have tested the response of Africanized honeybees and European honeybees to alarm pheromone by positioning a dark cloth near the entrance of hives and releasing an artificial pheromone. Africanized bees flew out of a hive and attacked the cloth faster than European bees and they plunged six to eight times as many stingers into the cloth.

Considering differences among honeybees leads us into the study of animal behavior. Scientists investigate proximate causes of behavior—the genetic and physiological factors that explain how a behavior is carried out. They also investigate ultimate causes—the factors that favored evolution of a behavior. Interactions among animals and between animals and the environment are the focus of **ecology**. Behavioral ecology, our focus here, is one facet of that field. Chapters that follow consider the study of populations, ecosystems, communities, and the biosphere.

Figure 25.1 Africanized honeybees on guard duty. If threatened, they will release an alarm pheromone that stimulates hivemates to join an attack.

ecology Study of organism–environment interactions.
pheromone Intraspecific chemical communication signal.

Animal Communication
Animal communication behavior can only evolve if communication benefits both signal senders and signal receivers. Animals communicate by a variety of mechanisms. Predators sometimes take advantage of the communication behavior of their prey.

Mating and Parenting
Males and females maximize their reproductive success in different ways. Females tend to be choosier about mates than males. Monogamy is rare among animals, but it tends to correlate with parental care by both parents. More often, the female provides parental care.

Costs and Benefits of Social Living
Life in social groups has reproductive benefits and costs. Not all environments favor the evolution of such groups. Self-sacrificing behavior has evolved among a few kinds of animals that live in large family groups.

> Much variation in behavior within or among species results from inherited differences. In a few instances, scientists have even pinpointed genes responsible for the variation.
< Link to Polygenic inheritance 14.2

How Genes Can Influence Behavior

Animal behavior requires a capacity to detect stimuli. A **stimulus** is some type of information about the environment that a sensory receptor detects and responds to. The structure of the nervous system determines the types of stimuli that an animal can detect and the types of responses it can make. Differences in genes that affect the structure and activity of the nervous system cause many differences in behavior.

Genes that affect metabolism or structural traits also influence behavior. For example, some African seedcrackers routinely eat large seeds and others focus on small seeds (Section 17.6). The birds that focus on small seeds do so because their genes specify a beak structure that allows them to easily open small seeds but not larger ones.

Genetic Variation Within a Species

One way to investigate the genetic basis of behavior is to examine behavioral and genetic differences among members of a single species. Studies designed to determine the roots of the difference in defensive behavior between Africanized honeybees and European honeybees fall into this category.

Stevan Arnold's study of snake feeding behavior is another example. Garter snakes that live in coastal forests of the Pacific Northwest prefer to eat the banana slugs that are common on the forest floor (Figure 25.2). Farther inland, there are no banana slugs. Garter snakes that live

Figure 25.2 Coastal garter snake dining on a banana slug. The snake is genetically predisposed to recognize slugs as prey. By contrast, snakes from inland regions where there are no banana slugs ignore slugs when experimenters offer them.

inland prefer to eat fishes and tadpoles. The food preference is inborn. When Arnold offered newborn garter snakes a slug as their first meal, offspring of coastal snakes ate it, but offspring of inland snakes ignored it.

Arnold hypothesized that inland snakes lack the genetically determined ability to associate the scent of slugs with food. He predicted that if coastal garter snakes were crossed with inland snakes, the resulting offspring would make an intermediate response to slugs. Results from his experimental crosses confirmed this prediction. We do not know which gene or genes underlie this difference.

We know more about the genetic basis of differences in foraging behavior among fruit fly larvae. The wingless, wormlike larvae crawl about eating yeast that grows on decaying fruit. In wild fruit fly populations, about 70 percent of the fly larvae are "rovers," which means they tend to move around a lot as they feed. Rovers often leave one patch of food to seek another (Figure 25.3A). The remaining 30 percent of larvae are "sitters"; they tend to move little once they find a patch of yeast (Figure 25.3B). When food is absent, rovers and sitters move the same amount, so both are equally energetic.

The proximate cause of the difference in larval behaviors is a difference in alleles of a gene called *foraging*. Flies with a dominant allele of this gene have the rover phenotype. Sitters are homozygous for the recessive allele. The *foraging* gene encodes an enzyme involved in learning about olfactory cues. Rovers make more of the enzyme than sitters.

The ultimate cause of the behavioral variation in larval foraging behavior is natural selection that arises from competition for food. In experimental populations with

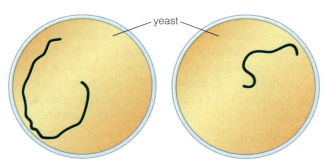

yeast

A Rovers (genotype *FF* or *Ff*) move often as they feed. When a rover's movements on a petri dish filled with yeast are traced for 5 minutes, the trail is relatively long.

B Sitters (genotype *ff*) move little as they feed. When a sitter's movements on a petri dish filled with yeast are traced for 5 minutes, the trail is relatively short.

Figure 25.3 Genetic polymorphism for foraging behavior in fruit fly larvae. When a larva is placed in the center of a yeast-filled plate, its genotype at the *foraging* locus influences whether it moves a little or a lot while it feeds. *Black* lines show a representative larva's path.

B PET scan of a monogamous prairie vole's brain with many receptors for the hormone oxytocin (*red*).

C PET scan of a promiscuous prairie vole's brain with few hormone receptors for oxytocin.

Figure 25.4 Studying the genetic roots of mating and bonding behavior. Voles (*Microtus*) are small rodents in which closely related species vary in their mating and bonding behavior, and in the number and distribution of receptors for the hormone oxytocin.

limited food, both rovers and sitters are most likely to survive to adulthood when their foraging type is rare. A rover does best when surrounded by sitters, and vice versa. Presumably, when there are lots of larvae of one type they all compete for food in the same way. Under these circumstances, a fly that behaves differently than the majority is at an advantage. As a result, natural selection maintains both alleles of the *foraging* gene.

Genetic Variation Among Species

Comparing behavior of related species can sometimes clarify the genetic basis of a behavior. For example, studies of voles (Figure 25.4) have revealed that inherited differences in the number and distribution of certain hormone receptors influence mating and bonding behavior.

Differences in mating behavior between closely related species of vole make these animals ideal subjects for study of this behavior. Most voles, like most mammals, are promiscuous. However, some voles form lifelong, largely monogamous relationships. For example, in prairie voles, a permanent social bond forms after a night of repeated matings. The hormone oxytocin plays a central role in a female prairie vole's bonding. When females who are part of an established pair are injected with a chemical that interferes with oxytocin action, they dump their long-term partners and mate with other males.

Females of promiscuous vole species may behave as they do because they are less influenced by oxytocin than prairie voles. When researchers compared the brains of promiscuous and monogamous species, they found a striking difference in the number and distribution of oxytocin receptors (Figure 25.4B,C). Monogamous prairie voles have many oxytocin receptors in part of the brain associated with social learning. Promiscuous mountain voles have far fewer of these receptors.

In male voles, variations in the distribution of receptors for another hormone (arginine vasopressin, or AVP) correlate with bonding tendency. Compared to males of promiscuous vole species, males of monogamous species

have more AVP receptors. When scientists isolated the prairie vole gene for the AVP receptor and used a virus to transfer copies of this gene into the forebrain of male mice (which are naturally promiscuous), the genetically modified mice gave up their playboy ways. They now preferred a female with whom they had already mated over an unfamiliar female. These results confirmed the role of AVP in fostering monogamy among male rodents.

Human Behavior Genetics

Nearly all human behavioral traits have a polygenic basis and are influenced by the environment. Keep this in mind when you see headlines touting discovery of a gene "for" thrill-seeking behavior, alcoholism, or some other human trait. Generally, such discoveries show a small statistical correlation between a particular allele and an increased tendency toward a particular behavior.

Insights from studies of animal behavior sometimes inspire human studies. For example, investigators are looking at the role oxytocin plays in human social behavior. Impaired oxytocin production or reception may contribute to autism, a disorder in which a person has trouble making social and emotional attachments. Scientists are collecting data about oxytocin receptor genes of autistic children and their unaffected siblings. The aim of the genetic study is to determine whether particular alleles of the oxytocin receptor gene raise the likelihood of autism. Oxytocin is also being tested as a treatment for autism.

---

stimulus Environmental cue that a sensory receptor responds to.

Take-Home Message How do genes affect behavior?

❯ Many genes shape the nervous system. Such genes affect traits such as an animal's ability to sense and respond to stimuli.

❯ Genetic differences affect species-specific behavior, and also differences in behavior among individuals of a species.

❯ Most human behaviors are complex, polygenic traits. Studies of genetic differences can shed light on predispositions to particular behaviors.

> Some behaviors are inborn and can be performed without any practice.
> Most behaviors can be modified as a result of experience.

Instinctive Behavior

All animals are born with the capacity for **instinctive behavior**—an innate response to a specific and usually simple stimulus. For example, a newborn coastal garter snake behaves instinctively when it attacks a banana slug in response to the slug's scent.

The life cycle of the cuckoo bird provides several examples of instinct at work. This European bird is a brood parasite, an animal that lays its eggs in another animal's nest. A female cuckoo lays her eggs in the nests of other birds. The newly hatched cuckoo is blind, but contact with an egg laid by its foster parent stimulates an instinctive response. That hatchling maneuvers the egg onto its back, then shoves it out of the nest (Figure 25.5A). This behavior removes any potential competition for the foster parent's attention.

A cuckoo's egg-dumping response is a **fixed action pattern**: a series of instinctive movements, triggered by a specific stimulus, that—once started—continues to completion without the need for further cues. Such fixed behavior has survival advantages when it permits a rapid reaction to an important stimulus. However, a fixed response to a simple stimulus has limitations. For example, the cuckoo's foster parent responds to the simple stimulus of a chick's gaping mouth with the fixed action pattern of parental feeding behavior (Figure 25.5B). It disregards a cuckoo's unusual large size and coloration.

Figure 25.5 Instinctive behavior.
A Young cuckoo shoves its foster parent's eggs out of the nest.
B The foster parent feeds the cuckoo chick in response to one simple cue: a gaping mouth.

Time–Sensitive Learning

Learned behavior is behavior that is altered by experience. Some instinctive behavior can be modified with learning. A garter snake's initial strikes at prey are instinctive, but the snake learns to avoid dangerous or unpalatable prey. Learning may occur throughout an animal's life, or be restricted to a critical period.

Imprinting is a form of learning that occurs during a genetically determined time period early in life. For example, baby geese learn to follow the large object that bends over them in response to their first peep (Figure 25.6). With rare exceptions, this object is their mother. When mature, the geese will seek out a sexual partner that is similar to the imprinted object.

A genetic capacity to learn, combined with actual experiences in the environment, shapes most forms of behavior. For example, a male songbird has an inborn capacity to recognize his species' song when he hears older males singing it. The young male uses these overheard songs as a guide to fill in details of his own song. Males reared with no model or exposed only to songs of other species often sing a simplified version of their species' song.

Many birds can only learn the details of their species-specific song during a limited period early in life. For example, a male white-crowned sparrow will not sing normally if he does not hear a male "tutor" of his own species during his first 50 or so days. Hearing a same-species tutor later in life will not influence his singing.

Most birds must also practice their song to perfect it. In one experiment, researchers temporarily paralyzed throat muscles of zebra finches who were beginning to sing. After being temporarily unable to practice, these birds never mastered their song. In contrast, temporary

Figure 25.6 Nobel laureate Konrad Lorenz with geese that imprinted on him. The inset photograph shows a more typical imprinting pattern.

paralysis of throat muscles in very young birds or adults did not impair later song production. Thus, in this species, there is a critical period for song practice, as well as for song learning.

Conditioned Responses

Nearly all animals are lifelong learners. Most learn to associate certain stimuli with rewards and others with negative consequences.

With classical conditioning, an animal's involuntary response to a stimulus becomes associated with a stimulus that accompanies it. In the most famous example, Ivan Pavlov rang a bell whenever he fed a dog. Eventually, the dog's reflexive response—increased salivation—was elicited by the sound of the bell alone. Taste aversion is a natural variant of classical conditioning. When an animal becomes nauseous after eating a new food, it avoids that food in the future.

With operant conditioning, an animal modifies its voluntary behavior in response to consequences of that behavior. This type of learning was first described in lab animals. A rat that presses a lever and is rewarded with a food pellet becomes more likely to press the lever again. A rat that receives a shock when it enters a particular area will quickly learn to avoid that area. In nature as well, animals learn to repeat behaviors that provide food or mating opportunities and to avoid those that cause pain.

Other Types of Learned Behavior

With **habituation**, an animal learns by experience not to respond to a stimulus that has neither positive nor negative effects. For example, pigeons in cities learn not to flee from the large numbers of people who walk past them.

Many animals learn about the landmarks in their environment and form a sort of mental map. This map may be put to use when the animal needs to return home. For example, a fiddler crab foraging up to 10 meters (30 feet) away from its burrow is able to scurry straight home when it perceives a threat.

Many animals also learn the details of their social landscape; they learn to recognize mates, offspring, or competitors by appearance, calls, odor, or some combination of cues. For example, two male lobsters that meet up for the first time will fight (Figure 25.7). Later, they will recognize one another by scent and behave accordingly, with the loser actively avoiding the winner.

---

fixed action pattern Series of instinctive movements elicited by a simple stimulus and carried out with little variation once begun.
habituation Learning not to respond to a repeated stimulus.
imprinting Learning that can occur only during a specific interval in an animal's life.
instinctive behavior An innate response to a simple stimulus.
learned behavior Behavior that is modified by experience.

Figure 25.7 Social learning. Two male lobsters battling at their first meeting. Later, the loser will remember the scent of the winner and avoid him. Without another meeting, memory of the defeat lasts up to two weeks.

Figure 25.8 Results of observational learning. A marmoset opens a container using its teeth. After watching one individual successfully perform this maneuver, other marmosets used the same technique. Analysis of videos of their movements showed that the observers closely imitated the behavior they had seen earlier.

With imitation learning, an animal copies the behavior it observes in another individual. Primates often learn new skills in this manner. For example, Ludwig Huber and Bernhard Voelkl allowed marmoset monkeys to watch another marmoset open a container to get the treat inside. Some marmosets used their hands to open the container, others used their teeth. When the observing monkeys were later given a similar container, they used either their hands or their teeth depending on which technique they had observed (Figure 25.8).

25.4 Adaptive Behavior

> Scientists use experiments to test hypotheses about the adaptive value of a behavior.
< Link to Adaptive traits 16.4

The fact that a particular behavior exists does not mean that it is an adaptive trait (Section 16.4). The behavior may have been adaptive under previous conditions, or it could be a consequence of another trait. Thus, behavior researchers often test hypotheses about the adaptive value of a behavior.

For example, starlings and many other birds place sprigs of aromatic plants such as wild carrot in their nest (Figure 25.9A). Larry Clark and Russell Mason hypothesized that nest decorating behavior increases a bird's reproductive success by reducing the number of bloodsucking mites that feed on nestlings. Some plant chemicals deter insect reproduction or maturation. The researchers tested their hypothesis by replacing natural starling nests with artificial ones. Some replacement nests had wild carrot sprigs and others were sprig-free. After the starling chicks had fledged, Clark and Mason counted the mites left behind in nests. As they expected, sprig-free nests had more mites (Figure 25.9B).

In other studies, researchers have found that aromatic plants encourage chick growth and health even when they do not affect parasite numbers. Thus, nest decorating behavior may be adaptive for a variety of reasons.

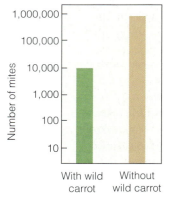

Figure 25.9 Starling nest decorating behavior.

A Starling with nesting material.

B Results of an experiment to test the effect of wild carrot sprigs on the number of mites in nests. Nests with wild carrots had significantly fewer mites than those with no greenery.

Take-Home Message What makes a behavior adaptive?

> A behavior is adaptive if it enhances reproductive success.

> Biologists can use experiments to test hypotheses about the adaptive value of a behavior.

25.5 Communication Signals

> Communication signals play a role in both cooperative and competitive behavior.

Communication signals are chemical, acoustical, visual, or tactile cues that transmit information from one member of a species to another. A communication signal evolves and persists only if it benefits both the signal sender and the signal receiver.

Pheromones are chemical cues. Signal pheromones cause a rapid shift in the receiver's behavior. The honeybee alarm pheromone is an example, as are sex attractants that help males and females of many species find each other. Priming pheromones cause longer-term responses, as when a chemical dissolved in the urine of male mice triggers ovulation in females of the same species.

Bird song is an acoustical signal used to attract mates. Territorial animals often use calls or song to advertise their presence to potential competitors. Alarm calls are emitted in response to a threat. Some convey a surprising amount of information. A prairie dog makes one call

Figure 25.10 Visual signals. **A** Male baboon showing his teeth in a threat display. **B** Penguins engaged in a courtship display. **C** A wolf's play bow tells another wolf that behavior that follows is play, not aggression.

A

B

When bee moves straight up comb, recruits fly straight toward the sun.

C

When bee moves straight down comb, recruits fly to source directly away from the sun.

D

When bee moves to right of vertical, recruits fly at 90° angle to right of the sun.

E

Figure 25.11 Animated Honeybee dances, an example of a tactile display. **A** Bees that have visited a source of food close to their hive return and perform a round dance on the hive's vertically oriented honeycomb. Bees that maintain contact with the dancer fly out and search for food near the hive.

B A bee that visits a feeding source more than 100 meters (110 yards) from her hive performs a *waggle* dance. The orientation of an abdomen-waggling dancer in the straight run of her dance informs other bees about the direction of the food.

C If the food is in line with the sun, the dancer's waggling run proceeds straight up the honeycomb. **D** If food is in the opposite direction from the sun, the dancer's waggle run is straight down. **E** If food is 90 degrees to the right of the direction of the sun, the waggle run is 90 degrees to the right of vertical.

The speed of the dance and the number of waggles in the straight run provide information about distance to the food. When food is 200 meters away, a bee dances much faster, and with more waggles per straight run, compared with a dance inspired by a food source that is 500 meters away.

when it sees an eagle and a different call when it sights a coyote. Upon hearing the call, other prairie dogs respond appropriately: they either dive into burrows (to escape an eagle's attack) or stand erect (to spot the coyote).

An acoustical signal's characteristics are related to its adaptive value. Courtship-related signals are easily localized. A singing male songbird benefits when a female finds him. By contrast, alarm calls often have properties that make it harder to pinpoint the caller's position.

A male baboon's threat display is a visual signal that communicates readiness to fight a rival (Figure 25.10A). Most bird courtship involves coordinated visual signaling (Figure 25.10B). Selection favors unambiguous signals, so movements that serve as signal often become exaggerated. Body form may evolve in concert with movements, as when bright-colored feathers enhance a courtship display.

Honeybees use tactile signals to communicate the location of food. A honeybee worker who finds food returns to the hive and moves in a defined pattern, jostling a crowd of other bees that surround her. The signals give other bees information about the distance and the direction of the food source (Figure 25.11).

The same signal sometimes functions in more than one context. For example, dogs and wolves solicit play behavior with a play bow (Figure 25.10C). Without the visual cue, a signal receiver may construe behaviors that follow as aggressive or sexual—but not playful.

Predators sometimes tap into signaling systems of their prey. For example, male tungara frogs attract females with complex calls. Frog-eating bats use these calls to zero in on the caller. When bats are near, male frogs make fewer and simpler calls. The subdued signal is a trade-off between locating a partner for mating and the need for immediate survival. As another example, fireflies attract mates by producing flashes of light in a characteristic pattern. Some female fireflies prey on males of other species. When a predatory female sees the flash from a male of the prey species, she flashes back as if she were a female of his own species. When he approaches ready to mate, she captures and eats him.

---

communication signal Chemical, acoustical, visual, or tactile cue that is produced by one member of a species and detected and responded to by other members of the same species.

Take-Home Message **What are animal communication signals?**

❯ A chemical, visual, acoustical, or tactile communication signal transfers information from one individual to another of the same species. Both signaler and receiver benefit from the transfer.

❯ Signals can draw attention of predators as well as intended receivers. Features of signals reflect a balance between the benefit of sending information and the potential cost of signaling.

❯ Mating and parenting behaviors reflect often opposing selective pressures operating on males and females.
❮ Link to Sexual selection 17.7

Mating Behavior

Males or females of a species often compete for access to mates, and many are choosy about their partners. Both situations lead to sexual selection. This microevolutionary process favors characteristics that provide a competitive advantage in attracting and retaining mates.

Male animals produce many small sperm, and females produce fewer, larger eggs. Thus a male can produce more offspring than a female. Males usually maximize their reproductive success by mating as many times as possible. Mating with more than one male can benefit a female by increasing the genetic diversity of her offspring. However, given that her number of potential offspring is more limited, she has more at stake in each offspring. Thus females are more likely than males to turn down opportunities to breed with a low-quality individual.

Female hangingflies only breed with males who provide an energy-laden meal. A male hunts and kills a moth or some other insect. Then he releases a sex pheromone that attracts females to him and his "nuptial gift" (Figure 25.12A). Only after a female has been eating for five minutes or so does she start to accept sperm from her partner. Even after mating begins, a female can break off from her suitor, if she finishes her meal. If she does end the mating, she will seek out a new male and his sperm will replace the first male's. Thus, the larger a male's gift, the

longer a female will spend eating it and the greater the chance that his sperm will end up fertilizing her eggs.

A female fiddler crab assesses both a male and his real estate. The crabs live along many sandy shores. One of the male's two claws is enlarged; it often accounts for more than half his total body weight (Figure 25.12B). During their breeding season, hundreds of males dig mating burrows in close proximity to one another. Each male stands at the entrance to his burrow, waving his oversized claw. Female crabs stroll along, checking out males. If a female likes what she sees, she inspects her suitor's burrow. Only when a burrow has the right location and dimensions does she mate with its owner and lay eggs in his burrow. One particularly fussy female was observed to visit 106 burrows in an hour. Burrow location and size are important because they affect larval development.

By contrast, male sage grouse converge at a **lek**, a communal display ground that serves only as a dance floor. With tail feathers erect, the males stamp their feet and emit booming calls by puffing and deflating big neck pouches (Figure 25.12C). Each female mates with a single male, but popular males mate many times.

In species in which females cluster around a necessary resource, males may hold a **territory**, a region that they occupy and defend from others. The territory holder mates with all the females in his territory. Lions, elk, elephant seals, and bison (Figure 25.13) have this sort of mating behavior.

When females choose among a group of displaying males or males defend a mating territory, nearly

A Male hangingfly dangling a moth as a nuptial gift for a potential mate.

B Male fiddler crabs (*top*) wave their one enlarged claw to a attract a female (*bottom*).

C Male sage grouse gather on a communal display ground, where they dance, puff out their neck, and make booming calls.

Figure 25.12 How to impress a female.

all females mate but many males do not. Such systems impose strong selective pressure for mating-related traits on males. The result is often **sexual dimorphism**, a difference in morphology between the sexes. The large claw of the male fiddler crab, inflatable neck pouch of a male sage grouse, and the massive body size of a male bison are examples of sexually dimorphic traits.

With the exception of birds, monogamy is rare in animals. An estimated 3 percent of mammals are socially monogamous; they pair bond and cooperate to raise their young. However, genetic analysis nearly always turns up evidence of mixed paternity in offspring of some pairs. Both males and females engage in outside matings.

Parental Care

Parental behavior requires an investment of time and energy that might otherwise be invested in reproducing again. Evolution of parental care requires that the benefit to a parent in terms of increased offspring survival be greater than the cost in lost reproductive opportunity.

In mammals, young are usually born in a relatively helpless state, so parental care is essential. Typically, the female is the sole caregiver (Figure 25.14A). Males can leave immediately after mating and reproduce again while a females is left holding eggs or developing young inside her body. Also, males who care for young may unknowingly end up protecting another male's offspring—an investment with no genetic payoff.

Males seldom serve as sole caregivers. The midwife toad is an interesting exception. A male holds strings of fertilized eggs around his legs until the eggs hatch (Figure 25.14B). A female will mate with multiple males.

Most birds are monogamous, and parents cooperate in caring for young (Figure 25.14C). Sage grouse and other birds in which females alone care for the young tend to hatch when they are more fully developed and require less care.

---

lek Of some birds, a communal mating display area for males.
sexual dimorphism Distinct male and female phenotypes.
territory Region that an animal or animals occupy and defend against competitors.

Take-Home Message How does natural selection affect mating and parental care?

> Each individual's behavior tends to maximize his or her own reproductive success.

> In many species, a few males monopolize mating opportunities either by enticing females to choose them or by defending a territory with many females.

> Monogamy is rare and even socially monogamous species often mate with nonpartners. Male parental care is associated with monogamy.

Figure 25.13 Male bison locked in combat during the breeding season. A few males will mate with many females. Some males will not mate at all.

A Female grizzlies care for their cub for as long as two years. The male takes no part in its upbringing.

B A male midwife toad carries developing eggs.

C A pair of Caspian terns cooperate in the care of their chick.

Figure 25.14 Caring for offspring.

› Many animals benefit by clustering in groups, but social living also has costs.

‹ Link to Culture 24.9

Defense Against Predators

In some groups, cooperative responses to predators reduce the net risk to all. Multiple individuals can be on the alert for predators, join a counterattack, or engage in more effective defenses.

Birds, monkeys, meerkats, prairie dogs, and many other animals that live in social groups make alarm calls in response to a predator (Figure 25.15A). An alarm call typically provokes such a flurry of activity in the group that the predator ends up looking elsewhere for food.

Sawfly caterpillars that feed in groups on branches benefit by the group's coordinated response to predatory birds. When a bird approaches, the caterpillars all rear up and vomit partly digested eucalyptus leaves (Figure 25.15B). Predatory birds tend to avoid the wiggling mass of fluid-exuding caterpillars, preferring instead to attack individuals. In one experiment, birds offered caterpillars one at a time ate an average of 5.6. When offered a cluster of twenty caterpillars, the birds ate an average of 4.1.

Whenever animals cluster, some individuals shield others from predators (Figure 25.15C). Preference for the center of a group can create a **selfish herd**, in which individuals hide behind one another. Selfish-herd behavior occurs in bluegill sunfishes. A male sunfish builds a nest by scooping out a depression in mud on the bottom of a lake, and females lay eggs in these nests. Snails and fishes prey on the eggs. Competition for the safest sites is greatest near the center of a group, with large males taking the

innermost locations. Smaller males cluster around them and bear the brunt of the egg predation. Even so, small males are better off nesting at the edge of the group than on their own.

Improved Feeding Opportunities

Many mammals, including wolves, lions, wild dogs, and chimpanzees, live in social groups and cooperate in hunts (Figure 25.16). Are cooperative hunters more efficient than solitary ones? Often, no. In one study, researchers observed that a solitary lion catches prey about 15 percent of the time. Two lions cooperatively hunting catch prey twice as often as a solitary lion, but having to share the spoils of the hunt means the amount of food per lion is the same. When more lions join a hunt, the success rate per lion falls. Wolves have a similar pattern. Among carnivores that hunt cooperatively, hunting success is not the major advantage of group living. Individuals hunt together, but they also fend off scavengers, care for one another's young, and protect territory together.

Group living also allows transmission of cultural traits, or behaviors learned by imitation. For example, chimpanzees make and use simple tools by stripping leaves from branches. They use thick sticks to make holes in a termite mound, then insert long, flexible "fishing sticks" into the holes (Figure 25.17). The long stick agitates the termites, which attack and cling to it. Chimps withdraw the stick

---

dominance hierarchy Social system in which resources and mating opportunities are unequally distributed within a group.
selfish herd Temporary group that forms when individuals cluster to minimize their individual risk of predation.

A Black-tailed prairie dog giving an alarm call.

B Clump of Australian sawfly caterpillars regurgitating fluid.

C Musk oxen adults forming a ring of horn. Young are protected in the center of the ring.

Figure 25.15 Group defenses.

Figure 25.16 Members of a pack of wolves (*Canis lupus*). Wolves cooperate in hunting, caring for the young, and defending territory. Benefits are not distributed equally. Most of the time, only the highest-ranking individuals, the alpha male and alpha female, breed.

and lick off termites. Different groups of chimpanzees use slightly different tool-shaping and termite-fishing methods. Youngsters learn by imitating the adults.

Dominance Hierarchies

Many animals that live in permanent groups form a **dominance hierarchy**. In this type of social system, dominant animals get a greater share of resources and breeding opportunities than subordinate ones. Typically, dominance is established by physical confrontation. In most wolf packs, one dominant male breeds with one dominant female. The other members of the pack are nonbreeding brothers and sisters, aunts and uncles. All hunt and carry food back to individuals that guard the young in their den.

Why would a subordinate give up resources and often breeding privileges? Challenging a strong individual can be dangerous, as is living on one's own. Subordinates get their chance to reproduce by outliving a dominant peer.

Regarding the Costs

Relatively few animals spend the bulk of their time in social groups. Why? In most habitats, the costs of social living outweigh the benefits. Individuals that live in denser groups compete more for resources (Section 26.5). Penguins and many other seabirds form dense breeding colonies in which competition for space and food is intense (Figure 25.18). Given the opportunity, a pair of breeding herring gulls will cannibalize the eggs and even the chicks of their neighbors. Large social groups also attract more predators, and individuals that live in dense groups are at a higher risk of parasites and contagious diseases that jump from host to host.

Figure 25.17 Chimpanzees (*Pan troglodytes*) using sticks as tools for extracting tasty termites from a nest. This behavior is learned by imitation.

Figure 25.18 A crowded penguin breeding colony.

Take-Home Message What are the benefits and costs of living in a group?

> Living in a social group can provide benefits, as through cooperative defenses or shielding against predators.

> Group living has costs such as increased competition and increased vulnerability to infections.

❯ Extreme cases of sterility and self-sacrifice have evolved in only two groups of insects and one group of mammals. How are genes of the nonreproducers perpetuated?

❮ Link to Insects 23.13

Social Insects

True social insects include honeybees, termites, and ants. All of these eusocial insects stay together for generations in a group that has a division of labor. Groups include permanently sterile workers that care cooperatively for the offspring of just a few breeding individuals. Such workers often are highly specialized in form and function (Figure 25.19).

A queen honeybee is the only fertile female in her hive (Figure 25.20A). She is larger than her worker daughters, partly because of her enlarged ovaries. The queen honeybee secretes a pheromone that makes all other female bees in the hive sterile. The 30,000 to 50,000 sterile female workers feed larvae, clean and maintain the hive, and construct honeycomb from waxy secretions. Workers also gather the nectar and pollen that feeds the colony. They guard the hive and will sacrifice themselves to repel any intruders.

Male stingless drones are produced seasonally and subsist on food gathered by their worker sisters. Each day, drones fly in search of a mate. The occasional lucky one will meet a virgin queen on her one flight away from a colony. He dies after mating. The queen mates with many males, then uses their stored sperm for years.

Like honeybees, termites live in enormous family groups with a queen who specializes in egg production (Figure 25.20B). Unlike the honeybee hive, a termite mound holds sterile males and females. A king supplies the female with sperm. Winged reproductive termites of both sexes develop seasonally.

Social Mole–Rats

Sterility and extreme self-sacrifice are uncommon in vertebrates. The only eusocial mammals are two species of African mole-rat. The best studied is *Heterocephalus glaber*, the naked mole-rat. Clans of this nearly hairless rodent live in burrows in dry parts of East Africa.

A reproducing female dominates the clan and mates with one to three males (Figure 25.20C). Nonbreeding members live to protect and care for the "queen" and "king" (or kings) and their offspring. Sterile diggers excavate tunnels and chambers. When a digger finds an edible root, it hauls a bit back to the main chamber and chirps. Its chirps recruit others, which help carry food back to the chamber. In this way, the queen, her mates, and her young offspring get fed. Other sterile helpers guard the colony. When a predator appears, they chase and attack it at great risk to themselves.

Evolution of Altruism

A sterile worker in a social insect colony or a naked mole-rat clan shows **altruistic behavior**, which is behavior that enhances another individual's reproductive success at the altruist's expense. How did this behavior evolve? According to William Hamilton's **theory of inclusive fitness**, genes associated with altruism are selected if they lead to behavior that promotes the reproductive success of an altruist's closest relatives.

A sexually reproducing, diploid parent caring for offspring is not helping exact genetic copies of itself. Each of its gametes, and each offspring, inherits one-half of its genes. Other individuals of the social group that have the same ancestors also share genes. Siblings (brothers or sisters) are as genetically similar as a parent and offspring. Nephews and nieces share about one-fourth of their uncle's genes.

A Australian honeypot ant worker, a living container for the colony's food reserves.

B Army ant soldier with formidable mandibles.

C Eyeless soldier termite. It bombards intruders with a stream of sticky goo from its nozzle-shaped head.

Figure 25.19 Specialized ways of serving and defending the colony. All of these individuals are sterile.

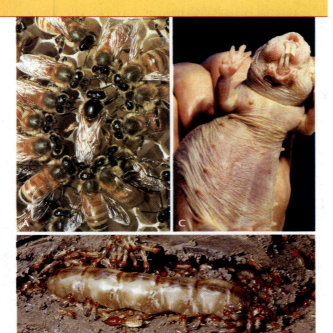

Figure 25.20 Three queens. **A** Queen honeybee with sterile daughters. **B** A termite queen dwarfs her offspring and mate. **C** A naked mole-rat queen.

Sterile workers promote genes for "self-sacrifice" through behavior that can benefit very close relatives. In honeybee, termite, and ant colonies, sterile workers assist fertile relatives with whom they share genes. A guard bee dies after she stings, but her sacrifice preserves copies of her genes in her hivemates.

Inbreeding increases the genetic similarity among relatives and may play a role in mole-rat sociality. A clan is highly inbred as a result of many generations of sibling, mother–son, and father–daughter matings. Researchers are now searching for other factors that select for eusocial behavior in mole-rats. According to one hypothesis, arid habitats and patchy food sources favor mole-rat genes that contribute to cooperation in digging burrows, searching for food, and fending off competitors for resources.

altruistic behavior Behavior that benefits others at the expense of the individual.
theory of inclusive fitness Genes associated with altruism can be advantageous if the expense of this behavior to the altruist is outweighed by the reproductive success of relatives.

An Aggressive Defense (revisited)

When a European queen bee mates with an Africanized drone, her worker offspring are just as aggressive as workers in a pure Africanized colony. In contrast, a cross between an Africanized queen and a European drone yields workers with an intermediate level of aggression. Unfortunately, European queen–Africanized male pairings occur far more frequently than the reciprocal cross. Africanized males outcompete European males for matings.

How Would You Vote? Africanized honeybees continue to increase their range. Should study of their genetics be a high priority? See CengageNow for details, then vote online (cengagenow.com).

25.9 ## 25.9 Evolution and Human Behavior

❯ Evolutionary forces shaped and continue to influence human behavior—but humans alone can make moral choices about their actions.
❮ Link to Philosophy of science 1.9

Many people resist the idea of analyzing the evolutionary basis of human behavior. A common fear is that an objectionable behavior will be defined as "adaptive." To evolutionary biologists, however, "adaptive" does not mean "morally right." It simply means a behavior increases reproductive success. Scientific studies do not address moral issues (Section 1.9).

For example, infanticide is morally repugnant, but it happens in many animal groups and all human cultures. In her book on maternal behavior, primatologist Sarah Blaffer Hrdy cites a study of a village in Papua New Guinea in which parents killed about 40 percent of the newborns. Hrdy argues that when resources and social support are hard to come by, a mother who is unlikely to be able to raise a newborn to adulthood can reduce her costs by killing it. The mother can then allocate her limited resources to her other offspring.

Is such behavior appalling? Yes. Can considering the possible evolutionary advantages of the behavior help us prevent it? Perhaps. An analysis of the conditions under which infanticide occurs tells us this: When mothers lack the resources they need to care for their children, they are more likely to harm them. We as a society can act on such information.

Take-Home Message How can altruistic behavior evolve?

❯ Altruistic behavior may arise and persist when altruists help relatives survive and reproduce. The altruist's actions increase the likelihood that copies of its genes for altruism will be passed to the next generation.

Take-Home Message What do studies of evolution tell us about human behavior?

❯ Evolutionary analysis of human behavior can tell us why some behaviors occur in many cultures. It does not tell us anything about the morality of a behavior.

Summary

Section 25.1 Scientists study the proximate causes of animal behavior, such as differing responses to **pheromones**. They also investigate a behavior's ultimate cause, its evolutionary roots. The study of environmental effects on behavior are one facet of **ecology**.

Section 25.2 Behavior refers to coordinated responses that an animal makes to **stimuli**. It starts with genes that influence the development and activity of the nervous, endocrine, and muscular systems. Studies of behavioral differences within a species or among closely related species can shed light on the proximate and ultimate causes of a behavior.

Section 25.3 **Instinctive behavior** can occur without a prior experience. A **fixed action pattern** is an instinctive response to a simple stimulus.

Learned behavior arises in response to experience. **Imprinting** is time-sensitive learning. With **habituation**, an animal learns to disregard certain stimuli. Animals also form mental maps, learn the identity of other individuals, develop conditioned responses, and imitate observed behaviors.

Section 25.4 A behavior that has a genetic basis is subject to evolution by natural selection. Experiments can test hypotheses about the current adaptive value of a specific behavior.

Section 25.5 Chemical, acoustical, or tactile **communication signals** are meant to change the behavior of individuals of the same species. Evolution influences properties of signals, as when courtship calls are easily localized, but alarm calls are not. Predators sometimes take advantage of the communication system of their prey.

Section 25.6 Sexual selection favors traits that give an individual a competitive edge in attracting mates. Both males and females benefit by mating with multiple partners, but males tend to be less choosy about mates than females. When large numbers of females cluster in a defensible area, males compete with one another to control the areas. Males may display for females at a **lek** or hold a **territory** with resources that females need. Mating systems in which some males father most offspring favor **sexual dimorphism**. Monogamy is rare; individuals of socially monogamous species often "cheat."

Parental care has reproductive costs in terms of reduced frequency of reproduction. It is adaptive when benefits to a present set of offspring offset the costs. Male parental care is typically associated with monogamy.

Section 25.7 Some animals come together as a **selfish herd** in response to the threat of predation. Others live in social groups and benefit by cooperating in predator detection, defense, and rearing the young. With a **dominance hierarchy,** resource and mating opportunities are distributed unequally among group members. Species that live in large groups incur costs, including increased disease and parasitism, and more intense competition for resources.

Section 25.8 Ants, termites, and some other insects as well as mole-rats are eusocial: They live in colonies with overlapping generations and have a reproductive division of labor. Most colony members do not reproduce; they assist their relatives instead.

The **theory of inclusive fitness** states that **altruistic behavior** can be perpetuated when altruistic individuals help their reproducing relatives. Altruistic individuals help perpetuate the genes that lead to their altruism by promoting reproductive success of close relatives that also carry copies of these genes.

Section 25.9 A human behavior that is adaptive in the evolutionary sense may still be judged by society to be morally wrong. Science does not address morality.

Self-Quiz   Answers in Appendix III

1. Genes affect the behavior of individuals by _____ .
   a. influencing the development of nervous systems
   b. affecting how individuals respond to hormones
   c. determining which stimuli can be detected
   d. all of the above

2. Stevan Arnold offered slug meat to newborn garter snakes from different populations to test his hypothesis that the snakes' response to slugs _____ .
   a. was shaped by indirect selection
   b. is an instinctive behavior
   c. is based on pheromones
   d. is adaptive

3. A behavior is defined as adaptive if it _____ .
   a. varies among individuals of a population
   b. occurs without prior learning
   c. increases an individual's reproductive success
   d. is widespread across a species

4. The honeybee dance transmits information about _____ by way of tactile signals.
   a. predators   c. location of food
   b. mating opportunities   d. amount of honey

5. A _____ is a chemical that conveys information between individuals of the same species.
   a. pheromone   c. hormone
   b. neurotransmitter   d. all of the above

Spread of Africanized Honeybees Honeybees disperse by forming new colonies. An old queen leaves the hive along with a group of workers. These bees find a new nest site, and set up a new hive. Meanwhile, at the old hive, a new queen emerges, mates, and takes over. A new hive can be several kilometers from the old one.

Africanized honeybees form new colonies more often than European ones, a trait that contributes to their spread. Africanized bees also spread by taking over existing hives of European bees. In addition, in areas where European and Africanized hives coexist, European queens are more likely to mate with Africanized males, thus introducing Africanized traits into the colony. Figure 25.21 shows the counties in the United States where Africanized honeybees became established between 1990 and 2009.

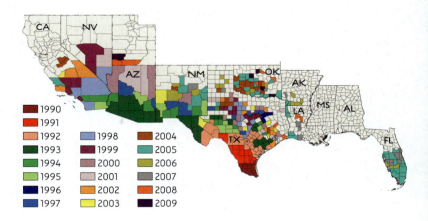

Figure 25.21 The spread of Africanized honeybees in the United States, from 1990 through 2009. The USDA adds a county to this map only when the state officially declares bees in that county Africanized. Bees can be identified as Africanized on the basis of morphological traits or analysis of their DNA.

1. Where in the United States did Africanized bees first become established?

2. In what states did Africanized bees first appear in 2005?

3. Why is it likely that human transport of bees contributed to the spread of Africanized honeybees to Florida?

6. In what group of animals are monogamy and cooperative care of the young by two parents most common?

7. List two possible benefits of living in a group.

8. All honeybee workers are _____ .
   a. male   c. female
   b. sterile   d. b and c

9. Eusocial insects _____ .
   a. live in extended family groups
   b. include termites, honeybees, and ants
   c. show a reproductive division of labor
   d. all of the above

10. Helping other individuals at a reproductive cost to oneself might be adaptive if those helped are _____ .

11. Match the terms with their most suitable description.
    ___ fixed action   a. time-dependent form of
        pattern   learning
    ___ altruism   b. communal display area
    ___ habituation   c. learning to ignore a stimulus
    ___ lek   d. defended area with resources
    ___ territory   e. assisting another individual
    ___ imprinting   at one's own expense
    ___ dominance   f. unequal distribution of
        hierarchy   benefits within group
    ___ classical   g. involuntary response becomes
        conditioning   tied to a stimulus
                           h. series of instinctive movements
                              triggered by a simple stimulus

Additional questions are available on **CENGAGENOW**.

Critical Thinking

1. For billions of years, the only bright objects in the night sky were stars or the moon. Night-flying moths use them to navigate a straight line. Today, the instinct to fly toward bright objects causes moths to exhaust themselves fluttering around streetlights and banging against brightly lit windowpanes. This behavior clearly is not adaptive, so why does it persist?

2. A female chimpanzee engages in sex only during her fertile period, which is advertised by a swelling of her external genitals. Bonobos, the closest relatives of the chimpanzees, are more sexually active than chimpanzees. Female bonobos mate even when they are not fertile. In both species, mating is promiscuous; each female mates with many males. By one hypothesis, female promiscuity may help prevent male infanticide. Chimpanzee males have been observed to kill infants in the wild, but this behavior has never been observed among bonobos.

   Explain why it would be disadvantageous for a male to kill the infant of a female with whom he had sex. How might the bonobo female's increased sexual activity lower likelihood of male infanticide still further?

Animations and Interactions on **CENGAGENOW**:
> Honeybee dance language.

Earlier chapters explored the evolutionary history and genetic nature of populations, including those of humans (Sections 1.4, 16.4, 17.2, 24.11). Now you will consider ecological factors that limit population growth. Again, science cannot address social issues—in this case, how exponential growth affects human lives (1.9). It can only help explain how ecological conditions and events sustain a population's growth, or put a stop to it.

Key Concepts

The Vital Statistics
Ecologists explain population growth in terms of population size, density, distribution, and number of individuals in different age categories. They have methods of estimating population size and density in the field.

Exponential Rates of Growth
A population's size and reproductive base influence its rate of growth. As long as births exceed deaths, a population will grow exponentially. Each generation will be larger than the preceding one.

26 Population Ecology

Canada geese (*Branta canadensis*) were hunted to near extinction in the late 1800s. In the early 1900s, federal laws and international treaties were put in place to protect them and other migratory birds. In recent decades, the number of geese in the United States has soared. For example, Michigan had about 9,000 birds in 1970 and today has more than 300,000. These plant-eating birds often congregate at golf courses and parks (Figure 26.1). They are considered pests because they produce slimy, green feces that soil shoes and stain clothes. Goose feces that gets into water adds nutrients that encourage bacterial and algal growth.

Canada geese also pose a hazard to air traffic. In January of 2009, both engines of a US Airways flight failed shortly after the plane took off from New York's La Guardia airport. The quick-thinking pilot managed to land the plane in the nearby Hudson River, where boats safely unloaded all 155 people aboard. After the crash, investigators determined that the engine failures occurred after Canada geese were sucked into both engines.

Controlling the number of Canada geese poses a challenge, because several different Canada goose populations spend time in the United States. A **population** is a group of organisms of the same species that occupy a particular area. Members of a population breed with one another more than they breed with members of other populations. In the past, nearly all Canada geese seen in the United States were migratory. (A migration is a round-trip between regions, usually in response to expected shifts or gradients in environmental resources.) The geese nested in northern Canada, flew to the United States to spend the winter, then returned to Canada. The common name of the species reflects this tie to Canada.

Most Canada geese still migrate, but some populations have lost this trait. Canada geese breed where they were raised, and nonmigratory birds are generally descendants of geese deliberately introduced to a park or hunting preserve. During the win-

Figure 26.1 Goose troubles. *Opposite page*, a park in Oakland, California, overrun by Canada geese. *Above*, US Airways Flight 1549 floats in the Hudson River after collisions with geese incapacitated both of its engines.

ter, migratory birds often mingle with nonmigratory ones. For example, a bird that breeds in Canada and flies to Virginia for the winter finds itself beside geese that have never left Virginia.

Life is more difficult for migratory geese than for nonmigratory ones. Flying hundreds of miles to and from a northern breeding area takes lots of energy and is dangerous. Compared to a migratory bird, one that stays put can devote more energy to producing young. If the nonmigrant lives in a suburban or urban area, it also benefits from an unnatural abundance of food (grass) and an equally unnatural lack of predators. Not surprisingly, the biggest increases in Canada geese have been among nonmigratory birds that live where humans are plentiful.

In 2006, increasing complaints about Canada geese led the U.S. Fish and Wildlife Service to encourage wildlife managers to look for ways to reduce nonmigratory Canada goose populations, without unduly harming migratory birds. To do so, these biologists need to know which traits characterize each goose population, as well as how populations interact with one another, with other species, and with their physical environment. This sort of information is the focus of the field of population ecology.

---

population Group of organisms of the same species that live in the same area and interbreed.

Limits on Increases in Size
Density dependent factors such as competition for resources lead to logistic growth. A population grows exponentially at first, then growth slows as the number approaches the environment's carrying capacity.

Patterns of Survival and Reproduction
Life history traits such as age at first reproduction and number of offspring per reproductive event vary and are shaped by natural selection. Adaptive life history traits are those that maximize an individual's lifetime reproductive success.

The Human Population
Human populations have sidestepped historical limits to growth by way of global expansion into new habitats, cultural traits, and technological innovations. But no population can expand indefinitely and we are starting to push our limits.

> Ecological factors affect the size, density, distribution, and age structure of a population.
❮ Links to Sampling error 1.8, Social groups 25.7

Studying population ecology often involves the use of **demographics**—statistics that describe a population. The demographics of a population often change over time.

Population Size

Population size is the number of individuals in a population. It is often impractical to actually count individuals, so biologists frequently use sampling techniques to estimate population size.

Plot sampling estimates the total number of individuals in an area on the basis of direct counts in a small portion of that area. For example, ecologists might estimate the number of grass plants in a grassland or the number of clams in a mudflat by measuring the number of individuals in several 1 meter by 1 meter square plots. To estimate the total population size, scientists multiply the average number of individuals in the sample plots by the number of these plots that can fit in the area where the population lives. Estimates derived from plot sampling are most accurate when the organisms under study are not very mobile and conditions across the area they occupy are more or less uniform.

Scientists use **mark–recapture sampling** to estimate the population size of mobile animals. Scientists capture animals and mark them with a unique identifier of some sort (**Figure 26.2**), then release them. After allowing a sufficient time to pass for the marked individuals to meld back into the population, the scientists capture animals again. The proportion of marked animals in the second sample is taken to be representative of the proportion marked in the whole population. For example, if 100 deer are captured and marked, and 50 of them are recaptured, scientists assume that they initially marked half of the population. Thus the total population is estimated at 200.

Information about the traits of individuals in a sample plot or capture group can be used to infer properties of the population as a whole. For example, if half the recaptured geese are of reproductive age, half of the population is assumed to share this trait. This sort of extrapolation is based on the assumption that the individuals in the sample are representative of the general population in terms of age, sex, and other traits under investigation.

Population Density and Distribution

Population density is the number of individuals per unit area or volume. Examples of population density include the number of dandelions per square meter of lawn or the number of euglenas per milliliter of pond water.

Figure 26.2 Florida Key deer marked for a population study.

Population distribution describes the location of individuals relative to one another. Members of a population may be clumped together, be an equal distance apart, or be distributed randomly.

Clumped Distribution Most often members of a population have a clumped distribution; they are closer to one another than would be predicted by chance alone. A patchy distribution of resources encourages clumping. For example, Canada geese tend to congregate in places with suitable food such as grass and a nearby body of water. A cool, damp, north-facing slope may be covered with ferns, whereas an adjacent drier south-facing slope has none.

Limited dispersal ability increases the likelihood of a clumped distribution. As the saying goes, the nut does not fall far from the tree. Asexual reproduction is another source of clusters. It produces colonies of coral and vast stands of King's holly clones.

---

age structure Of a population, the number of individuals in each of several age categories.
demographics Statistics that describe a population.
mark–recapture sampling Method of estimating population size of mobile animals by marking individuals, releasing them, then checking the proportion of marks among individuals recaptured at a later time.
plot sampling Method of estimating population size of organisms that do not move much by making counts in small plots, and extrapolating from this to the number in the larger area.
population density Number of individuals per unit area.
population distribution Where individuals are clumped, uniformly dispersed, or randomly dispersed in an area.
population size Total number of individuals in a population.
reproductive base Of a population, all individuals who are of reproductive age or younger.

A Clumped distribution of schooling squirrel fish

B Near-uniform distribution of nesting seabirds

C Random distribution of dandelions

Figure 26.3 Population distribution patterns.

Finally, as Section 25.7 explained, many animals benefit by grouping together, as when geese travel in flocks and fish swim in schools (Figure 26.3A).

Near–Uniform Distribution Competition for resources can produce a near uniform distribution, with individuals more evenly spaced than would be expected by chance. Creosote bushes in deserts of the American Southwest typically grow in this pattern. Competition for limited water among the root systems keeps the plants from growing in close proximity. Similarly, seabirds in breeding colonies often show a near uniform distribution. Each bird aggressively repels others that get within reach of its beak as it sits atop its nest (Figure 26.3B).

Random Distribution Members of a population become distributed randomly when environmental resources are uniformly distributed, and proximity to others neither benefits nor harms individuals. For example, when the wind-dispersed seeds of dandelions land on the uniform environment of a suburban lawn, dandelion plants grow in a random pattern (Figure 26.3C). Wolf spider burrows are also randomly distributed relative to one another. When seeking a burrow site, the spiders neither avoid one another nor seek one another out.

Age Structure

The **age structure** of a population refers to the number of individuals in various age categories. Individuals are frequently grouped as pre-reproductive, reproductive, or post-reproductive. Those in the pre-reproductive category

have the capacity to produce offspring when mature. Together with individuals in the reproductive group, they make up the population's **reproductive base**.

Effects of Scale and Timing

The scale of the area sampled and the timing of a study can influence the observed demographics. For example, seabirds are spaced almost uniformly at a nesting site, but the nesting sites are clumped along a shoreline. The birds crowd together during the breeding season, but disperse when breeding is over.

Wildlife managers use demographic information to decide how best to manage populations. For example, to set up a plan for management of nonmigratory Canada geese, wildlife managers began by evaluating the size, density, and distribution of nonmigratory populations. Based on this information, the U.S. Fish and Wildlife Service decided to allow destruction of some eggs and nests, and increased hunting opportunities during times when migratory Canada geese are least likely to be present.

> **Take-Home Message** **What characteristics do we use to describe a population and what factors affect these characteristics?**
>
> › Each population has characteristic demographics, such as its size, density, distribution pattern, and age structure.
>
> › Characteristics of the population as a whole are often inferred on the basis of a study of the traits of a smaller subsample.
>
> › Environmental conditions and interactions among individuals influence a population's demographics, which often change over time.

> With exponential growth, the size of the population grows faster and faster over time.

> A population will grow exponentially as long as the birth rate is greater than the death rate.

‹ Link to Binary fission 19.6

Gains and Losses in Population Size

Populations continually fluctuate in size. The number of individuals is increased by births and **immigration**, the arrival of new residents that previously belonged to another population. The number is decreased by deaths and **emigration**, the departure of individuals who take up permanent residence elsewhere.

In many animals, young of one or both sexes leave the area where they were born and breed elsewhere. For example, young freshwater turtles typically emigrate from their parental population and become immigrants at another pond some distance away.

By contrast, seabirds typically breed where they were born. However, some individuals may emigrate and end up at breeding sites more than a thousand kilometers away. The tendency of individuals to emigrate to a new breeding site is usually related to resource availability and crowding. As resources decline and crowding increases, the likelihood of emigration rises.

From Zero to Exponential Growth

If we set aside the effects of immigration and emigration, we can define **zero population growth** as an interval dur-ing which the number of births is balanced by an equal number of deaths. As a result, population size remains unchanged, with no net increase or decrease in the number of individuals.

We can measure births and deaths in terms of rates per individual, or per capita. *Capita* means head, as in a head count. Subtract a population's per capita death rate (d) from its per capita birth rate (b) and you have the **per capita growth rate**, or r:

$$
\begin{array}{ccccc}
r & = & b & - & d \\
\text{(per capita} & & \text{(per capita} & & \text{(per capita} \\
\text{growth rate)} & & \text{birth rate)} & & \text{death rate)}
\end{array}
$$

Imagine 2,000 mice living in the same cornfield. If 1,000 mice are born each month, then the birth rate is 0.5 births per mouse per month (1,000 births/2,000 mice). If 200 mice die one way or another each month, then the death rate is 200/2,000 or 0.1 deaths per mouse per month. Thus r is 0.5 – 0.1 or 0.4 per mouse per month.

As long as r remains constant and greater than zero, **exponential growth** will occur: Population size will increase by the exact same proportion of its total in every successive time interval. We can calculate population growth (G) for each interval based on the per capita growth rate and the number of individuals (N):

$$
\begin{array}{ccccc}
G & = & r & \times & N \\
\text{(population growth} & & \text{(per capita} & & \text{(number of} \\
\text{per unit time)} & & \text{growth rate)} & & \text{individuals)}
\end{array}
$$

Figure 26.4 Animated Exponential growth in hypothetical population of mice with a per capita rate of growth (r) of 0.4 per mouse per month and a population size of 2,000.

	Starting Population Size	Net Monthly Increase	New Population Size
$G = r \times$	2,000 =	800	2,800
$r \times$	2,800 =	1,120	3,920
$r \times$	3,920 =	1,568	5,488
$r \times$	5,488 =	2,195	7,683
$r \times$	7,683 =	3,073	10,756
$r \times$	10,756 =	4,302	15,058
$r \times$	15,058 =	6,023	21,081
$r \times$	21,081 =	8,432	29,513
$r \times$	29,513 =	11,805	41,318
$r \times$	41,318 =	16,527	57,845
$r \times$	57,845 =	23,138	80,983
$r \times$	80,983 =	32,393	113,376
$r \times$	113,376 =	45,350	158,726
$r \times$	158,726 =	63,490	222,216
$r \times$	222,216 =	88,887	311,103
$r \times$	311,103 =	124,441	435,544
$r \times$	435,544 =	174,218	609,762
$r \times$	609,762 =	243,905	853,667
$r \times$	853,667 =	341,467	1,195,134

A Increases in size over time. Note that the net increase becomes larger with each generation.

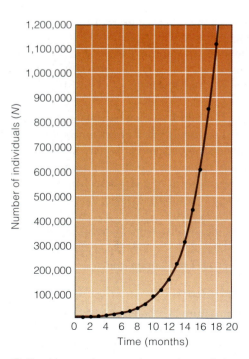

B Graphing numbers over time produces a J-shaped curve.

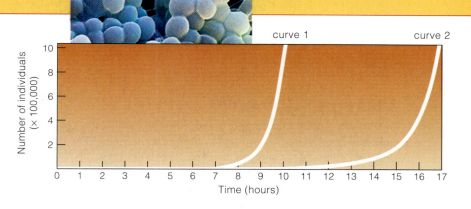

Figure 26.5 Effect of deaths on the rate of increase for two hypothetical populations of bacteria. Plot the population growth for bacterial cells that reproduce every half hour and you get growth curve 1. Next, plot the population growth of bacterial cells that divide every half hour, with 25 percent dying between divisions, and you get growth curve 2. Deaths slow the rate of increase, but as long as the birth rate exceeds the death rate, exponential growth will continue.

After one month, 2,800 mice are in the field (Figure 26.4A). A net increase of 800 fertile mice has made the reproductive base larger. They all reproduce, so the population size expands, for a net increase of 0.4 × 2,800 = 1,120. Population size is now 3,920. At this growth rate, the number of mice would rise from 2,000 to more than 1 million in under two years! Plot the increases against time and you end up with a curve, as shown in Figure 26.4B. Such a J-shaped curve is evidence of exponential growth.

With exponential growth, the number of new individuals added increases each generation, although the per capita growth rate stays the same. Exponential population growth is analogous to the compounding of interest on a bank account. The annual interest *rate* stays fixed, yet every year the *amount* of interest paid increases. Why? The annual interest paid into the account adds to the size of the balance, and the next interest payment will be based on that balance.

In exponentially growing populations, *r* is like the interest rate. Although *r* remains constant, population growth accelerates as the population size increases. When 6,000 individuals reproduce, population growth is three times higher than it was when there were only 2,000 reproducers. With exponential growth, each generation is larger than the prior one.

As another example, think of a single bacterium in a culture flask. After thirty minutes, the cell divides in two. Those two cells divide, and so on every thirty minutes. If no cells die between divisions, then the population size

will double in every interval—from 1 to 2, then 4, 8, 16, 32, and so on. The time it takes for a population to double in size is its doubling time.

After 9–1/2 hours, or nineteen doublings, there are more than 500,000 cells. Ten hours (twenty doublings) later, there are more than a million. Curve 1 in Figure 26.5 is a plot of this increase.

Suppose 25 percent of the descendant cells die every thirty minutes. It now requires about seventeen hours, not ten, for that population to reach 1 million. Deaths slow the rate of increase but do not stop exponential growth (curve 2 in Figure 26.5). Exponential growth will continue as long as birth rates exceed death rates—as long as *r* is greater than zero.

What Is the Biotic Potential?

The growth rate for a population under ideal conditions is its **biotic potential**. This is a theoretical value that would hold if shelter, food, and other essential resources were unlimited and there were no predators or pathogens. Factors that affect biotic potential include the age at which reproduction begins, how long individuals remain reproductive, and the number of offspring that are produced by each reproductive event. Microbes have some of the highest biotic potentials, whereas large-bodied mammals have some of the lowest.

Populations seldom reach their biotic potential because of the effects of limiting factors, a topic we discuss in detail in the next section.

---

biotic potential Maximum possible population growth rate under optimal conditions.
emigration Movement of individuals out of a population.
exponential growth A population grows by a fixed percentage in successive time intervals; the size of each increase is determined by the current population size.
immigration Movement of individuals into a population.
per capita growth rate For some interval, the added number of individuals divided by the initial population size.
zero population growth Interval in which births equal deaths.

Take-Home Message **What determines the size of a population and its growth rate?**

› The size of a population depends on its rates of births, deaths, immigration, and emigration.

› Subtract the per capita death rate from the per capita birth rate to get *r*, the per capita growth rate of a population. As long as *r* is constant and greater than zero, a population will grow exponentially. With exponential growth, the number of individuals increases at an ever accelerating rate.

› The biotic potential of a species is its maximum possible population growth rate under optimal conditions.

> Many complex interactions take place within and between populations in nature, and it is not always easy to identify all the factors that can restrict population growth.

Environmental Limits on Growth

Most of the time, a population cannot fulfill its biotic potential because of environmental limits. That is why, although a female sea star can produce 2,500,000 eggs a year, oceans do not overflow with sea stars.

Figure 26.6 One example of a limiting factor. **A** Wood ducks build nests only inside hollows of specific dimensions. With the clearing of old-growth forests, the access to natural cavities of the correct size and position is now a limiting factor on wood duck population size. **B** Artificial nesting boxes are being placed in preserves to help ensure the health of wood duck populations.

Figure 26.7 Idealized S-shaped curve characteristic of logistic growth. After a rapid growth phase (time B to C), growth slows and the curve flattens as carrying capacity is reached (time C to D).

In the real world, growth curves vary more, as when a change in the environment lowers carrying capacity (time D to E). That happened to the human population of Ireland in the mid-1800s. Late blight, a disease caused by a water mold, destroyed the potato crop that was the mainstay of Irish diets (Section 20.7).

An essential resource that is in short supply acts as a **limiting factor** for population growth. Food, mineral ions, refuge from predators, and safe nesting sites are examples (Figure 26.6). Many factors can potentially limit population growth. Even so, in any environment, one essential factor will kick in first, and it will act as the brake on population growth.

To get a sense of the limits on growth, start again with a bacterial cell in a culture flask, where you can control the variables. First, enrich the culture medium with glucose and other nutrients required for bacterial growth. Next, let many generations of cells reproduce.

Initially, growth will be exponential. Then it slows, and population size remains relatively stable. After a brief stable period, population size plummets until all the bacterial cells are dead. What happened? The larger population required more nutrients. In time, nutrient levels declined, and the cells could no longer divide. Even after cell division stopped, existing cells kept on taking up and using nutrients. Eventually, when the nutrient supply was exhausted, the last cells died out.

Even if you kept freshening the nutrient supply, the population would still eventually collapse. Like other organisms, bacteria generate metabolic wastes. Over time, accumulation of this waste would pollute the habitat and halt growth. Adding nutrients simply substitutes one limiting factor for another. All natural populations run up against limits eventually.

Carrying Capacity and Logistic Growth

Carrying capacity refers to the maximum number of individuals of a population that a given environment can sustain indefinitely. Ultimately, it means that the *sustainable* supply of resources determines population size. We can use the pattern of **logistic growth**, shown in Figure 26.7, to reinforce this point. By this pattern, a small population starts growing slowly in size, then it grows rapidly, then its size levels off as the carrying capacity is reached.

Logistic growth plots out as an S-shaped curve, as shown in Figure 26.7 (time A to C). In equation form:

population growth per unit time	=	maximum per capita population growth rate	×	number of individuals	×	proportion of resources not yet used

Two Categories of Limiting Factors

Factors that affect population growth fall into two categories: density dependent and density independent. **Density–dependent factors** decrease birth rates or increase death rates, and they come into play or worsen with crowding. Competition among members of a population for limited resources leads to density-dependent

Figure 26.8 Overshoot and crash. A reindeer herd introduced to a small island in 1944 increased in size exponentially, then crashed when a low food supply was coupled with an especially cold and snowy winter in 1963–64.

effects, as does infectious disease. Pathogens and parasites spread more easily when hosts are crowded than when host population density is low. The logistic growth pattern results from the effects of density-dependent factors on population size.

Density-independent factors also decrease births or increase deaths, but crowding does not influence the likelihood of their occurrence, or the magnitude of their effects. Fires, snowstorms, earthquakes, and other natural disasters affect crowded and uncrowded populations alike. For example, in December 2004, a powerful tsunami (a giant wave caused by an earthquake) hit Indonesia and killed about 250,000 people. The degree of crowding did not make the tsunami any more or less likely to happen, or to strike any particular island.

Density-dependent and density-independent factors can interact to determine the fate of a population. As an example, consider what happened after a herd of 29 reindeer was introduced to St. Matthew Island off the coast of Alaska in 1944 (Figure 26.8). Biologist David Klein

visited the island in 1957 and found 1,350 well-fed reindeer munching on lichens. In 1963, Klein returned to the island and counted 6,000 reindeer. The population had soared far above the island's carrying capacity. Lichens had become sparser and the average body size of the reindeer had decreased. When Klein returned again in 1966, bleached-out reindeer bones littered the island and only 42 reindeer were alive. Only one was a male; it had abnormal antlers, which made it unlikely to reproduce. There were no fawns. Klein figured out that thousands of reindeer had starved to death during the winter of 1963–1964. The winter had been unusually harsh, in both temperature and amount of snow. The reindeer had already been in poor condition because of the increased competition for food, so most starved when deep snow covered their food source. By the 1980s, there were no reindeer on the island at all.

---

carrying capacity Maximum number of individuals of a species that an environment can sustain.
density–dependent factor Factor that limits population growth and has a greater effect in dense populations than less dense ones.
density–independent factor Factor that limits population growth and arises regardless of population density.
limiting factor A necessary resource, the depletion of which halts population growth.
logistic growth A population grows slowly, then increases rapidly until it reaches carrying capacity and levels off.

Take-Home Message How do environmental factors affect population growth?

❯ Carrying capacity is the maximum number of individuals of a population that can be sustained indefinitely by the resources in a given environment.

❯ With logistic growth, population growth is fastest during times of low density, then it slows as the population approaches carrying capacity.

❯ The effects of density-dependent factors such as disease cause a logistic growth pattern. Density-independent factors such as natural disasters also affect population size.

> Age at maturity, the number of offspring produced per reproductive event, and life span affect population growth. Natural selection influences these life history traits.
❮ Link to Natural selection 16.4

Table 26.1 Life Table for an Annual Plant Cohort*

Age Interval (days)	Survivorship (number surviving at start of interval)	Number Dying During Interval	Death Rate (number dying/ number surviving)	"Birth" Rate During Interval (number of seeds from each plant)
0–63	996	328	0.329	0
63–124	668	373	0.558	0
124–184	295	105	0.356	0
184–215	190	14	0.074	0
215–264	176	4	0.023	0
264–278	172	5	0.029	0
278–292	167	8	0.048	0
292–306	159	5	0.031	0.33
306–320	154	7	0.045	3.13
320–334	147	42	0.286	5.42
334–348	105	83	0.790	9.26
348–362	22	22	1.000	4.31
362–	0	0	0	0
		996		

* *Phlox drummondii*; data from W. J. Leverich and D. A. Levin, 1979.

Table 26.2 Life Table for Humans in the United States (based on 2006 conditions)

Age Interval	Number at Start of Interval	Number Dying During Age Interval	Life Expectancy at Start of Interval	Reported Live Births
0–1	100,000	671	77.7	
1–5	99,329	113	77.2	
5–10	99,216	69	73.3	
10–15	99,147	81	68.4	6,396
15–20	99,065	318	63.4	415,262
20–25	98,747	494	58.6	1,080,437
25–30	98,253	495	53.9	1,181,899
30–35	97,759	546	49.2	950,258
35–40	97,213	718	44.4	498,616
40–45	96,495	1,098	39.7	105,539
45–50	95,397	1,647	35.2	6,480
50–55	93,750	2,398	30.7	494
55–60	91,352	3,295	26.5	
60–65	88,057	4,806	22.4	
65–70	83,251	6.591	18.5	
70–75	76,661	9,329	14.9	
75–80	67,331	13,130	11.6	
80–85	54,201	16,396	98.7	
85–90	37,806	16,906	6.4	
90–95	20,900	12,908	4.6	
95–100	7,992	6,255	3.2	
100+	1,737	1,737	2.3	

Patterns of Survival and Reproduction

Biologists refer to the reproduction-related events that occur between birth and death as a **life history pattern**. Traits that influence this pattern include the age at which an organism begins to reproduce, how often it reproduces, the number of offspring produced per reproductive event, and the length of the reproductive stage.

We can study life history traits within a population by recording what happens to a specific **cohort**, a group of individuals born at more or less the same time. Table 26.1 shows a life table for a cohort of 1,000 annual plants.

Human life tables are usually not based on a real cohort. Instead, information about current conditions is used to predict life expectancy. Table 26.2 shows the life expectancy for people in the United States based on conditions in the United States during 2006. The table also shows reported births for this year.

Information about age-specific death rates can also be summarized by a **survivorship curve**, a plot that shows how many members of a cohort remain alive over time. Three types of survivorship curves are common.

A type I curve indicates survivorship is high until late in life. Populations of large animals that bear one or, at most, a few offspring at a time and provide extended parental care show this pattern (Figure 26.9A). For example, a female elephant has one calf at a time and cares for it for several years. Type I curves are typical of human populations with access to good health care.

A type II curve indicates that death rates do not vary much with age (Figure 26.9B). In lizards, small mammals, and large birds, old individuals are about as likely to die of disease or predation as young ones.

A type III curve indicates that the death rate for a population peaks early in life. It is typical of species that produce many small offspring and provide little or no parental care. Figure 26.9C shows how the curve plummets for sea urchins. Sea urchin larvae are soft and tiny, so fish, snails, and sea slugs devour most of them before protective hard parts can develop. A type III curve is common for marine invertebrates, insects, fishes, fungi, and for annual plants.

cohort Group of individuals born during the same interval.
life history pattern A set of traits related to growth, survival, and reproduction such as life span, age-specific mortality, age at first reproduction, and number of breeding events.
***K*-selection** Individuals who produce offspring capable of out-competing others for limited resources have a selective advantage; occurs when a population is near carrying capacity.
***r*-selection** Individuals who produce the maximum number offspring as quickly as possible have a selective advantage; occurs when population density is low and resources are abundant.
survivorship curve Graph showing the decline in numbers of a cohort over time.

Figure 26.9 Three generalized survivorship curves. Graphs show the number of members of a cohort still alive as age increases.

›› Figure It Out What type of survivorship curve do phlox plants (Table 26.1) have?　*Answer: Type III, with high mortality early in life*

Allocating Reproductive Investment

Natural selection influences the timing of reproduction and how much a parent invests in each offspring. The most adaptive reproductive strategy is that which maximizes a parent's lifetime reproductive success.

Some species such as bamboo and Pacific salmon reproduce once, then die. Others such as oak trees, mice, and humans reproduce repeatedly. A one-shot strategy is evolutionarily favored when an individual is unlikely to have a second chance to reproduce. For Pacific salmon, reproduction requires a life-threatening journey from the sea to a stream. For bamboo, environmental conditions that favor reproduction occur only sporadically.

The most effective reproductive strategy can vary with population density. Reproduction involves trade-offs between offspring quality and quantity. At a low population density, there is little competition for resources and *r*-selection predominates: Individuals who turn resources into offspring fastest have a selective advantage. Species that colonize new habitats or are adapted to habitats where environmental changes frequently reduce population size tend to have a small body size and a short generation time.

By contrast, when a stable environment allows population density to near carrying capacity, individuals who can outcompete others for limited resources have the selective edge. Under these circumstances, *K*-selection predominates: Individuals who invest more resources in building their own body and produce fewer, but larger, more competitive offspring, leave the most descendants.

Life history traits of fruit flies and opportunistic weeds such as dandelions reflect mainly *r*-selection, whereas those of large mammals and oak trees reflect mainly *K*-selection. Most organisms are subject to both types of selection at different times and have intermediate traits.

A Elephants have type I survivorship, with low mortality until old age.

B Snowy egrets are a type II population, with a fairly constant death rate.

C Sea urchins are type III populations. Mortality is high for tiny soft-bodied larvae and in old age, but healthy spiny adults have low mortality.

Take-Home Message **How do researchers study and describe life history patterns?**

› Tracking a group of same-aged individuals from birth to death reveals patterns of reproduction and survivorship. This data can be summarized in life tables or survivorship curves.

› Different environmental conditions and population densities can favor different reproductive strategies.

> Evolution is an ongoing process. Organisms continually adapt to environmental changes, as when shifts in predation pressure alter life history traits.

‹ Link to Gene flow 17.9

Effect of Predation on Guppies

A long-term study by the evolutionary biologists John Endler and David Reznick illustrates the effect of predation on guppy life history traits. Endler and Reznick studied populations of guppies (*Poecilia reticulata*), small fishes that live in shallow freshwater streams (Figure 26.10) in the mountains of Trinidad.

For their study sites, the scientists focused on streams with many small waterfalls. These waterfalls serve as nat-

ural barriers that prevent guppies in one part of a stream from moving easily to another. As a result, each stream holds several populations of guppies that have very little gene flow between them (Section 17.9).

The waterfalls also keep guppy predators from moving from one part of the stream to another. The main guppy predators are killifishes and cichlids. The two differ in size and prey preferences. The killifish is relatively small and preys mostly on immature guppies. It ignores the larger adults. The cichlids are bigger fish. They tend to pursue mature guppies and pass on the small ones. Some parts of the streams hold one type of predator but not the other. Thus, different guppy populations face different predation pressures.

A *Right*, guppy that shared a stream with killifishes (*below*).

B *Right*, guppy that shared a stream with cichlids (*below*).

Figure 26.10 **A,B** Guppies and two guppy eaters, a killifish **A** and a cichlid **B**. **C** Biologist David Reznick contemplating interactions among guppies and their predators in a freshwater stream in Trinidad.

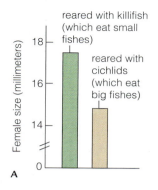

reared with killifish (which eat small fishes)

reared with cichlids (which eat big fishes)

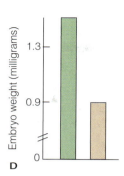

A Female size (millimeters)

B Male size (millimeters)

C Brood interval (days)

D Embryo weight (milligrams)

Figure 26.11 Experimental evidence of natural selection among guppy populations subject to different predation pressures. Compared to the guppies raised with killifish (*green* bars), guppies raised with cichlids (*tan* bars) differed in body size and in the length of time between broods.

Reznick and Endler discovered that guppies in streams with cichlids grow faster and are smaller at maturity than guppies in streams with killifish. Also, guppies hunted by cichlids reproduce earlier, have more offspring at a time, and breed more frequently (Figure 26.11).

Were these differences in life history traits genetic, or did some environmental variation cause them? To find out, the biologists collected guppies from both cichlid- and killifish-dominated streams. They reared the groups in separate aquariums under identical predator-free conditions. Two generations later, the groups continued to show the differences observed in natural populations. The researchers concluded that differences between guppies preyed on by different predators have a genetic basis.

Reznick and Endler hypothesized that the predators act as selective agents that influence guppy life history patterns. They made a prediction: If life history traits evolve in response to predation, then these traits will change when a population is exposed to a new predator that favors different prey traits.

To test their prediction, they found a stream region above a waterfall that had killifish but no guppies or cichlids. They brought in some guppies from a region below the waterfall where there were cichlids but no killifish. At the experimental site, the guppies that had previously lived only with cichlids were now exposed to killifish. The control site was the downstream region below the waterfall, where relatives of the transplanted guppies still coexisted with cichlids.

Reznick and Endler revisited the stream over the course of eleven years and thirty-six generations of guppies. They monitored traits of guppies above and below the waterfall. The recorded data showed that guppies at the upstream experimental site were evolving. Exposure to a previously unfamiliar predator caused changes in their rate of growth, age at first reproduction, and other life history traits. By contrast, guppies at the control site showed no such changes. Reznick and Endler concluded that life history traits in guppies can evolve rapidly in response to the selective pressure exerted by predation.

Effect of Overfishing on Atlantic Cod

The evolution of life history traits in response to predation pressure is not merely of theoretical interest. It has economic importance. Just as guppies evolved in response to predators, a population of Atlantic codfish (*Gadus morhua*) evolved in response to human fishing pressure. Atlantic codfish can grow quite large (Figure 26.12). However, the fishing pressure on the North Atlantic population increased from the mid-1980s to early 1990s. As it did, the age of sexual maturity shifted. The frequency of fast-maturing fish that reproduced while still young and smaller increased. Such individuals were at a selective advantage because both commercial fisherman and sports fishermen preferentially caught and kept the larger fish.

Figure 26.12 Fishermen with a prized catch, a large Atlantic codfish.

In 1992, the declining numbers of cod caused the Canadian government to ban cod fishing in some areas. That ban, and later restrictions, came too late to stop the Atlantic cod population from crashing. The population still has not recovered.

Looking back, it is clear that life history changes were an early sign that the North Atlantic cod population was in trouble. Had biologists recognized the sign, they might have been able to save the fishery and protect the livelihood of more than 35,000 fishers and associated workers. Ongoing monitoring of the life history data for other economically important fishes may help prevent similar disastrous crashes in the future.

Take-Home Message What effect does predation have on life history traits?

❯ When predators prefer large prey, individual prey who reproduce when still small and young are at a selective advantage. When predators focus on small prey, fast growing-individuals have the selective advantage.

> Human population size surpassed 6.8 billion in 2009. Take a look now at what the number means.

‹ Link to Human dispersal 24.11

A History of Human Population Growth

For most of its history, the human population grew very slowly. The growth rate began to increase about 10,000 years ago, and during the past two centuries, it soared (Figure 26.13). Three trends promoted the large increases. First, humans were able to migrate into new habitats and expand into new climate zones. Second, humans developed new technologies that increased the carrying capacity of existing habitats. Third, humans sidestepped some limiting factors that restrain growth of other species.

Early humans evolved in the dry woodlands of Africa, then moved into the savannas. Bands of hunter–gatherers moved out of Africa about 2 million years ago. By 40,000 years ago, their descendants were established in much of the world (Section 24.11).

Few species can expand into such a broad range of habitats, but the early humans had large brains that allowed them to develop the necessary skills. They learned how to start fires, build shelters, make clothing, manufacture tools, and cooperate in hunts. With the advent of language, knowledge of such skills did not die with the individual. Compared to most species, humans have a greater capacity to disperse fast over long distances and to become established in a variety of environments.

The invention of agriculture about 11,000 years ago provided a more dependable food supply. A pivotal factor was the domestication of wild grasses, including species ancestral to modern wheat and rice.

Infectious diseases are density-dependent factors that helped dampen human population growth. In the mid-1300s, one-third of Europe's population was lost to a pandemic known as the Black Death. Waterborne diseases such as cholera were widespread. In the past 200 years, improved sanitation and medical advances such as vaccines and antibiotics cut the death toll from disease. Births increasingly outpaced deaths, and exponential growth accelerated.

In the middle of the eighteenth century, people learned to harness the energy of fossil fuels, starting with coal. Fuel-driven, mechanized agriculture provided food to sustain the ever larger population.

In sum, by learning to grow food, controlling disease agents, and tapping into fossil fuels—a concentrated source of energy—the human population has sidestepped many factors that previously limited its rate of increase.

Viewed in a historical context, the rate of population growth in the past two centuries has been astonishing. It took more than 100,000 years for the human population size to reach 1 billion. It took just 123 years to reach 2 billion, 33 more to reach 3 billion, 14 more to reach 4 billion, and then 13 more to get to 5 billion. It took only 12 more years to arrive at 6 billion.

Estimated population

10,000 years ago	5 million
By 1804	1 billion
By 1927	2 billion
By 1960	3 billion
By 1974	4 billion
By 1987	5 billion
By 1999	6 billion
Projected for 2050	8.9 billion

domestication of plants, animals 9000 B.C. (about 11,000 years ago)

agriculturally based urban societies

beginning of industrial, scientific revolutions

number of individuals (billions)

Figure 26.13 Growth curve (*red*) for the world human population. To check the current world and U.S. population estimates, visit the U.S. Census web site at www.census.gov/main/www/popclock.html.

Figure 26.14 Animated Age structure diagrams for the world's three most populous countries. The width of each bar represents the number of individuals in a 5-year age group. *Green* bars represent people in their pre-reproductive years. The *left* side of each chart indicates males; the *right* side, females.

›› Figure It Out Which country has the largest number of men in the 45 to 49 age group?

Answer: China

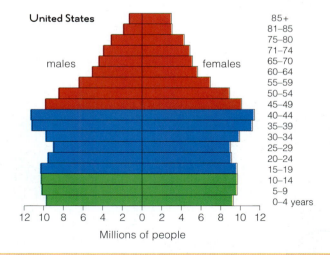

Fertility Rates and Future Growth

Most governments now acknowledge that population growth, resource depletion, pollution, and quality of life are interconnected. Many offer family planning programs. The United Nations Population Division estimates that globally, more than 60 percent of married women use family planning methods.

The **total fertility rate** of a human population is the average number of children born to a woman during her reproductive years. In 1950, the worldwide total fertility rate averaged 6.5. By 2008, it had declined to 2.6. It remains above the **replacement fertility rate**—the average number of children a woman must bear to replace herself with one daughter of reproductive age. At present, the replacement rate is 2.1 for developed countries and as high as 2.5 in some developing countries. (It is higher in developing countries because more daughters die before reaching the age of reproduction.) A population grows as long as total fertility rate exceeds the replacement rate.

World population is expected to reach 8.9 billion by 2050, and possibly to decline as the century ends. China and India already have more than one billion people apiece. Together they hold 38 percent of the world population. Next in line is the United States, with 307 million.

Age structure diagrams show the distribution of individuals among age groups. **Figure 26.14** shows the age structure in the three most populous countries. Notice the size of the age groups that will reproduce during the next fifteen years. The broader the base of an age structure diagram, the greater the proportion of young people, and the greater the expected growth. Government policies that favor couples who have only one child have helped China to narrow its pre-reproductive base.

Even if every couple now alive decides to bear no more than two children, world population growth will not slow for many years, because 1.9 billion people are about to come of reproductive age. More than one-third of the world population is in the broad pre-reproductive base.

---

replacement fertility rate Average number of children women of a population must bear to replace themselves with a daughter of reproductive age.
total fertility rate Average number of children the women of a population bear over the course of a lifetime.

Take-Home Message What is the history of human population growth and what are prospects for future growth?

❯ Through expansion into new habitats, improved agriculture, and technological innovations, the human population has temporarily skirted environmental resistance to growth.

❯ Global population growth has slowed somewhat, but large numbers of young people ensure that numbers will continue to rise for the foreseeable future.

> High population growth is correlated with low levels of economic development and low per capita consumption of resources.

Development and Demographics

Demographic factors vary among countries, with the most highly developed countries having the lowest birth rates and infant mortality, and the highest life expectancy (Figure 26.15). The **demographic transition model** describes how changes in population growth often unfold in four stages of economic development (Figure 26.16). Living conditions are harshest during the preindustrial stage, before technological and medical advances become widespread. Birth and death rates are both high, so the growth rate is low **1**.

Next, in the transitional stage, industrialization begins. Food production and health care improve. The death rate drops fast, but the birth rate declines more slowly **2**. As a result, the population growth rate increases rapidly.

During the industrial stage, when industrialization is in full swing, the birth rate declines. People move from the country to cities, where birth control is available and couples tend to want smaller families. The birth rate moves closer to the death rate, and the population grows less rapidly **3**.

In the postindustrial stage, a population's growth rate becomes negative. The birth rate falls below the death rate, and population size slowly decreases **4**.

The United States is in the industrial stage. Developing countries such as Mexico are in the transitional stage, with people continuing to migrate from agricultural regions to cities. Japan and some countries in Europe are in the postindustrial stage.

The demographic transition model is based on analysis of what happened when western Europe and North America industrialized. Whether it can accurately predict changes in modern developing countries remains to be seen. Less developed countries now receive aid from highly developed countries, but must also compete against these countries in a global market.

Runaway population growth is problematic, but negative growth also poses challenges. Some countries in Europe, some members of the former Soviet Union, and Japan currently have negative population growth. Their birth rate has fallen below their death rate. Negative

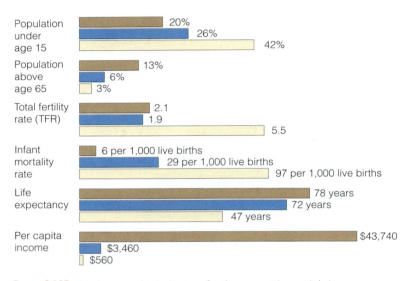

Population under age 15	20% / 26% / 42%
Population above age 65	13% / 6% / 3%
Total fertility rate (TFR)	2.1 / 1.9 / 5.5
Infant mortality rate	6 per 1,000 live births / 29 per 1,000 live births / 97 per 1,000 live births
Life expectancy	78 years / 72 years / 47 years
Per capita income	$43,740 / $3,460 / $560

Figure 26.15 Key demographic indicators for three countries, mainly in 2006. The United States (*brown* bar) is highly developed, Brazil (*blue* bar) is moderately developed, and Nigeria (*beige* bar) is less developed.

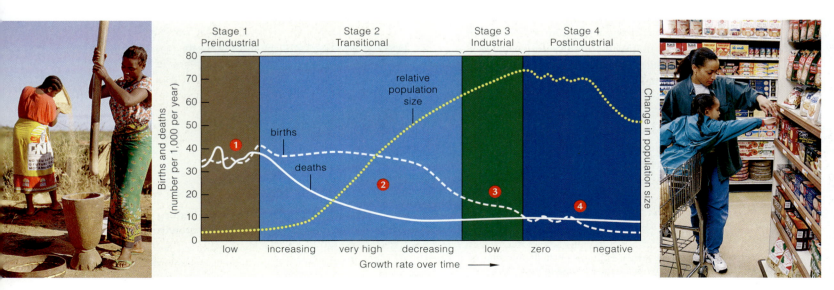

Figure 26.16 Animated Demographic transition model for changes in population growth rates and sizes, correlated with long-term changes in economy.

Table 26.3	Ecological Footprints*
Country	Hectares per Capita
United States	9.4
Canada	7.1
United Kingdom	5.3
France	4.9
Japan	4.9
Russia	3.7
Mexico	3.4
Brazil	2.4
China	2.1
India	0.9
World Average	2.7
* Global Footprint Network, 2008 data	

growth produces a population with more old people than young. In Japan, people over 65 currently make up about 20 percent of the population. The aging of a population has social implications because older individuals have traditionally been supported by a larger, younger workforce.

Development and Consumption

What is Earth's carrying capacity for humans? There is no simple answer to this question. For one thing, we cannot predict what new technologies may arise or their effects. For another, different types of societies require different amounts of resources to sustain them. On a per capita basis, people in highly developed countries use far more resources than those in less developed countries, and they also generate more waste and pollution.

Ecological footprint analysis is one widely used method of measuring and comparing resource use. An **ecological footprint** is the amount of Earth's surface required to support a particular level of development and consumption in a sustainable fashion. It includes the amount of area required to grow crops, graze animals, produce forest products, catch fish, hold buildings, and take up carbon emissions.

In 2008, the per capita global footprint for the human population was 2.7 hectares, or about 6.5 acres (Table 26.3). The world's two most populous countries, China and India, were below that average, whereas the per capita footprint of the United States was more than three times the average. In other words, the life-style of an average person in the United States requires three times as much of Earth's sustainable resources as the life-style of

---

demographic transition model Model describing the changes in human birth and death rates that occur as a region becomes industrialized.
ecological footprint Area of Earth's surface required to sustainably support a particular level of development and consumption.

A Honking Mess (revisited)

When a plane crashes or is forced to land unexpectedly, investigators from the Federal Aviation Agency try to determine what went wrong. In 2009, they were asked to find out why both engines of a passenger plane stopped, forcing the plane to land in the Hudson River. The pilot had reported a bird strike, and the plane had bits of feather, bone, and muscle in its wing flaps and its engines (*upper right*). Samples of this tissue were sent to the Smithsonian Institute, which analyzed the DNA. Unique sequences in the DNA identified the tissue in both engines as Canada goose. One engine had female goose DNA. The other had both male and female DNA, indicating that at least two birds entered it.

Researchers were even able to tell which population of geese these unlucky birds belonged to. As a feather develops, it incorporates hydrogen that the bird took up from its environment. The mix of hydrogen isotopes in the environment varies with latitude, so the isotope mix in a feather provides information about where that feather developed. The isotope mix in the feather bits from the engines indicated that these bird were migratory; their feathers had developed in Canada, not in New York.

How would you vote? One way to decrease the number of nonmigratory Canada geese is by encouraging hunting when migratory birds are unlikely to be present. Would you support relaxing hunting restrictions in regions where Canada geese have become pests? See CengageNow for details, then vote online (cengagenow.com).

---

an average world citizen. It requires more than ten times the resources of a person in India.

The United States is unlikely to lower its resource consumption to match that of India. In fact, billions of people in India, China, and other less developed nations dream that one day they or their offspring will enjoy the same type of life-style as the average American.

Ecological footprint analysis tells us that, with current technology, Earth may not have enough resources to make those dreams come true. For everyone now alive to live like an average American, four times the sustainable resources available on Earth would be required. Such analysis also suggests that the human population may already be living beyond its ecological means. If we divide the Earth equally, each of us now has 2.1 hectares available to support us sustainably, but our average per capita ecological footprint is 2.7 hectares.

Take-Home Message How does economic development affect population growth and resource consumption?

> Historically, population growth has slowed as countries have industrialized. The same trend may eventually slow growth in currently developing countries.

> The most highly developed countries have relatively low growth rates, but they consume a disproportionately large share of Earth's resources.

> The current global level of resource use is unsustainable, and the pressure on resources is expected to rise.

Summary

Section 26.1 A **population** is a group of individuals of the same species that live in the same area and tend to interbreed. Canada geese in the United States include migratory populations and resident ones.

Section 26.2 Demographics are statistics used to describe a population. We often estimate **population size** by using a sampling method such as **plot sampling** or **mark-recapture sampling**. Other demographics include **population density** and **population distribution**. Most populations have a clumped distribution. **Age structure** describes the proportion of individuals in each age category. The size of the **reproductive base** affects population growth.

Section 26.3 A population's per capita birth rate minus its per capita death rate gives us *r*, the **per capita growth rate**. When birth rate and death rate are equal, there is **zero population growth**.

A population in which *r* is greater than zero undergoes **exponential growth**. The increase in size in any interval is determined by the equation $G = r \times N$, where *G* is population growth and *N* is the number of individuals. With exponential growth, a graph of population size against time produces a J-shaped growth curve. The maximum possible exponential growth rate under optimal conditions is the population's **biotic potential**.

Emigration and **immigration** of individuals can also affect population size.

Section 26.4 The **carrying capacity** is the maximum number of individuals of a given population that can be sustained indefinitely by resources in their environment. **Density-dependent factors** are conditions or events that can lower reproductive success and that get worse with crowding. Disease and competition for food are examples. **Density-independent factors** are conditions or events that can lower reproductive success, but their effect does not vary with crowding. Carrying capacity varies among environments and over time. A population may temporarily overshoot its carrying capacity, then crash.

Density-dependent factors lead to **logistic growth**: a population begins growing exponentially, then growth levels off as **limiting factors** begin to come into play. With logistic growth, population size over time plots out as an S-shaped curve.

Sections 26.5, 26.6 The time to maturity, number of reproductive events, number of offspring per event, and life span are aspects of a **life history pattern**. Such patterns can be studied by following a **cohort**, a group of individuals born at the same time. Three types of **survivorship curves** are common: a high death rate late in life, a constant death rate at all ages, or a high death rate early in life. Life histories have a genetic basis and are subject to natural selection. At low population density, **r-selection** favors quickly producing as many offspring as possible. At a higher population density, **K-selection** favors investing more time and energy in fewer, higher-quality offspring. Most populations have a mixture of both *r*-selected and *K*-selected traits.

Section 26.7 The human population has surpassed 6.8 billion. Expansion into new habitats and agriculture allowed early increases. Medicine and technology have allowed greater increases. Today, the global **total fertility rate** is declining, but it remains above the **replacement fertility rate**. The very broad pre-reproductive base will cause numbers to increase for at least sixty years.

Section 26.8 The **demographic transition model** predicts economic development will slow population growth. Negative population growth increases the proportion of elderly in a population. World resource consumption will probably continue to rise because a highly developed nation has a much larger **ecological footprint** than a developing one. However, with current technology, Earth does not have enough resources to support the existing population in the style of developed nations.

Self-Quiz   Answers in Appendix III

1. Most commonly, individuals of a population show a _____ distribution through their habitat.

2. The rate at which population size grows or declines depends on the rate of _____ .
   a. births   c. immigration   e. a and b
   b. deaths   d. emigration   f. all of the above

3. Suppose 200 fish are marked and released in a pond. The following week, 200 fish are caught and 100 of them have marks. How many fish are in this pond?

4. A population of worms is growing exponentially in a compost heap. Thirty days ago there were 400 worms and now there are 800. How many worms will there be thirty days from now, assuming conditions remain constant?

5. For a given species, the maximum rate of increase per individual under ideal conditions is its _____ .
   a. biotic potential   c. environmental resistance
   b. carrying capacity   d. density control

6. _____ is a density-independent factor that influences population growth.
   a. Resource competition   c. Predation
   b. Infectious disease   d. Harsh weather

7. A life history pattern for a population is a set of adaptive traits such as _____ .
   a. longevity   c. age at reproductive maturity
   b. fertility   d. all of the above

Data Analysis Activities

Monitoring Iguana Populations In 1989, Martin Wikelski began a long-term study of marine iguana populations in the Galápagos Islands. He marked the iguanas on two islands—Genovesa and Santa Fe—and collected data on how their body size, survival, and reproductive rates varied over time. The iguanas eat algae and have no predators, so deaths typically result from food shortages, disease, or old age. His studies showed that the iguana populations decline during El Niño events, when water surrounding the islands heats up.

In January 2001, an oil tanker ran aground and leaked a small amount of oil into the waters near Santa Fe. Figure 26.17 shows the number of marked iguanas that Wikelski and his team counted just before the spill and about a year later.

1. Which island had more marked iguanas at the time of the first census?
2. How much did the population size on each island change between the first and second census?
3. Wikelski concluded that changes on Santa Fe were the result of the oil spill, rather than a factor common to both islands. How would the census numbers be different from those he observed if an adverse event had affected both islands?

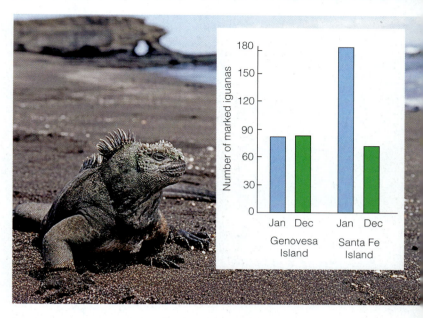

Figure 26.17 Shifting numbers of marked marine iguanas on two Galápagos islands. An oil spill occurred near Santa Fe just after the January 2001 census (*blue* bars). A second census was carried out in December 2001 (*green* bars).

8. The human population is now over 6.8 billion. It reached 6 billion in _____ .
   a. 2007   b. 1999   c. 1802   d. 1350

9. Compared to the less developed countries, the highly developed ones have a higher _____ .
   a. death rate   c. total fertility rate
   b. birth rate   d. resource consumption rate

10. An increase in infant mortality will _____ a population's replacement fertility rate.
   a. raise   b. lower   c. not affect

11. Species that usually colonize empty habitats are more likely to show traits that are favored by _____ .
   a. *r*-selection   b. *K*-selection

12. All members of a cohort are the same _____ .
   a. sex   b. size   c. age   d. weight

13. Match each term with its most suitable description.
   ___ carrying capacity   a. maximum rate of increase per individual under ideal conditions
   ___ exponential growth   b. population growth plots out as an S-shaped curve
   ___ biotic potential   c. maximum number of individuals sustainable by the resources in a given environment
   ___ limiting factor   d. population growth plots out as a J-shaped curve
   ___ logistic growth   e. essential resource that restricts population growth when scarce

Additional questions are available on CENGAGE**NOW**.

Critical Thinking

1. When researchers moved guppies from pools with cichlids that eat large guppies to pools with killifish that eat small ones, life history traits were not the only traits that changed. Over generations, the males became more colorful. Why do you think this change occurred?

2. Each summer, a giant saguaro cactus produces tens of thousands of tiny black seeds. Most die, but the few that land in a sheltered spot sprout the following spring. The saguaro is a slow-growing CAM plant (Section 6.8). After fifteen years, it may be only knee high, and it will not flower for another fifteen years. It may live for 200 years. Saguaros share their habitat with annuals such as poppies, which sprout, form seeds, and die in just a few weeks. Speculate on how these different life histories can both be adaptive in the same desert environment.

3. Age structure diagrams for two hypothetical populations are shown at *right*. Describe the growth rate of each population and discuss the current and future social and economic problems that each is likely to face.

Animations and Interactions on CENGAGE**NOW**:
› Mark–recapture sampling; Exponential growth; Logistic growth; Age structure; Demographic transition.

‹ Links to Earlier Concepts

In this chapter you will revisit biogeography and take a closer look at global patterns in species richness (Sections 16.2, 16.4). You will be reminded of the effects of pathogens (19.3) and how species interactions lead to natural selection (17.4, 17.13, 21.7). You will see examples of field experiments (1.7) and apply what you know about populations and factors affecting their growth (26.2–26.4).

Key Concepts

Community Characteristics

A community consists of all species in a habitat. A habitat's history, its biological and physical characteristics, and interactions among species in the habitat affect the number of species in the community and their relative abundance.

Forms of Species Interactions

Commensalism, mutualism, competition, predation, and parasitism are interspecific interactions. They influence the population size of participating species, which in turn influences the community's structure.

27 Community Ecology

Red imported fire ants, *Solenopsis invicta*, are native to South America. They arrived in the United States in the 1930s, probably as stowaways on a cargo ship. The ants spread out from the Southeast and in time established colonies as far west as California and as far north as Kansas and Delaware.

Like most ants, *S. invicta* nests in the ground (Figure 27.1). Accidentally step on one of these nests, and you will quickly realize your mistake. Like honeybees, fire ants defend their nest against perceived threats by stinging. Venom injected by the stinger causes a burning sensation, and results in formation of reddened bumps that are slow to heal.

S. invicta nests are an annoyance to people, but they pose a greater threat to wildlife. For example, researchers have demonstrated that the presence of *S. invicta* decreases the number of bobwhites (a type of quail) and vireos (a type of songbird). The ants feed on the birds' eggs and on nestlings. *S. invicta* also disrupts populations of Texas horned lizards. The invasive ants outcompete native ants that are the lizard's main source of food. The lizard cannot eat *S. invicta*. In addition, *S. invicta* can prey on the lizard's eggs.

Invicta means "invincible" in Latin, and *S. invicta* lives up to its name. Pesticides do not slow the spread of this species and may facilitate invasions by wiping out native ant populations. To fight the ants, scientists have turned to biological controls. A biological control agent is a natural enemy of a pest species. It can decrease pest numbers by preying on, infecting, or parasitizing the pest.

Biological control of *S. invicta* involves phorid flies, insects that prey on *S. invicta* in their native habitat. Phorid flies are parasitoids, specialized parasites that kill their host in a rather gruesome way. A female fly pierces the cuticle of an adult ant, then lays an egg in the ant's soft tissues. The egg hatches into a larva, which grows and then eats its way through the tissues to the ant's head. After the larva gets big enough, it secretes an

Figure 27.1 Red imported fire ants and their foes. *Opposite*, nest mounds of fire ants in a Texas pasture. **A** Phorid fly. This fly uses the hooked extension on its abdomen to insert a fertilized egg into an ant's thorax. **B** Parasitized ant that lost its head after a developing fly larva moved into it. The larva will undergo metamorphosis to an adult inside the detached head.

enzyme that makes the ant's head fall off. The fly larva develops into an adult within the shelter of the detached ant head.

Several phorid fly species have been introduced in various southern states. The flies seem to be surviving, reproducing, and increasing their range. They are not expected to kill off all *S. invicta* in affected areas. Rather, the hope is that the flies will reduce the density of invading colonies.

Ecologists are also exploring other options. They are testing effects of imported pathogenic fungi or protists that infect *S. invicta* but not native ants. Another idea is to introduce a parasitic South American ant that invades *S. invicta* colonies and kills the egg-laying queens.

Species interactions such as those between ants and flies or fungi are the focus of community ecology. A **community** is all the species that live in a region. As you will see, species interactions and disturbances can shift community structure (the types of species and their relative abundances) in small and large ways, some predictable, and others unexpected.

---

community All species that live in a particular region.

Long–Term Change in Communities
The array of species in a community changes over time, although the exact outcome of these changes is difficult to predict. When a new community forms, the early-arriving species often alter the habitat in a way that facilitates their own replacement.

Species Effects on Community Stability
Removing a species from a community or adding one to it can have a dramatic effect on other species. Some species are adapted to disturbances and a change in the frequency of disturbances can affect their number.

Global Patterns in Community Structure
Biogeographers identify regional patterns in species distribution. They have shown that tropical regions hold the greatest number of species and that characteristics of islands can be used to predict how many species an area will hold.

> Community structure refers to the number and relative abundances of species in a habitat. It changes over time.
❮ Link to Coevolution 17.13

The type of place where a species normally lives is its **habitat**, and all species living in a habitat constitute a community. Communities often are nested one inside another. For example, we find a community of microbial organisms inside the gut of a termite. That termite is part of a larger community of organisms living on a fallen log. The log-dwellers are part of a larger forest community.

Even communities that are similar in scale differ in their species diversity. There are two components to species diversity. The first, species richness, refers to the number of species. The second is species evenness, or the relative abundance of each species. For example, a pond that has five fish species in nearly equal numbers has a higher species diversity than a pond with one abundant fish species and four rare ones.

Community structure is dynamic. The array of species and their relative abundances change over time. Communities change over a long time span as they form and then age. They also change suddenly as a result of natural or human-induced disturbances.

Physical factors such as climate and resource availability influence community structure. Habitats that are nutrient-poor and have extreme conditions support fewer species than those that are nutrient-rich and have moderate conditions.

Species interactions also influence the types of species in a community and their relative abundances. In some cases, the effect is indirect. For example, when songbirds

Table 27.1 Direct Two-Species Interactions		
Type of Interaction	Effect on Species 1	Effect on Species 2
Commensalism	Helpful	None
Mutualism	Helpful	Helpful
Interspecific competition	Harmful	Harmful
Predation, herbivory parasitism, parasitoidism	Helpful	Harmful

eat caterpillars, the birds indirectly benefit the trees that the caterpillars feed on, while directly reducing the abundance of caterpillars.

We can group direct species interactions by their effects on both participants (Table 27.1). For example, **commensalism** helps one species and has no effect on the other. Commensal ferns live attached to the trunk or branches of a tree (Figure 27.2). Having a perch in the light benefits the fern, and the tree is unaffected. Relationships are considered commensal when one species benefits, and the other neither benefits nor is harmed by the relationship. Should evidence of either effect come to light, the relationship is reclassified.

The other types of species interaction are discussed in detail in sections that follow. Here we will simply note that interactions can be mutually beneficial, mutually harmful, or benefit one species while harming the other.

Species interactions may be fleeting or a long-term relationship. **Symbiosis** means "living together" and biologists use the term to refer to a relationship in which two species have a prolonged close association.

Regardless of whether one species helps or harms another, two species that interact closely for generations can coevolve. As Section 17.13 explained, coevolution is an evolutionary process in which each species acts as a selective agent that shifts the range of variation in the other.

commensalism Species interaction that benefits one species and neither helps nor harms the other.
habitat Type of environment in which a species typically lives.
symbiosis One species lives in or on another in a commensal, mutualistic, or parasitic relationship.

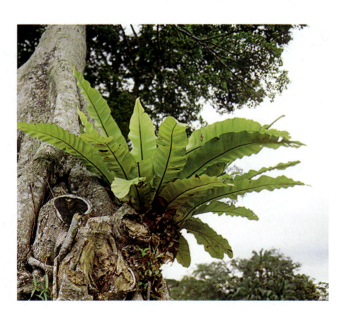

Figure 27.2 A tree with a commensal fern. The fern benefits by growing on the tree, which is unaffected by the presence of the fern.

Take-Home Message What factors affect the types and abundances of species in a community?

> The types and abundances of species in a community are affected by physical factors such as climate and by species interactions.

> A species can be benefited, harmed, or unaffected by its interaction with another species.

> In a mutualistic interaction, two species benefit by taking advantage of one another.

< Links to Endosymbiosis 18.6, Lichens and mycorrhizae 22.6, Pollinators 21.7

Mutualism is an interspecific interaction that benefits both species. For example, birds, insects, bats, and other animals pollinate flowering plants. Pollinators eat nectar and pollen. In return, they transfer pollen between plants of the same species. Similarly, many animals that eat fruits disperse the seeds inside the fruits to new sites.

In some mutualisms, neither species can complete its life cycle without the other. Yucca plants and the moths that pollinate them have such an interdependence (Figure 27.3). In other cases, the mutualism is helpful but not a life-or-death requirement. Most plants, for example, have more than one pollinator.

Mutualistic microorganisms help plants obtain the nutrients they require. Nitrogen-fixing bacteria living on roots of legumes such as peas provide the plant with extra nitrogen. Mycorrhizal fungi living in or on plant roots enhance a plant's mineral uptake. Other fungi interact with photosynthetic algae or bacteria in lichens.

There is some conflict between the partners in all mutualisms. For example, in a lichen, the fungus would benefit by getting as much sugar as possible from its photosynthetic partner. The photosynthesizer would benefit by keeping as much sugar as possible for its own use. In each species, natural selection will favor individuals who minimize their cost and maximize their benefit.

For some mutualists, the main benefit is defense. A sea anemone has stinging cells (nematocysts), so most fishes avoid its tentacles. However, an anemone fish can nestle among the tentacles in safety (Figure 27.4). A mucus layer shields the anemone fish from stings, and the tentacles keep it safe from predatory fish. The anemone fish repays its partner by chasing off the few fishes that are able to eat sea anemone tentacles.

Finally, think back on an idea introduced in Section 18.6. According to the theory of endosymbiosis, mutualistic aerobic bacteria that lived in early eukaryotic cells evolved into mitochondria. Cyanobacteria living inside eukaryotic cells evolved into chloroplasts by a similar process. Thus, all eukaryotes are the product of ancient mutualisms.

mutualism Species interaction that benefits both species.

Figure 27.3 Mutualism on a rocky slope of the high desert in Colorado.

Only one yucca moth species pollinates plants of each *Yucca* species; each moth cannot complete its life cycle with any other plant. The moth matures when yucca flowers blossom. The female has specialized mouthparts that collect and roll sticky pollen into a ball. She flies to another flower and pierces its ovary, where seeds will form and develop, and lays eggs inside. As the moth crawls out, she pushes a ball of pollen onto the flower's pollen-receiving platform.

After pollen grains germinate, they give rise to pollen tubes, which grow through the ovary tissues and deliver sperm to the plant's eggs. Seeds develop after fertilization.

Meanwhile, moth eggs develop into larvae that eat a few seeds, then gnaw their way out of the ovary. Seeds that larvae do not eat give rise to new yucca plants.

Figure 27.4 The sea anemone *Heteractis magnifica* has a mutualistic association with the pink anemone fish (*Amphiprion perideraion*). This tiny but aggressive fish chases away predatory butterfly fishes that would bite off tips of anemone tentacles. The fish cannot survive and reproduce without the protection of an anemone. The anemone does not need a fish to protect it, but it does better with one.

Take-Home Message What is a mutualism?

> A mutualism is a species interaction in which each species benefits by associating with the other.

> In each species, individuals who maximize the benefit they receive while minimizing their costs will be favored.

> Resources are limited and individuals of different species often compete for access to them.
> Competition adversely affects both species.
< Links to Darwin's theory 16.4, Directional selection 17.5

As Charles Darwin clearly understood, competition for resources among individuals of the same species can be intense and leads to evolution by natural selection. **Inter-specific competition**, competition among members of different species, is not usually as intense. The requirements of two species might be similar, but they are never as close as they are for members of one species.

Each species has a unique set of ecological requirements and roles that we refer to as its **ecological niche**. Both physical and biological factors define the niche. Aspects of an animal's niche include the temperature range it can tolerate, the species it eats, and the places it can breed. A description of a flowering plant's niche would include its soil, water, light, and pollinator requirements. The more similar the niches of two species are, the more intensely the species will compete.

Competition takes two forms. With interference competition, one species actively prevents another from accessing some resource. As an example, one species of scavenger will often chase another away from a carcass (Figure 27.5). As another example, some plants use chemical weapons against potential competition. Aromatic chemicals that ooze from tissues of sagebrush plants, black walnut trees, and eucalyptus trees seep into the soil around these plants. The chemicals prevent other kinds of plants from germinating or growing.

In exploitative competition, species do not interact directly; each reduces the amount of resources available to the other by using that resource. For example, deer and blue jays both eat acorns in oak forests. The more acorns the birds eat, the fewer there are for the deer.

Effects of Competition

Deer and blue jays share a fondness for acorns, but each also has other sources of food. Any two species differ in their resource requirements. Species compete most intensely when the supply of a shared resource is the main limiting factor for both (Section 26.4).

In the 1930s, G. Gause conducted experiments with two species of ciliated protozoans (*Paramecium*) that compete for the same prey: bacteria. He cultured the *Paramecium* species separately and together (Figure 27.6). Within weeks, population growth of one species outpaced the other, which went extinct.

Experiments by Gause and others are the basis for the concept of **competitive exclusion**: Whenever two species require the same limited resource to survive or reproduce, the better competitor will drive the less competitive species to extinction in that habitat.

When resource needs of competitors are not exactly the same, species can coexist, but competition reduces their numbers. For example, Nelson Hairston showed the effects of competition between two species of salamanders that coexist in woodlands. He created experimental plots by removing all members of one species or the other. Another group of plots served as unaltered controls. After five years, the number of salamanders in both types of single-species plots had increased, whereas the number in control plots was unchanged. Hairston concluded that where these salamanders coexist, competitive interactions suppress the population growth of both.

A Golden eagle and a red fox face off over a moose carcass.

Figure 27.5 Interspecific competition among scavengers.

B In a dramatic demonstration of interference competition, the eagle attacks the fox with its talons. After this attack, the fox retreated, leaving the eagle to exploit the carcass.

Paramecium

Figure 27.6 Animated Results of competitive exclusion between two *Paramecium* species that compete for the same food. When the two species were grown together, *P. aurelia* drove *P. caudatum* to extinction.

Resource Partitioning

Resource partitioning is an evolutionary process by which species become adapted to use a shared limiting resource in a way that minimizes competition. For example, three species of annual plants are common in abandoned fields. All require water and nutrients, but each is adapted to meet these needs in a slightly different way (Figure 27.7). This variation allows the plants to coexist.

Similarly, eight species of woodpecker coexist in Oregon forests. All feed on insects and nest in hollow trees, but the details of their foraging behavior and nesting preferences vary. Differences in nesting time also help reduce competitive interactions.

Resource partitioning occurs because directional selection occurs when species with similar requirements share a habitat and compete for a limiting resource. In each species, those individuals who differ most from the competing species have the least competition and thus leave the most offspring. Over generations, directional selection leads to **character displacement**: The range of variation for one or more traits is shifted in a direction that lessens the intensity of competition for a limiting resource.

A recently observed directional shift in the beak dimensions of the medium ground finch (*Geospiza fortis*) shows how competition can result in character displacement. The medium ground finch was the only seed-eating bird on one of the Galápagos islands until some large ground finches (*G. magnirostris*) arrived. When a drought caused a food shortage, the medium ground finches with the largest beaks found themselves at a selective disadvantage. These big-beaked, medium-sized birds had to compete

with the large new arrivals for the few big seeds. Small-beaked birds had no competition for the smaller seeds they favor. As a result, the large-beaked medium finches were more likely to die of starvation. The differential deaths caused a shift in the average beak size of the medium finch population; beaks became significantly shorter and shallower.

bristly foxtail

Figure 27.7 Resource partitioning among three annual plant species in a plowed but abandoned field.

Roots of each species take up water and mineral ions from a different soil depth. This difference reduces competition among the species and allows them to coexist.

Indian mallow   smartweed

character displacement Outcome of competition between two species; similar traits that lead to competition become dissimilar.
competitive exclusion Process whereby two species compete for a limiting resource, and one drives the other to local extinction.
ecological niche All of a species' requirements and roles in an ecosystem.
interspecific competition Competition between two species.
resource partitioning Species become adapted in different ways to access different portions of a limited resource; allows species with similar needs to coexist.

> Predation and herbivory are short-term interactions in which one species obtains nutrients and energy by feeding on another.

< Links to Ricin 9.1, Coevolution 17.13

Predator and Prey Abundance

With **predation**, one species (the predator) captures, kills, and digests another species (the prey). The abundance of prey species in a community affects how many predators it can support. The number of predators reduces the number of prey, but the extent of this effect depends partly on how the predator responds to changes in prey density.

With some predators, such as web-spinning spiders, the proportion of prey killed is constant, so the number killed in any given interval depends solely on prey density. As the number of flies in an area increases, more and more become caught in a web.

More often, the number of prey killed depends in part on the time it takes predators to capture, eat, and digest prey. As prey density increases, the rate of kills rises steeply at first because there are more prey to catch. Eventually, the rate of increase slows, because a predator is exposed to more prey than it can handle at one time. A wolf that just killed a caribou will not hunt another until it has eaten and digested the first one.

Predator and prey populations sometimes rise and fall in a cyclical fashion. For example, Figure 27.8 shows historical data for the numbers of lynx and their main prey, the snowshoe hare. Both populations rise and fall over an approximately ten-year cycle, with predator abundance lagging behind prey abundance. Field studies indicate that lynx numbers fluctuate mainly in response to hare numbers. However, the size of the hare population is affected by the abundance of the hare's food as well as the number of lynx. Hare populations continued to rise and fall even when predators were experimentally excluded from areas.

Coevolution of Predators and Prey

Predator and prey exert selection pressure on one another. Suppose a mutation gives members of a prey species a more effective defense. Over generations, directional selection will cause this mutation to spread through the prey population. If some members of a predator population have a trait that makes them better at thwarting the improved defense, these predators and their descendants will be at an advantage. Thus, predators exert selection pressure that favors improved prey defense, which in turn exerts selection pressure on predators, and so it goes over many generations.

You have already learned about some defensive adaptations. Many prey species have hard or sharp parts that make them difficult to eat. Think of a snail's shell or a sea urchin's spines. Others contain chemicals that taste bad or sicken predators. Most defensive toxins in animals come from the plants they eat. For example, a monarch butterfly caterpillar takes up chemicals from the milkweed plant that it feeds on. A bird that later eats the butterfly will be sickened by these chemicals.

Well-defended prey often have warning coloration, a conspicuous color pattern that predators learn to avoid. For example, many species of stinging wasps and bees have a pattern of black and yellow stripes (Figure 27.9A). The similar appearance of bees and wasps is an example of **mimicry**, an evolutionary pattern in which one species comes to resemble another. The bees and wasps benefit from their similar appearance. The more often a predator is stung by a black and yellow striped insect, the less likely it is to attack a similar insect.

In another type of mimicry, prey masquerade as a species that has a defense that they lack. For example, some

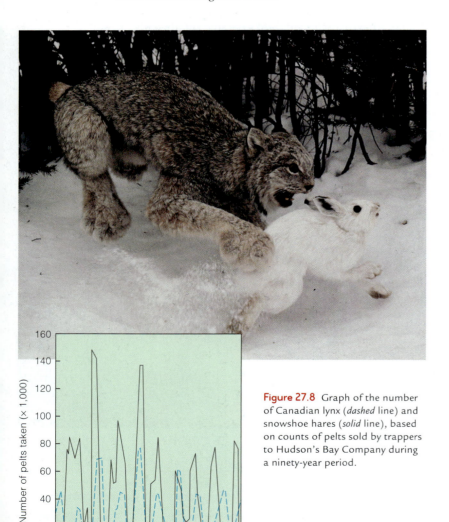

Figure 27.8 Graph of the number of Canadian lynx (*dashed* line) and snowshoe hares (*solid* line), based on counts of pelts sold by trappers to Hudson's Bay Company during a ninety-year period.

A Wasp that can inflict a painful sting.

B Fly, with no stinger, mimics wasp color pattern.

Figure 27.9 Mimicry. Some edible insect species resemble toxic or unpalatable species that are not closely related.

flies, which cannot sting, resemble stinging bees or wasps (Figure 27.9B). The fly benefits when predators avoid it after a run-in with the better-defended look-alike species.

Another adaptation to predation is a last-chance trick that can startle an attacking predator. Section 1.7 described how eye spots and a hissing sound protect some butterflies. Similarly, a lizard's tail may detach from the body and wiggle a bit as a distraction. Skunks squirt a foul-smelling, irritating repellent.

Many prey have camouflage—a form, patterning, color, and often behavior that allows them to blend into their surroundings and avoid detection (Figure 27.10A).

Predators too benefit from camouflage (Figure 27.10B,C). Other predator adaptations include sharp teeth and claws that can pierce protective hard parts. Speedy prey select for faster predators. For example, the cheetah, the fastest land animal, can run 114 kilometers per hour (70 mph). Its preferred prey, Thomson's gazelles, run 80 kilometers per hour (50 mph).

Coevolution of Herbivores and Plants

With **herbivory**, an animal eats a plant or plant parts. The number and type of plants in a community can influence the number and type of herbivores present.

Two types of defenses have evolved in response to herbivory. Some plants have adapted to withstand and recover quickly from herbivory. For example, prairie grasses are seldom killed by native grazers such as bison. The grasses have a fast growth rate and store enough resources in roots to grow back lost shoots.

Other plants have traits that deter herbivory. Such traits include spines or thorns, tough leaves that are difficult to chew or digest, and chemicals that taste bad or sicken herbivores. Ricin, the toxin made by castor bean

B Frilly pink body parts of a flower mantis hide it from insect prey attracted to the real flowers.

A Reedlike defensive posture of a bittern. The bird even sways in the wind like a reed.

C Fleshy protrusions give a predatory scorpionfish the appearance of an algae-covered rock. Algae-eating fish that come close for a nibble end up as prey.

Figure 27.10 Camouflage in prey and predators.

plants (Section 9.1), makes herbivores ill. Caffeine in coffee beans and nicotine in tobacco leaves are evolved defenses against insects.

The existence of plant defenses favors herbivores capable of overcoming those defenses. For example, eucalyptus leaves contain toxins that make them poisonous to most mammals, but not to koalas. Specialized liver enzymes allow koalas to break down toxins made by a few eucalyptus species.

---

herbivory An animal feeds on plant parts.
mimicry A species evolves traits that make it more similar in appearance to another species.
predation One species captures, kills, and eats another.

Take-Home Message **What are the evolutionary effects of predation and herbivory?**

> Predation benefits predators and harms prey. Predator numbers sometimes fluctuate in response to prey availability.

> Predators can coevolve with their prey. Herbivores coevolve with the plants that they eat.

> With mimicry, one species benefits by its resemblance to another species.

> Some species benefit by withdrawing nutrients from other species, or by tricking them into providing parental care.
< Links to Lamprey 24.3, Cuckoo 25.3

Parasitism

With **parasitism**, one species (the parasite) benefits by feeding on another (the host), without immediately killing it. Endoparasites such as parasitic roundworms live and feed inside their host (Figure 27.11A). An ectoparasite such as a tick feeds while attached to a host's external surface (Figure 27.11B).

A parasitic life-style has evolved in members of a diverse variety of groups. Bacterial, fungal, protistan, and invertebrate parasites feed on vertebrates. Lampreys (Section 24.3) attach to and feed on other fish. There are even a few parasitic plants that withdraw nutrients from other plants (Figure 27.12).

Parasites usually do not kill their host immediately. In terms of evolutionary fitness, killing a host too fast is bad for the parasite. Ideally, a host will live long enough to give a parasite time to produce some offspring. The

A Endoparasitic roundworms feeding in the intestine of a host pig.

B Ectoparasitic ticks attached to and sucking blood from a finch.

Figure 27.11 Parasites inside and out.

Figure 27.12 Dodder (*Cuscuta*), also known as strangleweed or devil's hair. This parasitic flowering plant has almost no chlorophyll. Leafless stems wrap around a host plant, and modified roots absorb water and nutrients from the host's vascular tissue.

longer the host survives, the more parasite offspring can be produced. Thus parasites with less-than-fatal effects on hosts are at a selective advantage.

Although parasites typically do not kill their hosts, they can still have an important impact on a host population. Many parasites are pathogens; they cause disease in hosts. Even when a parasite does not cause obvious symptoms, its presence can weaken a host so it is more vulnerable to predation or less attractive to potential mates. Some parasites cause their host to become sterile. Others shift the sex ratio among their host's offspring.

Adaptations to a parasitic life-style include traits that allow the parasite to locate hosts and to feed undetected. For example, ticks that feed on mammals or birds will move toward a source of heat and carbon dioxide, which may be a potential host. A chemical in tick saliva acts as a local anesthetic, preventing the host from noticing the feeding tick. Endoparasites often have traits that help them evade a host's immune defenses.

Among hosts, traits that minimize the negative effects of parasites confer a selective advantage. For example, the allele that causes sickle-cell anemia persists at high levels in some human populations because having one copy of the allele increases the odds of surviving malaria. Grooming and preening behavior are adaptations that minimize the impact of ectoparasites. Some animals produce chemicals that interfere with parasite activity. For example, crested auklets (a type of seabird) produce a citrus-scented secretion that repels ticks.

Strangers in the Nest

With **brood parasitism**, one egg-laying species benefits by having another raise its offspring. The European cuckoos described in Section 25.3 are brood parasites, as are North American cowbirds (Figure 27.13). Not having to invest in

Figure 27.13 A cowbird with its foster parent. A female cowbird minimizes the cost of parental care by laying her eggs in the nests of other bird species.

Figure 27.14 Biological control agent: a commercially raised parasitoid wasp about to deposit a fertilized egg in an aphid. This wasp reduces aphid populations. After the egg hatches, a wasp larva devours the aphid from the inside.

parental care allows a female cowbird to produce a large number of eggs, in some cases as many as thirty in a single reproductive season.

The presence of brood parasites decreases the reproductive rate of the host species and favors host individuals that can detect and eject foreign young. Some brood parasites counter this host defense by producing eggs that closely resemble those of the host species. In cuckoos, different subpopulations have different host preferences and egg coloration. Females of each subpopulation lay eggs that closely resemble those of their preferred host.

Brood parasitism also evolved in some bee species. Females of these species lay eggs in the nest of a different bee species. Larvae of the parasitic bee eat food stored by the host for its larvae, as well as the host's larvae.

Parasitoids

Parasitoids are insects that lay their eggs in other insects. Larvae hatch, develop in the host's body, eat its tissue, and eventually kill it. The fire ant–killing phorid flies described in Section 27.1 are an example. As many as 15 percent of all insects may be parasitoids.

Parasitoids reduce the size of a host population in two ways. First, as the parasitoid larvae grow inside their host, they withdraw nutrients and prevent it from reproducing. Second, the presence of these larvae eventually leads to the death of the host.

---

brood parasitism One egg-laying species benefits by having another raise its offspring.
parasitism Relationship in which one species withdraws nutrients from another species, without immediately killing it.
parasitoid An insect that lays eggs in another insect, and whose young devour their host from the inside.

Biological Controls

Parasites and parasitoids are commercially raised and released in target areas as biological control agents. They are promoted as an environmentally friendly alternative to pesticides. Most chemical pesticides kill a wide variety of insects. By contrast, biological control agents usually target a limited number of species. The phorid flies that attack fire ants and the parasitoid wasp in **Figure 27.14** are examples of biological control agents.

For a species to be an effective biological control agent, it must be adapted to take advantage of a specific host species and to survive in that species' habitat. The ideal biological control agent is good at finding hosts, has a population growth rate comparable to the host's, and has offspring that are good at dispersing.

Introducing a species into a community as a biological control does entail some risks. The introduced parasites sometimes go after nontargeted species in addition to, or instead of, those species they were expected to control. For example, parasitoids were introduced to Hawaii to control stink bugs that feed on some Hawaiian crops. Instead, the parasitoids decimated the population of koa bugs, Hawaii's largest native bug. Introduced parasitoids have also been implicated in ongoing declines of many native Hawaiian butterfly and moth populations.

Take-Home Message What are the effects of parasites, brood parasites, and parasitoids?

> Parasites reduce the reproductive rate of host individuals by withdrawing nutrients from them.

> Brood parasites reduce the reproductive rate of hosts by tricking them into caring for young that are not their own.

> Parasitoids reduce the number of host organisms by preventing reproduction and eventually killing the host.

> The array of species in a community depends on physical factors such as climate, biological factors such as which species arrived first, and the frequency of disturbances.
< Links to Mosses 21.3, Lichens 22.6

Successional Change

Species composition of a community can change over time. Species often alter the habitat in ways that allow other species to come in and replace them. We call this type of change ecological succession.

The process of succession starts with the arrival of **pioneer species**, species that are opportunistic colonizers of new or newly vacated habitats. Pioneer species have high dispersal rates, grow and mature fast, and produce many offspring. Later, other species replace the pioneers. Then the replacements are replaced, and so on.

Primary succession is a process that begins when pioneer species colonize a barren habitat with no soil, such as a new volcanic island or land exposed by the retreat of a glacier (Figure 27.15). The earliest pioneers to colonize a new habitat are often mosses and lichens (Sections 21.3 and 22.6), which are small, have a brief life cycle, and can tolerate intense sunlight, extreme temperature changes, and little or no soil. Some hardy annual flowering plants with wind-dispersed seeds are also typical pioneers.

Pioneers help build and improve the soil. In doing so, they often set the stage for their own replacement. Many pioneer species partner with nitrogen-fixing bacteria, so they can grow in nitrogen-poor habitats. Seeds of later species find shelter inside mats of the pioneers. Organic wastes and remains accumulate and, by adding volume and nutrients to soil, this material helps other species take hold. Later successional species often shade and eventually displace earlier ones.

In **secondary succession**, a disturbed area within a community recovers. If improved soil is still present, secondary succession can occur fast. It commonly occurs in abandoned agricultural fields and burned forests.

Factors That Influence Succession

When the concept of ecological succession was first developed in the late 1800s, it was thought to be a predictable and directional process. Physical factors such as climate, altitude, and soil type were considered to be the main determinants of which species appeared in what order during succession. Also by this view, succession culminates in a "climax community," an array of species that will persist over time and will be reconstituted in the event of a disturbance.

Ecologists now realize that the species composition of a community changes frequently, and in unpredictable ways. Communities do not journey along a well-worn path to some predetermined climax state.

Random events can determine the order in which species arrive in a habitat and thus affect the course of succession. The arrival of a certain species may make it easier or more difficult for others to take hold. As an example, surf grass can grow along a shoreline only if algae have

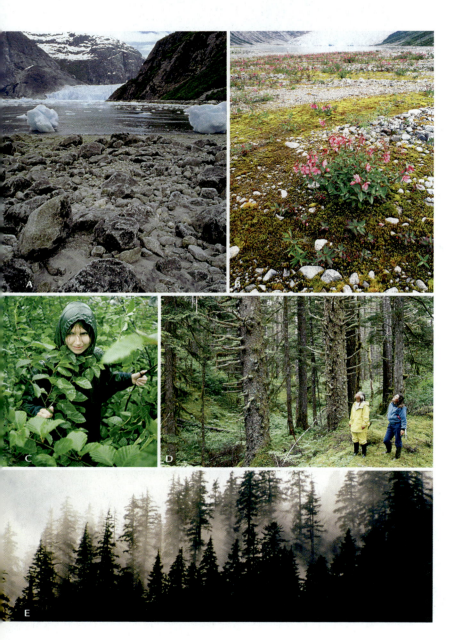

Figure 27.15 In Alaska's Glacier Bay region, one observed pathway of primary succession. **A** As a glacier retreats, meltwater leaches minerals from the rocks and gravel left behind. **B** Pioneer species include lichens, mosses, and some flowering plants such as mountain avens, which associate with nitrogen-fixing bacteria. Within 20 years, alder, cottonwood, and willow seedlings take hold. Alders also have nitrogen-fixing symbionts. **C** Within 50 years, the alders form dense, mature thickets. **D** After 80 years, western hemlock and spruce crowd out alders. **E** In areas deglaciated for more than a century, tall Sitka spruce dominate.

Figure 27.16 A natural laboratory for succession after the 1980 Mount Saint Helens eruption. **A** The community at the base of this volcano in the Cascades was destroyed. **B** In less than a decade, pioneer species had sprouted. **C** Twelve years later, seedlings of a dominant species, Douglas firs, took hold.

already colonized that area. The algae act as an anchoring site for the grass. In contrast, when sagebrush gets established in a dry habitat, chemicals it secretes into the soil keep most other plants out.

Ecologists had an opportunity to investigate these factors after the 1980 eruption of Mount Saint Helens leveled about 600 square kilometers (235 square miles) of forest in Washington State (Figure 27.16). Ecologists recorded the natural pattern of colonization and carried out experiments in plots inside the blast zone. The results showed that the presence of some pioneers helped other, later-arriving plants become established. Other pioneers kept the same late arrivals out.

Disturbances also influence species composition in communities. According to the **intermediate disturbance hypothesis**, species richness is greatest in communities where disturbances are moderate in their intensity or frequency. In such habitats, there is enough time for new colonists to arrive and become established but not enough for competitive exclusion to cause extinctions:

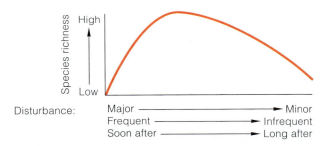

In short, the modern view of succession holds that the species composition of a community is affected by (1) physical factors such as soil and climate, (2) chance events such as the order in which species arrive, and (3) the extent of disturbances in the habitat. Because the second and third factors often vary even between two geographically close regions, it is generally difficult to predict exactly what any given community will look like in the future.

---

intermediate disturbance hypothesis Species richness is greatest in communities where disturbances are moderate in their intensity or frequency.
pioneer species Species that can colonize a new habitat.
primary succession A new community becomes established in an area where there was previously no soil.
secondary succession A new community develops in a site where the soil that supported an old community remains.

Take-Home Message **What is ecological succession?**

❯ Succession is a process in which one array of species replaces another over time. It can occur in a barren habitat (primary succession), or a region where a community previously existed (secondary succession).

❯ Physical factors affect succession, but so do species interactions and disturbances. As a result, the course that succession will take in any particular community is difficult to predict.

> ❯ The loss or addition of even one species may destabilize the number and abundances of species in a community.
> ❯ Some species adapted to being disturbed are at a competitive disadvantage if the disturbance does not occur.

The Role of Keystone Species

A **keystone species** is a species that has a disproportionately large effect on a community relative to its abundance. Robert Paine was the first to describe the effect of a keystone species after his experiments on the rocky shores of California's coast. Species in the rocky intertidal zone withstand pounding surf by clinging to rocks. A rock to cling to is a limiting factor. Paine set up control plots with the sea star *Pisaster ochraceus* and its main prey—chitons, limpets, barnacles, and mussels. Then he removed all sea stars from his experimental plots.

Sea stars prey mainly on mussels. With sea stars gone from experimental plots, mussels took over, crowding out seven other species of invertebrates. Paine concluded that sea stars are a keystone species. They normally keep the number of prey species in the intertidal zone high by preventing competitive exclusion by mussels.

The impact of a keystone species can vary between habitats that differ in their species arrays. For example, Jane Lubchenco discovered that periwinkle snails (Figure 27.17A) can increase or decrease the diversity of algal species, depending on the habitat. In tidepools, periwinkles prefer to eat a certain alga that can outgrow other algal species. By keeping the best algal competitor in check, periwinkles help other, less competitive algal species survive (Figure 27.17B). On rocks of the lower intertidal zone, a tough red alga dominates. Here, periwinkles preferentially graze on competitively weaker algae, giving the dominant competitor an added advantage (Figure 27.17C). Thus, periwinkles promote species diversity in tidepools but reduce it on intertidal rocks.

Keystone species need not be predators. For example, beavers can be a keystone species. These large rodents cut down trees by gnawing through their trunks. The beaver then uses the felled trees to build a dam, thus creating a deep pool where a shallow stream would otherwise exist. By altering the physical conditions in a section of the stream, the beaver affects the types of fish and aquatic invertebrates that can live there.

Adapting to Disturbance

In communities repeatedly subjected to a particular type of disturbance, individuals that withstand or benefit from that disturbance have a selective advantage. For example, some plants in areas subject to periodic fires produce seeds that germinate only after a fire. Seedlings of these plants benefit from the lack of competition for resources. Other plants have an ability to resprout quickly after a fire (Figure 27.18).

Because different species respond differently to fire, the frequency of fire affects competitive interactions. When naturally occurring fires are suppressed, species adapted to fire lose their competitive edge. They can be overgrown by species that devote all of their energy to

A Periwinkle snails (*Littorina littorea*) are a keystone species that affects the number of algal species in different ways in different tidal habitats.

B In tidepools, periwinkles eat the alga that is the most effective competitor and thus promote survival of less competitive algal species that would otherwise be overgrown.

C On rocks exposed only at high tide, periwinkles eat algae that are less effective competitors, thus enhancing the growth of the most competitive species.

Figure 27.17 Effect of competition and predation in an intertidal zone.

Figure 27.18 Adapted to disturbance. Some woody shrubs, such as this toyon, resprout from their roots after a fire. In the absence of occasional fire, toyons are outcompeted and displaced by species that grow faster but are less fire resistant.

A Kudzu native to Asia is overgrowing trees across the southeastern United States.

B Gypsy moths native to Europe and Asia feed on oaks through much of the United States.

C Nutrias native to South America are abundant in freshwater marshes of the Gulf States.

Figure 27.19 Three exotic species that are altering natural communities in the United States. To learn more about invasive species in the United States, visit the National Invasive Species Information Center online at www.invasivespeciesinfo.gov.

growing shoots, rather than storing some in underground parts in readiness for a spurt of regrowth.

Some species are especially sensitive to disturbances to the environment. These **indicator species** are the first to do poorly when conditions change, so they can provide an early warning of environmental degradation. For example, a decline in a trout population can be an early sign of problems in a stream, because trout are highly sensitive to pollutants and cannot tolerate low oxygen levels.

Species Introductions

The arrival of a new species in a community can also cause dramatic changes. When you hear someone speaking enthusiastically about exotic species, you can safely bet the speaker is not an ecologist. An **exotic species** is a resident of an established community that dispersed from its home range and became established elsewhere. Unlike most imports, which never take hold outside the home range, an exotic species becomes a permanent member of its new community.

More than 4,500 exotic species have become established in the United States. An estimated 25 percent of Florida's plant and animal species are exotics. In Hawaii, 45 percent are exotic. Some species were brought in for use as food crops, to brighten gardens, or to provide textiles. Other species, including red imported fire ants, arrived as stowaways along with cargo from distant regions.

One of the most notorious exotic species is a vine called kudzu (*Pueraria lobata*). Native to Asia, it was introduced to the American Southeast as a food for grazers and to control erosion. It quickly became an invasive weed. Kudzu overgrows trees, telephone poles, houses, and almost everything else in its path (Figure 27.19A).

Gypsy moths (*Lymantria dispar*) are an exotic species native to Europe and Asia. They entered the northeastern United States in the mid-1700s and now range into the Southeast, Midwest, and Canada. Gypsy moth caterpillars (Figure 27.19B) preferentially feed on oaks. Loss of leaves to gypsy moths weakens trees, making them less efficient competitors and more susceptible to parasites.

Large semiaquatic rodents called nutrias (*Myocastor coypus*) were imported from South America for their fur, then released into the wild in the 1940s. In states bordering the Gulf of Mexico, nutrias now thrive in freshwater marshes (Figure 27.19C). Their voracious appetite threatens the native vegetation. In addition, their burrowing contributes to marsh erosion and damages levees, increasing the risk of flooding.

---

exotic species A species that evolved in one community and later became established in a different one.
indicator species Species that is especially sensitive to disturbance and can be monitored to assess the health of a habitat.
keystone species A species that has a disproportionately large effect on community structure.

Take-Home Message What types of changes alter community structure?

❯ Removing a keystone species can alter the diversity of species in a community.

❯ Changing the frequency of a disturbance can favor some species over others.

❯ Introducing an exotic species can threaten native species.

❯ The richness and relative abundances of species differ from one region of the world to another.

❮ Link to Biogeography 16.2

Biogeography is the study of how species are distributed in the natural world (Section 16.2). **Species richness**, the number of species in an area, correlates with differences in sunlight, temperature, rainfall, and other factors that vary with latitude, elevation, or water depth. The history of a habitat also affects species richness.

Latitudinal Patterns

Perhaps the most striking pattern of species richness corresponds with distance from the equator. For most major plants and animal groups, the number of species is greatest in the tropics and declines from the equator to the poles. Figure 27.20 illustrates one example of this pattern.

Consider just a few factors that help bring about this latitudinal pattern and maintain it. First, tropical latitudes intercept more intense sunlight and receive more rainfall, and their growing season is longer. Thus resource availability tends to be greater and more reliable in the tropics than elsewhere. One result is specialized interrelationships not possible where species are active for shorter periods. Second, tropical communities have been evolving for a long time. Some temperate communities did not start forming until the end of the last ice age. Third, species richness may be self-reinforcing. The number of tree species in tropical forests is much greater than in comparable forests at higher latitudes. Where more plant species compete and coexist, more species of herbivores can coexist, partly because no single herbivore species can overcome all chemical defenses of all plants. In addition, more predators and parasites can evolve in response to more kinds of prey and hosts. The same principles apply to tropical reefs.

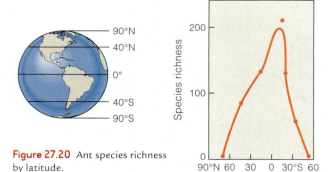

Figure 27.20 Ant species richness by latitude.

❯❯ **Figure It Out** Houston, Texas, is at 30 degrees north latitude and Minneapolis, Minnesota, is at 45 degrees north latitude. Where would you expect to find more ant species?

Answer: Houston

Island Patterns

Islands are natural laboratories for population studies. They have also been laboratories for community studies. For instance, in the mid-1960s volcanic eruptions formed a new island 33 kilometers (21 miles) from the coast of Iceland. The island was named Surtsey (Figure 27.21A). Bacteria and fungi were early colonists. The first vascular plant became established on the island in 1965. Mosses appeared two years later and thrived (Figure 27.21B). The first lichens were found five years after that. The rate of arrivals of new vascular plants picked up after a seagull colony became established in 1986 (Figure 27.21C).

The number of species on Surtsey will not continue increasing forever. How many species will there be when the number levels off? The **equilibrium model of island biogeography** addresses this question. According to this model, the number of species living on any island reflects a balance between immigration rates for new species and extinction rates for established ones. The distance between an island and a mainland source of colonists

Figure 27.21 Island colonizations. **A** Surtsey, a volcanic island, at the time of its formation. **B** Surtsey in 1983. **C** Graph of the number of vascular plant species found in yearly surveys. Seagulls began nesting on the island in 1986.

affects immigration rates. An island's size affects both immigration rates and extinction rates.

Consider first the **distance effect**: Islands far from a source of colonists receive fewer immigrants than those closer to a source. Most species cannot disperse very far, so they will not turn up far from a mainland.

Species richness is also shaped by the **area effect**: Big islands tend to support more species than small ones. More colonists will happen upon a larger island simply by virtue of its size. Also, big islands are more likely to offer a variety of habitats, such as high and low elevations. These options make it more likely that a new arrival will find a suitable habitat. Finally, big islands can support larger populations of species than small islands. The larger a population, the less likely it is to become locally extinct as the result of some random event.

Figure 27.22 illustrates how interactions between the distance effect and the area effect can influence the equilibrium number of species on islands.

Robert H. MacArthur and Edward O. Wilson first developed the equilibrium model of island biogeography in the late 1960s. Since then the model has been modified and its use has been expanded to help scientists think about habitat islands—natural settings surrounded by a "sea" of habitat that has been disturbed by humans. Many parks and wildlife preserves fit this description. Island-based models can help estimate the size of an area that must be set aside as a protected reserve to ensure survival of a species or a community.

---

area effect Larger islands have more species than small ones.
distance effect Islands close to a mainland have more species than those farther away.
equilibrium model of island biogeography Model that predicts the number of species on an island based on the island's area and distance from the mainland.
species richness Of an area, number of species.

Fighting Foreign Fire Ants (revisited)

Now that you have learned a bit about how communities and ecosystems work, let's take a second look at those red imported fire ants (RIFAs) we discussed at the beginning of this chapter. These ants are an introduced predator that is having a negative effect on competing ants and on prey species. RIFAs did not evolve in North America, so there are few predators, parasites, or pathogens to hold them in check. Freed from these natural restrictions, RIFAs have an easier time outcompeting native ants. As RIFA populations soar, populations of organisms that are their prey decline.

Global climate change is expected to help RIFAs extend their range. For example, RIFAs require mild winters and relatively high rainfall to do well. With winters becoming milder, the ants are expected to move farther north and ascend to higher altitudes. If current climate predictions are correct, the amount of habitat available to red imported fire ants in the United States is expected to increase by about 20 percent before the close of this century.

How Would You Vote? Increased global trade and faster ships are contributing to a rise in the rate of species introductions into North America. Faster ships mean shorter trips, increasing the likelihood that pests will survive a voyage. Wood-eating insects from Asia turn up with alarming frequency in the wood of packing crates and spools for steel wire. Some of these insects, such as the Asian long-horned beetle, now pose a serious threat to North America's forests. Currently, only a small fraction of the crates imported into the United States are inspected to see if they contain exotic species. Should inspections increase? See CengageNow for details, then vote online (cengagenow.com).

Take-Home Message **What are some biogeographic patterns in species richness?**

> Generally, species richness is highest in the tropics and lowest at the poles. Tropical habitats have conditions that more species can tolerate, and tropical communities have often been evolving for longer than temperate ones.

> When a new island forms, species richness rises over time and then levels off. The size of an island and its distance from a colonizing source influence its species richness.

Figure 27.22 Animated Island biodiversity patterns.

Distance effect: Species richness on islands of a given size declines as distance from a source of colonists rises. *Green* circles are values for islands less than 300 kilometers from the colonizing source. *Orange* triangles are values for islands more than 300 kilometers (190 miles) from a source of colonists.

Area effect: Among islands the same distance from a source of colonists, larger islands tend to support more species than smaller ones.

›› Figure It Out Which is likely to have more species, a 100-km² island more than 300 km from a colonizing source or a 500-km² island less than 300 km from a colonist source?

Answer: The 500-km² island

Summary

 Section 27.1 All the species that live in an area constitute a **community**. Introduction of a new species to a community can have negative effects on the species that evolved in that community.

 Section 27.2 Each species occupies a certain **habitat** characterized by physical and chemical features and by the array of other species living in it. Species interactions affect community structure. **Commensalism** is a common interaction in which one species benefits and the other is unaffected. A **symbiosis** is an interaction in which one species lives in or on another.

 Section 27.3 In a **mutualism**, both species benefit from an interaction. Some mutualists cannot complete their life cycle without the interaction. Mutualists who maximize their own benefits while limiting the cost of cooperating are at a selective advantage.

 Section 27.4 A species' roles and requirements define its unique **ecological niche**. With **interspecific competition**, two species with similar resource requirements are harmed by one another's presence. When one competitor drives the other to local extinction, we call the process **competitive exclusion**. Competing species with similar requirements become less similar when directional selection causes **character displacement**. This change in traits allows **resource partitioning**.

 Section 27.5 Predation benefits a predator at the expense of the prey it captures, kills, and eats. Predators and their prey exert selective pressures on one another. Evolved prey defenses include camouflage and **mimicry**. Prey have traits that allow them to overcome prey defenses. With **herbivory**, an animal eats plant or algae. Plants have traits that discourage herbivory or that allow a quick recovery from it.

 Section 27.6 Parasitism involves feeding on a host without killing it. **Brood parasites** steal parental care from another species. **Parasitoids** are insects that lay eggs on a host insect, then larvae devour the host.

 Section 27.7 Ecological succession is the sequential replacement of one array of species by another over time. **Primary succession** happens in new habitats. **Secondary succession** occurs in disturbed ones. The first species of a community are **pioneer species**. Their presence may help, hinder, or not affect later colonists.

Modern models of succession emphasize the unpredictability of the outcome as a result of chance events, ongoing changes, and disturbance. The **intermediate disturbance** hypothesis predicts that a moderate level of disturbance keeps a community diverse.

 Section 27.8 Keystone species are especially important in determining the composition of a community. An **indicator species** can provide information about the health of the environment. The removal of a keystone species or introduction of an **exotic species** can alter community structure.

 Section 27.9 Species richness, the number of species in a given area, varies with latitude. The **equilibrium model of island biogeography** predicts the number of species that an island will sustain based on the **area effect** and the **distance effect**. Scientists can use this model to predict the number of species that habitat islands such as parks can sustain.

Self-Quiz   Answers in Appendix III

1. The type of place where a species typically lives is called its _____ .
   a. niche   c. community
   b. habitat   d. population

2. Which cannot be a symbiosis?
   a. mutualism   c. commensalism
   b. parasitism   d. interspecific competition

3. Lizards that eat flies they catch on the ground and birds that catch and eat flies in the air are engaged in _____ competition.
   a. exploitative   d. interspecific
   b. interference   e. both a and d
   c. intraspecific   f. both b and c

4. _____ can lead to resource partitioning.
   a. Mutualism   c. Commensalism
   b. Parasitism   d. Interspecific competition

5. Match the terms with the most suitable descriptions.
   ___ mutualism   a. one free-living species feeds
   ___ parasitism   on another and usually kills it
   ___ commensalism   b. two species interact and both
   ___ predation   benefit by the interaction
   ___ interspecific   c. two species interact and one
        competition   benefits while the other is
                             neither helped nor harmed
                          d. one species feeds on another
                             but usually does not kill it
                          e. two species access a resource

6. A tick is a(n) _____ .
   a. brood parasite   b. ectoparasite   c. endoparasite

7. By a currently favored hypothesis, species richness of a community is greatest between physical disturbances of _____ intensity or frequency.
   a. low   c. high
   b. intermediate   d. variable

Data Analysis Activities

Biological Control of Fire Ants Ant-decapitating phorid flies are just one of the biological control agents used to battle imported fire ants. Researchers have also enlisted the help of a fungal parasite that infects the ants and slows their production of off-spring. An infected colony dwindles in numbers and eventually dies out.

Are these biological controls useful against imported fire ants? To find out, USDA scientists treated infested areas with either traditional pesticides or pesticides plus biological controls (both flies and the parasite). The scientists left some plots untreated as controls. Figure 27.23 shows the results.

1. How did population size in the control plots change during the first four months of the study?
2. How did population size in the two types of treated plots change during this same interval?
3. If this study had ended after the first year, would you conclude that biological controls had a major effect?
4. How did the two types of treatment (pesticide alone versus pesticide plus biological controls) differ in their longer-term effects? Which is most effective?

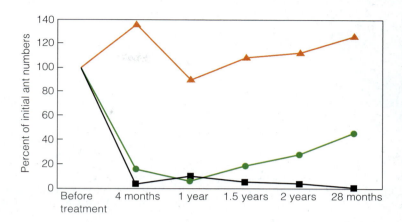

Figure 27.23 Effects of two methods of controlling red imported fire ants. The graph shows the numbers of red imported fire ants over a 28-month period. *Orange* triangles represent untreated control plots. *Green* circles are plots treated with pesticides alone. *Black* squares are plots treated with pesticide and biological control agents (phorid flies and a fungal parasite).

---

8. _____ species are the first to colonize a new habitat.

9. Growth of a forest in an abandoned corn field is an example of _____ .
    a. primary succession   c. secondary succession
    b. resource partitioning   d. competitive exclusion

10. Species richness is greatest in communities _____ .
    a. near the equator   c. near the poles
    b. in temperate regions   d. that recently formed

11. If you remove a species from a community, the population size of its main _____ is likely to increase.
    a. parasite   b. competitor   c. predator

12. Mammals are least likely to be _____ .
    a. mutualists   c. brood parasites
    b. commensal   d. predators

13. Herbivory benefits _____ .
    a. the herbivore   c. both a and b
    b. the plant   d. neither a nor b

14. Match the terms with the most suitable descriptions.
    ___ area effect   a. greatly affects other species
    ___ pioneer   b. first species established in a
        species   new habitat
    ___ indicator   c. more species on large islands
        species   than small ones at same distance
    ___ keystone   from the source of colonists
        species   d. species that is especially sensitive
    ___ exotic   to changes in the environment
        species   e. allows competitors to coexist
    ___ resource   f. often outcompete, displace native
        partitioning   species of established community

Additional questions are available on **CENGAGENOW.**

Critical Thinking

1. With antibiotic resistance rising, researchers are looking for ways to reduce use of antibiotics. Some cattle once fed antibiotic-laced food now get probiotic feed instead, with cultured bacteria that can establish or bolster populations of helpful bacteria in the animal's gut. The idea is that if a large population of beneficial bacteria is in place, then the harmful bacteria cannot become established or thrive. Which ecological principle is guiding this research?

2. Phasmids are herbivorous insects that mimic leaves or sticks, as shown at the *right*. Most rest during the day, and feed at night. If they do move in daytime, they do so very slowly. If disturbed, a phasmid will fall to the ground and lie motionless. Speculate on selective pressures that could have shaped phasmid morphology and behavior. Suggest an experiment to test a hypothesis about how its appearance or behavior may be adaptive.

Animations and Interactions on **CENGAGENOW:**
› Competitive exclusion; Island biodiversity patterns.

<< Links to Earlier Concepts

This chapter returns to the concept of a one-way flow of energy in nature (Sections 1.3, 5.2) and looks at nitrogen-fixing microbes, soil erosion, and leaching (19.7) in the context of nitrogen cycling. You will be reminded of properties of chemical bonds (2.4) and the pathways of photosynthesis (6.4) and aerobic respiration (7.2). You will see how slow movements of Earth's crust (16.7) influence the cycling of some nutrients.

Key Concepts

Organization of Ecosystems
A one-way flow of energy and the cycling of raw materials among species maintain an ecosystem. Nutrients and energy are transferred in a stepwise fashion through food chains that interconnect as complex food webs.

Biogeochemical Cycles
In a biogeochemical cycle, a nutrient moves relatively slowly among its environmental reservoirs. The reservoirs may include air, water, and rocks. Nutrients move more quickly into, through, and out of food webs.

28 Ecosystems

28.1 Too Much of a Good Thing

All organisms require certain elements to build their bodies and carry out metabolic processes. Phosphorus is one of these essential nutrients. It is part of ATP, phospholipids, nucleic acids, and other biological compounds. Plants meet their phosphorus requirement by taking up dissolved phosphates in soil water. Animals get phosphorus by eating plants or by eating other animals. Thus phosphorus taken up from the environment passes from one organism to another. When a plant or animal dies and decays, the phosphorus in its body returns to the environment. As a result, phosphorus moves continually from the environment, through organisms, and back to the environment.

All organisms in a community together with their environment constitute an ecosystem, and the study of nutrient cycles in ecosystems is one aspect of the science of ecology. Ecology is not the same as environmentalism, which is advocacy for protection of the environment. However, environmentalists often point to the results of ecological studies when drawing attention to a particular environmental concern.

As one example, environmentalists in many parts of the United States are pushing for laws that ban high-phosphorus laundry detergents, dish detergents, and lawn fertilizers. Their desire to reduce phosphate content of these products is based on evidence that phosphorus enrichment of water can disrupt aquatic ecosystems (Figure 28.1).

The addition of nutrients to an aquatic ecosystem is called **eutrophication**. It can occur slowly by natural processes, or fast as a result of human actions. Phosphorus is often a limiting factor for aquatic producers. A sudden addition of phosphorus removes this limitation and allows a population explosion of algae and cyanobacteria. The resulting algal bloom (Section 20.1) clouds water and threatens other aquatic species.

We began using phosphate-rich products when we had no idea of their effects beyond cleaner homes and greener lawns. Today, we confront the simple truth that the daily actions of

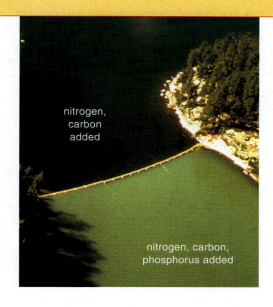

Figure 28.1 *Above*, Results of a field experiment in which different types of nutrients were added to two artificially separated portions of a lake. Including phosphorus in the mix (lower region) caused rapid overgrowth of algae.

Opposite page, Graphic from a Washington State Ecology Department campaign designed to remind people that lawn fertilizer often ends up in lakes.

millions of individuals can disrupt nutrient cycles that have been operating since long before humans existed. Our species has a unique capacity to shape the environment to our liking. In doing so, we have become major players in the global flows of energy and nutrients even before we fully comprehend how ecosystems work. Decisions we make today about environmental issues are likely to shape the quality of life and the environment far into the future.

---

eutrophication Nutrient enrichment of an aquatic ecosystem.

The Water Cycle
Most of Earth's water is in its oceans. Only a tiny fraction is fresh water. Evaporation, condensation, precipitation, and flow of rivers and streams move water. Water also plays a role in other nutrient cycles by transporting soluble forms of those nutrients.

The Carbon Cycle
Most of Earth's carbon is tied up in rocks, but organisms take carbon up from water or the air. Carbon dioxide is one of the atmospheric greenhouse gases that help keep Earth's surface warm. Increasing carbon dioxide in the air is the most likely cause of climate change.

Nitrogen and Phosphorus Cycles
Plants take up dissolved forms of nitrogen and phosphorus from soil water. Nitrogen is abundant in air, but only certain bacteria can use the gaseous form. Phosphorus has no major gaseous form; most of it is in rocks.

> All energy captured by the producers in an ecosystem eventually returns to the environment as heat.
> Nutrients taken up by producers are returned to the environment by decomposers, then taken up again.
❮ Links to Acquiring energy and nutrients 1.3, Energy transfers 5.2, Photosynthesis 6.4, Modes of nutrition 19.5

Each ecosystem is an array of organisms and a physical environment, all interacting through a one-way flow of energy and a cycling of nutrients (Figure 28.2). It is an open system, because it requires ongoing inputs of energy to persist. Most ecosystems also gain nutrients from and lose nutrients to other ecosystems.

Primary Producers and Production

An ecosystem runs on the energy captured by **primary producers**. These autotrophs get energy and carbon from a nonliving source to build organic compounds. In most ecosystems, photoautotrophs such as plants, algae, and photosynthetic bacteria are the primary producers. In some dark environments such as deep-sea vent ecosystems, chemoautrophs fill this role.

Primary production is the rate at which producers capture and store energy. Daylength, temperature, and the availability of nutrients, including phosphates, affect producer growth and so influence primary production. As a result, primary production can vary seasonally within an ecosystem and also differs among ecosystems. Per unit area, primary production on land tends to be higher than that in the oceans. However, because oceans cover about 70 percent of Earth's surface, marine producers contribute nearly half of the global net primary production.

The Roles of Consumers

As Section 1.3 explained, **consumers** get energy and carbon by feeding on tissues and remains of producers and one another. We describe consumers by their diets. Herbivores eat plants. Carnivores eat the flesh of animals. Parasites live inside or on a living host and feed on its tissues. Omnivores eat both animals and plants. **Detritivores** such as earthworms eat small bits of decaying organic matter, or detritus. **Decomposers** feed on wastes and remains, breaking them down into inorganic building blocks. Bacteria, archaeans, and fungi serve as decomposers.

---

consumer Organism that obtains energy and carbon by feeding on tissues, wastes, or remains of other organisms.
decomposer Organism that feeds on biological remains and breaks organic material down into its inorganic subunits.
detritivore Consumer that feed on small bits of organic material.
primary producer An organism that obtains energy and nutrients from inorganic sources; an autotroph.
primary production The rate at which an ecosystem's producers capture and store energy.

Figure 28.2 Animated One-way flow of energy (*yellow* arrows) and nutrient cycling (*blue* arrows) in the most common type of ecosystem. All light energy that enters the system eventually returns to the environment as heat energy that is not reused. By contrast, nutrients are continually recycled.

Energy Flow and Nutrient Cycling

Energy captured by producers is converted to bond energy in organic molecules. This energy is released by metabolic reactions that give off heat. The flow of energy through living organisms is a one-way process because heat energy is not recycled—producers do not use heat energy to form chemical bonds.

In contrast, nutrients are cycled within an ecosystem. Producers take up hydrogen, oxygen, and carbon from inorganic sources such as the air and water. They also take up dissolved nitrogen, phosphorus, and other necessary minerals. Nutrients move from producers into the consumers who eat them. Decomposition returns nutrients to the environment, from which producers take them up again.

Take-Home Message What factors characterize ecosystems?

> An ecosystem is a community of autrophic producers and heterotrophic consumers that interact with one another and the environment by a one-way flow of energy and a cycling of materials.

> Ecosystems vary in their primary production. Marine ecosystems generally have a lower primary production per unit area than land ecosystems.

› Food chains describe how energy and materials are transferred from one organism to another. Inefficient transfers constrain food chain length.

‹ Links to Energy transfers 5.2, Lignin 4.11, Ectotherms and endotherms 24.6

The hierarchy of feeding relationships within an ecosystem is the ecosystem's trophic structure (*troph* means nourishment). All organisms at the same **trophic level** are the same number of transfers away from the energy input into that system. The primary producers are at the first trophic level. The primary consumers that eat them are at the second trophic level, and so on.

A **food chain** is a sequence of steps by which some energy captured by primary producers is transferred to higher trophic levels. For example, in one tallgrass prairie food chain, energy flows from grasses to grasshoppers, to sparrows, and finally to coyotes (Figure 28.3). At the first trophic level in this food chain, grasses and other plants are the primary producers. At the second trophic level, grasshoppers are primary consumers. At the third trophic level, sparrows that eat grasshoppers are second-level consumers. At the fourth trophic level, coyotes that eat sparrows are third-level consumers.

Energy captured by producers usually passes through no more than four or five trophic levels. Even in ecosystems with many species, the number of participants in each food chain is limited. The length of food chains is constrained by the inefficiency of energy transfers. Only 5 to 30 percent of the energy in tissues of an organism at one trophic level ends up in tissues of an organism at the next trophic level.

Why isn't all the energy an organism takes in available to the organism that eats it? First, not all energy that an organism takes in is used to build body parts. Energy an organism invests in producing offspring or loses as metabolic heat is not available to a consumer. The more heat an organism produces, the less energy goes into building tissues. Thus, food chains that involve ectotherms are typically longer than those that involve endotherms.

Second, energy in some body parts cannot be accessed by most consumers. For example, few herbivores have the ability to break down the lignin that reinforces the bodies of most woody plants. Similarly, many animals have some energy tied up in a difficult-to-digest internal or external skeleton. Hair, feathers, and fur also resist digestion. A coyote that eats a sparrow does not benefit from energy that the bird invested in building bones and feathers.

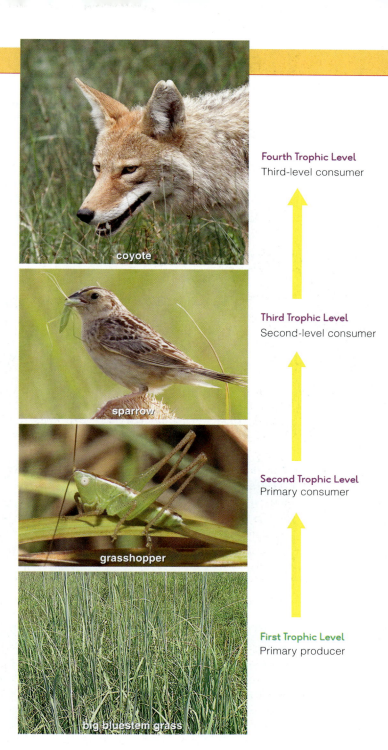

Fourth Trophic Level
Third-level consumer

Third Trophic Level
Second-level consumer

Second Trophic Level
Primary consumer

First Trophic Level
Primary producer

Figure 28.3 One food chain in a Kansas tallgrass prairie.

---

food chain Description of who eats whom in one path of energy in an ecosystem.
trophic level Position of an organism in a food chain.

Take-Home Message What is a food chain?

› A food chain is a description of a series of trophic interactions among the members of an ecosystem.

› Primary producers are the first trophic level of a food chain. Energy is transferred to primary consumers, and then to higher-level consumers.

› The inefficiency of energy transfers limits the number of steps in food chains. Only a small fraction of the energy that an organism takes in can be accessed by an organism that consumes it.

❯ Each food web consists of cross-connecting food chains. Its structure reflects environmental constraints and the inefficiency of energy transfers among trophic levels.

An organism that participates in one food chain usually has a role in many others as well. The food chains of an ecosystem cross-connect as a **food web**. Figure 28.4 shows a small sampling of the participants in an arctic food web.

Nearly all food webs include two types of interconnecting food chains. In **grazing food chains**, energy stored in producer tissues flows to herbivores, which often are relatively large animals such as mammals. In a **detrital food chain**, energy in producers flows to detritivores, which tend to be smaller animals such as worms or insects.

In most land ecosystems, the bulk of the energy that gets stored in producer tissues moves through detrital

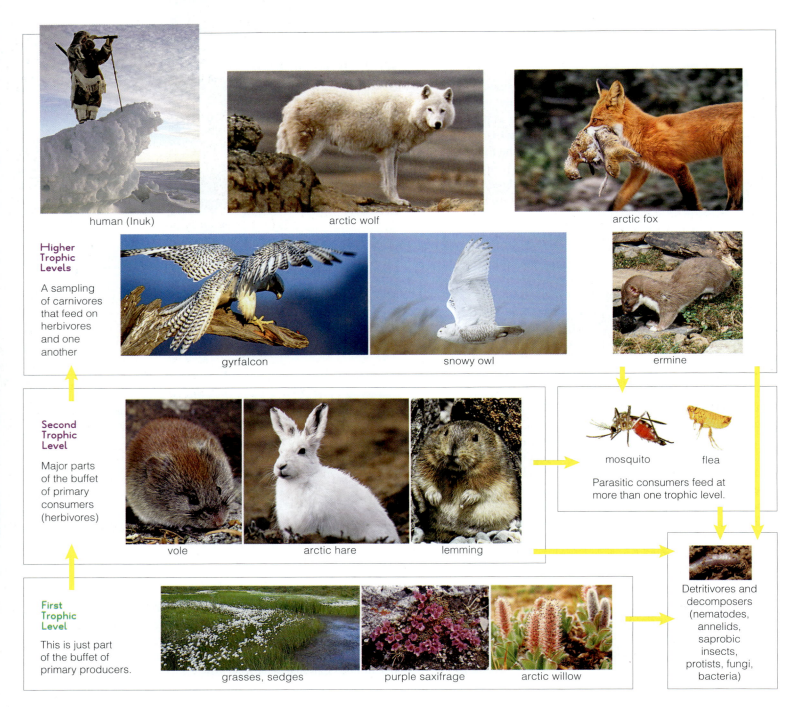

Higher Trophic Levels

A sampling of carnivores that feed on herbivores and one another

human (Inuk) arctic wolf arctic fox

gyrfalcon snowy owl ermine

Second Trophic Level

Major parts of the buffet of primary consumers (herbivores)

mosquito flea

Parasitic consumers feed at more than one trophic level.

vole arctic hare lemming

First Trophic Level

This is just part of the buffet of primary producers.

grasses, sedges purple saxifrage arctic willow

Detritivores and decomposers (nematodes, annelids, saprobic insects, protists, fungi, bacteria)

Figure 28.4 Animated Some organisms in an Arctic food web. Arrows point from eater to eaten.

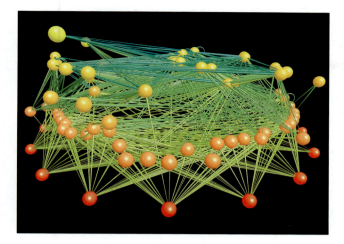

Figure 28.5 Computer model for a land food web in East River Valley, Colorado. Balls signify species. Their colors identify trophic levels, with producers (coded *red*) at the bottom and top predators (*yellow*) at top. The connecting lines thicken, as they go from an eaten species to the eater.

food chains. For example, in an arctic ecosystem, voles, lemmings, and hares eat some living plant parts. However, far more plant material ends up as detritus. Bits of dead plant material sustain detritivores such as nematodes and soil-dwelling insects.

Understanding food webs helps ecologists predict how ecosystems will respond to change. Neo Martinez and his colleagues constructed the food web diagram shown in Figure 28.5. By comparing different food webs, Martinez realized that trophic interactions connect species more closely than people thought. On average, each species in a food web was typically two links away from all other species. Ninety-five percent of species were within three links of one another, even in large communities with many species. As Martinez concluded in his paper on these findings, "Everything is linked to everything else." He cautioned that the extinction of any species in a food web has a potential impact on many other species.

---

detrital food chain Food chain in which energy is transferred directly from producers to detritivores.
food web Set of cross-connecting food chains.
grazing food chain Food chain in which energy is transferred from producers to grazers (herbivores).

Take-Home Message What have ecologists learned about food webs?

> Two types of food chains connect in most food webs. Tissues of living producers are the base for grazing food chains. Producer remains are the base for detrital food chains.

> Even in complex ecosystems, trophic interactions link each species in a food web with many others.

> Ecological pyramid diagrams dramatically illustrate the inefficiency of transfers between trophic levels.

A food web diagram is one way of depicting the trophic interactions among species in an ecosystem. Ecological pyramid diagrams are another. A biomass pyramid shows the amount of organic material in the bodies of organisms at each trophic level at a specific time. An energy pyramid shows the amount of energy that flows through each trophic level in a given interval. Figure 28.6 shows ecological pyramids for one freshwater spring ecosystem in Florida.

Most commonly, primary producers make up most of the biomass in a pyramid, and top carnivores make up very little. The Florida ecosystem has lots of aquatic plants but very few gars (a top predator in this ecosystem). Similarly, if you walk through a prairie, you will see more grams of grass than grams of coyote.

An energy pyramid is always broadest at the bottom. This is why people promote a vegetarian diet by touting the ecological benefits of "eating lower on the food chain." They are referring to the energy losses in transfers between plants, livestock, and humans. When a person eats a plant food, he or she gets most of the calories in that food. When the plant food is used to grow livestock, only a small percentage of the food's calories ends up in meat a person can eat. Thus, feeding a population of meat-eaters requires far greater crop production than sustaining a population of vegetarians.

A Biomass pyramid (grams per square meter)

top carnivores (gar and bass) — 1.5
carnivores (smaller fishes, invertebrates) — 11
herbivores (plant-eating fishes, invertebrates, turtles) — 37
producers (algae and aquatic plants) — 809
5 detritivores (crayfish) and decomposers (bacteria)

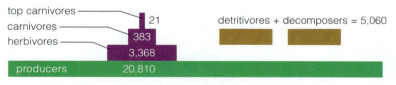

B Energy flow pyramid (kilocalories per square meter per year)

top carnivores — 21
carnivores — 383
herbivores — 3,368
producers — 20,810
detritivores + decomposers = 5,060

Figure 28.6 Ecological pyramids for Silver Springs, an aquatic ecosystem in Florida.

Take-Home Message What are ecological pyramids?

> Ecological pyramids depict the distribution of materials and energy among the trophic levels of an ecosystem.

28.6 Biogeochemical Cycles

> Elements essential to life move between a community and its environment in a biogeochemical cycle.

< Link to Plate tectonics 16.7

In a **biogeochemical cycle**, an essential element moves from one or more environmental reservoirs, through the biological component of an ecosystem, and then back to the reservoirs (Figure 28.7). Depending on the element, environmental reservoirs may include Earth's rocks and sediments, waters, and atmosphere.

Chemical and geologic processes move elements to, from, and among environmental reservoirs. For example, elements locked in rocks can become part of the atmosphere as a result of volcanic activity. As one of Earth's crustal plates moves under another, rocks on the sea floor can be uplifted so they become part of a land mass. On land, the rocks are exposed to the erosive forces of wind and rain. As the rocks are slowly broken down, elements in them enter rivers, and eventually seas. Compared to the movement of elements among organisms of an ecosystem, the movement of elements among nonbiological reservoirs is far slower. Processes such as erosion and uplifting operate over thousands or millions of years.

---

biogeochemical cycle A nutrient moves among environmental reservoirs and into and out of food webs.

Take-Home Message What is a biogeochemical cycle?

> A biogeochemical cycle is the slow movement of a nutrient among environmental reservoirs and into, through, and out of food webs.

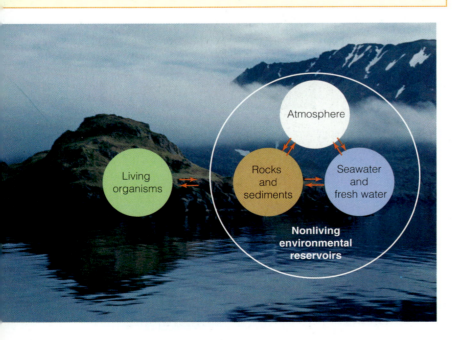

Figure 28.7 Generalized biogeochemical cycle. For all nutrients, the cumulative amount in all environmental reservoirs far exceeds the amount in living organisms.

28.7 The Water Cycle

> Water makes up the bulk of all organisms and serves as a transport medium for many soluble nutrients.

< Link to Properties of water 2.5

How and Where Water Moves

Most of Earth's water (97 percent) is in its oceans (Table 28.1). The **water cycle** moves water from the ocean to the atmosphere, onto land, and back to the oceans (Figure 28.8). Sunlight energy drives the water cycle by causing evaporation, the conversion of liquid water to water vapor. Water vapor that enters the cool upper layers of the atmosphere condenses into droplets, forming clouds. When droplets get large and heavy enough, they fall as precipitation—as rain, snow, or hail.

Oceans cover about 70 percent of Earth's surface, so most rainfall returns water directly to the oceans. On land, we define a **watershed** as an area in which all precipitation drains into a specific waterway. A watershed may be as small as a valley that feeds a stream, or as large as the 5.88 million square kilometers of the Amazon River Basin. The Mississippi River Basin watershed includes 41 percent of the continental United States.

Most precipitation that enters a watershed seeps into the ground. Some of this water remains between soil particles as **soil water**. Plant roots can tap into this water source. Soils differ in water holding capacity, with clay-rich soils holding the most water and sandy soils the least. Water that drains through soil layers often collects in **aquifers**. These natural underground reservoirs consist of porous rock layers that hold water. **Groundwater** is water in soil and aquifers. Water that falls on impermeable rock or on saturated soil becomes **runoff**: It flows over the ground into streams. The flow of groundwater and surface water slowly returns water to oceans.

The movement of water causes the movement of other nutrients. Carbon, nitrogen, and phosphorus all have soluble forms that can be moved from place to place by flowing water. As water trickles through soil, it carries nutrient particles from topsoil into deeper soil layers. As a stream flows over limestone, water slowly dissolves the rock and carries carbonates back to the seas where the limestone formed. On a less natural note, runoff from heavily fertilized lawns and agricultural fields carries dissolved phosphates and nitrates into streams and lakes.

Limited Fresh Water

The vast majority of Earth's water is too salty to drink or irrigate crops. If all Earth's water filled a bathtub, the amount of fresh water that could be used each year without decreasing the overall supply would fill a teaspoon. In addition, most fresh water is frozen as ice.

Groundwater supplies drinking water to about half of the United States population. Water overdrafts from

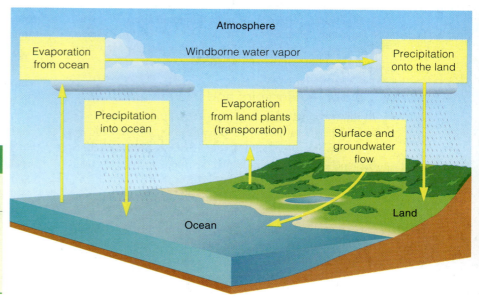

Figure 28.8 **Animated** The water cycle. Water moves from the ocean to the atmosphere, land, and back. The arrows identify processes that move water.

Table 28.1 Environmental Water Reservoirs	
Reservoir	Volume (10³ cubic kilometers)
Ocean	1,370,000
Polar ice, glaciers	29,000
Groundwater	4,000
Lakes, rivers	230
Atmosphere (water vapor)	14

aquifers are now common; water is drawn from aquifers faster than natural processes replenish it. When too much fresh water is withdrawn from a coastal aquifer, salt water moves in and replaces it. Figure 28.9 shows areas of aquifer depletion and saltwater intrusion in the United States.

Overdrafts have reduced the volume of water in the Ogallala aquifer by about half. This aquifer stretches from South Dakota to Texas and supplies irrigation water for about 20 percent of the nation's crops. For the past thirty years, withdrawals have exceeded replenishment by a factor of ten.

Desalinization, the removal of salt from seawater, can help increase freshwater supplies. However, the process is expensive because it is energy intensive. Most likely, desalinization will never be cost-effective for widespread use in the United States. In addition, the process produces mountains of waste salts that must be disposed of.

A more practical approach to meeting freshwater needs involves more efficient use of this limited resource. In the United States, about 80 percent of the water withdrawn for human use ends up irrigating agricultural fields. Using the most efficient irrigation methods available would go a long way toward ensuring that future generations have an adequate and safe water supply.

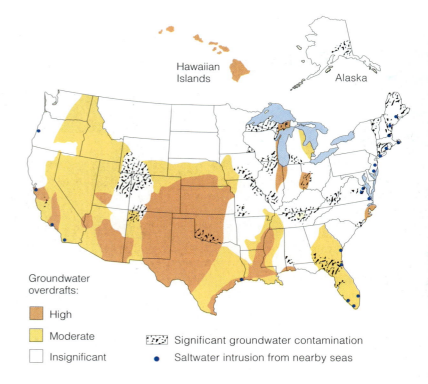

Figure 28.9 Groundwater troubles in the United States.

aquifer Porous rock layer that holds some groundwater.
groundwater Soil water and water in aquifers.
runoff Water that flows over soil into streams.
soil water Water between soil particles.
water cycle Movement of water among Earth's oceans, atmosphere, and the freshwater reservoirs on land.
watershed Land area that drains into a particular stream or river.

Take-Home Message What is the water cycle and how do human activities affect it?

❯ Water moves slowly from the world ocean—the main reservoir—through the atmosphere, onto land, then back to the ocean.

❯ Fresh water makes up only a tiny portion of the global water supply. Excessive water withdrawals threaten many sources of drinking water.

> After water, carbon is the most abundant substance in living things. Most of it is in rocks, but it enters food webs as gaseous carbon dioxide or bicarbonate dissolved in water.
> ⟨ Link to Photosynthesis 6.4

Carbon Reservoirs and Movements

In the **carbon cycle**, natural processes move carbon among Earth's atmosphere, oceans, soils, and into and out of food webs (Figure 28.10). It is an **atmospheric cycle**, a biogeochemical cycle in which a gaseous form of the element plays a significant role.

On land, plants take up carbon dioxide from the atmosphere and incorporate it into their tissues when they carry out photosynthesis ❶. Plants and most other land organisms release carbon dioxide into the atmosphere by the process of aerobic respiration ❷.

On an annual basis, the greatest flow of carbon between nonbiological reservoirs takes place between the ocean and atmosphere. The ocean holds 38,000–40,000 gigatons of dissolved carbon, primarily in the form of bicarbonate (HCO_3^-) and carbonate (CO_3^{2-}) ions. The air holds about 750 gigatons of carbon, mainly in the form of carbon dioxide (CO_2).

Bicarbonate ions form when atmospheric carbon dioxide dissolves in water ❸. Aquatic producers take up bicarbonate and convert it to carbon dioxide for use in photosynthesis. As on land, aquatic organisms carry out aerobic respiration and release carbon dioxide ❹.

The single greatest reservoir of carbon is Earth's rocks, which contain about 50 million gigatons of it. Sedimentary rocks such as limestone form over millions of years when sediments containing carbon-rich shells of marine organisms became compacted ❺. The carbon-rich rocks become part of land ecosystems when movements of crustal plates uplift portions of the sea floor. However, carbon locked in rocks does not move in significant amounts into the air or water and it is not available to producers, so it has little effect on ecosystems.

Plant remains in soil hold about 1,600 gigatons of carbon, about twice that in the atmosphere. Fossil fuels such

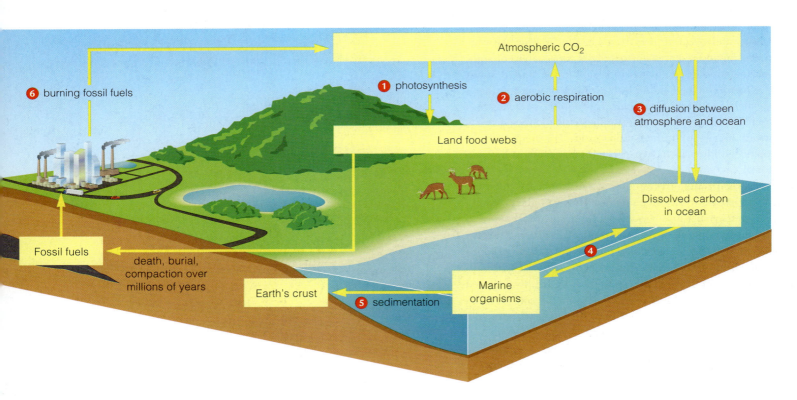

Figure 28.10 Animated The carbon cycle. Most carbon is in Earth's crust, where it is largely unavailable to living organisms.

❶ Carbon enters land food webs when plants take up carbon dioxide from the air for use in photosynthesis.

❷ Carbon returns to the atmosphere as carbon dioxide when plants and other land organisms carry out aerobic respiration.

❸ Carbon diffuses between the atmosphere and the ocean. Bicarbonate forms when carbon dioxide dissolves in seawater.

❹ Marine producers take up bicarbonate for use in photosynthesis, and marine organisms release carbon dioxide from aerobic respiration.

❺ Many marine organisms incorporate carbon into their shells. After they die, these shells become part of the sediments. Over time, the sediments become carbon-rich rocks such as limestone and chalk in Earth's crust.

❻ Burning of fossil fuels derived from the ancient remains of plants puts additional carbon dioxide into the atmosphere.

Figure 28.11 Animated Greenhouse effect.

1 Earth's atmosphere reflects some sunlight energy back into space.

2 More light energy reaches and warms Earth's surface.

3 Earth's warmed surface emits heat energy. Some of this energy escapes through the atmosphere into space. But some is absorbed and then emitted in all directions by greenhouse gases. The emitted heat warms Earth's surface and lower atmosphere.

》 Figure It Out Do greenhouse gases reflect heat energy toward the Earth?

Answer: No. The gases absorb heat energy, then reemit it in all directions.

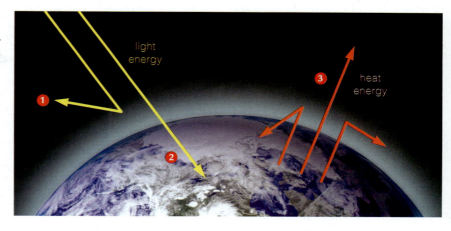

light energy

heat energy

as coal and gas are the ancient remains of such organic material. Fossil fuels hold an estimated 5,000 gigatons of carbon. Until recently, this carbon, like the carbon in rocks, had little impact on ecosystems. We now withdraw 4 to 5 gigatons of carbon from fossil fuel reservoirs each year **6** . Burning this fuel, cutting down forests, and other activities adds more carbon to the air than can dissolve in the oceans. Each year, only about 2 percent of the extra carbon that we put into the atmosphere dissolves in ocean water. The rest of it remains in the atmosphere.

Carbon, the Greenhouse Effect, and Global Warming

Atmospheric carbon dioxide helps keep the Earth warm enough for life. In what is known as the **greenhouse effect**, sunlight heats Earth's surface, then carbon dioxide and other "greenhouse gases" absorb some heat radiating from that surface and reradiate the heat toward Earth (**Figure 28.11**). Without the greenhouse effect, heat from Earth's surface would escape into space, leaving the planet cold and lifeless.

Scientists can use bubbles of air trapped in ancient glaciers to determine atmospheric conditions when the ice first formed. This and other evidence indicates that the current atmospheric carbon dioxide is the highest in 420,000 years and is climbing.

Given the greenhouse effect, we would predict that increases in the atmospheric concentration of carbon dioxide and other greenhouse gases would raise the tem-

perature of Earth's surface. Evidence supports this prediction. We are in the midst of a period of **global climate change**, with a trend toward rising temperature and shifts in other climate patterns.

Earth's climate has varied greatly over its long history. During ice ages, much of the planet was covered by glaciers. Other periods were warmer than the present, and tropical plants and coral reefs thrived at what are now cool latitudes. Scientists can correlate past large-scale temperature changes with shifts in Earth's orbit, which varies in a regular fashion over 100,000 years, and Earth's tilt, which varies over 40,000 years. Changes in solar output and volcanic eruptions also affect Earth's temperature. However, most scientists do not see evidence that these factors play a major role in the current temperature rise.

In 2007, the Intergovernmental Panel on Climate Change reviewed the results of many scientific studies related to climate change. The panel of hundreds of scientists from all over the world concluded that the warming in recent decades is very likely due to a human-induced increase in atmospheric greenhouse gases. Unlike other possible factors, the rise in greenhouse gases correlates well with the rise in temperature.

We return to the evidence of global climate change and its effects on natural systems in Section 30.8.

---

atmospheric cycle Biogeochemical cycle in which a gaseous form of an element plays a significant role.

carbon cycle Movement of carbon, mainly between the oceans, atmosphere, and living organisms.

global climate change A currently ongoing rise in average temperature that is altering climate patterns around the world.

greenhouse effect Warming of Earth's lower atmosphere and surface as a result of heat trapped by greenhouse gases.

Take-Home Message How does carbon move between reservoirs, and what are the effects of releasing excess carbon?

> Most of Earth's carbon is in rocks, but little carbon moves out of this reservoir.

> Oceans and the soil hold more carbon than the air. Carbon flows continually between these three reservoirs and into and out of food webs.

> Burning fossil fuels, cutting forests, and other activities add more carbon to the air than the ocean naturally absorbs.

> Carbon dioxide is a greenhouse gas. Rising levels of these gases correlate with rising global temperature. This indicates to scientists that human induced increases in these gases are the most likely cause of recent global climate change.

> Gaseous nitrogen abounds in the atmosphere, but only bacteria can make it available to other organisms.
❮ Links to Acids 2.6, Nitrogen fixation 19.7

Nitrogen Reservoirs and Movements

Nitrogen moves in an atmospheric cycle known as the **nitrogen cycle** (Figure 28.12). The main reservoir is the atmosphere, which is about 80 percent nitrogen. Triple covalent bonds hold two atoms of nitrogen gas together (N_2, or $N \equiv N$). Plants cannot use gaseous nitrogen because they cannot break these bonds. Only certain bacteria can carry out **nitrogen fixation**. They break the bonds in N_2, and use the nitrogen atoms to form ammonia, which is ionized in water as ammonium (NH_4^+) ❶. Nitrogen-fixing cyanobacteria live in aquatic habitats, soil, and as components of lichens. Another group of nitrogen-fixing bacteria forms nodules on the roots of peas and other legumes.

Plants take up ammonium from soil water ❷ and use it in metabolic reactions. Consumers get nitrogen by eating plants or one another. Bacterial and fungal decomposers that break down wastes and remains return ammonium to the soil ❸.

Nitrification is a two-step process that converts ammonium to nitrates ❹. First, ammonia-oxidizing bacteria and

archaeans convert ammonium to nitrite (NO_2^-), then bacteria convert nitrites to nitrates (NO_3^-). Like ammonium, nitrates are taken up and used by producers ❺.

Ecosystems lose nitrogen by **denitrification**. Denitrifying bacteria that use nitrate for energy produce nitrogen gas that escapes into the atmosphere ❻.

Human Effects on the Nitrogen Cycle

Manufactured ammonia fertilizers improve crop yields, but also modify soil chemistry. These fertilizers increase the concentration of hydrogen ions, as well as nitrogen. As a result of the increased acidity, nutrient ions bound to soil particles get replaced by hydrogen ions, and nutrients essential for plant growth trickle away in soil water. Nitrogen fertilizer also causes problems by running off and polluting aquatic habitats.

Burning fossil fuels releases nitrous oxide, which is a greenhouse gas and contributes to acid rain (Section 2.6). Nitrogen deposition in acid rain has the same negative effects on soil as overfertilizing.

---

denitrification Bacteria convert nitrates or nitrites to gaseous forms of nitrogen.
nitrification Bacteria convert ammonium to nitrates.
nitrogen cycle Movement of nitrogen among the atmosphere, soil, and water, and into and out of food webs.
nitrogen fixation Bacteria use nitrogen gas to form ammonia.

Take-Home Message How does nitrogen cycle in ecosystems?

> Nitrogen-fixing bacteria convert gaseous nitrogen to ammonium that plants can use. Other bacteria, archaeans, and fungi convert ammonium to nitrites and nitrates. Still other bacteria convert these soluble forms back to nitrogen gas.

> Use of manufactured fertilizers and nitrogen oxides from vehicle emissions add excess nitrogen to ecosystems.

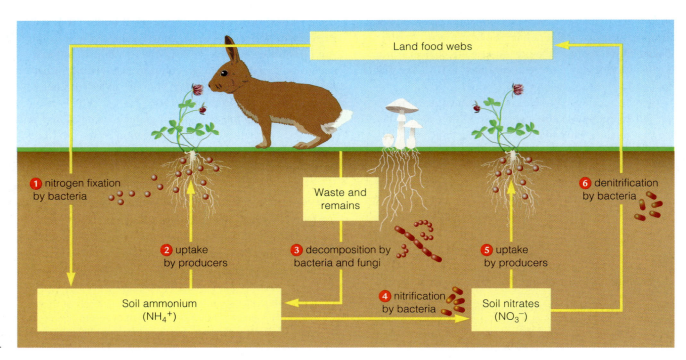

Figure 28.12 Animated Nitrogen cycle in a land ecosystem.

Land food webs

❶ nitrogen fixation by bacteria
❷ uptake by producers
Waste and remains
❸ decomposition by bacteria and fungi
❺ uptake by producers
❻ denitrification by bacteria
❹ nitrification by bacteria
Soil ammonium (NH_4^+)
Soil nitrates (NO_3^-)

28.10 The Phosphorus Cycle

> Unlike carbon and nitrogen, phosphorus seldom occurs as a gas. It moves in a sedimentary cycle.
< Link to Plate tectonics 16.7

Most of Earth's phosphorus is bonded to oxygen as phosphate (PO_4^{3-}), an ion that occurs in rocks and sediments. In the **phosphorus cycle**, phosphorus passes quickly through food webs as it moves from land to ocean sediments, then slowly back to land (Figure 28.13). There is no commonly occurring gaseous form of phosphorus, so the phosphorus cycle is a **sedimentary cycle**.

Weathering and erosion move phosphates from rocks into soil, lakes, and rivers ❶. Leaching and runoff carry dissolved phosphates to the ocean ❷. Here, most phosphorus comes out of solution and settles as deposits along continental margins ❸. Slow movements of Earth's crust can uplift these deposits onto land ❹, where weathering releases phosphates from rocks once again.

The biological portion of the phosphorus cycle begins when producers take up phosphorus. Land plants take up dissolved phosphate from the soil water ❺. Land animals get phosphates by eating the plants or one another. Phosphorus returns to the soil in wastes and remains ❻.

In the seas, phosphorus enters food webs when producers take up phosphate dissolved in seawater ❼. As on land, wastes and remains replenish the supply ❽.

---

phosphorus cycle Movement of phosphorus among Earth's rocks and waters, and into and out of food webs.
sedimentary cycle Biochemical cycle in which the atmosphere plays little role and rocks are the major reservoir.

Too Much of a Good Thing (revisited)

Water treatment systems can remove phosphates from detergents out of wastewater before it is discharged, but the additional treatment adds to the cost. Phosphate-rich runoff from lawns usually gets into waterways without going through a treatment plant. Thus, the most effective and economical way to keep aquatic ecosystems healthy is to avoid the use of phosphate-rich products when substitutes are available. Detergents that get their cleaning power from enzymes can replace phosphate-based products and most lawns will thrive even with low-phosphate fertilizer.

How Would You Vote? Is banning high–phosphate products the best way to prevent phosphate–related eutrophication, or are consumer education programs a better option? See CengageNow for details, then vote online (cengagenow.com).

Phosphorus is often a limiting factor for plant growth, so phosphate-rich droppings from seabird or bat colonies are collected and used as fertilizer. Phosphate-rich rock is also mined for this purpose. When fertilizer, detergent, or sewage with lots of phosphate gets into water, it can cause the type of eutrophication described in Section 28.1.

Take-Home Message How does phosphorus cycle in ecosystems?

> Rocks are the main phosphorus reservoir. Weathering puts phosphates into water. Producers take up dissolved phosphates.

> Phosphate-rich rock or wastes can be used as fertilizer; however, excess phosphate from these or other sources can cause eutrophication if they get into aquatic habitats.

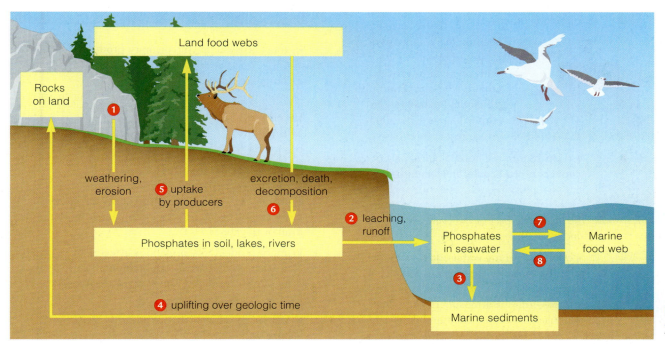

Land food webs

Rocks on land

❶ weathering, erosion

❺ uptake by producers

excretion, death, decomposition ❻

Phosphates in soil, lakes, rivers

❷ leaching, runoff

❹ uplifting over geologic time

Phosphates in seawater

❼

❽

Marine food web

❸

Marine sediments

Figure 28.13 Animated
The phosphorus cycle.

Summary

Section 28.1 Inorganic substances such as phosphorus serve as nutrients for living organisms. Excessive inputs of nutrients as a result of human activities can alter ecosystem dynamics, as when phosphate from detergents or fertilizers causes **eutrophication**.

Section 28.2 There is a one-way flow of energy into and out of an ecosystem, and a cycling of materials among the organisms within it. All ecosystems have inputs and outputs of energy and nutrients.

Primary producers convert energy from an inorganic source (usually light) into chemical bond energy. **Primary production**, the rate at which producers capture and store energy, can vary over time and between locations.

Consumers feed on producers or one another. For example, **detritivores** eat small bits of organic remains and **decomposers** break wastes and remains into their inorganic components.

Section 28.3 A **food chain** shows one path of energy and nutrient flow among organisms. Each organism in a food chain is at a different **trophic level**, with the primary producer being the first level and consumers at higher levels. Efficiency of energy transfers from one trophic level to the next is always low, so most food chains have only four or five links.

Section 28.4 A **food web** consists of interconnecting food chains. Most food webs include both **grazing food chains**, in which herbivores eat producers, and **detrital food chains**, in which producers die and are consumed by detritivores. Because of connections through food webs, a change that affects one species in an ecosystem will have effects on many others.

Section 28.5 Energy pyramids and biomass pyramids depict how energy and organic compounds are distributed among organisms. All energy pyramids are largest at their base.

Section 28.6 In the nonbiological part of a **biogeochemical cycle**, an element moves among environmental reservoirs such as Earth's atmosphere, rocks, and waters. In the biological part of the cycle, elements move through food webs, then return to the environment.

Section 28.7 In the **water cycle**, evaporation, condensation, and precipitation move water from its main reservoir—the oceans—into the atmosphere, onto land, then back to oceans. A **watershed** is a region in which all water drains into the same stream or river. Water that falls onto land may become part of the **groundwater**; it may be become **soil water** or be stored in an **aquifer**. Alternatively, it may become **runoff**. The water cycle helps move soluble forms of other nutrients.

Section 28.8 The main reservoir for carbon is rocks, but the **carbon cycle** moves carbon mainly among seawater, the air, soils, and living organisms in an **atmospheric cycle**. Carbon dioxide contributes to the **greenhouse effect**. Greenhouse gases keep Earth's surface warm enough to support life. However, as a result of fossil fuel consumption and other human activities, the levels of these gases are increasing. The increase correlates with and is the most likely cause of **global climate change**.

Section 28.9 The **nitrogen cycle** is an atmospheric cycle. Air is the main reservoir for N_2, a gaseous form of nitrogen that plants cannot use. Plants can take up and use ammonium that bacteria produce by **nitrogen fixation**. Fungi and bacteria that act as decomposers add ammonium derived from remains to the soil. Plants also use nitrates that some bacteria produce from ammonium through **nitrification**. Nitrogen is returned to the air by bacteria that carry out **denitrification** of nitrates. Humans add extra nitrogen to ecosystems by using synthetic fertilizer and by burning fossil fuels, which releases nitrous oxide.

Section 28.10 The **phosphorus cycle** is a **sedimentary cycle** with no significant atmospheric component. Phosphorus from rocks dissolves in water and is taken up by producers. Phosphate-rich rocks and deposits of bird droppings are mined for use as fertilizer.

Self-Quiz Answers in Appendix III

1. In most ecosystems, the primary producers use energy from _____ to build organic compounds.
   a. inorganic chemicals   c. heat
   b. sunlight   d. lower trophic levels

2. Decomposers are commonly _____ .
   a. fungi   c. plants
   b. bacteria   d. a and b

3. Organisms at the first trophic level _____ .
   a. capture energy from a nonliving source
   b. are eaten by organisms at higher trophic levels
   c. are shown at the bottom of an energy pyramid
   d. all of the above

4. Primary productivity on land is affected by _____ .
   a. nutrient availability   c. temperature
   b. amount of sunlight   d. all of the above

5. A(n) _____ is an autotroph.
   a. primary producer   c. detritivore
   b. herbivore   d. top carnivore

Data Analysis Activities

Rising Atmospheric Carbon To assess the impact of human activity on the carbon dioxide level in Earth's atmosphere, it helps to take a long view. One useful data set comes from deep core samples of Antarctic ice. The oldest ice core that has been fully analyzed dates back a bit more than 400,000 years. Air bubbles trapped in the ice provide information about the gas content in Earth's atmosphere at the time the ice formed. Combining ice core data with more recent direct measurements of atmospheric carbon dioxide—as in **Figure 28.14**—can help scientists put current changes in the atmospheric carbon dioxide into historical perspective.

1. What was the highest carbon dioxide level between 400,000 B.C. and 0 A.D.?
2. During this period, how many times did carbon dioxide reach a level comparable to that measured in 1980?
3. The industrial revolution occurred around 1800. How much did carbon dioxide levels change in the 800 years prior to this event? In 175 years after it?
4. Did carbon dioxide levels rise more between 1800 and 1975 or between 1980 and 2007?

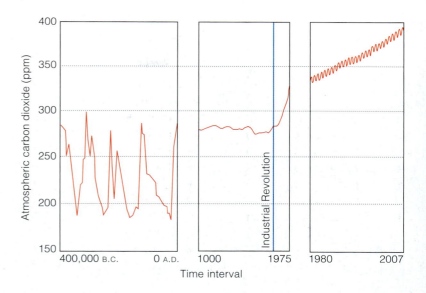

Figure 28.14 Changes in atmospheric carbon dioxide levels (in parts per million). Direct measurements began in 1980. Earlier data are based on ice cores.

6. Most of earth's fresh water is _____ .
   a. in lakes and streams   c. frozen as ice
   b. in aquifers and soil   d. in bodies of organisms

7. Earth's largest carbon reservoir is _____ .
   a. the atmosphere   c. seawater
   b. sediments and rocks   d. living organisms

8. Carbon is released into the atmosphere by _____ .
   a. photosynthesis   c. burning fossil fuels
   b. aerobic respiration   d. b and c

9. Greenhouse gases _____ .
   a. help keep Earth's surface warm enough for life
   b. are released by natural and human activities
   c. are at higher levels than they were 100 years ago
   d. all of the above

10. The _____ cycle is a sedimentary cycle.
    a. phosphorus   c. nitrogen
    b. carbon   d. water

11. Earth's largest phosphorus reservoir is _____ .
    a. the atmosphere   c. sediments and rocks
    b. bird droppings   d. living organisms

12. Plants obtain _____ by taking it up from the air.
    a. nitrogen   c. phosphorus
    b. carbon   d. a and b

13. Nitrogen fixation converts _____ to _____ .
    a. nitrogen gas; ammonia   c. ammonia; nitrates
    b. nitrates; nitrites   d. nitrites; nitrogen oxides

14. Which holds the least carbon?
    a. soils   b. the air   c. seawater   d. rocks

15. Match each term with its most suitable description.
    ___ carbon dioxide   a. contains triple bond
    ___ bicarbonate   b. product of nitrogen fixation
    ___ ammonium   c. marine carbon source
    ___ nitrogen gas   d. greenhouse gas

Additional questions are available on **CENGAGENOW**.

Critical Thinking

1. Marguerite has a vegetable garden in Maine. Eduardo has one in Florida. List the variables that could cause differences in the primary production of these gardens.

2. Where does your water come from? A well, a reservoir? Beyond that, what area is included within your watershed and what are the current flows? Visit the *Science in Your Watershed* site at water.usgs.gov/wsc and find the answers to these questions.

3. Why do all organisms require both phosphorus and nitrogen? List some of the molecules common to all life that contain these essential elements.

4. Rather than using fertilizer, a farmer may rotate crops, planting legumes one year, then another crop, then legumes again. Explain how crop rotation keeps soil fertile.

Animations and Interactions on **CENGAGENOW**:
❭ Energy flow and nutrient cycling; Food webs; Water cycle; Carbon cycle; Greenhouse effect; Nitrogen cycle; Phosphorus cycle.

‹ Links to Earlier Concepts

In this chapter, you will consider the biosphere as a whole (Section 1.2), revisit properties of water (2.5), and see how carbon-fixing pathways (6.8), herbivory (27.5), and fire (27.8) affect plant distribution. We compare regional primary production (28.2) and learn more about the effects of global warming (28.8), eutrophication (28.1), and morphological convergence (16.8).

Key Concepts

Air Circulation Patterns

Air circulation starts with latitudinal differences in energy inputs from the sun. Movement of air from the equator toward poles is affected by Earth's rotation and gives rise to major surface winds and latitudinal patterns in rainfall.

Ocean Circulation Patterns

Heating of the tropics also sets ocean waters in motion. As water circulates, it carries and releases heat, and so affects the climate on land. Interactions between oceans, air, and land affect coastal climates.

29 The Biosphere

29.1 Effects of El Niño

Professional surfer Ken Bradshaw has ridden a lot of waves, but one stands out in his memory. In January of 1998, he found himself off the coast of Hawaii riding the biggest wave he had ever seen (Figure 29.1). It towered more than 12 meters (39 feet) high and gave him the ride of a lifetime.

The giant wave was one manifestation of an **El Niño**, a recurring climate event in which equatorial waters of the eastern and central Pacific Ocean warm above their average temperature. The term El Niño means "baby boy" and refers to Jesus; it was first used by Peruvian fishermen to describe local weather changes and a shortage of fishes that occurred in some years around Christmas. Scientists now know that during an El Niño, marine currents interact with the atmosphere in ways that influence weather patterns worldwide.

An El Niño affects marine food webs along eastern Pacific coasts. As unusually warm water flows toward these coasts, it displaces currents that would otherwise bring up nutrients from the deep. Without these nutrients, marine primary producers decline in numbers. The dwindling producer populations and warming water cause a decrease in populations of small, cold-water fishes such as anchovies, as well as the larger consumers that rely on them. During the 1997–1998 El Niño, about half of the sea lions on the Galápagos Islands starved to death. California's population of northern fur seals also suffered a sharp decline.

Rainfall patterns shift during an El Niño. During the winter of 1997–1998, torrential rains caused flooding and landslides along eastern Pacific coasts, while Australia and Indonesia suffered from drought-driven crop failures and raging wildfires. A El Niño typically brings cooler, wetter weather to the American Gulf states, and reduces the likelihood of hurricanes.

An El Niño typically persists for 6 to 18 months. It may be followed by an interval in which the temperature of the eastern Pacific remains near its average, or by a La Niña. During a

Figure 29.1 El Niño. *Opposite*, Ken Bradshaw rides a wave more than 12 meters (39 feet) high, during the most powerful El Niño of the past century. *Above*, interactions among Earth's oceans and atmosphere give rise to an El Niño and to other climate patterns.

La Niña, eastern Pacific waters become cooler than average. As a result, the west coast of the United States gets little rainfall and the likelihood of hurricanes in the Atlantic increases.

The previous three chapters have considered three levels of biological organization—populations, communities, and ecosystems. With this discussion of El Niño/La Niña, we invite you to move to the next level, to consider the factors that influence the properties of the biosphere. The **biosphere** includes all places where we find life on Earth. Many organisms live in the hydrosphere: the ocean, ice caps, and other bodies of water, liquid and frozen. Others live on and in sediments and soils of the lithosphere: Earth's outer, rocky layer. Still others lift off into the lower region of the atmosphere: the gases and airborne particles that envelop Earth.

biosphere All regions of Earth where organisms live.
El Niño Periodic warming of equatorial Pacific waters and the associated shifts in global weather patterns.
La Niña Periodic cooling of equatorial Pacific waters and the associated shifts in global weather patterns.

Land Biomes
A biome consists of geographically separated regions that have a similar climate and soils, and so support similar types of vegetation. Biomes include deserts, grasslands, chaparral, various types of forests, and tundra.

Freshwater Ecosystems
Lakes have gradients of light and temperature. They undergo succession, changing over time. In temperate zones, their waters mix in response to seasonal changes in temperature. Rivers vary along their length in their properties, and in the organisms they contain.

Coastal and Marine Ecosystems
Productivity is high in coastal wetlands, on coral reefs, and in the ocean's upper, sunlit water. Life also thrives in the ocean's deeper, darker waters and on the sea floor, especially on undersea mountains and at hydrothermal vents.

❯ The amount of solar energy that reaches Earth's surface varies from place to place and with the season.

❯ Latitudinal changes in sunlight give rise to air circulation patterns that determine regional rainfall.

Climate refers to average weather conditions, such as cloud cover, temperature, humidity, and wind speed, over time. Regional climates differ because many factors that influence winds and ocean currents, such as intensity of sunlight, the distribution of land masses and seas, and elevation, vary from place to place.

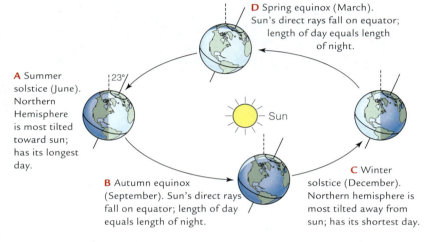

D Spring equinox (March). Sun's direct rays fall on equator; length of day equals length of night.

A Summer solstice (June). Northern Hemisphere is most tilted toward sun; has its longest day.

23°

Sun

B Autumn equinox (September). Sun's direct rays fall on equator; length of day equals length of night.

C Winter solstice (December). Northern hemisphere is most tilted away from sun; has its shortest day.

Figure 29.2 Earth's tilt and yearly rotation around the sun cause seasonal effects. The 23° tilt of Earth's axis causes the Northern Hemisphere to receive more intense sunlight and have longer days in summer than in winter.

A

B

Figure 29.3 Variation in intensity of solar radiation with latitude. For simplicity, we depict two equal parcels of incoming radiation on an equinox, a day when incoming rays are perpendicular to Earth's axis.

Rays that fall on high latitudes **A** pass through more atmosphere (*blue*) than those that fall near the equator **B**. Compare the length of the *green* lines. Atmosphere is not to scale.

Also, energy in the rays that fall at the high latitude is spread over a greater area than energy that falls on the equator. Compare the length of the *red* lines.

Seasonal Effects

Each year, Earth rotates around the sun in an elliptical path (Figure 29.2). Seasonal changes in daylength and temperature arise because Earth's axis is not perpendicular to the plane of this ellipse, but rather is tilted about 23 degrees. In June, when the Northern Hemisphere is angled toward the sun, it receives more intense sunlight and has longer days than the Southern Hemisphere (Figure 29.2A). In December, the opposite occurs (Figure 29.2C). Twice a year—on spring and autumn equinoxes—Earth's axis is perpendicular to incoming sunlight. On these days, every place on Earth has 12 hours of daylight and 12 hours of darkness (Figure 29.2B,D).

In each hemisphere, the degree of seasonal change in daylength increases with latitude. At 25° north or south of the equator, the longest daylength is a bit less than 14 hours. By contrast, 60° north or south of the equator, the longest daylength is nearly 19 hours.

Air Circulation and Rainfall

On any given day, equatorial regions get more sunlight energy than higher latitudes for two reasons (Figure 29.3). First, fine particles of dust, water vapor, and greenhouse gases absorb some solar radiation or reflect it back into space. Because sunlight traveling to high latitudes passes through more atmosphere to reach Earth's surface than light traveling to the equator, less energy reaches the ground. Second, energy in an incoming parcel of sunlight is spread out over a smaller surface area at the equator than at the higher latitudes. As a result, Earth's surface warms more at the equator than at the poles.

Knowing about two properties of air can help you understand how regional differences in surface warming give rise to global air circulation and rainfall patterns. First, as air warms, it becomes less dense and rises. Hot air balloonists take advantage of this effect when they take off from the ground by heating the air inside their balloon. Second, warm air can hold more water than cooler air. This is why you can "see your breath" in cold weather; water vapor in warm exhaled air condenses into droplets when exposed to the cool external environment.

The global air circulation pattern begins at the equator, where intense sunlight warms the air and causes evaporation from the ocean. The result is an upward movement of warm, moist air (Figure 29.4A). As the air from the equator rises to higher altitudes, it cools and flows north and south, releasing moisture as rain that supports tropical rain forests.

By the time the air has moved to 30° north or south of the equator, it has given up most moisture and cooled off, so it sinks back toward Earth's surface (Figure 29.4B). Many of the world's great deserts, including the Sahara, are about 30° from the equator.

Idealized Pattern of Air Circulation

D At the poles, cold air sinks and moves toward lower latitudes.

C Air rises again at 60° north and south, where air flowing poleward meets air coming from the poles.

B As the air flows toward higher latitudes, it cools and loses moisture as rain. At around 30° north and south latitude, the air sinks and flows north and south along Earth's surface.

A Warmed by energy from the sun, air at the equator picks up moisture and rises. It reaches a high altitude, and spreads north and south.

Major Winds Near Earth's Surface

E Major winds near Earth's surface do not blow directly north and south because of the effects of Earth's rotation. Winds deflect to the right of their original direction in the Northern Hemisphere and to the left in the Southern Hemisphere.

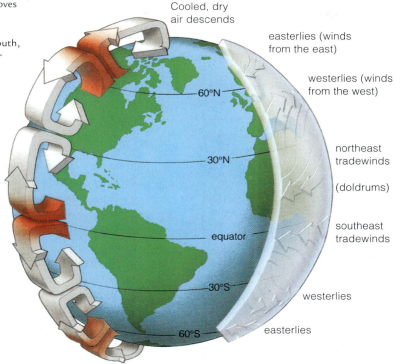

Cooled, dry air descends

easterlies (winds from the east)

westerlies (winds from the west)

60°N

30°N

northeast tradewinds

(doldrums)

equator

southeast tradewinds

30°S

westerlies

60°S

easterlies

Figure 29.4 Animated Global air circulation patterns.

As air continues flowing along Earth's surface toward the poles, it again picks up heat and moisture. At a latitude of about 60°, warm, moist air rises again, losing moisture as it does so (Figure 29.4C). The resulting rains support temperate zone forests.

Cold, dry air descends near the poles (Figure 29.4D). Precipitation is sparse, and polar deserts form.

Surface Wind Patterns

Major wind patterns arise as air in the lower atmosphere moves continually from latitudes where air is sinking toward those where air is rising. Earth's rotation affects the apparent trajectory of these winds. Air masses are not attached to Earth's surface, so the Earth spins beneath them, moving fastest at the equator and most slowly at the poles. Thus, as an air mass moves away from the equator, the speed at which the Earth rotates beneath it continually slows. As a result, major winds trace a curved path relative to the Earth's surface (Figure 29.4E). In the

Northern Hemisphere, winds curve toward the right; in the Southern Hemisphere, they curve toward the left. For example, between 30° north and 60° north, surface air traveling toward the North Pole is deflected right, or toward the east. Winds are named for the direction from which they blow, so the prevailing winds in the United States are westerlies.

Winds blow most consistently between regions where air rises. In the regions where the air is rising, winds can be intermittent, as in the doldrums near the equator.

Take-Home Message **How does sunlight affect climate?**

❯ Equatorial regions receive more intense sunlight than higher latitudes.

❯ Sunlight drives the rise of moisture-laden air at the equator. The air cools as it moves north and south, releasing rains that support tropical forests. Deserts form where cool, dry air descends. Sunlight energy also drives moisture-laden air aloft at 60° north and south latitude. This air gives up moisture as it flows toward the equator or the pole.

❯ Air flow in the lower atmosphere toward latitudes where air is rising and away from latitudes where it is sinking creates major surface winds. These winds trace a curved path relative to Earth's surface because of Earth's rotation.

---

climate Average weather conditions in a region over a long time period.

> The ocean is a continuous body of water that covers more than 71 percent of Earth's surface. Its water moves in currents that distribute nutrients through marine ecosystems.
< Link to Properties of water 2.5

Ocean Currents

Latitudinal variations in sunlight affect ocean temperature and set major currents in motion. At the equator, where vast volumes of water warm and expand, the sea level is about 8 centimeters (3 inches) higher than at either pole. The volume of water in this "slope" is enough to start sea surface water moving toward the poles. As the water moves, it gives up heat energy to the air above it.

Enormous volumes of water flow as ocean currents. The directional movement of surface currents is shaped by the force of major winds, Earth's rotation, and topography. Surface currents circulate clockwise in the Northern Hemisphere and counterclockwise in the Southern Hemisphere (Figure 29.5).

Swift, deep, and narrow currents of nutrient-poor water flow away from the equator along the east coast of continents. Along the east coast of North America, warm water flows north, as the Gulf Stream. Slower, shallower, broader currents of cold water parallel the west coast of continents and flow toward the equator.

Ocean currents affect climate. For example, Pacific Northwest coasts are cool and foggy in summer because the cold California current chills the air, and water condenses out of the cooled air as droplets. As another example, Boston and Baltimore are warm and muggy in summer because air masses pick up heat and moisture from the warm Gulf Stream, then flow over these cities.

Regional Effects

Mountains, valleys, and other surface features of the land affect climate. Suppose you track a warm air mass after it picks up moisture off California's coast. It moves inland as wind from the west, and piles up against the Sierra Nevada, a high mountain range that parallels the coast. The air cools as it rises in altitude and loses moisture as rain (Figure 29.6). The result is a **rain shadow**, a semiarid or arid region of sparse rainfall on the leeward side of high mountains. *Leeward* is the side facing away from the wind. The Himalayas, Andes, Rockies, and other great mountain ranges cause vast rain shadows.

Figure 29.5 Animated Major climate zones correlated with surface currents and surface drifts of the world ocean. Warm surface currents start moving from the equator toward the poles, but prevailing winds, Earth's rotation, gravity, the shape of ocean basins, and landforms influence the direction of flow. Water temperatures, which differ with latitude and depth, contribute to regional differences in air temperature and rainfall.

Figure 29.6 Animated Rain shadow effect. On the side of mountains facing away from prevailing winds, rainfall is light. *Black* numbers signify annual precipitation, in centimeters, averaged on both sides of the Sierra Nevada, a mountain range. *White* numbers signify elevations, in meters.

Differences in the heat capacity of water and land give rise to coastal breezes. In the daytime, land warms faster than water. As air over land warms and rises, cooler offshore air moves in to replace it (Figure 29.7A). After sundown, land cools more than the water, so the breezes reverse direction (Figure 29.7B).

Differential heating of water and land also causes **monsoons**, which are winds that change direction seasonally. For example, the continental interior of Asia heats up in the summer, so air rises above it. Moist air from over the warm Indian Ocean to the south moves in to replace the rising air, and this north-blowing wind delivers heavy rains. In the winter, the continental interior is cooler than the ocean. As a result, cool, dry wind blowing from the north toward southern coasts causes a seasonal drought.

Proximity to an ocean moderates climate. For example, Seattle, Washington, is slightly farther north than Minneapolis, Minnesota, but Seattle has much milder winters. Air over Seattle draws heat from the adjacent Pacific Ocean, a heat source not available to Minneapolis.

A In afternoon; the land is warmer than the sea, so the breeze blows onto shore.

B In the evening, the sea is warmer than the land; the breeze blows out to sea.

Figure 29.7 Coastal breezes.

Take-Home Message How do ocean currents arise and affect regional climates?

❯ Surface ocean currents are set in motion by latitudinal differences in solar radiation. Currents are affected by winds and by Earth's rotation.

❯ The collective effects of air masses, oceans, and landforms determine regional temperature and moisture levels.

monsoon Wind that reverses direction seasonally.
rain shadow Dry region downwind of a coastal mountain range.

> Similar communities often evolve in widely separated regions as a result of similar environmental factors.

‹ Links to Carbon-fixing pathways 6.8, Morphological convergence 16.8

Differences Between Biomes

Biomes are areas of land characterized by their climate and type of vegetation (Figure 29.8). A biome is discontinuous; most biomes include widely separated areas on different continents. For example, the temperate grassland biome includes areas of North American prairie, South African veld, South American pampa, and Eurasian steppe. In all of these regions, the main vegetation is grasses and other nonwoody flowering plants.

Rainfall and temperature are the main determinants of the type of biome in a given region. Desert biomes get the least annual rainfall, grasslands and shrublands get more, and forests get the most. Deserts occur where temperatures soar the highest and tundra where they drop the lowest.

Soils also influence biome distribution. Soils consist of a mixture of mineral particles and varying amounts of humus. Water and air fill spaces between soil particles. Properties of soils vary depending on the types, proportions, and compaction of particles. Deserts have sandy or gravelly, fast-draining soil with little topsoil. Topsoil tends to be deepest in natural grasslands, where it can be more than one meter thick. For this reason, grasslands are often converted to agricultural uses.

Climate and soils affect primary production, so primary production varies among biomes (Figure 29.9).

Similarities Within a Biome

Unrelated species living in widely separated parts of a biome often have similar body structures that arose by the process of morphological convergence (Section 16.8). For example, cactuses with water-storing stems live in

---

biome Discontinuous region characterized by its climate and dominant vegetation.

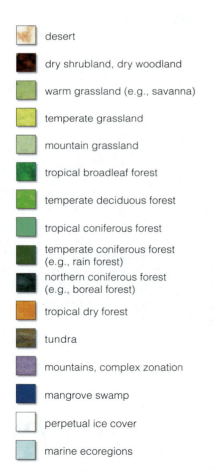

desert

dry shrubland, dry woodland

warm grassland (e.g., savanna)

temperate grassland

mountain grassland

tropical broadleaf forest

temperate deciduous forest

tropical coniferous forest

temperate coniferous forest (e.g., rain forest)

northern coniferous forest (e.g., boreal forest)

tropical dry forest

tundra

mountains, complex zonation

mangrove swamp

perpetual ice cover

marine ecoregions

Figure 29.8 Global distribution of major categories of biomes and marine ecoregions.

North American deserts and euphorbs with water-storing stems live in African deserts. Cactuses and euphorbs do not share an ancestor with a water-storing stem. This feature evolved independently in the two groups as a result of similar selection pressures. Similarly, an ability to carry out C4 photosynthesis evolved independently in grasses growing in warm grasslands on different continents. C4 photosynthesis is more efficient than the more common C3 pathway under hot, dry conditions.

Take-Home Message What factors shape marine communities?

❯ Biomes are vast expanses of land dominated by distinct kinds of plants that support characteristic communities. They vary in their primary productivity.

❯ Evolution often produces similar solutions to environmental challenges in different regions of a biome.

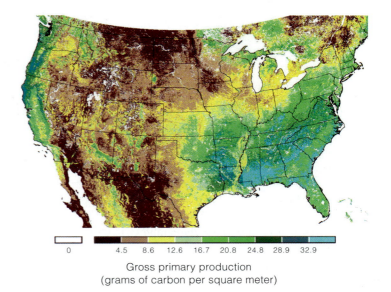

Gross primary production
(grams of carbon per square meter)

0   4.5   8.6   12.6 16.7 20.8 24.8 28.9 32.9

Figure 29.9 Remote satellite monitoring of gross primary productivity across the United States. The differences roughly correspond with variations in soil types and moisture.

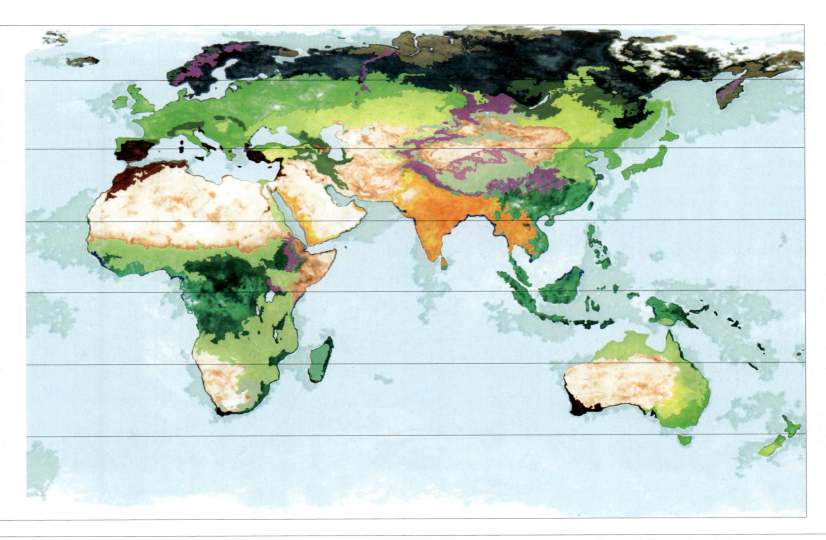

> Low rainfall shapes the desert biome.
< Links to Carbon-fixing pathways 6.8, Atacama Desert 18.1, Nitrogen fixation 19.7

Deserts

Desert Locations and Conditions

Deserts receive an average of less than 10 centimeters (4 inches) of rain per year. They cover about one-fifth of Earth's land surface and many are located at about 30° north and south latitude, where global air circulation patterns cause dry air to sink. Rain shadows also reduce rainfall. For example, Chile's Atacama desert is on the leeward side of the Andes. Similarly, the Himalayas prevent rain from falling in China's Gobi desert.

Lack of rainfall keeps the humidity in deserts low. With little water vapor to block rays, intense sunlight reaches the ground and heats it. At night, the lack of insulating water vapor in the air allows the temperature to fall fast. As a result, deserts tend to have larger daily temperature shifts than other biomes.

Desert soils have little topsoil (Figure 29.10), the layer most important for plant growth. The soils often are somewhat salty, because rain that falls usually evaporates before seeping into the ground. Rapid evaporation allows any salt in rainwater to accumulate at the soil surface.

O horizon:
Pebbles, little organic matter

A horizon:
Shallow, poor soil

B horizon:
Evaporation causes salt buildup; leaching removes nutrients

C horizon:
Rock fragments from uplands

Figure 29.10 Desert soil profile.

Adaptations to Desert Conditions

Despite their harsh conditions, most deserts support some plant life. Diversity is highest in regions where moisture is available in more than one season (Figure 29.11).

Many desert plants have adaptations that reduce water loss. For example, some have spines or hairs (Figure 29.12A). In addition to detering herbivory, these structures reduce water loss by trapping some water and keeping the humidity around the stomata high. Where rains fall seasonally, some plants reduce their water loss by leafing out after a rain, then dropping their leaves when dry conditions return (Figure 29.12B).

Some desert plants store water in their tissues during the wet season, for use in drier times. For example, the stem of the barrel cactus in Figure 29.12A has a spongy pulp that holds water. The cactus stem swells after a rain, then shrinks as the plant uses up its store of water.

Woody desert shrubs such as mesquite and creosote have extensive, efficient root systems that take up the little water that is available. Mesquite roots have been found as deep as 60 meters (197 feet) beneath the soil surface.

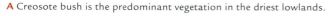

A Creosote bush is the predominant vegetation in the driest lowlands.

B A greater variety of plants survive in uplands, which are a bit wetter and cooler.

Figure 29.11 Vegetation in Arizona's Sonoran Desert.

A Barrel cactuses are covered by spines that reduce evaporative water loss. The cactus is a CAM plant.

B Ocitillo, a desert shrub, grows leaves on its stems after a rain, then sheds them when conditions become dry again.

Figure 29.12 Perennials adapted to desert conditions.

Figure 29.13 Mojave Desert after the rains. Annual poppies sprout, flower, produce seeds, and die within weeks beneath slow-growing perennial cactuses.

Alternative carbon-fixing pathways also help desert plants conserve water. Cactuses, agaves, and euphorbs are CAM plants. They open their stomata only at night when temperature declines.

Most deserts contain a mix of annuals and perennials (Figure 29.13). The annuals are adapted to desert life by a life cycle that allows them to sprout and reproduce in the short time that the soil is moist.

Animals also have adaptations that allow them to conserve water. For example, the highly efficient kidneys of a desert kangaroo rat minimize its water needs. Most desert animals are not active at the height of the daytime heat (Figure 29.14).

The Crust Community

In many deserts, the soil is covered by a desert crust, a community that can include cyanobacteria, lichens, mosses, and fungi. The organisms secrete organic molecules that glue them and the surrounding soil particles together. The crust benefits members of the larger desert community in important ways. Its cyanobacteria fix nitrogen and make ammonia available to plants. The crust also holds soil particles in place. When the fragile connections within the desert crust are broken, the soil can blow away. Negative effects of such disturbance are exacerbated when windblown soil buries healthy crust in an undisturbed area, killing additional crust organisms and allowing more soil to take flight.

A The Sonoran desert tortoise spends much of its life inactive. In hot summer months, it ventures out of its burrow only in cool mornings to feed. It hibernates during the cold winter, when little food is available.

B Lesser long-nosed bats spend spring and summer in the Sonoran Desert, where they avoid the daytime heat by resting in caves.

Figure 29.14 Two Sonoran desert animals.

---

desert Biome with little rain and low humidity; plants that have water-storing and conserving adaptations predominate.

> ## Take-Home Message **What are deserts?**
>
> ❯ A desert gets very little rain and has low humidity. There is plenty of sunlight, but the lack of water prevents most plants from surviving here.
>
> ❯ The plants that dominate in deserts have adaptations that allow them to reduce water lost by transpiration, store water, or access water deep below ground.
>
> ❯ Desert animals often spend the day inactive, sheltering from the heat.
>
> ❯ Desert soils are held in place by a community of organisms that form a desert crust. Disruption of this crust allows wind to strip away soil.

> Perennial grasses adapted to fire and to grazing are the main plants in grasslands.
< Link to Adaptations to herbivory 27.5

Temperate grasslands and tropical savannahs

Figure 29.16 African savanna, a tropical grassland with scattered trees. Africa's savannas are famous for their abundant wildlife such as this immense herd of wildebeests.

Grasslands form in the interior of continents between deserts and temperate forests. Grassland soils are rich, with deep topsoil. Annual rainfall is enough to keep desert from forming, but not enough to support woodlands. Low-growing grasses and other nonwoody plants tolerate strong winds, sparse and infrequent rain, and intervals of drought. Growth tends to be seasonal. Constant trimming by grazers, along with periodic fires, keeps trees and most shrubs from taking hold.

Temperate Grasslands

Temperate grasslands are warm in summer, but cold in winter. Annual rainfall is 25 to 100 centimeters (10–40 inches), with rains throughout the year. Grass roots extend profusely through the thick topsoil and help hold it in place, preventing erosion by the constant winds. North America's grasslands are shortgrass and tallgrass prairies (Figure 29.15).

During the 1930s, much of the shortgrass prairie of the American Great Plains was plowed to grow wheat. The strong winds, a prolonged drought, and unsuitable farm-

ing practices turned much of the region into what the newspapers of that time called the Dust Bowl.

Tallgrass prairie has somewhat richer topsoil and slightly more frequent rainfall than in the shortgrass prairie. Before the arrival of Europeans, it covered about 140 million acres, mostly in Kansas. Nearly all tallgrass prairie has now been converted to cropland. The Tallgrass Prairie National Preserve was created in 1996 to protect the little that remains.

Savannas

Savannas are broad belts of grasslands with scattered shrubs and trees. They lie between tropical forests and deserts. Temperatures are warm year-round, but rainfall is seasonal. Africa has the most extensive savanna; it covers about half the continent. African savanna supports enormous herds of hoofed grazers (Figure 29.16), and the predators such as lions that feed on them.

---

grassland Biome in the interior of continents where grasses and nonwoody plants adapted to grazing and fire predominate.

Take-Home Message What are grasslands?

> Grasslands are biomes dominated by grasses and other nonwoody plants that are adapted to fire and to grazers.

A horizon: Alkaline, deep, rich in humus

B horizon: Percolating water enriches layer with calcium carbonates

A Prairie soil profile.

B Tallgrass prairie in Kansas. See Figure 42.3 for a food chain.

C Bison grazing in South Dakota's shortgrass prairie.

Figure 29.15 North America's temperate grasslands.

❭ Regions with cool, rainy winters and hot, dry summers support dry shrublands and woodlands.
❬ Link to Adaptation to fire 27.8

Chaparral

Dry shrublands receive 25 to 60 centimeters (10–24 inches) of rain annually. We see them in South Africa, Greece, Italy, Chile, and California, where they are known as **chaparral**. California has about 6 million acres of chaparral (Figure 29.17A,B).

Rains occur seasonally, and lightning-sparked fires sometimes sweep through shrublands during the dry season. In California, where homes are often built near chaparral, the fires frequently cause property damage (Figure 29.17C). Foliage of many chaparral shrubs has oils that deter herbivores and also make the plant highly flammable. However, the plants are adapted to occasional fires. Some grow back from root crowns after a fire. Seeds of other chaparral species germinate only after they are exposed to heat or smoke, ensuring that the seeds sprout only when young seedlings face little competition.

---

chaparral Biome of dry shrubland in regions with hot, dry summers and cool, rainy winters.

Figure 29.18 Oak woodland in northern California.

Dry woodlands prevail where the seasonal rainfall is 40 to 100 centimeters (16–40 inches). Examples include the eucalyptus forests of Australia, and California's oak forests (Figure 29.18). Acorns produced by the oaks are an important seasonal source of food for birds and mammals in this community.

Take-Home Message **What are dry shrublands and woodlands?**

❭ Dry shrublands and woodlands form in areas with a hot, dry season and a cool, rainy one. Plants in these biomes are adapted to seasonal drought and to fire.

❭ Dry shrublands, known as chaparral, get less water than dry woodlands.

A Chaparral is California's most extensive natural community.

B Evergreen shrubs with leathery leaves predominate. Most are less than 2 meters (6 feet) high.

C Fire on chaparral-covered hills above Malibu. Most fires are now started by people, rather than lightning.

Figure 29.17 California's chaparral.

> Broadleaf (angiosperm) trees are the main plants in temperate and tropical forests.
< Links to Deforestation 21.1, Primary production 28.2

Semi–Evergreen and Deciduous Forests

Semi-evergreen forests occur in the humid tropics of Southeast Asia and India. The forests include broadleaf (angiosperm) trees that retain leaves year-round, and deciduous broadleaf trees. A deciduous plant sheds leaves annually, prior to a season when cold or dry conditions would not favor growth. In semi-evergreen forests, deciduous trees shed their leaves at the start of the dry season.

Where less than 2.5 centimeters (1 inch) of rain falls in the dry season, tropical deciduous forests form. In tropical deciduous forests, most trees shed leaves at the start of the dry season.

Temperate deciduous forests form in the Northern Hemisphere in parts of eastern North America, western and central Europe, and parts of Asia, including Japan. In these regions, 50 to 150 centimeters (about 20–60 inches) of precipitation falls throughout the year. Winters are cool and summers are warm.

Growth of temperate deciduous forests is seasonal. Leaves often turn color before dropping in autumn (Figure 29.19). Winters are cold and trees remain dormant while water is locked in snow and ice. In the spring, when conditions again favor growth, deciduous trees flower and put out new leaves. Also during the spring, leaves that were shed the prior autumn decay to form a rich humus. Rich soil and a somewhat open canopy that lets sunlight through allows understory plants to flourish.

The temperate deciduous forests of North America are the most species-rich examples of this biome. Different tree species characterize different regions of these forests. For example, Appalachian forests include mainly oaks, whereas beeches and maples dominate Ohio's forests. Animals in North American deciduous forests include grazing deer, seed-eating squirrels and chipmunks, as well as omnivores such as raccoons, opossums, and black bears. Native predators such as wolves and mountain lions have been largely eliminated.

Tropical Rain Forests

Tropical rain forests of evergreen broadleaf trees form between latitudes 10° north and south in equatorial Africa, the East Indies, Southeast Asia, South America, and Central America. Rain falls throughout the year for an annual total of 130 to 200 centimeters (50 to 80 inches).

Regular rains, combined with an average temperature of 25°C (77°F) and little variation in daylength, allows photosynthesis to continue year-round. Of all land biomes, tropical forests have the greatest primary production. Per unit area, they remove more carbon from the atmosphere than other forests or grasslands.

Tropical rain forest is the most structurally complex and species-rich biome. The forest has a multilayer structure (Figure 29.20). Its broadleaf trees can stand 30 meters (100 feet) tall. The trees often form a closed canopy that prevents most sunlight from reaching the forest floor.

Temperate deciduous forest

O horizon:
Scattered litter

A horizon:
Rich in organic matter above humus layer unmixed with minerals

B horizon:
Accumulated minerals leached from above

C horizon:
Poorly weathered rocks

Figure 29.19 North American temperate deciduous forest. Forest soil profile at *right*.

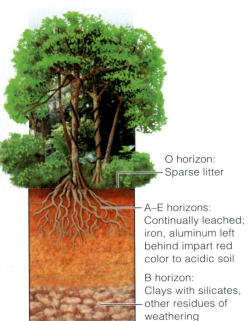

O horizon:
Sparse litter

A–E horizons:
Continually leached;
iron, aluminum left
behind impart red
color to acidic soil

B horizon:
Clays with silicates,
other residues of
weathering

Figure 29.20 Tropical rain forest. Forest soil profile at *right*.

Tropical
rain forest

Vines and epiphytes (plants that grow on another plant, but do not withdraw nutrients from it) thrive in the shade beneath the canopy.

Trees of topical rain forests shed leaves continually, but decomposition and mineral cycling happen so fast in this warm, moist environment that litter does not accumulate. Soils are highly weathered, heavily leached, and are very poor nutrient reservoirs.

Deforestation is an ongoing threat to tropical rain forests. Tropical forests are located in developing countries with fast-growing human populations who look to the for-

est as a source of lumber, fuel, and potential agricultural land. As human populations expand, more and more trees fall to the ax.

Deforestation in any region leaves fewer trees to remove carbon dioxide from the atmosphere. In rain forests, it also causes the extinction of species found nowhere else in the world. Compared to other land biomes, tropical rain forests have the greatest variety and numbers of insects, as well as the most diverse collection of birds and primates. This great species diversity means many species are affected by the loss of any amount of forest. Among the potential losses are species with chemicals that could save human lives. Two chemotherapy drugs, vincristine and vinblastine, were extracted from the rosy periwinkle, a low-growing plant native to Madagascar's rain forests. Today, these drugs help fight leukemia, lymphoma, breast cancer, and testicular cancer. No doubt other similarly valuable species live in the rain forests and will go extinct before we learn how they can help us.

---

temperate deciduous forest Northern Hemisphere biome in which the main plants are broadleaf trees that lose their leaves in fall and become dormant during cold winters.
tropical rain forest Highly productive and species-rich biome in which year-round rains and warmth support continuous growth of evergreen broadleaf trees.

Take-Home Message **What are broadleaf forests?**

❭ Temperate broadleaf forests grow in the Northern Hemisphere where cold winters prevent year-round growth. Trees lose their leaves in autumn, then remain dormant during the winter.

❭ Year-round warmth and rains support tropical rain forests, the most productive, structurally complex, and species-rich land biome.

> Conifers withstand harsher conditions than broadleaf trees, so they grow farther north and at higher altitudes.
< Link to Conifers 21.6

Conifers (evergreen trees with seed-bearing cones) are the main plants in coniferous forests. Conifer leaves are typically needle-shaped, with a thick cuticle and stomata that are sunk below the leaf surface. These adaptations help conifers conserve water during drought or times when the ground is frozen. As a group, conifers tolerate poorer soils and drier habitats than deciduous trees.

The most extensive land biome is the coniferous forest that sweeps across northern Asia, Europe, and North America (Figure 29.21A). It is known as **boreal forest**, or taiga, which means "swamp forest" in Russian. The conifers are mainly pine, fir, and spruce. Most rain falls in the summer, and little evaporates into the cool summer air. Winters are long, cold, and dry. Moose are dominant grazers in this biome.

Also in the Northern Hemisphere, montane coniferous forests extend southward through the great mountain ranges (Figure 29.21B). Spruce and fir dominate at the highest elevations. At lower elevations, the mix becomes firs and pines.

Conifers also dominate temperate lowlands along the Pacific coast from Alaska into northern California. These coniferous forests hold the world's tallest trees, Sitka spruce to the north and coast redwoods to the south.

We find other conifer-dominated ecosystems in the eastern United States. About a quarter of New Jersey is pine barrens, a mixed forest of pitch pines and scrub oaks that grow in sandy, acidic soil. Pine forest covers about one-third of the Southeast. Fast-growing loblolly pines dominate these forests and are a major source of lumber and wood pulp. The pines can survive periodic fires that kill most hardwood species. When fires are suppressed, hardwoods outcompete pines.

boreal forest Extensive high-latitude forest of the Northern Hemisphere; conifers are the predominant vegetation.

Take-Home Message What are coniferous forests?

> Conifers prevail across the Northern Hemisphere's high-latitude forests, at high elevations, and in temperate regions with nutrient-poor soils.

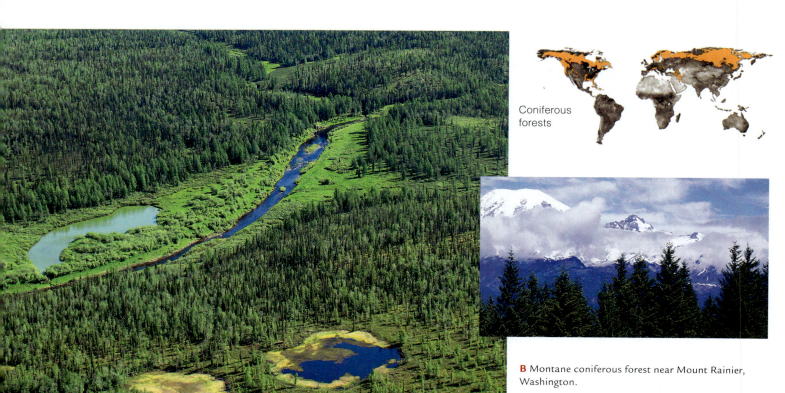

Coniferous forests

A Boreal forest (taiga) in Siberia.

B Montane coniferous forest near Mount Rainier, Washington.

Figure 29.21 Coniferous forests.

> Low-growing, cold-tolerant plants have only a brief growing season on the tundra.
< Link to Carbon and global warming 28.8

Arctic tundra

Arctic Tundra

Arctic tundra forms between the polar ice cap and the belts of boreal forests in the Northern Hemisphere. Most is in northern Russia and Canada. Arctic tundra is Earth's youngest biome, having first appeared about 10,000 years ago when glaciers retreated at the end of the last ice age.

Conditions in this biome are harsh; snow blankets arctic tundra for as long as nine months of the year. Annual precipitation is usually less than 25 centimeters (10 inches), but cold temperature keeps the snow that does fall from melting. During a brief summer, plants grow fast under the nearly continuous sunlight (Figure 29.22). Lichens and shallow-rooted, low-growing plants are the producers for food webs that include voles, arctic hares, caribou, arctic foxes, wolves, and brown bears. Enormous numbers of migratory birds nest here in the summer.

Only the surface layer of tundra soil thaws during summer. Below that lies **permafrost**, a frozen layer 500 meters (1,600 feet) thick in places. Permafrost acts as a barrier that prevents drainage, so the soil above it remains perpetually waterlogged. The cool, anaerobic conditions in this soil slow decay, so organic remains can build up. Organic matter in the permafrost makes the arctic tundra one of Earth's greatest stores of carbon.

As global temperatures rise, the amount of frozen soil that melts each summer is increasing. With warmer temperatures, much of the snow and ice that would otherwise reflect sunlight is disappearing. As a result, newly exposed dark soil absorbs heat from the sun's rays, which encourages more melting.

Alpine Tundra

Alpine tundra occurs at high altitudes throughout the world (Figure 29.23). Even in the summer, some patches of snow persist in shaded areas, but there is no permafrost. The alpine soil is well drained, but thin and nutrient-poor. As a result, primary productivity is low. Grasses, heaths, and small-leafed shrubs grow in patches where soil has accumulated to a greater depth. These low-growing plants can withstand the strong winds that discourage the growth of trees.

Figure 29.22 Arctic tundra in the summer. Permafrost underlies the soil.

Figure 29.23 Alpine tundra. Low-growing, hardy plants at a high altitude in Washington's Cascade range.

alpine tundra Biome of low-growing, wind-tolerant plants adapted to high-altitude conditions.
arctic tundra Highest-latitude Northern Hemisphere biome, where low, cold-tolerant plants survive with only a brief growing season.
permafrost Continually frozen soil layer that lies beneath arctic tundra and prevents water from draining.

Take-Home Message What is tundra?

> Arctic tundra prevails at high latitudes, where short, cold summers alternate with long, cold winters.
> Alpine tundra prevails in high, cold mountains across all latitudes.

> Gradients in light penetration, temperature, and dissolved gases affect the distribution of life in aquatic habitats.
> Links to Properties of water 2.5, Eutrophication 28.1, Indicator species 27.8

With this section, we turn our attention to Earth's waters. We begin here with freshwater systems, continue to coasts, then dive into the oceans.

Lakes

A lake is a body of standing fresh water. If sufficiently deep, it will have zones that differ in their physical characteristics and species composition (Figure 29.24). Nearest shore is the littoral zone. Here, sunlight penetrates all the way to the lake bottom and aquatic plants are primary producers. A lake's open waters include an upper, well-lit limnetic zone, and a profundal zone where light does not penetrate. Primary producers in the limnetic zone are members of the phytoplankton, a group of photosynthetic microorganisms that includes green algae, diatoms, and cyanobacteria. They serve as food for zooplankton, which are tiny consumers such as copepods. In the profundal zone, there is not enough light for photosynthesis, so consumers here depend on food produced above. Debris that drifts down feeds detritivores and decomposers.

Nutrient Content and Succession Like a habitat on land, a lake undergoes succession; it changes over time. A newly formed lake is oligotrophic: deep, clear, and nutrient-poor, with low primary productivity (Figure 29.25). Over time, the lake becomes eutrophic. Again, eutrophication refers to processes, either natural or artificial, that enrich a body of water with nutrients. Increased nutrients allow producers to grow and productivity rises.

limit of effective light penetration

Figure 29.24 Animated Lake zonation. A lake's littoral zone extends all around the shore to a depth where rooted aquatic plants stop growing. Its limnetic zone is the open water where light penetrates and photosynthesis occurs. Below the limnetic zone is the cooler, dark water of the profundal zone.

Figure 29.25 An oligotrophic lake. Crater Lake in Oregon is a collapsed volcano that filled with snow melt. It began filling about 7,700 years ago; from a geologic standpoint, it is a young lake.

Seasonal Changes Temperate zone lakes undergo seasonal changes that affect primary productivity. Unlike most substances, water is not most dense in its solid state (ice). As water cools, its density increases until it reaches 4°C (39°F). Below this temperature, any additional cooling decreases water's density—which is why ice floats on water (Section 2.5). In an ice-covered lake, water just under the ice is near freezing and at its lowest density. The densest (4°C) water is at the bottom (Figure 29.26A).

In spring, winds cause vertical currents that lead to a spring overturn, during which oxygen-rich water in the surface layers moves down and nutrient-rich water from the lake's depths moves up (Figure 29.26B). After the spring overturn, longer days and the dispersion of nutrients through the water encourage primary productivity.

In summer, a lake has three layers (Figure 29.26C). The upper layer is warm and oxygen-rich. Below this is a thermocline, a thin layer where temperature falls rapidly. Beneath the thermocline is the coolest water. The upper and lower waters on either side of this boundary do not mix. As a result, decomposers deplete oxygen dissolved near the lake bottom, and nutrients near the lake bottom cannot escape into surface waters. Nutrient shortages limit growth and production declines.

In autumn, the lake's upper waters cool, the thermocline vanishes, and a fall overturn occurs (Figure 29.26D). Oxygen-rich water moves down while nutrient-rich water moves up. The overturn in the fall brings nutrients to the surface and favors a brief burst of primary productivity. However, unlike the spring overturn, it does not lead to sustained production because decreasing light and temperature slows photosynthesis. Primary productivity will not peak again until after the next spring overturn.

A Winter. Ice covers the thin layer of slightly warmer water just below it. Densest (4°C) water is at bottom. Winds do not affect water under the ice, so there is little circulation.

B Spring. Ice thaws. Upper water warms to 4°C and sinks. Winds blow across the lake causing currents that help overturn water, bringing nutrients up from the bottom.

C Summer. Sun warms the upper water, which floats on a thermocline, a layer across which temperature changes abruptly. Waters above and below the thermocline do not mix.

D Fall. Upper water cools and sinks, thus causing the thermocline to disappear. Vertical currents can now mix waters that remained separate during summer.

Figure 29.26 Season changes in a temperate zone lake.

Streams and Rivers

Flowing-water ecosystems start as freshwater springs or seeps. As they flow downslope, they grow and merge. Rainfall, snowmelt, geography, altitude, and shade cast by plants affect flow volume and temperature. The composition of the rocks under flowing water can affect the water's solute concentrations. Because water in different parts of a river moves at different speeds, contains different solutes, and differs in temperature, the species composition of a river varies along its length (Figure 29.27).

The Importance of Dissolved Oxygen

The dissolved oxygen content of water is one of the most important factors affecting aquatic organisms. More oxygen dissolves in cooler, fast-flowing water than in warmer, still water. When water temperature increases or water becomes stagnant, aquatic species that have high oxygen needs suffocate.

In freshwater habitats, aquatic larvae of mayflies and stoneflies are the first invertebrates to disappear when oxygen levels fall. These insect larvae are active predators that demand considerable oxygen, so they serve as indicator species. Gilled snails disappear, too. Declines in populations of invertebrates can have cascading effects on the fishes that feed on them. Fishes can also be more directly affected. Trout and salmon are especially intolerant of low oxygen. Carp (including koi and goldfish) are among the most tolerant; they survive even in tepid algae-rich ponds and in tiny goldfish bowls.

No fishes can survive when the oxygen content of water falls below 4 parts per million. Leeches thrive as most competing invertebrates disappear. In waters with the lowest oxygen concentration, annelids called sludge worms (*Tubifex*) often are the only animals. The worms

Figure 29.27 Variation in properties along the length of a river. Smoothly flowing water (*left*), holds less oxygen than water that mixes with air as it runs over rocks (*right*).

are colored red by large amounts of hemoglobin. A high affinity for oxygen is an adaptation that allows these worms to exploit low-oxygen habitats where predators and competition for food are scarce.

Take-Home Message What factors affect life in freshwater habitats?

> Lakes have gradients in light, dissolved oxygen, and nutrients.

> Primary productivity varies with a lake's age and—in temperate zones—with the season.

> Rivers move nutrients into and out of ecosystems. Characteristics such as temperature and nutrient content typically vary along the length of a river.

> Species differ in their dissolved oxygen needs. Cold, fast-moving water holds more oxygen than still, warm water.

> Near the coasts of continents and islands, concentrations of nutrients support some of the world's most productive aquatic ecosystems.
< Link to Food chains 28.3

Coastal Wetlands

An **estuary** is an enclosed coastal region where seawater mixes with nutrient-rich fresh water from rivers and streams (Figure 29.28A). Water inflow continually replenishes nutrients, so estuaries are often highly productive. Primary producers include algae and other types of phytoplankton, along with plants that tolerate submergence at high tide. Detrital food chains predominate. Estuaries are marine nurseries; many larval and juvenile invertebrates and fishes live in them.

In tidal flats at tropical latitudes, we find nutrient-rich mangrove wetlands. Mangroves are salt-tolerant woody plants that live in sheltered areas along tropical coasts. The plants have prop roots that extend out from the trunk (Figure 29.28B). Specialized cells at the surface of some roots allow gas exchange.

Rocky and Sandy Shores

Organisms that live along ocean shores are adapted to withstand the force of the waves and repeated tidal changes. Many species are underwater during high tide, but are exposed to the air when the tide is low.

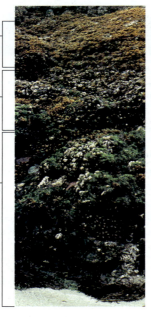

intertidal zone's upper littoral; submerged only at highest tide of lunar cycle

midlittoral; submerged at each highest regular tide and exposed at lowest tide

lower littoral; exposed only at low tide of lunar cycle

Figure 29.29 Vertical zonation in the intertidal zone.

Tide height varies over an approximately monthly cycle as a result of the position of the moon and Earth. Biologists divide a shoreline into three zones (Figure 29.29). The *upper* littoral zone is submerged only at the highest tide of a lunar cycle. It holds the fewest species. The *mid*littoral zone is submerged during the highest regular tide and exposed at the lowest tide. The *lower* littoral zone, exposed only during the lowest tide of the lunar cycle, holds the most diverse collection of species.

Along rocky shores, where waves prevent detritus from piling up, algae that cling to rocks are the

A South Carolina salt marsh dominated by marsh grass (*Spartina*).

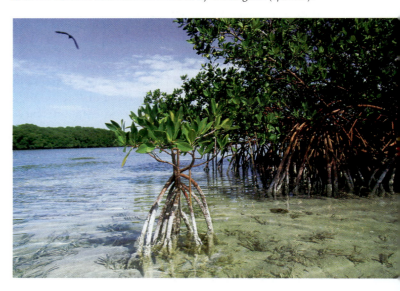

B Florida mangrove wetland.

Figure 29.28 Two types of coastal wetlands.

producers in grazing food chains. In contrast, waves that continually rearrange loose sediments along sandy shores make it difficult for algae to take hold. Here, detrital food chains start with organic debris from land or offshore.

Take-Home Message What are some coastal ecosystems?

> Estuaries are highly productive areas where fresh water and seawater mix.

> Mangrove wetlands form along tropical tidal flats.

> Grazing food chains predominate on rocky shores, and detrital food chains on sandy shores.

estuary A highly productive ecosystem where nutrient-rich water from a river mixes with seawater.

> Coral reefs are highly diverse—and highly threatened—marine ecosystems.

‹ Link to Corals 23.5

Coral reefs are wave-resistant formations that consist primarily of calcium carbonate secreted by generations of coral polyps. Reef-forming corals live mainly in shallow, clear, warm waters between latitudes 25° north and 25° south. About 75 percent of all coral reefs are in the Indian and Pacific oceans. A healthy reef is home to living corals and a huge number of other species (Figure 29.30). Biologists estimate that about a quarter of all marine fish species are associated with coral reefs.

The largest existing reef, Australia's Great Barrier Reef, parallels Queensland for 2,500 kilometers (1,550 miles), and is the largest example of biological architecture. Scientists estimate it began forming about 600,000 years ago. Today it is a string of reefs, some 150 kilometers (95 miles) across. The Great Barrier Reef supports about 500 coral species, 3,000 fish species, 1,000 kinds of mollusks, and 40 kinds of sea snakes.

Photosynthetic dinoflagellates live as symbionts inside the tissues of all reef-building corals (Section 23.5). The dinoflagellates find protection in the coral's tissues and provide the coral polyp with oxygen and sugars that it depends on. When stressed, coral polyps expel the dinoflagellates. Because the dinoflagellates give a coral its color, expelling these protists turns the coral white, an event called **coral bleaching**. When a coral is stressed for more than a short time, the dinoflagellate population in a coral's tissues cannot rebound and the coral dies, leaving only its bleached hard parts behind (Figure 29.31).

The incidence of coral bleaching events has been increasing. Rising sea temperatures and sea level associated with global climate change most likely play a role. People also stress reefs by discharging sewage and other pollutants into coastal waters, by causing erosion that clouds water with sediments, and by destructive fishing practices. Fishing nets break pieces off corals. Fishermen hoping to capture reef fishes for the pet trade use explosives or sodium cyanide to stun the fishes, and destroy corals in the process. Invasive species also threaten reefs. Hawaiian reefs are threatened by exotic algae, including several species imported for cultivation during the 1970s.

Assaults on reefs are taking a huge toll. For example, the Indo-Pacific region, the global center for reef diversity, lost about 3,000 square kilometers (1,160 square miles) of living coral reef each year between 1997 and 2003.

coral bleaching A coral expels its photosynthetic dinoflagellate symbionts in response to stress and becomes colorless.
coral reef Highly diverse marine ecosystem centered around reefs built by living corals that secrete calcium carbonate.

Figure 29.30 Healthy coral reef near Fiji. The coral gets its color from pigments of symbiotic dinoflagellates that live in the its tissues and supply it with sugars.

Figure 29.31 "Bleached" reef near Australia. The coral skeletons shown here belong mainly to staghorn coral (*Acropora*), a genus especially likely to undergo coral bleaching.

Take-Home Message **What are coral reefs, how are they threatened, and why is reef loss a matter of concern?**

> Coral reefs form by the action of living corals that lay down a calcium carbonate skeleton. Photosynthetic dinoflagellates in the coral's tissues are necessary for the coral's survival.

> Rising water temperature, pollutants, fishing, and exotic species contribute to loss of reefs.

> Declines in coral reefs will affect the enormous number of fishes and invertebrate species that make their home on or near the reefs.

> From its upper, brightly lit waters to the hydrothermal vents on its deep, dark floor, the ocean is filled with life.
‹ Link to Modes of nutrition 19.5

As in fresh water, gradients of light and temperature affect the distribution of marine life. We divide the ocean into two regions: pelagic and benthic provinces (Figure 29.32).

The Pelagic Province

The ocean's open waters are the **pelagic province**. This province includes the water over continental shelves and the more extensive waters farther offshore. In the ocean's upper, bright waters, phytoplankton such as single-celled algae and bacteria are the primary producers, and grazing food chains predominate.

Depending on the region, some light may penetrate as far as 1,000 meters (more than a half mile) beneath the sea surface. Below that, organisms live in continual darkness, and organic material that drifts down from above serves as the basis of detrital food chains.

The Benthic Province

The **benthic province** is the ocean bottom, its rocks and sediments. Species richness is greatest on continental shelves (the underwater edges of continents). The benthic province also includes largely unexplored species-rich

Figure 29.33 Seamounts.
A Computer model of three seamounts on the sea floor off the coast of Alaska. Patton Seamount, at the rear, stands 3.6 kilometers (about 2 miles) high.
B Flytrap anemone, from Davidson Seamount near the California coast.

regions, including seamounts and the regions around hydrothermal vents.

Seamounts are undersea mountains that stand 1,000 meters or more tall, but are still below the sea surface (Figure 29.33A). They attract large numbers of fishes and are home to many marine invertebrates (Figure 29.33B). Like islands, seamounts often are home to species that evolved there and are found nowhere else.

The abundance of life at seamounts makes them attractive to commercial fishing vessels. Fishes and other organisms are often harvested by trawling, a fishing technique in which a large net is dragged along the bottom, capturing everything in its path. The process is ecologically devastating; trawled areas are stripped bare of life, and the silt stirred up by the giant, weighted nets suffocates filter-feeders in adjacent areas.

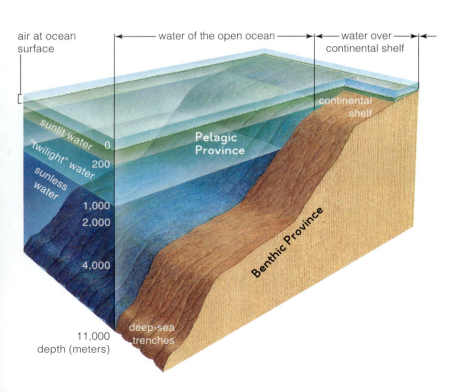

Figure 29.32 Animated Oceanic zones. Dimensions of the zones are not to scale.

---

benthic province The ocean's sediments and rocks.
hydrothermal vent Place where hot, mineral-rich water streams out from an underwater opening in Earth's crust.
pelagic province The ocean's waters.
seamount An undersea mountain.

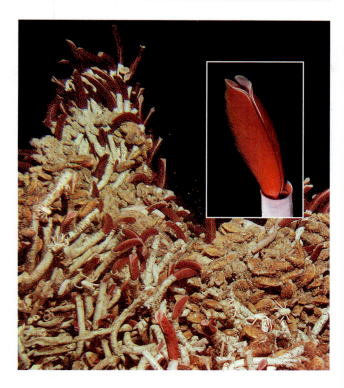

Figure 29.34 Members of a hydrothermal vent community. Tube worms (annelids) and crabs are among the consumers. The producers are chemoautotrophic bacteria and archaeans.

At **hydrothermal vents**, superheated water rich in dissolved minerals spews out from an opening on the ocean floor. The water is seawater that seeped into cracks in the ocean floor at the margins of tectonic plates and was heated by geothermal energy. Where this mineral-rich water mixes with the cold deep-sea water, minerals settle out and form extensive deposits. Chemoautotrophic bacteria and archaeans that obtain energy by removing electrons from minerals are the primary producers for food webs that include diverse invertebrates, including large numbers of tube worms (Figure 29.34).

Take-Home Message What factors shape marine communities?

> In the pelagic province's upper waters, photosynthesis supports grazing food chains. In deeper, darker waters of this province, organisms feed mainly on detritus that drifts down from above.

> The benthic province has pockets of high species diversity at undersea mountains (seamounts) and near hydrothermal vents. A hydrothermal vent ecosystem does not run on energy from the sun; the producers are chemoautotrophs rather than photoautotrophs.

Effects of El Niño (revisited)

An El Niño event can have surprising effects. For example, it can increase the number of cases of cholera. During the 1997–1998 El Niño event, 30,000 cases of cholera were reported in Peru alone, compared with only 60 cases from January to August in 1997.

The bacterium, *Vibrio cholerae*, shown at the *right*, causes cholera. Severe diarrhea is the main symptom. During a cholera outbreak, bacteria-contaminated feces often enter the water supply. People who drink or wash food with the tainted water become infected.

Marine biologist Rita Caldwell figured out how a change in the ocean temperature can affect the incidence of cholera in humans. Caldwell discovered that copepods, a type of small crustacean, serve as a reservoir for cholera-causing bacteria between disease outbreaks. During an El Niño, the rise in the temperature of surface waters causes a rise in abundance of the phytoplankton that the copepods feed on, so the number of disease-carrying copepods increases. When Caldwell analyzed records of sea surface temperature and cholera outbreaks in the Bay of Bengal, she found that reports of cholera cases rise four to six weeks after an increase in water temperature. Caldwell continues to research how the environment influences the incidence of cholera. She also investigates simple, inexpensive methods that people can use to avoid the disease (Figure 29.35).

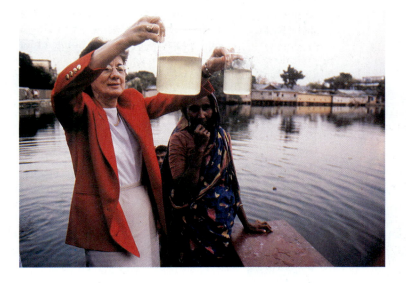

Figure 29.35 Rita Caldwell examining filtered and unfiltered water. She advised women in Bangladesh to use sari cloth as a filter to remove disease-carrying copeods from the water. The copepod hosts are too big to pass through the thin cloths, which can be rinsed in clean water, sun-dried, and used again. This inexpensive, simple method has cut cholera outbreaks by half.

How Would You Vote? Is supporting studies of El Niño and the other long-term climate cycles a good use of government funds? See CengageNow for details, then vote online (cengagenow.com).

Summary

Section 29.1 Interactions among Earth's air and waters, as when oceans warm during an **El Niño** and then cool during a **La Niña**, affect weather throughout the **biosphere**. Such events alter ocean currents and rainfall patterns that influence primary production, thus affecting biological communities.

Sections 29.2, 29.3 Global air circulation patterns affect **climate** and the distribution of communities. The patterns are set into motion when sunlight heats tropical regions more than higher latitudes. Ocean currents distribute heat energy worldwide and influence weather patterns. Interactions between ocean currents, air currents, and landforms determine where regional phenomena such as **rain shadows** or **monsoons** occur.

Section 29.4 Biomes are discontinuous areas characterized by a particular type of vegetation. Differences in climate, elevation, and soil properties affect the distribution of biomes.

Sections 29.5–29.7 Deserts form around latitudes 30° north and south and in rain shadows. **Grasslands** dominated by plants adapted to fire and grazing form in the somewhat moister interior of midlatitude continents. Shrubby, fire-adapted **chaparral** is common in California and other coastal regions with hot, dry summers and cool, wet winters.

Section 29.8 Broadleaf trees are angiosperms. Those in **temperate deciduous forests** shed their leaves all at once just before a cold winter that prevents growth. Broadleaf trees in **tropical rain forests** can grow year-round, and these enormously productive forests are home to a large number of species.

Sections 29.9, 29.10 Conifers withstand cold and drought better than broadleaf trees and dominate Northern Hemisphere high latitude **boreal forests**. Farther north in this hemisphere, **arctic tundra** overlies **permafrost. Alpine tundra** occurs at high altitudes.

Section 29.11 Lakes undergo succession, becoming more nutrient-rich over time. In temperate zone lakes, seasonal variation in water circulation patterns affects productivity. Properties of a river vary along its length. Fast-moving, cool water holds the most oxygen.

Sections 29.12, 29.13 Nutrient-rich freshwater mixes with seawater in an **estuary**. In tropical regions, mangrove wetlands form on tidal flats. Along rocky shores, algae form the base for grazing food chains.

On sandy shores, detrital food chains predominate. All tidal zones show a vertical zonation, with most diversity in the region submerged during all but the lowest tides.

The accumulated calcium carbonate skeletons of coral polyps form **coral reefs** in shallow tropical seas, especially in the Indo-Pacific Ocean. Many marine species associate with reefs. When stressed, a coral may eject its photosynthetic symbionts, an event called **coral bleaching**.

Section 29.14 Life exists throughout the ocean. In the **pelagic province**, diversity is highest in sunlit waters, where grazing food chains predominate. Detritus forms the base for food chains in deeper, darker waters. **Seamounts** are regions of high diversity in the **benthic province**. At **hydrothermal vents**, chemoautotrophic bacteria and archaeans are the producers; they get energy from mineral-rich seawater emitted from the vent.

Self-Quiz   Answers in Appendix III

1. The Northern Hemisphere is most tilted toward the sun in _____ .
   a. spring   b. summer   c. autumn   d. winter

2. Which latitude will have the most hours of daylight on the summer solstice?
   a. 0° (the equator)   c. 45° north
   b. 30° north   d. 60° north

3. Warm air _____ and it holds _____ water than cold air.
   a. sinks; less   c. sinks; more
   b. rises; less   d. rises; more

4. A rain shadow is a reduction in rainfall _____ .
   a. on the inland side of a coastal mountain range
   b. during an El Niño event
   c. that results from global warming

5. The Gulf Stream is a current that flows _____ along the eastern coast of the United States.
   a. north to south   b. south to north

6. _____ have a deep layer of humus-rich topsoil.
   a. Deserts   c. Rain forests
   b. Grasslands   d. Seamounts

7. Biome distribution depends on _____ .
   a. climate   c. soils
   b. elevation   d. all of the above

8. Grasslands most often are found _____ .
   a. at 30° north and south   c. in interior of continents
   b. at high altitudes   d. all of the above

9. Permafrost underlies _____ .
   a. arctic tundra   c. boreal forest
   b. alpine tundra   d. all of the above

10. Warm, still water holds _____ oxygen than cold, fast-flowing water.

Data Analysis Activities

Changing Sea Temperatures In an effort to predict El Niño or La Niña events in the near future, the National Oceanographic and Atmospheric Administration collects information about sea surface temperature (SST) and atmospheric conditions. Scientists compare monthly temperature averages in the eastern equatorial Pacific Ocean to historical data and calculate the difference (degree of anomaly) to see if El Niño conditions, La Niña conditions, or neutral conditions are developing. El Niño is a rise in the average SST above 0.5°C. A decline of the same amount is La Niña. **Figure 29.36** shows data for nearly 39 years.

1. When did the greatest positive temperature deviation occur during this time period?
2. What type of event, if any, occurred during the winter of 1982–1983? What about the winter of 2001–2002?
3. During a La Niña event, less rain than normal falls in the American West and Southwest. In the time interval shown, what was the longest interval without a La Niña event?
4. What type of conditions were in effect in the fall of 2007 when California suffered severe wildfires?

Figure 29.36 Sea surface temperature anomalies (differences from the historical mean) in the eastern equatorial Pacific Ocean. A rise above the dashed *red* line is an El Niño event, a decline below the *blue* line is La Niña.

---

11. Chemoautotrophic bacteria and archaeans are the primary producers for food webs _____ .
    a. in mangrove wetlands   c. on coral reefs
    b. at seamounts   d. at hydrothermal vents

12. Corals rely on symbiotic _____ for sugars.
    a. fungi   c. dinoflagellates
    b. bacteria   d. green algae

13. What biome borders on boreal forest?
    a. savanna   c. tundra
    b. taiga   d. chaparral

14. Unrelated species living in geographically separated parts of a biome may resemble one another as a result of _____ .
    a. competitive interactions
    b. morphological convergence
    c. morphological divergence
    d. coevolution

15. Match the terms with the most suitable description.
    ___ tundra   a. broadleaf forest near equator
    ___ chaparral   b. partly enclosed by land; where
    ___ desert   fresh water and seawater mix
    ___ savanna   c. African grassland with trees
    ___ estuary   d. low-growing plants at
    ___ boreal forest   high latitudes or elevations
    ___ prairie   e. dry shrubland
    ___ tropical rain   f. at latitudes 30° north and south
        forest   g. mineral-rich, superheated
    ___ hydrothermal   water supports communities
        vents   h. conifers dominate
                        i. North American grassland

Additional questions are available on **CENGAGENOW**.

Critical Thinking

1. On April 26, 1986, a meltdown occurred at the Chernobyl nuclear power plant in Ukraine. Nuclear fuel burned for nearly ten days and released 400 times more radioactive material than the atomic bomb that dropped on Hiroshima. Winds carried radioactive fallout around the globe. By 1998, the rate of thyroid abnormalities in children living downwind from the site was nearly seven times as high as for those upwind; their thyroid gland concentrated the iodine radioisotopes. Chernoboyl is at 51° north latitude. In what direction did the major winds carry the fallout after the accident?

2. Owners of off-road recreational vehicles would like increased access to government-owned deserts. Some argue that it is the perfect place for off-roaders because "There's nothing there." Do you agree?

3. Rita Caldwell, the scientist who discovered why cholera outbreaks often occur during an El Niño, is concerned that global climate change could increase the incidence of this disease. By what mechanism might global warming cause an increase in cholera outbreaks?

Animations and Interactions on **CENGAGENOW**:
❯ Earth's tilt and the seasons; Air circulation patterns; Major ocean currents; Rain shadow effect; Coastal breezes; Biome distribution; Oceanic zones.

❮ **Links to Earlier Concepts**

This chapter considers the causes of an ongoing mass extinction (Section 16.1) in light of human population growth (26.7). We look again at effects of species introductions (27.8), deforestation (21.1, 29.8), coral bleaching (29.13), and mercury pollution (2.1). You will draw on your knowledge of pH (2.6), the ozone layer (18.5), trophic levels (28.3), and water and nutrient cycles (28.7–28.9).

Key Concepts

An Extinction Crisis
Extinction is a natural process, but human activities have increased the frequency of extinction events. Species go extinct when they are adversely affected by habitat destruction, degradation, and fragmentation. The extent of species losses is not fully known.

Harmful Land Uses
Plowing under grasslands and cutting down forests can have long-term and long-range effects. Loss of plants allows soil erosion, raises soil temperature, and affects rainfall patterns in ways that make it difficult to restore plant cover.

30 Human Effects on the Biosphere

30.1 A Long Reach

We began this book with the story of biologists who ventured into a remote forest in New Guinea, and their excitement at the many previously unknown species that they encountered (Section 1.1). At the far end of the globe, a U.S. submarine surfaced in Arctic waters and discovered polar bears hunting on the ice-covered sea. The polar bears were about 445 kilometers (270 miles) from the North Pole and 805 kilometers (500 miles) from the nearest land (Figure 30.1).

Even such seemingly remote regions are no longer beyond the reach of human explorers—and human influence. You already know that the temperature of Earth's atmosphere and seas are rising. In the Arctic, unusually warm temperatures are affecting the seasonal cycle of sea ice melting and formation. In recent years, sea ice has begun to thin and to break up earlier in the spring and to form later in the fall. A decrease in the persistence of sea ice is bad news for polar bears. They can only reach their main prey—seals—by traveling across ice. A longer ice-free period means less time for bears to feed. In addition, an earlier seasonal ice melt raises the risk that polar bears hunting far from land will become stranded and unable to return to solid ground before the ice thaws.

Polar bears face other threats as well. They are top predators and their tissues contain a surprisingly high amount of mercury and organic pesticides. The pollutants entered the water and air far away, in more temperate regions. Winds and ocean currents deliver them to polar realms. Contaminants also travel north in the tissues of migratory animals such as seabirds that spend their winters in temperate regions and nest in the Arctic.

In places less remote than the Arctic, human activities have a more direct effect. As we cover more and more of the world with our dwellings, factories, and farms, less appropriate habitat remains for other species. We also put species at risk by competing with them for resources, overharvesting them, and introducing non-native competitors.

Figure 30.1 Polar bears on ice. *Opposite page*, a polar bear investigates an American submarine that surfaced in ice-covered Arctic waters.

It would be presumptuous to think that we alone have had a profound impact on the world of life. As long ago as the Proterozoic, photosynthetic cells were irrevocably changing the course of evolution by enriching the atmosphere with oxygen. Over life's existence, the evolutionary success of some groups ensured the decline of others. What is new is the increasing pace of change and the capacity of our own species to recognize and affect its role in this increase.

A century ago, Earth's physical and biological resources seemed inexhaustible. Now we know that many practices put into place when humans were largely ignorant of how natural systems operate take a heavy toll on the biosphere. The rate of species extinctions is on the rise and many types of biomes are threatened. These changes, the methods scientists use to document them, and the ways that we can address them, are the focus of this chapter.

Effects of Pollutants
Some pollutants fall to Earth as acid rain that can harm aquatic communities and soils. Other pollutants disrupt metabolism when they accumulate to high levels in animal bodies. Still other pollutants harm the protective ozone layer or contribute to global climate change.

Conserving Biodiversity
Earth's biodiversity is the product of billions of years of evolution, and our well-being depends on sustaining it. Conservation biologists help us prioritize which areas to protect first by assessing which are the most threatened and most biodiverse.

Reducing Negative Impacts
Extraction of fuel and other resources harms the environment. Individual actions that minimize the use of resources and energy can help reduce the threats to the health of the planet and to biodiversity.

> Extinction is a natural process, but human activities are raising the rate at which other species disappear.

< Links to Mass extinction 16.1, Codfish population crash 26.6, Human population growth 26.7, Species interactions and introductions 27.8, Nitrogen cycle 28.9

The rising size of the human population and increasing industrialization have far-reaching effects on the biosphere. We begin by discussing effects on individual species, then turn to the wider impacts.

Threatened and Endangered Species

Extinction, like speciation, is a natural process. Species arise and become extinct on an ongoing basis. The rate of extinction picks up dramatically during a mass extinction, when many kinds of organisms in many different habitats become extinct in a relatively short period. We are currently in the midst of such an event. Unlike historical mass extinctions, this one does not stem from some natural catastrophe such as an asteroid impact. Rather, this mass extinction is the outcome of the success of a single species—humans—and their effects on Earth.

An **endangered species** is a species that faces extinction in all or part of its range. A **threatened species** is one that is likely to become endangered in the near future. Keep in mind that not all rare species are threatened or endangered. Some species have always been uncommon. A species is considered endangered when one or more of its populations have declined or are declining.

Causes of Species Declines

When European settlers first arrived in North America, they found between 3 and 5 billion passenger pigeons. In the 1800s, commercial hunting caused a steep decline in the bird's numbers. The last time anyone saw a wild passenger pigeon was 1900—and he shot it. The last captive bird died in 1914.

We are still overharvesting species. The crash of the Atlantic codfish population, described in Section 40.6, is one recent example. Another is the fate of the white abalone, a gastropod mollusk native to kelp forests off the coast of California (Figure 30.2A). Heavy harvesting of this species during the 1970s reduced the population to about 1 percent of its original size. In 2001, it became the first invertebrate to be listed as endangered by the United States Fish and Wildlife Service. Although some white abalone remain in the wild, the population density is too low for effective reproduction. The species' only hope for survival is a program of captive breeding. If this program succeeds, individuals can be reintroduced to the wild.

Overharvest directly reduces population size, but we also affect species by altering their habitat. Each species requires a specific type of habitat, and any degradation,

A White abalone   **B** Panda

Figure 30.2 Some threatened and endangered species. To learn about others, visit the International Union for Conservation of Nature and Natural Resources web site at www.iucnredlist.org/.

fragmentation, or destruction of that habitat reduces population numbers.

An **endemic species**, one that is confined to the limited area in which it evolved, is more likely to go extinct than a species with a more widespread distribution. For example, giant pandas are endemic to China's bamboo forests (Figure 30.2B). As China's human population soared, bamboo was cut for building materials and to make room for farms. As the bamboo forests disappeared, so did the pandas. Their numbers, which may once have been as high as 100,000, have fallen to 1,000 or so animals in the wild.

Similarly, logging of forests in the southeastern United States reduced numbers of ivory-billed woodpeckers endemic to these forests (Figure 30.2C). The birds are listed as endangered, but may be extinct. Other species endangered by logging of their habitat include Costa Rica's harlequin frog (Figure 30.2D) and the gorillas of the Congo (Figure 30.2E).

Habitat loss also affects plants. Plowing of North American prairies threatens prairie fringed orchids (Figure 30.2F). Destruction of the Indonesian rain forest has put all species of *Rafflesia*, the plants with world's largest flowers, on the endangered species list (Figure 30.2G).

Human activities that degrade, rather than destroy, habitats also endanger species. Texas blind salamanders (Figure 30.2H) are among the species endemic to the Edwards aquifer, a series of water-filled, underground limestone formations. Excessive withdrawals of water, along with water pollution, threatens species in the aquifer. Similarly, development of beaches interferes with reproduction of sea turtles such as leatherbacks that lay their eggs on these shores (Figure 30.2I). As a final example, manatees (Figure 30.2J) in Florida's waters are injured by motor boat propellers and get tangled in fishing line.

Deliberate or accidental species introductions also can threaten a species (Section 27.8). Rats that reached islands by stowing away on ships attack and endanger many ground-nesting birds that evolved in the absence of egg-eating ground predators. Exotic species also cause

C Ivory-billed woodpecker

D Harlequin frog

E Gorilla

F Prairie fringed orchid

G *Rafflesia*

H Texas blind salamander

I Leatherback sea turtle

J Manatee

K Koa bug

problems by outcompeting native ones. After European brown trout and eastern brook trout were introduced into California's mountain streams for sport fishing, populations of the native golden trout declined.

The koa bug (Figure 30.2K) is the victim of a biological control program gone awry. This Hawaiian native is threatened by an exotic parasitoid fly introduced to attack another bug that is an agricultural pest. Unfortunately, the fly prefers koa bugs to its intended target.

The decline or loss of one species can endanger others. For example, running buffalo clover and the buffalo that grazed on it were once common in the Midwest. The plants thrived in the open woodlands where soil was enriched by buffalo droppings and periodically disturbed by the animals' hooves. Buffalo helped to disperse the clover's seeds. When buffalo were hunted to near extinction, buffalo clover populations declined.

Like most endangered species, buffalo clover faces a number of simultaneous threats. In addition to the loss of the buffalo, the clover is threatened by conversion of its habitat to housing developments, competition from introduced plants, and attacks by introduced insects.

The Unknown Losses

In November of 2009, the International Union for Conservation of Nature and Natural Resources (IUCN) reported that of the 47,677 species they had assessed, 36 percent were threatened or endangered. We do not know the level of threat for the vast majority of the approximately 1.8 million named species, or for the countless millions of species yet to be discovered and named.

Endangered species listings have historically focused on vertebrates. Scientists have only recently begun to consider the threats to invertebrates and to plants. Our impact on protists and fungi is largely unknown, and the IUCN does not address threats to bacteria or archaeans.

Microbiologist Tom Curtis is among those making a plea for increased research on microbial ecology and microbial diversity. He argues that we have barely begun to comprehend the vast number of microbial species and to understand their importance. Curtis writes, "I make no apologies for putting microorganisms on a pedestal above all other living things. For if the last blue whale choked to death on the last panda, it would be disastrous but not the end of the world. But if we accidentally poisoned the last two species of ammonia-oxidizers, that would be another matter. It could be happening now and we wouldn't even know . . . " Ammonia-oxidizing bacteria play an essential role in the nitrogen cycle by converting ammonia in wastes and remains to nitrites. Without them wastes would pile up and plants would not have the nitrogen they need to grow.

---

endangered species A species that faces extinction in all or a part of its range.
endemic species Species that remains restricted to the area where it evolved.
threatened species Species likely to become endangered in the near future.

Take-Home Message How do humans threaten other species?

❯ Species often decline when humans destroy or fragment natural habitat by converting it to human use, or degrade it through pollution or withdrawal of an essential resource.

❯ Humans also directly cause declines by overharvesting species.

❯ Global travel and trade can introduce exotic species that harm native ones.

❯ The number of endangered species remains largely unknown.

> Around the globe, alteration of habitats as a result of farming, grazing, and logging contributes to species loss.
< Links to Deforestation 21.1, Tropical forest loss 29.8

With this section, we begin a survey of some of the ways that human activities threaten species by destroying or degrading habitat.

Desertification

Deserts naturally expand and contract over time as climate conditions vary. However, humans sometimes convert a dry grassland or woodland to desert, a process called **desertification**. Desertification occurs when plowing or grazing removes plants and so exposes topsoil to wind erosion. Plants cannot thrive where topsoil has blown away.

In a positive feedback cycle, drought encourages desertification, which worsens drought. Fewer plants means less transpiration. Because less water enters the atmosphere, local rainfall decreases.

Drought and poor agricultural practices are currently expanding Africa's Sahara Desert south into the Sahel region. Effects of this expansion are felt as far away as the southern United States and the Caribbean, where dust blown across the Atlantic eventually alights (Figure 30.3). Desertification also threatens China's northwestern region. Dust clouds that often darken skies above Beijing are a symptom of expansion of the Gobi Desert.

Deforestation

The amount of forested land is currently stable or increasing in North America, Europe, and China, but tropical forests continue to disappear at an alarming rate. Deforestation has detrimental effects beyond the immediate destruction of forest organisms. For example, deforestation encourages flooding because water runs off into

Figure 30.3 One symptom of ongoing desertification. Increasing amounts of dust now blow from Africa's Sahara Desert into and across the Atlantic Ocean.

streams, rather than being taken up by tree roots. Deforestation also raises risk of landslides in hilly areas. Tree roots tend to stabilize the soil. When they are removed, waterlogged soil becomes more likely to slide.

Deforested areas also become nutrient-poor. Figure 30.4 shows results of an experiment in which scientists deforested a region in New Hampshire and monitored the nutrients in runoff. Deforestation caused a spike in loss of essential soil nutrients such as calcium.

Like desertification, deforestation affects local weather. Temperatures in forests are cooler than in adjacent nonforested areas because trees shade the ground and transpiration causes evaporative cooling. When forest is cut down, daytime temperatures rise and reduced transpiration results in less rainfall.

Once a tropical forest is logged, the resulting nutrient losses and drier, hotter conditions can make it impossible for tree seeds to germinate or for seedlings to survive. Thus, deforestation can be difficult to reverse.

Because forests take up and store huge amounts of carbon, ongoing forest losses also contribute to global climate change.

---

desertification Conversion of dry grassland to desert.

A plot of forest was stripped of vegetation as an experiment.

After experimental deforestation, calcium levels in runoff increased sixfold (*purple*). An undisturbed control plot in the same forest showed no similar increase during this time (*light blue*).

Figure 30.4 Animated Experimental deforestation in the Hubbard Brook watershed.

Take-Home Message What are the effects of desertification and deforestation?

> With desertification, excess plowing or grazing causes soil to blow away. With fewer plants, rainfall declines.

> Deforestation increases flooding and loss of soil nutrients, raises the local temperature, and decreases rainfall. Changes in soil and temperature produced by deforestation make it difficult for new trees to become established.

30.4 Acid Rain

❯ Acid deposition contributes to forest destruction and degrades aquatic habitats.
❮ Link to pH 2.6

Pollutants are natural or man-made substances released into soil, air, or water in greater than natural amounts. The presence of a pollutant disrupts the physiological processes of organisms that evolved in its absence, or that are adapted to lower levels of it.

Sulfur dioxides and nitrogen oxides are examples of common air pollutants. Most sulfur dioxide comes from coal-burning power plants. Vehicles and power plants that burn gas and oil emit nitrogen oxides. Synthetic nitrogen-rich fertilizers also contribute nitrogen oxides to the atmosphere.

In dry weather, airborne sulfur and nitrogen oxides coat dust particles. Dry acid deposition occurs when the coated dust falls to the ground. Wet acid deposition, or **acid rain**, occurs when pollutants combine with water and fall as acidic precipitation. The pH of unpolluted rainwater is about 5 (Section 2.6). Acid rain can be ten times more acidic (Figure 30.5).

Acid rain that falls on or drains into waterways, ponds, and lakes affects aquatic organisms. Low pH prevents fish eggs from developing and kills adult fish. In New York's Adirondack Mountains, acid rain has left more than 200 lakes free of fish (Figure 30.5A).

When acid rain falls on forests it burns tree leaves and alters the composition of soils. As acidic water drains through the soil, positively charged hydrogen ions displace positively charged nutrient ions such as calcium, causing nutrient loss. The acidity also causes soil particles to release metals such as aluminum that can harm plants. The combination of poor nutrition and exposure to toxic aluminum weakens trees, making them more susceptible to insects and pathogens, and thus more likely to die (Figure 30.5B). Effects are most pronounced at higher elevations where trees are frequently exposed to clouds of acidic droplets.

---

acid rain Low pH rain that forms when sulfur dioxide and nitrogen oxides mix with water vapor in the atmosphere.
pollutant A substance that is released into the environment by human activities and interferes with the function of organisms that evolved in the absence of the substance or with lower levels.

Take-Home Message What is acid rain?

❯ Acid rain contains nitric or sulfuric acid. It forms when pollutants mix with water vapor, and it harms plants and aquatic organisms.

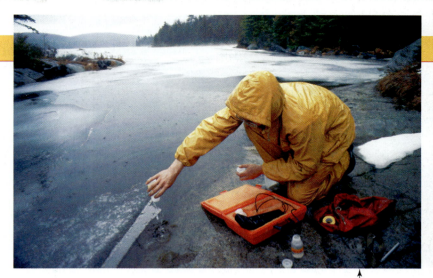

A Biologist testing the pH of water in New York's Woods Lake. In 1979, the lake's pH was 4.8. Since then, experimental addition of calcite to soil around the lake has raised the pH to more than 6.

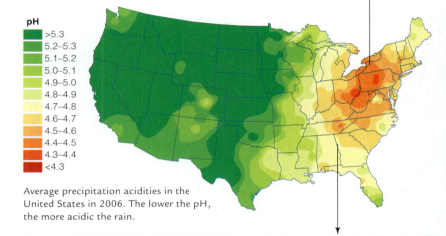

pH
- >5.3
- 5.2–5.3
- 5.1–5.2
- 5.0–5.1
- 4.9–5.0
- 4.8–4.9
- 4.7–4.8
- 4.6–4.7
- 4.5–4.6
- 4.4–4.5
- 4.3–4.4
- <4.3

Average precipitation acidities in the United States in 2006. The lower the pH, the more acidic the rain.

B Dying trees in Great Smoky Mountains National Park, where acid rain harms leaves and causes loss of nutrients from soil.

Figure 30.5 Animated Acid rain.

❯❯ **Figure It Out** Is rain more acidic on the East Coast or the West Coast?

Answer: The East Coast

Biological Effects of Chemical Pollutants

> Chemicals released by human activities can accumulate in the bodies of organisms and interfere with metabolism.
< Link to Mercury 2.1

Accumulation and Magnification

Industrial processes and the burning of fossil fuels put a vast array of chemical pollutants into Earth's air, water, and soils. Some of these pollutants can build up in the tissues of organisms. By the process of **bioaccumulation**, an organism's tissues store a pollutant taken up from the environment, causing the amount in the body to increase over time. The ability of some plants to bioaccumulate toxic substances makes them useful in phytoremediation of polluted soils.

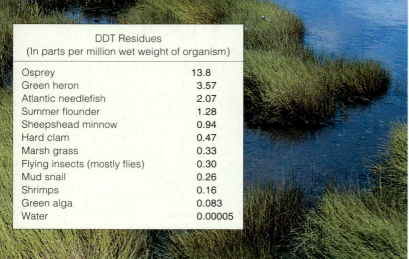

DDT Residues (In parts per million wet weight of organism)	
Osprey	13.8
Green heron	3.57
Atlantic needlefish	2.07
Summer flounder	1.28
Sheepshead minnow	0.94
Hard clam	0.47
Marsh grass	0.33
Flying insects (mostly flies)	0.30
Mud snail	0.26
Shrimps	0.16
Green alga	0.083
Water	0.00005

Figure 30.6 Biological magnification in an estuary on Long Island, New York, as reported in 1967 by George Woodwell, Charles Wurster, and Peter Isaacson. The *upper* photo shows an osprey, a top predator in this ecosystem.

In animals, hydrophobic chemical pollutants ingested or absorbed across the skin tend to accumulate in fatty tissues. Because the amount of pollutant in an animal's body increases over time, longer-lived species tend to be more affected by fat-soluble pollutants than shorter-lived ones. Within each species, older individuals tend to have a higher pollutant load than younger ones.

Pollutant concentration in tissues also varies with trophic level. By the process of **biological magnification**, the concentration of a chemical increases as the pollutant moves up a food chain. Figure 30.6 provides data documenting the biological magnification of DDT (a pesticide) in a salt marsh ecosystem during the 1960s. Notice that the concentration of DDT in osprey tissues was 276,000 times higher than that in the water. As a result of bioaccumulation and biological magnification, even seemingly low environmental concentrations of pollutants can have major detrimental effects on a species. During the 1960s, ospreys were driven toward extinction by widespread use of DDT, because this chemical interferes with their egg shell formation. A ban on DDT use in the United States has allowed the species to recover.

Bioaccumulation and magnification of other pollutants such as methylmercury continue to pose a threat to wildlife and to human health (Section 2.1).

Point and Nonpoint Sources

Some pollutants come from a few easily identifiable sites, or point sources. A factory that discharges pollutants into the air or water is a point source. Pollutants that come from point sources are the easiest to control: Identify the sources, and you can take action there.

Dealing with pollution from nonpoint sources is more challenging. Such pollution stems from widespread release of a pollutant. For example, oil that drips from vehicles is a nonpoint source of water pollution. Rain washes accumulated oil from pavement into natural waterways or storm drains. In many places, water that flows into storm drains is discharged into a body of water without any treatment. Halting nonpoint source pollution is difficult because it requires large numbers of people to alter their activity.

bioaccumulation An organism accumulates increasing amounts of a chemical pollutant in its tissues over the course of its lifetime.
biological magnification A chemical pollutant becomes increasingly concentrated as it moves up through food chains.

Take-Home Message How do chemical pollutants become concentrated in tissues?

> Some chemical pollutants can accumulate to high levels over an organism's lifetime. Such chemicals also become more concentrated as they move up food chains.

> Improperly disposing of trash can threaten water supplies and harm wildlife.

< Link to Groundwater 28.7

Six billion people use and discard a lot of stuff. Where does all the waste go? Historically, unwanted material was simply buried in the ground or dumped out at sea. Trash was out of sight, and also out of mind. We now know that chemicals in buried trash can contaminate groundwater, as when lead from discarded batteries seeps into the ground. Waste that gets into oceans harms marine life. For example, seabirds often eat floating bits of plastic and feed them to their chicks, with deadly results (Figure 30.7).

In 2006, the United States generated 251 million tons of garbage, which averages out to 2.1 kilograms (4.6 pounds) per person per day. By weight, about a third of that material was recycled, but there is plenty of room for improvement. Two-thirds of plastic soft drink bottles and three-quarters of glass bottles were not recycled. Nonrecycled trash now gets burned in high-temperature incinerators or placed in engineered landfills lined with material that minimizes the risk of groundwater contamination. No solid municipal waste can legally be dumped at sea.

Nevertheless, plastic and other garbage constantly enters our coastal waters. Foam cups and containers from fast-food outlets, plastic shopping bags, plastic water bottles, and other material discarded as litter ends up in storm drains. From there it is carried to streams and rivers that can convey it to the sea. A seawater sample taken near the mouth of the San Gabriel River in southern California had 128 times as much plastic as plankton by weight.

Once in the ocean, trash can persist for a surprisingly long time. Components of a disposable diaper will last for more than 100 years, as will fishing line. A plastic bag will be around for more than 50 years, and a cigarette filter for more than 10.

To reduce the impact of plastic trash, choose more durable objects over disposable ones, and avoid buying plastic when other, less environmentally harmful alternatives exist. If you use plastic, be sure to recycle or dispose of it properly.

Figure 30.7 One effect of discarded plastic in a marine environment.

A Recently deceased Laysan albatross chick, dissected to reveal the contents of its gut.

B Scientists found more than 300 pieces of plastic inside the bird. One of the pieces had punctured its gut wall, resulting in its death. The chick was fed the plastic by its parents, who gathered the material from the ocean surface, mistaking it for food.

Take-Home Message What are the ecological effects of discarding trash?

> Trash, especially plastics, often ends up in oceans where it harms marine life.

> You can minimize your environmental impact by avoiding disposable plastic goods and by recycling.

> Ozone is said to be "good up high, but bad nearby." It forms a protective layer in the upper atmosphere, but is a harmful pollutant in air near the ground.

< Links to UV as mutagen 9.6, Ozone layer formation 18.5

Depletion of the Ozone Layer

In the upper layers of the atmosphere, between 17 and 27 kilometers (10.5 and 17 miles) above sea level, the ozone (O_3) concentration is so great that scientists refer to this region as the **ozone layer**. The ozone layer benefits living organisms by absorbing most ultraviolet (UV) radiation from incoming sunlight. UV radiation, remember, damages DNA and causes mutations (Section 9.6).

In the mid-1970s, scientists noticed that Earth's ozone layer was thinning. Its thickness had always varied a bit with the season, but now the average level was declining steadily from year to year. By the mid-1980s, the spring ozone thinning over Antarctica was so pronounced that people were referring to the lowest-ozone region as an "ozone hole" (Figure 30.8A).

A Ozone levels in the upper atmosphere in September 2007, the Antarctic spring.

Purple indicates the least ozone, with *blue*, *green*, and *yellow* indicating increasingly higher levels.

Check the current status of the ozone hole at NASA's web site (http://ozonewatch .gsfc.nasa.gov/).

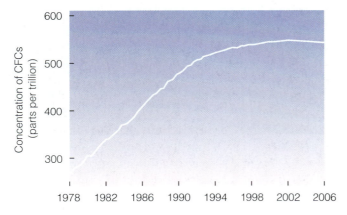

B Concentration of CFCs in the upper atmosphere. These pollutants destroy ozone. A worldwide ban on CFCs has successfully halted the rise in CFC concentration.

Figure 30.8 Animated Ozone and CFCs.

Declining ozone quickly became an international concern. With a thinner ozone layer, people would be exposed to more UV radiation, the main cause of skin cancers. Higher UV levels also harm wildlife, which do not have the option of avoiding sunlight. In addition, exposure to higher than normal UV levels affects plants and other producers, slowing the rate of photosynthesis and the release of oxygen into the atmosphere.

Chlorofluorocarbons, or CFCs, are the main ozone destroyers. These odorless gases were once widely used as propellants in aerosol cans, as coolants, and in solvents and plastic foam. In response to the potential threat posed by ozone thinning, countries worldwide agreed in 1987 to phase out the production of CFCs and other ozone-destroying chemicals. As a result of that agreement (the Montreal Protocol), the concentrations of CFCs in the atmosphere are no longer rising dramatically (Figure 30.8B). However, CFCs break down quite slowly, so scientists expect them to remain at a level that significantly impairs the ozone layer for several decades.

Near-Ground Ozone Pollution

Near the ground, where ozone levels are naturally low, ozone is considered a pollutant. Ozone irritates the eyes and respiratory tracts of humans and wildlife. It also interferes with plant growth.

Ground-level ozone forms when nitrogen oxides and volatile organic compounds released by burning or evaporating fossil fuels are exposed to sunlight. Warm temperatures speed the reaction. Thus, ground-level ozone tends to vary daily (being higher during the daytime) and seasonally (being higher during the summer).

You can help reduce ozone pollution by avoiding actions that put fossil fuels or their combustion products into the air at times that favor ozone production. On hot, sunny, still days, postpone filling your gas tank or using gasoline-powered appliances such as lawn mowers until the evening, when there is less sunlight to power the conversion of pollutants to ozone.

---

ozone layer High atmospheric layer with a high concentration of ozone (O_3) that prevents much ultraviolet radiation from reaching Earth's surface.

Take-Home Message How do human activities affect ozone levels?

> Certain synthetic chemicals destroy ozone in the upper atmosphere's protective ozone layer. This layer serves as a protective shield against UV radiation.

> Evaporation and burning of fossil fuels increases the amount of ozone in the air near the ground, where ozone is considered a harmful pollutant.

❯ Climate change is the most widespread threat to habitats. Among other effects, it melts ice, causing sea level to rise.
❮ Links to Greenhouse gases 28.8, Coral bleaching 29.13

Ongoing climate change affects ecosystems worldwide. How is the global climate changing? Most notably, average temperatures are increasing (Figure 30.9). Warming is more pronounced at temperate and polar latitudes than at the equator.

A rising temperature raises sea level. Water expands as it is heated, and heating also melts sea ice and glaciers (Figure 30.10). Together, thermal expansion and the addition of meltwater from glaciers cause sea level to rise. In the past century, the sea level has risen about 20 centimeters (8 inches). As a result, some coastal wetlands are disappearing underwater.

The warming climate is having widespread effects on biological systems. Temperature changes are important cues for many temperate zone species. Warmer than normal springs are causing deciduous trees to put leaves out earlier, and spring-blooming flowers to flower earlier. Animal migration times and breeding seasons are also shifting. Species arrays in biological communities are changing as warmer temperatures allow some species

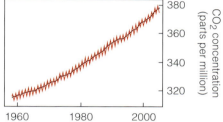

Figure 30.9 Changing global average temperature and atmospheric carbon dioxide. The large graph illustrates the increase in average temperature between 1850 and 2005.

The small graph shows the rise in atmospheric carbon dioxide during the last 45 years of this period. Carbon dioxide is a greenhouse gas.

A 1941 photo of Muir Glacier in Alaska.

B 2004 photo of the same region.

Figure 30.10 Melting glaciers, one sign of a warming world. Water from melting glaciers contributes to rising sea level.

to shift their range to higher latitudes or elevations. Of course, not all species can move or spread quickly, and warmer temperatures are expected to drive some of these species to extinction. For example, warming of tropical waters is already stressing reef-building corals and increasing the frequency of coral bleaching events.

Global warming is just one aspect of global climate change. Because the temperature of the land and seas affects evaporation, winds, and currents, many weather patterns are expected to change as temperature continues to rise. For example, warmer temperatures are correlated with extremes in rainfall patterns: periods of drought interrupted by unusually heavy rains. Also, warmer seas tend to increase the intensity of hurricanes.

As Section 28.8 explained, the most widely accepted explanation for global climate change is the rising levels of greenhouse gases such as carbon dioxide. Reducing greenhouse gas emissions will be a challenge. Fossil fuel combustion is the single biggest source of greenhouse gas emissions. Fossil fuel use is still rising as large nations such as China and India become increasingly industrialized. However, efforts are under way to increase the efficiency of processes that require fossil fuels, shift to alternative energy sources such as solar and wind power, and develop innovative ways to store carbon dioxide.

Take-Home Message How is the global climate changing, and what are some biological effects of the changes?

❯ The rise in global temperature is causing the sea level to rise, and is affecting weather patterns.

❯ These changes are altering the range of some species, threatening others with extinction, and disrupting the structure of biological communities.

❯ Conservation biologists survey and seek ways to protect the world's existing biodiversity.

The Value of Biodiversity

Every nation has several forms of wealth: material wealth, cultural wealth, and biological wealth. Biological wealth is called **biodiversity**. We measure a region's biodiversity at three levels: the genetic diversity within species, species diversity, and ecosystem diversity. Biodiversity is currently declining at all three levels, in all regions.

Conservation biology addresses these declines. The goals of this relatively new field of biology are (1) to survey the range of biodiversity, and (2) to find ways to maintain and use biodiversity to benefit human populations. The aim is to conserve biodiversity by encouraging people to value it and use it in ways that do not destroy it.

Why should we protect biodiversity? From a selfish standpoint, doing so is an investment in our future. Healthy ecosystems are essential to the survival of our species. Other organisms produce the oxygen we breathe and the food we eat. They remove waste carbon dioxide from the air and decompose and detoxify wastes. Plants take up rain and hold soil in place, preventing erosion and reducing the risk of flooding. Compounds in wild species can serve as medicines. Wild relatives of crop plants are reservoirs of genetic diversity that plant breeders draw on to protect and improve crops.

In addition, there are ethical reasons to preserve biodiversity. As we have emphasized many times, all living species are the result of an ongoing evolutionary process that stretches back billions of years. Each species has a unique combination of traits. The extinction of a species removes its unique collection of traits from the world of life forever.

Setting Priorities

Protecting biological diversity is often a tricky proposition. Even in developed countries, people often oppose environmental protections because they fear such measures will have adverse economic consequences. However, taking care of the environment can make good economic sense. With a bit of planning, people can both preserve and profit from their biological wealth.

Conservation biologists can help us make the difficult choices about which regions should be targeted for protection first. These biologists identify **hot spots**, places that are home to species found nowhere else and are under great threat of destruction. Once identified, hot spots can take priority in worldwide conservation efforts.

On a broader scale, conservation biologists define ecoregions, which are land or aquatic regions characterized by climate, geography, and the species found within them. The most widely used ecoregion system was developed by conservation scientists of the World Wildlife Fund. These scientists defined 867 distinctive land ecoregions. Figure 30.11 shows the locations and conservation status of ecoregions that are considered the top priority for conservation efforts.

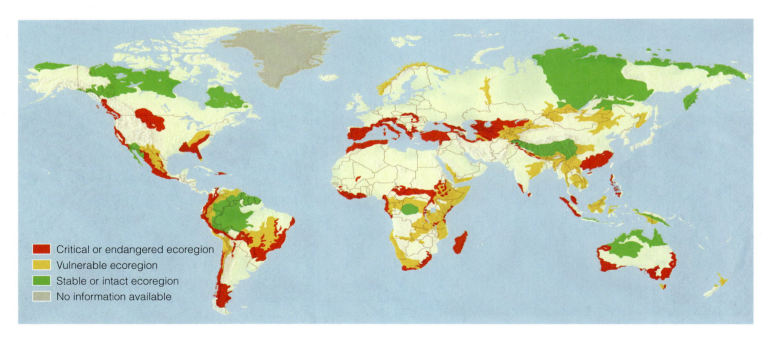

Legend:
- Critical or endangered ecoregion
- Vulnerable ecoregion
- Stable or intact ecoregion
- No information available

Figure 30.11 The location and conservation status of the land ecoregions deemed most important by the World Wildlife Fund.

The Klamath-Siskiyou forest in southwestern Oregon and northwestern California is one of North America's endangered ecoregions. It is home to many rare conifers. Two endangered birds, the northern spotted owl and the marbled murrelet, nest in old-growth parts of the forest, and endangered coho salmon breed in streams that run through the forest. Logging threatens all of these species.

By focusing on hot spots and critical ecoregions rather than on individual endangered species, scientists hope to maintain ecosystem processes that naturally sustain biological diversity.

Preservation and Restoration

Worldwide, many ecologically important regions have been protected in ways that benefit local people. The Monteverde Cloud Forest in Costa Rica is one example. During the 1970s, George Powell was studying birds in this forest, which was rapidly being cleared. Powell decided to buy part of the forest as a nature sanctuary. His efforts inspired individuals and conservation groups to donate funds, and much of the forest is now protected as a private nature reserve (Figure 30.12). The reserve's plants and animals include more than 100 mammal species, 400 bird species, and 120 species of amphibians and reptiles. It is one of the few habitats left for jaguars and ocelots. A tourism industry centered on the reserve provides economic benefits to local people.

Sometimes, an ecosystem is so damaged, or there is so little of it left, that conservation alone is not enough to sustain biodiversity. **Ecological restoration** is work designed to bring about the renewal of a natural ecosystem that has been degraded or destroyed, fully or in part.

Restoration work in Louisiana's coastal wetlands is an example. More than 40 percent of the coastal wetlands in the United States are in Louisiana. The marshes are an ecological and economic treasure, but they are in trouble. Dams and levees built upstream of the marshes keep back sediments that would normally replenish sediments lost to the sea. Channels cut through the marshes for oil exploration and production have encouraged erosion, and the rising sea level threatens to flood the existing plants. Since the 1940s, Louisiana has lost an area of marshland the size of Rhode Island. Restoration efforts now under way aim to reverse some of those losses (Figure 30.13).

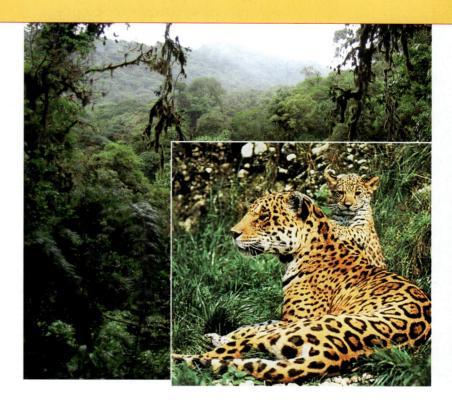

Figure 30.12 A conservation success story. Costa Rica's Monteverde Cloud Forest Reserve protects an area of tropical rain forest that is home to many endangered species, including jaguars (*inset photo*).

Figure 30.13 Ecological restoration in Louisiana's Sabine National Wildlife Refuge. In places where marshland has become open water, sediments are barged in and marsh grasses are planted on them. The squares are new sediment with marsh grass.

---

biodiversity Of a region, the genetic variation within its species, variety of species, and variety of ecosystems.
conservation biology Field of applied biology that surveys biodiversity and seeks ways to maintain and use it.
ecological restoration Actively altering an area in an effort to restore or create a functional ecosystem.
hot spot Threatened region with great biodiversity that is considered a high priority for conservation efforts.

Take-Home Message How do we sustain biodiversity?

> Biodiversity is the genetic diversity of individuals of a species, the variety of species, and the variety of ecosystems.

> Conservation biologists identify threatened regions with high biodiversity and prioritize which should be first to receive protection.

> Through ecological restoration, we re-create or renew a biologically diverse ecosystem that has been destroyed or degraded.

❯ The cumulative effects of individual actions will determine the health of our planet.

Ultimately, the health of our planet depends on our ability to recognize that the principles of energy flow and of resource limitation, which govern the survival of all systems of life, do not change. We must take note of these principles and find a way to live within our limits. The goal is living sustainably, which means meeting the needs of the present generation without reducing the ability of future generations to meet their own needs.

Promoting sustainability begins with recognizing the environmental consequences of one's own life-style. People in industrial nations use enormous quantities of resources, and the extraction, delivery, and use of these resources has negative effects on biodiversity. In the United States, the size of the average family has declined since the 1950s, while the size of the average home has doubled. All of the materials used to build and furnish those larger homes come from the environment. For example, an average new home contains about 500 pounds of copper in its wiring and plumbing.

Where does copper come from? Like most other mineral elements used in manufacturing, most copper is mined from the ground (Figure 30.14). Surface mining strips an area of vegetation and soil, creating an ecological

Figure 30.15 Volunteers restoring the Little Salmon River in Idaho so that salmon can migrate upstream to their breeding grounds.

dead zone. It puts dust into the air, creates mountains of rocky waste, and can contaminate nearby waterways.

Globalization makes it difficult to know the source of the raw materials in products you buy. Resource extraction in developing countries is often carried out under regulations that are less strict or less stringently enforced than those in the United States, so the environmental impact is even greater.

Figure 30.14 Effects of resource extraction. Bingham copper mine near Salt Lake City, Utah. This open pit mine is 4 kilometers (2.5 miles) wide and 1,200 meters (0.75 miles) deep, the largest man-made excavation on Earth.

Nonrenewable mineral resources are also used in electronic devices such as phones, computers, televisions, and MP3 players. Constantly trading up to the newest device may be good for the ego and the economy, but it is bad for the environment. Reducing consumption by fixing existing products is a sustainable resource use, as is recycling. Obtaining nonrenewable materials by recycling reduces the need for extraction of those resources, and it also helps keep materials out of landfills.

Reducing your energy use is another way to promote sustainability. Fossil fuels such as petroleum, natural gas, and coal supply most of the energy used by developed countries. You already know that burning these nonrenewable fuels contributes to global warming and acid rain. In addition, extracting and transporting these fuels have negative impacts. Oil harms many species when it leaks from pipelines or from ships.

Renewable energy sources do not produce greenhouse gases, but they have their own drawbacks. For example, dams in rivers of the Pacific Northwest generate renewable hydroelectric power, but they also prevent endangered salmon from returning to streams above the dam to breed. Similarly, wind turbines can harm birds and bats. Panels used to collect solar energy are made using nonrenewable mineral resources, and manufacturing the panels generates pollutants.

In short, all commercially produced energy has negative environmental impacts, so the best way to minimize your impact is to use less energy. Buy energy-efficient appliances, use fluorescent bulbs instead of incandescent ones, do not overheat or overcool rooms, and turn out unneeded lights. Walking, bicycling, and using public transportation are energy-efficient alternatives to driving. Shopping locally and purchasing locally produced goods also reduces energy use.

If you want to make more of a difference, learn about the threats to ecosystems in your own area. Support efforts to preserve and restore local biodiversity. Many ecological restoration projects are supervised by trained biologists but carried out primarily through the efforts of volunteers (Figure 30.15). Keep in mind that unthinking actions of billions of individuals are the greatest threat to biodiversity. Each of us may have little impact on our own, but our collective behavior, for good or for bad, will determine the future of the planet.

Take-Home Message What can individuals do to reduce their harmful impact on biodiversity?

> Resource extraction and usage have side effects that threaten biodiversity.

> You can save energy and other resources by reducing energy consumption and recycling and reusing materials.

■ A Long Reach (revisited)

The Arctic is not a separate continent like Antarctica, but rather a region that encompasses the northernmost parts of several continents. Eight countries, including the United States, Canada, and Russia, control parts of the Arctic and have rights to its extensive oil, gas, and mineral deposits. Until recently, ice sheets covered the Arctic Ocean, making it difficult for ships to move to and from the Arctic land mass, but those sheets are breaking up as a result of global climate change (Figure 30.16). At the same time, ice that covered the Arctic land mass is melting. These changes will make it easier for people to remove minerals and fossil fuels from the Arctic. With the world supply of fuel and minerals dwindling, pressure to exploit Arctic resources is rising. However, conservationists warn that extracting these resources will harm Arctic species such as the polar bear that are already threatened by global climate change.

Arctic perennial sea ice in 1979

Arctic perennial sea ice in 2003

Figure 30.16 Declining Arctic perennial ice.

How Would You Vote? The Arctic has extensive deposits of minerals and fossil fuel, but tapping into these resources might pose a risk to species already threatened by global climate change. Should the United States exploit its share of the Arctic resources or advocate for protection of the region? See CengageNow for details, then vote online (cengagenow.com).

Summary

Section 30.1 Human activities affect even remote places such as the Arctic. Polar bears in the Arctic have pollutants in their bodies and are threatened by thinning sea ice, one effect of global climate change.

Section 30.2 We are in the midst of a human-caused mass extinction, with numbers of **threatened species** and **endangered species** rising. **Endemic species** are especially likely to be at risk. Overharvest, species introductions, and habitat destruction, degradation, and fragmentation can push species toward extinction.

We know about only a tiny fraction of the species that are currently under threat.

Section 30.3 Overplowing or overgrazing can cause **desertification**. Both desertification and deforestation affect soil properties and rainfall. Changes caused by deforestation are especially difficult to reverse.

Section 30.4 Burning fossil fuels releases sulfur dioxide and nitrogen oxides. These **pollutants** mix with water vapor in the air, then fall to earth as **acid rain**. The resulting increase in the acidity of soils and waters can sicken or kill organisms.

Section 30.5 With **bioaccumulation**, a pollutant taken up from the environment becomes stored in an organism's tissues. With **biological magnification**, a pollutant increases in concentration as it is passed up a food chain. As a result of bioaccumulation and biological magnification, the tissues of an organism can have a far higher concentration of a pollutant than the environment.

Section 30.6 Human populations discard large amounts of trash that is burned, buried, or dumped in seas. Trash that washes into or is dumped into oceans poses a threat to marine life.

Section 30.7 The **ozone layer** in the upper atmosphere protects against incoming UV radiation. Chemicals called CFCs were banned when they were found to cause thinning of the ozone layer. Near the ground, where the ozone concentration is naturally low, ozone emitted as a result of fossil fuel use is considered a pollutant. It irritates animal respiratory tracts and interferes with photosynthesis by plants.

Section 30.8 Global climate change is causing glaciers to melt, thus raising the sea level. It also is affecting the range of species, allowing some to move into higher elevations or latitudes. Other species such as corals are showing signs of temperature-related stress.

Global climate change is also expected to further alter rainfall patterns and the intensity of hurricanes.

Sections 30.9, 30.10 Genetic diversity, species diversity, and ecosystem diversity are components of **biodiversity**, which is declining in all regions. **Conservation biology** involves surveying biodiversity and developing strategies to protect it and allow its sustainable use. Because resources are limited, biologists often focus on **hot spots**, where many unique species are under threat. They also attempt to ensure that portions of all ecoregions are protected. When an ecosystem has been totally or partially degraded, **ecological restoration** can help restore biodiversity.

Individuals can help sustain biodiversity by limiting their energy use and material consumption. Reuse and recycling help prevent destructive practices required to extract resources.

Self-Quiz   Answers in Appendix III

1. A(n) _____ species has population levels so low it is at great risk of extinction in the near future.
   - a. endemic
   - b. endangered
   - c. threatened
   - d. indicator

2. Species are threatened by habitat _____ .
   - a. fragmentation
   - b. degradation
   - c. destruction
   - d. all of the above

3. Deforestation _____ the amount of minerals that run off from soil.
   - a. increases   b. decreases   c. has no effect on

4. Sulfur dioxide released by coal-burning power plants contributes to _____ .
   - a. ozone destruction
   - b. sea level rise
   - c. acid rain
   - d. desertification

5. As a result of _____ , an old animal usually has more pollutants in its body than a young one.
   - a. bioaccumulation
   - b. biological magnification

6. With biological magnification, a _____ will have the highest pollutant load.
   - a. producer
   - b. primary consumer
   - c. secondary consumer
   - d. top-level consumer

7. An increase in the size of the ozone hole would be expected to _____ .
   - a. increase skin cancers
   - b. reduce respiratory disorders
   - c. both a and b

8. True or false? All pollutants are synthetic chemicals.

9. Global climate change is causing _____ .
   - a. a decrease in sea level
   - b. glacial melting
   - c. acid rain
   - d. all of the above

10. The Montreal Protocol banned use of _____ , which contribute(s) to ozone depletion.
    - a. DDT
    - b. CFCs
    - c. fossil fuels
    - d. sulfur dioxides

Data Analysis Activities

Arctic PCB Pollution Winds carry chemical contaminants produced and released at temperate latitudes to the Arctic, where the chemicals enter food webs. Top carnivores in arctic food webs, including polar bears and people, end up with high concentrations of these chemicals in their body. Arctic people who eat a lot of local wildlife tend to have unusually high levels of industrial chemicals called polychlorinated biphenyls, or PCBs, in their body. The Arctic Monitoring and Assessment Programme studies the effects of these industrial chemicals on the health and reproduction of Arctic people. Figure 30.17 shows how sex ratio at birth varies with average maternal PCB levels among people native to the Russian Arctic.

1. Which sex was more common in offspring of women with less than 1 microgram per milliliter of PCB in serum?
2. At what PCB concentrations were women more likely to have daughters?
3. In some villages in Greenland, nearly all recent newborns are female. Would you expect PCB levels in those villages to be above or under 4 micrograms per milliliter?

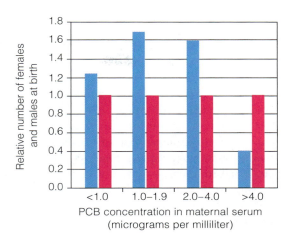

Figure 30.17 Effect of maternal PCB concentration on sex ratio of newborns in human populations native to the Russian Arctic. *Blue* bars indicate the relative number of males born per one female (*pink* bars).

---

11. A highly threatened region that is home to many unique species is a(n) _____ .
    a. ecoregion   c. hot spot
    b. biome   d. community

12. Biodiversity refers to _____ .
    a. genetic diversity   c. ecosystem diversity
    b. species diversity   d. all of the above

13. Restoring a marsh that has been damaged by human activities is an example of _____ .
    a. biological magnification c. ecological restoration
    b. bioaccumulation   d. globalization

14. Individuals help sustain biodiversity by _____ .
    a. reducing resource consumption
    b. reusing materials
    c. recycling materials
    d. all of the above

15. Match the terms with the most suitable description.
    ___ hot spot   a. good up high; bad nearby
    ___ ozone   b. tree loss alters rainfall pattern
    ___ biodiversity   and is difficult to reverse
    ___ acid rain   c. can increase dust storms
    ___ endemic   d. evolved in one region and
         species   remains there
    ___ biological   e. coal-burning is major cause
         magnification f. results in highest pollution
    ___ global climate   level at top trophic level
         change   g. has lots of threatened species
    ___ deforestation   h. cause of rising sea level
    ___ desertification   i. genetic, species, and ecosystem
                  diversity

Additional questions are available on **CENGAGENOW**.

Critical Thinking

1. In one seaside community in New Jersey, the U.S. Fish and Wildlife Service suggested trapping and removing feral cats (domestic cats that live in the wild). The goal was to protect some endangered wild birds (plovers) that nested on the town's beaches. Many residents were angered by the proposal, arguing that the cats have as much right to be there as the birds. Do you agree? Why or why not?

2. Burning fossil fuel puts excess carbon dioxide into the atmosphere, but deforestation and desertification also affect carbon dioxide concentration. Explain why a global decrease in the amount of vegetation is contributing to the rise in carbon dioxide.

3. The magnitude of acid rain's effects can be influenced by the properties of the rock that the rain runs over. Acid rain is least likely to significantly acidify lakes in regions where the bedrock consists of calcium carbonate–rich limestone or marble. How does the presence of these rocks mitigate the effects of acid rain?

4. Two arctic marine mammals that live in the same waters differ in the level of pollutants in their bodies. Bowhead whales have a lower pollutant load than ringed seals. What are some factors that might explain this difference?

Animations and Interactions on **CENGAGENOW**:
❭ Acid rain; Ozone and CFCs.

Appendix I. Classification System

This revised classification scheme is a composite of several that microbiologists, botanists, and zoologists use. The major groupings are agreed upon, more or less. However, there is not always agreement on what to name a particular grouping or where it might fit within the overall hierarchy. There are several reasons why full consensus is not possible at this time.

First, the fossil record varies in its completeness and quality. Therefore, the phylogenetic relationship of one group to other groups is sometimes open to interpretation. Today, comparative studies at the molecular level are firming up the picture, but the work is still under way. Also, molecular comparisons do not always provide definitive answers to questions about phylogeny. Comparisons based on one set of genes may conflict with those comparing a different part of the genome. Or comparisons with one member of a group may conflict with comparisons based on other group members.

Second, ever since the time of Linnaeus, systems of classification have been based on the perceived morphological similarities and differences among organisms. Although some original interpretations are now open to question, we are so used to thinking about organisms in certain ways that reclassification often proceeds slowly.

A few examples: Traditionally, birds and reptiles were grouped in separate classes (Reptilia and Aves); yet there are compelling arguments for grouping the lizards and snakes in one group and the crocodilians, dinosaurs, and birds in another. Many biologists still favor a six-kingdom system of classification (archaea, bacteria, protists, plants, fungi, and animals). Others advocate a switch to the more recently proposed three-domain system (archaea, bacteria, and eukarya).

Third, researchers in microbiology, mycology, botany, zoology, and other fields of inquiry inherited a wealth of literature, based on classification systems that have been developed over time in each field of inquiry. Many are reluctant to give up established terminology that offers access to the past.

For example, botanists and microbiologists often use *division*, and zoologists *phylum*, for taxa that are equivalent in hierarchies of classification.

Why bother with classification frameworks if we know they only imperfectly reflect the evolutionary history of life? We do so for the same reasons that a writer might break up a history of civilization into several volumes, each with a number of chapters. Both are efforts to impart structure to an enormous body of knowledge and to facilitate retrieval of information from it. More importantly, to the extent that modern classification schemes accurately reflect evolutionary relationships, they provide the basis for comparative biological studies, which link all fields of biology.

Bear in mind that we include this appendix for your reference purposes only. Besides being open to revision, it is not meant to be complete. Names shown in "quotes" are polyphyletic or paraphyletic groups that are undergoing revision. For example, "reptiles" comprise at least three and possibly more lineages.

The most recently discovered species, as from the mid-ocean province, are not listed. Many existing and extinct species of the more obscure phyla are also not represented. Our strategy is to focus primarily on the organisms mentioned in the text or familiar to most students. We delve more deeply into flowering plants than into bryophytes, and into chordates than annelids.

RELATIONSHIPS AMONG THE THREE DOMAINS

As a general frame of reference, note that almost all bacteria and archaeans are microscopic in size. Their DNA is typically concentrated in a nucleoid (a region of cytoplasm), not in a membrane-bound nucleus. They reproduce by binary fission or budding, transfer genes by conjugation, and none are multicellular.

Bacteria and Archaeans have historically been grouped together as "prokaryotes." However, some microbiologists now advocate avoiding the use of this term. They point out that bacteria and archaeans do not constitute a monophyletic group, and that the two lineages differ in many structural and genetic traits. In deference to this sentiment, we use the term "prokaryote" only in discussion of historical classification systems.

Use of the term "eukaryote" remains noncontroversial. The eukaryotes are a monophyletic group thought to be descended from an archaean ancestor. Eukaryotic cells are typically larger than those of archaeans or bacteria and they have a DNA-enclosing nucleus and other membrane-bound organelles such as endoplasmic reticulum. Mitosis and meiosis occur only in eukaryotes.

Eukaryotes include a diverse array of single-celled organisms, as well as all multicellular species.

DOMAIN BACTERIA — Kingdom Bacteria

DOMAIN ARCHAEA — Kingdom Archaea

DOMAIN EUKARYA — Kingdom Protista · Kingdom Fungi · Kingdom Plantae · Kingdom Animalia

DOMAIN OF BACTERIA

KINGDOM BACTERIA

The largest, and most diverse group of microorganisms. Includes photosynthetic autotrophs, chemosynthetic autotrophs, and heterotrophs. Includes many important pathogens of vertebrates.

PHYLUM AQIFACAE   Most ancient branch of the bacterial tree. Gram-negative, mostly aerobic chemoautotrophs, mainly of volcanic hot springs. *Aquifex*.

PHYLUM DEINOCOCCUS-THERMUS   Gram-positive, heat-loving chemoautotrophs. *Deinococcus* is the most radiation resistant organism known. *Thermus* occurs in hot springs and near hydrothermal vents.

PHYLUM CHLOROFLEXI   Green nonsulfur bacteria. Gram-negative bacteria of hot springs, freshwater lakes, and marine habitats. Act as nonoxygen-producing photoautotrophs or aerobic chemoheterotrophs. *Chloroflexus*.

PHYLUM ACTINOBACTERIA   Gram-positive, mostly aerobic heterotrophs in soil, freshwater and marine habitats, and on mammalian skin. *Propionibacterium, Actinomyces, Streptomyces*.

PHYLUM CYANOBACTERIA   Gram-negative, oxygen-releasing photoautotrophs mainly in aquatic habitats. They have chlorophyll *a* and photosystem I. Includes many nitrogen-fixing genera. *Anabaena, Nostoc, Oscillatoria*.

PHYLUM CHLOROBIUM   Green sulfur bacteria. Gram-negative nonoxygen-producing photosynthesizers, mainly in freshwater sediments. *Chlorobium*.

PHYLUM FIRMICUTES   Gram-positive walled cells and the cell wall-less mycoplasmas. All are heterotrophs. Some survive in soil, hot springs, lakes, or oceans. Others live on or in animals. *Bacillus, Clostridium, Heliobacterium, Lactobacillus, Listeria, Mycobacterium, Mycoplasma, Streptococcus*.

PHYLUM CHLAMYDIAE   Gram-negative intracellular parasites of birds and mammals. *Chlamydia*.

PHYLUM SPIROCHETES   Free-living, parasitic, and mutualistic gram-negative spring-shaped bacteria. *Borelia, Pillotina, Spirillum, Treponema*.

PHYLUM PROTEOBACTERIA   The largest bacterial group. Includes photoautotrophs, chemoautotrophs, and heterotrophs; free-living, parasitic, and colonial groups. All are gram-negative.

Class Alphaproteobacteria. *Agrobacterium, Azospirillum, Nitrobacter, Rickettsia, Rhizobium*.

Class Betaproteobacteria. *Neisseria*.

Class Gammaproteobacteria. *Chromatium, Escherichia, Haemopilius, Pseudomonas, Salmonella, Shigella, Thiomargarita, Vibrio, Yersinia*.

Class Deltaproteobacteria. *Azotobacter, Myxococcus*.

Class Epsilonproteobacteria. *Campylobacter, Helicobacter*.

DOMAIN OF ARCHAEA

KINGDOM ARCHAEA

Single cells that are evolutionarily between eukaryotic cells and the bacteria. Most are anaerobes. None are photosynthetic. Originally discovered in extreme habitats, they are now known to be widely dispersed. Compared with bacteria, the archaea have a distinctive cell wall structure and unique membrane lipids, ribosomes, and RNA sequences. Some are symbiotic with animals, but none are known to be animal pathogens.

PHYLUM EURYARCHAEOTA   Largest archean group. Includes extreme thermophiles, halophiles, and methanogens. Others are abundant in the upper waters of the ocean and other more moderate habitats. *Methanocaldococcus, Nanoarchaeum*.

PHYLUM CRENARCHAEOTA   Includes extreme theromophiles, as well as species that survive in Antarctic waters, and in more moderate habitats. *Sulfolobus, Ignicoccus*.

PHYLUM KORARCHAEOTA   Known only from DNA isolated from hydrothermal pools. As of this writing, none have been cultured and no species have been named.

DOMAIN OF EUKARYOTES

KINGDOM "PROTISTA"

A collection of single-celled and multicelled lineages, which does not constitute a monophyletic group. Some biologists consider the groups listed below to be kingdoms in their own right.

PARABASALIA   Parabasalids. Flagellated, single-celled anaerobic heterotrophs with a cytoskeletal "backbone" that runs the length of the cell. There are no mitochondria, but a hydrogenosome serves a similar function. *Trichomonas, Trichonympha*.

DIPLOMONADIDA   Diplomonads. Flagellated, anaerobic single-celled heterotrophs that do not have mitochondria or Golgi bodies and do not form a bipolar spindle at mitosis. May be one of the most ancient lineages. *Giardia*.

EUGLENOZOA   Euglenoids and kinetoplastids. Free-living and parasitic flagellates. All with one or more mitochondria. Some photosynthetic euglenoids with chloroplasts, others heterotrophic. *Euglena, Trypanosoma, Leishmania*.

RHIZARIA   Formaminiferans and radiolarians. Free-living, heterotrophic amoeboid cells that are enclosed in shells. Most live in ocean waters or sediments. *Pterocorys, Stylosphaera*.

ALVEOLATA   Single cells having a unique array of membrane-bound sacs (alveoli) just beneath the plasma membrane.

Ciliata. Ciliated protozoans. Heterotrophic protists with many cilia. *Paramecium, Didinium*.

Dinoflagellates. Diverse heterotrophic and photosynthetic flagellated cells that deposit cellulose in their alveoli. *Gonyaulax, Gymnodinium, Karenia, Noctiluca*.

Apicomplexans. Single-celled parasites of animals. A unique microtubular device is used to attach to and penetrate a host cell. *Plasmodium*.

STRAMENOPHILA   Stramenophiles. Single-celled and multicelled forms; flagella with tinsel-like filaments.

Oomycotes. Water molds. Heterotrophs. Decomposers, some parasites. *Saprolegnia, Phytophthora, Plasmopara*.

Chrysophytes. Golden algae, yellow-green algae, diatoms, coccolithophores. Photosynthetic. *Emiliania, Mischococcus*.

Phaeophytes. Brown algae. Photosynthetic; nearly all live in temperate marine waters. All are multicellular. *Macrocystis, Laminaria, Sargassum, Postelsia*.

RHODOPHYTA   Red algae. Mostly photosynthetic, some parasitic. Nearly all marine, some in freshwater habitats. Most multicellular. *Porphyra, Antithamion*.

CHLOROPHYTA   Green algae. Mostly photosynthetic, some parasitic. Most freshwater, some marine or terrestrial. Single-celled, colonial, and multicellular forms. Some biologists place the chlorophytes and charophytes with the land plants in a kingdom called the Viridiplantae. *Acetabularia, Chlamydomonas, Chlorella, Codium, Udotea, Ulva, Volvox*.

CHAROPHYTA   Photosynthetic. Closest living relatives of plants. Include both single-celled and multicelled forms. Desmids, stoneworts. *Micrasterias, Chara, Spirogyra*.

AMOEBOZOA   True amoebas and slime molds. Heterotrophs that spend all or part of the life cycle as a single cell that uses pseudopods to capture food. *Amoeba, Entoamoeba* (amoebas), *Dictyostelium* (cellular slime mold), *Physarum* (plasmodial slime mold).

KINGDOM FUNGI

Nearly all multicelled eukaryotic species with chitin-containing cell walls. Heterotrophs, mostly saprobic decomposers, some parasites. Nutrition based upon extracellular digestion of organic matter and absorption of nutrients by individual cells. Multicelled species form absorptive mycelia and reproductive structures that produce asexual spores (and sometimes sexual spores).

PHYLUM CHYTRIDIOMYCOTA   Chytrids. Primarily aquatic; saprobic decomposers or parasites that produce flagellated spores. *Chytridium.*

PHYLUM ZYGOMYCOTA   Zygomycetes. Producers of zygospores (zygotes inside thick wall) by way of sexual reproduction. Bread molds, related forms. *Rhizopus, Philobolus.*

PHYLUM ASCOMYCOTA   Ascomycetes. Sac fungi. Sac-shaped cells form sexual spores (ascospores). Most yeasts and molds, morels, truffles. *Saccharomycetes, Morchella, Neurospora, Claviceps, Candida, Aspergillus, Penicillium.*

PHYLUM BASIDIOMYCOTA   Basidiomycetes. Club fungi. Most diverse group. Produce basidiospores inside club-shaped structures. Mushrooms, shelf fungi, stinkhorns. *Agaricus, Amanita, Craterellus, Gymnophilus, Puccinia, Ustilago.*

"IMPERFECT FUNGI"   Sexual spores absent or undetected. The group has no formal taxonomic status. If better understood, a given species might be grouped with sac fungi or club fungi. *Arthobotrys, Histoplasma, Microsporum, Verticillium.*

"LICHENS"   Mutualistic interactions between fungal species and a cyanobacterium, green alga, or both. *Lobaria, Usnea.*

KINGDOM PLANTAE

Most photosynthetic with chlorophylls *a* and *b*. Some parasitic. Nearly all live on land. Sexual reproduction predominates.

BRYOPHYTES (NONVASCULAR PLANTS)

Small flattened haploid gametophyte dominates the life cycle; sporophyte remains attached to it. Sperm are flagellated; require water to swim to eggs for fertilization.

PHYLUM HEPATOPHYTA   Liverworts. *Marchantia.*

PHYLUM ANTHOCEROPHYTA   Hornworts.

PHYLUM BRYOPHYTA   Mosses. *Polytrichum, Sphagnum.*

SEEDLESS VASCULAR PLANTS

Diploid sporophyte dominates, free-living gametophytes, flagellated sperm require water for fertilization.

PHYLUM LYCOPHYTA   Lycophytes, club mosses. Small single-veined leaves, branching rhizomes. *Lycopodium, Selaginella.*

PHYLUM MONILOPHYTA

Subphylum Psilophyta. Whisk ferns. No obvious roots or leaves on sporophyte, very reduced. *Psilotum.*

Subphylum Sphenophyta. Horsetails. Reduced scalelike leaves. Some stems photosynthetic, others spore-producing. *Calamites* (extinct), *Equisetum.*

Subphylum Pterophyta. Ferns. Large leaves, usually with sori. Largest group of seedless vascular plants (12,000 species), mainly tropical, temperate habitats. *Pteris, Trichomanes, Cyathea* (tree ferns), *Polystichum.*

SEED-BEARING VASCULAR PLANTS

PHYLUM CYCADOPHYTA   Cycads. Group of gymnosperms (vascular, bear "naked" seeds). Tropical, subtropical. Compound leaves, simple cones on male and female plants. Plants usually palm-like. Motile sperm. *Zamia, Cycas.*

PHYLUM GINKGOPHYTA   Ginkgo (maidenhair tree). Type of gymnosperm. Motile sperm. Seeds with fleshy layer. *Ginkgo.*

PHYLUM GNETOPHYTA   Gnetophytes. Only gymnosperms with vessels in xylem and double fertilization (but endosperm does not form). *Ephedra, Welwitchia, Gnetum.*

PHYLUM CONIFEROPHYTA   Conifers. Most common and familiar gymnosperms. Generally cone-bearing species with needle-like or scale-like leaves. Includes pines (*Pinus*), redwoods (*Sequoia*), yews (*Taxus*).

PHYLUM ANTHOPHYTA   Angiosperms (the flowering plants). Largest, most diverse group of vascular seed-bearing plants. Only organisms that produce flowers, fruits. Some families from several representative orders are listed:

BASAL FAMILIES

Family Amborellaceae. *Amborella.*
Family Nymphaeaceae. Water lilies.
Family Illiciaceae. Star anise.

MAGNOLIIDS

Family Magnoliaceae. Magnolias.
Family Lauraceae. Cinnamon, sassafras, avocados.
Family Piperaceae. Black pepper, white pepper.

EUDICOTS

Family Papaveraceae. Poppies.
Family Cactaceae. Cacti.
Family Euphorbiaceae. Spurges, poinsettia.
Family Salicaceae. Willows, poplars.
Family Fabaceae. Peas, beans, lupines, mesquite.
Family Rosaceae. Roses, apples, almonds, strawberries.
Family Moraceae. Figs, mulberries.
Family Cucurbitaceae. Squashes, melons, cucumbers.
Family Fagaceae. Oaks, chestnuts, beeches.
Family Brassicaceae. Mustards, cabbages, radishes.
Family Malvaceae. Mallows, okra, cotton, hibiscus, cocoa.
Family Sapindaceae. Soapberry, litchi, maples.
Family Ericaceae. Heaths, blueberries, azaleas.
Family Rubiaceae. Coffee.
Family Lamiaceae. Mints.
Family Solanaceae. Potatoes, eggplant, petunias.
Family Apiaceae. Parsleys, carrots, poison hemlock.
Family Asteraceae. Composites. Chrysanthemums, sunflowers, lettuces, dandelions.

MONOCOTS

Family Araceae. Anthuriums, calla lily, philodendrons.
Family Liliaceae. Lilies, tulips.
Family Alliaceae. Onions, garlic.
Family Iridaceae. Irises, gladioli, crocuses.
Family Orchidaceae. Orchids.
Family Arecaceae. Date palms, coconut palms.
Family Bromeliaceae. Bromeliads, pineapples.
Family Cyperaceae. Sedges.
Family Poaceae. Grasses, bamboos, corn, wheat, sugarcane.
Family Zingiberaceae. Gingers.

KINGDOM ANIMALIA

Multicelled heterotrophs, nearly all with tissues and organs, and organ systems, that are motile during part of the life cycle. Sexual reproduction occurs in most, but some also reproduce asexually. Embryos develop through a series of stages.

PHYLUM PORIFERA   Sponges. No symmetry, tissues.

PHYLUM PLACOZOA   Marine. Simplest known animal. Two cell layers, no mouth, no organs. *Trichoplax.*

PHYLUM CNIDARIA   Radial symmetry, tissues, nematocysts.
Class Hydrozoa. Hydrozoans. *Hydra, Obelia, Physalia, Prya.*
Class Scyphozoa. Jellyfishes. *Aurelia.*
Class Anthozoa. Sea anemones, corals. *Telesto.*

PHYLUM PLATYHELMINTHES   Flatworms. Bilateral, cephalized; simplest animals with organ systems. Saclike gut.

Class Turbellaria. Triclads (planarians), polyclads. *Dugesia.*
Class Trematoda. Flukes. *Clonorchis, Schistosoma.*
Class Cestoda. Tapeworms. *Diphyllobothrium, Taenia.*

PHYLUM ROTIFERA   Rotifers. *Asplancha, Philodina.*

PHYLUM MOLLUSCA   Mollusks.

Class Polyplacophora. Chitons. *Cryptochiton, Tonicella.*

Class Gastropoda. Snails, sea slugs, land slugs. *Aplysia, Ariolimax, Cypraea, Haliotis, Helix, Liguus, Limax, Littorina.*

Class Bivalvia. Clams, mussels, scallops, cockles, oysters, shipworms. *Ensis, Chlamys, Mytelus, Patinopectin.*

Class Cephalopoda. Squids, octopuses, cuttlefish, nautiluses. *Dosidiscus, Loligo, Nautilus, Octopus, Sepia.*

PHYLUM ANNELIDA   Segmented worms.

Class Polychaeta. Mostly marine worms. *Eunice, Neanthes.*

Class Oligochaeta. Mostly freshwater and terrestrial worms, many marine. *Lumbricus* (earthworms), *Tubifex.*

Class Hirudinea. Leeches. *Hirudo, Placobdella.*

PHYLUM NEMATODA   Roundworms. *Ascaris, Caenorhabditis elegans, Necator* (hookworms), *Trichinella.*

PHYLUM ARTHROPODA

Subphylum Chelicerata. Chelicerates. Horseshoe crabs, spiders, scorpions, ticks, mites.

Subphylum Crustacea. Shrimps, crayfishes, lobsters, crabs, barnacles, copepods, isopods (sowbugs).

Subphylum Myriapoda. Centipedes, millipedes.

Subphylum Hexapoda. Insects and sprintails.

PHYLUM ECHINODERMATA   Echinoderms.

Class Asteroidea. Sea stars. *Asterias.*
Class Ophiuroidea. Brittle stars.
Class Echinoidea. Sea urchins, heart urchins, sand dollars.
Class Holothuroidea. Sea cucumbers.
Class Crinoidea. Feather stars, sea lilies.
Class Concentricycloidea. Sea daisies.

PHYLUM CHORDATA   Chordates.

Subphylum Urochordata. Tunicates, related forms.

Subphylum Cephalochordata. Lancelets.

CRANIATES

Class Myxini. Hagfishes.

VERTEBRATES (SUBGROUP OF CRANIATES)

Class Cephalaspidomorphi. Lampreys.

Class Chondrichthyes. Cartilaginous fishes (sharks, rays, skates, chimaeras).

Class "Osteichthyes." Bony fishes. Not monophyletic (sturgeons, paddlefish, herrings, carps, cods, trout, seahorses, tunas, lungfishes, and coelocanths).

TETRAPODS (SUBGROUP OF VERTEBRATES)

Class Amphibia. Amphibians. Require water to reproduce.
Order Caudata. Salamanders and newts.
Order Anura. Frogs, toads.
Order Apoda. Apodans (caecilians).

AMNIOTES (SUBGROUP OF TETRAPODS)

Class "Reptilia." Skin with scales, embryo protected and nutritionally supported by extraembryonic membranes.

Subclass Anapsida. Turtles, tortoises.

Subclass Lepidosaura. *Sphenodon,* lizards, snakes.

Subclass Archosaura. Crocodiles, alligators.

Class Aves. Birds. In some classifications birds are grouped in the archosaurs.

Order Struthioniformes. Ostriches.

Order Sphenisciformes. Penguins.

Order Procellariiformes. Albatrosses, petrels.

Order Ciconiiformes. Herons, bitterns, storks, flamingoes.

Order Anseriformes. Swans, geese, ducks.

Order Falconiformes. Eagles, hawks, vultures, falcons.

Order Galliformes. Ptarmigan, turkeys, domestic fowl.

Order Columbiformes. Pigeons, doves.

Order Strigiformes. Owls.

Order Apodiformes. Swifts, hummingbirds.

Order Passeriformes. Sparrows, jays, finches, crows, robins, starlings, wrens.

Order Piciformes. Woodpeckers, toucans.

Order Psittaciformes. Parrots, cockatoos, macaws.

Class Mammalia. Skin with hair; young nourished by milk-secreting mammary glands of adult.

Subclass Prototheria. Egg-laying mammals (monotremes; duckbilled platypus, spiny anteaters).

Subclass Metatheria. Pouched mammals or marsupials (opossums, kangaroos, wombats, Tasmanian devils).

Subclass Eutheria. Placental mammals.

Order Edentata. Anteaters, tree sloths, armadillos.

Order Insectivora. Tree shrews, moles, hedgehogs.

Order Chiroptera. Bats.

Order Scandentia. Insectivorous tree shrews.

Order Primates.

Suborder Strepsirhini (prosimians). Lemurs, lorises.

Suborder Haplorhini (tarsioids and anthropoids).

Infraorder Tarsiiformes. Tarsiers.

Infraorder Platyrrhini (New World monkeys).

Family Cebidae. Spider monkeys, howler monkeys, capuchin.

Infraorder Catarrhini (Old World monkeys and hominoids).

Superfamily Cercopithecoidea. Baboons, macaques, langurs.

Superfamily Hominoidea. Apes and humans.

Family Hylobatidae. Gibbon.

Family "Pongidae." Chimpanzees, gorillas, orangutans.

Family Hominidae. Existing and extinct human species (*Homo*) and humanlike species, including the australopiths.

Order Lagomorpha. Rabbits, hares, pikas.

Order Rodentia. Most gnawing animals (squirrels, rats, mice, guinea pigs, porcupines, beavers, etc.).

Order Carnivora. Carnivores (wolves, cats, bears, etc.).

Order Pinnipedia. Seals, walruses, sea lions.

Order Proboscidea. Elephants, mammoths (extinct).

Order Sirenia. Sea cows (manatees, dugongs).

Order Perissodactyla. Odd-toed ungulates (horses, tapirs, rhinos).

Order Tubulidentata. African aardvarks.

Order Artiodactyla. Even-toed ungulates (camels, deer, bison, sheep, goats, antelopes, giraffes, etc.).

Order Cetacea. Whales, porpoises.

Appendix II. Annotations to A Journal Article

This journal article reports on the movements of a female wolf during the summer of 2002 in northwestern Canada. It also reports on a scientific process of inquiry, observation and interpretation to learn where, how and why the wolf traveled as she did. In some ways, this article reflects the story of "how to do science" told in section 1.5 of this textbook. These notes are intended to help you read and understand how scientists work and how they report on their work.

1 ARCTIC

2 VOL. 57, NO. 2 (JUNE 2004) P. 196–203

3 Long Foraging Movement of a Denning Tundra Wolf

4 Paul F. Frame,[1,2] David S. Hik,[1] H. Dean Cluff,[3] and Paul C. Paquet[4]

5 (Received 3 September 2003; accepted in revised form 16 January 2004)

6 **ABSTRACT** Wolves (*Canis lupus*) on the Canadian barrens are intimately linked to migrating herds of barren-ground caribou (*Rangifer tarandus*). We deployed a Global Positioning System (GPS) radio collar on an adult female wolf to record her movements in response to changing caribou densities near her den during summer. This wolf and two other females were observed nursing a group of 11 pups. She traveled a minimum of 341 km during a 14-day excursion. The straight-line distance from the den to the farthest location was 103 km, and the overall minimum rate of travel was 3.1 km/h. The distance between the wolf and the radio-collared caribou decreased from 242 km one week before the excursion to 8 km four days into the excursion. We discuss several possible explanations for the long foraging bout.

7 *Key words:* wolf, GPS tracking, movements, *Canis lupus*, foraging, caribou, Northwest Territories

8 **RÉSUMÉ** Les loups (*Canis lupus*) dans la toundra canadienne sont étroitement liés aux hardes de caribous des toundras (*Rangifer tarandus*). On a équipé une louve adulte d'un collier émetteur muni d'un système de positionnement mondial (GPS) afin d'enregistrer ses déplacements en réponse au changement de densité du caribou près de sa tanière durant l'été. On a observé cette louve ainsi que deux autres en train d'allaiter un groupe de 11 louveteaux. Elle a parcouru un minimum de 341 km durant une sortie de 14 jours. La distance en ligne droite de la tanière à l'endroit le plus éloigné était de 103 km, et la vitesse minimum durant tout le voyage était de 3,1 km/h. La distance entre la louve et le caribou muni du collier émetteur a diminué de 242 km une semaine avant la sortie à 8 km quatre jours après la sortie. On commente diverses explications possibles pour ce long épisode de recherche de nourriture.

Mots clés: loup, repérage GPS, déplacements, *Canis lupus*, recherche de nourriture, caribou, Territoires du Nord-Ouest

Traduit pour la revue *Arctic* par Nésida Loyer.

9 Introduction

Wolves (*Canis lupus*) that den on the central barrens of mainland Canada follow the seasonal movements of their main prey, migratory barren-ground caribou (*Rangifer tarandus*) (Kuyt, 1962; Kelsall, 1968; Walton et al., 2001). However, most wolves do not den near caribou calving grounds, but select sites farther south, closer to the tree line (Heard and Williams, 1992). Most caribou migrate beyond primary wolf denning areas by mid-June and do not return until mid-to-late July (Heard et al., 1996; Gunn et al., 2001). Conse- quently, caribou density near dens is low for part of the summer.

During this period of spatial separation from the main caribou herds, wolves must either search near **10** the homesite for scarce caribou or alternative prey (or both), travel to where prey are abundant, or use a combination of these strategies.

Walton et al. (2001) postulated that the travel of tundra wolves outside their normal summer ranges is **11** a response to low caribou availability rather than a pre-dispersal exploration like that observed in terri- torial wolves (Fritts and Mech, 1981; Messier, 1985). The authors postulated this because most such travel was directed toward caribou calving grounds. We report details of such a long-distance excursion by a breeding female tundra wolf wearing a GPS radio collar. We discuss the relationship of the excursion to movements of satellite-collared caribou (Gunn et al., 2001), supporting the hypothesis that tundra wolves make directional, rapid, long-distance movements in response to seasonal prey availability.

[1] Department of Biological Sciences, University of Alberta, Edmonton, Alberta T6G 2E9, Canada
[2] Corresponding author: pframe@ualberta.ca
[3] Department of Resources, Wildlife, and Economic Development, North Slave Region, Government of the Northwest Territories, P.O. Box 2668, 3803 Bretzlaff Dr., Yellowknife, Northwest Territories X1A 2P9, Canada; Dean_Cluff@gov.nt.ca
[4] Faculty of Environmental Design, University of Calgary, Calgary, Alberta T2N 1N4, Canada; current address: P.O. Box 150, Meacham, Saskatchewan S0K 2V0, Canada

196

---

1 Title of the journal, which reports on science taking place in Arctic regions.

2 Volume number, issue number and date of the journal, and page numbers of the article.

3 Title of the article: a concise but specific description of the subject of study—one episode of long-range travel by a wolf hunting for food on the Arctic tundra.

4 Authors of the article: scientists working at the institutions listed in the footnotes below. Note #2 indicates that P. F. Frame is the *corresponding author*—the person to contact with questions or comments. His email address is provided.

5 Date on which a draft of the article was received by the journal editor, followed by date one which a revised draft was accepted for publication. Between these dates, the article was reviewed and critiqued by other scientists, a process called peer review. The authors revised the article to make it clearer, according to those reviews.

6 ABSTRACT: A brief description of the study containing all basic elements of this report. First sentence summarizes the *background* material. Second sentence encapsulates the *methods* used. The rest of the paragraph sums up the *results*. Authors introduce the main *subject* of the study—a female wolf (#388) with pups in a den—and refer to later *discussion* of possible explanations for her behavior.

7 Key words are listed to help researchers using computer databases. Searching the databases using these key words will yield a list of studies related to this one.

8 RÉSUMÉ: The French translation of the abstract and key words. Many researchers in this field are French Canadian. Some journals provide such translations in French or in other languages.

9 INTRODUCTION: Gives the background for this wolf study. This paragraph tells of known or suspected wolf behavior that is important for this study. Note that (a) major species mentioned are always accompanied by scientific names, and (b) statements of fact or *postulations* (claims or assumptions about what is likely to be true) are followed by references to studies that established those facts or supported the postulations.

10 This paragraph focuses directly on the wolf behaviors that were studied here.

11 This paragraph starts with a statement of the *hypothesis* being tested, one that originated in other studies and is supported by this one. The hypothesis is restated more succinctly in the last sentence of this paragraph. This is the *inquiry* part of the scientific process—asking questions and suggesting possible answers.

Figure 1. Map showing the movements of satellite radio-collared caribou with respect to female wolf 388's summer range and long foraging movement, in summer 2002.

12 This map shows the study area and depicts wolf and caribou locations and movements during one summer. Some of this information is explained below.

13 STUDY AREA: This section sets the stage for the study, locating it precisely with latitude and longitude coordinates and describing the area (illustrated by the map in Figure 1).

14 Here begins the story of how prey (caribou) and predators (wolves) interact on the tundra. Authors describe movements of these nomadic animals throughout the year.

15 We focus on the denning season (summer) and learn how wolves locate their dens and travel according to the movements of caribou herds.

13 Study Area

Our study took place in the northern boreal forest–low Arctic tundra transition zone (63° 30' N, 110° 00' W; Figure 1; Timoney et al., 1992). Permafrost in the area changes from discontinuous to continuous (Harris, 1986). Patches of spruce (*Picea mariana, P. glauca*) occur in the southern portion and give way to open tundra to the northeast. Eskers, kames, and other glacial deposits are scattered throughout the study area. Standing water and exposed bedrock are characteristic of the area.

14 *Details of the Caribou-Wolf System*

The Bathurst caribou herd uses this study area. Most caribou cows have begun migrating by late April, reaching calving grounds by June (Gunn et al., 2001;

Figure 1). Calving peaks by 15 June (Gunn et al., 2001), and calves begin to travel with the herd by one week of age (Kelsall, 1968). The movement patterns of bulls are less known, but bulls frequent areas near calving grounds by mid-June (Heard et al., 1996; Gunn et al., 2001). In summer, Bathurst caribou cows generally travel south from their calving grounds and then, parallel to the tree line, to the northwest. The rut usually takes place at the tree line in October (Gunn et al., 2001). The winter range of the Bathurst herd varies among years, ranging through the taiga and along the tree line from south of Great Bear Lake to southeast of Great Slave Lake. Some caribou spend the winter on the tundra (Gunn et al., 2001; Thorpe et al., 2001).

In winter, wolves that prey on Bathurst caribou do **15** not behave territorially. Instead, they follow the herd throughout its winter range (Walton et al., 2001; Musiani, 2003). However, during denning (May–

16 Other variables are considered—prey other than caribou and their relative abundance in 2002.

17 METHODS: There is no one scientific method. Procedures for each and every study must be explained carefully.

18 Authors explain when and how they tracked caribou and wolves, including tools used and the exact procedures followed.

19 This important subsection explains what data were calculated (average distance ...) and how, including the software used and where it came from. (The calculations are listed in Table 1.) Note that the behavior measured (traveling) is carefully defined.

20 RESULTS: The heart of the report and the *observation* part of the scientific process. This section is organized parallel to the Methods section.

21 This subsection is broken down by periods of observation. Pre–excursion period covers the time between 388's capture and the start of her long–distance travel. The investigators used visual observations as well as telemetry (measurements taken using the global positioning system (GPS)) to gather data. They looked at how 388 cared for her pups, interacted with other adults, and moved about the den area.

Table 1. Daily distances from wolf 388 and the den to the nearest radio-collared caribou during a long excursion in summer 2002.

Date (2002)	Mean distance from caribou to wolf (km)	Daily distance from closest caribou to den
12 July	242	241
13 July	210	209
14 July	200	199
15 July	186	180
16 July	163	162
17 July	151	148
18 July	144	137
19 July[1]	126	124
20 July	103	130
21 July	73	130
22 July[2]	40	110
23 July[2]	9	104
29 July[3]	16	43
30 July	32	43
31 July	28	44
1 August	29	46
2 August[4]	54	52
3 August	53	53
4 August	74	74
5 August	75	75
6 August	74	75
7 August	72	75
8 August	76	75
9 August	79	79

[1] Excursion starts.
[2] Wolf closest to collared caribou.
[3] Previous five days' caribou locations not available.
[4] Excursion ends.

August, parturition late May to mid-June), wolf movements are limited by the need to return food to the den. To maximize access to migrating caribou, many wolves select den sites closer to the tree line than to caribou calving grounds (Heard and Williams, 1992). Because of caribou movement patterns, tundra denning wolves are separated from the main caribou herds by several hundred kilometers at some time during summer (Williams, 1990:19; Figure 1; Table 1).

Muskoxen do not occur in the study area (Fournier and Gunn, 1998), and there are few moose there (H.D. Cluff, pers. obs.). Therefore, alternative prey for wolves includes waterfowl, other ground-nesting birds, their eggs, rodents, and hares (Kuyt, 1972; Williams, 1990:16; H.D. Cluff and P.F. Frame, unpubl. data). During 56 hours of den observations, we saw no ground squirrels or hares, only birds. It appears that the abundance of alternative prey was relatively low in 2002.

Methods

Wolf Monitoring

We captured female wolf 388 near her den on 22 June 2002, using a helicopter net-gun (Walton et al., 2001). She was fitted with a releasable GPS radio collar (Merrill et al., 1998) programmed to acquire locations at 30-minute intervals. The collar was electronically released (e.g., Mech and Gese, 1992) on 20 August 2002. From 27 June to 3 July 2002, we observed 388's den with a 78 mm spotting scope at a distance of 390 m.

Caribou Monitoring

In spring of 2002, ten female caribou were captured by helicopter net-gun and fitted with satellite radio collars, bringing the total number of collared Bathurst cows to 19. Eight of these spent the summer of 2002 south of Queen Maud Gulf, well east of normal Bathurst caribou range. Therefore, we used 11 caribou for this analysis. The collars provided one location per day during our study, except for five days from 24 to 28 July. Locations of satellite collars were obtained from Service Argos, Inc. (Landover, Maryland).

Data Analysis

Location data were analyzed by ArcView GIS software (Environmental Systems Research Institute Inc., Redlands, California). We calculated the average distance from the nearest collared caribou to the wolf and the den for each day of the study.

Wolf foraging bouts were calculated from the time 388 exited a buffer zone (500 m radius around the den) until she re-entered it. We considered her to be traveling when two consecutive locations were spatially separated by more than 100 m. Minimum distance traveled was the sum of distances between each location and the next during the excursion.

We compared pre- and post-excursion data using Analysis of Variance (ANOVA; Zar, 1999). We first tested for homogeneity of variances with Levene's test (Brown and Forsythe, 1974). No transformations of these data were required.

Results

Wolf Monitoring

Pre-Excursion Period: Wolf 388 was lactating when captured on 22 June. We observed her and two other females nursing a group of 11 pups between 27 June and 3 July. During our observations, the pack consisted of at least four adults (3 females and 1 male) and 11 pups. On 30 June, three pups were moved to a location 310 m from the other eight and cared for by an uncollared female. The male was not seen at the den after the evening of 30 June.

Before the excursion, telemetry indicated 18 foraging bouts. The mean distance traveled during these bouts was 25.29 km (± 4.5 SE, range 3.1–82.5 km). Mean greatest distance from the den on foraging

198   *P.F. Frame, et al*

Figure 2. Details of a long foraging movement by female wolf 388 between 19 July and 2 August 2002. Also shown are locations and movements of three satellite radio-collared caribou from 23 July to 21 August 2002. On 23 July, the wolf was 8 km from a collared caribou. The farthest point from the den (103 km distant) was recorded on 27 July. Arrows indicate direction of travel.

bouts was 7.1 km (± 0.9 SE, range 1.7–17.0 km). The average duration of foraging bouts for the period was 20.9 h (± 4.5 SE, range 1–71 h).

The average daily distance between the wolf and the nearest collared caribou decreased from 242 km on 12 July, one week before the excursion period, to 126 km on 19 July, the day the excursion began (Table 1).

Excursion Period: On 19 July at 2203, after spending 14 h at the den, 388 began moving to the northeast and did not return for 336 h (14 d; Figure 2). Whether she traveled alone or with other wolves is unknown. During the excursion, 476 (71%) of 672 possible locations were recorded. The wolf crossed the southeast end of Lac Capot Blanc on a small land bridge, where she paused for 4.5 h after traveling for 19.5 h (37.5

km). Following this rest, she traveled for 9 h (26.3 km) onto a peninsula in Reid Lake, where she spent 2 h before backtracking and stopping for 8 h just off the peninsula. Her next period of travel lasted 16.5 h (32.7 km), terminating in a pause of 9.5 h just 3.8 km from a concentration of locations at the far end of her excursion, where we presume she encountered caribou. The mean duration of these three movement periods was 15.7 h (± 2.5 SE), and that of the pauses, 7.3 h (± 1.5). The wolf required 72.5 h (3.0 d) to travel a minimum of 95 km from her den to this area near caribou (Figure 2). She remained there (35.5 km2) for 151.5 h (6.3 d) and then moved south to Lake of the Enemy, where she stayed (31.9 km^2) for 74 h (3.1 d) before returning to her den. Her greatest distance from the den, 103 km, was recorded 174.5 h (7.3 d) after the excursion

22 The key in the lower right–hand corner of the map shows areas (shaded) within which the wolves and caribou moved, and the dotted trail of 388 during her excursion. From the results depicted on this map, the investigators tried to determine when and where 388 might have encountered caribou and how their locations affected her traveling behavior.

23 The wolf's excursion (her long trip away from the den area) is the focus of this study. These paragraphs present detailed measurements of daily movements during her two–week trip—how far she traveled, how far she was from collared caribou, her time spent traveling and resting, and her rate of speed. Authors use the phrase "minimum distance traveled" to acknowledge they couldn't track every step but were measuring samples of her movements. They knew that she went at least as far as they measured. This shows how scientists try to be exact when reporting results. Results of this study are depicted graphically in the map in Figure 2.

Appendix II

24 Post-excursion measurements of 388's movements were made to compare with those of the pre-excursion period. In order to compare, scientists often use *means*, or averages, of a series of measurements—mean distances, mean duration, etc.

25 In the comparison, authors used statistical calculations (F and df) to determine that the differences between pre- and post-excursion measurements were *statistically insignificant*, or close enough to be considered essentially the same or similar.

26 As with wolf 388, the investigators measured the movements of caribou during the study period. The areas within which the caribou moved are shown in Figure 2 by shaded polygons mentioned in the second paragraph of this subsection.

27 This subsection summarizes how distances separating predators and prey varied during the study period.

28 DISCUSSION: This section is the *interpretation* part of the scientific process.

29 This subsection reviews observations from other studies and suggests that this study fits with patterns of those observations.

30 Authors discuss a prevailing *theory* (CBFT) which might explain why a wolf would travel far to meet her own energy needs while taking food caught closer to the den back to her pups. The results of this study seem to fit that pattern.

---

began, at 0433 on 27 July. She was 8 km from a collared caribou on 23 July, four days after the excursion began (Table 1).

The return trip began at 0403 on 2 August, 318 h (13.2 d) after leaving the den. She followed a relatively direct path for 18 h back to the den, a distance of 75 km.

The minimum distance traveled during the excursion was 339 km. The estimated overall minimum travel rate was 3.1 km/h, 2.6 km/h away from the den and 4.2 km/h on the return trip.

24 **Post-Excursion Period:** We saw three pups when recovering the collar on 20 August, but others may have been hiding in vegetation.

Telemetry recorded 13 foraging bouts in the post-excursion period. The mean distance traveled during these bouts was 18.3 km (+ 2.7 SE, range 1.2–47.7 km), and mean greatest distance from the den was 7.1 km (+ 0.7 SE, range 1.1–11.0 km). The mean duration of these post-excursion foraging bouts was 10.9 h (+ 2.4 SE, range 1–33 h).

When 388 reached her den on 2 August, the distance to the nearest collared caribou was 54 km. On 9 August, one week after she returned, the distance was 79 km (Table 1).

Pre- and Post-Excursion Comparison

25 We found no differences in the mean distance of foraging bouts before and after the excursion period (F = 1.5, df = 1, 29, p = 0.24). Likewise, the mean greatest distance from the den was similar pre- and post-excursion (F = 0.004, df = 1, 29, p = 0.95). However, the mean duration of 388's foraging bouts decreased by 10.0 h after her long excursion (F = 3.1, df = 1, 29, p = 0.09).

26 *Caribou Monitoring*

Summer Movements: On 10 July, 5 of 11 collared caribou were dispersed over a distance of 10 km, 140 km south of their calving grounds (Figure 1). On the same day, three caribou were still on the calving grounds, two were between the calving grounds and the leaders, and one was missing. One week later (17 July), the leading radio-collared cows were 100 km farther south (Figure 1). Two were within 5 km of each other in front of the rest, who were more dispersed. All radio-collared cows had left the calving grounds by this time. On 23 July, the leading radio-collared caribou had moved 35 km farther south, and all of them were more widely dispersed. The two cows closest to the leader were 26 km and 33 km away, with 37 km between them. On the next location (29 July), the most southerly caribou were 60 km

200 *P.F. Frame, et al*

farther south. All of the caribou were now in the areas where they remained for the duration of the study (Figure 2).

A Minimum Convex Polygon (Mohr and Stumpf, 1966) around all caribou locations acquired during the study encompassed 85 119 km².

Relative to the Wolf Den: The distance from the **27** nearest collared caribou to the den decreased from 241 km one week before the excursion to 124 km the day it began. The nearest a collared caribou came to the den was 43 km away, on 29 and 30 July. During the study, four collared caribou were located within 100 km of the den. Each of these four was closest to the wolf on at least one day during the period reported.

28 **Discussion**

Prey Abundance

Caribou are the single most important prey of tundra **29** wolves (Clark, 1971; Kuyt, 1972; Stephenson and James, 1982; Williams, 1990). Caribou range over vast areas, and for part of the summer, they are scarce or absent in wolf home ranges (Heard et al., 1996). Both the long distance between radio-collared caribou and the den the week before the excursion and the increased time spent foraging by wolf 388 indicate that caribou availability near the den was low. Observations of the pups' being left alone for up to 18 h, presumably while adults were searching for food, provide additional support for low caribou availability locally. Mean foraging bout duration decreased by 10.0 h after the excursion, when collared caribou were closer to the den, suggesting an increase in caribou availability nearby.

Foraging Excursion

One aspect of central place foraging theory (CPFT) **30** deals with the optimality of returning different-sized food loads from varying distances to dependents at a central place (i.e., the den) (Orians and Pearson, 1979). Carlson (1985) tested CPFT and found that the predator usually consumed prey captured far from the central place, while feeding prey captured nearby to dependants. Wolf 388 spent 7.2 days in one area near caribou before moving to a location 23 km back towards the den, where she spent an additional 3.1 days, likely hunting caribou. She began her return trip from this closer location, traveling directly to the den. While away, she may have made one or more successful kills and spent time meeting her own energetic needs before returning to the den. Alternatively, it may have taken several attempts to make a kill, which she then fed on before beginning her return trip. We do not know if she returned food to the pups, but such behavior would be supported by CPFT.

31 Other workers have reported wolves' making long round trips and referred to them as "extraterritorial" or "pre-dispersal" forays (Fritts and Mech, 1981; Messier, 1985; Ballard et al., 1997; Merrill and Mech, 2000). These movements are most often made by young wolves (1–3 years old), in areas where annual territories are maintained and prey are relatively sedentary (Fritts and Mech, 1981; Messier, 1985). The long excursion of 388 differs in that tundra wolves do not maintain annual territories (Walton et al., 2001), and the main prey migrate over vast areas (Gunn et al., 2001).

Another difference between 388's excursion and those reported earlier is that she is a mature, breeding female. No study of territorial wolves has reported reproductive adults making extraterritorial movements in summer (Fritts and Mech, 1981; Messier, 1985; Ballard et al., 1997; Merrill and Mech, 2001). However, Walton et al. (2001) also report that breeding female tundra wolves made excursions.

Direction of Movement

32 Possible explanations for the relatively direct route 388 took to the caribou include landscape influence and experience. Considering the timing of 388's trip and the locations of caribou, had the wolf moved northwest, she might have missed the caribou entirely, or the encounter might have been delayed.

A reasonable possibility is that the land directed 388's route. The barrens are crisscrossed with trails worn into the tundra over centuries by hundreds of thousands of caribou and other animals (Kelsall, 1968; Thorpe et al., 2001). At river crossings, lakes, or narrow peninsulas, trails converge and funnel towards and away from caribou calving grounds and summer range. Wolves use trails for travel (Paquet et al., 1996; Mech and Boitani, 2003; P. Frame, pers. observation). Thus, the landscape may direct an animal's movements and lead it to where cues, such as the odor of caribou on the wind or scent marks of other wolves, may lead it to caribou.

33 Another possibility is that 388 knew where to find caribou in summer. Sexually immature tundra wolves sometimes follow caribou to calving grounds (D. Heard, unpubl. data). Possibly, 388 had made such journeys in previous years and killed caribou. If this were the case, then in times of local prey scarcity she might travel to areas where she had hunted successfully before. Continued monitoring of tundra wolves may answer questions about how their food needs are met in times of low caribou abundance near dens.

Caribou often form large groups while moving **34** south to the tree line (Kelsall, 1968). After a large aggregation of caribou moves through an area, its scent can linger for weeks (Thorpe et al., 2001:104). It is conceivable that 388 detected caribou scent on the wind, which was blowing from the northeast on 19–21 July (Environment Canada, 2003), at the same time her excursion began. Many factors, such as odor strength and wind direction and strength, make systematic study of scent detection in wolves difficult under field conditions (Harrington and Asa, 2003). However, humans are able to smell odors such as forest fires or oil refineries more than 100 km away. The olfactory capabilities of dogs, which are similar to wolves, are thought to be 100 to 1 million times that of humans (Harrington and Asa, 2003). Therefore, it is reasonable to think that under the right wind conditions, the scent of many caribou traveling together could be detected by wolves from great distances, thus triggering a long foraging bout.

Rate of Travel

Mech (1994) reported the rate of travel of Arctic **35** wolves on barren ground was 8.7 km/h during regular travel and 10.0 km/h when returning to the den, a difference of 1.3 km/h. These rates are based on direct observation and exclude periods when wolves moved slowly or not at all. Our calculated travel rates are assumed to include periods of slow movement or no movement. However, the pattern we report is similar to that reported by Mech (1994), in that homeward travel was faster than regular travel by 1.6 km/h. The faster rate on return may be explained by the need to return food to the den. Pup survival can increase with the number of adults in a pack available to deliver food to pups (Harrington et al., 1983). Therefore, an increased rate of travel on homeward trips could improve a wolf's reproductive fitness by getting food to pups more quickly.

Fate of 388's Pups

Wolf 388 was caring for pups during den observa- **36** tions. The pups were estimated to be six weeks old, and were seen ranging as far as 800 m from the den. They received some regurgitated food from two of the females, but were unattended for long periods. The excursion started 16 days after our observations, and it is improbable that the pups could have traveled the distance that 388 moved. If the pups died, this would have removed parental responsibility, allowing the long movement.

Our observations and the locations of radio-collared caribou indicate that prey became scarce in

31 Here our authors note other possible explanations for wolves' excursions presented by other investigators, but this study does not seem to support those ideas.

32 Authors discuss possible reasons for why 388 traveled directly to where caribou were located. They take what they learned from earlier studies and apply it to this case, suggesting that the lay of the land played a role. Note that their description paints a clear picture of the landscape.

33 Authors suggest that 388 may have learned in traveling during previous summers where the caribou were. The last two sentences suggest ideas for future studies.

34 Or maybe 388 followed the scent of the caribou. Authors acknowledge difficulties of proving this, but they suggest another area where future studies might be done.

35 Authors suggest that results of this study support previous studies about how fast wolves travel to and from the den. In the last sentence, they speculate on how these observed patterns would fit into the theory of evolution.

36 Authors also speculate on the fate of 388's pups while she was traveling. This leads to . . .

37 Discussion of cooperative rearing of pups and, in turn, to speculation on how this study and what is known about cooperative rearing might fit into the animal's strategies for survival of the species. Again, the authors approach the broader theory of evolution and how it might explain some of their results.

38 And again, they suggest that this study points to several areas where further study will shed some light.

39 In conclusion, the authors suggest that their study supports the hypothesis being tested here. And they touch on the implications of increased human activity on the tundra predicted by their results.

40 ACKNOWLEDGEMENTS: Authors note the support of institutions, companies and individuals. They thank their reviewers ad list permits under which their research was carried on.

41 REFERENCES: List of all studies cited in the report. This may seem tedious, but is a vitally important part of scientific reporting. It is a record of the sources of information on which this study is based. It provides readers with a wealth of resources for further reading on this topic. Much of it will form the foundation of future scientific studies like this one.

the area of the den as summer progressed. Wolf 388 may have abandoned her pups to seek food for herself. However, she returned to the den after the excursion, where she was seen near pups. In fact, she foraged in a similar pattern before and after the excursion, suggesting that she again was providing for pups after her return to the den.

37 A more likely possibility is that one or both of the other lactating females cared for the pups during 388's absence. The three females at this den were not seen with the pups at the same time. However, two weeks earlier, at a different den, we observed three females cooperatively caring for a group of six pups. At that den, the three lactating females were observed providing food for each other and trading places while nursing pups. Such a situation at the den of 388 could have created conditions that allowed one or more of the lactating females to range far from the den for a period, returning to her parental duties afterwards. However, the pups would have been weaned by eight weeks of age (Packard et al., 1992), so nonlactating adults could also have cared for them, as often happens in wolf packs (Packard et al., 1992; Mech et al., 1999).

Cooperative rearing of multiple litters by a pack could create opportunities for long-distance foraging movements by some reproductive wolves during summer periods of local food scarcity. We have recorded multiple lactating females at one or more tundra wolf dens per year since 1997. This reproductive strategy may be an adaptation to temporally and **38** spatially unpredictable food resources. All of these possibilities require further study, but emphasize both the adaptability of wolves living on the barrens and their dependence on caribou.

39 Long-range wolf movement in response to caribou availability has been suggested by other researchers (Kuyt, 1972; Walton et al., 2001) and traditional ecological knowledge (Thorpe et al., 2001). Our report demonstrates the rapid and extreme response of wolves to caribou distribution and movements in summer. Increased human activity on the tundra (mining, road building, pipelines, ecotourism) may influence caribou movement patterns and change the interactions between wolves and caribou in the region. Continued monitoring of both species will help us to assess whether the association is being affected adversely by anthropogenic change.

40 ## Acknowledgements

This research was supported by the Department of Resources, Wildlife, and Economic Development, Government of the Northwest Territories; the Department of Biological Sciences at the University of Alberta; the Natural Sciences and Engineering Research Council of Canada; the Department of Indian and Northern Affairs Canada; the Canadian Circumpolar Institute; and DeBeers Canada, Ltd. Lorna Ruechel assisted with den observations. A. Gunn provided caribou location data. We thank Dave Mech for the use of GPS collars. M. Nelson, A. Gunn, and three anonymous reviewers made helpful comments on earlier drafts of the manuscript. This work was done under Wildlife Research Permit – WL002948 issued by the Government of the Northwest Territories, Department of Resources, Wildlife, and Economic Development.

41 ## References

BALLARD, W.B., AYRES, L.A., KRAUSMAN, P.R., REED, D.J., and FANCY, S.G. 1997. Ecology of wolves in relation to a migratory caribou herd in northwest Alaska. Wildlife Monographs 135. 47 p.

BROWN, M.B., and FORSYTHE, A.B. 1974. Robust tests for the equality of variances. Journal of the American Statistical Association 69:364–367.

CARLSON, A. 1985. Central place foraging in the red-backed shrike (Lanius collurio L.): Allocation of prey between forager and sedentary consumer. Animal Behaviour 33:664–666.

CLARK, K.R.F. 1971. Food habits and behavior of the tundra wolf on central Baffin Island. Ph.D. Thesis, University of Toronto, Ontario, Canada.

ENVIRONMENT CANADA. 2003. National climate data information archive. Available online: http://www.climate.weatheroffice.ec.gc.ca/Welcome_e.html

FOURNIER, B., and GUNN, A. 1998. Musk ox numbers and distribution in the NWT, 1997. File Report No. 121. Yellowknife: Department of Resources, Wildlife, and Economic Development, Government of the Northwest Territories. 55 p.

FRITTS, S.H., and MECH, L.D. 1981. Dynamics, movements, and feeding ecology of a newly protected wolf population in northwestern Minnesota. Wildlife Monographs 80. 79 p.

GUNN, A., DRAGON, J., and BOULANGER, J. 2001. Seasonal movements of satellite-collared caribou from the Bathurst herd. Final Report to the West Kitikmeot Slave Study Society, Yellowknife, NWT. 80 p. Available online: http://www.wkss.nt.ca/HTML/08_ProjectsReports/PDF/Seasonal MovementsFinal.pdf

HARRINGTON, F.H., and ASA, C.S. 2003. Wolf communication. In: Mech, L.D., and Boitani, L., eds. Wolves: Behavior, ecology, and conservation. Chicago: University of Chicago Press. 66–103.

HARRINGTON, F.H., MECH, L.D., and FRITTS, S.H. 1983. Pack size and wolf pup survival: Their relationship under varying ecological conditions. Behavioral Ecology and Sociobiology 13:19–26.

HARRIS, S.A. 1986. Permafrost distribution, zonation and stability along the eastern ranges of the cordillera of North America. Arctic 39(1):29–38.

HEARD, D.C., and WILLIAMS, T.M. 1992. Distribution of wolf dens on migratory caribou ranges in the Northwest

202 *P.F. Frame, et al*

Territories, Canada. Canadian Journal of Zoology 70:1504–1510.

HEARD, D.C., WILLIAMS, T.M., and MELTON, D.A. 1996. The relationship between food intake and predation risk in migratory caribou and implication to caribou and wolf population dynamics. Rangifer Special Issue No. 2:37–44.

KELSALL, J.P. 1968. The migratory barren-ground caribou of Canada. Canadian Wildlife Service Monograph Series 3. Ottawa: Queen's Printer. 340 p.

KUYT, E. 1962. Movements of young wolves in the Northwest Territories of Canada. Journal of Mammalogy 43:270–271.

———. 1972. Food habits and ecology of wolves on barren-ground caribou range in the Northwest Territories. Canadian Wildlife Service Report Series 21. Ottawa: Information Canada. 36 p.

MECH, L.D. 1994. Regular and homeward travel speeds of Arctic wolves. Journal of Mammalogy 75:741–742.

MECH, L.D., and BOITANI, L. 2003. Wolf social ecology. In: Mech, L.D., and Boitani, L., eds. Wolves: Behavior, ecology, and conservation. Chicago: University of Chicago Press. 1–34.

MECH, L.D., and GESE, E.M. 1992. Field testing the Wildlink capture collar on wolves. Wildlife Society Bulletin 20:249–256.

MECH, L.D., WOLFE, P., and PACKARD, J.M. 1999. Regurgitative food transfer among wild wolves. Canadian Journal of Zoology 77:1192–1195.

MERRILL, S.B., and MECH, L.D. 2000. Details of extensive movements by Minnesota wolves (*Canis lupus*). American Midland Naturalist 144:428–433.

MERRILL, S.B., ADAMS, L.G., NELSON, M.E., and MECH, L.D. 1998. Testing releasable GPS radiocollars on wolves and white-tailed deer. Wildlife Society Bulletin 26:830–835.

MESSIER, F. 1985. Solitary living and extraterritorial movements of wolves in relation to social status and prey abundance. Canadian Journal of Zoology 63:239–245.

MOHR, C.O., and STUMPF, W.A. 1966. Comparison of methods for calculating areas of animal activity. Journal of Wildlife Management 30:293–304.

MUSIANI, M. 2003. Conservation biology and management of wolves and wolf-human conflicts in western North America. Ph.D. Thesis, University of Calgary, Calgary, Alberta, Canada.

ORIANS, G.H., and PEARSON, N.E. 1979. On the theory of central place foraging. In: Mitchell, R.D., and Stairs, G.F., eds. Analysis of ecological systems. Columbus: Ohio State University Press. 154–177.

PACKARD, J.M., MECH, L.D., and REAM, R.R. 1992. Weaning in an arctic wolf pack: Behavioral mechanisms. Canadian Journal of Zoology 70:1269–1275.

PAQUET, P.C., WIERZCHOWSKI, J., and CALLAGHAN, C. 1996. Summary report on the effects of human activity on gray wolves in the Bow River Valley, Banff National Park, Alberta. In: Green, J., Pacas, C., Bayley, S., and Cornwell, L., eds. A cumulative effects assessment and futures outlook for the Banff Bow Valley. Prepared for the Banff Bow Valley Study. Ottawa: Department of Canadian Heritage.

STEPHENSON, R.O., and JAMES, D. 1982. Wolf movements and food habits in northwest Alaska. In: Harrington, F.H., and Paquet, P.C., eds. Wolves of the world. New Jersey: Noyes Publications. 223–237.

THORPE, N., EYEGETOK, S., HAKONGAK, N., and QITIRMIUT ELDERS. 2001. The Tuktu and Nogak Project: A caribou chronicle. Final Report to the West Kitikmeot/Slave Study Society, Ikaluktuuttiak, NWT. 160 p.

TIMONEY, K.P., LA ROI, G.H., ZOLTAI, S.C., and ROBINSON, A.L. 1992. The high subarctic forest-tundra of northwestern Canada: Position, width, and vegetation gradients in relation to climate. Arctic 45(1):1–9.

WALTON, L.R., CLUFF, H.D., PAQUET, P.C., and RAMSAY, M.A. 2001. Movement patterns of barren-ground wolves in the central Canadian Arctic. Journal of Mammalogy 82:867–876.

WILLIAMS, T.M. 1990. Summer diet and behavior of wolves denning on barren-ground caribou range in the Northwest Territories, Canada. M.Sc. Thesis, University of Alberta, Edmonton, Alberta, Canada.

ZAR, J.H. 1999. Biostatistical analysis. 4th ed. New Jersey: Prentice Hall. 663 p.

Appendix III. Answers to Self-Quizzes and Genetics Problems

Italicized numbers refer to relevant section numbers

Chapter 1

1.	atoms	*1.2*
2.	cell	*1.2*
3.	Animals	*1.4*
4.	energy, nutrients	*1.3*
5.	Homeostasis	*1.3*
6.	d	*1.3*
7.	reproduction	*1.3*
8.	d	*1.3*
9.	a	*1.2*
	d	*1.3*
	e	*1.4*
10.	a	*1.2*
	d	*1.3*
	f	*1.4*
11.	b	*1.9*
12.	b	*1.6*
13.	c	*1.2*
	b	*1.5*
	d	*1.9*
	e	*1.6*
	a	*1.6*
	f	*1.8*

Chapter 2

1.	False, H lacks neutrons	*2.2*
2.	d	*2.2*
3.	b	*2.2*
4.	b	*2.3*
5.	a	*2.3*
6.	a	*2.4*
7.	polar covalent	*2.4*
8.	c	*2.5*
9.	H+ and OH−	*2.5*
10.	d	*2.6*
11.	a	*2.6*
12.	c	*2.6*
13.	b	*2.5*
14.	c	*2.5*
	b	*2.2*
	d	*2.3*
	h	*2.2*
	g	*2.5*
	f	*2.3*
	e	*2.3*
	a	*2.3*

Chapter 3

1.	c	*3.2*
2.	four	*3.2*
3.	a	*3.2*
4.	e	*3.3, 3.7*
5.	c	*3.4*
6.	False	*3.4*
7.	b	*3.4*
8.	starch, glycogen, cellulose	*3.3*
9.	e	*3.4*
10.	d	*3.5, 3.7*
11.	d	*3.6*
12.	d	*3.7*
13.	a	*3.7*
14.	a, amino acid	*3.5*
	b, carbohydrate	*3.3*
	c, polypeptide	*3.5*
	d, fatty acid	*3.4*

15.	c	*3.5*
	e	*3.7*
	f	*3.4*
	d	*3.7*
	a	*3.3*
	b	*3.4*
16.	g	*3.5*
	b	*3.5*
	d	*3.7*
	f	*3.7*
	a	*3.4*
	c	*3.4*
	e	*3.3*
	i	*3.5*
	h	*3.3*

Chapter 4

1.	cell	*4.2*
2.	c	*4.2*
3.	False	*4.6*
4.	c	*4.4*
5.	c	*4.5*
6.	c	*4.4*
7.	lysosomes	*4.8*
8.	c, b, d, a	*4.8*
9.	lipids and proteins	*4.8*
10.	False. Some secrete a covering	*4.5*
11.	a	*4.4*
12.	d	*4.11*
13.	a	*4.11*
14.	c	*4.9*
	f	*4.9*
	a	*4.5*
	e	*4.8*
	d	*4.8*
	b	*4.8*

Chapter 5

1.	c	*5.2*
2.	a	*5.3*
3.	b	*5.2*
4.	d	*5.4*
5.	c	*5.3, 5.4*
6.	temperature, pH, salt concentration are all correct	*5.7*
7.	e	*5.5*
8.	more, less	*5.6*
9.	gases, small nonpolar molecules, and water are all correct	*5.7*
10.	b	*5.7*
11.	a	*5.6*
12.	e	*5.8*
13.	c	*5.3*
	d	*5.4*
	e	*5.2*
	b	*5.3*
	a	*5.4*
	f	*5.6*
	g	*5.7*
	i	*5.7*
	h	*5.8*

Chapter 6

1.	autotroph: weed; heterotrophs: cat, bird, caterpillar	*6.1*

2.	carbon dioxide; sunlight	*6.1*
3.	a	*6.2*
4.	a	*6.4*
5.	d	*6.5*
6.	b	*6.5*
7.	b	*6.5*
8.	c	*6.4, 6.7*
9.	b	*6.7*
10.	e	*6.7*
11.	f	*6.7*
12.	f	*6.7*
	h	*6.7*
	g	*6.5*
	d	*6.5, 6.6*
	e	*6.8*
	b	*6.1*
	a	*6.2*
	c	*6.1*

Chapter 7

1.	False (plants use aerobic respiration too)	*7.2*
2.	d	*7.3*
3.	a	*7.2*
4.	c	*7.3*
5.	b	*7.4*
6.	e	*7.3–7.5*
7.	b	*7.4*
8.	c	*7.5*
9.	c	*7.6*
10.	b	*7.6*
11.	c	*7.5*
12.	d	*7.7*
13.	f	*7.6*
14.	b	*7.3*
	c	*7.6*
	a	*7.4*
	d	*7.5*
15.	b	*7.4*
	d	*7.3*
	a	*7.3, 7.6*
	c	*7.5*
	f	*7.5*
	e	*7.4*

Chapter 8

1.	b	*8.2*
2.	centromere	*8.2*
3.	d	*8.2*
4.	c	*8.4*
5.	d	*8.4*
6.	c	*8.4*
7.	a	*8.6*
8.	d	*8.6*
9.	3′—CCAAAGAAGTTCTCT—5′	*8.6*
10.	d	*8.7*
11.	c	*8.4*
12.	d	*8.7*
13.	b	*8.2*
14.	therapeutic cloning	*8.7*
15.	d	*8.3*
	b	*8.1, 8.7*
	a	*8.4*
	f	*8.2*
	e	*8.6*
	g	*8.6*
	c	*8.2*

Chapter 9

1.	c	*9.2*
2.	b	*9.3*
3.	d	*9.3*
4.	a	*9.2*
5.	c	*9.2*
6.	c	*9.4*
7.	a	*9.4*
8.	15 amino acids	*9.4*
9.	a	*9.3*
10.	a	*9.5*
11.	d	*9.5*
12.	a	*9.4*
13.	c	*9.4*
14.	f	*9.6*
15.	c	*9.4*
	g	*9.3*
	e	*9.5*
	a	*9.4*
	f	*9.4*
	d	*9.3*
	b	*9.6*

Chapter 10

1.	d	*10.2*
2.	d	*10.2, 10.3*
3.	d	*10.2*
4.	transcription factors	*10.2*
5.	b	*10.2*
6.	h	*10.2*
7.	c	*10.3*
8.	d	*10.4*
9.	d	*10.2*
10.	b	*10.4*
11.	c	*10.4*
12.	c	*10.3*
13.	d	*10.3*
14.	b	*10.5*
15.	f	*10.4*
	a	*10.4*
	b	*10.5*
	e	*10.4*
	c	*10.2*
	d	*10.2*

Chapter 11

1.	d	*11.2*
2.	b	*11.2*
3.	d	*11.2*
4.	e	*11.2*
5.	b	*11.3*
6.	c	*11.3*
7.	b	*11.2*
8.	a	*11.2*
9.	interphase, prophase, metaphase, anaphase, telophase	*11.3*
10.	c	*11.2*
11.	a	*11.2*
12.	The EGF receptor gene (*EGFR*), as well as *BRCA1*, *BRCA2*, and *53BP1* are all named in this chapter	*11.5, 11.6*
13.	b	*11.6*

14.	c	*11.4*
	f	*11.3*
	a	*11.6*
	g	*11.4*
	b	*11.4*
	e	*11.6*
	d	*11.3*
15.	d, b, c, a	*11.3*

Chapter 12

1.	b	*12.1*
2.	c	*12.2*
3.	d	*12.2*
4.	b	*12.2*
5.	b	*12.4*
6.	prophase I	*12.4*
7.	c	*12.3, 12.6*
8.	a	*12.2*
9.	e	*12.4*
10.	sister chromatids are still attached	*12.3*
11.	b	*12.4, 12.6*
12.	b	*12.2*
	c	*12.3*
	a	*12.2*
	e	*12.5*
	d	*12.2*

Chapter 13

1.	b	*13.2*
2.	a	*13.2*
3.	offspring	*13.3*
4.	b	*13.3*
5.	c	*13.3*
6.	a	*13.3*
7.	b	*13.3*
8.	d	*13.4*
9.	c	*13.4*
10.	c	*13.5*
11.	b	*13.5*
12.	continuous variation	*13.6*
13.	b	*13.4*
	d	*13.3*
	a	*13.2*
	c	*13.2*

Chapter 14

1.	b	*14.2*
2.	b	*14.2*
3.	a	*14.2*
4.	b	*14.3*
5.	False	*14.4*
6.	c	*14.4*
7.	d	*14.4*
8.	Y-linked dominant inheritance	*14.4, 14.6*
9.	X from mom and dad	*14.6*
10.	d	*14.6*
11.	b	*14.6*
12.	d	*14.6*
13.	True	*14.6*
14.	c	*14.6*
15.	c	*14.6*
	e	*14.5*
	f	*14.6*
	b	*14.5*
	a	*14.2*
	d	*14.6*

Chapter 15

#	Answer	Section
1.	c	15.2
2.	a	15.2
3.	b	15.2
4.	b	15.3
5.	c	15.3
6.	DNA sequencing	15.4
7.	b	15.4
8.	d	15.3, 15.5
9.	d	15.5
10.	b	15.10
	d	15.2
	e	15.10
	f	15.7
11.	c	15.6
12.	b	15.10
13.	a	15.3
	d	15.2
	c	15.2
	e	15.4
	b	15.2
14.	c	15.5
	g	15.7
	d	15.3
	e	15.10
	b	15.1
	a	15.6
	f	15.6

Chapter 16

#	Answer	Section
1.	b	16.2
2.	d	16.3, 16.4
3.	e	16.5
4.	Gondwana	16.7
5.	c	16.2, 16.8
6.	d	16.5
7.	a	16.2, 16.3
	c	16.7
	e	16.7
	f	16.3
	g	16.1
	h	16.7
8.	65.5	16.6
9.	Archaean Eon	16.6
10.	a	16.8
11.	d	16.8
12.	e	16.4
	a	16.2
	d	16.9
	h	16.5
	c	16.8
	b	16.3
	f	16.8
	g	16.4

Chapter 17

#	Answer	Section
1.	populations	17.2
2.	b	17.2
3.	a	17.2
4.	c	17.2, 17.3
5.	allopatric speciation	17.11
6.	a, d	17.6
7.	b, c	17.6
8.	b, e	17.5
9.	d	17.7
10.	e	17.7
11.	b	17.9
12.	c	17.14
13.	b	17.14
14.	c	17.9
	e	17.4
	g	17.2
	b	17.8
	f	17.13
	a	17.13
	d	17.14
	h	17.14

Chapter 18

#	Answer	Section
1.	c	18.2, 18.5
2.	c	18.4
3.	a	18.4
4.	a	18.4
5.	c	18.5
6.	b	18.6
7.	electrodes	18.3
8.	d	18.5
9.	c	18.5
10.	RNA	18.4
11.	a	18.5
12.	d	18.7
13.	c	18.6
14.	DNA	18.4
15.	f	18.2
	c	18.4
	d	18.5
	a	18.5
	b	18.7
	e	18.7

Chapter 19

#	Answer	Section
1.	c	19.2
2.	b	19.4
3.	c	19.2
4.	RNA	19.2
5.	c	19.3
6.	False	19.6
7.	one	19.5
8.	c	19.7
9.	b	19.7
10.	d	19.7
11.	a	19.7
12.	a	19.5
13.	c	19.3, 19.7
14.	a	19.8
15.	e	19.8
	f	19.7
	b	19.2
	a	19.4
	g	19.6
	h	19.8
	d	19.5
	c	19.8

Chapter 20

#	Answer	Section
1.	c	20.3
2.	d	20.4
3.	b	20.3
4.	c	20.5
5.	d	20.3
6.	d	20.9
7.	d	20.8
8.	b	20.6
9.	c	20.7
10.	d	20.7
11.	c	20.8
12.	a	20.9
13.	a	20.3
14.	a	20.8
15.	d	20.3
	g	20.6
	a	20.3
	b	20.7
	f	20.7
	h	20.8
	e	20.8
	c	20.9

Chapter 21

#	Answer	Section
1.	c	21.2
2.	a	21.4
3.	b	21.3
4.	c	21.4
5.	a	21.5
6.	e	21.3, 21.4, 21.6
7.	b	21.5
8.	True	21.4
9.	c	21.6, 21.7
10.	a	21.3, 21.4
11.	a	21.2
12.	d	21.8
13.	b	21.8
14.	c	21.3
	d	21.4
	a	21.6
	b	21.7
15.	c	21.5
	i	21.2
	a	21.2
	b	21.2
	e	21.7
	f	21.7
	d	21.4
	g	21.4
	h	21.5

Chapter 22

#	Answer	Section
1.	c	22.2
2.	a	22.2
3.	a	22.2
4.	c	22.4
5.	d	22.5
6.	hypha	22.2
7.	c	22.5
8.	b	22.5
9.	a	22.3
10.	b	22.6
11.	a	22.3, 22.6
12.	d	22.3, 22.7
13.	d	22.7
14.	b	22.7
15.	e	22.2
	b	22.2
	a	22.3
	f	22.4
	g	22.5
	c	22.6
	e	22.6

Chapter 23

#	Answer	Section
1.	True	23.2
2.	d	23.2
3.	a	23.5
4.	a	23.6
5.	b	23.13
6.	a	23.8
7.	b	23.11
8.	a	23.2
9.	c	23.10, 23.13
10.	b	23.15
11.	closed	23.7, 23.8
12.	b	23.3
	j	23.3
	d	23.4
	i	23.5
	c	23.6
	a	23.2, 23.9
	g	23.7
	e	23.10
	f	23.8
	h	23.15

Chapter 24

#	Answer	Section
1.	c	24.2
2.	d	24.2
3.	a	24.3
4.	c	24.2, 24.3
5.	d	24.3, 24.4
6.	c	24.5
7.	f	24.5
8.	a	24.5, 24.6
9.	c	24.7
10.	f	24.2, 24.8–24.10
11.	c	24.10, 24.11
12.	True	24.6
13.	b	24.2
	i	24.3
	g	24.4
	f	24.9
	c	24.7
	d	24.8
	a	24.8
	h	24.8
	e	24.9

Chapter 25

#	Answer	Section
1.	d	25.2
2.	b	25.2
3.	c	25.4
4.	c	25.5
5.	a	25.1
6.	birds	25.6
7.	protection from predators; greater ability to protect territory in a group	25.7
8.	d	25.8
9.	d	25.8
10.	close relatives	25.8
11.	h	25.3
	e	25.8
	c	25.3
	b	25.6
	d	25.6
	a	25.3
	f	25.7
	g	25.3

Chapter 26

#	Answer	Section
1.	clumped	26.2
2.	f	26.3
3.	400	26.2
4.	1600	26.3
5.	a	26.3
6.	d	26.4
7.	d	26.5
8.	b	26.7
9.	d	26.8
10.	a	26.7
11.	a	26.5
12.	c	26.5
13.	c	26.4
	d	26.3
	a	26.3
	e	26.4
	b	26.4

Chapter 27

#	Answer	Section
1.	b	27.2
2.	d	27.2
3.	e	27.4
4.	c	27.4
5.	b	27.3
	d	27.6
	c	27.2
	a	27.5
	e	27.4
6.	b	27.6
7.	b	27.7
8.	Pioneer	27.7
9.	c	27.7
10.	a	27.9
11.	b	27.4
12.	c	27.6
13.	a	27.5
14.	c	27.9
	b	27.7
	d	27.8
	a	27.8
	f	27.8
	e	27.4

CHAPTER 28

#	Answer	Section
1.	b	28.2
2.	d	28.2
3.	d	28.3
4.	d	28.2
5.	a	28.2
6.	c	28.7
7.	b	28.8
8.	d	28.8
9.	d	28.8
10.	a	28.10
11.	c	28.10
12.	b	28.8
13.	a	28.9
14.	b	28.8
15.	d	28.8
	c	28.8
	b	28.9
	a	28.9

Chapter 29

#	Answer	Section
1.	b	29.2
2.	d	29.2
3.	d	29.2
4.	a	29.3
5.	b	29.3
6.	b	29.4
7.	d	29.4
8.	c	29.6
9.	a	29.10
10.	less	29.11
11.	d	29.14
12.	c	29.13
13.	c	29.10
14.	b	29.4
15.	d	29.10
	e	29.7
	f	29.5
	c	29.6
	b	29.12
	h	29.9
	i	29.6
	a	29.8
	g	29.14

Chapter 30

#	Answer	Section
1.	b	30.2
2.	d	30.2
3.	a	30.3
4.	c	30.4
5.	a	30.5
6.	d	30.5
7.	a	30.7
8.	False	30.4
9.	b	30.8
10.	b	30.7
11.	c	30.9
12.	d	30.9
13.	c	30.9
14.	d	30.10
15.	g	30.9
	a	30.7
	i	30.9
	e	30.4
	d	30.2
	f	30.5
	h	30.8
	b	30.3
	c	30.3

Chapter 13: Genetics Problems

1. yellow

2. a. *AB*
   b. *AB, aB*
   c. Ab, ab
   d. *AB, Ab, aB, ab*

3. a. All offspring will be AaBB.
   b. 1/4 *AABB*   (25% each genotype)
      1/4 *AABb*
      1/4 *AaBB*
      1/4 *AaBb*
   c. 1/4 *AaBb*   (25% each genotype)
      1/4 *Aabb*
      1/4 *aaBb*
      1/4 *aabb*
   d. 1/16 *AABB*   (6.25% of genotype)
      1/8 *AaBB*   (12.5%)
      1/16 *aaBB*   (6.25%)
      1/8 *AABb*   (12.5%)
      1/4 *AaBb*   (25%)
      1/8 *aaBb*   (12.5%)
      1/16 *AAbb*   (6.25%)
      1/8 *Aabb*   (12.5%)
      1/16 *aabb*   (6.25%)

4. monohybrid cross

5. Incomplete dominance; a mating of two *Mm* cats yields 1/2 *Mm*, 1/4 *MM*, and 1/4 *mm*. Because *MM* is lethal, the probability that any one kitten among the survivors will be heterozygous is 2/3.

6. a. Both parents are heterozygous (*Aa*), Children with the albino phenotype are homozygous recessive (*aa*); unaffected children may be homozygous dominant (*AA*) or heterozygous (*Aa*).
   b. All are homozygous recessive (*aa*).
   c. Homozygous recessive (*aa*) father, and heterozygous (*Aa*) mother. The child with the albino phenotype is homozygous recessive (*aa*), the unaffected children, heterozygous (*Aa*).

Chapter 14: Genetics Problems

1. Autosomal dominant. One of the first generation females is unaffected, so the allele cannot be carried on the X chromosome. The autosomal allele cannot be recessive, because if it were, both parents would be homozygous for the allele, so all of their children would be affected.

2. a. With respect to X-linked alleles, a male can produce only one type of gamete. Half of the gametes he produces carry an X chromosome and half carry a Y chromosome. All the gametes that carry the X chromosome will have the X-linked allele.
   b. A female homozygous for an X-linked allele produces only one kind of gamete.
   c. Fifty percent of the gametes of a female who is heterozygous for an X-linked allele carry one of the two alleles at that locus; the other fifty percent carry its partner allele for that locus.

3. a. 25%; b. 25%; c. 50%

4. By translocation, one of the individual's three chromosomes 21 may have become attached to the end of a chromosome 14. The individual would still have 46 chromosomes, but his or her somatic cells would have the translocated chromosome 21 in addition to two normal chromosomes 21.

5. 50%.

Appendix IV. Periodic Table of the Elements

The symbol for each element is an abbreviation of its name. Some symbols for elements are abbreviations for their Latin names. For instance, Pb (lead) is short for *plumbum*; the word "plumbing" is related—ancient Romans made their water pipes with lead.

Elements in each vertical column of the table behave in similar ways. For instance, all of the elements in the far right column of the table are inert gases; they do not interact with other atoms. In nature, such elements occur only as solitary atoms.

Appendix V. Molecular Models

A molecule's structure can be depicted by different kinds of molecular models. Such models allow us to visualize different characteristics of the same structure.

Structural models show how atoms in a molecule connect to one another:

methane   glucose

In such models, each line indicates one covalent bond: Double bonds are shown as two lines; triple bonds as three lines. Some atoms or bonds may be implied but not shown. For example, carbon ring structures such as those of glucose and other sugars are often represented as polygons. If no atom is shown at the corner of a polygon, a carbon atom is implied. Hydrogen atoms bonded to one of the atoms in the carbon backbone of a molecule may also be omitted:

glucose   glucose

Ball-and-stick models show the relative sizes of the atoms and their positions in three dimensions:

methane   glucose

All types of covalent bonds (single, double, or triple) are shown as one stick. Typically, the elements in such models are coded in standardized colors:

carbon   hydrogen   oxygen   nitrogen

Space-filling models show the outer boundaries of the atoms in three dimensions:

methane   glucose

A model of a large molecule can be quite complex if all the atoms are shown. This space-filling model of hemoglobin is an example:

To reduce visual complexity, other types of models omit individual atoms. Surface models of large molecules can show features such as an active site crevice (Figure 5.10). In this surface model of hemoglobin, you can see two heme groups (red) nestled in pockets of the protein:

Large molecules such as proteins are often shown as ribbon models. Such models highlight secondary structure such as coils or sheets. In this ribbon model of hemoglobin, you can see the four coiled polypeptide chains, each of which folds around a heme group:

Such structural details are clues to how a molecule functions. Hemoglobin is the main oxygen carrier in vertebrate blood. Oxygen binds at the hemes, so one hemoglobin molecule can hold four molecules of oxygen.

The Amino Acids

Neutral, nonpolar side group

glycine (gly)

alanine (ala)

valine (val)

isoleucine (ile)

leucine (leu)

phenylalanine (phe)

proline (pro)

methionine (met)

Neutral, polar side group

serine (ser)

threonine (thr)

tyrosine (tyr)

tryptophan (trp)

asparagine (asn)

glutamine (gln)

cysteine (cys)

Acidic side group

aspartic acid (asp)

glutamic acid (glu)

Basic side group

lysine (lys)

arginine (arg)

histidine (his)

ENERGY-REQUIRING STEPS OF GLYCOLYSIS

(two ATP invested)

ENERGY-RELEASING STEPS OF GLYCOLYSIS

(four ATP produced)

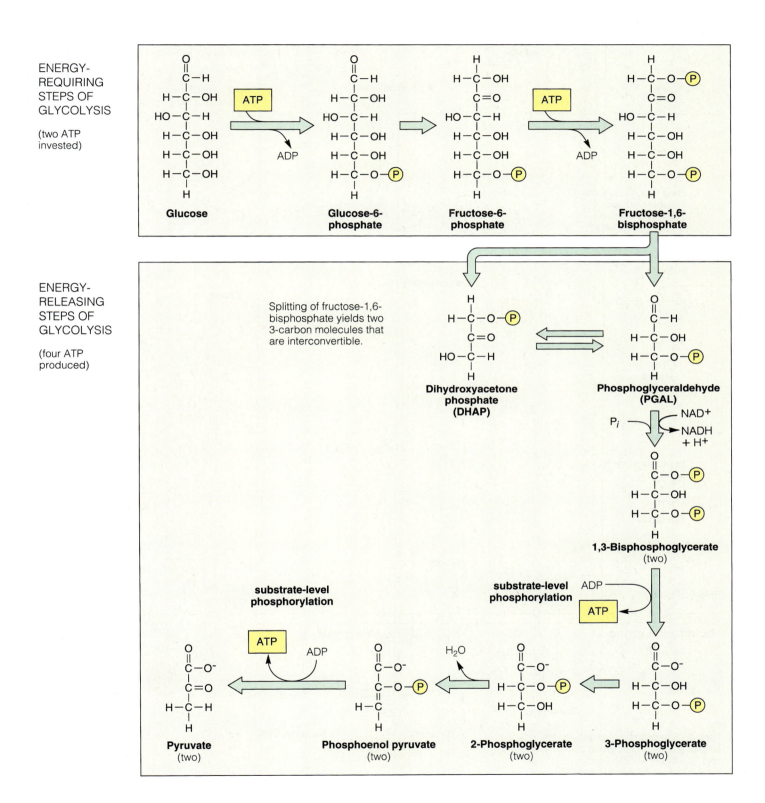

Splitting of fructose-1,6-bisphosphate yields two 3-carbon molecules that are interconvertible.

Figure A Glycolysis, ending with two 3-carbon pyruvate molecules for each 6-carbon glucose molecule entering the reactions. The *net* energy yield is two ATP molecules (two invested, four produced).

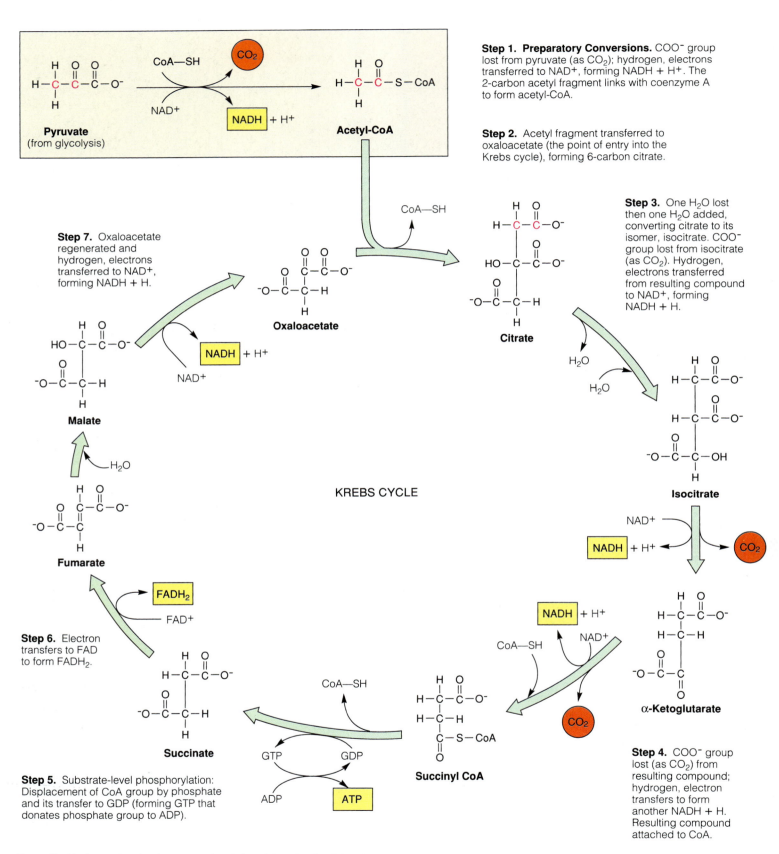

Step 1. Preparatory Conversions. COO^- group lost from pyruvate (as CO_2); hydrogen, electrons transferred to NAD^+, forming $NADH + H^+$. The 2-carbon acetyl fragment links with coenzyme A to form acetyl-CoA.

Step 2. Acetyl fragment transferred to oxaloacetate (the point of entry into the Krebs cycle), forming 6-carbon citrate.

Step 3. One H_2O lost then one H_2O added, converting citrate to its isomer, isocitrate. COO^- group lost from isocitrate (as CO_2). Hydrogen, electrons transferred from resulting compound to NAD^+, forming $NADH + H$.

Step 4. COO^- group lost (as CO_2) from resulting compound; hydrogen, electron transfers to form another $NADH + H$. Resulting compound attached to CoA.

Step 5. Substrate-level phosphorylation: Displacement of CoA group by phosphate and its transfer to GDP (forming GTP that donates phosphate group to ADP).

Step 6. Electron transfers to FAD to form $FADH_2$.

Step 7. Oxaloacetate regenerated and hydrogen, electrons transferred to NAD^+, forming $NADH + H$.

KREBS CYCLE

Figure B Krebs cycle, also known as the citric acid cycle. *Red* identifies carbon atoms entering the cyclic pathway (by way of acetyl-CoA) and leaving (by way of carbon dioxide). These cyclic reactions run twice for each glucose molecule that has been degraded to two pyruvate molecules.

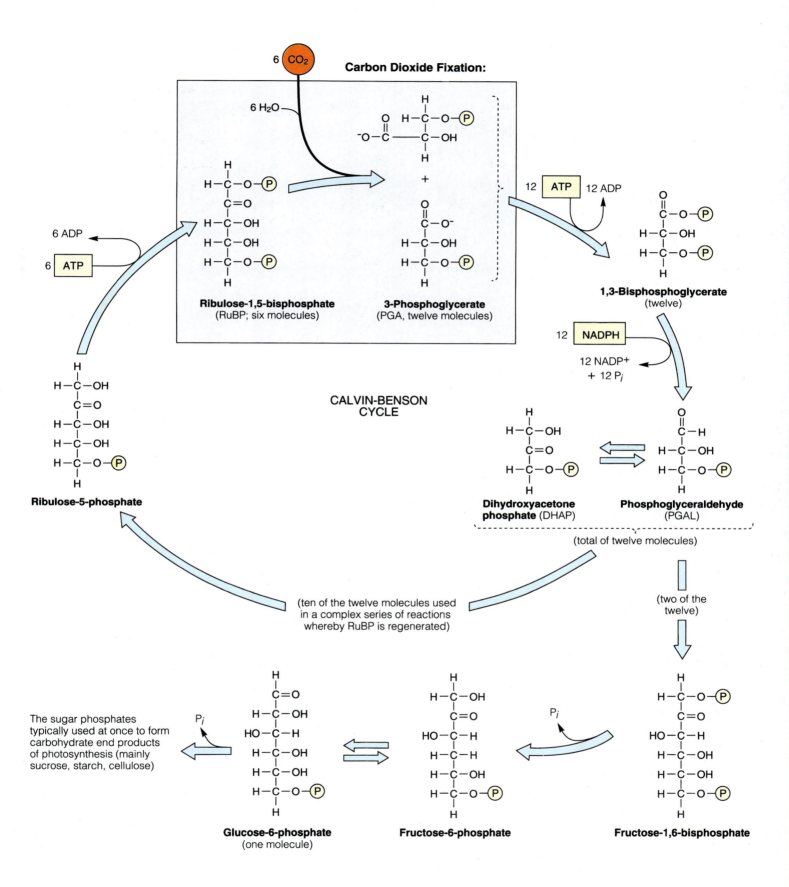

Figure C Calvin–Benson cycle of the light-independent reactions of photosynthesis.

Noncyclic photophosphorylation

light

photosystem II

cytochrome b_6f complex

plastocyanin

light

ferredoxin

to Calvin–
Benson cycle

NADPH ATP

ATP
synthase

ADP + P_i

stroma

thylakoid
compartment

plastoquinone

photosystem I

NADP+

ferredoxin NADP+
reductase

Cyclic photophosphorylation

cytochrome b_6f complex

ferredoxin

light

ATP
synthase

ATP

ADP + P_i

stroma

thylakoid
compartment

plastocyanin

photosystem I

The arrangement of electron transfer chain components in
highly folded thylakoid membranes maximizes the efficiency
of ATP production. ATP synthases are positioned only on the
outer surfaces of the thylakoid stacks, in contact with the
stroma and its supply of NADP+ and ADP.

stroma

thylakoid
compartment

ATP synthase   cytochrome b_6f complex
photosystem I   photosystem II

Figure D   Electron transfer in the light-dependent reactions of photosynthesis.
Members of the electron transfer chains are densely packed in thylakoid membranes;
electrons are transferred directly from one molecule to the next. For clarity, we show
the components of the chains widely spaced.

Appendix VII. A Plain English Map of the Human Chromosomes

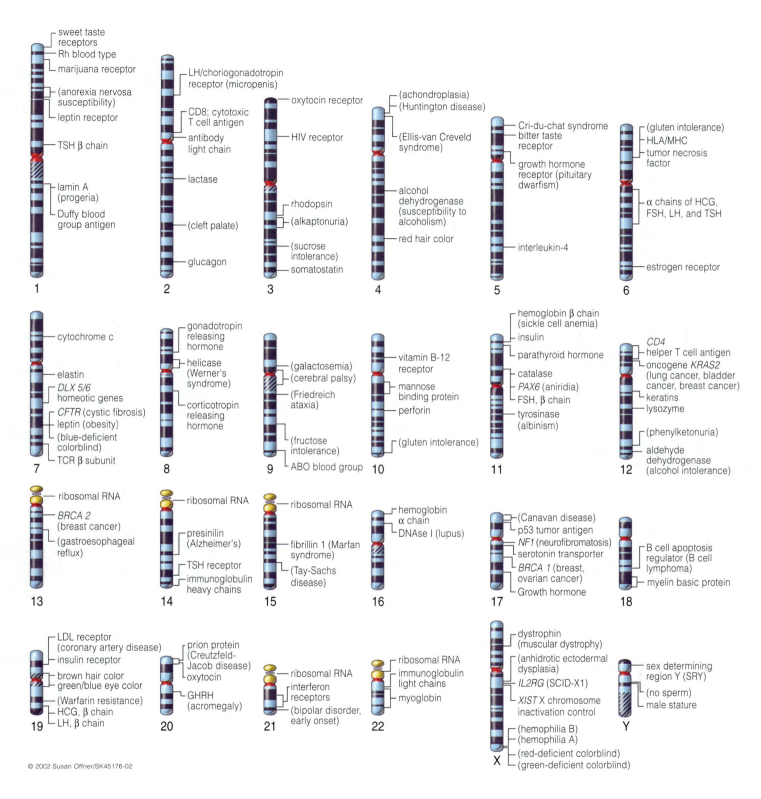

1
- sweet taste receptors
- Rh blood type
- marijuana receptor
- (anorexia nervosa susceptibility)
- leptin receptor
- TSH β chain
- lamin A (progeria)
- Duffy blood group antigen

2
- LH/choriogonadotropin receptor (micropenis)
- CD8; cytotoxic T cell antigen
- antibody light chain
- lactase
- (cleft palate)
- glucagon

3
- oxytocin receptor
- HIV receptor
- rhodopsin
- (alkaptonuria)
- (sucrose intolerance)
- somatostatin

4
- (achondroplasia)
- (Huntington disease)
- (Ellis-van Creveld syndrome)
- alcohol dehydrogenase (susceptibility to alcoholism)
- red hair color

5
- Cri-du-chat syndrome
- bitter taste receptor
- growth hormone receptor (pituitary dwarfism)
- interleukin-4

6
- (gluten intolerance)
- HLA/MHC
- tumor necrosis factor
- α chains of HCG, FSH, LH, and TSH
- estrogen receptor

7
- cytochrome c
- elastin
- DLX 5/6 homeotic genes
- CFTR (cystic fibrosis)
- leptin (obesity)
- (blue-deficient colorblind)
- TCR β subunit

8
- gonadotropin releasing hormone
- helicase (Werner's syndrome)
- corticotropin releasing hormone

9
- (galactosemia)
- (cerebral palsy)
- (Friedreich ataxia)
- (fructose intolerance)
- ABO blood group

10
- vitamin B-12 receptor
- mannose binding protein
- perforin
- (gluten intolerance)

11
- hemoglobin β chain (sickle cell anemia)
- insulin
- parathyroid hormone
- catalase
- PAX6 (aniridia)
- FSH, β chain
- tyrosinase (albinism)

12
- CD4
- helper T cell antigen
- oncogene KRAS2 (lung cancer, bladder cancer, breast cancer)
- keratins
- lysozyme
- (phenylketonuria)
- aldehyde dehydrogenase (alcohol intolerance)

13
- ribosomal RNA
- BRCA 2 (breast cancer)
- (gastroesophageal reflux)

14
- ribosomal RNA
- presinilin (Alzheimer's)
- TSH receptor
- immunoglobulin heavy chains

15
- ribosomal RNA
- fibrillin 1 (Marfan syndrome)
- (Tay-Sachs disease)

16
- hemoglobin α chain
- DNAse I (lupus)

17
- (Canavan disease)
- p53 tumor antigen
- NF1 (neurofibromatosis)
- serotonin transporter
- BRCA 1 (breast, ovarian cancer)
- Growth hormone

18
- B cell apoptosis regulator (B cell lymphoma)
- myelin basic protein

19
- LDL receptor (coronary artery disease)
- insulin receptor
- brown hair color
- green/blue eye color
- (Warfarin resistance)
- HCG, β chain
- LH, β chain

20
- prion protein (Creutzfeld-Jacob disease)
- oxytocin
- GHRH (acromegaly)

21
- ribosomal RNA
- interferon receptors
- (bipolar disorder, early onset)

22
- ribosomal RNA
- immunoglobulin light chains
- myoglobin

X
- dystrophin (muscular dystrophy)
- (anhidrotic ectodermal dysplasia)
- IL2RG (SCID-X1)
- XIST X chromosome inactivation control
- (hemophilia B)
- (hemophilia A)
- (red-deficient colorblind)
- (green-deficient colorblind)

Y
- sex determining region Y (SRY)
- (no sperm)
- male stature

Haploid set of human chromosomes. The banding patterns characteristic of each type of chromosome appear after staining with a reagent called Giemsa. The locations of some of the 20,065 known genes (as of November, 2005) are indicated. Also shown are locations that, when mutated, cause some of the genetic diseases discussed in the text.

180°   90°   0°

45°

Mohns Ridge

Reykjanes Ridge
7.8

North

2.3

American

Aleutian Trench

Juan de Fuca Plate

Azores F.Z.

45°

0°

Plate

2.3

2.5

Pacific

Caribbean Plate

Mid-Atlantic Ridge

Plate

Cocos Plate

8.6

7.0

5.0

East Pacific Rise

African

15.1

3.5

Mid-Atlantic Ridge

Plate

South

0°

15.1

Nazca

American

3.5

Plate

Peru-Chile Trench

Plate

Tonga Trench

45°

5.9

Chile Ridge

9.4

1.4

Sandwich Plate

Antarctic

Scotia Plate

Plate

45°

90°   0°

90°   0°

Appendix VIII.
Restless Earth—Life's Changing
Geologic Stage

This NASA map summarizes the tectonic and volcanic activity of Earth during the past 1 million years. The reconstructions at far right indicate positions of Earth's major land masses through time.

Actively-spreading ridges and transform faults

1.4   Total spreading rate, cm/year

Major active fault or fault zone; dashed where nature, location, or activity uncertain

Normal fault or rift; hachures on downthrown side

Reverse fault (overthrust, subduction zones); generalized; barbs on upthrown side

Volcanic centers active within the last one million years; generalized. Minor basaltic centers and seamounts omitted.

Middle Miocene. Polar regions again iced over, as in Cambrian. All land masses are assuming their current distribution

10 mya

65 mya

Cretaceous into Tertiary. Extinction of dinosaurs; rise of mammals

Pangea —

240 mya

Permian into Triassic. Vast swamp forests (eventual coal source); seed plants evolve

370 mya

Devonian. Jawed fishes evolve, diversify; ancestors of amphibians invade land

420 mya

Silurian. Sea level rises, diverse marine life; plants, invertebrates invade land

540 mya

Cambrian. Fragments of Rodinia, the first supercontinent. Major adaptive radiations in equatorial seas; icy polar regions

Appendix VIII

Appendix IX. Units of Measure

Length

1 kilometer (km) = 0.62 miles (mi)
1 meter (m) = 39.37 inches (in)
1 centimeter (cm) = 0.39 inches

To convert	multiply by	to obtain
inches	2.25	centimeters
feet	30.48	centimeters
centimeters	0.39	inches
millimeters	0.039	inches

Area

1 square kilometer = 0.386 square miles
1 square meter = 1.196 square yards
1 square centimeter = 0.155 square inches

Volume

1 cubic meter = 35.31 cubic feet
1 liter = 1.06 quarts
1 milliliter = 0.034 fluid ounces = 1/5 teaspoon

To convert	multiply by	to obtain
quarts	0.95	liters
fluid ounces	28.41	milliliters
liters	1.06	quarts
milliliters	0.03	fluid ounces

Weight

1 metric ton (mt) = 2,205 pounds (lb) = 1.1 tons (t)
1 kilogram (kg) = 2.205 pounds (lb)
1 gram (g) = 0.035 ounces (oz)

To convert	multiply by	to obtain
pounds	0.454	kilograms
pounds	454	grams
ounces	28.35	grams
kilograms	2.205	pounds
grams	0.035	ounces

Temperature

Celcius (°C) to Fahrenheit (°F):
$$°F = 1.8 \ (°C) + 32$$

Fahrenheit (°F) to Celsius:
$$°C = \frac{(°F - 32)}{1.8}$$

	°C	°F
Water boils	100	212
Human body temperature	37	98.6
Water freezes	0	32

Appendix X. A Comparative View of Mitosis in Plant and Animal Cells

For step-by-step description of the stages of mitosis, refer to Figure 11.5.

Mitosis in a generalized animal cell. For simplicity, only two chromosomes are shown.

| Prophase | Metaphase | Anaphase | Telophase |

Mitosis in a white-fish cell.

| Prophase | Metaphase | Anaphase | Telophase |

Mitosis in a lily cell.

| Prophase | Metaphase | Anaphase | Telophase |

Glossary

acid Substance that releases hydrogen ions in water. **32**

acid rain Low pH rain that forms when sulfur dioxide and nitrogen oxides mix with water vapor in the atmosphere. **491**

activation energy Minimum amount of energy required to start a reaction. **79**

activator Regulatory protein that increases the rate of transcription when it binds to a promoter or enhancer. **152**

active site Pocket in an enzyme where substrates bind and a reaction occurs. **80**

active transport Energy-requiring mechanism in which a transport protein pumps a solute across a cell membrane against its concentration gradient. **87**

adaptation (adaptive trait) A heritable trait that enhances an individual's fitness. **242**

adaptive radiation A burst of genetic divergences from a lineage gives rise to many new species. **276**

adhering junction Cell junction composed of adhesion proteins; anchors cells to each other and to extracellular matrix. Role in cell signaling. **69**

adhesion protein Membrane protein that is a component of adhering junctions. **56**

aerobic Involving or occurring in the presence of oxygen. **108**

aerobic respiration Oxygen-requiring pathway that breaks down carbohydrates to produce ATP. Occurs in stages: glycolysis; acetyl-CoA formation and the Krebs cycle; electron transfer phosphorylation. **108**

age structure Of a population, the number of individuals in each of several age categories. **415**

alcoholic fermentation Anaerobic carbohydrate breakdown pathway that produces ATP and ethanol. Begins with glycolysis; end reactions regenerate NAD$^+$ so glycolysis can continue. **116**

algal bloom Population explosion of tiny aquatic producers such as dinoflagellates. **311**

allele frequency Abundance of a particular allele among members of a population. **259**

alleles Forms of a gene that encode slightly different versions of the gene's product. **176**

allopatric speciation Speciation pattern in which a physical barrier that separates members of a population ends gene flow between them. **272**

allosteric Describes a region of an enzyme other than the active site that can bind regulatory molecules. **82**

alpine tundra Biome of low-growing, wind-tolerant plants adapted to high-altitude conditions. **477**

alternation of generations Of land plants and some algae, a life cycle in which both haploid and diploid multicelled bodies form. **318**

alternative splicing RNA processing event in which some exons are removed or joined in various combinations. **141**

altruistic behavior Behavior that benefits others at the expense of the individual. **408**

alveolate Member of a protist lineage having small sacs beneath the plasma membrane; dinoflagellate, ciliate, or apicomplexan. **314**

amino acid Small organic compound that is a subunit of proteins. Consists of a carboxyl group, an amine group, and a characteristic side group (R), all typically bonded to the same carbon atom. **44**

amniote Vertebrate whose embryos develop within a fluid enclosed within a membrane (the amnion); a reptile (including birds) or mammal. **379**

amniote egg Egg with internal membranes that allow the amniote embryo to develop away from water. **384**

amoeba Single-celled protist that extends pseudopods to move and to capture prey. **320**

amoebozoan Shape-shifting heterotrophic protist with no pellicle or cell wall; an amoeba or slime mold. **320**

amphibian Tetrapod with a three-chambered heart and scaleless skin; typically develops in water, then lives on land as an air-breathing carnivore. **382**

anaerobic Occurring in the absence of oxygen. **108**

analogous structures Similar body structures that evolved separately in different lineages. **251**

anaphase Stage of mitosis during which sister chromatids separate and move to opposite spindle poles. **166**

aneuploidy A chromosome abnormality in which an individual's cells carry too many or too few copies of a particular chromosome. **212**

angiosperms Largest seed plant lineage. Only group that makes flowers and fruits. **336**

animal Multicelled consumer with unwalled cells; develops through a series of stages and moves about during part or all of the life cycle. **8, 354**

annelid Segmented worm with a coelom, complete digestive system, and closed circulatory system. **362**

antenna Of some arthropods, sensory structure on the head that detects touch and odors. **367**

antioxidant Substance that prevents molecules from reacting with oxygen. **81**

apicomplexan Single-celled protist that lives as a parasite inside animal cells. **315**

aquifer Porous rock layer that holds some groundwater. **454**

arachnid Land-dwelling arthropod with four pairs of walking legs; a spider, scorpion, mite, or tick. **368**

archaean Member of a group of single-celled microorganisms that superficially resemble bacteria, but are genetically and structurally distinct. **8**

arctic tundra Highest-latitude Northern Hemisphere biome, where low, cold-tolerant plants survive with only a brief growing season. **477**

area effect Larger islands have more species than small ones. **445**

arthropod Invertebrate with jointed legs and a hard exoskeleton that is periodically molted. **367**

artificial selection Selective breeding of animals by humans. **242**

asexual reproduction Reproductive mode by which offspring arise from a single parent only. **165**

astrobiology The scientific study of life's origin and distribution in the universe. **283**

atmospheric cycle Biogeochemical cycle in which a gaseous form of an element plays a significant role. **456**

atom Particle that is a fundamental building block of all matter. **4, 24**

atomic number Number of protons in the atomic nucleus; determines the element. **24**

ATP Adenosine triphosphate. Nucleotide that consists of an adenine base, a five-carbon ribose sugar, and three phosphate groups. Often functions as an energy carrier that couples endergonic with exergonic reactions in cells. **47, 79**

ATP/ADP cycle Process in which cells regenerate ATP. ADP forms when ATP loses a phosphate group, then ATP forms again as ADP gains a phosphate group. **79**

australopiths Collection of now-extinct hominid lineages; some may be ancestral to humans. **390**

autosome Any chromosome other than a sex chromosome. **125**

autotroph Organism that makes its own food using carbon from inorganic molecules such as

CO_2 and energy from the environment; a producer. 93

bacterial chromosome Circle of double-stranded DNA that resides in the bacterial cytoplasm. 302

bacteriophage Virus that infects bacteria. 127, 298

bacterium Single-celled organism belonging to the Domain Bacteria; cells are typically walled and do not contain a nucleus. 8

balanced polymorphism Maintenance of two or more alleles for a trait at high frequency in a population as a result of natural selection against homozygotes. 267

basal body Organelle that develops from a centriole. 67

base Substance that accepts hydrogen ions in water. 32

base–pair substitution Type of mutation in which a single base-pair changes. 147

bell curve Bell-shaped curve; typically results from graphing frequency versus distribution for a trait that varies continuously. 198

benthic province The ocean's sediments and rocks. 482

big bang theory Model describing formation of the universe as a nearly instant distribution of matter through space. 284

bilateral symmetry Having paired structures so the right and left halves are mirror images. 354

binary fission Method of asexual reproduction that divides one bacterial or archaean cell into two identical descendant cells. 303

bioaccumulation An organism accumulates a chemical pollutant in its tissues over the course of its lifetime. 492

biodiversity Of a region, the genetic variation within its species, variety of species, and variety of ecosystems. 8, 496

biofilm Community of microorganisms living within a shared mass of slime. 59

biogeochemical cycle Movement of a chemical nutrient among environmental reservoirs and into and out of food webs. 454

biogeography Study of patterns in the geographic distribution of species and communities. 238

biological magnification A chemical pollutant becomes increasingly concentrated as it moves up through food chains. 492

biology The scientific study of life. 3

biomarker Molecule produced only by a specific type of cell. 289

biome Discontinuous region characterized by its climate and dominant vegetation. 468

biosphere All regions of Earth where organisms live. 5, 463

biotic potential Maximum possible population growth rate under optimal conditions. 417

bipedalism Standing and walking on two legs. 389

bird An animal with feathers. 386

bivalve Mollusk with a hinged two-part shell. 364

bony fish Fish with a lung or swim bladder and a skeleton consisting largely of bone. 381

boreal forest Extensive high-latitude forest of the Northern Hemisphere; conifers are the predominant vegetation; also known as tiaga. 476

bottleneck Reduction in population size so severe that it reduces genetic diversity. 268

brood parasitism One egg-laying species benefits by having another raise its offspring. 438

brown alga Multicelled marine protist with a brown accessory pigment in its chloroplasts. 317

bryophyte Member of an early-evolving plant lineage with a gametophyte-dominant life cycle; a moss, liverwort, or hornwort. 328

buffer Set of chemicals that can keep the pH of a solution stable by alternately donating and accepting ions that contribute to pH. 33

C3 plant Type of plant that uses only the Calvin–Benson cycle to fix carbon. 102

C4 plant Type of plant that minimizes photorespiration by fixing carbon twice, in two cell types. 102

Calvin–Benson cycle Light-independent reactions of photosynthesis; cyclic carbon-fixing pathway that forms sugars from CO_2. 101

CAM plant Type of C4 plant that conserves water by fixing carbon twice, at different times of day. 102

cancer Disease that occurs when a neoplasm physically and metabolically disrupts body tissues. 151, 170

carbohydrate Molecule that consists primarily of carbon, hydrogen, and oxygen atoms in a 1:2:1 ratio. 40

carbon cycle Movement of carbon, mainly between the oceans, atmosphere, and living organisms. 456

carbon fixation Process by which carbon from an inorganic source such as carbon dioxide gets incorporated into an organic molecule. 101

carpel Floral reproductive structure that produces female gametophytes; a sticky or hairlike stigma together with an ovary and a style. 336

carrying capacity Maximum number of individuals of a species that an environment can sustain. 418

cartilaginous fish Jawed fish with a skeleton of cartilage. 380

catalysis The acceleration of a reaction rate by a molecule that is unchanged by participating in the reaction. 80

catastrophism Now-abandoned hypothesis that catastrophic geologic forces unlike those of the present day shaped Earth's surface. 240

cDNA DNA synthesized from an RNA template by the enzyme reverse transcriptase. 221

cell Smallest unit that has the properties of life; at minimum, consists of plasma membrane, cytoplasm, and DNA. 5, 52

cell cortex Reinforcing mesh of cytoskeletal elements under a plasma membrane. 66

cell cycle A series of events from the time a cell forms until its cytoplasm divides. 164

cell junction Structure that connects a cell to another cell or to extracellular matrix. 69

cell plate After nuclear division in a plant cell, a disk-shaped structure that forms a cross-wall between the two new nuclei. 168

cell theory Theory that all organisms consist of one or more cells, which are the basic unit of life; all cells come from division of preexisting cells; and all cells pass hereditary material to their offspring. 53

cell wall Semirigid but permeable structure that surrounds the plasma membrane of some cells. 58

cellular slime mold Amoeba-like protist that feeds as a single predatory cell; joins with others to form a multicellular spore-bearing structure when conditions are unfavorable. 320

central vacuole Fluid-filled vesicle in many plant cells. 63

centriole Barrel-shaped organelle from which microtubules grow. 67

centromere Constricted region in a eukaryotic chromosome where sister chromatids are attached. 124

cephalization A concentration of nerve and sensory cells at the head end. 355

cephalopod Predatory mollusk with a closed circulatory system; moves by jet propulsion. 365

chaparral Biome of dry shrubland in regions with hot, dry summers and cool, rainy winters. 473

character Quantifiable, heritable characteristic or trait. 278

character displacement Outcome of competition between two species; similar traits that result in competition become dissimilar. 435

charge Electrical property. Opposite charges attract and like charges repel. 24

chelicerates Arthropod subgroup with specialized feeding structures (chelicerae) and no antennae. 368

chemical bond An attractive force that arises between two atoms when their electrons interact. 27

chlamydias Tiny round bacteria that are intracellular parasites of eukaryotic cells. 305

chlorophyll _a_ Main photosynthetic pigment in plants. 94

chloroplast Organelle of photosynthesis in the cells of plants and many protists. 65, 97

choanoflagellate Member of the protist group most closely related to animals. **356**

chordate Animal with an embryo that has a notochord, dorsal nerve cord, pharyngeal gill slits, and a tail that extends beyond the anus. For example, a lancelet or a vertebrate. **378**

chromosome A structure that consists of DNA and associated proteins; carries part or all of a cell's genetic information. **124**

chromosome number The sum of all chromosomes in a cell of a given type. **124**

chytrid Fungus that makes flagellated spores. **345**

ciliate Single-celled, heterotrophic protist with many cilia. **315**

cilium Short, movable structure that projects from the plasma membrane of some eukaryotic cells. **67**

clade A species or group of species that share a set of characters; ideally, a monophyletic group. **278**

cladistics Method of determining evolutionary relationships by grouping species into clades based on shared characters. **278**

cladogram Evolutionary tree that shows a network of evolutionary relationships among clades. **278**

cleavage furrow In a dividing animal cell, the indentation where cytoplasmic division will occur. **168**

climate Average weather conditions in a region over a long time period. **464**

clone Genetically identical copy of an organism. **123**

cloning vector A DNA molecule that can accept foreign DNA, be transferred to a host cell, and get replicated in it. **221**

closed circulatory system Circulatory system in which blood flows through a continuous system of vessels, and substances are exchanged across the walls of the smallest vessels. **362**

club fungi Fungi that have septate hyphae and produce spores by meiosis in club-shaped cells. **347**

cnidarian Radially symmetrical invertebrate with two tissue layers; uses tentacles with unique stinging structures (nematocysts) to capture food. **358**

coal Fossil fuel formed over millions of years by compaction and heating of plant remains. **332**

codominant Refers to two alleles that are both fully expressed in heterozygous individuals. **196**

codon In mRNA, a nucleotide base triplet that codes for an amino acid or stop signal during translation. **142**

coelom Body cavity lined with tissue derived from mesoderm. **355**

coenzyme An organic cofactor. **81**

coevolution The joint evolution of two closely interacting species; each species is a selective agent for traits of the other. **277**

cofactor A metal ion or an organic molecule that associates with an enzyme and is necessary for its function. **81**

cohesion Tendency of molecules to resist separating from one another. **31**

cohort Group of individuals born during the same interval. **420**

commensalism Species interaction that benefits one species and neither helps nor harms the other. **432**

communication signal Chemical, acoustical, visual, or tactile cue that is produced by one member of a species and detected and responded to by other members of the same species. **402**

community All populations of all species that live a particular region. **5, 431**

comparative morphology Study of body plans and structures among groups of organisms. **239**

competitive exclusion Process whereby two species compete for a limiting resource, and one drives the other to local extinction. **434**

compound Type of molecule that has atoms of more than one element. **27**

concentration The number of molecules or ions per unit volume of a solution. **32, 84**

concentration gradient Difference in concentration between adjoining regions of fluid. **84**

condensation Process by which enzymes build large molecules from smaller subunits; water also forms. **39**

conifer Gymnosperm with nonmotile sperm and woody cones; for example, a pine. **334**

conjugation Mechanism of gene exchange in which one bacterial or archaean cell passes a plasmid to another. **303**

conservation biology Field of applied biology that surveys biodiversity and seeks ways to maintain and use it. **496**

consumer Organism that gets energy and nutrients by feeding on tissues, wastes, or remains of other organisms; a heterotroph. **6, 450**

continuous variation In a population, a range of small differences in a shared trait. **198**

contractile vacuole In freshwater protists, an organelle that collects and expels excess water. **313**

control group In an experiment, group of individuals who are not exposed to the independent variable that is being tested. **13**

coral bleaching A coral expels its photosynthetic dinoflagellate symbionts in response to stress and becomes colorless. **481**

coral reef Highly diverse marine ecosystem centered around reefs built by living corals that secrete calcium carbonate. **481**

covalent bond Chemical bond in which two atoms share a pair of electrons. **28**

craniate Chordate with a braincase; a hagfish or a vertebrate. **378**

critical thinking Judging the quality of information before allowing it to guide one's thoughts or actions. **12**

crossing over Process in which homologous chromosomes exchange corresponding segments during prophase I of meiosis. **181**

crustaceans Mostly marine arthropod group with two pairs of antennae; for example, shrimp, crabs, lobsters, and barnacles. **369**

culture Learned behavior patterns transmitted among members of a group and between generations. **389**

cuticle Secreted covering at a body surface. **68, 326**

cyanobacteria Photosynthetic, oxygen-producing bacteria. **304**

cycad Tropical or subtropical gymnosperm with flagellated sperm, palmlike leaves, and fleshy seeds. **334**

cytokinesis Cytoplasmic division. **168**

cytoplasm Semifluid substance enclosed by a cell's plasma membrane. **52**

cytoskeleton Dynamic framework of protein filaments that support, organize, and move eukaryotic cells and their internal structures. **66**

data Experimental results. **13**

decomposer Organism that feeds on biological remains and breaks organic material down into its inorganic subunits. **450**

deductive reasoning Using a general idea to make a conclusion about a specific case. **12**

deletion Mutation in which one or more base pairs are lost from a chromosome. **146, 210**

demographics Statistics that describe the traits of a population. **414**

demographic transition model Model describing the changes in human birth and death rates that occur as a region becomes industrialized. **426**

denature To unravel the shape of a protein or other large biological molecule. **46**

denitrification Bacteria convert nitrates or nitrites to gaseous forms of nitrogen. **458**

density–dependent factor Factor that limits population growth and has a greater effect in dense populations than less dense ones. **418**

density–independent factor Factor that limits population growth and arises regardless of population density. **419**

dependent variable In an experiment, variable that is presumably affected by the independent variable being tested. **13**

desert Biome with little rain and low humidity; plants that have water-storing and conserving adaptations predominate. **470**

desertification Conversion of dry grassland to desert. **490**

detrital food chain Food chain in which energy is transferred directly from producers to detritivores. **452**

detritivore Consumer that feeds on small bits of organic material. **450**

deuterostomes Lineage of bilateral animals in which the second opening on the embryo surface develops into a mouth. **355**

development Multistep process by which the first cell of a new individual becomes a multicelled adult. **7**

diatom Single-celled photosynthetic protist with brown accessory pigments in its choroplasts and a two-part silica shell. **317**

differentiation The process by which cells become specialized. **152**

diffusion Net movement of molecules or ions from a region where they are more concentrated to a region where they are less so. **84**

dihybrid cross Breeding experiment in which individuals identically heterozygous for two genes are crossed. The frequency of traits among the offspring offers information about the dominance relationships between the paired alleles. **194**

dikaryotic Having two genetically distinct nuclei ($n + n$). **344**

dinoflagellate Single-celled, aquatic protist with cellulose plates and two flagella; may be heterotrophic or photosynthetic. **314**

dinosaur Reptile lineage abundant in the Jurassic to Cretaceous; now extinct with the exception of birds. **384**

diploid Having two of each type of chromosome characteristic of the species ($2n$). **124**

directional selection Mode of natural selection in which phenotypes at one end of a range of variation are favored. **262**

disruptive selection Mode of natural selection that favors forms of a trait at the extremes of a range of variation; intermediate forms are selected against. **265**

distance effect Islands close to a mainland have more species than those farther away. **445**

DNA Deoxyribonucleic acid. Nucleic acid that carries hereditary information about traits; consists of two nucleotide chains twisted in a double helix. **7, 47**

DNA cloning Set of procedures that uses living cells to make many identical copies of a DNA fragment. **220**

DNA library Collection of cells that host different fragments of foreign DNA, often representing an organism's entire genome. **222**

DNA ligase Enzyme that seals gaps in double-stranded DNA. **131**

DNA polymerase DNA replication enzyme. Uses a DNA template to assemble a complementary strand of DNA. **130**

DNA profiling Identifying an individual by analyzing the unique parts of his or her DNA. **227**

DNA repair mechanism Any of several processes by which enzymes repair damaged DNA. **131**

DNA replication Process by which a cell duplicates its DNA before it divides. **130**

DNA sequence Order of nucleotide bases in a strand of DNA. **130**

DNA sequencing Method of determining the order of nucleotides in DNA. **224**

dominance hierarchy Social system in which resources and mating opportunities are unequally distributed within a group. **406**

dominant Refers to an allele that masks the effect of a recessive allele paired with it. **191**

dosage compensation Theory that X chromosome inactivation equalizes gene expression between males and females. **156**

duplication Repeated section of a chromosome. **210**

echinoderms Invertebrates with hardened plates and spines embedded in the skin or body, and a water–vascular system. **373**

ecological footprint Area of Earth's surface required to sustainably support a particular level of development and consumption. **426**

ecological niche All of a species' requirements and roles in an ecosystem. **434**

ecological restoration Actively altering an area in an effort to restore or create a functional ecosystem. **497**

ecology Study of interactions among organisms and their environment. **397**

ecosystem A community interacting with its environment. **5**

ectoderm Outermost tissue layer of an animal embryo. **354**

ectotherm Animal that controls its internal temperature by altering its behavior; for example, a fish or a lizard. **385**

egg Mature female gamete, or ovum. **182**

electron Negatively charged subatomic particle that occupies orbitals around an atomic nucleus. **24**

electronegativity Measure of the ability of an atom to pull electrons away from other atoms. **27**

electron transfer chain Array of enzymes and other molecules that accept and give up electrons in sequence, thus releasing the energy of the electrons in usable increments. **83**

electron transfer phosphorylation Process in which electron flow through electron transfer chains sets up a hydrogen ion gradient that drives ATP formation. **98**

electrophoresis Technique that separates DNA fragments by size. **224**

element A pure substance that consists only of atoms with the same number of protons. **24**

El Niño Periodic warming of equatorial Pacific waters and the associated shifts in global weather patterns. **463**

embryophyte Member of the land plant clade. **326**

emergent property A characteristic of a system that does not appear in any of the system's component parts. **4**

emerging disease A disease that was previously unknown or has recently begun spreading to a new region. **300**

emigration Movement of individuals out of a population. **416**

endangered species A species that faces extinction in all or a part of its range. **488**

endemic disease Disease that persists at a low level in a region or population. **300**

endemic species Species that remains restricted to the area where it evolved. **488**

endergonic Type of reaction that requires a net input of free energy to proceed. **78**

endocytosis Process by which a cell takes in a small amount of extracellular fluid by the ballooning inward of its plasma membrane. **88**

endoderm Innermost tissue layer of an animal embryo. **354**

endomembrane system Series of interacting organelles (endoplasmic reticulum, Golgi bodies, vesicles) between the nucleus and plasma membrane. **62**

endoplasmic reticulum (ER) Organelle that is a continuous system of sacs and tubes extending from the nuclear envelope. Rough ER is studded with ribosomes; smooth ER is not. **62**

endoskeleton Internal skeleton made up of hardened components such as bones. **378**

endosperm Nutritive triploid tissue in angiosperm seeds. **336**

endospore Resistant resting stage of some soil bacteria. **305**

endosymbiosis One species lives and reproduces inside another. **290**

endotherm Animal controls its internal temperature by adjusting its metabolism; for example, a bird or mammal. **385**

energy The capacity to do work. **6, 76**

enhancer Binding site in DNA for proteins that enhance the rate of transcription. **152**

entropy Measure of how much the energy of a system is dispersed. **76**

enzyme Compound (usually a protein) that speeds a reaction without being changed by it. **39**

epidemic Disease outbreak limited to one region. **300**

epiphyte Plant that grows on another plant but does not harm it. **331**

epistasis Effect in which a trait is influenced by the products of multiple genes. **197**

equilibrium model of island biogeography Model that predicts the number of species on an island based on the island's area and distance from the mainland. **444**

estuary A highly productive ecosystem where nutrient-rich water from a river mixes with seawater. **480**

eudicots Largest lineage of angiosperms; includes herbaceous plants, woody trees, and cacti. **336**

eugenics Idea of deliberately improving the genetic qualities of the human race. **232**

euglenoid Flagellated protozoan with multiple mitochondria; may be heterophic or have chloroplasts decended from algae. **313**

eukaryote Organism whose cells characteristically have a nucleus. **8**

eutrophication Nutrient enrichment of an aquatic ecosystem. **449**

evaporation Transition of a liquid to a gas. **31**

evolution Change in a line of descent. **240**

evolutionary tree Type of diagram that summarizes evolutionary relationships among a group of species. **278**

exaptation Adaptation of an existing structure for a completely different purpose; a major evolutionary novelty. **276**

exergonic Type of reaction that ends with a net release of free energy. **78**

exocytosis Process by which a cell expels a vesicle's contents to extracellular fluid. **88**

exon Nucleotide sequence that is not spliced out of RNA during processing. **141**

exoskeleton Of some invertebrates, hard external parts that muscles attach to and move. **367**

exotic species A species that evolved in one community and later became established in a different one. **443**

experiment A test designed to support or falsify a prediction. **13**

experimental group In an experiment, group of individuals who are exposed to an independent variable. **13**

exponential growth A population grows by a fixed percentage in successive time intervals; the size of each increase is determined by the current population size. **416**

extinct Refers to a species that has been permanently lost. **276**

extracellular matrix (ECM) Complex mixture of cell secretions that supports cells and tissues. Also has roles in cell signaling. **68**

extreme halophile Organism adapted to life in a highly salty environment. **306**

extreme thermophile Organism adapted to life in a very high-temperature environment. **306**

fat Lipid that consists of a glycerol molecule with one, two, or three fatty acid tails. **42**

fatty acid Organic compound that consists of a chain of carbon atoms with an acidic carboxyl group at one end. Carbon chain of saturated types has single bonds only; that of unsaturated types has one or more double bonds. **42**

feedback inhibition Mechanism by which a change that results from some activity decreases or stops the activity. **82**

fermentation An anaerobic pathway by which cells harvest energy from carbohydrates to produce ATP. **109**

fertilization Fusion of two gametes to form a zygote. **177**

first law of thermodynamics Energy cannot be created or destroyed. **76**

fitness Degree of adaptation to an environment, as measured by an individual's relative genetic contribution to future generations. **242**

fixed Refers to an allele for which all members of a population are homozygous. **268**

fixed action pattern Series of instinctive movements elicited by a simple stimulus and carried out with little variation once begun. **400**

flagellated protozoan Protist belonging to an entirely or mostly heterotrophic lineage with no cell wall and one or more flagella. **313**

flagellum Long, slender cellular structure used for motility. **58**

flatworm Bilaterally symmetrical invertebrate with organs but no body cavity; for example, a planarian or tapeworm. **360**

flower Specialized reproductive shoot of a flowering plant. **336**

fluid mosaic Model of a cell membrane as a two-dimensional fluid of mixed composition. **56**

food chain Description of who eats whom in one path of energy in an ecosystem. **451**

food web Set of cross-connecting food chains. **452**

foraminiferan Heterotrophic single-celled protist with a porous calcium carbonate shell and long cytoplasmic extensions. **314**

fossil Physical evidence of an organism that lived in the ancient past. **239**

founder effect Change in allele frequencies that occurs when a small number of individuals establish a new population. **268**

fruit Mature ovary of a flowering plant, often with accessory parts; encloses a seed or seeds. **336**

functional group A group of atoms bonded to a carbon of an organic compound; imparts a specific chemical property to the molecule. **38**

fungus Eukaryotic heterotroph with cell walls of chitin; obtains nutrients by digesting food outside the body and absorbing them. **8, 344**

gamete Mature, haploid reproductive cell; e.g., an egg or a sperm. **177**

gametophyte A haploid, multicelled body that produces gametes in the life cycle of land plants and some protists. **182, 318**

gap junction Cell junction that forms a channel across the plasma membranes of adjoining animal cells. **69**

gastropod Mollusk in which the lower body is a broad "foot." **364**

gene Part of a DNA base sequence; specifies an RNA or protein product. **138**

gene expression Process by which the information in a gene becomes converted to an RNA or protein product. **139**

gene flow The movement of alleles into and out of a population. **269**

gene pool All of the alleles of all of the genes in a population; a pool of genetic resources. **259**

gene therapy The transfer of a normal or modified gene into an individual with the goal of treating a genetic defect or disorder. **232**

genetic code Complete set of sixty-four mRNA codons. **142**

genetic drift Change in allele frequencies in a population due to chance alone. **268**

genetic engineering Process by which deliberate changes are introduced into an individual's genome. **228**

genetic equilibrium Theoretical state in which a population is not evolving. **259**

genetically modified organism (GMO) Organism whose genome has been modified by genetic engineering. **228**

genome An organism's complete set of genetic material. **222**

genomics The study of genomes. **226**

genotype The particular set of alleles carried by an individual. **191**

genus A group of species that share a unique set of traits; also the first part of a species name. **10**

geologic time scale Chronology of Earth's history. **246**

germ cell Diploid reproductive cell that gives rise to haploid gametes by meiosis. **177**

ginkgo Deciduous gymnosperm with flagellated sperm, fan-shaped leaves, and fleshy seeds. **334**

global climate change A rise in temperature and shifts in other climate patterns. **457**

glomeromycete Fungus with hyphae that grow inside the wall of a plant root cell. **345**

glycolysis Set of reactions in which glucose or another sugar is broken down to two pyruvate for a net yield of two ATP. **109**

gnetophyte Shrubby or vinelike gymnosperm, with nonmotile sperm; for example, *Ephedra*. **334**

Golgi body Organelle that modifies polypeptides and lipids; also sorts and packages the finished products into vesicles. **63**

Gondwana Supercontinent that existed before Pangea, more than 500 million years ago. **249**

Gram staining Process used to prepare bacterial cells for microscopy, and to distinguish groups based on cell wall structure. **304**

grassland Biome in the interior of continents where grasses and nonwoody plants adapted to grazing and fire predominate. **472**

grazing food chain Food chain in which energy is transferred from producers to grazers (herbivores). **452**

green alga Photosynthetic protist that deposits cellulose in its cell wall, stores sugars as starch, and has chloroplasts containing chlorophylls *a* and *b*. **318**

greenhouse effect Warming of Earth's lower atmosphere and surface as a result of heat trapped by greenhouse gases. **457**

groundwater Soil water and water in aquifers. **454**

growth In multicelled species, an increase in the number, size, and volume of cells. **7**

growth factor Molecule that stimulates mitosis. **169**

gymnosperm Seed plant that does not make flowers or fruits; for example, a conifer. **334**

habitat Type of environment in which a species typically lives. **432**

habituation Learning not to respond to a repeated stimulus. **401**

hagfish Jawless fish with a skull case but no backbone. **380**

half–life Characteristic time it takes for half of a quantity of a radioisotope to decay. **245**

haploid Having one of each type of chromosome characteristic of the species. **177**

herbivory An animal feeds on plant parts. **437**

hermaphrodite Animal that produces both eggs and sperm, either simultaneously or at different times in its life. **357**

heterotroph Organism that obtains energy and carbon from organic compounds assembled by other organisms; a consumer. **93**

heterozygous Having two different alleles of a gene. **191**

histone Type of protein that structurally organizes eukaryotic chromosomes. **124**

homeostasis Set of processes by which an organism keeps its internal conditions within tolerable ranges. **7**

homeotic gene Type of master gene; its expression controls formation of specific body parts during development. **154**

hominid Human or extinct humanlike species. **388**

homologous chromosomes Chromosomes with the same length, shape, and set of genes. **164**

homologous structures Similar body parts that evolved in a common ancestor. **250**

homozygous Having identical alleles of a gene. **191**

horizontal gene transfer Transfer of genetic material among existing individuals. **303**

hot spot Threatened region with great biodiversity that is considered a high priority for conservation efforts. **496**

humans Members of the genus *Homo*. **391**

hybrid The offspring of a cross between two individuals that breed true for different forms of a trait; a heterozygous individual. **191**

hydrocarbon Compound or region of one that consists only of carbon and hydrogen atoms. **38**

hydrogen bond Attraction that forms between a covalently bonded hydrogen atom and another atom taking part in a separate covalent bond. **29**

hydrolysis Process by which an enzyme breaks a molecule into smaller subunits by attaching a hydroxyl group to one part and a hydrogen atom to the other. **39**

hydrophilic Describes a substance that dissolves easily in water. **30**

hydrophobic Describes a substance that resists dissolving in water. **30**

hydrostatic skeleton Of soft-bodied invertebrates, a fluid-filled chamber that muscles exert force against, redistributing the fluid. **358**

hydrothermal vent Rocky, underwater opening where mineral-rich water heated by geothermal energy streams out. **285, 483**

hypertonic Describes a fluid that has a high overall solute concentration relative to another fluid. **84**

hypha Component of a fungal mycelium; a filament made up of cells arranged end to end. **344**

hypothesis Testable explanation of a natural phenomenon. **12**

hypotonic Describes a fluid that has a low overall solute concentration relative to another fluid. **84**

immigration Movement of individuals into a population. **416**

imprinting Learning that can occur only during a specific interval in an animal's life. **400**

inbreeding Mating among close relatives. **269**

incomplete dominance Condition in which one allele is not fully dominant over another, so the heterozygous phenotype is between the two homozygous phenotypes. **196**

independent variable Variable that is controlled by an experimenter in order to explore its relationship to a dependent variable. **13**

indicator species Species that is especially sensitive to disturbance and can be monitored to assess the health of a habitat. **443**

induced–fit model Substrate binding to an active site improves the fit between the two. **80**

inductive reasoning Drawing a conclusion based on observation. **12**

inheritance Transmission of DNA from parents to offspring. **7**

insects Most diverse arthropod group; members have six legs, two antennae, and, in some groups, wings. **370**

insertion Mutation in which one or more base pairs become inserted into DNA. **146**

instinctive behavior An innate response to a simple stimulus. **400**

intermediate disturbance hypothesis Species richness is greatest in communities where disturbances are moderate in their intensity or frequency. **441**

intermediate filament Stable cytoskeletal element structurally supports cells and tissues. **66**

interphase In a eukaryotic cell cycle, the interval between mitotic divisions when a cell grows, roughly doubles the number of its cytoplasmic components, and replicates its DNA. **164**

interspecific competition Competition between two species. **434**

intron Nucleotide sequence that intervenes between exons and is excised during RNA processing. **141**

inversion Structural rearrangement of a chromosome in which part of it becomes oriented in the reverse direction. **210**

invertebrate General term for an animal that does not have a backbone. **353**

ion Charged atom. **30**

ionic bond Type of chemical bond in which a strong mutual attraction forms between ions of opposite charge. **28**

isotonic Describes two fluids with identical solute concentrations. **84**

isotopes Forms of an element that differ in the number of neutrons their atoms carry. **25**

karyotype Image of an individual's complement of chromosomes arranged by size, length, shape, and centromere location. **125**

key innovation An evolutionary adaptation that gives its bearer the opportunity to exploit a particular environment more efficiently or in a new way. **276**

keystone species A species that has a disproportionately large effect on community structure. **442**

kinetic energy The energy of motion. **76**

knockout An experiment in which a gene is deliberately inactivated in a living organism. **154**

Krebs cycle Cyclic pathway that, along with acetyl–CoA formation, breaks down two pyruvate to carbon dioxide for a net yield of two ATP and many reduced coenzymes. Part of aerobic respiration pathway. **113**

K–selection Individuals who produce offspring capable of outcompeting others for limited resources have a selective advantage; occurs when a population is near carrying capacity. **421**

lactate fermentation Anaerobic carbohydrate breakdown pathway that produces ATP and lactate. **116**

lamprey Jawless fish with a backbone of cartilage. **380**

lancelet Invertebrate chordate that has a fishlike shape and retains the defining chordate traits into adulthood. **378**

La Niña Periodic cooling of equatorial Pacific waters and the associated shifts in global weather patterns. **463**

larva Preadult stage in some animal life cycles. **357**

law of independent assortment During meiosis, members of a pair of genes on homologous chromosomes tend to be distributed into gametes independently of other gene pairs. **194**

law of nature Generalization that describes a consistent natural phenomenon for which there is incomplete scientific explanation. **18**

law of segregation The two members of each pair of genes on homologous chromosomes end up in different gametes during meiosis. **193**

learned behavior Behavior that is modified by experience. **400**

lek Of some birds, a communal mating display area for males. **404**

lethal mutation Mutation that drastically alters phenotype; causes death. **258**

lichen Composite organisms, consisting of a fungus and a single-celled alga or a cyanobacterium. **348**

life history pattern A set of traits related to growth, survival, and reproduction such as age-specific mortality, life span, age at first reproduction, and number of breeding events. **420**

light–dependent reactions First stage of photosynthesis; a set of reactions that convert light energy to chemical energy of ATP and NADPH. **97**

light–independent reactions Second stage of photosynthesis; a set of reactions that use ATP and NADPH to assemble sugars from water and CO_2. **97**

lignin Material that stiffens cell walls of vascular plants. **68, 326**

limiting factor A necessary resource, the depletion of which halts population growth. **418**

lineage Line of descent. **240**

linkage group All genes on a chromosome. **195**

lipid Fatty, oily, or waxy organic compound. **42**

lipid bilayer Structural foundation of cell membranes; double layer of lipids arranged tail-to-tail. **56**

lobe–finned fish Fish with fleshy fins that contain bones. **381**

locus Location of a gene on a chromosome. **191**

logistic growth A population grows slowly, then increases rapidly until it reaches carrying capacity and levels off. **418**

lysogenic pathway Bacteriophage replication path in which viral DNA becomes integrated into the host's chromosome and is passed to the host's descendants. **298**

lysosome Enzyme-filled vesicle that functions in intracellular digestion. **63**

lytic pathway Bacteriophage replication pathway in which a virus immediately replicates in its host and kills it. **298**

mammal Animal with hair or fur; females secrete milk from mammary glands. **387**

mark–recapture sampling Method of estimating population size of mobile animals by marking individuals, releasing them, then checking the proportion of marks among individuals recaptured at a later time. **414**

marsupial Mammal in which young are born at an early stage and complete development in a pouch on the mother's surface. **387**

mass extinction Simultaneous loss of many lineages from Earth. **237**

mass number Total number of protons and neutrons in the nucleus of an element's atoms. **25**

master gene Gene encoding a product that affects the expression of many other genes. **154**

megaspore Haploid spore formed in ovule of seed plants; develops into an egg-producing gametophyte. **332**

meiosis Nuclear division process that halves the chromosome number. Basis of sexual reproduction. **176**

mesoderm Middle tissue layer of a three-layered animal embryo. **354**

messenger RNA (mRNA) A type of RNA that carries a protein-building message. **138**

metabolic pathway Series of enzyme-mediated reactions by which cells build, remodel, or break down an organic molecule. **82**

metabolism All the enzyme-mediated chemical reactions by which cells acquire and use energy as they build and break down organic molecules. **39**

metamorphosis Dramatic remodeling of body form during the transition from larva to adult. **367**

metaphase Stage of mitosis at which the cell's chromosomes are aligned midway between poles of the spindle. **166**

metastasis The process in which cancer cells spread from one part of the body to another. **171**

methanogen Organism that produces methane gas as a metabolic by-product. **306**

microevolution Change in allele frequencies in a population or species. **259**

microfilament Reinforcing cytoskeletal element that consists of actin subunits. **66**

microspore Walled, haploid spore formed in pollen sacs of seed plants; develops into a sperm-producing gametophyte (a pollen grain). **332**

microtubule Cytoskeletal element involved in cellular movement; hollow filament of tubulin subunits. **66**

mimicry A species evolves traits that make it similar in appearance to another species. **436**

mitochondrion Double-membraned organelle that produces ATP by aerobic respiration in eukaryotes. **64**

mitosis Nuclear division mechanism that maintains the chromosome number. Basis of body growth and tissue repair in multicelled eukaryotes; also asexual reproduction in some plants, animals, fungi, and protists. **164**

mixture An intermingling of two or more types of molecules. **27**

model Analogous system used for testing hypotheses. **13**

molecule Two or more atoms joined by chemical bonds. **4, 27**

mollusk Invertebrate with a reduced coelom and a mantle; such as a bivalve, gatropod, or cephalopod. **364**

molting Periodic shedding of an outer body layer or part. **366**

monocots Lineage of angiosperms with one seed leaf (cotyledon); includes grasses, orchids, and palms. **336**

monohybrid cross Breeding experiment in which individuals identically heterozygous for one gene are crossed. The frequency of traits among the offspring offers information about the dominance relationship between the alleles. **192**

monomers Molecules that are subunits of polymers. **39**

monophyletic group An ancestor and all of its descendants. **278**

monotreme Egg-laying mammal. **387**

monsoon Wind that reverses direction seasonally. **467**

morphological convergence Evolutionary pattern in which similar body parts evolve separately in different lineages. **251**

morphological divergence Evolutionary pattern in which a body part of an ancestor changes in its descendants. **250**

moss Nonvascular plant with a leafy green gametophyte and an attached, dependent sporophyte consisting of a capsule on a stalk. **329**

motor protein Type of energy-using protein that interacts with cytoskeletal elements to move the cell's parts or the whole cell. **66**

multiple allele system Gene for which three or more alleles persist in a population. **196**

multiregional model Model that postulates that *H. sapiens* populations in different regions evolved from *H. erectus* in those regions. **392**

mutation Permanent change in the sequence of DNA. **131**

mutualism Species interaction that benefits both species. **348, 433**

mycelium Mass of threadlike filaments (hyphae) that make up the body of a multicelled fungus. **344**

mycorrhiza Mutually beneficial partnership between a fungus and a plant root. **348**

naturalist Person who observes life from a scientific perspective. **238**

natural selection A process in which environmental pressures result in the differential survival and reproduction of individuals of a population who vary in the details of shared, heritable traits. **243, 261**

nematocyst Stinging organelle unique to cnidarians. **358**

neoplasm An accumulation of abnormally dividing cells. **169**

nerve net Of cnidarians, a mesh of nerve cells with no central control organ; allows movement and other behavior. **358**

neutral mutation A mutation that has no effect on survival or reproduction. **258**

neutron Uncharged subatomic particle in the atomic nucleus. **24**

niche See ecological niche.

nitrification Bacteria convert ammonium to nitrates. **458**

nitrogen cycle Movement of nitrogen among the atmosphere, soil, and water, and into and out of food webs. **458**

nitrogen fixation Incorporation of nitrogen gas into ammonia. **304, 458**

nondisjunction Failure of sister chromatids or homologous chromosomes to separate during nuclear division. **212**

normal flora Normally harmless or beneficial microorganisms that typically live in or on a host body. **304**

notochord Stiff rod of connective tissue that runs the length of the body in chordate larvae or embryos. **378**

nuclear envelope A double membrane that constitutes the outer boundary of the nucleus. **61**

nucleic acid Single- or double-stranded chain of nucleotides joined by sugar–phosphate bonds; for example, DNA, RNA. **47**

nucleic acid hybridization Base-pairing between DNA or RNA from different sources. **222**

nucleoid Region of cytoplasm where the DNA is concentrated in a bacterium or archaean. **58, 302**

nucleolus In a cell nucleus, a dense, irregularly shaped region where ribosomal subunits are assembled. **61**

nucleoplasm Viscous fluid enclosed by the nuclear envelope. **61**

nucleosome A length of DNA wound around a spool of histone proteins. **124**

nucleotide Monomer of nucleic acids. Consists of a five-carbon sugar, a nitrogen-containing base, and phosphate groups. **47**

nucleus Of an atom, the core region occupied by protons and neutrons. **24** Of a eukaryotic cell, an organelle with two membranes that holds the cell's DNA. **8, 52**

nutrient Substance that an organism needs for growth and survival, but cannot make for itself. **6**

oncogene Gene that has the potential to transform a normal cell into a tumor cell. **170**

open circulatory system System in which blood leaves vessels and seeps through tissues before returning to the heart. **364**

operator Part of an operon; a DNA binding site for a repressor. **158**

operon Group of genes together with a promoter–operator DNA sequence that controls their transcription. **158**

organ In multicelled organisms, a grouping of tissues engaged in a collective task. **5**

organelle Structure that carries out a specialized metabolic function inside a cell. **52**

organic Type of compound that consists primarily of carbon and hydrogen atoms. **38**

organism Individual that consists of one or more cells. **5**

organ system In multicelled organisms, set of organs engaged in a collective task that keeps the body functioning properly. **5**

osmosis The diffusion of water across a selectively permeable membrane in response to a differing overall solute concentration. **85**

osmotic pressure Amount of turgor that prevents osmosis into cytoplasm or other hypertonic fluid. **85**

ovary Of flowering plants, the enlarged base of a carpel, inside which one or more ovules form and eggs are fertilized. **336**

ovule Of seed plants, reproductive structure in which egg-producing female gametophyte develops; after fertilization, matures into a seed. **332**

ozone layer High atmospheric layer with a great concentration of ozone (O_3); prevents much ultraviolet radiation from reaching Earth's surface. **494**

pandemic Outbreak of disease that affects many separate regions and poses a serious threat to human health. **300**

Pangea Supercontinent that formed about 237 million years ago and broke up about 152 million years ago. **248**

parapatric speciation Speciation model in which differing selection pressures lead to divergences within a single population. **275**

parasitism Relationship in which one species withdraws nutrients from another species, without immediately killing it. **438**

parasitoid An insect that lays eggs in another insect, and whose young devour their host from the inside. **438**

passive transport Mechanism by which a concentration gradient drives the movement of a solute across a cell membrane through a transport protein. Requires no energy input. **86**

pathogen Disease-causing agent. **300**

pattern formation Process by which a complex body forms from local processes during embryonic development. **155**

PCR See polymerase chain reaction.

peat Carbon-rich moss remains; can be dried for use as fuel. **329**

pedigree Chart showing the pattern of inheritance of a trait through generations in a family. **204**

pelagic province The ocean's waters. **482**

pellicle Layer of proteins that gives shape to many unwalled, single-celled protists. **313**

peptide bond A bond between the amine group of one amino acid and the carboxyl group of another. Joins amino acids in proteins. **44**

per capita growth rate For some interval, the added number of individuals divided by the initial population size. **416**

periodic table Tabular arrangement of the known elements by atomic number. **24**

permafrost Continually frozen soil layer that lies beneath arctic tundra and prevents water from draining. **477**

peroxisome Enzyme-filled vesicle that breaks down amino acids, fatty acids, and toxic substances. **63**

pH A measure of the number of hydrogen ions in a fluid. **32**

phagocytosis "Cell eating"; an endocytic pathway by which a cell engulfs particles such as microbes or cellular debris. **88**

phenotype An individual's observable traits. **191**

pheromone Intraspecific chemical communication signal. **397**

phloem Plant vascular tissue that distributes sugars through sieve tubes. **326**

phospholipid A lipid with a phosphate group in its hydrophilic head, and two nonpolar fatty acid tails; main constituent of eukaryotic cell membranes. **43**

phosphorus cycle Movement of phosphorus among Earth's rocks and waters, and into and out of food webs. **459**

phosphorylation Addition of a phosphate group to a molecule; occurs by the transfer of a phosphate group from a donor molecule such as ATP. **79**

photoautotroph Photosynthetic autotroph. **108**

photolysis Process by which light energy breaks down a molecule. **98**

photorespiration In plants, reaction in which rubisco attaches oxygen instead of carbon dioxide to ribulose bisphosphate. Reduces sugar production by photosynthesis because it diverts carbon atoms from the Calvin–Benson cycle. **102**

photosynthesis Metabolic pathway by which most autotrophs capture light energy and use it to make sugars from CO_2 and water. **6, 93**

photosystem Cluster of pigments and proteins that converts light energy to chemical energy in photosynthesis. **98**

phylogeny Evolutionary history of a species or group of species. **278**

pigment An organic molecule that can absorb light of certain wavelengths. **94**

pilus Protein filament that projects from the surface of some bacterial cells. **58, 302**

pioneer species Species that can colonize a new habitat. **440**

placental mammal Mammal in which a mother and her embryo exchange materials by means of an organ called the placenta. **387**

placozoan The simplest modern animal known, with an asymmetrical flat body, four types of cells, and a small genome. **356**

planarian Free-living freshwater flatworm. **360**

plankton Community of tiny drifting or swimming organisms. **314**

plant A multicelled, typically photosynthetic autotroph. **8**

plasma membrane A cell's outermost membrane. **52**

plasmid Of many bacteria and archaeans, a small ring of nonchromosomal DNA replicated independently of the chromosome. **58, 220, 303**

plasmodesmata Cell junctions that connect the cytoplasm of adjacent plant cells. **69**

plasmodial slime mold Protist that feeds as a multinucleated mass; forms a spore-bearing structure when enviromental conditions become unfavorable. **320**

plastid An organelle that functions in photosynthesis or storage; e.g., chloroplast, amyloplast. **65**

plate tectonics Theory that Earth's outer layer of rock is cracked into plates, the slow movement of which rafts continents to new locations over geologic time. **249**

pleiotropic Refers to a gene whose product influences multiple traits. **197**

plot sampling Method of estimating population size of organisms that do not move much by making counts in small plots, and extrapolating from this to the number in the larger area. **414**

polarity Any separation of charge into distinct positive and negative regions. **29**

pollen grain Male gametophyte of a seed plant. **327**

pollen sac Of seed plants, reproductive structure in which sperm-bearing gametophytes (pollen grains) develop. **332**

pollination Arrival of a pollen grain on the egg-bearing part of a seed plant. **332**

pollinator Animal that moves pollen, thus facilitating pollination. **336**

pollutant A substance that is released into the environment by human activities and interferes with the function of organisms that evolved in the absence of the substance or with lower levels. **491**

polymer Molecule that consists of multiple monomers. **39**

polymerase chain reaction (PCR) Method that rapidly generates many copies of a specific section of DNA. **222**

polypeptide Chain of amino acids linked by peptide bonds. **44**

polyploid Having three or more of each type of chromosome characteristic of the species. **212**

population A group of organisms of the same species who live in a specific location and breed with one another more often than they breed with members of other populations. **5, 258, 413**

population density Number of individuals per unit area. **414**

population distribution Where individuals are clumped, uniformly dispersed, or randomly dispersed in an area. **414**

population size Total number of individuals in a population. **414**

potential energy Stored energy. **77**

predation One species captures, kills, and eats another. **344, 436**

prediction Statement, based on a hypothesis, about a condition that should exist if the hypothesis is correct. **12**

primary producer In an ecosystem, an organism that captures energy from an inorganic source and stores it as biomass; first trophic level. **450**

primary production The rate at which an ecosystem's producers capture and store energy. **450**

primary succession A new community becomes established in an area where there was no soil. **440**

primary wall The first cell wall of young plant cells. **68**

primate Mammal having grasping hands with nails; includes prosimians, monkeys, apes, and hominids such as humans. **388**

primer Short, single strand of DNA designed to hybridize with a particular nucleotide sequence. **223**

prion Infectious protein. **46**

probability The chance that a particular outcome of an event will occur. **17**

probe Short fragment of DNA labeled with a tracer; designed to hybridize with a nucleotide sequence of interest. **222**

producer Organism that makes its own food using energy and simple raw materials from the environment; an autotroph. **6**

product A molecule that remains at the end of a reaction. **78**

promoter In DNA, a sequence to which RNA polymerase binds. **140**

prophase Stage of mitosis in which chromosomes condense and become attached to a newly forming spindle. **166**

protein Organic compound that consists of one or more chains of amino acids (polypeptides). **44**

proteobacteria Largest bacterial lineage. **304**

protist General term for a eukaryote that is not a fungus, animal, or plant. **8, 311**

protocell Membranous sac that contains interacting organic molecules; hypothesized to have formed prior to the earliest life forms. **286**

proton Positively charged subatomic particle that occurs in the nucleus of all atoms. **24**

proto-oncogene Gene that can become an oncogene. **170**

protostomes Lineage of bilateral animals in which the first opening on the embryo surface develops into a mouth. **355**

pseudocoelom Unlined body cavity. **355**

Punnett square Diagram used to predict the genetic and phenotypic outcome of a cross. **192**

pyruvate Three-carbon organic compound that is the end product of glycolysis. **109**

radial symmetry Having parts arranged around a central axis, like spokes around a wheel. **354**

radioactive decay Process by which atoms of a radioisotope emit energy and/or subatomic particles when their nucleus spontaneously disintegrates. **25**

radioisotope Isotope with an unstable nucleus. **25**

radiolarian Heterotrophic single-celled protist with a porous silica shell, long cytoplasmic extensions. **314**

radiometric dating Method of estimating the age of a rock or fossil by measuring the content and proportions of a radioisotope and its daughter elements. **245**

rain shadow Dry region downwind of a coastal mountain range. **466**

ray-finned fish Fish with fins supported by thin rays derived from skin; member of the most diverse lineage of fishes. **381**

reactant Molecule that enters a reaction. **78**

reaction Process of chemical change. **78**

receptor protein Plasma membrane protein that binds to a particular substance outside of the cell. **57**

recessive Refers to an allele with an effect that is masked by a dominant allele on the homologous chromosome. **191**

recognition protein Plasma membrane protein that identifies a cell as belonging to self (one's own body); e.g., MHC marker. **56**

recombinant DNA A DNA molecule that contains genetic material from more than one organism. **220**

red alga Photosynthetic protist that deposits cellulose in its cell wall, stores sugars as starch, and has chloroplasts containing chlorophyll *a* and red pigments (phycobilins). **318**

redox reaction Oxidation–reduction reaction in which one molecule accepts electrons (it becomes reduced) from another molecule (which becomes oxidized). Also called electron transfer. **83**

replacement fertility rate Average number of children women of a population must bear to replace themselves with a daughter of reproductive age. **425**

replacement model Model for origin of *H. sapiens*; humans evolved in Africa, then migrated to different regions and replaced the other hominids that lived there. **392**

repressor Regulatory protein that blocks transcription. **152**

reproduction Processes by which parents produce offspring. **7**

reproductive base Of a population, all individuals who are of reproductive age or younger. **415**

reproductive cloning Technology that produces genetically identical individuals; e.g., SCNT. **132**

reproductive isolation Absence of gene flow between populations; always part of speciation. **270**

reptile Amniote subgroup that includes lizards, snakes, turtles, crocodilians, and birds. **384**

resource partitioning Species become adapted in different ways to access different portions of a limited resource; allows species with similar needs to coexist. **435**

restriction enzyme Type of enzyme that cuts specific nucleotide sequences in DNA. **220**

reverse transcriptase A viral enzyme that uses mRNA as a template to make a strand of cDNA. **221**

rhizoid Threadlike structure that anchors a bryophyte. **329**

rhizome Stem that grows horizontally along or under the ground. **330**

ribosomal RNA (rRNA) A type of RNA that is a component of ribosomes. **138**

ribosome Organelle of protein synthesis. **58**

ribozyme RNA that functions as an enzyme. **287**

RNA Ribonucleic acid; e.g., mRNA, rRNA, tRNA. **47**

RNA polymerase Enzyme that carries out transcription. **140**

RNA world Hypothetical early interval when RNA served as the genetic information. **287**

roundworm Unsegmented worm with a pseudocoel and a cuticle that is molted as the animal grows. **366**

r–selection Individuals who produce the maximum number offspring as quickly as possible have a selective advantage; occurs when population density is low and resources are abundant. **421**

rubisco Ribulose bisphosphate carboxylase. Carbon-fixing enzyme of the Calvin–Benson cycle. **101**

runoff Water that flows over soil into streams. **454**

sac fungi Most diverse fungal group; sexual reproduction produces ascospores inside a saclike structure (an ascus). **346**

salt Compound that releases ions other than H^+ and OH^- when it dissolves in water. **30**

sampling error Difference between results derived from testing an entire group of events or individuals, and results derived from testing a subset of the group. **16**

saprobe Organism that feeds on wastes and remains. **344**

science Systematic study of the observable world. **12**

scientific method Systematically making, testing, and evaluating hypotheses. **13**

scientific theory Hypothesis that has not been disproven after many years of rigorous testing. **18**

SCNT See somatic cell nuclear transfer.

seamount An undersea mountain. **482**

secondary metabolite Chemical that has no known role in an organism's normal metabolism; often deters predation. **339**

secondary succession A new community develops in a site where the soil that supported an old community remains. **440**

secondary wall Lignin-reinforced wall that forms inside the primary wall of a plant cell. **68**

second law of thermodynamics Energy tends to disperse spontaneously. **76**

sedimentary cycle Biochemical cycle in which the atmosphere plays little role and rocks are the major reservoir. **459**

seed Embryo sporophyte of a seed plant packaged with nutritive tissue inside a protective coat. **327**

seedless vascular plant Plant such as a fern or horsetail that has vascular tissue and disperses by producing spores. **330**

segmentation Having a body composed of similar units that repeat along its length. **355**

selfish herd Temporary group that forms when individuals cluster to minimize their individual risk of predation. **406**

sex chromosome Member of a pair of chromosomes that differs between males and females. **125**

sexual dimorphism Having distinct male and female phenotypes. **404**

sexual reproduction Reproductive mode by which offspring arise from two parents and inherit genes from both. **175**

sexual selection Mode of natural selection in which some individuals outreproduce others of a population because they are better at securing mates. **266**

shell model Model of electron distribution in an atom. **26**

short tandem repeats In chromosomal DNA, sequences of 4 or 5 bases repeated multiple times in series. **227**

sister chromatid One of the two attached DNA molecules of a duplicated eukaryotic chromosome. **124**

sister groups The two lineages that emerge from a node on a cladogram. **278**

soil water Water between soil particles. **454**

solute A dissolved substance. **30**

solvent Liquid that can dissolve other substances. **30**

somatic Relating to the body. **176**

somatic cell nuclear transfer (SCNT) Method of reproductive cloning in which genetic material is transferred from an adult somatic cell into an unfertilized, enucleated egg. **132**

sorus Cluster of spore-producing capsules on a fern leaf. **330**

speciation One of several processes by which new species arise. **270**

species A type of organism. Designated by genus name together with specific epithet. **10**

species richness Of an area, number of species. **444**

specific epithet Second part of a species name. **10**

sperm Mature male gamete. **182**

spindle Dynamically assembled and disassembled network of microtubules that moves chromosomes during nuclear division. **166**

spirochetes Bacteria that resemble a stretched-out spring. **305**

sponge Aquatic invertebrate that has no tissues or organs and filters food from the water. **357**

sporophyte Diploid, spore-producing body in the life cycle of land plants and some algae. **182, 318**

stabilizing selection Mode of natural selection in which intermediate forms of a trait are favored over extreme forms. **264**

stamen Floral reproductive structure that produces male gametophytes; in most plants it consists of a pollen-producing anther on the tip of a filament. **336**

stasis Evolutionary pattern in which a lineage persists with little or no change over evolutionary time. **276**

statistically significant Refers to a result that is statistically unlikely to have occurred by chance alone. **17**

steroid Type of lipid with four carbon rings and no fatty acid tails. **43**

stimulus Form of energy that a sensory receptor detects. **398**

stoma Opening across a plant's cuticle and epidermis; can be opened for gas exchange or closed to prevent water loss. **326**

stomata Closeable gaps defined by guard cells on plant surfaces; when open, they allow water vapor and gases to diffuse across the epidermis. **102**

stroma Semifluid matrix between the thylakoid membrane and the two outer membranes of a chloroplast. **97**

stromatolite Dome-shaped structures composed of layers of bacterial cells and sediments. **288**

substrate A molecule that is specifically acted upon by an enzyme. **80**

substrate-level phosphorylation A reaction that transfers a phosphate group from a substrate directly to ADP, thus forming ATP. **110**

surface-to-volume ratio A relationship in which the volume of an object increases with the cube of the diameter, but the surface area increases with the square. **52**

survivorship curve Graph showing the decline in numbers of a cohort over time. **420**

symbiosis One species lives in or on another in a commensal, mutualistic, or parasitic relationship. **432**

sympatric speciation Pattern in which speciation occurs in the absence of a physical barrier to gene flow. **274**

taiga See boreal forest.

taxon Linnaean category; a grouping of organisms. Each consists of a group of the next lower taxon. **10**

taxonomy The science of naming and classifying species. **10**

telophase Stage of mitosis during which chromosomes arrive at the spindle poles and decondense, and new nuclei form. **166**

temperate deciduous forest Northern Hemisphere biome in which the main plants are broadleaf trees that lose their leaves in fall and become dormant during cold winters. **474**

temperature Measure of molecular motion. **31**

territory Region that an animal or animals occupy and defend against competitors. **404**

testcross Method of determining genotype in which an individual of unknown genotype is crossed with one that is known to be homozygous recessive. **192**

tetrapod Vertebrate with four legs, or a descendant thereof; an amphibian, reptile (including birds), or mammal. **379**

theory of inclusive fitness Genes associated with altruism can be advantageous if the expense of

this behavior to the altruist is outweighed by the reproductive success of relatives. **408**

theory of uniformity Idea that gradual repetitive processes occurring over long time spans shaped Earth's surface. **240**

therapeutic cloning The use of SCNT to produce human embryos for research purposes. **133**

threatened species Species likely to become endangered in the near future. **488**

thylakoid membrane A chloroplast's highly folded inner membrane system; molecules embedded in it carry out the light-dependent reactions of photosynthesis. **97**

tight junctions Arrays of fibrous proteins that join the plasma membranes of adjacent cells in epithelial tissue; collectively, they prevent fluids from leaking between the cells. **69**

tissue In multicelled organisms, specialized cells organized in a pattern that allows them to perform a collective function. **5**

total fertility rate Average number of children the women of a population bear over the course of a lifetime. **425**

toxin Chemical that is made by one organism and harms another. **311**

tracer A molecule that has been labeled with a detectable substance. **25**

transcription Process by which an RNA is assembled from nucleotides using the base sequence of a gene as a template. **138**

transcription factor Regulatory protein that influences transcription; e.g., an activator or repressor. **152**

transfer RNA (tRNA) A type of RNA that delivers amino acids to a ribosome during translation. **138**

transgenic Refers to a genetically modified organism that carries a gene from a different species. **228**

translation Process by which a polypeptide chain is assembled from amino acids in the order specified by an mRNA. **139**

translocation In genetics, chromosomal mutation in which a broken piece gets reattached in the wrong location. **210**

transport protein Protein that passively or actively assists specific ions or molecules across a membrane. **57**

transposable element Segment of DNA that can spontaneously move to a new location in a chromosome. **147**

triglyceride A fat with three fatty acid tails. **42**

trophic level Position of an organism in a food chain. **451**

tropical rain forest Highly productive, species-rich biome in which year-round rains and warmth support continuous growth of evergreen broadleaf trees. **474**

trypanosome Parasitic flagellate with a single mitochondrion and a membrane-encased flagellum. **313**

tumor A neoplasm that forms a lump. **170**

tunicate Invertebrate chordate that loses its defining chordate traits during the transition to adulthood. **378**

turgor Pressure that a fluid exerts against a wall, membrane, or other structure that contains it. **85**

vacuole A fluid-filled organelle that isolates or disposes of waste, debris, or toxic materials. **63**

variable In an experiment, a characteristic or event that differs among individuals or over time. **13**

vascular plant Plant that has xylem and phloem; a seedless vascular plant, gymnosperm, or angiosperm. **326**

vascular tissue Internal pipelines of xylem and phloem in the body of a vascular plant. **326**

vector Animal that carries a pathogen from one host to the next. **300**

vertebrate Animal with a backbone. **353, 378**

vesicle Small, membrane-enclosed, saclike organelle; different kinds store, transport, or degrade their contents. **62**

viroid Small noncoding RNA that can infect and cause disease in plants. **301**

virus Noncellular, infectious particle of protein and nucleic acid; replicates only in a host cell. **298**

water cycle Movement of water among Earth's oceans, atmosphere, and freshwater reservoirs on land. **454**

water mold Heterotrophic protist that grows as nutrient-absorbing filaments. **317**

watershed Land area that drains into a particular stream or river. **454**

water-vascular system Of echinoderms, a system of fluid-filled tubes and tube feet that function in locomotion. **373**

wavelength Distance between the crests of two successive waves of light. **94**

wax Water-repellent mixture of lipids with long fatty acid tails bonded to long-chain alcohols or carbon rings. **43**

X chromosome inactivation Shutdown of one of the two X chromosomes in the cells of female mammals. **156**

xenotransplantation Transplantation of an organ from one species into another. **231**

xylem Complex vascular tissue of plants; its tracheids and vessel members distribute water and mineral ions. **326**

zero population growth Interval in which births equal deaths. **416**

zygote Cell formed by fusion of two gametes; the first cell of a new individual. **177**

zygote fungus Fungus that forms a zygospore during sexual reproduction. **345**

Art Credits & Acknowledgments

TABLE OF CONTENTS **Page vi** from top, © Raymond Gehman/ Corbis; © Bill Beatty/ Visuals Unlimited. **Page vii** left, © Dylan T. Burnette and Paul Forscher; middle left, Hemoglobin models: PDB ID: 1GZX; Paoli, M., Liddington, R., Tame, J., Wilkinson, A., Dodson, G.; Crystal structure of T state hemoglobin with oxygen bound at all four haems. *J.Mol.Biol.*, v256, pp. 775–792, 1996; middle right, Kristian Peters, http://commons.wikimedia.org/wiki/User:Fabelfroh; right, Lilly M/ http://commons.wikimedia.org. **Page viii** middle bottom, © Maria Samsonova and John Reinitz; bottom, Michael Clayton/ University of Wisconsin, Department of Botany. **Page ix** from left, Francis Leroy, Biocosmos/ Science Photo Library/ Photo Researchers, Inc.; © Michael Stuckey/ Comstock, Inc.; Photo courtesy of The Progeria Research Foundation; Photo courtesy of MU Extension and Agricultural Information. **Page x** from top, USGS; Courtesy of Hopi Hoekstra, University of California, San Diego; © Jack Jeffrey Photography. **Page xi** from left, Courtesy of the University of Washington; SciMAT/ Photo Researchers, Inc.; Dr. Richard Feldmann/ National Cancer Institute; © Kim Taylor/ Bruce Coleman, Inc./ Photoshot; Photo USDA. **Page xii** from top, Eye of Science/ Photo Researchers, Inc.; © Wim van Egmond/ Visuals Unlimited; CDC/ Piotr Naskrecki; © Kenneth Garrett/National Geographic Image Collection. **Page xiii** from left, © Kevin Schafer/Tom Stack & Associates; © Alexander Wild; © Jacques Langevin/ Corbis Sygma. **Page xiv** from top, E.R. Degginger/ Photo Researchers, Inc.; © D. A. Rintoul; © George H. Huey/ Corbis. **Page xv** from left, © John Easley, www.johneasley.com; © Gary Head; Image © Lee Prince, Used under license from Shutterstock.com.

PREFACE **Page xvi** Clockwise from top left, Courtesy East Carolina University; Photo by Peggy Greb/ USDA; Photo courtesy of The Progeria Research Foundation.

CHAPTER 1 **Key Concepts** from left, California Poppy, © 2009, Christine M. Welter; Dr. Marina Davila Ross, University of Portsmouth, Courtesy of Allen W. H. Bé and David A. Caron; © Roger W. Winstead, NC State University; © Adrian Vallin. **1.1** page 2, © Steve Richards; page 3, © Bruce Beehler/ Conservation International. **1.3** (3–4) © Umberto Salvagnin, www.flickr.com/photos/kaibara; (5) California Poppy, © 2009, Christine M. Welter; (6) Lady Bird Johnson Wildflower Center; (7) Michael Szoenyi/ Photo Researchers, Inc.; (8) Photographers Choice RF/ SuperStock; (9) © Sergei Krupnov, www.flickr.com/photos/7969319@N03; (10) © Mark Koberg Photography; (11) NASA. **1.4** above, © Victoria Pinder, http://www.flickr.com/photos/vixstarplus. **1.5** Dr. Marina Davila Ross, University of Portsmouth. **1.6** (a) Clockwise from top left, Dr. Richard Frankel; SPL/ Photo Researchers, Inc.; © Susan Barnes, Tony Brian, David Parker/ SPL/ Photo Researchers, Inc.; www.zahnarzt-stuttgart.com; SciMAT/ Photo Researchers, Inc.; (b) left, ArchiMeDes; right, © Dr. Harald Huber, Dr. Michael Hohn, Prof. Dr. K.O.Stetter, University of Regensburg, Germany; (c) Protists, left, © Lewis Trusty/ Animals Animals; right, clockwise from top left, M I Walker/ Photo Researchers, Inc.; Courtesy of Allen W. H. Bé and

David A. Caron; © Emiliania Huxleyi photograph, Vita Pariente, scanning electron micrograph taken on a Jeol T330A instrument at Texas A&M University Electron Microscopy center; Oliver Meckes/ Photo Researchers, Inc.; © Carolina Biological Supply Company; Plants, left © John Lotter Gurling / Tom Stack & Associates; right © Edward S. Ross; Fungi, left, © Robert C. Simpson/ Nature Stock; right, © Edward S. Ross; Animals, from left, © Martin Zimmerman, Science, 1961, 133:73–79, © AAAS.; Thomas Eisner, Cornell University; © Tom & Pat Leeson, Ardea London Ltd.; © Pixtal/ SuperStock. **1.7** from left, Joaquim Gaspar; Bogdan; Opiola Jerzy; Ravedave; Luc Viatour. **1.9** From Meyer, A., Repeating Patterns of Mimicry. PLoS Biology Vol. 4, No. 10, e341 doi:10.1371/journal.pbio.0040341. **Page 12** left, © Bruce Beehler/ Conservation International; right, Volker Steger/ SPL/ Photo Researchers, Inc.; **1.10** from left, Cape Verde National Institute of Meteorology and Geophysics and the U.S. Geological Survey; © Roger W. Winstead, NC State University; Photo by Scott Bauer, USDA/ARS. **1.11** © Superstock. **1.12** (a) © Matt Rowlings, www.eurobutterflies.com; (b) © Adrian Vallin; (c) © Antje Schulte. **1.13** © Gary Head. **Page 16** right, © Bruce Beehler/ Conservation International. **1.14** © Adrian Vallin. **Page 18** © Raymond Gehman/ Corbis. **1.15** above, Photo courtesy of Dr. Robert Zingg/ Zoo Zurich; below, Courtesy East Carolina University. **Page 20** Section 1.1, © Bruce Beehler/ Conservation International; Section 1.2, California Poppy, © 2009, Christine M. Welter; Section 1.3, © Victoria Pinder, http://www.flickr.com/photos/vixstarplus; Section 1.4, Courtesy of Allen W. H. Bé and David A. Caron ; Section 1.5, From Meyer, A., Repeating Patterns of Mimicry. PLoS Biology Vol. 4, No. 10, e341 doi:10.1371/journal.pbio.0040341; Section 1.6, © Bruce Beehler/ Conservation International; Section 1.7, © Adrian Vallin; Section 1.8, © Gary Head; Section 1.9, © Raymond Gehman/ Corbis. **Page 21** Scientific Paper; Adrian Vallin, Sven Jakobsson, Johan Lind and Christer Wiklund, Proc. R. Soc. B (2005 272, 1203, 1207). Used with permission of The Royal Society and the author.

CHAPTER 2 **Key Concepts** from left, © Michael S. Yamashita/ Corbis; © Bill Beatty/ Visuals Unlimited; © Herbert Schnekenburger; W. K. Fletcher/ Photo Researchers, Inc. **2.1** page 22, © Kim Westerskov, Photographer's Choice/ Getty Images; page 23, © Michael Grecco/ Picture Group. **2.4** Brookhaven National Laboratory. **Page 26** © Michael S. Yamashita/ Corbis. **2.7** (a) © Bill Beatty/ Visuals Unlimited. **2.12** © Herbert Schnekenburger. **2.13** left, © Vicki Rosenberg, www.flickr.com/photos/roseofredrock. **2.14** © JupiterImages Corporation. **2.15** W. K. Fletcher/ Photo Researchers, Inc. **Page 33** © Jared C. Benedict. **Page 34** Section 2.1, © Kim Westerskov, Photographer's Choice/ Getty Images; Section 2.3, © Michael S. Yamashita/ Corbis; Section 2.4, © Bill Beatty/ Visuals Unlimited; Section 2.5, © Herbert Schnekenburger; Section 2.6, W. K. Fletcher/ Photo Researchers, Inc.

CHAPTER 3 **Key Concepts** from left, © JupiterImages Corporation; Tim Davis/ Photo Researchers, Inc.; © JupiterImages Corporation. **3.1 page 36** © ThinkStock/ SuperStock. **3.5** left, © JupiterImages Corporation. **3.6** (c) © JupiterImages Corporation.

3.11 right, Tim Davis/ Photo Researchers, Inc. **Page 43** Kenneth Lorenzen. **3.14** #5 right, © JupiterImages Corporation. **Page 43** This lipoprotein image was made by Amy Shih and John Stone using VMD and is owned by the Theoretical and Computational Biophysics Group, NIH Resource for Macromolecular Modeling and Bioinformatics, at the Beckman Institute, University of Illinois at Urbana-Champaign. Labels added to the original image by book author. **3.15** (a) © Lily Echeverria/ Miami Herald; (b) Sherif Zaki, MD PhD, Wun-Ju Shieh, MD PhD; MPH/ CDC. **Page 47** © JupiterImages Corporation. **Page 48** Section 3.1, 3.3, 3.5 © JupiterImages Corporation; Section 3.6, Sherif Zaki, MD PhD, Wun-Ju Shieh, MD PhD; MPH/ CDC.

CHAPTER 4 **Key Concepts** from left, Astrid Hanns-Frieder michler/ SPL/ Photo Researchers, Inc.; © JupiterImages Corporation; Courtesy of © Roberto Kolter Lab, Harvard Medical School; © Martin W. Goldberg, Durham University, UK. **4.1** page 50 © JupiterImages Corporation; page 51, Stephanie Schuller/ Photo Researchers, Inc. **4.2** from left, Astrid Hanns-Frieder michler/SPL/ Photo Researchers, Inc.; Courtesy of Allen W. H. Bé and David A. Caron ; M I Walker/ Photo Researchers, Inc.; © Dr. Dennis Kunkel/ Visuals Unlimited; © Wim van Egmond/ Visuals Unlimited. **4.5** (a) left, © JupiterImages Corporation; (b) right, Geoff Tompkinson/ Science Photo Library/ Photo Researchers, Inc. **4.6** (a–b, d–e) © Jeremy Pickett-Heaps, School of Botany, University of Melbourne; (c) © Prof. Franco Baldi. **4.7** Virus, CDC; Bacteria, Tony Brian, David Parker/ SPL/ Photo Researchers, Inc.; Eukaryotic cells, Jeremy Pickett-Heaps, School of Botany, University of Melbourne; Louse, Edward S. Ross; Ladybug, Image © Dole, Used under license from Shutterstock.com; Goldfish, Image © Ultrashock, Used under license from Shutter stock.com; Butterfly, From Meyer, A., Repeating Patterns of Mimicry. PLoS Biology Vol. 4, No. 10, e341 doi:10.1371/journal.pbio.0040341; Guinea pig, Image © Sascha Burkard, Used under license from Shutterstock.com; Dog, Image © Pavel Sazonov, Used under license from Shutterstock.com; Human, Jupiter Images Corporation; Giraffe, © Ingram Publishing, SuperStock; Whale, Jupiter Images Corporation; Tree, Courtesy of © Billie Chandler. **4.10** (a) Rocky Mountain Laboratories, NIAID, NIH; (b) © R. Calentine/ Visuals Unlimited; (c) ArchiMeDes; (d–e) © K.O. Stetter & R. Rachel, Univ. Regensburg. **4.12** Courtesy of © Roberto Kolter Lab, Harvard Medical School. **4.13** (a) © Dr. Gopal Murti/ Photo Researchers, Inc.; (b) M.C. Ledbetter, Brookhaven National Laboratory. **4.14** right, © Kenneth Bart. **4.15** left, © Martin W. Goldberg, Durham University, UK. **4.16** left, © Kenneth Bart; (#1,3) Don W. Fawcett/ Visuals Unlimited; (#4) Micrograph, Gary Grimes. **4.17** left, Keith R. Porter. **4.18** top, Kristian Peters, http://commons.wikimedia.org/wiki/User:Fabelfroh; bottom, Dr. Jeremy Burgess/ SPL/ Photo Researchers, Inc. **4.19** below, © Dylan T. Burnette and Paul Forscher. **4.21** (a) Dow W. Fawcett/ Photo Researchers, Inc.; (b) © Mike Abbey/ Visuals Unlimited. **4.22** upper right, Don W. Fawcett/ Photo Researchers, Inc. **4.23** © George S. Ellmore. **4.24** (c) Russell Kightley/ Photo Researchers, Inc. **4.25** © ADVANCELL (Advanced In Vitro

Cell Technologies; S.L.) www.advancell.com. **Page 71**, Section 4.1 & 4.2, © Wim van Egmond/ Visuals Unlimited; Section 4.3, © JupiterImages Corporation. **Page 72** Section 4.5, © K.O. Stetter & R. Rachel, Univ. Regensburg; Section 4.6 & 4.7, © Kenneth Bart; Section 4.10, © Dylan T. Burnette and Paul Forscher; Section 4.11, © ADVANCELL (Advanced In Vitro Cell Technologies; S.L.) www.advancell.com. **4.27** From "Tissue & Cell", Vol. 27, pp.421–427, Courtesy of Bjorn Afzelius, Stockholm University. **Page 73**, Critical Thinking, P.L. Walne and J. H. Arnott, Planta, 77:325–354, 1967.

CHAPTER 5 **Key Concepts** from left, © Martin Barraud/ Stone/ Getty Images; © JupiterImages Corporation; Andrew Lambert Photography/ Science Photo Library/ Photo Researchers, Inc.; Biology Media/ Photo Researchers, Inc. **5.1** page 74, © BananaStock/ SuperStock. **5.2** © Martin Barraud/ Stone/ Getty Images. **5.4** © JupiterImages Corporation. **5.10** Hemoglobin models: PDB ID: 1GZX; Paoli, M., Liddington, R., Tame, J., Wilkinson, A., Dodson, G.; Crystal structure of T state hemoglobin with oxygen bound at all four haems. *J.Mol.Biol.*, v256, pp. 775–792, 1996. **5.12** © Scott McKiernan/ ZUMA Press. **5.13** Perennou Nuridsany / Photo Researchers, Inc. **5.15** Lisa Starr, after David S. Goodsell, RCSB, Protein Data Bank. **5.16** (a) © JupiterImages Corporation. **Page 84** Andrew Lambert Photography/ Science Photo Library/ Photo Researchers, Inc. **5.19** (b–d) Reproduced from *The Journal of Cell Biology*, 1976, vol. 70, pp. 193 by copyright permission of the Rockefeller Univeristy Press. **5.20** PDB files from NYU Scientific Visualization Lab. **5.21** After: David H. MacLennan, William J. Rice, and N. Michael Green, "The Mechanism of Ca2+ Transport by Sarco (Endo) plasmic Reticulum Ca2+-ATPases." *JBC* Volume 272, Number 46, Issue of November 14, 1997, pp. 28815–28818. **5.24** © R.G.W. Anderson, M.S. Brown and J.L. Goldstein. Cell 10:351 (1977). **5.25** (a, right) Biology Media/ Photo Researchers, Inc. **Page 89** © BananaStock/ SuperStock. **Page 90** Section 5.1, © BananaStock/ SuperStock; Section 5.2, © Martin Barraud/ Stone/ Getty Images; Section 5.5, © JupiterImages Corporation; Section 5.6, Andrew Lambert Photography/ Science Photo Library/ Photo Researchers, Inc.; Section 5.8, Biology Media/ Photo Researchers, Inc.

CHAPTER 6 **Key Concepts** from left, Kristian Peters, http://commons.wikimedia.org/wiki/User:Fabelfroh; © Bill Boch/ FoodPix/ Jupiter Images. **6.1** page 92, Photo by Peggy Greb/ USDA; page 93, © Roger W. Winstead, NC State University. **6.3** above, © Larry West/ FPG/ Getty Images. **6.4** (a) Jason Sonneman. **6.5** above, Kristian Peters, http://commons.wikimedia.org/wiki/User:Fabelfroh. **6.10** (a) Courtesy of John S. Russell, Pioneer High School; (c) © Bill Boch/ FoodPix/ Jupiter Images; (d) Lisa Starr. **6.11** © JupiterImages Corporation. **Page 103** Photo by Peggy Greb/ USDA. **Page 104** Section 6.1, Photo by Peggy Greb/ USDA; Section 6.4, Kristian Peters, http://commons.wikimedia.org/wiki/User:Fabelfroh; Section 6.8 © Bill Boch/ FoodPix/ Jupiter Images. **Page 105** Critical Thinking, © E.R. Degginger.

CHAPTER 7 **Key Concepts** from left, © Dr. Dennis Kunkel/ Visuals Unlimited; © Lois Ellen Frank/ Corbis. **7.1** page 106, Courtesy of © Louise Chalcraft-Frank and FARA; page 107, Professors P. Motta and T Naguro/ SPL/ Photo Researchers, Inc. **7.7** right, SPL/ Photo Researchers, Inc. **7.10** (b–c) © Ben Fink/ Foodpix/ Jupiter Imges; (d) © Dr. Dennis

Kunkel/ Visuals Unlimited. **7.11** (b) Courtesy of © William MacDonald, M.D.; (c) © Randy Faris/ Corbis; (d) Lilly M/ http://commons.wikimedia.org. **7.12** left, © Lois Ellen Frank/ Corbis. **Page 119** top, Professors P. Motta and T Naguro/ SPL/ Photo Researchers, Inc.; Section 7.1, Courtesy of © Louise Chalcraft-Frank and FARA. **Page 120** Section 7.6, © Dr. Dennis Kunkel/ Visuals Unlimited; Section 7.7 © Lois Ellen Frank/ Corbis. **7.13** (a) Steve Gschmeissner/ Photo Researchers, Inc.; (b) © Images Paediatr Cardiol.

CHAPTER 8 **Key Concepts** from left, Andrew Syred/ Photo Researchers, Inc., Courtesy of Cyagra, Inc., www.cyagra.com. **8.1** page 122, © James Symington; page 123, Ben Glass, courtesy of © BioArts International. **8.2** inset, Andrew Syred/ Photo Researchers, Inc. **8.3** © University of Washington Department of Pathology. **8.6** (a) below, Eye of Science/ Photo Researchers, Inc. **8.8** C. Barrington Brown, 1968 J. D. Watson; left, A C. Barrington Brown © 1968 J. D. Watson; right, PDB ID: 1BBB; Silva, M. M., Rogers, P. H., Arnone, A.: A third quaternary structure of human hemoglobin A at 1.7-≈ resolution. *J Biol Chem* 267 pp. 17248 (1992). **8.13, 8.14** Courtesy of Cyagra, Inc., www.cyagra.com. **Page 133** Ben Glass, courtesy of © BioArts International. **Page 134** Section 8.1, Ben Glass, courtesy of © BioArts International; Section 8.2, Andrew Syred/ Photo Researchers, Inc.; Section 8.7, Andrew Syred/ Photo Researchers, Inc. **8.15** above, Journal of General Physiology 36(1), September 20, 1952: "Independent Functions of viral Protein and Nucleic Acid in Growth of Bacteriophage;" below, © JupiterImages Corporation.

CHAPTER 9 **Key Concepts** from left, O. L. Miller, © John W. Gofman and Arthur R. Tamplin. From Poisoned Power: The Case Against Nuclear Power Plants Before and After Three Mile Island, Rodale Press, PA, 1979. **9.1** page 136, Vaughan Fleming/ SPL/ Photo Researchers, Inc. **9.5** O. L. Miller. **9.12** inset, © Kiseleva and Donald Fawcett/ Visuals Unlimited. **9.13** (e) Dr. Gopal Murti/ SPL/ Photo Researchers, Inc. **9.14** (a) © John W. Gofman and Arthur R. Tamplin. From Poisoned Power: The Case Against Nuclear Power Plants Before and After Three Mile Island, Rodale Press, PA, 1979. **Page 148** Vaughan Fleming/ SPL/ Photo Researchers, Inc.; Section 9.3, O. L. Miller; Section 9.6, © John W. Gofman and Arthur R. Tamplin. From Poisoned Power: The Case Against Nuclear Power Plants Before and After Three Mile Island, Rodale Press, PA, 1979.

CHAPTER 10 **Key Concepts** from left, From the collection of Jamos Werner and John T. Lis; © Maria Samsonova and John Reinitz; © Jose Luis Riechmann. **10.1** page 150, Courtesy of Robin Shoulla and Young Survival Coalition; page 151, From the archives of www.breastpath.com, courtesy of J.B. Askew, Jr., M.D., P.A. Reprinted with permission, copyright 2004 Breastpath.com. **10.4** From the collection of Jamos Werner and John T. Lis. **10.5** (a–b) David Scharf/ Photo Researchers, Inc.; (c) Eye of Science/ Photo Researchers, Inc.; (d) Courtesy of the Aniridia Foundation International, www.aniridia.net; (e) M. Bloch. **10.6** (a–e) © Maria Samsonova and John Reinitz; (f) © Jim Langeland, Jim Williams, Julie Gates, Kathy Vorwerk, Steve Paddock and Sean Carroll, HHMI, University of Wisconsin-Madison. **10.7** © Dr. William Strauss. **10.9** (a) below, Juergen Berger, Max Planck Institute for Developmental Biology, Tuebingen, Germany; (b) © Jose Luis Riechmann. **Page 159** left, ©

Lowe Worldwide, Inc. as Agent for National Fluid Milk Processor Promotion Board; right, Courtesy of Robin Shoulla and Young Survival Coalition. **Page 160** Section 10.1, From the archives of www.breastpath.com, courtesy of J.B. Askew, Jr., M.D., P.A. Reprinted with permission, copyright 2004 Breastpath.com; Section 10.2, From the collection of Jamos Werner and John T. Lis; Section 10.3, © Maria Samsonova and John Reinitz; Section 10.4, © Dr. William Strauss.

CHAPTER 11 **Key Concepts**, from left © Carolina Biological Supply Company/ Phototake; Michael Clayton/ University of Wisconsin, Department of Botany; From "Expression of the epidermal growth factor receptor (EGFR) and the phosphorylated EGFR in invasive breast carcinomas." http://breast-cancer research.com/content/10/3/R49. **11.1** page 162, Dr. Paul D. Andrews/ University of Dundee; page 163, Courtesy of the family of Henrietta Lacks. **11.4** © Carolina Biological Supply Company/ Phototake. **11.5** left (all), Michael Clayton/ University of Wisconsin, Department of Botany; right (all) Ed Reschke. **11.8** © Phillip B. Carpenter, Department of Biochemistry and Molecular Biology, University of Texas — Houston Medical School. **11.9** From "Expression of the epidermal growth factor receptor (EGFR) and the phosphorylated EGFR in invasive breast carcinomas." http://breast-cancer research.com/content/10/3/R49. **11.11** (a) © Ken Greer/Visuals Unlimited; (b) Biophoto Associates/Science Source/Photo Researchers; (c) James Stevenson/ Photo Researchers, Inc. **Page 171** above, Dr. Paul D. Andrews/ University of Dundee; below, Courtesy of the family of Henrietta Lacks. **Page 172** Section 11.1, Courtesy of the family of Henrietta Lacks; Section 11.2, © Carolina Biological Supply Company/ Phototake; Section 11.3, Michael Clayton/ University of Wisconsin, Department of Botany; Section 11.5, From "Expression of the epidermal growth factor receptor (EGFR) and the phosphorylated EGFR in invasive breast carcinomas." http://breast-cancer research.com/content/10/3/R49; Section 11.6, James Stevenson/ Photo Researchers, Inc. **11.12** Courtesy of © Dr. Thomas Ried, NIH and the American Association for Cancer Research. **Page 173**, Critical Thinking, © micrograph, D. M. Phillips/ Visuals Unlimited.

CHAPTER 12 **Key Concepts** from left, Image courtesy of Carl Zeiss MicroImaging, Thornwood, NY; With thanks to the John Innes Foundation Trustees, computer enhanced by Gary Head; Francis Leroy, Biocosmos/ Science Photo Library/ Photo Researchers, Inc. **12.1** page 174, © JupiterImages Corporation; page 175, Susumu Nishinaga / Photo Researchers, Inc. **12.2** Image courtesy of Carl Zeiss MicroImaging, Thornwood, NY. **12.5** upper, With thanks to the John Innes Foundation Trustees, computer enhanced by Gary Head. **12.8** right, Courtesy of © Billie Chandler. **12.10** Francis Leroy, Biocosmos/ Science Photo Library/ Photo Researchers, Inc. **Page 185** AP/ Wide World Photos. **Page 186** Section 12.1, © JupiterImages Corporation; Section 12.2, Image courtesy of Carl Zeiss MicroImaging, Thornwood, NY; Section 12.3, With thanks to the John Innes Foundation Trustees, computer enhanced by Gary Head; Section 12.5, Francis Leroy, Biocosmos/ Science Photo Library/ Photo Researchers, Inc. **12.12** above, Reprinted from *Current Biology*, Vol 13, (Apr 03), Authors Hunt, Koehler, Susiarjo, Hodges, Ilagan, Voigt, Thomas, Thomas and Hassold, Bisphenol A Exposure Causes Meiotic Aneuploidy in the Female Mouse, pp. 546–553, © 2003 Cell Press. Published by Elsevier Ltd. With permission from Elsevier.

CHAPTER 13 **Key Concepts** from left, © The Moravian Museum, Brno; © Michael Stuckey/ Comstock, Inc. **13.1** page 188, clockwise from top left, Courtesy of © Steve & Ellison Widener and Breathe Hope, http://breathehope.tamu.edu; Courtesy of © The Family of Savannah Brooke Snider; Courtesy of © The Cody Dieruf Benefit Foundation, www.breathinisbelievin.org; Courtesy of © Bobby Brooks and The Family of Jeff Baird; Courtesy of The family of Benjamin Hill, reprinted with permission of © Chappell/Marathonfoto; Courtesy of © the family of Brandon Herriott. **13.2** (a) Jean M. Labat/ Ardea, London. **Page 190**, © The Moravian Museum, Brno. **13.6** right, At Tip, © George Lepp/ Corbis. **13.9** above, © David Scharf/ Peter Arnold, Inc. **13.10** (a) © JupiterImages Corporation. **13.11** dogs, left and right, © Michael Stuckey/ Comstock, Inc.; center, Bosco Broyer, photograph by Gary Head. **13.12** Courtesy of the family of Haris Charalambous and the University of Toledo. **13.13** below, Courtesy of Ray Carson, University of Florida News and Public Affairs. **13.14** left, © JupiterImages Corporation; right, © age photostock/ SuperStock. **Page 198** right, from top, © Frank Cezus/ FPG/ Getty Images; speakslowly, http://commons.wikimedia.org/wiki/File:Heterochromia.jpg; i love images/ Jupiter Images; © Georg Tanner, http://www.flickr.com/photos/gtanner; © Michael Prince/ Corbis; © Medioimages/Photodisc/ Jupiter Images. **13.15** © Dr. Christian Laforsch. **Page 199** © Gary Gaugler/ The Medical File/ Peter Arnold Inc. **Page 200** Section 13.1, Courtesy of © The Cody Dieruf Benefit Foundation, www.breathinisbelievin.org; Section 13.2, © The Moravian Museum, Brno; Section 13.5, © Michael Stuckey/ Comstock, Inc.; Section 13.6, © Dr. Christian Laforsch. **Page 201** © Leslie Faltheisek.

CHAPTER 14 **Key Concepts** from left, Courtesy of Irving Buchbinder, DPM, DABPS, Community Health Services, Hartford CT; Photo courtesy of The Progeria Research Foundation; © Lennart Nilsson/ Bonnierforlagen AB. **14.1** page 202, © Gary Roberts/ worldwidefeatures.com; page 203, Richard A. Sturm, Molecular genetics of human pigmentation diversity, Human Molecular Genetics, 2009 Apr 15;18(R1):R9–17, by permission of Oxford University Press. **14.2** (a) right, Courtesy of Irving Buchbinder, DPM, DABPS, Community Health Services, Hartford CT; (c) © Steve Uzzell. **14.4** (a) © Newcastle Photos and Ivy & Violet Broadhead and family; (b) Photo courtesy of The Progeria Research Foundation. **14.5** (b) © Rick Guidotti/ Positive Exposure; (c) © Conner's Way Foundation, www.connersway.com. **14.7** (a–b) Photo by Gary L. Friedman, www.FriedmanArchives.com. **14.13** Left, L. Willatt, East Anglian Regional Genetics Service / Photo Researchers, Inc; right, Ciarra, photo by © Michelle Harmon. **14.14** (a) http://en.wikipedia.org/wiki/File:Embryo_at_14_weeks_profile.jpg; (b) © Howard Sochurek/ The Medical File/ Peter Arnold, Inc. **14.15** © Lennart Nilsson/ Bonnierforlagen AB. **14.16** © Fran Heyl Associates/ Jacques Cohen, computer-enhanced by © Pix Elation. **Page 215** © Gary Roberts/ worldwidefeatures.com. **Page 216** Section 14.1, © Gary Roberts/ worldwidefeatures.com; Section 14.2, Courtesy of Irving Buchbinder, DPM, DABPS, Community Health Services, Hartford CT; Section 14.3, Photo courtesy of The Progeria Research Foundation; Section 14.7, © Lennart Nilsson/ Bonnierforlagen AB.

CHAPTER 15 **Key Concepts** from left, Professor Stanley Cohen/ SPL/ Photo Researchers, Inc.; Patrick Landmann/ Photo Researchers, Inc.; Photo courtesy of MU Extension and Agricultural Infomation; © Corbis/ SuperStock. **15.1** page 218, © 2009 Jupiter Images Corporation/ Fotosearch; page 219, Courtesy of Affymetrix. **15.3** (a) Professor Stanley Cohen/ SPL/ Photo Researchers, Inc.; (b) with permission of © QIAGEN, Inc. **15.8** left, Volker Steger/ SPL/ Photo Researchers, Inc.; right, Patrick Landmann/ Photo Researchers, Inc. **15.10** left, The Sanger Institute. Wellcome Images; right, Wellcome Trust Sanger Institute. **15.11** Raw STR data courtesy of © Orchid Cellmark, www.orchidcellmark.com. **Page 228** Photo Courtesy of Systems Biodynamics Lab, P.I. Jeff Hasty, UCSD Department of Bioengineering, and Scott Cookson. **15.12** (d) © Lowell Georgis/ Corbis; (e) © Keith V. Wood. **15.13** The Bt and Non-Bt corn photos were taken as part of field trial conducted on the main campus of Tennessee State University at the Institute of Agricultural and Environmental Research. The work was supported by a competitive grant from the CSREES, USDA titled "Southern Agricultural Biotechnology Consortium for Underserved Communities, (2000–2005)." Dr. Fisseha Tegegne and Dr. Ahmad Aziz served as Principal and Co-principal Investigators respectively to conduct the portion of the study in the State of Tennessee. **15.14** (a) © Adi Nes, Dvir Gallery Ltd.; (b) Photo courtesy of MU Extension and Agricultural Infomation; (c) Transgenic goat produced using nuclear transfer at GTC Biotherapeutics. Photo used with permission. **15.15** Courtesy of © Dr. Jean Levit. The Brainbow technique was developed in the laboratories of Jeff W. Lichtman and Joshua R. Sanes at Harvard University. This image has received the Bioscape imaging competition 2007 prize. **15.16** © Jeans for Gene Appeal. **Page 233** left, © Corbis/ SuperStock; right, © 2009 Jupiter Images Corporation/ Fotosearch; Section 15.1, Courtesy of Affymetrix; Section 15.2, Professor Stanley Cohen/ SPL/ Photo Researchers, Inc. **Page 234** Section 15.4, Patrick Landmann/ Photo Researchers, Inc.; Section 15.5, Wellcome Trust Sanger Institute; Section 15.6–15.9, Photo courtesy of MU Extension and Agricultural Infomation; Section 15.10 © Corbis/ SuperStock. **15.17** Courtesy of Dr. S. Thuret, Kings College London.

CHAPTER 16 **Key Concepts** from left, © Wolfgang Kaehler/ Corbis; Courtesy of Stan Celestian/ Glendale Comunity College Earth Science Image Archive; USGS; © Taro Taylor, www.flickr.com/photos/tjt195. **16.1** page 236, © 2009 Mike Park/ www.flickr.com/photos/38869446@N00; page 237, © David A. Kring, NASA/ Univ. Arizona Space Imagery Center. **16.2** (a) © Earl & Nazima Kowall/ Corbis; (b–c) © Wolfgang Kaehler/ Corbis. **16.3** © Edward S. Ross. **16.4** (a) © Dr. John Cunningham/ Visuals Unlimited; (b) © Gary Head. **Page 239** Courtesy of Daniel C. Kelley, Anthony J. Arnold, and William C. Parker, Florida State University Department of Geological Science. **16.5** above, © Gordon Chancellor. **16.6** above, Painting by George Richmond; right, Cambridge University Library. **16.7** (a) © John White; (b) 2004 Arent. **16.8** © Down House and The Royal College of Surgeons of England. **Page 244** top, Courtesy of Stan Celestian/ Glendale Comunity College Earth Science Image Archive. **16.9** (a) W. B. Scott (1894); (b) Doug Boyer in P. D. Gingerich et al. (2001) © American Association for Advancement of Science; (c) © P. D. Gingerich and M. D. Uhen (1996), © University of Michigan. Museum of Paleontology; bottom © P. D. Gingerich, University of Michigan. Museum of Paleontology. **16.10** (b) © Photodisc/ Getty Images. **Page 245** , right, Courtesy of Stan Celestian/ Glendale Community College Earth Science Image Archive. **16.11** (c) © Michael Pancier. **16.12** left, USGS. **16.13** (a–e) After A.M. Ziegler, C.R. Scotese, and S.F. Barrett, "Mesozoic and Cenozoic Paleogeographic Maps," and J. Krohn and J. Sundermann (Eds.), Tidal Frictions and the Earth's Rotation II, Springer-Verlag, 1983. **16.15** (a) © Taro Taylor, www.flickr.com/photos/tjt195; (b) © JupiterImages Corporation; (c) © Linda Bingham. **16.16** from left, © Lennart Nilsson/ Bonnierforlagen AB; Courtesy of Anna Bigas, IDIBELL-Institut de Recerca Oncologica, Spain; From "Embryonic staging system for the short-tailed fruit bat, Carollia perspicillata, a model organism for the mammalian order Chiroptera, based upon timed pregnancies in captive-bred animals" C.J. Cretekos et al., Developmental Dynamics Volume 233, Issue 3, July 2005, Pages: 721–738. Reprinted with permission of Wiley-Liss, Inc. a subsidiary of John Wiley & Sons, Inc.; Courtesy of Prof. Dr. G. Elisabeth Pollerberg, Institut für Zoologie, Universität Heidelberg, Germany; USGS. **16.17** (a) © Jürgen Berger, Max-Planck-Institut für Developmental Biology, Tübingen; (b) © Visuals Unlimited. **16.18** Courtesy of Ann C. Burke, Wesleyan University. **Page 253** © David A. Kring, NASA/ Univ. Arizona Space Imagery Center. **Page 254** Section 16.1, © 2009 Mike Park/ www.flickr.com/photos/38869446@N00; Section 16.2, © Wolfgang Kaehler/ Corbis; Section 16.3, Painting by George Richmond; Section 16.4, © Down House and The Royal College of Surgeons of England; Section 16.5, Courtesy of Stan Celestian/ Glendale Comunity College Earth Science Image Archive; Section 16.6, © Michael Pancier; Section 16.7, USGS; Section 16.8, © Taro Taylor, www.flickr.com/photos/tjt195; Section 16.9, Courtesy of Ann C. Burke, Wesleyan University. **Page 255** Lawrence Berkeley National Laboratory.

CHAPTER 17 **Key Concepts** from left, © Alan Solem; Courtesy of Hopi Hoekstra, University of California, San Diego; © Jack Jeffrey Photography; © Jeremy Thomas/ Natural Visions; © Jack Jeffrey Photography. **17.1** page 256, © Rollin Verlinde/ Vilda; page 257, © Reuters NewMedia, Inc./ Corbis. **17.2** (a) © Alan Solem; (b) from left, © Roderick Hulsbergen/ http://www.photography.euweb.nl; all others, © JupiterImages Corporation. **17.6** J. A. Bishop, L. M. Cook. **17.7** (a) © James "Bo" Insogna/ Photos.com Plus; (b–d) Courtesy of Hopi Hoekstra, University of California, San Diego. **17.9** above, Peter Chadwick/ Science Photo Library/ Photo Researchers, Inc. **17.11** © Thomas Bates Smith. **17.12** (a) © Ingo Arndt/ Nature Picture Library; (b) © Bruce Beehler; (c) Courtesy of Gerald Wilkinson. **17.13** (a,b) After Ayala and others; (c) © Michael Freeman/ Corbis. **17.14** (a,b) Adapted from S. S. Rich, A. E. Bell, and S. P. Wilson, "Genetic drift in small populations of Tribolium," Evolution 33:579–584, Fig. 1, p. 580, 1979. Used by permission of the publisher; below, Photo by Peggy Greb/ USDA. **17.15** © Dr. Victor A. McKusick. **17.16** (a) © David Neal Parks; (b) © W. Carter Johnson. **Page 270** lvin E. Staffan/ Photo Researchers, Inc. **17.18** (a) Courtesy of Dr. James French; (b) Courtesy of Ron Brinkmann, www.flickr.com/photos/ronbrinkmann; (c) © David Goodin. **17.19** G. Ziesler/ ZEFA. **17.20** right, © Arthur Anker. **17.21** Po'ouli, Bill Sparklin/ Ashley Dayer; All others, © Jack Jeffrey Photography. **17.23** Kevin Bauman, www.african-cichlid.com. **17.24** top, Courtesy of The Virtual

Fossil Museum, www.fossilmuseum.net; bottom, © Mark Erdman. **17.26** (a) © David Simcox; (b) © Jeremy Thomas/ Natural Visions. **17.28** (a–b) © Jack Jeffrey Photography; (c) Bill Sparklin/ Ashley Dayer. **Page 279** © Rollin Verlinde/ Vilda. **Page 280** Section 17.1, © Rollin Verlinde/ Vilda; Section 17.2–3, © Alan Solem; Section 17.4–6, Courtesy of Hopi Hoekstra, University of California, San Diego; Section 17.7, © Bruce Beehler; Section 17.8–9, © Dr. Victor A. McKusick; Section 17.10, Alvin E. Staffan/ Photo Researchers, Inc.; Section 17.11, © Jack Jeffrey Photography; Section 17.12, Kevin Bauman, african-cichlid.com; Section 17.13, © Jeremy Thomas/ Natural Visions; Section 17.14, © Jack Jeffrey Photography.

CHAPTER 18 **Key Concepts** from left, Painting by William K. Hartmann; From Hanczyc, Fujikawa, and Szostak, Experimental Models of Primitive Cellular Compartments: Encapsulation, Growth, and Division;www.sciencemag.org Science 24 October 2003; 302;529, Figure 2, page 619. Reprinted with permission of the authors and AAAS; Bruce Runnegar, NASA Astrobiology Institute. **18.1** page 282, © Jeff Hester and Paul Scowen, Arizona State University, and NASA; page 283, Photo by Julio Betancourt/ U.S. Geological Survey. **18.2** Painting by William K. Hartmann. **18.5** Courtesy of the University of Washington. **18.7** © Micheal J. Russell, Scottish Universities Environmental Research Centre. **18.8** (a) © Janet Iwasa; (b) From Hanczyc, Fujikawa, and Szostak, Experimental Models of Primitive Cellular Compartments: Encapsulation, Growth, and Division;www.sciencemag.org Science 24 October 2003; 302;529, Figure 2, page 619. Reprinted with permission of the authors and AAAS; (c) Photo by Tony Hoffman, courtesy of David Deamer. **18.9** (a) © Stanley M. Awramik; (b) Russell Kightley/ Science Photo Library/ Photo Researchers, Inc.; (c) © University of California Museum of Paleontology; (d) Chase Studios/ Photo Researchers, Inc.; (e) Courtesy of © Department of Industry and Resources, Western Australia. **18.10** (a–b) Bruce Runnegar, NASA Astrobiology Institute; (c) © N.J. Butterfield, University of Cambridge. **18.12** (a) Photo provided by M.G. Klotz, Department of Biology, University of Louisville, KY, USA (http://mgkmicro.com); (b) Courtesy of John Fuerst, University of Queensland. originally published in Archives of Microbiology vol 175, p 413–429 (Lindsay MR, Webb RI, Strous M, Jetten MS, Butler MK, Forde RJ, Fuerst JA. Cell compartmentalisation in planctomycetes: novel types of structural organisation for the bacterial cell.Arch Microbiol. 2001 Jun;175(6):413–29]. **18.13** (a) CNRI/ Photo Researchers, Inc.; (b) © Robert Trench, Professor Emeritus, University of British Columbia; above, © Courtesy of Isao Inouye, Institute of Biological Sciences, University of Tsukuba. **Page 294**, NASA/JPL. **Page 294** Section 18.1, Photo by Julio Betancourt/ U.S. Geological Survey; Section 18.3, Courtesy of the University of Washington; Section 18.4, © Janet Iwasa; Section 18.5, Chase Studios/ Photo Researchers, Inc.; Section 18.6, © Courtesy of Isao Inouye, Institute of Biological Sciences, University of Tsukuba.

CHAPTER 19 **Key Concepts** from left, Dr. Richard Feldmann/ National Cancer Institute; SciMAT/ Photo Researchers, Inc.; © Dr. Dennis Kunkel/ Visuals Unlimited; CDC/ Janice Haney Car; © Savannah River Ecology Laboratory. **19.1** page 296, © Lennart Nilsson/ Bonnierforlagen AB; page 297, Micrographs Z. Salahuddin, National Institutes of Health. **19.2** (b) left, after Stephen L. Wolfe; (c) Dr. Richard Feldmann/ National Cancer Institute; (d) Russell Knightly/ Photo Researchers, Inc. **19.3** top left, Science Photo Library/ Photo Researchers, Inc. **19.5** © WHO, Pierre-Michel Virot, photographer. **Page 300** Sercomi/ Photo Researchers, Inc. **Page 301** left, CDC/ C. S. Goldsmith and A. Balish; right, Electron Microscopy Laboratory, Agricultural Research Service, U. S. Department of Agriculture. **19.7** top, L. Santo. **19.8** © Dr. Dennis Kunkel / Visuals Unlimited. **19.9** (a) © P. W. Johnson and J. MeN. Sieburth, Univ. Rhode Island/ BPS; (b) © Dr. Manfred Schloesser, Max Planck Institute for Marine Microbiology; (c) SciMAT/ Photo Researchers, Inc. **19.10** (a) CDC/ Janice Haney Car; (b) CDC/ Bruno Coignard, M.D.; Jeff Hageman, M.H.S. **19.11** (a) Stem Jems/ Photo Researchers, Inc.; (b) CDC/ James Gathany. **Page 306** © Courtesy Jack Jones, Archives of Microbiology, Vol. 136, 1983, pp. 254–261. Reprinted by permission of Springer-Verlag. **19.13** (a) © Dr. W. Michaelis/ Universitat Hamburg; (b) Courtesy of Benjamin Brunner; (c) © Savannah River Ecology Laboratory. **Page 308** Section 19.1, Micrographs Z. Salahuddin, National Institutes of Health; Section 19.3, © WHO, Pierre-Michel Virot, photographer; Section 19.7, © P. W. Johnson and J. MeN. Sieburth, Univ. Rhode Island/ BPS; Section 19.8, © Dr. W. Michaelis/ Universitat Hamburg. **19.14** © Photodisc/ Getty Images.

CHAPTER 20 **Key Concepts**, from left © Dr. David Phillips/ Visuals Unlimited; © Wim van Egmond/ Visuals Unlimited; © Lewis Trusty/ Animals Animals; © Wim van Egmond/ Visuals Unlimited; M I Walker/ Photo Researchers, Inc. **20.1** page 310, © Peter M. Johnson - www.flickr.com/photos/pmjohnso; page 311, © Dr. David Phillips/ Visuals Unlimited. **20.2** (a) Astrid Hanns-Frieder michler/SPL/ Photo Researchers, Inc.; (b) © Dr. David Phillips/ Visuals Unlimited; (c) Science Museum of Minnesota; (d) D. P. Wilson/ Eric & David Hosking; (e) © Jeffrey Levinton, State University of New York, Stony Brook. **Page 313**, © Dr. Stan Erlandsen, University of Minnesota. **20.3** (a) © Dr. Dennis Kunkel/ Visuals Unlimited; (b) Oliver Meckes/ Photo Researchers, Inc.. **20.5** (a) Courtesy of Allen W. H. Bé and David A. Caron ; (b) © Ric Ergenbright/ Corbis; (c) © John Clegg/ Ardea, London. **20.6** © Frank Borges Llosa/ www.frankley.com; inset, © Wim van Egmond/ Visuals Unlimited. **20.7** (a) Gary W. Grimes and Steven L'Hernault. **20.8** Based on Fig. 1 from *Genetic linkage and association analyses for trait mapping in Plasmodium falciparum*, by Xinzhuan Su, Karen Hayton & Thomas E. Wellems, *Nature Reviews Genetics* 8, 497–506 (July 2007). **Page 317**, Heather Angel. **20.9** (a) © Wim van Egmond/ Visuals Unlimited; (b) Image courtesy of Woodstream Corporation. **20.10** (a) © Lewis Trusty/ Animals Animals. **20.11** below left, © PhotoDisc/ Getty Images. **Page 318**, Courtesy of Professeur Michel Cavalla. **20.12** (a) Courtesy of © Brian P. Piasecki and Carolyn D. Silflow; (b) © Kim Taylor/ Bruce Coleman, Inc./ Photoshot; (c) © Lawson Wood/ Corbis; (d) © Wim van Egmond/ Visuals Unlimited. **20.13** M I Walker/ Photo Researchers, Inc. **20.14** (a) © Edwards S. Ross; (b) © Courtesy of www.hiddenforest.co.nz. **20.15** below, © Carolina Biological Supply Company. **20.16** Larry Brand. **Page 321**, Woods Hole Oceanographic Institution. **Page 322** Section 20.1, © Peter M. Johnson - www.flickr.com/photos/pmjohnso; Section 20.4, Courtesy of Allen W. H. Bé and David A. Caron ; Section 20.5–6, © Wim van Egmond/ Visuals Unlimited; Section 20.7,

Heather Angel; Section 20.8, D. P. Wilson/ Eric & David Hosking; Section 20.9, © Courtesy of www.hiddenforest.co.nz. **Page 323**, Critical Thinking, Courtesy Brian Duval.

CHAPTER 21 **Key Concepts** from left, Courtesy of © Professor T. Mansfield; National Park Services, Paul Stehr-Green; © A. & E. Bomford/ Ardea, London; © Dave Cavagnaro/ Peter Arnold, Inc.; © Sanford/ Agliolo/ Corbis. **21.1** page 324, T. Kerasote/ Photo Researchers, Inc.; page 325, © William Campbell/ TimePix/ Getty Images. **21.2** right, Courtesy of © Professor T. Mansfield. **21.4** Courtesy of © Christine Evers. **21.5** left, © Jane Burton/ Bruce Coleman Ltd. **21.6** © Fred Bavendam/ Peter Arnold, Inc. **21.7** (a) National Park Services, Paul Stehr-Green; (b) © Wayne P. Armstrong, Professor of Biology and Botany, Palomar College, San Marcos, California. **21.8** © University of Wisconsin-Madison, Department of Biology, Anthoceros CD. **21.9** (a) © Martin LaBar, www.flickr.com/photos/martinlabar; (b) Courtesy of BOCATEC Sales and Rent GmbH & Co. KG. **21.10** (a) © William Ferguson; (b) Courtesy of © Christine Evers. **21.11** right, © A. & E. Bomford/ Ardea, London. **21.12** (a) © S. Navie; (b) David C. Clegg/ Photo Researchers, Inc.; (c) © Klein Hubert/ Peter Arnold, Inc. **Page 332** © Reprinted with permission from Elsevier. **21.14** © Jeri Hochman and Martin Hochman, Illustration by Zdenek Burian. **21.16** © Karen Carr Studio/ www.karencarr.com. **21.18** (a) © Dave Cavagnaro/ Peter Arnold, Inc.; (b) © M. Fagg, Australian National Botanic Gardens; (c) Michael P. Gadomski/ Photo Researchers, Inc.; (d) © E. Webber/ Visuals Unlimited; (e) © Gerald & Buff Corsi/ Visuals Unlimited. **21.19** far left © Robert Potts, California Academy of Sciences; left upper, © Robert & Linda Mitchell Photography; lower, R. J. Erwin/ Photo Researchers, Inc. **Page 336** Courtesy of © Christine Evers. **21.22** © Sanford/ Agliolo/ Corbis. **21.23** (a–b) Photo USDA; (c) Courtesy of Linn County, Oregon Sheriff's Office. **21.24** R. Bieregaard/Photo Researchers, Inc. **Page 340** Section 21.1, © William Campbell/ TimePix/ Getty Images; Section 21.4, © Martin LaBar, www.flickr.com/photos/martinlabar; Section 21.6, © Robert & Linda Mitchell Photography; Section 21.7, Courtesy of © Christine Evers; Section 21.8, Photo USDA. **21.25** right, © Clinton Webb.

CHAPTER 22 **Key Concepts** from left, Garry T. Cole, University of Texas, Austin/ BPS; © Ed Reschke; © Dr. Dennis Kunkel/ Visuals Unlimited; © Robert C. Simpson/ Nature Stock; Gary Head. **Page 342** © Robert C. Simpson/ Nature Stock. **Page 343** Photo by Yue Jin/ USDA. **22.1** (a) Photo by Scott Bauer/ USDA; (b) © Robert C. Simpson/ Nature Stock. **22.2** Garry T. Cole, University of Texas, Austin/ BPS. **22.3** top, © Ed Reschke. **22.4** © Dr. Mark Brundrett, The University of Western Australia. **22.5** (a) above, © Michael Wood/ mykob.com; below, © North Carolina State University, Department of Plant Pathology; (b) © Bill Beatty/ Visuals Unlimited; (c) © agefotostock/ SuperStock. **22.6** N. Allin and G. L. Barron. **Page 346** right, © Dr. Dennis Kunkel/ Visuals Unlimited. **22.7** top left, Dr. J. O'Brien, USDA Forest Service; all others, © Robert C. Simpson/ Nature Stock. **22.8** below, Eye of Science/ Photo Researchers, Inc ; art, After T. Rost, et al., *Botany*, Wiley 1979. **22.9** (a) Gary Head; (b) After Raven, Evert, and Eichhorn, *Biology of Plants*, 4th ed., Worth Publishers, New York, 1986; (c) © Mark E. Gibson/ Visuals Unlimited. **22.10** (a) © Gary Braasch; (b) © F. B. Reeves. **22.11** © Harry Regin. **Page 349** right, above, Courtesy of D. G. Schmale III Courtesy of D. G. Schmale III; below, Dr. P. Marazzi/ Photo Researchers, Inc. **Page 350** Section 22.1, Photo by Yue Jin/ USDA; Section 22.2, Garry T. Cole, University of Texas, Austin/ BPS; Section 22.3, © Ed Reschke; Section 22.4, © North Carolina State University, Department of Plant Pathology; Section 22.5, Eye of Science/ Photo Researchers, Inc.; Section 22.6, © Mark E. Gibson/ Visuals Unlimited; Section 22.7, Dr. P. Marazzi/ Photo Researchers, Inc. **22.12** After graph from www.pfc.foresty.ca.

CHAPTER 23 **Key Concepts** from left © The Natural History Museum (London); © Wim van Egmond/ Visuals Unlimited; Courtesy of Karen Swain, North Carolina Museum of Natural Sciences; © Fred Bavendam/ Minden Pictures. **Page 352** © K.S. Matz. **Page 353** © Callum Roberts, University of York. **23.4** (a–b) David Patterson, courtesy micro*scope/http://microscope.mbl.edu/ (c) © 2003 Ana Signorovitch. **23.5** © The Natural History Museum (London). **23.6** (a) After Eugene Kozloff; (b) Image © ultimathule, Used under license from Shutterstock.com. **23.7** (a) Marty Snyderman/ Planet Earth Pictures; (b) © Dr. Sally Leys/ University of Alberta. **23.8** (b) © Wim van Egmond/ Visuals Unlimited; (c) © Brandon D. Cole/ Corbis. **23.10** After T. Storer, et al., *General Zoology*, Sixth Edition. **23.11** (a) © Jeffrey L. Rotman/ Corbis; (b) © Kim Taylor/ Bruce Coleman, Ltd.; (c) A.N.T./ Photo Researchers, Inc. **23.12** © Cory Gray. **23.14** right, Andrew Syred/ SPL/ Photo Researchers, Inc. **23.15** (a) © J. Solliday/ BPS; center art, adapted from Rasmussen, "Ophelia," Vol. 11, in Eugene Kozloff, *Invertebrates*, 1990; (b) © Jon Kenfield/ Bruce Coleman Ltd. **23.16** © J. A. L. Cooke/ Oxford Scientific Films. **23.17** left art, from Solomon, 8th edition, p. 624, figure 29-4; right, Courtesy of © Christine Evers. **23.19** left, Danielle C. Zacherl with John McNulty. **23.20** (a) © B. Borrell Casals/ Frank Lane Picture Agency/ Corbis; (b) © Frank Park/ ANT Photo Library; (c) © Dave Fleetham/Tom Stack & Associates. **23.21** (a) © Joe McDonald/Corbis; (b) Courtesy of © Christine Evers; (c) © Alex Kirstitch. **23.22** (a) © Alex Kirstitch; (b) J. Grossauer/ ZEFA. **23.24** (a) Courtesy of © Emily Howard Staub and The Carter Center; (b) Sinclair Stammers/ SPL/ Photo Researchers, Inc. **23.25** (a) © Jane Burton/ Bruce Coleman, Ltd.; (b) NOAA; (c) Courtesy of © Christine Evers; (d) Photo by Peggy Greb/ USDA. **23.26** (a) Image © Eric Isselée, Used under license from Shutterstock.com; (b) © Angelo Giampiccolo; (c) © Frans Lemmens/ The Image bank/ Getty Images; (d) Eye of Science/ Photo Researchers, Inc.; (e) Andrew Syred/ Photo Researchers, Inc. **23.27** (a) © David Tipling/ Photographer's Choice/ Getty Images; (b) Herve Chaumeton/ Agence Nature; (c) © Peter Parks/ Imagequestmarine.com. **23.28** After D.H. Milne, *Marine Life and the Sea*, Wadsworth, 1995. **23.30** CDC/ Piotr Naskrecki. **23.32** (a,h) Courtesy of © Christine Evers; (b–c,i) © Edward S. Ross (d) Alvin E. Staffan/ Photo Researchers, Inc.; (e) © Joseph L. Spencer; (f) John Alcock, Arizona State University; (g) © D. A. Rintoul; (j) © Mark Moffett/ Minden Pictures; (k) CDC/ Harvard University, Dr. Gary Alpert; (l) Courtesy of Karen Swain, North Carolina Museum of Natural Sciences. **23.33** Gregory G. Dimijian, M.D./ Photo Researchers, Inc. **23.34** Photo by Scott Bauer/ USDA. **23.35** Photo by James Gathany, Centers for Disease Control. **Page 373** © Walter Deas/ Seaphot Limited/ Planet Earth Pictures. **23.36** (a) © Herve Chaumeton/ Agence Nature; (c) © Fred Bavendam/

Minden Pictures; (d) © Jan Haaga, Kodiak Lab, AFSC/NMFS. **Page 374** Section 23.1, © Callum Roberts, University of York; Section 23.3, © 2003 Ana Signorovitch; Section 23.4, Marty Snyderman/ Planet Earth Pictures; Section 23.5, © Jeffrey L. Rotman/ Corbis; Section 23.6, © Cory Gray; Section 23.7, © J. Solliday/ BPS; Section 23.8, © B. Borrell Casals/ Frank Lane Picture Agency/ Corbis; Section 23.9, Sinclair Stammers/ SPL/ Photo Researchers, Inc.; Section 23.10–14, CDC/ Harvard University, Dr. Gary Alpert; Section 23.15, © Herve Chaumeton/ Agence Nature. **23.37** right, Jane Burton/ Bruce Coleman, Ltd.

CHAPTER 24 **Key Concepts** from left, Peter Parks/ Oxford Scientific Films/ Animals Animals; © Wernher Krutein/ photovault.com; © Bill M. Campbell, MD; © Z. Leszczynski/ Animals Animals; © Jean-Paul Tibbles, "Book of Life", Ebury Press. **Page 376** © P. Morris/ Ardea London. **Page 377** above, © James Reece, Nature Focus, Australian Museum; below, With permission of the Australian Museum. **24.1** right, Runk & Schoenberger / Grant Heilman, Inc. **24.2** (a) Peter Parks/ Oxford Scientific Films/ Animals Animals; (b) Redrawn from *Living Invertebrates*, V. & J. Pearse and M. & R. Buchsbaum. The Boxwood Press, 1987. Used by permission; (c) © California Academy of Sciences. **24.4** (a) © Brandon D. Cole/ Corbis; (b) Heather Angel/ Natural Visions. **24.5** Adapted from A.S. Romer and T.S. Parsons, *The Vertebrate Body*, Sixth Edition, Saunders, 1986. **24.6** (a) Jonathan Bird/ Oceanic Research Group, Inc.; (b) © Gido Braase/ Deep Blue Productions. **24.7** (a) from E. Solomon, L. Berg, and D.W. Martin, *Biology*, Seventh Edition, Thomson Brooks/Cole; (b) Robert & Linda Mitchell Photography; (c) Mark Dixon, NOAA; (d) Patrice Ceisel/ © 1986 John G. Shedd Aquarium. **24.8** © Wernher Krutein/ photovault.com. **24.9** (a–c) © P. E. Ahlberg; (d) © Alfred Kamajian. **24.10** (a) © Bill M. Campbell, MD; (b,c) Adapted from A.S. Romer and T.S. Parsons, The Vertebrate Body, Sixth Edition, Saunders, 1986. **24.11** (a) Stephen Dalton/ Photo Researchers, Inc.; (b) © David M. Dennis/ Tom Stack & Associates, Inc. **24.12** © John Serraro/ Visuals Unlimited. **24.13** © Pieter Johnson. **24.16** © Karen Carr Studio/ www.karencarr.com. **24.17** (a) © Z. Leszczynski/ Animals Animals; (b) © Kevin Schafer/Corbis; (c) © Kevin Schafer/ Tom Stack & Associates. **Page 385** © S. Blair Hedges, Pennsylvania State University. **24.18** (a) © Gerard Lacz/ ANTPhoto.com.au. **24.19** http://en.wikipedia.org/ wiki/User:Jmgarg1 'Creating awareness of Indian Flora & Fauna' Image Resource of thousands of my images of Birds, Butterflies, Flora etc. (arranged alphabetically & place-wise): http://commons.wiki media.org/wiki/Category:J.M.Garg. For learning about Indian Flora, visit/ join google e-group-Indiantreepix:http://groups.google.co.in/group/ indiantreepix?hl=en. **24.20** After M. Weiss and A. Mann, *Human Biology and Behavior*, 5th Edition, HarperCollins, 1990. **24.21** (a) Jean Phillipe Varin/ Jacana/ Photo Researchers, Inc.; (b) Jack Dermid; (c) © Sandy Roessler/ FPG/ Getty Images. **24.22** © Kenneth Garrett/ National Geographic Image Collection. **24.24** above, © Rod Williams/ www .bciusa.com. **24.25** (a) © Bone Clones®, www.bone clones.com; (b) © Gary Head. **24.26** (a) © AFP/ Getty Images; (b–e) Pascal Goetgheluck/ Photo Researchers, Inc. **24.27** (a) © Dr. Donald Johanson, Institute of Human Origins; (b) © Louise M. Robbins; (c) © Kenneth Garrett/ National Geographic Image Collection. **24.28** © Jean-Paul Tibbles, "Book

of Life", Ebury Press. **24.29** John Reader/ Photo Researchers, Inc. **24.30** top, © Pascal Goetgheluck/ Photo Researchers, Inc.; others, Peter Brown. **24.32** © Christopher Scotese, PALEOMAP Project. **Page 394** Section 24.1, © James Reece, Nature Focus, Australian Museum; Section 24.2, Runk & Schoenberger / Grant Heilman, Inc.; Section 24.3, Robert & Linda Mitchell Photography; Section 24.4, © John Serraro/ Visuals Unlimited; Section 24.5, © Karen Carr Studio/ www.karencarr.com; Section 24.6, © Kevin Schafer/Corbis; Section 24.7, © Gerard Lacz/ ANTPhoto.com.au; Section 24.8, © Sandy Roessler/ FPG/ Getty Images; Section 24.9, © Kenneth Garrett/ National Geographic Image Collection; Section 24.10, Pascal Goetgheluck/ Photo Researchers, Inc.

CHAPTER 25 **Key Concepts** from left, Reprinted from Trends in Neuroscience, Vol. 21, Issue 2, 1998, L.J.Young, W. Zuoxin, T.R. Insel, "Neuroendocrine bases of monogamy", Pages 71–75, ©1998, with permission from Elsevier Science; © Eric Hosking; © Steve Kaufman/ Corbis; © Alexander Wild. **25.1** © Scott Camazine. **25.2** © Christine Majul/ www .flickr.com/photos/kitkaphotogirl. **25.4** (a) © Robert M. Timm & Barbara L. Clauson, University of Kansas; (b–c) Reprinted from Trends in Neuroscience, Vol. 21, Issue 2, 1998, L.J.Young, W. Zuoxin, T.R. Insel, "Neuroendocrine bases of monogamy", Pages 71–75, ©1998, with permission from Elsevier Science. **25.5** (a) © Eric Hosking; (b) Stephen Dalton/ Photo Researchers, Inc. **25.6** © Nina Leen/ TimePix/ Getty Images; inset, © Robert Semeniuk/ Corbis. **25.7** © Professor Jelle Atema, Boston University. **25.8** © Bernhard Voelkl. **25.9** (a) © Robert Maier/ Animals Animals. **25.10** (a) © Tom and Pat Leeson, leesonphoto.com; (b) © Kevin Schafer/ Corbis; (c) © Monty Sloan, www.wolfphotography.com. **25.12** (a) © John Alcock, Arizona State University; (b) upper, © Pam Gardner, Frank Lane Picture Agency/ Corbis; lower, © Pam Gardner, Frank Lane Picture Agency/ Corbis; (c) © D. Robert Franz/ Corbis. **25.13** © Michael Francis/ The Wildlife Collection. **25.14** (a) John Conrad/ Corbis; (b) © B. Borrell Casals, Frank Lane Picture Agency/ Corbis; (c) © Steve Kaufman/ Corbis. **25.15** (a) © Tom and Pat Leeson, leesonphoto.com; (b) © John Alcock, Arizona State University; (c) © Paul Nicklen/ National Geographic/ Getty Images. **25.16** © Jeff Vanuga/ Corbis. **25.17** © Steve Bloom/ stevebloom.com. **25.18** © A. E. Zuckerman/ Tom Stack & Associates. **25.19** (a) © Australian Picture Library/ Corbis; (b) © Alexander Wild; (c) © Professor Louis De Vos. **25.20** (a) © Kenneth Lorenzen; (b) © Peter Johnson/ Corbis; (c) © Nicola Kountoupes/ Cornell University. **Page 409** © Lynda Richardson/ Corbis. **Page 410** Section 25.1, © Scott Camazine; Section 25.2, Reprinted from Trends in Neuroscience, Vol. 21, Issue 2, 1998, L.J.Young, W. Zuoxin, T.R. Insel, "Neuroendocrine bases of monogamy," Pages 71–75, ©1998, with permission from Elsevier Science; Section 25.3, Stephen Dalton/ Photo Researchers, Inc.; Section 25.4, © Robert Maier/ Animals Animals; Section 25.5, © Tom and Pat Leeson, leesonphoto.com; Section 25.6, © D. Robert Franz/ Corbis; Section 25.7, © A. E. Zuckerman/ Tom Stack & Associates; Section 25.8, © Kenneth Lorenzen.

CHAPTER 26 **Key Concepts** from left, © Cynthia Bateman, Bateman Photography; © G. K. Peck; © Wayne Bennett/ Corbis; © Don Mason/ Corbis. **26.1** page 412, Courtesy of @ Joel Pete; page 413, AP Images/ Steven Day. **26.2** © Cynthia Bateman, Bateman Photography. **26.3** (a) © Amos Nachoum/

Art Credits & Acknowledgments